OCCUPATIONAL SAFETY AND HYGIENE

PROCEEDINGS OF THE INTERNATIONAL SYMPOSIUM ON OCCUPATIONAL SAFETY AND HYGIENE, GUIMARÃES, PORTUGAL, 14–15 FEBRUARY 2013

Occupational Safety and Hygiene

Editors

Pedro M. Arezes
University of Minho, Guimarães, Portugal

João Santos Baptista
University of Porto, Porto, Portugal

Mónica P. Barroso, Paula Carneiro, Patrício Cordeiro &
Nélson Costa
University of Minho, Guimarães, Portugal

Rui B. Melo
University of Lisbon, Lisbon, Portugal

A. Sérgio Miguel
University of Minho, Guimarães, Portugal

Gonçalo Perestrelo
SMGP Consultores, Porto, Portugal

CRC Press
Taylor & Francis Group
Boca Raton   London   New York   Leiden

CRC Press is an imprint of the
Taylor & Francis Group, an **informa** business

A BALKEMA BOOK

CRC Press/Balkema is an imprint of the Taylor & Francis Group, an informa business

© 2013 Taylor & Francis Group, London, UK

Typeset by V Publishing Solutions Pvt Ltd., Chennai, India
Printed and bound in Great Britain by CPI Group (UK) Ltd, Croydon, CR0 4YY

Published by: CRC Press/Balkema
        P.O. Box 11320, 2301 EH Leiden, The Netherlands
        e-mail: Pub.NL@taylorandfrancis.com
        www.crcpress.com – www.taylorandfrancis.com

ISBN: 978-1-138-00047-6 (Hbk)
ISBN: 978-0-203-72965-6 (eBook)

Occupational Safety and Hygiene – Arezes et al. (eds)
© 2013 Taylor & Francis Group, London, ISBN 978-1-138-00047-6

Table of contents

Environmental ergonomics

Fire safety

Health monitoring and occupational medicine

Human and organisational factors

Occupational health and safety management systems

Risk assessment methods

Other

Occupational Safety and Hygiene – Arezes et al. (eds)
© 2013 Taylor & Francis Group, London, ISBN 978-1-138-00047-6

Foreword

The Portuguese Society of Occupational Safety and Hygiene (SPOSHO) organised on 14 and 15 February 2013 the 9th edition of the International Symposium on Occupational Safety and Hygiene—SHO 2013. Similarly to the past six years, the event was held in the main Auditorium of the School of Engineering at University of Minho in Guimarães. The 2013 edition covered the issues of Ergonomics and Physical Environment, Chemical and Biological Risk, Fire Safety and Prevention Management, which occurred in plenary sessions, as well as in parallel sessions of submitted works in more than 25 subjects covered by the event, and several sessions of posters.

More than 250 papers from more than 500 authors of 20 different countries were submitted, corresponding to an equal number of extended abstracts. These were reviewed by the international Scientific Committee (SC) of the Symposium, which has involved more than 100 specialists in the various scientific fields covered by the event. The authors of the papers presented at the SHO2013 were later invited to submit their full papers to a peer review process, including a review by 2 or 3 members of a restricted scientific committee. From this selection process, 109 full papers were accepted and are now included in this book.

We take this opportunity to thank our academic partners of the organisation of SHO2013, namely, the School of Engineering of the University of Minho, the School of Engineering of the University of Porto, the Faculty of Human Kinetics of the Technical University of Lisbon, the Polytechnic University of Catalonia and the Technical University of Delft. We also thank the scientific sponsorship of more than 20 academic and professional institutions and the official support of the Portuguese Authority for Working Conditions (ACT), as well as the valuable support of several Companies and Institutions including several media partners, which have contributed to the broad dissemination of the event.

The Editors,

Pedro M. Arezes
J. Santos Baptista
Mónica P. Barroso
Paula Carneiro
Patrício Cordeiro
Nélson Costa
Rui B. Melo
A. Sérgio Miguel
Gonçalo Perestrelo

The editors wish also to thank all the reviewers, listed below, which were involved in the process of reviewing and editing the papers included in this book.

A. Sérgio Miguel	Ana Colim	Anabela Simões
Ângela Malcata	Ashis Bhattacherjiee	Beata Mrugalska
Béda Barkokébas Jr.	Carla Barros	Celeste Jacinto
Celina Leão	Denis Coelho	Divo Quintela
Enrico Cagno	Eugénia Pinho	Fernando Amaral
Florentino Serranheira	Francisco Másculo	Francisco Rebelo
Hernâni Neto	Ignácio Castellucci	Ignácio Pavon
Isabel Lopes Nunes	Isabel Loureiro	J. Torres Da Costa
João Santos Baptista	João Ventura	Jorge Patrício

José D. Carvalhais
Luís Silva
M. Fernanda Rodrigues
Marino Menozzi
Matilde Rodrigues
Nélson Costa
Paula Carneiro
Paulo Vila-Real
Pere Sanz Gallén
Rui Azevedo
Sílvia Agostinho Silva
Susana Viegas

José Keating
Luiz Bueno
Marcelo Pereira
Mário Vaz
Miguel Tato Diogo
Olga Mayan
Paulo Barros de Oliveira
Pedro Arezes
Raquel R. Santos
Rui B. Melo
Sophia Piacenza
Thais Morata

Laura Bezerra Martins
Luiz Franz
Maria Antónia Gonçalves
Marta Santos
Mohammad Shahriari
Paul Swuste
Paulo Sampaio
Pedro Domingues
Ricardo Vasconcelos
Sérgio Sousa
Susana Costa
Timo Kauppinen

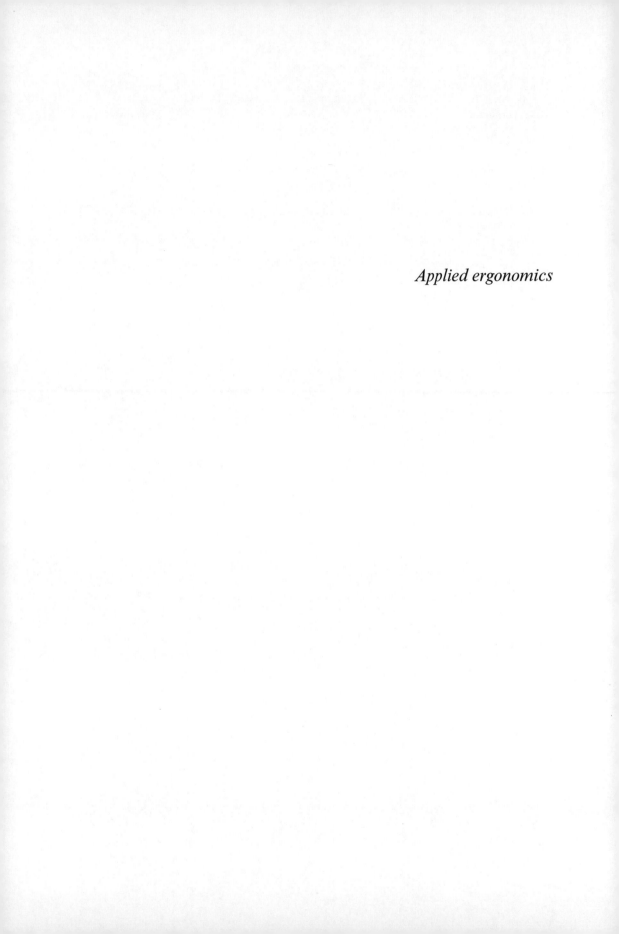

Applied ergonomics

Occupational Safety and Hygiene – Arezes et al. (eds)
© 2013 Taylor & Francis Group, London, ISBN 978-1-138-00047-6

Evaluation method of low back injuries in nursing: An ergonomic approach

R. Schlossmacher & F.G. Amaral

Production Engineering Department, Federal University of Rio Grande do Sul(UFRGS),
Porto Alegre, RS, Brazil

ABSTRACT: Studies have shown that nursing professionals are exposed to unfavorable working conditions. Ergonomic conditions of nursing daily activities are related to the characteristics of the occurrence of low back pain, this overload can be linked to three main factors: professional, organizational and personal. In this context, the objective of this study is evaluate the factors related to working conditions and the risk of low back injuries among nursing professionals through a case study in a geriatric clinic. As a result, the evaluation method allowed the generation of macro and micro organizational and structural indicators to assist in the management of occupational health and safety in clinics or similar. Thus we conclude that the characteristics interaction of the methods employed allow the knowledge of the work as a whole.

1 INTRODUCTION

Studies have shown that nurses are exposed to unfavorable working conditions. So there are many occupational hazards to which these professionals are undergoing during the labor exercise, and patient care is the task more arduous. These risks can be classified as: chemical, physical, biological, ergonomic and psychosocial (Simão et al., 2008; Lisboa et al., 2008; Smedley et al., 1995; Rosa, 2012; Cunha & Souza, 2008; Marziale, 2007). In this context, the ergonomic factors are those that will directly affect the relationship between task and worker (Cavassa, 1997). For nursing professionals those risks concerns: load (patient), lack of materials access, imposing unfavorable postures, excessive work demands, psychological factors, etc.

Ergonomic conditions of daily nursing tasks are related to the characteristics of low back pain occurrence, especially those concerning transport of patients (Cavassa, 1997). Some works undertaken a focus on the health of nurses showed that the low back is the region most affected when vertebral lesions are investigated, due to the peculiar work characteristics of this professional group (Alexandre, 1996; Rocha & Oliveira, 1998; Hoogendoorn et al., 2002; Magora, 1970; Rocha, 1997; Smedley, 1997).

The musculoskeletal overload may be due to three main factors: professional, organizational and personal (Malchaire, 2001). The musculoskeletal problems are related to the individual factors (age, health history, stress, etc.) and company (work organization as work cycles duration, breaks, structure, social climate, and biomechanical factors as repeatability, effort, etc.) (Aptel, 1993). These risk factors are not independent, in practice there is an interaction between them. In this context, the objective of this study is to evaluate the factors interactively to the overload indicators caused by working conditions related to the risk of low back injuries in nurses.

2 METHODS

The methodological approach is based on a case study research (Yin, 1994) and was applied in a clinical health care, implemented in 3 steps as shown in Figure 1.

3 RESULTS

3.1 *Step 1: Preliminary analysis*

Three female workers were interviewed collecting personal and occupational information. It can be inferred as general information that the study participants are not smokers and usually do not practice exercises in their free time. However, all workers reported low back pain symptoms in the last 12 months, 2 workers reported low back pain in the past seven days, claiming that this pain was

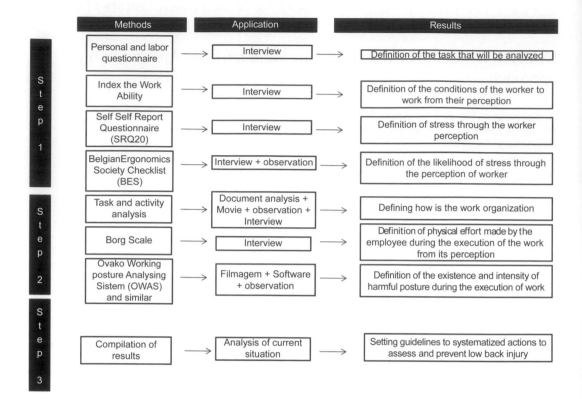

Methods	Application	Results

Step 1

Personal and labor questionnaire	Interview	Definition of the task that will be analyzed
Index the Work Ability	Interview	Definition of the conditions of the worker to work from their perception
Self Self Report Questionnaire (SRQ20)	Interview	Definition of stress through the worker perception
BelgianErgonomics Society Checklist (BES)	Interview + observation	Definition of the likelihood of stress through the perception of worker

Step 2

Task and activity analysis	Document analysis + Movie + observation + Interview	Defining how is the work organization
Borg Scale	Interview	Definition of physical effort made by the employee during the execution of the work from its perception
Ovako Working posture Analysing Sistem (OWAS) and similar	Filmagem + Software + observation	Definition of the existence and intensity of harmful posture during the execution of work

Step 3

| Compilation of results | Analysis of current situation | Setting guidelines to systematized actions to assess and prevent low back injury |

Figure 1.   Methodological approach.

linked to work conditions, and especially the handling activities with patients. Also, 2 workers perceived stronger effort in the low back during their work execution.

From the questionnaire of labor and individual characteristics application, the three workers answered the Work Ability Index (WAI), with the following results: 2 workers with great working capacity and functional working and 1 with good functional capacities. After, we applied the Self Report Questionnaire (SRQ20), but its application did not show decline of vital energy and depressive thoughts symptoms. For symptoms of depressed humor/ anxiety, 2 of respondents said they have been feeling sad lately, and 1 says she was nervous and worried easy. Somatic symptoms group had 50% positive responses to the question "sleep badly?" on the other hand "stomachache?" and "hand tremors?" showed 25% of positive responses

During the work observation, the Belgian Ergonomics Society checklist (BES) was applied. Its objective is to present the existence of unfavorable ergonomic conditions that may be related to occupational low back pain (Baucke, 2008). This consists on observational methodology and risk

identification to low back pain divided into four categories: driving transport vehicles, frequent displacements, unfavorable postures and charge maintenance. However, this study was applied only to the last three categories. The ap-plication showed the existence of unfavorable factors in the displacement mainly in the design of ramps during the change in level. There are two wooden ramps resting on stairs to transport wheelchair patients. In this case, due to its instability, the nursing techniques need flex the spine, keeping this region in static contraction during level change with patients.

For the maintenance of adopted postures, there is existence of unfavorable factors was observed in the activity of handling the patient in bed. In this activity, the nurses remain standing and unstable for a long time, when they need to sustain the patient, with the trunk flexed or sometimes combined with rotation and flexion. Some factors are aggravating the difficulty postural as poor access, lack of aid tools and work relationship and the need of furniture suitability that does not have mechanisms adapted to the workers psychophysical characteristics. Another aggravating factor is connected to overhead work and the inability to rest periods because patients are handled and removed by one hour uninterrupted, these features are related to the work organization, so contributing to the presence of low back pain risk.

The action of removing patient from the bed is characterized the worst workload, in this case there is an unfavorable factor for lifting conditions (patients are not compact, unstable and not easy to hold). Besides, the manual transport conditions (heavier than 25 kg and existence of path with level difference) and conditions to push and pull (no handle and the existence of physical obstacles and visual field). These situations can establish the presence of significant risk to manual handling load (patient).

3.2 *Step 2: Task analysis*

The target system consists on a geriatric clinic that aims to provide multidisciplinary care by promoting quality healthcare to their patients who are hospitalized and dependent on specific care.

Eighteen patients were admitted to the clinic, 9 were classified as level II, or need care that will increase the workload of the employees. The classification of patients' dependence follows the definition according to Resolution 283/05 ANVISA (2005) which is defined as grade I: independent, even requiring the use of self-help equipment, grade II: dependency within three self-care activity for life daily, such as: power, mobility, hygiene, without cognitive impairment and grade III dependency requiring assistance in all self-care activities for daily living and/or cognitively impaired.

Everyone should be removed from bed and cleaned at 8am o'clock by two professionals and stay in bed from 14 h to 15 h. Workers complain

about the inability to distribute this activity during the shift by creating a routine for patients. The discrepancies between task and activity are described in Table 1. During the execution of nursing technical work they were asked about the efforts made during their activities. In this case, it was reported the following information: 1 worker alleged effort equivalent to 2 on the Borg scale. The others, 2 and 3, rated their efforts as 6 and 5 respectively.

The REBA application, used to quantify the risk level, together with task analysis, indicated a high level of risk, so improvements actions are needed soon. The REBA was applied to the task of handling and hygiene of the patient in bed, because according to their own perception is the most painful task.

3.3 *Step 3: Compilation of results*

The problems existing in each task can be categorized (Moraes & Mont'Alvão, 2000). In task called Hygiene, with discrepancies between Task and Activity is possible to identify seven categories of problems: 1-interface (harmful postures resulting from inadequacies of sight/outlet information, wrap action/ranges, the positioning of components communication with damage to the muscular and skeletal systems); 2-Movement (overweight, travel distance of the load, frequency of lift movement or carry objects, disregard the recommended limits of manual handling of materials, with risks to the muscular and skeletal systems); 3-accessibility (poor accessibility, inadequate space for movement of wheelchairs, lack of support for use of equipment) 4-Biological (lack of hygiene and asepsis, allowing the proliferation of pathogens (bacteria

Table 1. Task versus activity.

Task	Task prescription	Activity	Task versus activity
Hygiene	– Put the patients' bedding and bath in own basket for washing. – Put dirty clothes with secretion in separate basket – Tie and put the used disposable diapers in locations indicated – It is mandatory to use disposable gloves for bathing and hygiene of patients	– Workers use disposable gloves and this is your only individual protection equipment, they put the used disposable diapers on the floor, only after completion of the patient's care, they discards disposable diapers in appropriate baskets	– Workers use disposable gloves, but does not do the discard of the clothes and disposable diapers used in own basket.
Withdrawal of bed	– Change the patients positions bedridden 2 after two hours	– All patients are withdrawn from the bed at predetermined time and this must be doing in the time of one hour. This work is done in pairs of nursing techniques.	– The exchanges of decubitus are not made every two hours as indicated in the prescription and yes at times linked to the feeding of patients.

and viruses), fungi and other microorganisms; 5-Operating (intense pace, repetitiveness and monotony); 6-Management (lack of participatory management, ignoring opinions and suggestions from employees centralization of decisions; lack of transparency in communication of decisions, lack of priorities and strategies policy positions and salaries coherent); 7-Instructional (7-disregard of the specific activities of the task). Task of Withdrawal of the bed provides the categories 1-Interface -2-Movement 3-Accessibility; 4-Operating, 5-Management (Table 1).

By analyzing results collected in the methods contemplated, it was possible to understand the dynamics of the work as a whole. The compilation of the results demonstrates the problematic work management of the nursing staff. The management presents itself inefficient to have two chiefs asking different tasks, according to the workers reports.

Another situation observed after comparing the data collected was the difficulty of work performing ac-cording to their description. This is due to the lack of task structure and the workplace, ie the layout is not organized for the specific job. Besides, the difficulties of movement in this environment increase the runtime of tasks, subjecting workers to unfavorable postures during work performance due to hard accesses. These were considered an important data or indicators that should be used for planning and maintaining a program for managing health and safety at work in clinics or similar organizations.

4 DISCUSSION

The aim of this paper was to assess the unfavorable work conditions related to low back injuries among nursing professionals. The methods were able to build an overview of the work taking into account the individual factors and the productive system. In this case, different from other evaluation methods, analyzing the characteristics of the existing work organization and the risk of job stress, the worker's individual characteristics and biomechanical risk. The system allowed the generation of organizational and structural indicators of company (macro and micro) to assist in the management of occupational health and safety within clinics or similar.

Its structure divided into three steps, as shown in figure 1, was based on interviews and observation of work during implementation and comparison of these data with the documents analysis has highlighted discrepancies between the described and actual work situation. These differences are indicators that demonstrate the lack of proper training, high team members turnover, and poor leadership reported in this case study.

The application was based on one evaluation methods of the work as a whole, supporting the ideas of Aptel (2003), which advocates understanding the dynamics of work organization and how it is expressed to understand musculoskeletal problems. This procedure is different of other evaluation methods presented in the literature. Thus, methods that constitute this study were selected from a literature review. They present the advantage in evaluating unfavorable factors that may cause work overload (physical and/or mental) perceived by the assessor or by the worker himself.

5 CONCLUSION

The methods used in this paper identified and evaluated work problems related to low back pain. The assessment approach allowed treat problems of the structure inherent to the specific problem and related in literature to musculoskeletal problems in nursing.

More studies should be conducted to increase the understanding of the interaction between the unfavorable work situations in order to better appreciate the existing problems in overload conditions as a whole on a basis of work organization. In addition, other methods can be used to evaluate in an interactive way specific factors of the nursing work in all its dimensions.

REFERENCES

Alexandre, N.M.C; Angerami, E.L.S; Moreira Filho, D.C. Dores nas costas e enfermagem. Rev. Esc. Enf.USP. Agosto 1996; v.30(2): 267–85.

Anvisa. Regulamento técnico de funcionamento para instituições de longa permanência para idosos. RDC 283/05, 2005.

Aptel, M. Étude dans une entreprise de montage d'appareils électroménagers des facteurs de risques professionnels du syndrome du canal carpien. INRS, 1993.

Baucke, O.J.S. Sistemática preventiva e participativa para avaliação ergonômica de quadros lombálgicos: o caso de uma indústria fabricante de dormitórios e cozinhas em MDF. Escola de Engenharia. Universidade Federal do Rio Grande do Sul [dissertação]. Porto Alegre, 2008.

Cavassa, C.R. Ergonomia y productividad. Limsa Noriega. 1997: 415.

Cunha, L.S; Souza N.V.D.O. Adaptações e improvisações no trabalho hospitalar e suas implicações na saúde do trabalhador de enfermagem. 3º Simpósio Nacional—Enfcuidar—O cuidar em saúde e enfermagem (Saberes e práticas de cuidar em enfermagem). Rio de Janeiro, 2008.

Hoogendoorn, W.E et al. High physical work load and low job satisfaction increase the risk of sickness absence due to low back pain: results of a prospective cohort study. OccupEnvironMed; 2002.

Lisboa, M.T.L; Salvador, R.S.P; Valente, G.L; Peres, M.A.A, Alvin, N.AT. Hábitos de vida e de saúde de trabalhadores – uma extensão do cuidado. 3° Simpósio Nacional—Enfcuidar—O cuidar em saúde e enfermagem (Saberes e práticas de cuidar em enfermagem). Rio de Janeiro,2008.

Magora, A. Investigation of the relation between low back pain and occupation.Ind.Med. 1970; v.39 (11): 31–37.

Malchaire, J. Évaluation et prévention des risques lombaires: Classification des méthodes. Médecine du Travail e Ergonomie. Volume XXXVIII. N°2, 2001.

Marziale, M.H.P; Rodrigues, C.M. A Produção científica sobre os acidentes de trabalho com material perfurocortantes entre trabalhadores de enfermagem. Rev Latino-americana de Enfermagem. Julho-agosto, 2002.

Moraes, A; Mont'Alvão, C. Ergonomia: conceitos e aplicações. 2 AB editor, 2ª edição, Rio de Janeiro, 2000.

Rocha, A.M. Fatores ergonômicos e traumáticos envolvidos com a ocorrência de dor nas costas em trabalhadores de enfermagem. (Dissertação de Mestrado). Belo Horizonte, Minas Gerais: Universidade Federal de Minas Gerais; 1997: 156.

Rocha, A.M; Oliveira, A.G.C. Estudo da dor nas costas em trabalhadores de enfermagem de um hospital universitário de Belo Horizonte. Revista Mineira de Enfermagem, 1998.

Rosa, L.R. Fatores Intervenientes no trabalho Coletivo dos profissionais de Enfermagem: Uma proposta de Ação. [Dissertação]. Porto Alegre: Universidade Federal do Rio Grande do Sul—Escola de Engenharia; 2010.

Simão, S.A.F et al. Acidentes de trabalho com perfurocortantes envolvendo profissionais de enfermagem de um hospital público do Rio de Janeiro.3° Simpósio Nacional—Enfcuidar—O cuidar em saúde e enfermagem (Saberes e práticas de cuidar em enfermagem). Rio de Janeiro,2008.

Smedley, J.P; Egger, C; Cooper; Coggon, D. Manual handling activities and risk of low back pain in nurses. Occupational and Environmental Medicine 1995; 52: 160–163.

Smedley, J. Prospective cohort study of predictors of incident low back pain in nurses.BMJ.April 1997; (314).

Yin, R.K. Case Study Research—Design and Methods. Sage Publications, 2ª ed. London, 1994.

Are dental students at risk of developing occupational musculoskeletal disorders?

M.E. Pinho & M.A. Vaz
FEUP, Porto, Portugal

P.M. Arezes
U. Minho, Guimarães, Portugal

J. Reis Campos
FMDUP, Porto, Portugal

ABSTRACT: The prevalence of musculoskeletal symptoms has been found high among dental practitioners and symptoms seem to appear soon after a short period of clinical practice. This study aims to investigate the 12 months prevalence of musculoskeletal symptoms among a cohort of Portuguese dental students, and find out if there is any significant association with socio-demographic variables. Results showed that the prevalence of musculoskeletal symptoms is high. 91.6% of the respondents reported musculoskeletal symptoms in at least one body region during the preceding 12 months and the most affected body regions were lower back and neck (67.0%), upper back (56.8%), shoulders (41.0%) and wrists/hands (33.3%). However, few significant associations were found between reported symptoms and the socio-demographic variables. Understanding the causes and mechanisms for the onset and progression of musculoskeletal disorders is a fundamental step for the design of effective prevention strategies.

1 INTRODUCTION

The clinical practice of dental medicine has been indicated as involving an increased risk of musculoskeletal disorders for its practitioners (Akesson et al., 1999, de Carvalho et al., 2009, Winkel & Westgaard, 2008), namely due to cumulative trauma (Andrews & Vigoren, 2002, Newell & Kumar, 2005, Rice et al., 1996).

High prevalence rates of musculoskeletal symptoms, such as pain, have been found among dental students (Al Rayes, 2012, Fahim, 2011, Rising et al., 2005, Smith et al., 2009, Thornton et al., 2008), indicating that symptoms may appear soon after a short period of clinical practice (Melis et al., 2004). They have also been pointed out at higher risk for developing musculoskeletal disorders than students of other disciplines (e.g. Psychology) (Melis et al., 2004). The situation among dental students is quite similar to that of dentists either in the prevalence of symptoms (Finsen et al., 1998, Marshall et al., 1997, Morse et al., 2010) or in the increased risk of musculoskeletal disorders (Akesson et al., 1999, de Carvalho et al., 2009, Winkel & Westgaard, 2008), particularly in neck, shoulder and lower back (Hagberg & Hagberg (1989) and van Doorn (1995), as cited by Rising et al., 2005).

The present study aims to investigate the 12 months prevalence of musculoskeletal symptoms among a cohort of Portuguese dental students, and find out if there is any significant association with socio-demographic variables.

2 MATERIAL AND METHODS

2.1 Subjects

One hundred and twenty six dental students (Table 1), enrolled in 1st, 4th and 5th class years at the Faculty of Dental Medicine of University of Porto, volunteered to fulfill the questionnaire.

Table 1.   The distribution [n (%)] of dental students participating in the survey.

Class year	Gender		Total
	Female	Male	
1st year	23	11	34 (27.0%)
4th year	20	8	28 (22.2%)
5th year	50	14	64 (50.8%)
Total	93 (73.8%)	33 (26.2%)	126 (100.0%)

The study was submitted to and approved by the Commission of Ethics of the Faculty of Dental Medicine of University of Porto.

2.2 *Instrument and administration procedure*

A self-administered questionnaire survey was carried out in April–May 2012. The questionnaire was based on the Standardized Nordic Questionnaire (SNQ) (Kuorinka et al., 1987). Socio-demographic information, concerning gender, age, height, weight, dominant hand, class year, clinical and/or pre-clinical practice hours per week, regular physical exercise practice, and physical exercise practice hours per week, was also asked for.

A pre-test was performed, in March 2012, using the e-mail to distribute the questionnaires, and the opinion of some Professors of the Faculty of Dental Medicine was asked for. In view of that, some adjustments were made. Afterwards, Professors distributed the questionnaires by students attending 1st, 4th and 5th year classes.

2.3 *Data analysis*

Microsoft Office Excel 2007 and IBM SPSS Statistics 20 were used in data analysis. Contingency Tables and Chi-Square Tests were performed to analyze the association between the prevalence of symptoms and the socio-demographic variables. Pearson Chi-Square (χ^2) and, whenever the cells of contingency tables had expected count less than 5, Fisher's Exact Test were used. For analysis purpose, continuous socio-demographic variables were dichotomized as follows: age (≤22 years old and >22), height (<170 cm and ≥170 cm), BMI (<25 and ≥25), clinical and/or pre-clinical practice hours per week (≤5 and >5) and physical exercise practice hours per week (≤2 and >2). The significance level was set at $\alpha = 0.05$ and, since the direction of the association was unknown in advance, two-tailed tests were used. Additionally, a bivariate analysis was performed and Spearman rank correlation coefficients (r_s) were computed to estimate the magnitude and direction of the association between variables. For the statistically significant associations found, and whenever possible, relative risk (RR), and 95% confidence intervals (95% CI) were computed to estimate the effect size.

3 RESULTS AND DISCUSSION

An overall response rate of 46.8% was achieved. Survey participants (126) were mostly female (73.8%), right-handed (93.7%) and a little less than half (46.8%) of them reported to practice regular physical exercise. Other demographic characteristics (age, height and BMI) are presented in Table 2.

91.6% (75.2% females and 24.8% males) of the respondents reported musculoskeletal symptoms in at least one body region during the preceding 12 months, although no statistically significant association has been found with any of the socio-demographic variables. The most reported symptoms in the previous 12 months, affected equally lower back and neck (67.0%), upper back (56.8%), shoulders (41.0%) and wrists/hands (33.3%), as shown in Table 3.

A higher prevalence of musculoskeletal symptoms, in the preceding 12 months, was found in female students for all body regions (Table 3), despite no statistically significant differences have been found between genders, except for upper back $\chi^2(1) = 6.782$, $p = 0.013$ and hips/thighs ($p = 0.036$).

Table 2. Participants' demographic characteristics.

Variable parameters	Gender		Total
	Female	Male	
Age (in years) (n = 121)			
Mean	22.4	22.1	22.3
SD*	3.3	2.4	3.1
Range	18–45	18–30	18–45
Height (in cm) (n = 126)			
Mean	165.3	177.5	168.2
SD*	6.3	6.2	8.2
Range	150–178	167–191	150–191
BMI (in Kg/m²) (n = 126)			
Mean	21.3	23.4	21.9
SD*	2.7	2.2	2.7
Range	15.1–29.2	19.0–28.9	15.2–29.2

* SD = standard deviation.
BMI = Body Mass Index.

Table 3. Self-reported musculoskeletal symptoms in preceding 12 months, by gender and body region.

Body region	N	Females		Males		Total	
		Yes	No	Yes	No	Yes	No
Neck	103	56	22	13	12	67.0%	33.0%
Shoulders	105	34	41	9	21	41.0%	59.0%
Elbows	111	5	80	1	25	5.4%	94.6%
Wrists/Hands	111	32	53	5	21	33.3%	66.7%
Upper Back*	111	54	31	9	17	56.8%	43.2%
Lower Back	103	52	21	17	13	67.0%	33.0%
Hips/Thighs*	111	13	72	0	26	11.7%	88.3%
Knees	111	16	69	2	24	16.2%	83.8%
Ankles/Feet	111	9	76	1	25	9.0%	91.0%

* Statistically significant at p < 0.05.

10

Female students reported symptoms in these body regions significantly more than their male colleagues (r_s = –0.247, p < 0.05 and r_s = –0.201, p < 0.05, respectively). The estimated risk for females students reporting upper back symptoms in preceding 12 months (RR: 1.835; 95% CI: 1.056–3.188) is almost twofold when compared with their male colleagues. These results partly support the findings of Rising et al., (2005) who did not find any statistically significant association between reported symptoms and gender.

The prevalence of symptoms in the preceding 12 months affecting at least one body region is very high (91.6%) and is quite similar to the results of other studies carried out among either dental students (Table 4) (Al Rayes, 2012, Smith et al., 2009) or dentists (Table 5) (Leggat & Smith, 2006, Lin et al., 2012), despite lower prevalence rates have been found either in students (61.0%) (Thornton et al., 2008) or in dentists (62.0%) (Alexopoulos et al., 2004). Similarities were also found in the 12 months prevalence of musculoskeletal symptoms by body region, as shown in Table 4, and the results are not very different from those found among dentists (Table 5).

A Swedish study (Rolander et al., 2005) found that body height and lower back pain are significantly correlated. Despite the fact that it is easily understandable that the "Taller individuals probably have to work with more neck flexion to compensate for their body height" (Rolander et al., 2005) and, therefore, more likely could be that taller students reported more musculoskeletal symptoms, in the present study, significant differences were only found between students' height and hips/thighs symptoms ($\chi^2(1)$ = 4.939, p = 0.036) reported in the preceding 12 months.

Table 4. Prevalence (%) of musculoskeletal symptoms among dental students, in the preceding 12 months.

Body region	Study			
	(1)	(2)	(3)	(4)
Any region	91.6	92.7	–	83.9
Neck	67.0	72.7	64.3	60.7
Shoulders	41.0	64.9	48.4	44.6
Elbows	5.4	12.3	5.6	12.5
Wrists/hands	33.3	25.6	42.1	37.5
Upper back	56.8	57.4	41.2	37.5
Lower back	67.0	68.0	57.9	46.4
Hips/thighs	11.7	14.3	11.9	16.1
Knees	16.2	14.8	26.2	28.6
Ankles/feet	9.0	20.4	12.7	17.9

(1) Current study.
(2) (Al Rayes, 2012).
(3) (Fahim, 2011).
(4) (Smith et al., 2009).

Table 5. Prevalence (%) of musculoskeletal symptoms among dentists, in the preceding 12 months.

Body region	Study			
	(1)	(2)	(3)	(4)
Any region	87.2	92.4	–	–
Neck	57.5	71.6	63.0	38.3
Shoulders	53.3	75.1	49.0	25.0
Elbows	13.0	27.4	17.0	6.7
Wrists/hands	33.7	41.1	42.0	6.7
Upper back	34.4	45.2	32.0	20.0
Lower back	53.7	66.5	63.0	55.0
Hips/thighs	12.6	14.7	21.0	5.0
Knees	18.9	13.2	22.0	10.0
Ankles/feet	11.6	13.2	18.0	5.0

(1) (Leggat & Smith, 2006).
(2) (Lin et al., 2012).
(3) (Palliser et al., 2005).
(4) (Ratzon et al., 2000).

Table 6. Prevalence (%) of musculoskeletal symptoms in the preceding 12 months, by students' class year and body region.

Body region	1st year	4th year	5th year
At least one	90.0	96.2	95.0
Neck	60.0	76.2	66.7
Shoulders	22.2	56.0	43.4
Elbows	10.7	3.8	3.5
Wrists/hands	28.6	34.6	35.1
Upper back	46.4	57.7	61.4
Lower back	73.9	70.8	62.5
Hips/thighs	7.1	15.4	12.3
Knees	21.4	15.4	14.0
Ankles/feet	10.7	3.8	10.5

Students below 170 cm height reported more musculoskeletal symptoms in that body region (r_s = –0.211, p < 0.05) and their estimated risk of hips/thighs symptoms (RR: 4.347; 95% CI: 1.010–18.704) is more than four times that of their taller colleagues. As only female students reported hips/thighs symptoms, and the prevalence rate increases from 1st (7.1%) to 4th class year (15.4%) (Table 6), a possible explanation to be investigated is their use of an inappropriate adjustment of the seat height and the biomechanical load due to the foot pressure exerted over the pedal. A 5-year follow-up study of symptoms carried out among female dental personnel performed by Akesson et al., (1999) may support this last hypothesis once they found higher 12 months prevalence rates of hips symptoms for dentists (23%) and dental hygienists (23%) than for dental assistants (15%).

11

The excess of body weight has been indicated as causing increased stress not only on bones and joints but also on connective-tissue structures (tendons and ligaments), thus increasing the risk of osteoarthritis and the potential for musculoskeletal injury, respectively (Wearing et al., 2006). However, in the present study, BMI was only found to be associated to the reported elbows symptoms (p = 0.002). Overweight students reported significantly more symptoms at elbows ($r_s = 0.409$, p < 0.01) and are over 15 times more likely to have elbows symptoms (RR: 15.077; 95% CI; 3.057–74.359) than their colleagues with BMI lower than 25. An increased risk of elbow symptoms was found in dental professionals (Akesson et al., 1999) which may be explained by the prolonged periods working with elevated arms (Akesson et al., 2000) and the biomechanical load associated (Finsen & Christensen, 1998). Working with elevated arms for over 10 hours per week is considered as "heavy work" (Schneider et al., 2010) and working with elevated arms over an average of 2 hours per day is common even for dental students from the 4th class year on. However, the highest prevalence rate of elbows symptoms was found in 1st year students (Table 6), raising the question that other causes, not related to the dental work, are probably in the origin of their symptoms. This may also be the case concerning the highest prevalence of upper back, knees and ankles/feet symptoms in 1st year students, as shown in Table 6.

Several studies have shown significant differences between class years in the prevalence of reported symptoms in the preceding 12 months (Hodson et al., 2008, Rising et al., 2005, Thornton et al., 2008). While in Hodson (2008), in general, the higher the class level the higher the prevalence of symptoms, others found an increasing trend from 1st trough the 3rd year and a reduction in the 4th (last year) (Rising et al., 2005, Thornton et al., 2008). Contrary to Thornton et al. (2008) who found that 3rd year students had the highest prevalence of symptoms in all body regions, the present study does not show a consistent pattern (Table 6). However, significant differences were found in shoulders symptoms reported by students of different class years ($\chi^2(2) = 6.389$, p = 0.041). The percentage of dental students in each class year reporting shoulders symptoms was higher in 4th year students (56.0%), followed by 5th year students (43.4%) and, then, 1st year students (22.2%). However, significant differences were only found in reported shoulders symptoms between 1st and 4th year students ($\chi^2(1) = 6.257$, p = 0.022). 4th year students reported significantly more symptoms ($r_s = 0.347$, p < 0.05) and the estimated risk of symptoms is more than twofold

of that of their 1st year colleagues (RR: 2.520; 95% CI: 1.148–5.534). These results seem to be more similar to those of Rising et al., (2005) that, although had found that the percentage of dental students reporting symptoms had increased with the number of years in dental school, did not find any statistically significant difference based on years in dental school.

Clinical and/or pre-clinical practice hours per week were only found associated with shoulders symptoms in the preceding 12 months ($\chi^2(1) = 4.666$, p = 0.038). Students practicing more than 5 hours per week reported more shoulders symptoms ($r_s = 0.216$, p < 0.05) and have an estimated risk (RR: 2.050; 95% CI; 0.976–4.304) of twice that of their colleagues practicing 5 or less hours per week. Not very differently, an Iranian study shows that hours of practice per day and years of work were not found associated to symptoms reported by dentists (Chamani et al., 2012), despite this may arise from the "healthy worker effect".

No significant differences in reported symptoms were found, concerning students' age, dominant hand, regular physical exercise practice, and physical exercise practice hours per week.

The study has several limitations. Apart from the bias inherent to the subjective method (self-administered questionnaire) used in the data collection process (Alexopoulos et al., 2006), some of the main limitations of the study are its cross-sectional design and the low response rate (46.8%) (Blatter & Bongers, 2002). One of the most likely causes for this may have been the voluntary participation. Also the method selected for the distribution of the questionnaire (administration procedure) may have affected the response rate, since the students who were not attending the classes, in which questionnaire was distributed, were not given the opportunity to participate. However, conversely, this method granted a higher response rate than that of the pre-test. Therefore, the reproducibility of the results must be interpreted carefully.

4 CONCLUSIONS

The prevalence of musculoskeletal symptoms seems to be high among Portuguese dental students and quite similar to the results of other studies among either dental students or dentists. However, only a few significant associations were found between the reported symptoms and the socio-demographic variables. On the other hand, unexpectedly, the prevalence of symptoms was found high among the 1st year students or even the highest among the three years under study. This seems to mean that musculoskeletal symptoms in dental students rather than being occupation-related may have other important underlying causes. Further research is needed in order to allow understanding the causes and mechanisms leading to the onset and progression of musculoskeletal disorders among dental practitioners as well as finding out more effective prevention strategies in addition to continuous education and training.

ACKNOWLEDGMENTS

The authors express their sincere gratitude to all those who contributed for the questionnaire improvement and to FMDUP Professors Helena Figueiral and João Paulo Dias for their valuable participation in data collection process. Utmost recognition is also due to the dental students participating in the pre-test and/or in the survey for their fundamental contribution.

REFERENCES

Akesson, I., Johnsson, B., Rylander, L., Moritz, U., & Skerfving, S. 1999. Musculoskeletal disorders among female dental personnel-clinical examination and a 5-year follow-up study of symptoms. *Int Arch Occup Environ Health* 72(6): 395–403.

Akesson, I., Schutz, A., Horstmann, V., Skerfving, S., & Moritz, U. 2000. Musculoskeletal symptoms among dental personnel; lack of association with mercury and selenium status, overweight and smoking. *Swedish Dental Journal* 24(1–2): 23–38.

Al Rayes, F. 2012. *The prevalence of musculoskeletal disorders (MSD) among dental students, general dental practitioners and dental specialists in Kuwait.* Unpublished M.S. dissertation. Tufts University School of Dental Medicine. Massachusetts, United States.

Alexopoulos, E., Stathi, I.-C., & Charizani, F. 2004. Prevalence of musculoskeletal disorders in dentists. *BMC Musculoskeletal Disorders* 5(1): 16.

Alexopoulos, E., Tanagra, D., Konstantinou, E., & Burdorf, A. 2006. Musculoskeletal disorders in shipyard industry: prevalence, health care use, and absenteeism. *BMC Musculoskeletal Disorders* 7: 88.

Andrews, N., & Vigoren, G. 2002. Ergonomics: muscle fatigue, posture, magnification, and illumination. *Compend Contin Educ Dent* 23(3): 261–266, 268, 270 passim; quiz 274.

Blatter, B.M., & Bongers, P.M. 2002. Duration of computer use and mouse use in relation to musculoskeletal disorders of neck or upper limb. *International Journal of Industrial Ergonomics* 30(4–5): 295–306.

Chamani, G., Zarei, M.R., Momenzadeh, A., Safizadeh, H., Rad, M., & Alahyari, A. 2012. Prevalence of Musculoskeletal Disorders among Dentists in Kerman, Iran. *Journal of Musculoskeletal Pain* 20(3): 202–207.

de Carvalho, M.V.D., Soriano, E.P., Caldas, A.D., Campello, R.I.C., de Miranda, H.F., & Cavalcanti, F.I.D. 2009. Work-Related Musculoskeletal Disorders Among Brazilian Dental Students. *Journal of Dental Education* 73(5): 624–630.

Fahim, A.E. 2011. Factors Affecting Musculoskeletal Disorders among Final Year Dental Students in Ismailia. *Egyptian Journal of Community Medicine* 29(1): 49–58.

Finsen, L., & Christensen, H. 1998. A biomechanical study of occupational loads in the shoulder and elbow in dentistry. *Clinical Biomechanics* 13(4–5): 272–279.

Finsen, L., Christensen, H., & Bakke, M. 1998. Musculoskeletal disorders among dentists and variation in dental work. *Applied Ergonomics* 29(2): 119–125.

Hodson, N.A., Pankhurst, C., & Linden, R.W.A. 2008. Incidence of musculoskeletal problems in different years of dental students. [Abstract]. *Journal of Dental Research* (87 C).

Kuorinka, I., Jonsson, B., Kilbom, A., Vinterberg, H., Biering Sørensen, F., Andersson, G., et al. 1987. Standardised Nordic questionnaires for the analysis of musculoskeletal symptoms. *Applied Ergonomics* 18(3): 233–237.

Leggat, P.A., & Smith, D.R. 2006. Musculoskeletal disorders self-reported by dentists in Queensland, Australia. *Australian Dental Journal* 51(4): 324–327.

Lin, T.-H., Liu, Y.C., Hsieh, T.-Y., Hsiao, F.-Y., Lai, Y.-C., & Chang, C.-S. 2012. Prevalence of and risk factors for musculoskeletal complaints among Taiwanese dentists. *Journal of Dental Sciences* 7(1): 65–71.

Marshall, E.D., Duncombe, L.M., Robinson, R.Q., & Kilbreath, S.L. 1997. Musculoskeletal symptoms in New South Wales dentists. *Australian Dental Journal* 42(4): 240–246.

Melis, M., Abou-Atme, Y.S., Cottogno, L., & Pittau, R. 2004. Upper body musculoskeletal symptoms in Sardinian dental students. *Journal of the Canadian Dental Association* 70(5): 306–310.

Morse, T., Bruneau, H., & Dussetschleger, J. 2010. Musculoskeletal disorders of the neck and shoulder in the dental professions. *Work: A Journal of Prevention, Assessment & Rehabilitation* 35(4): 419–429.

Newell, T.M., & Kumar, S. 2005. Comparison of instantaneous and cumulative loads on the low back and neck in orthodontists. *Clinical Biomechanics* 20(2): 130–137.

Palliser, C.R., Firth, H.M., Feyer, A.M., & Paulin, S.M. 2005. Musculoskeletal discomfort and work-related stress in New Zealand dentists. *Work and Stress* 19(4): 351–359.

Ratzon, N.Z., Yaros, T., Mizlik, A., & Kanner, T. 2000. Musculoskeletal symptoms among dentists in relation to work posture. *Work: A Journal of Prevention, Assessment & Rehabilitation* 15(3): 153–158.

Rice, V.J., Nindl, B., & Pentikis, J.S. 1996. Dental Workers, Musculoskeletal Cumulative Trauma, and Carpal Tunnel Syndrome, Who is at Risk? A Pilot Study. *International journal of occupational safety and ergonomics: JOSE* 2(3): 218–233.

13

Rising, D.W., Bennett, B.C., Hursh, K., & Plesh, O. 2005. Reports of body pain in a dental student population. *Journal of American Dental Association 136*(1): 81–86.

Rolander, B., Karsznia, A., Jonker, D., Oberg, T., & Bellner, A.-L. 2005. Perceived contra observed physical work load in Swedish dentists. *Work: A Journal of Prevention, Assessment & Rehabilitation 25*(3): 253–262.

Schneider, E., Irastorza, X., & European Agency for Safety and Health at Work. 2010. *OSH in figures: Work-related musculoskeletal disorders in the EU—Facts and figures.* Luxembourg: Publications Office of the European Union.

Smith, D.R., Leggat, P.A., & Walsh, L.J. 2009. Workplace Hazards Among Australian Dental Students. *Australian Dental Journal 54*(2): 186–188.

Thornton, L.J., Barr, A.E., Stuart-Buttle, C., Gaughan, J.P., Wilson, E.R., Jackson, A.D., et al. 2008. Perceived musculoskeletal symptoms among dental students in the clinic work environment. *Ergonomics 51*(4): 573–586.

Wearing, S.C., Hennig, E.M., Byrne, N.M., Steele, J.R., & Hills, A.P. 2006. Musculoskeletal disorders associated with obesity: a biomechanical perspective. *Obesity Reviews 7*(3): 239–250.

Winkel, J., & Westgaard, R.H. 2008. Risk factors of occupational MSDs and potential solutions: past, present and future. *HESA Newsletter 34.*

Occupational Safety and Hygiene – Arezes et al. (eds)
© 2013 Taylor & Francis Group, London, ISBN 978-1-138-00047-6

Popliteal height as a measure for classroom furniture selection: An exploratory analysis

H.I. Castellucci & M. Catalán
Universidad de Valparaíso, Chile

C. Viviani
Universidad Técnico Federico Santa María, Chile

J. Rojas
Pontificia Universidad Católica, Chile

P.M. Arezes
Universidade do Minho, Portugal

ABSTRACT: The aim of this study is to determine if popliteal height can be used as a better measure for classroom furniture selection than stature. The sample study consisted of 2068 volunteer subjects from the Valparaiso Region. Regarding the methodology, eight anthropometric measures were gathered, as well as six furniture dimensions from the Chilean standard. For the evaluation of classroom furniture a match criterion equation was defined. A match criterion equation were applied for both cases using both popliteal height and stature. Results show that using popliteal height presents a higher number of match in the seat height dimension and lower levels of match in other three furniture dimensions. Also in the transversal match, popliteal height presents a higher match level than stature. In conclusion, it is possible that popliteal height can be the most accurate anthropometric measure for classroom furniture selection.

1 INTRODUCTION

School work requires students to spend long hours sitting down, presenting a high risk of developing permanent postural changes. This situation causes an increased concern about the school classrooms, in particular about the study and design of school furniture suitable to the needs of the students and with appropriate dimensions according to the students' anthropometrics characteristics. In Chile, an important milestone in this increasing concern was the publication of the Chilean standard N°. 2566 (Instituto Nacional de Normalización, 2002), which determines the dimensions and characteristics of five different types of school furniture for the whole Chilean population.

It is important to mention that the Chilean standard N°. 2566, like most of the standards worldwide used for furniture selection, tend to use, as reference, the anthropometric dimension 'Stature' or 'Height' of the school children, assuming that all the other anthropometric characteristics will be also appropriate. However, other authors, such as Molenbroek et al. (2003), suggest that the furniture selection can be done more efficiently is the popliteal height is used instead of stature. The aim of this study is to determine if popliteal height can be used as a better measure for classroom furniture selection than stature.

2 MATERIALS AND METHOD

2.1 Sample

This cross-sectional study involved a representative group of students, with ages ranging from 5 to 19 years old (11.8 ± 3.56 mean), from basic and secondary schools in the Valparaiso Region, in Chile. Ten schools were randomly selected from a list given by the Regional Ministerial Secretary of Education and the selection used a cluster design regarding the three types of elementary school administrations in Chile, as well as of the economical background level of the corresponding students.

The sample study consisted of 2068 volunteer subjects (1153 male and 915 female). The study started after the written information about the study from the headmaster of the school, which were also followed by the collection of the written authorizations obtained from all parents and students.

2.2 Anthropometric measures

The anthropometric measurement were collected from the right side of the subjects while they were sitting in an erect position on a height-adjustable chair with a horizontal surface, with their legs flexed at a 90° angle, and with their feet flat on an adjustable footrest. During the measurement process, the subjects were without shoes, and wearing shorts and T-shirts.

All measurements were taken with a portable anthropometer (Holtain), exception made to subjects' stature, which was measure with an estadiometer.

The following anthropometric measures (ISO 7250., 1996) were considered and collected during this study:

Shoulder Height Sitting: vertical distance from subject's seated surface to the acromion.

Elbow Height Sitting: taken with a 90° angle elbow flexion, as the vertical distance from the bottom of the tip of the elbow (olecranon) to the subject's seated surface.

Thigh Thickness: the vertical distance from the highest uncompressed point of thigh to the subject's seated surface.

Buttock-Popliteal Length: taken with a 90° angle knee flexion as the horizontal distance from the posterior surface of the buttock to the popliteal surface.

Popliteal Height: measured with 90° knee flexion, as the vertical distance from the floor or footrest and the posterior surface of the knee (popliteal surface).

Subscapular Height: the vertical distance from the lowest point (inferior angle) of the scapula to the subject's seated surface.

Hip Width: the horizontal distance measured in the widest point of the hip in the sitting position.

Stature: determined as the vertical distance between the floor and the top of the head, and measured with the subject erect and looking straight ahead (Frankfort plane).

2.3 Furniture dimensions

The following furniture dimensions (with the corresponding description) were analysed:

Seat Height (SH): the vertical distance from the floor to the middle point of the front edge of the seat.

Seat Depth (SD): the distance from the back to the front of the sitting surface.

Seat Width (SW): the horizontal distance between the lateral edges of the seat.

Upper Edge of Backrest (UEB): the vertical distance between the middle points of the upper edge of the backrest and the top of the seat.

Seat to Desk Height (SDH): the vertical distance from the top of the front edge of the seat to the top of front edge of the desk (Seat Height +Seat to Desk Height = Desk height).

Seat to Desk Clearance: the vertical distance from the top of the front edge of the seat to the lowest structure point below the desk (Seat height + Seat to Desk clearance = Desk clearance).

2.4 Equation for mismatch criterion

Popliteal Height against Seat Height: evidence shows that PH should be higher than the SH but it does not have to be higher than four centimetres or 88% of the PH. This match criterion was determined using the criteria described by Gouvali and Boudolos (2006), but PH was modified according to a shoe height of 3 centimetres. Therefore, the match criterion was determined by the application of Equation (1):

$$(PH + 3) \cos 30° \leq SH \leq (PH + 3) \cos 5° \qquad (1)$$

Buttock-Popliteal Length against Seat Depth: to proper fit in the seat in order to be able to use the backrest of the seat to support the lumbar spine without compression of the popliteal surface, most researchers convey that SD should be designated for the fifth percentile of BPL distribution (Gutiérrez & Morgado, 2001; Helander, 1997; Khalil et al., 1993; Milanese & Grimmer, 2004; Occhipinti et al., 1993; Orborne, 1996; Pheasant, 1991). On the other hand, if the seat depth is considerably less than the BPL of the subjects, the thigh would not be supported enough. The match criterion was defined according to the equation (2) (Parcells et al., 1999):

$$0.80 \; BPL \leq SD \leq 0.95 \; BPL \qquad (2)$$

Hip Width against Seat Width: based on available evidence HW is the major criterion for SW, and is designated for the 95 percentile of HW distribution (Evans et al., 1988; Mondelo et al., 2000; Orborne, 1996; Sanders & McCormick., 1993). Also, Gouvali and Boudolos (2006), recommended that the seat width should be at least 10% (to accommodate hip breadth) and at the most 30% (for space economy) larger than the hip breadth. In this case the match criterion was:

$$HW \; 1.1 \leq SW \leq HW \; 1.3 \qquad (3)$$

Subscapular Height against Upper Edge of Backrest: UEB is considered appropriate when it is lower than SUH in order to permit trunk and arms movement. Also, is designated for the 5 percentile of SUH distribution or the smaller SUH (García-Acosta & Lange-Morales, 2007; Gutiérrez et al., 2001). As a result, the match criterion was one-way:

$$SUH \geq UEB \qquad (4)$$

Thigh Thickness against Seat to Desk Clearance: to be able to move the leg, the TT should be lower than the SDC in order to allow leg movement. SDC is considered appropriate when it is higher than TT (García-Acosta et al., 2007; Molenbroek et al., 2003). Castellucci et al. (2010) also proposed that the desk clearance should be 2 cm higher than TT. Considering these situations, the match criterion was defined according to the equation (5):

$$TT +2 < SDC \qquad (5)$$

Elbow Height Sitting against Seat to Desk Height: EHS is the major criterion for SDH (García-Acosta et al., 2007; Milanese et al., 2004; J.F.M. Molenbroek et al., 2003; Sanders et al., 1993). Parcells et al. (1999) also suggested that SDH depends on shoulder flexion and shoulder abduction angles. This match criterion was determined using the criteria described by Parcells:

$$EHS \leq SDH \leq 0.8517\ EHS + 0.1483\ HS \qquad (6)$$

2.5 Statistical analysis

All data were entered into Microsoft Office Excel 2007 and analysed using SPSS (v16.0). Categorical data were summarised using percentages and analysed using the chi-square test (cross table), with a 95% confidence interval, which was performed for testing the independence between the results of the equation for mismatch criterion (match/mismatch) and the two anthropometric measures used for furniture selection (stature/popliteal height).

3 RESULTS AND DISCUSSION

The following results presents the percentage of students who fit (or did not fit) using "Selection by Stature" (SBS) and "Selection by Popliteal height" (SBPH) as an indicator for classroom furniture selection.

Before the presentation of the results is important to mention that using SBPH 25 students will not fit in any of the furniture size. Also 5 students will be in the same situation using SBS.

Seat height should be considered as the starting point and the most important variable for the design of the classroom furniture (Molenbroek et al., 2003; Castellucci et al., 2010). From the

analysis of the data in Figure 1, it is possible to see when the furniture size is assigned by stature, 17% of students will find that the chair is too high and this percentage is only 1% of the students if we use SBPH. These students will not be able to support their feet in the floor, generating increase pressure on the tissue on the posterior surface of the knee. Also, 1% of the students will use a lower than needed chair if SBS is used. This result shows that there is dependence between seat height and the measure used for furniture selection ($X^2 = 384.16$; $p < 0.001$).

Figure 2 shows that too shallow seats for 29% and 23% of the students using SBPH and SBS respectively. This situation of mismatch produces that thighs would not be supported enough and would generate discomfort. In the other hand, using SBPH 3% of the students will use a too deep seat. A similar situation occurs with SBS. In this case, and to avoid the compression on the posterior surface of the knee, the students will place their buttocks forward on the edge of the seat (Panagiotopoulou et al., 2004), causing kyphotic postures. The difference between the two measures is statistically significant ($X^2 = 16.74$; $p < 0.001$).

Narrower seats exit for 29% and 28% of the students using SBPH and SBS respectively (Figure 3). This situation of mismatch produces mobility

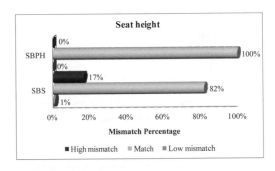

Figure 1. Percentages of match/mismatch in seat height.

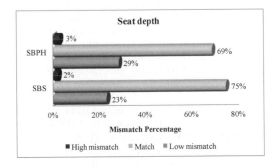

Figure 2. Percentages of match/mismatch in seat depth.

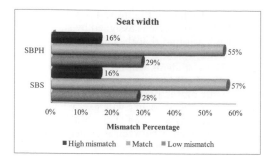

Figure 3. Percentages of match/mismatch in seat width.

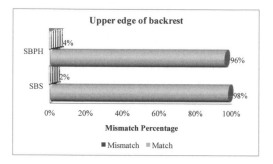

Figure 4. Percentages of match/mismatch in upper edge of backrest.

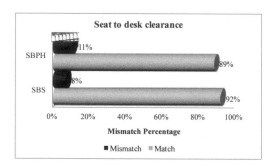

Figure 5. Percentages of match/mismatch in seat to desk clearance.

restrictions and discomfort (Evans et al., 1988; Helander, 1997; Orborne, 1996). Otherwise, 16% of the students will use a wider seat in both situations. The overall result shows that the levels of mismatch for this furniture dimension are 45% and 44% using SBPH and SBS respectively, these variables are statistically independent ($X^2 = 0.62$; p = 0.429).

After selecting furniture with SBPH and SBS, 4% and 2% of the students will find the backrest higher than SUH, respectively (Figure 4). This percentage of student will reduce the arm mobility due compression in the scapula (Gutiérrez et al., 2001; Orborne, 1996). The applied statistical analy-

sis shows that the variables are statistically dependent ($X^2 = 12.77$; p < 0.001).

Seat to Desk Clearance mismatched 11% and 8% of the students using SBPH and SBS, respectively (Figure 5). This situation of mismatch produces mobility constraint because of the contact of the thighs with the desk (Parcells et al., 1999; Sanders et al., 1993). As most of the analyzes, there are differences between the results from the two anthropometric measures used for furniture selection ($X^2 = 14.78$; p < 0.001).

There are no differences regarding the Seat to desk height dimension (Figure 6). The frequency of higher desks was more than 60% in both cases. These results were also independent between the variables ($X^2 = 0.483$; p = 0.486). In most of the cases, students are required to work with shoulder flexion and abduction or scapular elevation, causing more muscle work load, discomfort and pain in the shoulder region (Szeto et al., 2002). Moreover, this result is better than the 100% and 99% of mismatch from a previous study (Castellucci et al., 2010). Although remains a high percentages of mismatch for optimal use of the standard.

In order to improve the analysis, figure 7 shows the percentage of "transversal match" using SBS and SBPH. Match transversal is defined as the match in the cumulative of different furniture

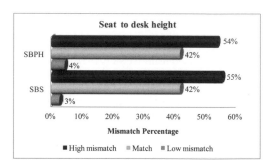

Figure 6. Percentages of match/mismatch in seat to desk height.

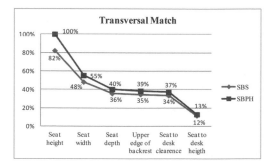

Figure 7. Percentages of transversal match.

dimensions. The first furniture dimension was seat height, as it was explained before, is the starting point and the most important variable for the design of the classroom furniture. After that, it was considered seat width and the other four furniture dimensions. The final results shows that only 13%, for SBS, and 12%, for SBPH, of the student will fit in the different furniture size presented in the Chilean standard N° 2566.

4 CONCLUSIONS

According to the obtained data it can be concluded that popliteal height is the most accurate anthropometric measure for classroom furniture selection. Furthermore, it can be stated that this measure can be also more efficient, as the measurement of popliteal high can be easier to measure than stature. Finally, it also seems that it will be important to analyse the Chilean Standard recommended values, since it seems that there is great percentage of mismatch between children characteristics and the furniture recommended values.

ACKNOWLEDGMENTS

This research was funded by The National Fund for Health Research and Development, Chilean Government, N° SA11I2105 (Fondo Nacional de Investigación y Desarrollo en Salud (FONIS), Gobierno de Chile).

REFERENCES

Castellucci, H.I., Arezes, P.M. & Viviani, C.A. 2010. Mismatch between classroom furniture and anthropometric measures in Chilean schools. *Applied Ergonomics*, 41(4): 563–568.

Evans, W.A., Courtney, A.J. & Fok, K.F. 1988. The design of school furniture for Hong Kong school children: an anthropometric case study. *Applied Ergonomics*, 19, 122–134.

García-Acosta, G. & Lange-Morales, K. 2007. Definition of sizes for the design of school furniture for Bogotá schools based on anthropometric criteria. *Ergonomics*, 50, 1626–1642.

Gouvali, M.K. & Boudolos, K. 2006. Match between school furniture dimensions and children's anthropometry. *Applied Ergonomics*, 37, 765–773.

Gutiérrez, M. & Morgado, P. 2001. *Guía de recomendaciones para el diseño del mobiliario escolar*. Chile: Ministerio de Educación and UNESCO.

Helander, M. 1997. Anthropometry in workstation design. In M. Helander (Ed.), *A Guide to the Ergonomics of Manufacturing*. (pp. 17–28). London: Taylor & Francis.

ISO, 1996. ISO 7250: Basic Human Body Measurements for Technological Design. *International Organization for Standardization*, Geneva, Switzerland.

Instituto nacional de normalización (Chile). 2002. *Norma Chilena 2566*. Mobiliario escolar—Sillas y mesas escolares - Requisitos dimensionales.

Khalil, T.M., Abdel-Moty, E.M., Rosomoff, R.S. & Rosomoff, H.L. 1993. *Ergonomics in Back Pain: A Guide to Prevention and Rehabilitation*. Van Nostrand Reihold, New York.

Milanese, S. & Grimmer, K. 2004. School furniture and the user population: an anthropometric perspective. *Ergonomics*, 47, 416–426.

Molenbroek, J.F.M., Kroon-Ramaekers, Y.M.T. & Snijders, C.J. 2003. Revision of the design of a standard for the dimensions of school furniture. *Ergonomics*, 46, 681–694.

Mondelo, P., Gregori, E. & Barrau, P. 2000. *Ergonomía 1: Fundamentos* (3rd ed.). México: Alfaomega Grupo Editor - UPC.

Occhipinti, E., Colombini, D., Molteni, G. & Grieco, A. 1993. Criteria for the ergonomic evaluation of work chairs. *Le Medicina del Lavoro*, 84(4), 274–285.

Orborne, D.J. 1996. *Ergonomics at Work: Human Factors in Design and Development* (third). John Wiley & Sons, Chihester.

Panagiotopoulou, G., Christoulas, K., Papanickolaou, A. & Mandroukas, K. 2004. Classroom furniture dimensions and anthropometric measures in primary school. *Applied Ergonomics*, 35, 121–128.

Parcells, C., Stommel, M. & Hubbard, R.P. 1999. Mismatch of classroom furniture and student body dimensions: empirical findings and health implications. *J. Adolesc. Health*, 24(4), 265–273.

Pheasant, S. 1991. Ergonomics, *Work and Health*. Macmillan, Hong Kong.

Sanders, M.S. & McCormick, E.J. 1993. Applied anthropometry, work-space design and seating. *Human Factors in Engineering and Design* (7th ed.). Singapore: McGraw-Hill.

Szeto, G., Straker, L. & Raine, S. 2002. A field comparison of neck and shoulder postures in symptomatic and asymptomatic office workers. *Applied Ergonomics*, 33, 75–84.

Occupational Safety and Hygiene – Arezes et al. (eds)
© *2013 Taylor & Francis Group, London, ISBN 978-1-138-00047-6*

Hierarchical classification of ergonomic methods for applications in current engineering practice

S. Fiserova
Faculty of Safety Engineering, VSB, Technical University of Ostrava, Ostrava, Czech Republic

ABSTRACT: The current ergonomic methodology enables applications related to the evaluation and design of work systems that take into account both basic aspects of each work system, namely the technical and human aspects. The article summarises the basic methodological principles of ergonomic activities in work systems. It focuses on the characteristic divisions of a large number (more than 300) of methods intended for the ergonomic evaluation and design of work systems that are contained in current technical publications. For effective applications of ergonomic activities in order to optimize work systems, the need for method classification understandable not only to ergonomists but also technicians and engineers appears. The proposed classification of ergonomic methods according to the hierarchy of obligation corresponds to the common approaches applied in real conditions of engineering practice and can contribute to an increase in the number of goal directed and successful ergonomic projects.

1 INTRODUCTION

In ergonomics, the design process, management system, methods, equipment and environment are always evaluated in relation to human abilities, capacity and human limit values. It is typical of a modern world that disciplines of present-day relevance and value are generally multi-, inter-and trans-disciplinary, and thus they cannot be defined simply (Charlton, 2001).

A considerable number of methods so far published, i.e. more than three hundred methods, including obligatory and recommended methods, are greatly indebted to special methodological division, and do not give technicians—engineers, who carry out the optimization of work systems, a sufficient overview of possibilities of applications of ergonomic evaluation methods. With reference to the great number of methods and variety in possibilities of their application, the need appears to summarise and classify systematically the methods, above all with regard to transparency, effectiveness and significance of required applications. Under the current conditions of engineering practice, the points of view of obligation and also economical, professional and time demands play a great role.

The significance of ergonomics increases with the efficiency of applications and on the basis of correctly used methods of design and evaluation of effectiveness of ergonomic solutions (Turekova, 2011). The ergonomic methodology deals with principles, procedures, methods and techniques that are at present used or can be used in analyses, designs and evaluations of machinery in relation to machinery-human interaction, analyses, designs and evaluations of work positions, work tasks, working conditions, work environment and work organisation.

2 METHODOLOGICAL PRINCIPLES OF ERGONOMIC ACTIVITIES IN WORK SYSTEMS

For managing successfully the process of a comprehensive ergonomic evaluation and for achieving the improvement or optimization of a work system, a system approach to solving is expected. The system approach shall be applied in cases of making ergonomic analyses for current as well as new situations. In evaluations and designs basic ergonomic principles shall be always utilized for the design of optimal working conditions with regard to well-being, safety and health of workers, including the development of existing skills and the acquirement of new skills, with taking technological and economic effectiveness and efficiency into account (Salvendy, 2005).

Accepted and valid international standards are orientated towards the evaluation and design of work systems and can be used for other areas of human activity as well. In the process of ergonomic evaluation and design, the major interactions between one or more people and the components of the work system, such as tasks, equipment, workspace and environment shall be dealt with.

The process of ergonomic design of a work system can be divided into the following phases:

– formulation of goals (analysis of requirements),
– analysis and allocation of functions,
– design concept,
– detailed design,
– realization, implementation and validation,
– evaluation.

All the phases contain a fixed methodological framework, e.g. in the form of the detailed design it is necessary to evaluate and design individual elements of the parts of which the work system is composed. In accordance with the relevant international standard, the detailed design should be carried out so that all related ergonomic knowledge and requirements may be respected. The work system design shall include the design of the following components:

– work organisation
– work tasks
– work
– work environment
– work equipment, hardware and software
– workspace and workstations.

Ergonomically designed work systems increase safety, effectiveness and performance, improve human working and living conditions and compensate unfavourable effects on human health and performance (Taylor, 2004). Interactions among the design of machinery, design of work task and job position are of key importance to the accomplishment of goals of ergonomic activities.

3 EXISTING CLASSIFICATIONS OF ERGONOMIC METHODS FOR THE EVALUATION AND DESIGN OF WORK SYSTEMS

A relatively large number of methods applied, verified and published (after the year 2000) abroad exist. It is more than 300 methods that are associated with the evaluation and assessment of ergonomic parameters both in already implemented systems and in design of new systems. Commented and applied methods are contained especially in foreign technical monographs and binding and recommended legal and technical standards—international and national ones. A considerable number of modern methods of ergonomic evaluation and design are usually divided by authors according to various methodological and ergonomically logical criteria. Those usually differ from the concept of binding and recommending standards covering this area. For engineering practice, technical divisions are insufficiently clear because they require a complex of knowledge of ergonomic problems. For the purpose of creation of a clear set of applicable ergonomic evaluation methods, it is necessary to approach the ergonomics of work systems as the ergonomics applied in specific conditions. The aim is to achieve analogy to basic processes of ergonomic design of work systems, which are analysis, synthesis, evaluation and subsequent ergonomic intervention in a closed cycle of continuous verification.

Common divisions of methods of ergonomic evaluation and design that are presented frequently in topical technical literature are as follows:

a) classification of methods according to the extension of the base of process of ergonomic project (Karwowski, 2003, 2006; Salvendy, 2006; Wilson 2005; Stanton, 2005),
b) classification of methods according to the conditions of application (Karwowski, 2003, 2006; Salvendy, 2006; Wilson 2005),
c) classification of methods according to the purpose of use (Stanton, 2005; Karwowski, 2006),
d) classification of methods according to ergonomic disciplines (Karwowski, 2003, 2006; Salvendy, 2006; Wilson 2005).

These divisions of methods of ergonomic evaluation and design are very beneficial in cases of sufficient knowledge of the problems of general ergonomics of work systems that however is not usual in conditions of current engineering practice. Technical literature, which deals in detail with ergonomic methodologies with focus on specific areas of processes of ergonomic design is not usually available to engineering practice.

The correctness of the selected method depends on many weighty aspects, but the basic precondition is, in addition to knowledge and sufficient understanding of the work system, a sufficient overview of methods, their properties and range of usage. The selection of the method should be in relation to the expected benefits from its use and to the quality and level of results obtained.

4 CLASSIFICATION OF ERGONOMIC METHODS ACCORDING TO THE HIERARCHY OF OBLIGATION

The great number of current methods of the ergonomic evaluation and design of work systems determine the need to systematise them so that they may be lucid and usable under the present-day modern conditions of engineering practice with high requirements for effectiveness and economic return of activities carried out.

After application of rules and principles of ergonomic evaluation and design and correct determination of priorities of ergonomic applications in work systems, subsequent dealing with the ergonomic approach is to be subject, also with reference to the availability of methods used in ergonomic activities for engineering practice, to a uniform hierarchy (Fiserova, 2011).

A proposal for the division of methods used in ergonomic activities for their application in engineering practice assumes and respects the hierarchy of obligation.

I. Methods obligatory and methods necessary for the application of obligatory requirements (i.e. minimum requirements for occupational safety and health).
II. Methods recommended, developing obligatory requirements and contained in international standards related to the ergonomics of work systems.
III. Methods supplementary, identifying and analysing specific areas of human activities in work systems.

Respecting the modern ergonomic principles, including the systematic classification and hierarchical analysis of selected methods can contribute to the wide use of specific ergonomic designs, which have been implemented so far only in a limited degree, in practice.

4.1 *Methods obligatory*

Methods obligatory and methods necessary for the application of obligatory requirements (i.e. minimum requirements for occupational safety and health) are methods used for health risk assessment are to be taken as the basis of all ergonomic activities in work systems. Applications of these methods that are obligatory in the range of valid legal regulations are intended especially for use in a real environment. The use of them is connected with a concrete work process with concrete work activities and tasks carried out in the framework of a characteristic working shift. Exposure assessment has become more or less an independent specialisation with its own international professional society ISEA (International Society of Exposure Assessment). Risk factors that may have, from the health point of view, a significant influence on the quality of working conditions are as follows: duct, chemical substances, noise, vibration, nonionizing radiation and electromagnetic field, physical load, work position, heat load, cold load, psychical load, visual load, work with biological agents, work under increased air pressure. The analysis of health risk factors at work is based on the systematic monitoring of all factors of work environment and working conditions from the point of view of human health load due to these factors and their possible harmful effects on occupational health and safety. It pre-

dicts a possibility of occurrence of occupational injuries, occupational diseases or other injuries to health associated with work and working conditions (e.g. diseases connected with work). Part of this activity is the evaluation of proposals for measures to limit or eliminate risks, including the check and evaluation of accepted measures. The assessment of all health risks, with some exceptions, rests on requirements for objectivity, which brings with it the fact that for the reproducibility and subsequent classification of results of evaluation it is necessary to accept the response of the "average healthy human" organism. The methods are for example in accordance with valid legal rules of European Union (Directives) as methods contained in the technical standards, under conditions determined as obligatory in the national legislations members states of European Union.

4.2 *Methods recommended*

The detailed evaluation of working conditions with a view to their optimization can be made according to valid international standards so that ergonomic activities may be in accordance with generally valid rules and principles of ergonomic design. Methods contained in international standards are recommended and take into account also criteria not dealt with by obligatory regulations. It is a case of analysis procedures with respecting differences e.g. anthropometric and biomechanical, in age, etc. Work systems designed to respect also requirements and recommendations of international standards increase safety, effectiveness and efficiency, improve human working and living conditions and diminish unfavourable effects on human health and work performed. For this reason, a good ergonomic design has a favourable influence on a work system and human reliability in this system.

The recommended methods that develop obligatory requirements and are contained in international standards are in the draft classification for applications in engineering practice divided into methods intended for the assessment of:

– Machinery risks
– Human physical performance
– Human mental performance
– Human performance with regard to physical, chemical and physiological working conditions
– Human sensory performance
– Ergonomic design of control centres
– Display units and clerical work

Recommended methods are for example contained in valid international standards (ISO, EN) and in informative parts of the obligatory standards (Karwowski, 2006).

4.3 Methods supplementary

The set goal in the area of optimization of work with regard to health protection, safety and reliability and also optimization of work efficiency, work quality and productivity can be achieved by applications of other suitable and selected methods that are contained neither in any regulations nor international/national standards. Their application follows from the requirements given by abilities, limits and requirements of all elements of a system (Taylor, 2004). The supplementary methods can be divided for example as the classification of methods according to the purpose (Stanton, 2005; Karwowski, 2003) of use into the following eleven categories:

– *Data collection methods*
  Data collection methods are used for the collection of specific data related to the system and scenarios. They are basic methods for designing and planning new systems and for evaluating currently operated systems. For example interviews, observations, questionnaires, simply checklists, etc. (Charlton, 2001; Stanton, 2005)

– *Task analysis methods*
  These methods are used for analyses of human position and human role in executing tasks and scenarios in systems. Analytical methods specify tasks and scenarios (e.g. working procedures, task contents) to individual steps, for human-machine, human-human (other persons) interactions. For example HTA, GOMS, VPA, TD, TTA etc. (Stanton, 2005).

– *Cognitive Task Analysis (CTA) methods*
  CTA methods are used for the description of yet not known sets of arrangement of activities and operations. They are used in the description of mental processes of system operators in the course of completing and making up operations to be performed and their sets. For example ACTA, CWT, CWA, CDM, etc. (Stanton, 2005; Karwowski, 2003).

– *Process charting methods*
  They are used for the graphic representation of tasks and processes by means of standardized symbols. The output of process charting methods and techniques can be a basis for the cognition and understanding of different sequences of tasks that are contained as part in the overall scenario—a detailed overview of work activities. Furthermore, they are used for the clarification of time schedules of operations that may occur and for the clarification of which technological aspects of the system and its relations are required. For example ETA, FTA, OSD, DAD, etc. (Charlton, 2001; Karwowski, 2003; Stanton, 2005).

– *Human Errors Identification (HEI) methods*
  Human error identification methods are designed for the prediction, identification of possible human errors in a work system, especially those that may occur in interaction with machinery. By the application of Human Reliability Analysis (HRA) methods is then carried out the quantification of cases of human failure in the system. For example SHERPA, HET, SPEAR, etc. (Charlton, 2001; Karwowski, 2003; Stanton, 2005).

– *Mental workload assessment methods*
  A mental workload represents a level of abilities of a human to satisfy requirements imposed on the human. Quite a lot of such methods exist and they can be used widely in the assessment of processes and also in the design of them. For example PTPM, STPM, NASA-TLX etc. (Charlton, 2001; Karwowski, 2003; Stanton, 2005).

– *Situation awareness assessment methods*
  Situation awareness assessment methods are used for the analysis of human preparedness for situations that may occur in a system. They are used for the determination of requirements for knowledge and abilities of operators and machinery operators and are also a confrontation with the determination of target requirements for system functionality and quality of management preparedness in relation to the corresponding comprehension of formulation of individual operations and their interrelations. They are also used for planning the overall layout of the system. These techniques are used for partial as well as comprehensive evaluations of mainly dynamic systems. For example CARS, SACRI, SARS, SPAM etc. (Stanton, 2005).

– *Interface analysis methods*
  Methods and techniques used for the analyses of interfaces in a system serve the evaluation and design and planning of requirements and functions of interconnections between specific elements of a system with a view to optimization, including the evaluation of e.g. employee satisfaction and consideration of employee opinion. For example special checklists-RULA, REBA, SUMI, Heuristic Evaluation etc. (Charlton, 2001; Karwowski, 2003; Stanton, 2005).

– *Design methods*
  They are the methods that are typically used in designing and planning new systems, activities and human factor relations in processes—of individuals, groups and sequences in the framework of large working teams. For example TCDS, SBD, etc. (Stanton, 2005).

– Performance time prediction methods

They are used for the determination of corresponding time requirements for work operations, tasks and activities, including the creation of designs of overall detailed overviews of work activities and scenarios. For example CPA, KLM, Timeline Analysis, etc. (Stanton, 2005).

– Team assessment methods

They are used for the assessment of performance of groups and teams for individual activities and also overall scenarios and work images. For such assessments, a whole series of aspects is usually specified and those are later evaluated and compared. Requirements and the level of intercommunication, awareness, co-decision-making, load and co-operation are assessed. For example BOS, CDA, DRX, SNA, etc. (Karwowski, 2003; Stanton, 2005).

Supplementary methods can be applied in any phase of ergonomic design; they do not replace either obligatory or recommended methods.

5 CONCLUSION

Requirements of modern developing companies for increasing work performance, quality and productivity and a related increase in competitiveness and better market positions are reasons for increasing the significance of correct ergonomic design and functionality of work systems.

In the framework of ergonomic evaluations of work systems, equal attention shall be paid to three basic aspects from the point of view of human position in a work system. They are the propositional aspect, task aspect and social aspect, and their interfaces. For these purposes, assuming the hierarchical process, the applications of methods intended for data collection, analytical methods and presentation methods are necessary. New technologies together with development lead on the one hand to the facilitation of some working procedures; on the other hand they bring growing requirements for the professional and also psychical abilities of employees to give expected and high-quality work performance. Engineering practice usually inclines towards extensive and detailed investigations of working conditions and subsequent ergonomic interventions provided that sufficient arguments on benefits and improvements are available. Only a correctly and reliably made ergonomic evaluation, utilizing the combination of methods that are suitable and verified tools, can lead to an effective and long-term ergonomic solution in the framework of work systems.

REFERENCES

Charlton, S., O´Brien, T.: *Handbook of Human Factors Testing and Evaluation,* 2001, LEA, ISBN: 0805832904.

Fiserova, S: *Applied ergonomic methodology for current engineering practice,* Communications-Scientific Letters of the University of Žilina, No. 2/2011, ISSN 1335-4205.

Karwowski, W.: *Handbook of Standards and Guidelines in Ergonomics and Human Factors,* IEA, UK, 2006, ISBN: 0-8058-4129-6.

Karwowski, W. Marras, W.S.: *Occupational Ergonomics,* CRC Press LLC, 2003, ISBN: 0-8493-1802-5.

Salvendy, G.: *Handbook of Human Factors and Ergonomics,* John Wiley & Sons, Inc., USA, 2006, ISBN: 0-471-44917-2.

Stanton et al.: *Handbook of Human Factors and Ergonomics Methods,* CRC Press, 2005, ISBN: 0-415-28700-6.

Taylor et al: *Enhancing Occupational Safety and Health,* Elsevier, UK, 2004, ISBN: 0-7506-6197-6.

Tureková, I., Kuracina, R. & Rusko, M.: *Manažment nebezpečných činností (Dangerous operation management),* Trnava, AlumniPress, 2011, ISBN 978-80-8096-139-8.

Wilson, J.R., Corlett, N.: *Evaluation of Human Work,* Taylor & Francis Group, CRC, 2005, ISBN: 0-415-26757-9.

Occupational Safety and Hygiene – Arezes et al. (eds)
© *2013 Taylor & Francis Group, London, ISBN 978-1-138-00047-6*

Comfort underwear, their implications for women's health in task performance

R.P. Alves & L.B. Martins
Federal University of Pernambuco, Brazil

S.B. Martins
Estadual University of Londrina, Paraná, Brazil

ABSTRACT: With the progressive inclusion of women in the public space of production, initiated during the two world wars, the demand for comfortable and better adapted female working clothes underwear grew. In contemporary times, it is understood that the design and production of lingerie by prêt-à-porter, has the challenge of creating and producing also underwear that suits to the different bodies of the users, to work clothes and the specificities of the activities that they perform. Therefore, in this study, we developed a research on comfort and usability in intimate apparel and its possible implications for women's health during the execution of tasks. It is based on the Oikos methodology of assessment of usability and comfort in clothing. The underwear analyzed presented features of discomfort, such as inadequate and restrictive mold materials with negative implications on the health of women, especially with the long use during the execution of tasks.

1 INTRODUCTION

Since the emergence of ready-to-wear garment, mass production has been directed to a standard public–young bodies (dynamic and slim). In addition, in their relation with fashion, industries, research and cultural references have been applied with greater emphasis on aesthetics. A fact that has often resulted in nice clothes visually, but with restrictions relating to comfort and to use. It is the case of some underwear.

In Brazil, from an economic point of view, this topic has been significant in terms of the internal market and for export. Only in the Intimate Fashion sector, in 2011, 3.4 thousand companies were accounted for (71% micro and small), which produced 808 million pieces and generated 167 thousand jobs. In the same year, Brazil exported 32.9 million dollars (BOLETIM ABNT, 2012).

It is understood, therefore, that the design and production of lingerie, as well as their use by different female bodies, especially during the execution of tasks, with repercussions on the health of working women, are an open laboratory for research in the context of design and ergonomics.

Ergonomics is presented in Martins and Martins (2012) as a useful tool for the development of clothing with shapes, combinations of materials and assembly techniques capable of contributing with the comfort and a good fit in static conditions

and human mobility. In other words, it seeks to adapt the garment design to the user. According to Santos (2009), this adaptation has as a starting point: 1) the process of construction of clothes; 2) the user as focus; 3) analysis of the user adaptation time to the new clothing.

Some works were carried out in this respect. In Brazil: 1) "Modeling under an ergonomic perspective" (GRAVE, 2004); 2) "Comfort in clothing: An interpretation of ergonomics–methodology evaluation of usability and comfort in garments" (Martins, 2005); "The fashion-clothing and the ergonomics of the hemiplegic" (GRAVE, 2010). In Portugal: 1) "The comfort in fine wool fabrics" (BROEGA, 2007); 2) "Optimising the design of surgical attire through the thermo physiological comfort" (BRAGA, 2008). And additionally, publications of the World Congress of Ergonomics (IEA 2012): 1) "Ergonomics, universal design and fashion" (MARTINS and MARTINS, 2012); 2) "Ergonomics principles to design clothing work for electrical workers in Colombia" (CASTILLO and CUBILLOS, 2012); among others.

Concerning the use of the garment there are the following types of comfort: 1) Thermo physiological—transfer of heat and water vapor through the clothing or textile materials; 2) Sensory—set of neural sensations caused by contact of a textile material with the skin; 3) Ergonomic—the capacity of the clothing to dress well and to allow the movements of the body; 4) Psycho-Aesthetic—is related to the subjective perception of aesthetic evaluation with reference in vision, touch, hearing and smell that contribute to the general well-being of the user.

Regarding clothing, comfort and usability as an ergonomics object of study should be project requirements. However, when it comes to clothing projective methodology in Brazil, at least in the academy, Montemezzo (2003) has been used, which includes five phases: 1) Preparation; 2) Generation of alternatives; 3) Evaluation and selection of proposals; 4) Realization; 5) Documentation. This author suggests that the knowledge generated through ergonomics should be used early in the project. However, their methodology does not address ergonomic aspects. This is what says Martins (2005), in his thesis that originated the OIKOS methodology evaluation of physical comfort and usability of the clothing. This contains a list of items and sub-items that indicate which aspects should be evaluated, such as: 1) Easy handling; 2) Easy maintenance; 3) Easy assimilation; 4) Security; 5) Indicators of usability (Jordan); and 6) Comfort—contact of the fabric with the skin and adjustment to the body.

It is known that some researches in the field of ergonomics include the work clothes and health, but ignore the interference of the undergarment, which positions itself between the uniforms and the body, for example.

Accordingly, the objectives set for this study were: 1) Develop research on usability and comfort in intimate apparel; 2) Describe their implications for women's health during the execution of tasks.

2 MATERIAL AND METHOD

It is important to emphasize that this study is in progress. We are raising the evaluation parameters and indicators of different levels of comfort in the use of underwear. Therefore, the data presented in the results are empirical and originated from exploratory research, conducted through a qualitative approach, which occurred at the level of human representations related to the object of study.

It is based on the Oikos methodology evaluation of comfort and usability of the clothing (Martins, 2005). At this stage of the work, we evaluate empirically the following items in it and some sub-items: 1) Easy handling: Easy to use, dress and undress, activate, hold, manipulate, the materials, finishing of trimmings and mobility during use; 2) Security: Resistance to microorganisms, that shall not affect the circulation or hurt the skin, mobility and if the fabric allows perspiration; 3) Indicators of usability: Consistency and compatibility with user functionality prioritization; 4) Comfort: Contact of the fabric with the skin (touch, abrasion, tenderness) and the adjustment of the garment to the body.

The following procedures were used: 1) Bibliographical and secondary data survey about the history of the body and underwear; 2) Direct observation and empirical survey of the interaction of intimate apparel with the body of adult users.

3 RESULTS AND DISCUSSION

Based on the empirical data, one can see that when thinking of designing intimate apparel it is necessary to study the body, its appearance, its location in space and time, as well as its role in the environment that accommodates it. Saltzman (2009) confirms that the body is the starting point and support for creation and construction of the clothes.

The concern with the appearance of garments has intensified since the Middle Ages with the rise of fashion and at the expense of welfare. The corset that emerged in the sixteenth century by the rigidity with which squeezed the female silhouette, was condemned by physicians. Then, with the successive developments of fashion, the focus turned to other underwear in order to better adapt to the clothes, especially since the entry of women into the labor market. This is the case of the bra, which in 1859 had characteristics of discomfort, setting precedents for its redesign.

In 1910 some brands of lingerie ads emphasized aspects of durability and comfort to attract consumers. However, largely reformulations are given as a function of the aesthetic fashion. In the 1920s, the bra pressed the bust to assign women an androgynous look, and in the 1950s the padded bra appeared, approaching the appearance of the feminine aesthetic body presented in the movies. From the 1960s the attention is directed to features of well-being: The cotton fiber was considered comfortable for the intimate apparel segment, the technological development has produced new synthetic fibers (micro-fiber), textile finishing and productive processes which expanded the options of materials and techniques for design and production of lingerie (NAZERETH, 2007).

Thus, if on one hand, one finds in the market underwear made with fabrics which have bacteriostatic and fungistatic finishing (treatment applied on the surface of the fabric that inhibits the growth of bacteria and fungi), such as the example of panties, on the other hand there are also those made with materials that, in the interaction with the body and with the climate, cause genital diseases.

In the research of Alves and Costa (2005), performed with gynecologists and microbiologists in the Metropolitan Region of Recife-PE, it was found that the synthetic textile fiber (polyester and polyamide) and restrictive models, used in the manufacture of panties, associated with the hot and humid weather of the region, was related to the proliferation of Candida and Trichomonas in female genitalia. Health professionals recommend the use of underwear made with natural fibers (cotton), molds that do not cause strong friction with the skin, to cleanse the underwear with neutral soap and change of underwear every six hours. However, such recommendations, according to those interviewed, encounter resistance from women who work outside the home environment.

The work done by Giongo and Heinrich (2010) on the perception of physical and psychological comfort for users of panties, based on the methodology of Linden, proposes that the evaluation of the perceived comfort in the use of products is mediated by the personal values of potential users, by size of hedonic experiences for them. This evaluation should be done starting from the characteristics of the product, the stimulus formula and the dominant reference for the user. The research revealed that physical comfort while wearing panties is disapproved of due to the psychological comfort, as long as the aesthetic requirements are met.

By analyzing the ergonomics of underwear (panties and bras) produced by industries prêt-à-porter, Silva et al. (2011) identified problems in modeling the bra, that instead of being designed based on human anatomy, had reference to the bulges in the market. According to users, the bulge did not allow a good accommodation for the breasts, leaving them open laterally. They also stated their preference for cotton panties in daily use.

Spaine (2008) emphasizes that ergonomics went on to be employed in the design of lingerie from the moment comfort was seen as a key feature of the clothing. That study sought to evaluate the ergonomic knowledge of the activities carried out in the underwear modeling stage, with the intention of identifying the form it being elaborated in the manufacturing industry. Four underwear companies were analyzed. In them, the need for adequate standard anthropometric measures table was identified, for the biotype of the target public. Combined with the lack of ergonomics knowledge in the development of modeling. Insufficient and inappropriate use of the product technical sheet was also identified, interfering in the productive process and undermining the quality and accuracy in all stages of production. The study proposes the use of standard anthropometric measurements for construction of models as a prerequisite in obtaining comfort characteristics and parameter for the correction of graduation processes of mold sizes.

As regards the adaptation of the undergarment to work clothes for the control and the health of workers during the execution of tasks, the following concepts, singled out by Castillo and Cubillos (2012), in the study that aimed to collect information of workers with experience as users of work clothes to establish the usability and the criteria for improvement or design recommendation: comfort, functionality and usability. The authors highlighted that the functionality is related to the perceived comfort by the worker in terms of thermal conditions, anthropometric and visual appearance.

For the development of design criteria of working clothes, according to Castillo and Cubillos (2012), the following should be considered: The task, the action, the context, the knowledge and the situation, to subsequently perform the steps of systematization, modeling and transfer rules (knowledge activities for the technical specifications of work clothes).

Thus, based on the Oikos methodology of Martins (2005), the following items must be taken into account: Easy handling, security, indicators of usability and comfort as the least met items in using underwear during work hours with negative implications on women's health. Such as the proliferation of fungi and bacteria in the genitalia, arising from the use of panties with restrictive molds and made with fabrics that are preventing the exchange of heat with the external environment. Plus bruises on the skin in the region of the chest and breasts, caused by: Modeling that do not carry the variety of breast size and its relationship with the width of the thorax; boning that extends beyond the fabric; bra strap that loses its elasticity and decreases support. It should be noted that such implications are compounded when the outfit features modeling and materials that amplify the problems cited.

4 FINAL REMARKS

In general the female underwear reviewed in this study showed characteristics of discomfort and limitations on use, with negative implications on the health of women, especially during the execution of tasks. It should be noted, therefore, that from the point of view of design, designing intimate apparel requires the convergence of multiple knowledges to create underwear adapted, in terms of materials, models, and stitching finishes, to feminine bodily needs and esthetic, in a manner that health and well-being is preserved.

REFERENCES

Alves, R.P., Costa, A.F.S., & Pires, D.A. (2005). *Identificação dos têxteis usados na confecção das roupas íntimas e a saúde da mulher.* In: Annals of the XVIII Brazilian Congress, VI Latin American Meeting and IX State Symposium of Home Economics. Francisco Beltrão: UNIOESTE/ABED ISBN 85-89441-25-3.

Boletim ABNT. (2012). *Pequenas notáveis.* In: Boletim ABNT no. 18 March. Available in: <http://portalmpe. abnt.org.br/bibliotecadearquivos/Biblioteca%20de%20 Documentos/Pequenas_Notaveis.pdf> Accessed on 20 April 2012.

Braga, I.M. da S. (2008). *Optimização do Design do Vestuário Cirúrgico através do Conforto Termofisiológico.* MSc thesis (Masters in Textile Design and Marketing from the University of Minho). Portugal.

Broega, A.C.L. (2007). *Contribuição para a definição de padrões de conforto de tecidos finos de lã.* Thesis (Ph.D in Textile Engineering—Physical textile industry). University of Minho—School of Engineering. Portugal: UM.

Castillo, J., Cubillos, A. (2012). *Ergonomics principles to design clothing work for electrical workers in Colombia,* 623–627. In: 18th World Congress on Ergonomics of the IEA. Recife-BR: IEA.

Giongo, Mariana A, Heinrich, D.P. (2010). *Avaliação da percepção de conforto pelas usuárias de calcinhas.* In. DAMT: Design, Art, Fashion and Technology. Org. Gisela Belluzzo e Jofre Silva. São Paulo: Edições Rosari, 389–397.

Grave, M.F. (2010). *A moda-vestuário e a ergonomia do hemiplégico.* São Paulo: Escrituras Editora.

_____ (2004). *A modelagem sob a ótica da ergonomia.* São Paulo: Zennex Publishing.

Martins, S.B. (2005). *O conforto no vestuário: uma interpretação da ergonomia:* metodologia de avaliação de usabilidade e conforto no vestuário. Thesis (Ph.D in Industrial Engineering from UFV). Florianópolis-SC.

Martins, S.B., Martins, L.B. (2012). *Ergonomics, design universal and fashion,* 4733–4738. In: 18th World Congress on Erongomics of the International Ergonomics Association. Recife-BR.

Montemezzo, M.C. de F.S. (2003). *Diretrizes metodológicas para o projeto de produtos de moda no âmbito acadêmico.* (Dissertation—Masters in Industrial Design from Sao Paulo State University, College of Architecture, Arts and Communication) Bauru-SP.

Nazareth, O. (2007). *Intimidade revelada.* São Paulo: Estúdio Substância: Olhares Editora.

Saltzman, A. (2009). *El cuerpo diseñado:* Sobre la forma em el proyecto de la vestimenta. Ed. 1. Buenos Aires: Paidós.

Santos, C. de S. (2009). *O corpo.* In: Sabrá, F. (Org.) Modelagem: tecnologia em produção do vestuário, 38–55. 1 ed. São Paulo: Estação das Letras e Cores.

Silva, M., Raphael, N., Keitty, S. & Alves, R.P. (2011). *Análise ergonômica de roupas íntimas femininas produzida na cidade de Caruaru-PE.* In: Annals of the XXI Brazilian Congress, IX Latin American Meeting and II Intercontinental Meeting of Home Economics. Recife-PE.

Spaine, P.A. de A., Maffei, S.T.A., Paschoarelli, L.C., Silva, J.C.P. da, Santos Filho, A.G. & Menezes, M.S. (2008). *O conhecimento ergonômico nas fases de modelagem de underwear: uma contribuição.* In: XV ABERGO. Porto Seguro-BA.

Occupational Safety and Hygiene – Arezes et al. (eds)
© 2013 Taylor & Francis Group, London, ISBN 978-1-138-00047-6

Finger tapping rates and the effects of various factors

A. İşeri & M. Ekşioğlu
Department of Industrial Engineering, Boğaziçi University, İstanbul, Turkey

ABSTRACT: In this study, finger tapping rate capacities of eight fingers were determined and the effects of gender, age, finger, hand preference, tapping period, smoking and exercise were investigated. A sample of 124 subjects consisted of 56 males and 68 females between ages of 18 and 84 were tested. All investigated factors, except hand preference, significantly affected finger tapping rate. Index and middle fingers, the fastest fingers, had the same tapping rates and this was true for both hands. All right hand fingers, except little finger, had higher tapping rates than the fingers of left hand. Right little finger had the same tapping rate of the left index and middle fingers. The rates generally decreased with increasing age. Tapping rate was approximately stable in the first 10 seconds; and gradually decreased up to about 30 seconds then somewhat stabilized afterwards. Nicotine intake increased the tapping speed significantly. Males had greater mean tapping rate, and the number of weekly exercise increased the tapping rate; however, these effects were marginal, and may be considered insignificant in practical applications.

1 INTRODUCTION

Finger tapping rate is the total number of finger taps in a specified period of time and it is commonly used as a psychomotor test for evaluating the patients who have neurologic problems like Parkinson, schizophrenia, Alzheimer, etc. (e.g., Dodrill and Troupin, 1979; Wing et al., 1984; Ott et al., 1995). Generally, it is measured as the number of taps of the index finger in 10 seconds (e.g., Schmidt et al., 2000; Cousins et al., 1997).

In the field of ergonomics, the tapping rate capacity of fingers can be used as a design parameter in various products. For example, in the optimization of keyboard layouts (e.g., Dvorak, 1936; Eggers et al., 2003; and Yin and Su, 2011), and in the designs of some musical devices, tools and machines that require high rates of finger tapping. In the keyboard optimization case, the mentioned studies distributed the number of key presses to the fingers according to some defined capacities of the fingers.

Finger tapping rate is affected by a number of factors. There are studies that attempted to identify these effects with some lacks. For instance, Cousins et al., (1997) studied the effects of age and found that older subjects performed fewer total finger taps than younger subjects.

Schmidt et al., (2000) investigated the effects of gender and hand preference and found that men performed faster and more regularly than women. They explained this phenomenon with the effects of male hormones (testosterone) on muscles. They also concluded that hand preference doesn't have a significant effect on overall tapping rate. However, in the same study they found that manual asymmetry (difference in tapping rate) between left and right hands are smaller in left-handed subjects.

Silver et al., (2002) studied the effects of smoking on finger tapping rate of schizophrenic patients and found that smoking is associated with faster central processing. Perkins et al., (1990), and Roth and Batting (1991) also found that nicotine administration can increase finger tapping rate in normal smokers. However, the mechanism by which nicotine improves central processing is unknown (Silver et al., 2002).

The study by Jackson (1953) found that tapping rates of index and middle fingers are not significantly different. However, tapping rates for the ring and little fingers were significantly lower than for those two.

As the review of the literature shows, most of these studies are limited to the index finger of the right hand. There are only a few studies that measured the tapping rates of all eight fingers (e.g.; Jackson, 1953). Sample sizes of the studies in the literature generally vary between 50 and 100. There is no study that investigated the effects of age, gender, fingers, time duration, hand preference, smoking and exercise on tapping rate in a single study, considering a wide spectrum of age range.

Therefore, the objectives of this study are set as follows: Determining tapping rate capacity of fingers that can be used as a parameter in ergonomic designs; and determining the finger, gender, tapping duration, hand preference and age effects on the finger tapping rate.

2 MATERIALS AND METHODS

2.1 Subjects

The sample of participants consisted of 56 male and 68 female adults (a total of 124) with a wide range of age spectrum (18 to 84 years). All subjects participated in this study were in good physical health, and had no records of upper extremity, neck or shoulder disorders or pain. A self-report survey tool is used to determine the health status of the subjects. Healthy subjects signed a consent form prior to the experiments.

Age range is divided by 10-year intervals. The distribution of participants by gender, age-group and hand preference can be seen in Table 1.

2.2 Experimental procedure

Prior to the experiments, some characteristics of the subjects such as gender, date of birth, dominant hand, daily cigarette usage, weekly exercise rate and occupation are recorded. In addition, several anthropometric variables (hand length and breadth, and middle finger length) were measured. A briefing was given to the subjects regarding the task and experimental procedure. A few trials were performed for the purpose of familiarization before the actual tests.

The experimental task consisted of tapping the specified key in the keyboard with the specified finger for a period of one minute with the maximum volitional tempo. The subject was seated in neutral sitting posture with keyboard at about elbow height while the thumb and bottom part of palm were supported. The associated key was the standard position on the home row of the specified finger for touch typing. For instance, for left little finger, subject typed on key A (on Q layout) repetitively with his/her maximum tempo for one minute. Experiments were carried out with all eight fingers except thumbs. The order of the eight fingers was randomized for each subject to reduce experimental error. Between the two test sessions, there was a 30-second rest time for recovering from the fatigue. In addition, to minimize fatigue effect, finger of one hand followed the finger of other hand: Such as one left hand finger, then one right hand finger, then one left hand finger, and so on.

A laptop computer was used as the test device. A computer program that is written in MatLab was installed into that computer, and during all experiments the same laptop computer was used. While subjects performed tapping with the specified finger at his/her maximum volitional speed, the same computer program counted and recorded the number of taps every five-second. There was no feedback to the subject about his/her speed during the experiments but encouraged to perform tapping as fast as possible.

3 RESULTS AND DISCUSSION

This study had two objectives: Determining tapping rate capacity of fingers, and to investigate the effects of considered factors on finger tapping rate. In order to achieve these goals, descriptive and inferential statistical analyses were carried out on the recorded data.

3.1 Descriptive statistics

The figures 1 to 3 show the boxplots of the tapping rates, which depict mean (circles), median, 1st and 3rd quartile values. Figure 1 indicates that tapping rate decreases nonlinearly with increasing age. An examination of Figure 2 shows that tapping rates of the right hand fingers are higher than the left hand fingers. Figure 3 depicts the male-female difference: Female tapping rate shows higher variation.

3.2 Factor analysis

A general linear model and post-hoc analyses using Tukey tests are performed to investigate the effects of the factors on tapping rate. In this model, the number of keystrokes in every 5-second period of 60 seconds is used as the dependent variable. That is, each of 1-minute experiments is divided into 5-second periods (a total of 12 periods in 60 s) to investigate the changes in the response from first period to the last period. The independent variables are age-group (6 levels), gender, finger, period (12 levels), hand preference, smoking and exercise. All the factors are categorized except smoking and exercise, which are taken as continuous variables. Smoking is taken as the number of cigarette

Table 1. The distribution of the participants by gender, age-group and hand preference.

| | Males | | Females | |
	Right handed	Left handed	Right handed	Left handed
18–29	10	1	13	1
30–39	10	2	9	1
40–49	11	1	11	1
50–59	9	0	12	1
60–69	8	0	11	0
Over 70	4	0	7	1
Total	52	4	63	5

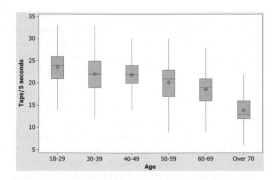

Figure 1. Boxplot of finger tapping rate by age.

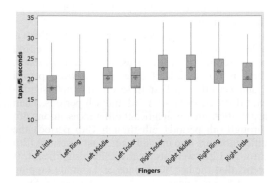

Figure 2. Boxplot of finger tapping rate by finger.

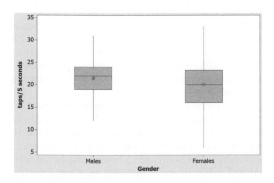

Figure 3. Boxplot of finger tapping rate by gender.

packages smoked daily and exercise is taken as the number of exercises weekly.

The results of the general linear model as taken from Minitab 16 statistical software package is shown in Table 2. According to the results, all the factors, except hand preference, have significant effect on finger tapping rate.

Following this, a general linear model analysis without hand preference factor is rerun and post-hoc analyses with Tukey tests are performed for further investigation. The results with the model coefficients

Table 2. General linear model results.

Source	DF	Seq SS	Adj SS	Adj MS	F	p
Age	5	88783	82489	16498	1761.8	<0.001
Finger	7	28793	28699	4100	437.8	<0.001
Period	11	5144	5112	465	49.6	<0.001
Smoking	1	2540	1586	1586	169.4	<0.001
Exercise	1	1709	1343	1343	143.5	<0.001
Gender	1	1045	1042	1042	111.3	<0.001
Hand preference	1	2	2	2	0.2	0.627
Error	11322	106019	106019	9		
Total	11349	234034				

$S = 3.06$ $R^2 = 54.7\%$ $R^2(adj) = 54.6\%$.

are given in Table 3. Using the model coefficients, tapping rates can be estimated as a function of the investigated factors. For instance, in order to estimate the tapping rate of a 64 years old woman's right index finger in the first 5 seconds (assuming she smokes two packages of cigarettes daily and does not exercise regularly), one can use the coefficients in Table 3 in such a way: $(19.67 - 0.32 - 1.71 + 1.97 + 1.29 + 2 * 0.97 + 0 * 0.22 = 22.84$ taps/5 seconds).

However as explained earlier, these coefficients are for estimating the tapping rates in periods of 5 seconds. For estimating the tapping rate as taps/second, the result should be divided by 5.

According to the statistical results, age-group is found one of the most defining factors in finger tapping. Overall, the finger tapping rate decreases nonlinearly with increasing age-group (from 18–29 to 70 and up). As seen in Table 3, there is a large difference between the groups (18–29) and (30–39). However, between the groups (30–39) and (40–49), this difference is insignificant. Moreover, after age of 50, tapping rate decreases rapidly. The rate of decrease is even faster after age of 70 years. The subjects within the age group of 70 years and above have 0.83 taps/sec lower than the subjects within (60–69) age-group.

Finger is found another important factor significantly affecting tapping rate. Index and middle fingers were the fastest fingers for both hands and the little fingers the slowest. There was no significant difference between index and middle fingers and this was true for both hands.

All right hand fingers, except little finger, had higher tapping rates than the fingers of left hand. The slowest finger of right hand, the little finger, had the same tapping rate of the fastest fingers of left hand, index and middle fingers (Table 3). It should be noted that for left hand dominant hand subjects, possibly the inverses of the results would be true, if their tapping data were analyzed separately.

Table 3. General linear model coefficients and post-hoc analysis.

Term	Coef	SE Coef	t	p	*Grouping
Constant	19.67	0.038	512.6	<0.01	
Gender					
Female	−0.32				A
Male	0.32	0.030	10.6	<0.01	B
Age					
18–29	3.73	0.062	60.4	<0.01	A
30–39	1.83	0.064	28.7	<0.01	B
40–49	1.85	0.061	30.1	<0.01	B
50–59	0.16	0.064	2.5	<0.01	C
60–69	−1.71	0.070	−24.5	<0.01	D
Over 70	−5.86				E
Finger					
Left little	−2.90	0.076	−38.0	<0.01	A
Left ring	−1.51				B
Left middle	−0.35	0.076	−4.6	<0.01	C
Left index	−0.11	0.076	−1.5	0.14	C
Right index	1.97	0.076	26.0	<0.01	D
Right middle	1.99	0.076	26.1	<0.01	D
Right ring	1.24	0.076	16.3	<0.01	E
Right little	−0.33	0.076	−4.4	<0.01	C
Period					
1	1.29	0.099	13.1	<0.01	A
2	1.22	0.097	12.6	<0.01	A
3	0.72	0.095	7.6	<0.01	B
4	0.25	0.095	2.6	0.03	C
5	−0.12	0.095	−1.3	0.35	C D
6	−0.20	0.095	−2.1	<0.01	C D E
7	−0.31	0.095	−3.3	<0.01	D E F
8	−0.44	0.094	−4.7	<0.01	D E F
9	−0.52	0.094	−5.5	<0.01	D E F
10	−0.53	0.095	−5.6	<0.01	D E F
11	−0.62	0.095	−6.5	<0.01	E F
12	−0.74				F
Smoking	0.97	0.074	13.1	<0.01	
Exercise	0.22	0.019	12.0	<0.01	

* Means that do not share a letter are significantly different with 95% confidence.

Although the absolute tapping rates obtained through the current study are somewhat different from the rates reported by Jackson (1953), the speed order of fingers are similar between the two studies.

Smoking is another factor significantly affecting tapping rate. The finger tapping rates increase with increased number of packages smoked per day. With every package of cigarette consumed daily,

there is an increase of 0.19 taps/sec (0.97 taps/5 seconds) in the finger tapping rate. This result supports the results of previous studies (Silver et al., 2002; Perkins et al., 1990; Roth & Batting, 1991) on the effect of smoking and nicotine on finger tapping rates.

Males are found to have statistically higher tapping rates than females. Although this effect is statistically significant, the difference, only 0.13 taps/ sec, may be considered insignificant in practical sense. This result is in agreement with the study by Schmidt et al., (2000). However, the magnitude of the difference in their study was higher, 0.32 taps/sec, between the right-hand dominant male and female university students. This difference in experimental results may partly arise from the characteristics of the participants.

For the effect of period factor, in general, there is a decreasing trend in the tapping rate from the first period to the last. However, as can be seen in Table 3, there is no significant difference between period 1 and period 2 (in first 10 seconds). Then, significant decrease occurs for 3rd and 4th periods. Then again, there are no significant differences among 4th through 6th periods. 5th though 10th periods the differences are also insignificant. Last 6 periods (7th through 12th) were also statistically insignificant. It seems that the last 30 seconds of 1 minute tapping task, there is a nearly steady rate of tapping. Since most of the tapping studies in the literature used a period of 10 seconds, the results obtained from this study cannot be used for comparisons.

Weekly exercise rate is also found statistically significant. The finger tapping rate increases with the increased number of weekly exercises. However, this difference (0.04 taps/second per number of exercises weekly) is too small in order to have a practical value.

Hand preference is found to be insignificant, so it is not considered in further analysis. In addition, with only 9 left-handed subjects, it is not feasible to perform an accurate statistical analysis. However, Schmidt et al., (2000) used a higher number of left-handed subjects, and concluded that hand preference does not have a significant effect on overall finger tapping rate.

3.2 *Tapping rate capacity of fingers*

In order to calculate the maximum volitional tapping rates for each finger, main effect coefficients of the fingers (constant + finger coefficient + covariate's coefficients * mean covariate values) in the general linear model (see Table 3) are used by normalizing the sum of these coefficients against 100%. The resulting parameters show the normalized tapping rates of each finger with respect to each other (Table 4).

Table 4. Normalized tapping rate capacity of fingers.

Finger	Normalized tapping rates (%)
Left little	10.8*
Left ring	11.6
Left middle	12.2
Left index	12.4
Right index	13.8
Right middle	13.7
Right ring	13.2
Right little	12.4

*Numbers add up to 100%.

These parameters, the normalized tapping rates, can be used in keyboard layout designs and similar equipment, musical instruments, tools, and jobs that require finger tapping. The ergonomic designs obviously should distribute the task loads to the fingers according to the tapping rate capacity of the fingers.

4 CONCLUSIONS

In this study, finger tapping rates of eight fingers, excluding thumbs, were determined and the effects of gender, age, tapping period, hand preference, smoking, and exercise were investigated. Based on the results, tapping rate capacity of each finger were calculated to be used as a parameter for various designs; specifically, in keyboard layout optimization.

Age and finger are found to be the most significant factors on tapping rate among the investigated factors. Smoking and tapping period were found as other important factors to be considered in estimating the tapping rates.

It is believed that taking into consideration of the findings of this study, the designs, which involve tapping rate capacity of fingers, will be improved. The results of this study can also be used as a reference in psychomotor tests.

ACKNOWLEDGEMENT

This study is supported and funded by the Scientific and Technological Research Council of Turkey (TÜBİTAK).

REFERENCES

Cousins, M.S., Corrow, C., Finn, M., Salamone J.D. 1998. Temporal Measures of Human Finger Tapping: Effects of Age. *Pharmacology Biochemistry and Behavior* 59: 445–449.

Dodrill, C.B., Troupin, A.S. 1975. Effects of repeated administrations of a comprehensive neuropsychological batter among chronic epileptics. *The Journal of Nervous and Mental Disease* 161: 185–190.

Dvorak, A., Merrick, N.L., Dealey, W.L, Ford, G.C. 1936. *Typewriting behavior*. NewYork: American Book Company.

Eggers, J., Feillet, D., Kehl, S., Wagner, M.O., Yannou, B. 2003. Optimization of the keyboard arrangement problem using an ant colony algorithm. *European Journal of Operational Research* 148 (3): 672–686.

Jackson, C.V. 1953. Differential finger tapping rates. *The Journal of Physiology* 122: 582–587.

Ott, B.R., Ellias, S.A., Lannon, M.C. 1995. Quantitative assessment of movement in Alzheimer's disease. *Journal of Geriatric Psychiatry and Neurology* 8: 71–76.

Perkins, K.A., Epstein, L.H., Stiller, R.L., Sexton, J.E., Debski, T.D., Jacob, R.G. 1990. Behavioral performance effects of nicotine in smokers and non-smokers. *Pharmacology Biochemistry and Behavior* 37: 11–15.

Roth, N., Batting, K., 1991. Effects of cigarette smoking upon frequencies of EEG alpha rhythm and finger tapping. *Psychopharmacology* 105: 186–190.

Schmidt, S.L., Oliveira, R.M., Krahe, T.E., Filgueiras, C.C. 2000. The effects of hand preference and gender on finger tapping performance asymmetry by the use of an infra-red light measurement device. *Neuropsychologia* 38: 529–534.

Silver, H., Shlomo, N., Hiemke, C., Rao, M.L., Ritsner, M., Modai, I. 2002. Schizophrenic patients who smoke have a faster finger tapping rate than non-smokers. *European Neuropsychopharmacology* 12: 141–144.

Wing, A.M., Keele, S., Margolin, D.I. 1984. Motor disorder and the timing of repetitive movements. *Annals of the New York Academy of Sciences* 423: 183–192.

Yin, P.Y., Su, E.P. 2011. Cyber swarm optimization for general keyboard arrangement problem. *International Journal of Industrial Ergonomics* 41(1): 43–52.

Occupational Safety and Hygiene – Arezes et al. (eds)
© *2013 Taylor & Francis Group, London, ISBN 978-1-138-00047-6*

Hand torque strength of female population of Turkey and the effects of various factors

M. Ekşioğlu & Z. Recep
Department of Industrial Engineering, Boğaziçi University, İstanbul, Turkey

ABSTRACT: This study aimed to establish the static hand torque strength norms of healthy adult female population of Turkey, and investigate the effects of handle type, posture, age-group, job-group and several anthropometric variables on hand torque strength. A sample of 257 female volunteers aged between 18 and 69 with roots from all seven regions of Turkey participated in the study. Maximum voluntary torque strengths of dominant hands were measured both in sitting and standing with four types of handles (cylindirical, circular, ellipsoid and key). Through statistical analysis descriptive values of torque strength and factor effects were determined. The results indicate that handle type, age-group and job-group significantly affect torque strength. The highest values were obtained with cylindrical handle and lowest with key handle. The oldest age-group (60–69) had significantly lower strength values than the remaining groups. Manual workers were stronger than non-manual workers. Marginally higher strength values were recorded in standing posture and with overweight subjects.

1 INTRODUCTION

The anthropometric and strength data of the population are fundamental and essential to the safe and usable design of products as well as safe and productive workplaces, equipment and tools in occupational settings (Norris and Wilson, 1997; Mital and Kumar, 1998; Ekşioğlu 2004; Ekşioğlu and Kızılaslan, 2008; Ekşioğlu, 2011). The strength evaluations are also necessary for predicting the capability of workers while performing a job requiring strength without incurring injurious strains (Chaffin, 1975; Mital and Kumar, 1998; Ekşioğlu 2004). Specifically, hand torque strength is an important parameter to be considered for the design and evaluation of a broad range of manual tasks which involve the tightening and loosening of fasteners, threaded parts or connectors, turning knobs, so on.

Hence, torque strength norms of populations need to be developed worldwide to determine the capacity and reference values for design and evaluation purposes. However, it is a known fact that there are many 'gaps' in the available strength data (e.g.; Peebles and Norris, 2003; Kroemer, 1970 and 1999). Kroemer (1970, 1999) emphasized that there is a gap in the data because strength studies have been generally based on highly selected groups and there is very little information about the force capabilities of women or of population in general. A thorough examination of available literature by the authors also supported this fact:

There are only a few studies on the establishment of hand torque strength norms of the world populations so far (e.g.; Peebles and Norris, 2003), and factors affecting hand torque strength need further investigations. The findings from the reviewed literature can be summarized as follows:

(i) males are stronger than females;
(ii) dominant hand applies greater torque;
(iii) strength increases with age throughout childhood, peaks in adulthood, and then starts to decrease from age around 50 years and above;
(iv) the hand torque strength is affected by torque direction, hand orientation, surface friction coefficient, handle size and type, grip type, knurling and indentations;
(v) torque strength is also affected by posture: in general, higher torque strengths are exerted in the standing posture compared with the sitting posture. For some studies this difference is insignificant in practical sense;
(vi) for cylindirical handles, maximum torque values are achieved at about 5 cm diameter; below and above this diameter, torque values decrease;
(vii) The effects of handle dimension, shape, surface finish and type of material on hand torque strength are worth to investigate.

Furthermore, while many studies investigated hand torque strength capabilities for various factors, none involved job factor as manual and non-manual classification. Relatively few studies have focused on relationship between body posture and hand torque strength and key handle was not studied before.

Therefore, this study is an attempt to address some of strength data 'gaps'. For the purpose, the study aimed to establish the maximum voluntary static (isometric) hand torque strength (TS) norms of healthy adult female population of Turkey; and investigate the effects of handle type, posture, age-group, job-group and several anthropometric measures on TS. The methodology and findings from this research, as well as detailed descriptions and results of the data collected, are presented and discussed.

2 MATERIALS AND METHOD

2.1 Participants

The sample of participants consisted of 257 healthy adult females aged between 18 and 69 years. All subjects, participated in this study were in good physical health, and had no records of upper extremity, neck or shoulder disorders or pain. A self-report survey is used to determine the health status of the subjects. Healthy subjects signed a consent form prior to the experiments. The participants were grouped into 10-year age bands and the distribution of the subjects across these age groups were nearly uniform with around 50 subjects in each age band.

All subjects were recruited from the major metropolitan city of İstanbul and its surrounding areas. It was assumed that İstanbul approximately represents the general population of Turkey since the population of the city is composed of people whose roots are from every region of Turkey. Sampling is made so that all seven regions and most ethnic groups of Turkey are represented in the sample. Stratified random sampling method is used dividing the population of Turkey into strata (age-group, job-group and seven geographical regions of Turkey based on family roots and birth place).

2.2 Equipment

Hand torque strengths were measured with the CAP-TT01-250 Digital Torque Tester (Electromatic Equipment Co., Inc., USA) which is a special high-capacity version calibrated for a range up to 2875 N-cm. The torque tester was custom-ordered from the manufacturer to meet the needs of this study, in terms of accomodating the handle types to be studied and required torque range.

Four handles that were used are: [cylindrical (diameter: 51 mm and length: 113 mm), circular (diameter: 60 mm and depth: 20 mm), ellipsoid (length of major axis: 55.6 mm, length of minor axis: 42 mm and z-axis: 35.4 mm), and key (width: 30.9 mm, length: 55.6 mm and depth: 3.9 mm)]. The grip surfaces of circular and ellipsoid handles are made of plastic, the grip surface of key handle is made of natural anodized coated aluminum covered with plastic, and the grip surface of cylindrical handle is made of black anodized coated aluminum. The circular handle has knurled grip and the remaining three handles have smooth grip surface.

Dominant hand grip strengths were measured by a standard Jamar handgrip dynamometer and body measurements were measured with an anthropometric kit.

For both sitting and standing posture adjustments, a remote controlled adjustable height table and an adjustable chair were used.

2.3 Procedure

Prior to the experiments, some characteristics of the subjects such as date of birth and occupation are recorded, and some anthropometric measures, including dominant hand maximum voluntary static grip strength, are measured. In addition, a briefing is given to the subjects regarding the required posture, experimental task, equipment and the test procedure. Experimental variables included TS of dominant hand as the response variable, and posture (sitting and standing) and handle type (4 types) as independent variables (a total of 8 test combinations). Randomized block design with subjects serving as blocks was used as the experimental design. All 8 test conditions were randomized within each block.

Maximum voluntary static torque strengths for sitting posture were measured following the standard testing position recommended by the American Society of Hand Therapists (Fess and Moran, 1981; Bohannon, 1991; Fess, 1992). For standing posture, the subject stood in front of the measuring device and adopted a free posture. For both sitting and standing, all handles were positioned at elbow height and oriented horizontally.

Each participant's dominant hand TS at each of 8 test conditions were measured following the static strength measurement protocol by Caldwell et al., (1974) at least twice using the CAP-TT01-250 Digital Torque Tester. A static twisting force was exerted with dominant hand in a clockwise direction by right handed people and counter clockwise direction around the long axis of forearm by left handed people on circular, ellipsoid and key handles. For cylindirical handle, torque direction was inward (counterclockwise) perpendicular to the long axis of forearm while forearm in fully pronated position.

If the difference between the two measurements was within 10%, the test for that experimental condition for that subject is ended; otherwise, the test was repeated as many times as needed to meet the 10% variation criterion. There was a minimum of 2-minute rest period between two trials. The maximum of the torque strength values of the trials (the mean torque value of the middle 3 sec of 5 sec torque exertion time as instructed by Caldwell et al., 1974) is taken as the TS for that subject at that test condition.

2.4 *Pilot study and sample size determination*

A pilot study is performed with randomly selected 50 participants to determine the necessary parameter values for the required minimum sample size calculations and to become familiar with the test procedures. The minimum sample size is estimated using the statistical power formula provided in "General requirements for establishing anthropometric databases" (ISO 15535:2006) for a 95% confidence interval for the 5th and 95th percentiles (since in most cases, anthropometric data for technological design are of interest at the 5th and 95th percentiles):

$$n \geq \left(3{,}000 \times \frac{CV}{\alpha}\right)^2 \ \& \ CV = \frac{s}{\bar{x}} \times 100$$

(where, n: sample size, CV: coefficient of variation, α: the percentage of relative accuracy desired, $\bar{x}$ and s: the mean and std. dev. of the sample of torque strength). Assuming a relative accuracy of 5%, and using the empirical means and standard deviations, the required minimum sample size is calculated as 151. However, more than 100 additional subjects were included in the study to increase the statistical power of the results (total sample size: 257).

A repeatability study was also carried out after actual data collection with a randomly selected sample of the participants that were part of the actual study. The differences, in general, were found insignificant, and thus the repeatability was assured.

3 RESULTS AND DISCUSSION

Statistical analyses of the collected torque strength data were performed by Minitab 16 statistical package. The demographic profile and anthropometric characteristics of 257 female subjects are summarized in Table 1. The mean, std. dev., range and percentile values of TS classified by five age groups (from 18 to 69 yrs by 10-year bands), job

(manual and non-manual) and four handle types are obtained for adult female population of Turkey. The figures 1 to 3 show the box plots of TS by handle type, age-group and job-group, respectively. Table 2 shows the TS percentiles by posture and handle type.

ANOVA results indicated that handle type, age-group, job-group, posture and body-weight group all have significant effect on TS ($p = 0.092$ for BMI,

Table 1. Characteristics of the female subjects (n = 257).

Measurements	$\bar{x} \pm s$	Min–Max
Age (years)	43.61 ± 14.26	19–69
Stature (cm)	161.29 ± 6.01	145–175
Body mass (kg)	69.62 ± 12.95	42–110
Hand length (cm)*	17.19 ± 0.84	14–19.2
Hand breadth (cm)	7.87 ± 0.47	6–9.1
Wrist circ. (cm)	16.76 ± 1.41	14–23.5
Forearm circ. (cm)	25.37 ± 2.59	17–34
Grip strength (N)	251.6 ± 68.9	78.5–70.7

*Length and strength values refer to dominant hand.

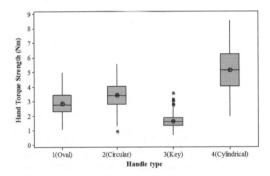

Figure 1.   Boxplots of torque strength by handle type.

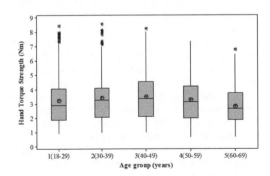

Figure 2.   Boxplots of torque strengths by age-group.

39

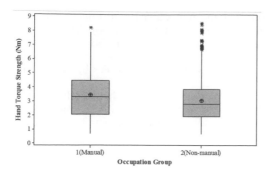

Figure 3. Box plot of torque strength by job-group.

Table 2. Dominant hand TS percentiles (in Nm) by posture and handle type for adult females of Turkey.

Handle	Sitting (%iles in Nm)				Standing (%iles in Nm)			
	5	50	95	sd	5	50	95	sd
Ellipsoid	1.74	2.62	3.96	0.71	1.76	2.95	4.31	0.78
Circular	1.79	3.33	4.86	0.86	2.03	3.49	5.00	0.88
Key	1.01	1.56	2.27	0.39	1.05	1.6	2.39	0.42
Cylindrical	2.97	4.98	6.94	1.30	3.15	5.26	7.56	1.38

and p < 0.001 for the remaining factors). Participants, on average, generated significantly higher TS values while standing compared to sitting posture, which is in agreement with previous studies.

Overweight group was found significantly stronger than normal weight group (There was only a few underweight subjects that were excluded from the analysis). However, the differences of both posture and body-weight effect are small to be considered significant for practical applications. Manual workers had statistically higher TS values than non-manual workers.

Tukey's test provided further details on the effects of handle-type and age-group on TS. Participants generated higher TS with cylindrical handle followed by circular, ellipsoid and the lowest with key handle. This was an anticipated result since circular handle allowed larger hand grasp area for power grip whereas key handle allowed only pinch grip torque application with thumb against index finger. On the other hand, the difference between circular and ellipsoid handle can partly be explained by the knurled surface of circular handle that allowed better grip.

Overall, the highest values of the hand torque strength obtained at the age-group of (40–49) and the lowest values at the age group of (60–69).

However, the TS remained relatively constant up to age-group of (40–49). If each of two job groups is examined separately, it can be seen that the (30–39) age-group is the strongest of non-manual group. On the other hand, age group of (40–49) is the strongest of manual group. This may be partly explained by the 'use or loose' principle of muscle: manual workers are expected to use their arm muscles at work more forcefully keeping them in shape, and thus resisting to the age-related muscle loss longer years.

Tukey tests indicated that only the oldest group (60–69) has significantly lower TS than the all other groups (p < 0.001).

The results showed significant correlations between the eight measurements, with Pearson product-moment correlation coefficients ranging from 0.277 to 0.872 (p < 0.001). Furthermore, grip strength and some of the anthropometric variables, such as forearm circumference and hand breadth were also significantly correlated with TS.

Direct comparisons with hand torque studies were very difficult to perform since experimental conditions vary among the studies (e.g.; handle type, size and shape). However, comparisons were made with studies that had somewhat similar conditions using t-tests. The results indicated that TS values obtained through this study were significantly higher than some study results and lower than some others (e.g.; Kim and Kim, 2000; Peebles and Norris, 2003; Seo et al., 2008).

4 CONCLUSIONS

Designing for hand torque strength, a designer must consider varying torque strength capabilities of all intended users. A knowledge of the torque strength of users when interacting with the product in question is therefore necessary for good designs.

Through this study, the static hand torque strength norms of adult female population of Turkey are established between the ages of 18 and 69 yrs. These norms can serve as reference in design for torque strength for female population of Turkey. Hence, one more 'gap' is filled in the ergonomics knowledgebase in the world.

Some of the handle types tested is unique to this study in the sense that the dimensions, type, shape, and surface finish are different from previous studies. Therefore, the torque strength data generated provide new data to be referred for future studies and for comparison purposes. Moreover, the authors believe that this is the first study that investigated the job effect (manual and non-manual) on torque strength.

REFERENCES

Bohannon, R.W. 1991. Hand grip dynamometers: issues relevant to application. Journal of Human Muscle Performance, 1: 16–36.

Caldwell, S.L., Chaffin D.B., Dukes-Dobos F.N., Kroemer K.H.E., Laubach L.L., Snook S.H. & Wasserman D.E. 1974. A proposed standard procedure for static muscle strength testing. *American Industrial Hygiene Association Journal*, 35(4), 201–206.

Chaffin, D.B. 1975. Ergonomics Guide for the Assessment of Human Static Strength. *American Industrial Hygiene Association Journal,* Vol. 36, pp. 505–511.

ISO 2006. General requirements for establishing anthropometric databases: EN ISO 15535:2006.

Ekşioğlu, M. 2004. Relative optimum grip span as a function of hand anthropometry. *International Journal of Industrial Ergonomics, 34(1)*, 1–12.

Ekşioğlu, M. & Kızılaslan, K. 2008. Steering-wheel grip force characteristics of drivers as a function of gender, speed, and road condition. *International Journal of Industrial Ergonomics,* 38 (3–4) 354–361.

Ekşioğlu, M. 2011. Endurance time of grip-force as a function of grip-span and arm posture. *International Journal of Industrial Ergonomics,* 41(5) 401–409.

Fess EE. & Moran C. 1981. Clinical assessment recommendations. Am Soc Hand Therapists.

Fess EE. 1992. Grip strength. In: Clinical assessment recommendations, 2nd ed. Am Soc Hand Therapists, Chicago, pp 41–45.

Kim, C.H. & Kim, T.K. 2000. Maximum Torque Exertion Capabilities of Korean at Varying Body Postures with Common Hand Tools. *Proceedings of the Human Factors and Ergonomics Society Annual Meeting,* v3, 157–160.

Kroemer, K.H.E. 1970. Human Strength: Terminology, Measurement and Interpretation of Data. *Human Factors*, 12(3), 297–313.

Kroemer, K.H.E 1999. Assessment of human muscle strength for engineering purposes: a review of the basics. *Ergonomics*, 42 (1), 74–93.

Mital. A. & Kumar, S. 1998. Human muscle strength deÞnitions, measurement, and usage: Part I -Guidelines for the Practitioner. *International Journal of Industrial Ergonomics,* 22, 101–121.

Norris, B. & Wilson, J.R. 1997. Designing safety into products. *Product Safety and Testing Group Institute for Occupational Ergonomics,* London.

Peebles, L. & Norris, B. 2003. Filling 'gaps' in strength data for design. *Applied Ergonomics,* 34, 73–88.

Portney, L.G. & Watkins M.P. 1993. *Foundation of clinical research: Application to practice,* Appleton & Lange (Norwalk, Conn.).

Seo, N.J., Armstrong, T.J., Ashton-Miller, J.A. Chaffin, D.B. 2007. The effect of torque direction and cylindrical handle diameter on the coupling between the hand and a cylindrical handle. *Journal of Biomechanics,* 40, 3236–3243.

Occupational Safety and Hygiene – Arezes et al. (eds)
© 2013 Taylor & Francis Group, London, ISBN 978-1-138-00047-6

Intervention in hydrometers industry: Recommendations for improvements based on ergonomic assessment

G.S. Ribeiro
Federal University of Rio de Janeiro, Rio de Janeiro, RJ, Brazil

C.R. Leite da Silva
State University of São Paulo, Bauru, SP, Brazil

F.P.C. Neves & L.B. Martins
Federal University of Pernambuco, Recife, PE, Brazil

ABSTRACT: With the aim of adapting working conditions to the skills, abilities and limitations of workers, this paper presents the results of an ergonomical intervention performed at an volumetric and velocimetric water meters factory, located in Recife, Pernambuco, Brazil. Based on a systemic approach which considers human-task-machine-environment interface, the results indicated that the factory exposes its workers to problems of physical-environmental, organizational, job, biomechanical and psychosocial order. Changes and adjustments in the surveyed employees' working conditions were proposed to guarantee comfort, safety and life quality to these people.

1 INTRODUCTION

The manual assemblage of equipment has been, profusely, cited as an occupation exposed to several ergonomic constraints, associated with the fractionation of the tasks, promoting monotony activities and high number of repetitive movements that can result in physical and psychic illnesses.

In assemblage systems, as in any subsystem manufacturing, there are factors that influence the way the operator performs activities, such as: Work place, work organization, layout, product design and employee training. However, generally, these factors aren't taken into account in an integrated way when there's a definition of tasks to workers, disregarding ergonomic aspects related to work and encouraging the involvement of insecurity, discomfort and illness.

It comes, then, from the assumption that better working conditions, resulting in comfort, health and safety, generate more satisfied workers and consequently better life quality.

It's in this context that there is the focus of this research, ie, the need to think about job from the precepts of ergonomics, understanding the work systemically, where one does not work without the other, solving the relationship human-machine-task-environment in order to create better working conditions for all. Dul and Weedmeester (2001) corroborate the idea by stating that ergonomics can help to solve

a large number of social problems related to health, safety, comfort and efficiency, and can contribute to preventing errors, improving performance.

With the aim of adapting working conditions to skills, abilities and limitations of workers from the systemic approach considering human interface task-machine-environment through ergonomics intervention, is intended to propose design requirements improvement to the problems found, in order to subsidize comfort, safety and quality of life for workers in an volumetric and velocimetric water meters factory.

2 MATERIALS AND METHODS

The conducted research was descriptive and exploratory, from literature and ergonomic intervention of human-machine-task system searches, proposed by Moraes and Mont'Alvão (2003), comprising the steps of assessment, diagnosis and projecting ergonomic.

The data collection was carried at a volumetric and velocimetric hydrometers assembler, located in the city of Recife, Pernambuco, northeastern Brazil. Particularly, in an organization sector that includes: (i) assembly of totalizers, (ii) recording of covers and closing, (iii) test and the totalizers.

On the occasion of the research, the company had thirty five employees, spread across all sectors of factory assembly. The surveyed area, at baseline, had eight people. Later, three more employees were hired. So a total of eleven people (six women and five men) were surveyed in the industry.

There are two types of factory layout investigated, which vary according to the demand of produced hydrometers. The hydrometers of models A, B and C are mounted in line, as shown in Figure 1.

Figure 2 illustrates the assembly process in the cell, which is used when mounting the model D.

2.1 Ergonomic appreciation phase

Identification of ergonomic constraints for employees step.

For this, there were: (i) unsystematic observations in real work situation, (ii) analysis of documentary of prevention program of environmental risk and control program of occupational health physician, (iii) open interviews and questionnaires structured with 100% of workers surveyed, which served as input for structuring the identification of plant, (iv) identification of ergonomic problems and grouping into categories, (v) formulation of ergonomic advice.

Figure 1. Workstation layout with in line.

Figure 2. Workstation layout in cell.

It was also applied the matrix of severity, urgency and trend (SUT), by Kepner and Tregoe (1976), in order to guide decisions and highlight the most complex problems to be prioritized in subsequent phase of research.

2.2 Ergonomic diagnosis phase

To further investigation of the encountered and prioritized problems, the occupational postures assumed by 100% of workers were analysed, relating them to the repetitiveness of tasks and expressions of discomfort or pain in body regions. Therefore, we carried out systematic observations of postures assumed, according to RULA spreadsheet, Rapid Upper Limb Assessment, by McAtamney and Corlett (1993). The physical costs of workers, with discomfort/pain diagram application, proposed by Corlett and Bishop (1976), and anthropometric studies with bidimensional anthropometric dummies were analysed.

For the analysis of assumed postures, systematic observations were made during visits in the industry installations, from the filming of the work in real-use situation and with a digital camcorder, for a period of 50 minutes, each job. Transferred and processed on computer, the movies were paused every 30 seconds and analysed from the RULA Online spreadsheet (www.rula.co.uk), also were identified the postures assumed for all body regions. Such analyzes encompassed two physical arrangements present in the sector under study.

A scale was used regarding discomfort/pain, graded 1 to 5, proposed by Corlett and Bishop (1976): 1, corresponds to a comfortable situation or no pain; 2, corresponds to light discomfort or pain; 3, corresponds to moderate pain or discomfort; 4, corresponds to severe pain or discomfort and; 5, corresponds to unbearable pain or discomfort.

To investigate whether the observed situation would result in a poor design of workstations and inadequate characteristics of workers, we proceeded to anthropometric studies of the situation found in the company. Thus, measurements of workstations were made; they were transferred to the program with CAD system, and studied from bidimensional anthropometric dummies, with representations in the human scale of the largest man (95 percentile) and smallest woman (5 percentile), based on Panero and Zelnik (2002).

2.3 Ergonomic design specifications phase

After phases of observation, records and diagnoses of ergonomic constraints to which workers were subjected highlighting the recognition of its causes, improvement suggestions to the encountered problems were proposed. It is worth mentioning that these proposals were in accordance with the strategic decisions of the organization, seeking greater life quality for workers.

3 RESULTS AND DISCUSSION

In accordance with the results obtained in the first phase of the research, the ergonomic appreciation, it was noticed that sector of assemblage totalizers offers plenty occupational hazards to workers.

Based on documentary analysis undertaken, the Prevention of Environmental Risks Program, the company had reported problems with noise and lighting.

The program reported the need for provision of personal protective equipment (PPE) to workers, to control damage from noise and lighting that does not meet the recommended levels by Brazilian standards, like NBR5413 and NR17.

The control program of occupational health directs procedures to be followed by the company, as admission examinations, periodic examinations, examinations of return to work in case of long absences, for cases where the employee changes function or in cases of termination. However, it needs to be clarified whether the recommendations are followed or not. Our analyzes found that aren't registered any accidents, number of lesions and human costs that adversely affect the health of workers, which could be interpreted as if organization did not offer the health risks of their workers. This situation differs from program data to mitigate environmental risks.

According to results obtained from unsystematic observations, interviews and questionnaires, there is occurrence of ergonomic constraints of physical-environmental, organizational, biomechanical and psychosocial levels.

The aspects related to the maintenance of occupational positions are noteworthy as inappropriate due to prolonged seated position and postures needing the workers to be outside their comfort, the remoteness of supplies and equipment required for totalizers assemblers and workstations with inadequate anthropometric scaling features for its users.

In both layouts of the assembly factory workstations studied, the SUT matrix showed flexion of the neck and upper limbs for long periods, frequent use of forceps type movement and noise while cleaning the base reached the maximum indices severity (5), urgency (5) and trend (5), total score of SUT = 125, for each of the elements of ergonomic demand pointed. Culminating in prioritizing and deepening of these items during the diagnosis ergonomic second phase of the research.

The results obtained with the application of RULA worksheet shows that the joints of workers which are overburdened due to the posture required

by the type of motion are: shoulder, neck, wrist and hand.

The jobs entail frequent evaluated shoulder elevation above 90 degrees, static contraction of the neck, causing muscle fatigue and pain and tension frames in character burning in the region of cervical paraspinal muscles and the upper trapezius. Postures of flexion and extension, radial and ulnar deviation and pincer movements of the joints of the wrist and hand, which, combined with the repetitiveness, are risk factors for the development of occupational diseases, such as tendonitis and carpal tunnel syndrome. Range of motion out of the comfort levels, mainly in relation to the shoulders (above 60 degrees), in ranges of objects and trigger buttons. Repetitive movements primarily of the wrist and hand, besides the maintenance of prolonged static postures of seated position and cervical flexion.

In cell layout, figure 3, the jobs most affected are those related to partial tasks, such as assembling and assembly of ogômetros or totalizers. Tasks involve assembly of subcomponents of the product, with greater repeatability of movements, because the tasks are less enriched when it comes to physical arrangement.

Regarding the layout (physical arrangement) in line, Figure 4 shows that the jobs that require more maintenance of inappropriate occupational

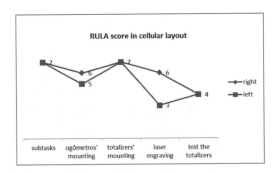

Figure 3.   RULA score in cell layout.

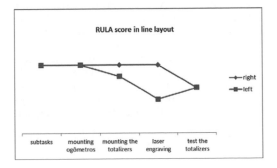

Figure 4.   RULA score in line layout.

workers positions are also related to the ogômetros partial assembly and assembly of totalizers. The assembly of subcomponents with little enriched tasks justifies the high scores.

When comparing two RULA worksheet scores in Figures 3 and 4, it is noticed that the assembly that exposes workers to greater risk factors is related to the D model, with cellular layout.

The application diagram discomfort/pain with workers studied corroborate the data obtained from the application of worksheet.

The results indicated that related workers felt pain from the start of work activities. And, along the occupational activities, the pain was intensified and had markedly elevated perception of its intensity, with strong and unbearable pain in neck, right shoulder and nape.

So with the analysis supporting the diagram of discomfort/pain, one realizes that many workers already start the workday feeling discomfort or pain. This situation can be justified by the task performed throughout the week, which, by exposing workers to ergonomic constraints, promotes the accumulation of pain from one day to another.

In all cases studied, the incidence of discomfort/pain is increasing throughout the workday, reaching unbearable levels, as can be seen in the job of finalizing the assembly of totalizers and recording cover.

The region of neck, upper limbs and back are the areas with the greatest number of complaints resulting from poorly sized workstation linked to repetitive tasks.

These findings were ratified with anthropometric study. For men the highest percentile, the accomplishment of the task was performed backed up on the bench, who possessed sharp areas. The elbow flexion remains above 90 degrees. Also, the footrest forcing the same to keep the hips and knees flexed above 90 degrees, the shock lower limbs against the lower part of the bench and keep the ankle spinal flexion Dornal to its full support.

In the case of the smallest percentile regarding women, we found problems in the scale of the post. To keep the shoulder in a neutral position and the forearm supported (with cutting area), the elbow is in flexion above 90 degrees, even with the shoulder associated with a high elbow extension, it can't achieve the objects before him. And the legs do not have support correctly, only the ends of the fingers.

Therefore, we can say that the work on the assembly sector of the factory totalizers above exposes its employees to various levels of ergonomic constraints. Being necessary changes and adjustments in the workplace to make-work safer, more comfortable and thereby increase the quality of life of these people and their work efficiency.

Among the proposed improvements presented, comprising the step of ergonomic design specification were addressed the broader issues, from organizational aspects of the physical and environmental conditions of the workstation. Among which stand out guidance on posture assumed with workers, purchase of furniture (chairs and benches) customizable, workstation redesign, with U-shaped bench, bringing the inputs and training for better performance and greater worker safety.

4 CONCLUSION

The results showed problems in the physical-environmental, organizational, and psychosocial biomechanics workstation investigated.

It was possible to understand the systemic nature of the job, being necessary to analyze the harmony between human-machine-environment-task when one does not work in perfect condition without the other.

More and more the trend reasserts itself holistic approach of systemic problems, emphasizing the organizational and psychosocial aspects, which are often suppressed by physiological aspects inherent in professional activity. So, are necessary adaptations and changes in working conditions of workers in the sector studied. In order to make-work safer, more comfortable and thereby increase the quality of life of these people and their work efficiency.

REFERENCES

Corlett, E.N.; Bishop, R.P. (1976). A tecnique for assessing postural discomfort. *Ergonomics*, v. 19, n. 2, p. 175–182.

Dul, J.; Weedmeester, B. (2001). *Ergonomia Prática*. São Paulo: Editora Afiliada.

Kepner, C.H.; Tregoe, B.B. (1976). *O Administrador Racional*: uma abordagem sistemática à solução de problemas e tomada de decisões. São Paulo: Atlas.

McAtamney, L.; Corlett, E.N. (1993). RULA: a survey method for the investigation of work-related upper limb disorders. *Applied Ergonomics*, v. 4, n. 2, p. 91–99.

Moraes, A.; Mont'Alvão, C. (2003). *Ergonomia*: Conceitos e Aplicações. Rio de Janeiro: iUsEr.

Panero, J.; Zelnik, M. (2002). *Dimensionamento humano para espaços interiores*. Barcelona: Gustavo Guili.

Occupational Safety and Hygiene – Arezes et al. (eds)
© *2013 Taylor & Francis Group, London, ISBN 978-1-138-00047-6*

The role of ergonomics in implementation of the social aspect of sustainability, illustrated with the example of maintenance

M. Jasiulewicz-Kaczmarek
Poznan University of Technology, Poznan, Poland

ABSTRACT: Traditionally, the scope of maintenance referred to production processes. It was widely agreed that the main aim of maintenance was the optimization of equipment availability at the lowest feasible cost. However, the shift from production paradigm to sustainable development resulted in a change of the maintenance paradigm towards of product life cycle management; as well as taking into account economic, environmental and social aspects. The many decisions made during the process of a technical object's design, production and operation directly influence the effect and outcome in the social dimension. Social dimension of sustainable maintenance means including safety and human factor aspects in every stage of product's life cycle, cooperation between a designer and maintenance staff, information exchange and continuous search for opportunities of improvement for all the parties involved in maintenance in a company.

1 INTRODUCTION

Sustainability is becoming a significant component of operational and competitive strategies in an increasing number of companies. The idea combines economic, environmental, and social aspects of activity of an organization. Social sustainability is implemented in concepts such as preventive occupational health and safety, human-centred design of work, empowerment, individual and collective learning, employee participation, and work-life balance. All these concepts aim to preserve or build up human capital, and they represent a conscious way to deal with human resources. This contemporary approach to running business caused changes in the way the functioning of basic processes and, more importantly, because of it support processes are perceived with some reference to production equipment maintenance.

Sustainable maintenance is a new challenge for enterprises realizing concepts of sustainable development. It can be defined as pro-active maintenance operations striving to provide balance in the social (welfare and satisfaction of maintenance operators), environmental and economic dimensions (Fig. 1).

The objective of this article is to show the practical aspects of implementation of the social dimension of sustainable development, in terms of safety and ergonomics. The subject of consideration are technical activities performed in the maintenance system.

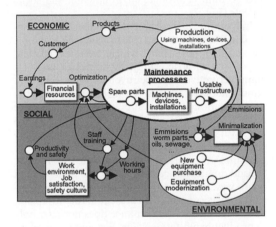

Figure 1. General model of interdependencies in sustainable maintenance (Jasiulewicz-Kaczmarek 2013).

2 WHAT IS THE MAINTENANCE?

According to BS EN 13306:2010, maintenance concerns the combination of all technical, administrative, and managerial actions during the life cycle of an item, intended to retain it in, or restore it to, a state in which it can perform the required function. The conclusion from the definition presented is the following: maintenance is not only purely technical tasks, such as disassembly and replacement of spare parts, lubrication, and repair. In practice, maintenance has a much wider spectrum involving numerous additional tasks. These include planning work/jobs, organizing work, as well as resources necessary to complete it and all the jobs f.ex.: the choice of appropriate tools, the choice of appropriate chemicals, preparing areas (evacuating non-involved personnel, traffic control and putting up signs), preparing machinery or areas for shut down, transporting spare parts (manually or in industrial vehicles), preparing the necessary safety precautions (for instance PPE, energy depletion, training), motivating people and controlling them. Maintenance is therefore a generic term for a variety of tasks in a very wide range of industrial sectors and in every type of working environment (Lind & Nenonen 2008).

Traditionally, the scope of maintenance is referred to production processes. It was widely agreed that the main aim of maintenance was the optimization of equipment availability at the lowest feasible cost. However, the shift from production paradigm to sustainable development resulted in a change of the maintenance paradigm towards of product life cycle management; as well as taking into account economic, environmental and social aspects (Fig. 2).

In literature maintenance function is usually analyzed in the context of its role in achieving economic and ecologic efficiency, which refers to efficient use of f.ex. electric energy, high level of products manufactured or competitiveness of a company. Simultaneously, it also contributes significantly to occupational safety and health. Maintenance influences the safety and health of workers in two ways. First, regular maintenance that is correctly planned and carried out is essential to keep both machines and work environment safe and reliable. Second, maintenance itself has to be performed in a safe way, with appropriate protection of maintenance workers and others present in the workplace. Such approach to maintenance does not refer only to product exploitation stage. Sustainable development prospective requires broader approach and taking the entire life cycle into consideration, from the design to disposal stage. During the design stage the basic characteristics of an object from mainte-

nance point of view is created—maintainability. Hence, safety and human factor/ergonomics as aspects of social dimension of sustainable maintenance should be analyzed in a holistic way and refer to all the aspects of technical object's life cycle.

3 MAINTENANCE, SAFETY, HUMAN FACTOR IN TECHNICAL OBJECT'S LIFE CYCLE

The system life cycle is a sequence of phases, each containing tasks, covering the total life of a system from the initial concept to decommissioning and disposal. The many decisions made during the process of a technical object's design, production and operation directly influence the effect and outcome in the social dimension. The maintenance managers hold all the instruments that allow the firm's technical service workers to participate in all phases of the life cycle of the machine and thus engage in the implementation of the social dimension of the strategy of the organization.

The first phase of the product's life cycle is its design. The fundamental rationale for the technical object design comprises, among others, a guarantee of operational reliability, as well as repair, diagnosis and maintenance adaptability. Therefore, the design phase exacts not only functional and safety requirements, but also ergonomics. Technical object's user and operator is a human being, thus if human factor issues are efficiently applied during the design stage, exposure to physical hazard and physiological and mental strain is likely to be decreased.

Elements such as maintainability, the ability to keep machinery in good condition by staff, have a direct relationship with the physical and psychological behavior of operators that undertake repairs or preventive maintenance tasks, besides the reduced efficiency and effectiveness of bad maintainability (in engineering, maintainability is defined as ease of maintaining technical objects, f.ex. isolate defects or their cause, correct defects or their cause, automate fault detection and isolation tasks whenever possible make future maintenance easier etc.).

In the literature some solutions for integration of maintainability criteria in product design stage can be found (Rajpal et al. 2006, Kuo & Chu 2005). A number of software tools have been developed for ergonomic design of machines, workplaces, and occupational devices. Coulibaly et al. 2008, presented an approach for maintainability and safety assessment in the design process using CAD model enriched with behavioral semantic data. Rui et al. 2009, classified the human factors requirement in the repair processes of products. These authors present a human-factor evaluation method for maintainability design with the VR-based environment.

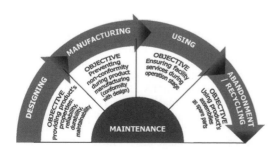

Figure 2.   Maintenance in product life cycle.

Their works give some rules to apply for the design for maintainability, but also characterize the problem. The solutions however should be searched for in cooperation between devices' designers and their final users, and specifically in cooperation between engineers and maintenance department staff. Expectations and requirements of final user concerning a device which is to be designed can be met only with knowledge on past experience with same or similar object maintaining. The knowledge can be obtained on one way, which is contact with people performing maintenance operations (service and maintenance staff). Their suggestions, remarks and initiatives should be the basic input data for design.

Contemporary approaches to maintenance, such as Total Productive Maintenance (TPM), consider this issue an important element of maintenance. The important pillar of TPM is so called "Early Equipment Management". One of the goals of that pillar is collecting information concerning opportunities for continuous improvement of exploited objects in terms of operational activities performed by operator and in terms of lubrication, inspection and repair performed by technical staff. Some examples of issues which are usually analyzed are presented in the table below (Table 1).

The tendency applied to machines design is striving for minimization of amount and range of service performed by humans. We would like to make next generations of machines as automatic as possible. However, independently from how complex and advanced automatic solutions will be applied and to what extend wireless technologies will be used, it is

impossible to eliminate human activities in maintenance ("zero maintenance"). There are and always will be some activities that need to performed by maintenance workers. Thus it is necessary to design a machine and its service to make it least physically and psychically inconvenient and troublesome.

Thus, the life cycle of the machinery should be considered during the design phase, and design features introduced to control unacceptable risks that may occur at different stages in its life. This should be done by eliminating hazards at source, f.ex. at:

- installation—to minimize hazards during installation, a large machine might be designed so that it is supplied in modules that can be placed in position by a crane. This avoids the need for installers to work at height or manually handle heavy items;
- maintenance—considerations should include providing ready access to areas identified during design requiring regular maintenance, such as cleaning, lubrication and adjustment; routine adjustments should be designed to be carried out with the machine stopped but without the need for removal of safeguards or dismantling of machine components; where frequent access is required, interlocked guards should be used; self-lubrication or central lubrication of parts should be considered if access is difficult, etc.

The next phase of the product's life cycle, which is influenced and participated by maintenance staff, is product exploitation. In every manufacturing company pressure on meeting predefined production goals is very high. It is transferred to requirements defined for maintenance and referring to providing high accessibility and proper parameters of machinery work. The question whether to produce or make some repairs is usually answered by managers with "to produce". The consequence of such decision is not only harmful for the production process (an results i.e. in increase of discordant products) but also for safety of process participants, f.ex. operators.

It seem quite a paradox that when a breakdown happens we can easily identify its cause—conservation that was not executed, but when everything goes according to the plan we do not appreciate preventive actions. That way of thinking is reflected in traditional approach to maintenance in which it is simply "a necessary evil". Wrong decisions and organizational culture are usually key factors for human mistakes when performing conservation works.

Despite continuous contradiction between production department goals and maintenance, maintenance as a support process is a very important stage of technical object's exploitation.

Starting from the decision on purchase of a new machine, it is necessary to take into consideration tasks it generates for maintenance department. Irrespective of its functioning or not, some technical

Table 1. Examples of perspectives for equipment weak points searching.

Area	Issues
Facilitating autonomous maintenance	Can cleaning and inspection be easier?
	Can lubrication be centralized so that lubricant is supplied at just one or two inlets per equipment unit?
Improving quality	Is diagnostic equipment easy to set up?
	Does it have visual displays?
Improving maintainability	Have equipment life data been collected, and is work in progress to extend equipment life?
	Can parts replacement be simplified?
	Are self-diagnostic functions built into the equipment?
	Can oil supply and oil changing be simplified?
Safety	Are interlocking methods safe?
	Are there safety fences around hazardous equipment?

services just have to be performed. The next activity is designing a workstation, a place where the object is to be installed. "The focus on design and planning of new installations or production systems as a goal for participatory ergonomics (PE) triggers a new question: How can workers and other workplace end-users participate in setting up measures for ergonomics, when the new workplace does not exist?" (Broberg et al. 2011).

After installation of a technical object and preliminary starting actual exploitation stage begins, which means that the object performs its functions according to its definition. However, realization of basic functions of a machine depend on performance in the right time, with use of proper tools and materials, according to recommendation of the producer of technical services.

Maintenance is not a routine activity. Working under time pressure (especially when shutdowns or high-priority repairs are involved), changing tasks and working environment are typical for maintenance operations. Studies indicate that industry maintenance workers might be especially at risk of contracting occupational diseases, too (AFIM 2007). This makes maintenance a high-risk activity.

Maintenance is rarely part of the production plan and is mostly considered a waste of time. As a result of this maintenance personnel works under stress to complete the tasks in a very limited time and under constant pressure, on pieces of machinery that are not always well maintained. Procedures are not followed, maintenance technicians often improvise and the quickest methods are used at the expense of their own safety. A consequence is that the maintenance schedule, when present, rarely includes testing the safety gear (for instance safety switches, light curtains, sirens, emergency buttons, signs and interlock switches) that are an integral part of the equipment, and which have a limited operational life. Analysis of the causes presented by Scroubelos (2011) leads to the conclusion that they all refer to organizational aspects. Thus, from safety point of view maintenance managers' actions should be focused on planning work and providing resources necessary for its execution (materials, tools, personal protection means, procedures, instructions etc.), motivation and building awareness, as well as consequent controlling if predefined schemes and instructions are followed and what are their results. Implementation of safety procedures and safe work systems should be supported with actions shaping safe behavior with trainings and information providing. System approach to safety issues, which consists of participation of employees and managers is the only way to create safety culture and include human factor/ergonomics in everyday routines. These elements can be easily

found in actions realized by companies according to *Behavior-based safety* approach (Jasiulewicz-Kaczmarek & Drożyner 2011).

When analyzing the product's life cycle from the social point of view, we reflect on how maintenance can contribute to worker's safety in the production processes. Generally the fact that maintenance operations involve some specific risks and that their operators are exposed to potential chemical, biological, physical (etc.) hazards that may influence their safety and health more than other workers is not taken into consideration.

Repair work requires sustained awkward postures of the back, neck, and shoulders as well as repetitive manipulations and awkward postures associated with hand tool use. Sustained awkward postures restrict blood flow and can cause muscle fatigue as well as place the employee at risk of developing Work-Related Musculoskeletal Disorders. Repeatedly performing tasks in such positions imposes increased stress on the muscles and joints. Employees are exposed to contact stress to the hands from using small tools, while a lack of task lighting can increase eye strain and induce awkward postures as employees try to adequately view a part (Fig. 3).

Opposite to what can be assumed, conservation is not only domain of maintenance staff. Maintenance is a role that can be executed by different operators in a principal or subsidiary way. Maintenance organizations are various and have undergone profound modifications (e.g. autonomous maintenance (AM); shared, integrated or specialized maintenance; subcontracting maintenance; and remote maintenance) leading for example to the allocation of maintenance tasks to production operators.

Figure 3.　"Ergonomics" in machine repair process.

Thus, maintenance operations may be carried out by a specialized operator, a user or an operator who is external to the company owning the items being maintained. Many actions are realized together by machines operators and maintenance staff and some are performed only by operators as so called autonomous maintenance. Autonomous maintenance is a particularly important pillar of TPM.

Implementation of the AM program requires preparing operators for active partnership with maintenance department staff not only in technology area (f.ex. inspection scheme), but also in safety and ergonomics area. Realizing the program of joining employees in autonomous maintenance and actions for realization of strategies such as Lean and TQM they usually benefit from "Kaizen events" system for initiatives introduction. Initiatives of employees are mostly focused on various aspects of work processes and production equipment. Solutions referred to equipment usually strive for providing better access to machine sub-assemblies, simpler inspection and repairs schemes, design of tools facilitating the works mentioned, "safety maps" development for machines for which hazards are visualized etc.

Participation of maintenance staff in work environment improvement is crucial as workers know problems connected with work performing best and when they are well prepared (instructions and trainings) they can and want to identify these problems. The results are improvement ideas concerning both work routine and machines and tools used in production process introduced by workers. When analysing the product's life cycle from the social point of view, we reflect on how maintenance can contribute to worker's safety in the production processes. Generally the fact that maintenance operations involve some specific risks and that their operators are exposed to potential chemical, biological, physical (etc.) hazards that may influence their safety and health more than other workers is not taken into consideration (Fig. 4).

Figure 4. Example of results of "Kaizen".

Participation of maintenance staff in work environment improvement is crucial as workers know problems connected with work performing best and when they are well prepared (instructions and trainings) they can and want to identify these problems. The results are improvement ideas concerning both work routine and machines and tools used in production process introduced by workers.

4 CONCLUSION

In recent years, maintenance has been the subject of fundamental change and is now regarded as an essential function within companies. However, maintenance-related risks continue to receive limited attention and little research has been devoted to the impact of maintenance on the safety of those who work in maintenance and their co-workers. It is essential to take a structured approach to maintenance, seeing it as a process rather than a task. The process should be analyzed in comprehensive way and include all the stages of technical object's life cycle. Social dimension of sustainable maintenance is including safety and human factor in every stage of product's life cycle, cooperation between designers and maintenance workers, exchange of information and continuous search for opportunities of improvement for all the parties involved in maintenance in a company.

REFERENCES

Association française des ingénieurs et techniciens de maintenance, *Guide national de la maintenance 2007*, Paris, AFIM, 2007.
BS EN 13306:2010 Maintenance. Maintenance terminology.
Broberg, O. et al. 2011. Participatory ergonomics in design processes: The role of boundary objects. *Applied Ergonomics* 42: 464–472.
Coulibaly, A. et al. 2008. Maintainability and safety indicators at design stage for mechanical products, *Computers in Industry* 59: 438–449.
Jasiulewicz-Kaczmarek M. & Drożyner P. 2011. Preventive and Pro-Active Ergonomics Influence on Maintenance Excellence Level, In: M.M. Robertson (eds), *Ergonomics and Health Aspects*: 49–58. LNCS 6779, Springer-Verlag.
Jasiulewicz-Kaczmarek, M. 2013. Sustainability: Orientation in Maintenance Management-Theoretical Background. In: P. Golinska et al. (eds.), *Eco-Production and Logistics. Emerging Trends and Business Practices*: 117–134. Berlin-Heidelberg: Springer-Verlag.
Kuo, C-F. & Chu, C-H. 2005. An online ergonomics evaluator for 3D product design. *Computers in Industry* 56(5): 479–792.
Lind, S. & Nenonen, S. 2008. Occupational risks in industrial maintenance, *Journal of Quality in Maintenance Engineering* 14(2): 194–204.

Rajpal, P.S. et al. 2006. An artificial neural network for modeling reliability, availability and maintainability of a repairable system, *Reliability Engineering & System Safety* 91(7): 809–819.

Rui, L. et al. 2009. The preliminary study on the human-factor evaluation system for maintainability design. Presiding of the International Conference on Intelligent Human-Machine Systems and Cybernetics 2: 107–112.

Scroubelos, G. 2011. Incidents in maintenance: their link to the tasks, special characteristics and proposed measures, In *Magazine-Healthy Workplaces. A European Campaign on Safe Maintenance 12:* 14–19. Luxembourg: European Agency for Safety and Health at Work.

Occupational Safety and Hygiene – Arezes et al. (eds)
© *2013 Taylor & Francis Group, London, ISBN 978-1-138-00047-6*

Postural changes and musculoskeletal disorders in workers with mental disabilities

F. Diniz-Baptista
Department of Biomedical Sciences, Universidad de León, León, Spain
ISLA, Higher Institute of Languages and Administration, Santarém, Portugal

ABSTRACT: This study's main goal was to identify postural changes and musculoskeletal disorders (MSDs) in workers with mental disabilities on a Protected Job Centre in Lisbon area. Were studied 36 workers of both sexes, aged from 26 to 50 years, assigned to the areas of; packaging cutlery, carpentry/joinery, stuffing, industrial laundry, ironing, gardening, and aid to patient transportation. Was applied the OWAS methodology for postural analysis, a survey questionnaire and a bodily pain map. The results showed a predominance of an exacerbated posture of the trunk in anterior inclination overload (73%), and high prevalence of MSDs, especially in the spine in 60% of cases: cervical (24%); lumbar (22%); and dorsal (14%). The results, point to a possible relation between the postural overload and MSDs when the workers they do tasks that require manual lifting with excessive loads, and prolonged exposure to incorrect postures with repetitive movements with biomechanical overload.

1 INTRODUCTION

Postural changes that young people and adults, and particularly active workers, develop throughout their lives, have led to the development of musculoskeletal disorders (MSDs) resulting from current living standards, and in particularly the activities of highly skilled labour, particularly those related with repetitive body movements, which mostly run on biomechanical overload (Kendall, McCreary and Provence 1995).

According the report of the European Agency for Safety and Health at Work (2000), workers from different professional areas suffer from MSDs, resulting from debilitating postures.

According to Codo and Almeida (1998) MSDs have become a global epidemic. Already in the 1980s the MSDs were reported as the most frequent cause of absenteeism across the world. The MSDs include all occupational injuries affecting tendons, synovial, nerves, muscles, fascia, ligaments, alone or associated, accompanied by degeneration or not of the involved tissues in those process, mainly affecting the upper limbs, and cervical, dorsal and lumbar spine regions.

To Granata and Marras (1995) low back pain usually occurs in the activities that are required lateral flexion movements of the trunk with association of rotations at high speed.

According to Kumar, Narayan and Zedka (1998) the rotations of the trunk at high speed are the third movement most run by individuals who have a symptomatology of low back pain.

The rotations of the trunk associated with work involving the use of strength with sustained loads normally produce an overload at musculoskeletal level particularly in the lumbar region, constituting clearly a risk factor for the appearance of low back pain (Amell *et al* 2000).

The localized pains at back level due the realization of work tasks in large musculoskeletal overload have in most cases aetiology related with inappropriate postures (Couto 1995).

The European Foundation for the Development and Improvement of Living and Working Conditions (1996) through a survey at European level on the working conditions, found that inappropriate postures when carrying out specific tasks associated with multiple risk factors existing in the workplace, have repercussion at spinal level with a big overhead biomechanics, with maintenance of incorrect postures for long periods, causing great physical and psychological suffering of the worker, becoming one of the leading causes of absenteeism to work, with high costs for society.

According the report of the European Agency for Safety and Health at Work (2000), under the title: *"Work—Related Neck and Upper Limb Musculoskeletal Disorders"*, across Europe, the various studies on MSDs have evidenced that workers who perform essentially manual work, either skilled or not, have a higher risk of contracting MSDs. The same report refer that 35% of women have more lesions at upper limbs level compared to men (30%) because those women do a type of work which in most cases are performed in repetitive overhead.

To Paoli and Damien (2000), in the European Union the workers who reported back pain, 25% are in the age range from 15 to 24 years old, 35% are situated in ages higher than 55 years old, and most of these have spent almost all time working in situations of constant risk at musculoskeletal level.

The present study aimed to analyse the prevalence of biomechanical overload and identify postural changes in *major* constraint, as well as the risk factors for the appearance of MSDs in workers with mental disabilities, in the course of carrying out their work tasks in a Protected Job Centre (PJC)[1], in Lisbon area.

2 MATERIALS AND METHOD

The cross-sectional analytical study was performed by observing the momentary dynamic process and identification of postural changes in the studied population of the PJC. The study involved the participation of all the PJC universe of 36 workers (n = 36) of both sexes, aged from 26 to 50 years, assigned to the tasks of; packing cutlery for aircrafts, carpentry/joinery, stuffing, industrial laundry. Ironing, gardening, and aid to patients transport.

For data collection was used the following instruments:

- Survey questionnaire applied to workers (n = 36) with the selected variables:
  Localization of the symptoms of body pain and discomfort; Frequency of postural events; Change in work habits; Working time reduction, and recurrence of symptoms;
- Map of body pain and discomfort applied to workers (n = 36);
- Data collection for postural analysis by OWAS method, through 30 direct observations in seven workers, randomly selected in the different above mentioned work tasks, in the PJC, through filming and photographic records.

Was used the program *Win-OWAS®* for analysis of biomechanical parameters by observation of the adopted postures in all above described tasks, and the program *Excel®* for statistical descriptive analysis of the variables indicated above.

1. CEP: Protected Job Centre is traditionally defined as a supervised work environment in a specialized institution that provides gainful employment to people with disabilities, with the objective of social integration of those workers.

3 RESULTS AND DISCUSSION

The participation of the studied universe (n = 36) was 100%.

The questionnaire reveals, in Figure 1, that 28% of workers operates in the industrial laundry, 25% in the area of packaging cutlery for aircrafts, and 22% in gardening, places where exists postural overload.

Regarding the time of practice, 61% of workers are working affects to their workplace for over ten years, 28% between five and ten years, and only 11% are working for more than five years.

In relation to the daily working hours, all workers fulfil with the full timetable of 8 hours per day, working five days a week, totalling 40 hours.

Regarding the exercise of another work activity, none of the workers perform any additional work beyond the tasks that are assigned to the CEP.

In the analysis of body discomfort and pain symptoms of Figure 2, reported by the total inquired, manifests itself great emphasis in the spinal region; 24% at the cervical level, 22% at lumbar level, and 14% at dorsal level. Respondents were unable to indicate the temporal dimension of symptoms.

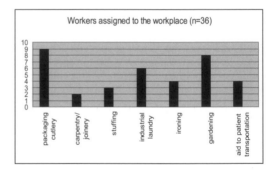

Figure 1. Distribution of the workers assigned to the workplace.

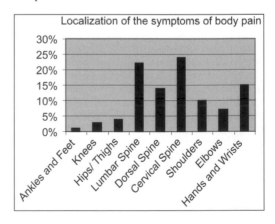

Figure 2. Distribution of the pain symptoms, and body discomfort.

It was found that 10% of workers use a standing posture, 8% use sitting posture, 11% use the rotation of the trunk standing, 12% use the trunk rotation seated, 13% use trunk flexion standing, 7% use trunk flexion seated, 8% use flexion with trunk rotation standing, 7% use flexion with trunk rotation seated, 4% use static posture prolonged seated, 6% use static posture prolonged standing, 9% use dynamic posture standing, and 5% use dynamic posture seated. We can observe in Figure 3, that 13% of all workers (n = 36) carrying out activities with a greater utilization of the trunk flexion in standing position that puts the spine and the related musculoskeletal structures at high risk of overloading.

When workers were asked about changes in work habits, Table 1, seeking a change to procedures in the tasks that perform, in order to modify the incorrect postures and diminish the painful symptoms, the response was negative, ie, 100% of workers do not wont to change their postural attitudes.

It was found, Table 2, that there was a recurrence of painful symptoms that had previously manifested in approximately 60% of workers.

In the analysis to the adopted postures by the workers during a workday of 8 hours in Figure 4, and by the 30 postural observations analyzed by the *Win-OWAS®* program, showed us a prevalence of 73% of trunk posture in flexion with overhead.

The results of this study revealed a high prevalence of postural disorders in the analyzed workers with mental disabilities. Being that overload occurrence it is at cervical level (24%), and lumbar level (22%), indicate that much of the work performed by those workers is done with great effort in those regions of the spine.

It was found that 13% of workers use primarily standing posture with trunk flexion for long time associated with repetitive movements.

The obtained data by OWAS method, revealing that 57% of the postures used, with greater emphasis on using anterior flexion of the spine and positioning of both arms below shoulder level, are the overloaded biomechanical postures more harmful to the musculoskeletal system, and should be modified in the shortest period of time. Was also observed that a daily load of 8 working hours performing repetitive and continuous tasks in biomechanical overload with accented postural changes, contribute to the significant musculoskeletal and physical wear,

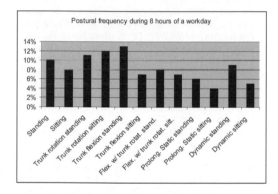

Figure 3. Distribution of the postural frequency during 8 hours of a workday.

Table 1. Distribution of changes in the work habits.

Changes	Workers (%)
Yes	0
No	100

Table 2. Distribution of recurrence of painful symptoms.

Recurrence of painful symptoms	(%)
Yes	58.34
No	41.66

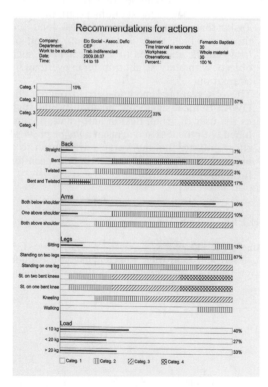

Figure 4. Results of biomechanical parameters of the postural analysis of all activities.

55

corroborating scientific work of Wisner (1997), and Kroemer and Grandjean (2005).

Workers with mental disabilities, analyzed by the present study, perform their activities mostly in static standing position without support of compensation, with incorrect postures, or moving, using muscle strength to lift and move heavy objects, or in sitting position in inadequate structures for the function they perform, doing flexions and trunk rotations repeatedly, caring heavy loads, with significant request of lumbar region with biomechanical overload and consequent disc compression.

4 CONCLUSIONS

According to the study results there was a possible relation between the postural overload and MSDs presented by workers with mental disabilities when they do tasks that require lifting and manual handling of excessive loads, and prolonged exposure to incorrect postures, repetitive movements with biomechanical overload.

From the analysis it was concluded:

- The worker with mental disability fits under the professionals with health risk activities;
- The biggest constraint postures are forward flexion of the trunk in the standing position;
- Tasks that require greater postural load are those in which there is caring and lifting of heavy loads, associated to a movement of such objects, with body weight transfer in overload by adaptation;
- The observational method carried out by the OWAS to most significant postural events, such as the user of trunk flexion and rotation combined with the legs and arms in a *major* physical effort indicate that the workers are subject to a biomechanical overload, indicating the need for future postural corrections.

REFERENCES

Amell, T.K.; Kumar, S.; Narayan, Y.; Gil Coury, H.C. (2000), Effect of Trunk Rotation and Arm Position on Gross Upper Extremity Adduction Strength and Muscular Activity, *Ergonomics*, Vol. 43, Issue 4, pp. 512–527.

Codo, W; Almeida, M.C.G. (1998), *LER – Lesões por Esforços Repetitivos*, 4ª Edição, São Paulo, Ed. Summus.

Couto, H.A. (1995), Ergonomia Aplicada ao Trabalho: Manual Técnico da Máquina Humana, Belo Horizonte, Ed. Ergo.

European Agency for Safety and Health at Work (2000), The State of Occupational Safety and Heath in the European Union – Pilot Study, Luxembourg, Office for Official Publications of the European Communities, pp. 262–271.

European Agency for Safety and Health at Work (2000), Work-Related Neck and upper Limb Musculoskeletal Disorders, *Facts issue 5*, Luxembourg, Office for Official Publications of the European Communities, pp. 1–2.

Fundação Europeia para a Melhoria das Condições de Vida e de Trabalho (1996), As Condições de Trabalho na União Europeia: *Segundo Inquérito Europeu sobre as Condições de Trabalho*, Edição resumida do Serviço das Publicações Oficiais das Comunidades Europeias, Luxembourg, pp. 1–8.

Granata, K.P.; Marras, W.S. (1995), EMG – Assisted Model of Biomechanical Trunk Loading during Free-Dynamic Lifting, *J. Biomechanics*, Vol. 20(11), pp. 1309–1317.

Grove, A.T. (1980), Geomorphic evolution of the Sahara and the Nile. In M.A.J. Williams & H. Faure (eds), *The Sahara and the Nile*: 21–35. Rotterdam: Balkema.

Jappelli, R.; Marconi, N. (1997), Recommendations and prejudices in the realm of foundation engineering in Italy: A historical review. In Carlo Viggiani (ed.), *Geotechnical engineering for the preservation of monuments and historical sites*; *Proc. intern. symp., Napoli, 3–4 October 1996*. Rotterdam: Balkema.

Johnson, H.L. (1965), Artistic development in autistic children. *Child Development* 65(1): 13–16.

Kendall, P.F; McCreary, E.K; Provence, P.G. (1995), *Músculos Provas e Funções*, São Paulo, Ed. Manole.

Kroemer, K.H.E; Grandjean, E. (2005), *Manual de Ergonomia: Adaptando o Trabalho ao Homem*, 5ª Ed., Porto Alegre, Ed. Artmed.

Kumar, S.; Narayan, Y.; Zedka, M. (1998), Strength in Combined Motions of Rotation and Flexion/Extension in Normal Young Adults, *Ergonomics*, Vol. 41, Issue 6, pp. 835–852.

Paoli, P.; Damien, M. (2000), Third European survey on working conditions 2000, *Eurofound*, Luxembourg, pp. 10–11.

Wisner, A. (1997), Por dentro do Trabalho: Ergonomia, Método e Técnica, São Paulo, FTD/Oboré.

Occupational Safety and Hygiene – Arezes et al. (eds)
© *2013 Taylor & Francis Group, London, ISBN 978-1-138-00047-6*

Analysis of interventions in the design of *potiguar* handicraft—Brazil

M.L. Leal
SEBRAE, Natal, Rio Grande do Norte, Brasil

M.C.W. Saldanha
Programa de Engenharia de Produção, Universidade Federal da Paraíba, João Pessoa, Paraíba, Brasil
GREPE-UFRN—Grupo de Extensão e Pesquisa em Ergonomia, Natal, Rio Grande do Norte, Brasil

ABSTRACT: The current article aims to analyze the interventions of product innovation in the handicraft production through design and it is based on the development of three design workshops given between 2005 and 2009 in the Craftsman Association of *Lajes Pintadas*-AALP, located in the municipality of *Lajes Pintadas*-RN, Brazil. The methodology used interactional and observational methods and techniques as well as document analysis. The actors involved in the social construction were the presidents of the association (currently and formerly), fifteen craftswomen from the Association and three design consultants who presented the design workshop at AALP. The results point to the need of involving the craftsmen since the moment of the planning until the final assessment of the workshop, and that the interaction bonds must be consolidated so that cooperative actions in the innovation process provide positive results in favour of the sustainability of craftsmen groups.

1 INTRODUCTION

Handicraft activity is an important expression of popular culture and a great development trigger. It is more often than not the only source of income for some families. The handicraft market in Brazil involves around 8.5 million people, out of which 3.5 million (40%) live in the northeast. It is estimated that the sector moves around R$ 28 billion a year, that represents 2.8% of Gross National Product— GNP (BNB, 2002; SEBRAE, 2006).

The quality of Brazilian handicraft has been standing out in the international market in recent times. According to the Agency of Exportation and Investment Promotion-APEX, Brazil exported R$ 1 million 410 thousand in handicraft (SEBRAE, 2008).

With the changes in the competitive parameters, craftsmen are at times forced to promote changes in their products through the use of design resources (Leal, 2011, Saldanha & Almeida, 2012). The projects of the Programme for promoting handicraft in *Rio Grande do Norte* supported 2551 craftsmen between 2005 and 2010.

However, even though craftsmen have the support of promotion institutions, cooperative and individual craftsmen associations, they find it difficult to keep the production, either because of the discontinued institutional support, difficulty in absorbing new technologies and knowledge or poor selling skills (Leal, 2011).

The current article aims to analyse the interventions of innovation in the *potiguar* (from the state of Rio Grande do Norte—Brazil) handicraft products through design using concepts and methods of ergonomics, based on three design workshops given between 2005 and 2009 in the Association of Craftsmen of *Lajes Pintadas*-AALP, located in the municipality of *Lajes Pintadas-RN*, Brazil.

2 MATERIALS AND METHODOLOGY

The research performed in 2010 used interactional methods and techniques, documental and observational analysis. The actors involved in the social construction were presidents of the association (former and current); fifteen craftswomen from the association, which corresponds to 65.22% of the members; three design consultants who conducted the workshops at the AALP in the years 2005, 2006 and 2009.

At first a global analysis was performed in the Association though four on-site visits lasting 8 hours each, and intending to get to know AALP and its craftswomen, the production process and to improve trust. At that moment interactional techniques were used (informal conversation, spontaneous and provoked verbalizations and questions to collect global data and also from the craftswomen population) as well as also open observations.

A conversational action was done with the first president of the Association, enabling the creation of a previous diagnosis, which along with the global and documental analysis led to the elaboration of research tools to be used in the collective analysis with the craftswomen.

In order to analyse the workshop, a conversation action was done with the presidents of the association (former and current), interviews with the consultants, analysis of the workshop reports and collective analysis with the craftswomen using slides with images of the workshops, self confrontation technique and a dynamic script of questions.

The collected information through conversational actions along with the collective analysis were transcribed and tables through the matrix of commentary inclusion (Vidal, 2003).

3 RESULTS AND DISCUSSIONS

3.1 Contextualization of handicraft work at AALP

The municipality of *Lajes Pintadas-RN-Brazil*, has approximately six thousand inhabitants and it is 134 km far from the capital of the state, *Natal-RN*. Handicraft work made of *Sisal* hemp appeared from the lack of alternatives for income generation and soon was perceived as a business opportunity. In the beginning of the year 2000, the craftswomen started a partnership with the Promotion Institution. In 2001 it was created an organization to represent the group, the Association of Craftsmen of *Lajes Pintadas*-AALP which has its own office and 23 craftswomen members.

The diagnosis performed in 2003 detected the following problems at the AALP (SEBRAE, 2003):

- Lack of control on the costs of production;
- Random calculation of retail price;
- Lack of main raw material—*sisal* hemp;
- Lack of production guidance;
- Lack of managerial control;
- Difficulty in the collective response to demands.

Some of the problems were solved through the empowerment performed by the Promotion Institution from 2000 to 2010, which consisted on consultancy in cooperativism, management, market access and three design workshops. In spite of the problems, the association was awarded twice with the *TOP 100 of SEBRAE* on a national level, contest sponsored by SEBRAE—Support Service to Micro and Small Enterprises, which awards the best of the year in several categories.

3.2 Craftswomen

At AALP, 23 craftswomen work with sisal hemp, 45.2% are married and 28.6% are single. On average they have three children. Age varies from 20 to 70 years old, 46% are aged between 20 to 40 years old. It is a group formed by relatively Young people when compared to other craftsmen group.

Only 05 craftswomen have a daily work routine at the association headquarters. 78% work from home due to difficulties in commuting and matching housework and craftwork.

Only three out of the fifteen craftswomen who took part in the collective analysis declared to create products or adapt the ones that already existed (incremental innovation), the others said that they preferred to copy.

3.3 Productive process

The main raw material-benefited sisal hemp- is acquired by the association. Each craftswoman is responsible for preparing the material used individually, beginning with the washing and drying of the material. As a follow-up the material is dyed in a collective way so that colours are uniform. After dying, the fibres are grouped, combed, separated in bunches, the tips are trimmed so that they all have the same size and look shiny.

The tools used in the process are: Manual loom, thick and thin needles and scissors. In the production of the pieces they use the fabric stitch (loom), closed stitch and the laced stitch (Fig. 1).

The fabric stitch in the pedal loom presents flexibility only in the direction of the sisal thread. It is normally used for rectangular pieces such as wallets and placemats. One of the negative points of such technique is the reduced size of the loom, which does not allow to produce bigger pieces, thus limiting the shape of the final product. On the other hand, once the thread is prepared in the loom, the technique becomes faster.

The technique of closed stitch allows to produce round pieces, basically *sousplat*, bags and decorative vases at times. Though it is considered an easier technique to be learned it is also the technique which demands more time to be produced, which affects the price of the product.

In the technique of laced stitch, the *sisal* hemp is grouped as a web and with the help of a paper shape it is marked a linear way for the stitch, where the thread and needle will pass (Rocha et al, 2008). With the laced stitch it is possible to use part of the wasted *sisal* in the production of hats, bags and piece mats. Though it is an easy and quick technique, craftswomen say that the market has some rejection to the products due to its durability for the thread loosens easily.

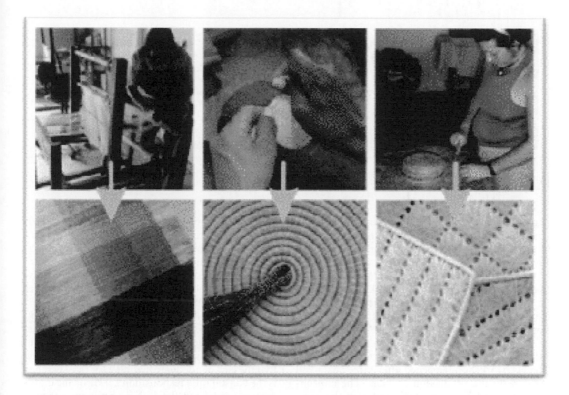

Figure 1. Techniques used in the craft of sisal products in AALP: fabric stitch, closed stitch and laced stitch.

Each technique presents certain limitations to the development of products, which must be taken into consideration by the consultants of design in the planning of the workshops, as it was reported by the craftswomen when they reinforced the need for the consultant to come earlier in order to become familiar with the techniques.

3.4 Design workshops at AALP

Concerning the innovation process in the products, the AALP craftswomen had the support of the Institution of Fomentation RN, through the design intervention from three design workshops in 2005 (bags), 2006 (utilities and decorative objects) and 2009 (bags) (Fig. 2).

Due to the amount of actions promoted by the Promotion Institution, the difficulty in finding consultants or the internal process of negotiations, many times the requests were considered in moments during which the craftswomen had no conditions to effectively take part for they had to produce pieces for certain times of the year. However, the dates and times of the workshops were kept. Each workshop had a 40 hour programme with activities during the morning and afternoon on weekdays.

3.4.1 Design workshops 2005

The workshop of 2005 (13th to the 17th of June) aimed to develop and produce bags. It was given by an industrial designer who did not visit the Association before the workshop but had Access to the products. According to the consultant, the methodology was participative and the craftswomen worked in a cooperative way. No ready pieces were presented, thus avoiding the copy and stimulating the creative process. It was taught how to finish using manual stitches for, as *sisal* is a natural fibre whose thickness is not even, each product needs individual finishing. According to the consultant, the most important aspect was the language used, for the more connected to the participants, the easier is the understanding and credibility. According to the craftswomen the workshop was useful.

3.4.2 Design workshop 2006

The workshop in 2006 (31/07 to 4/08) was given by an artist. The demand was to provide guidance to improve the products, solve colour composition issues and difficulties dying *sisal*. The consultant visited the association before the workshop and created a work plan. The methodology consisted on the development of practices of experimentation

Figure 2. Focus of the design workshops performed at the AALP in 2005, 2006 e 2009.

and experience exchange. They produced utilitarian and decorative pieces, which originated several innovative products—which led to the award TOP 100 given by National SEBRAE.

Despite the positive result, the craftswomen said that the market does not value that kind of product for the technique of closed stitch which was predominantly used is very time consuming in the production, which increases the final price of the pieces.

3.4.3 *Design workshop 2009*

The last workshop happened after three years and focused on the production of bags. The consultant, an industrial designer did not visit the association before the workshop but she had access to the reports of the previous workshops from 2005 and 2006. According the consultant, it was used a methodology of collective construction. The goal was to combine design and handicraft knowledge so that the craftswomen would participate actively in the process of creating, developing and improving products.

According to the craftswomen, only 50% of the products developed in the workshop were kept for some ideas were not adequate to the raw material (*sisal* hemp) and to the production techniques.

From the collective analysis of the craftswomen and the interview with the consultant it was observed that the craftswomen are used to working taking ready pieces as a model.

Thus, the expectation related to the workshop was that the consultant would present some ready pieces for them to copy, which made the creative process difficult for the workers thought themselves to be unable to create a product without having seen it ready beforehand.

3.5 *Design workshop analysis*

According the craftswomen, the workshop of 2005 was considered the most profitable, with 100% utilization of knowledge. The 2006 workshop had 75% success, because some products that use

closed point are unviable. The workshop in 2009 had a lower rating, 50% yield, resulting from lack of skill in dealing with the consultant group variability and technical, human and organizational.

Table 1 presents a summary of the positive and negative points about the workshops which occurred in 2005, 2006 and 2009.

Concerning the Promotion Institution and the consultants, the craftswomen (100%) pointed as the main problem the absence of a diagnosis by the consultant before the workshop. They (87%) believe that the lack of adequate planning for the workshop is the consultant's responsibility. 97% of the craftswomen considered the 40 hour program adequate but 87% said the daily class load was inadequate for they also had to cope with housework. As far as the methodology is concerned 100% of participants said they had difficulties in the beginning to understand technical language of design used by some consultants. Moreover, 87% of the craftswomen emphasized the long time waiting for having their demands for workshops accomplished, the time lapse between each workshop and the fact that they did not receive any copy of the design workshop report (100%) and also the lack of an assessment of the results after the workshop (93%).

The self-evaluation of the craftswomen demonstrated the lack of cooperation among them. They were unmotivated, probably due to the low and late financial return. 100% of the participants emphasized the non-commitment to the workshops due to the fact that housework is the priority for most of them and also the period when the workshops happened, which were moments with high demands of orders.

In the area of product innovation, 80% declared to prefer to copy products, for the creative process wastes too much raw material until the final product is reached. Besides that, they prefer to make products with shorter production time for they say clients prefer cheaper products.

In terms of having access to raw material, it seems to be a recurring issue.

Table 1. Positive and negative points in the design workshops at AALP.

Workshop	Positive points	Negative points
2005	– Active participation of the group. – Cooperation stimulates by the consultant. – Introduction of new techniques. – Improvement in the product finishing. – Improvement in the process of bag production. – Consultant skill to deal with the group. – Products with focus on the market.	– Lack of a diagnosis given by the consultant. – Only 01 sewing machine with a pedal. – Need of regulations for knotting the macramé. – Poor infrastructure (tables, equipment). – Difficulties in sewing. – Difficulty to obtain raw materials.
2006	– Diagnosis carried by the consultant. – Innovative products with high added value. – Experience Exchange and craftsmen self-esteem. – Guidance in the combination of colors. – Consultant skill to deal with the group.	– Poor infrastructure (inadequate machinery). – Difficulty to sew the sisal hemp (fabric stitch/loom). – Poor attendance of participants. – Long time for producing the pieces suggested. – High retail price of products.
2009	– Guidance in the combination of colors. – New technique for finishing the bag zippers. – Enough sewing machines. – Support equipment and adequate structure of association.	– Lack of diagnosis given by the consultant. – Period of the year in which the workshop happens. – Poor participation of the craftswomen. – Difficulty to find raw materials for the workshop. – Culture of copying. – Low self-esteem of craftswomen. – Lack of skills by the consultant.

Table 2. Evaluation of consultants with respect to the craftswomen, the AALP and Institution of Fomentation RN.

Problems identified by the consultants	
Craftswomen and AALP	Lack of cooperation and commitment Resistance to change Access to raw material and information Quality Control inefficient Lack motivation Dependence Institutions Development Lack of planning Precarious financial conditions Infrastructure precarious Culture of copying
Institution of Fomentation RN	Looking for immediate results Planning inappropriate actions Lack of diagnostic Lack methodology Internal bureaucracy

Table 2 presents the main problems identified by the consultants of workshops related to craftswomen, the AALP and Institution of Fomentation RN.

4 CONCLUSIONS

It has been verified that the absence of a diagnosis, which would point the demands, and characteristics of both products and institutions—a lack to be found in the operational flow of the Promotion Institution—does interfere directly in the planning of the workshop, mainly in the choice of methodology, techniques and products to be used.

With respect to product innovation, focus of design workshops, the consultants stated that craftsmen are resistant to change believe they are selling does not need to innovate, and that customers value more than the price the quality of the product. The craftsmen manifest preference for copies of products, depending on the cost of the process of creation, it is waste of raw material. Also, look for products with less complexity because customers prefer cheap products. The result of the inefficiency of the workshops and the "culture of copying" are standardized products, without identity and personality, devaluing the handicraft production *potiguar*.

It is necessary to improve not only the methodology for design intervention as well as the whole planning of actions in the Project of Handicraft in the Institution RN so that it is possible to reach the goals of the program, i.e.: to guarantee that the methodology used in the workshops leads to the autonomy of the craftsmen, making their activity independent from promotion institutions.

The results of this research allows to assert that for the process of innovation through design intervention to be satisfactory for everyone involved it is necessary to involve the craftsmen from the planning until the final assessment of the workshop, to consolidate bonds of interaction and to make the cooperative actions in the innovation process produce results in favour of the sustainability of the craftsmen groups.

FOMENTATION

CNPQ Projeto Cooperação Técnica—Processo: 620251/2008-5, SEBRAE-RN

REFERENCES

BNB. Banco do Nordeste do Brasil (2002) *Ações para o desenvolvimento do Artesanato do Nordeste*. Acesso Agosto 01, 2008, http://www.bnb.gov.br/Content/aplicacao/ Cadeas_Produtivas/Artesanato/gerados/art_publicacoes.asp.
Leal, M.L. (2011) Produção artesanal: análise do método de intervenção de design no artesanato potiguar sob o ponto de vista dos atores envolvidos no processo. *Dissertação Mestrado em Engenharia de Produção*. Natal: Universidade Federal do Rio Grande do Norte.
Rocha, F.B.A; Campos, M.C.; Pacheco, N.O.; Silveira, R.R; Falani, S;Y.A. (2008) Sisal em Tramas: o artesanato como alternativa de sustentabilidade. In: *Anais do XV Simpósio de Engenharia de Produção*. São Paulo.
Saldanha, M.C.W.; Almeida, J.D. (2012). Situated modeling in the drawing workshop for bobbin lace. *Work Journal* (Reading, MA), v. 41, p. 683–689.
SEBRAE. (2003) *Histórias de Sucesso: experiências empreendedoras*. Belo Horizonte: SEBRAE.
SEBRAE. (2006) *Programa de Desenvolvimento de Distritos Industriais: Uma experiência de Internacionalização de APLs*. 166p. Brasília: SEBRAE.
SEBRAE. (2008). *Artesanato: um negócio genuinamente brasileiro*. Vol 01. N° 01. Brasília: SEBRAE.
Vidal, M.C.R. (2003) *Guia para análise ergonômica do trabalho (AET) na empresa: uma metodologia realista, ordenada e sistematizada*. Rio de Janeiro: EVC.

Occupational Safety and Hygiene – Arezes et al. (eds)
© *2013 Taylor & Francis Group, London, ISBN 978-1-138-00047-6*

Prevalence of pain/discomfort on artisan fishing using rafts

A. Jaeschke

Grupo de Extensão e Pesquisa em Ergonomia, Universidade Federal do Rio Grande do Norte, Natal, Brasil

M.C.W. Saldanha

Programa de Pós-Graduação em Engenharia de Produção, Universidade Federal da Paraíba, João Pessoa, Paraíba, Brasil
GREPE, Grupo de Extensão e Pesquisa em Ergonomia, UFRN, Natal, Brasil

ABSTRACT: The current article analyses the activity of raftsmen in the capture expedition in the urban beach of *Ponta Negra, Natal-RN-Brazil* emphasizing the prevalence of pain/discomfort. Work Ergonomic Analysis-WEA was used as a reference methodology. In order to perform the modelling, observational and interactional methods were used. The diagram painful areas and Nordic Questionnaire were applied in search of reported of pain/discomfort of the raftsmen. Twenty-one raftsmen participated in the research (50% of the population). The artisan fishing using rafts represents a high musculoskeletal risk, joining physical effort, adopting forced postures with the rotation of the spine and movement repetition, mainly flexion extension of vertebral spine, aggravated by the demand of strength of the stabilizing muscles of the human body to keep the poise. It has been highlighted the high percentage of reports of musculoskeletal pain in the spine, and in both lower and upper limbs.

1 INTRODUCTION

In Brazil, sea fishing is responsible for 580 thousand tonnes of fish caught a year. In 2007, the north-eastern region caught 28.8% of national production, out of which 96.3% were from artisan fishing, which makes the fleet of that region the least industrialized of the country (Castello, 2010). Rio Grande do Norte (RN) has a 410 km coastline, 25 coastal towns, 97 fishing communities and an estimated contingent of 13,000 raftsmen who perform the fishing for subsistence and commerce. 28.5% of the State fleet are rafts and "paquetes" (smaller rafts) which represented 12.10% of the annual fish capture in 2007 (2174t) (IBAMA, 2008). The Colony of fish and aquiculture in Natal-RN has 381 registered vessels, out of which 22.80% (87) are rafts. In the beach of Ponta Negra, Natal-RN, the area of the current study, there are 31 rafts and 42 raftsmen (Saldanha et al, 2012).

Artisan fishing is the one that occurs through manual work performed by the raftsmen, using small vessels in the capture and in small scale. The raft is a secular vessel used in artisan fishing and named according to its dimensions. The small-sized ones are called "botes" or "catraias", measuring around 3.5 meters; the medium-sized ones are known as "paquetes", 4 to 5 meters length; the "jangada de alto" is the model that reaches 8 meters length (Araújo, 1985).

The current article analyses the activity of raftsmen in the capture expedition in the urban beach of Ponta Negra, Natal-RN-Brazil emphasizing the prevalence of pain/discomfort.

2 MATERIALS AND METHOD

Work Ergonomic Analysis-WEA was used as a reference methodology (Guèrin et al, 2001; Vidal, 2003). In order to perform the modelling, observational and interactional methods were used and also specific protocols. The observational methods include systematic open observational techniques and simulations and also the use of auxiliary resources to record and further analysis, such as photographic and video cameras. The methods and interactional techniques used were the following; social and economical questionnaire, conversational actions, spontaneous and provoked verbalizations, self-confrontations and workshops (collective analysis). The diagram of painful areas was applied (Corllet et al, 1980) and the Nordic Questionnaire for the analysis of musculoskeletal symptoms (Kourinka, 1986) in search of report of pain from the raftsmen. Twenty-one raftsmen participated in the research (50% of the population). For the statistic analysis of data it was used Microsoft Office Excel 2007 and the software R(version 2.11.1).

The analysis of the activity has enabled to obtain data which was submitted to constant restitutions and validations together with the raftsmen themselves, thus allowing a refined understanding of the raftsmen activity and their relationships with the aspects of physical demand and prevalence of pain/discomfort.

3 RESULTS AND DISCUSSION

3.1 Artisan fishing using rafts in the beach of Ponta Negra, Natal-RN-Brazil

3.1.1 Work population: Raftsmen

The raftsmen of Ponta Negra Beach are male, have incomplete primary education and their families have from 3 to 7 children. Most of their ages range from 40 to 49 years old (33.33%), from 30 to 39 and from 50 to 59 (28.57%). The youngest group (20 to 29 years old) has the smallest percentage (9.52%).

The learning of the job by the researched raftsmen started at an early age. 38.10% started between the age of 5 and 10 years old, and 42.86% of them from 11 to 15 years old. Only two raftsmen started fishing after 20 years old. 28.57% of the raftsmen are active between 21 and 30 years of age. The time of activity between 11 and 20 years; 31 to 40 years and over 40 years represent 23.81% of the raftsmen.

The index of experience, defined as the relation between the time of performing the fishing activity and the age of the fisherman in years, on a 0 to 1 scale, indicates that none of the fishermen in the research can be considered as a beginner (active time inferior to 25% of their age), 9.5% (2 raftsmen) have intermediate level of experience (active time between 26 and 50% of their age), 52.38% (11 raftsmen) are experienced (active time between 51 and 75% of their age) and 38.10% (8 raftsmen) are very experienced, that is, they have dedicated more than 75% of their age to the job (Table 1).

3.1.2 Rafts

The rafts of Ponta Negra, known as "paquetes", are medium-sized. They are built in marine plywood and wood, and have sail or engine propulsion. The dimensions of the rafts are varied, measuring from 3.6 to 5.14 m length by 1.4 to 1.7 m, weighing around 642 kg. They usually accommodate 2 crew members (master and assistant) who perform different tasks.

The use of Honda Stationary Engine is a recent innovation in the rafts. Its use began around 2005. According to reports by raftsmen, the use of the engine has brought improvements in the activity, among which a significant reduction in the sailing

Table 1. Index of raftsmen experience.

Experience index	Class	Raftsmen	
		Frequency	%
0.0–0.25	Beginner	00	0.0
0.26–0.50	Intermediate	02	9.52
0.51–0.75	Experienced	11	52.38
0.76–1	Very experienced	08	38.10

time, decrease in physical effort and less dependence on the wind, an indispensable factor when a sail is used. On the other hand, the use of fuel pollutes the ocean and increases the cost of the expedition.

3.1.3 Capture expeditions: Raft fishing activity

The fishing expeditions are made predominantly on Tuesdays and Saturdays. However, because of the weather and the ocean conditions along with the raftsmen physical resistance, three expeditions on average happen every week. Most of the fishing is made with a net. However, some raftsmen might use a line and a net, only the thread, "covos" or "manzuá" (fishing traps in the shape of a basket or a cage) besides free diving for fishing in some times of the year. Fishing with a line is also used while they wait for the nets to be hauled, being used as a regulation strategy against sleep and sleepiness.

The work strategy is directly linked to the kind of fishing: the "come and go", the "ice" fishing and, "with awaiting nets". The "come and go" fishing is predominant and happens throughout all the months of the year. The raftsmen go to the sea at dawn (2:00h.) and return in the morning (8:00h.) or they leave in the afternoon (14:00h) and return at night (21:00h.), depending on tidal conditions, moon phase, month of the year and the weather. In the "ice" fishing, raftsmen usually leave in the beginning of the day and they may spend until 24 hours fishing. In this kind of fishing they use ice to store the fish and the nets can be cast to the ocean more than once.

Fishing with "awaiting nets" occurs less frequently. In such mode, the nets are placed in specific points and the raftsmen go back to shore and then return to sea after a few hours or in the following day to pick up the nets.

The duration of one expedition varies from 3 h 30 min to 9 hours, depending on several factors: kind of expedition ("come and go" or "awaiting nets"), kind of propulsion used (engine or sail), sea conditions, location of fisheries, number of times the net is cast, amount of fish caught by number of nets cast and even the presence of seaweed in the net.

The first step of the capture expedition (Fig. 1) is the transport of the engine and the fishing tools from the residence of the raftsmen to the place where the rafts are docked at the beach (10–15 min.). Then the organization of the raft begins, they check items related to security and fuelling the engine (4–40 min.).

The transport of the raft (642 kg) is done by rolling it over two coconut trunks in small cycles. The cycle is finished once the roll with is placed

ahead reaches the centre of the boat and the one which was placed at the back is released and transported to the front. Raftsmen at the back then lean at the bow of the raft so that it raises its stern in order to reposition the roll. The raftsmen at the front the pulls the raft, controls its movements and speed, thus making it stop at the end of each cycle. Such process is repeated until they reach the sea, over a distance, which varies according to the tide. When the tide is low the distance to be followed is

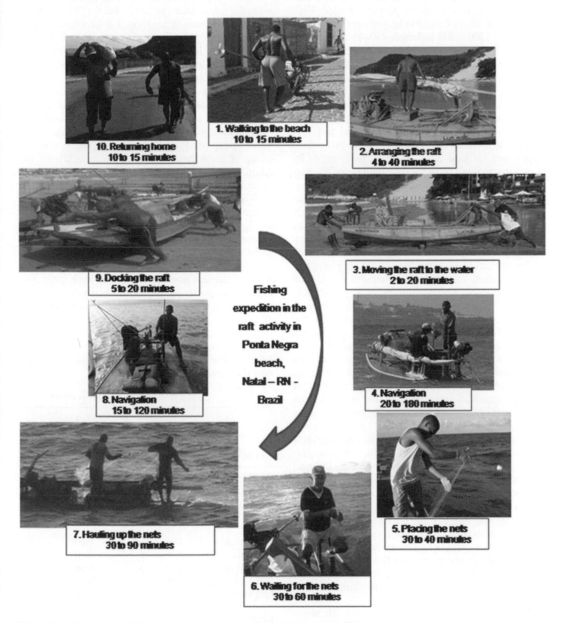

Figure 1.   Capture expedition in the raft activity in Ponta Negra, Natal-RN.

approximately 60 meters (14 cycles). In half tide it is around 30 meters (8 cycles) and in high tide from 2 to 9 meters (1 to 3 cycles), ranging from 2 to 20 minutes. The posture of raftsmen features the flexion and extension of the trunk associated with rotations and strength demand. 25% of the accidents that occur are due to the fishermen hitting the rolls or the rafts during the moving, which harm lower limbs. They can also cause damage to the hull of the vessel with risk of sinking.

The crew boards and starts the procedures that vary according to the type of propulsion used. They sail until the fishery (30–180 min.) where they cast the nets along 1700–2700 m length (30–40 min.). They anchor the raft and wait until it is time to haul the nets (30–60 min.). During that time, they fish with line.

Two raftsmen are in charge of hauling up the nets (30–90 min), one of them at the bow of the raft—he has to pull the net moving the raft against the tide for about 1,700 to 2,700 meters. The other one removes the fish from the net and puts the nets in the internal compartment of the raft, which forces him to bend down from 17 to 27 times by launching/hauling the nets. The hauling of the nets represents a high musculoskeletal risk, joining physical effort, adopting forced postures with the rotation of the spine and movement repetition, mainly flex extension of vertebral spine, was pointed by 52.39% of the fishermen as being the hardest task of the job. The physical strength is directly related to the amount of fish caught (2–200 kg). The physical exhaustion is increased when the net is cast more than once, which often happens in the "ice" fishing, or when there is little fish caught. Depending on the tidal movement, the ocean currents, the distance from the coast and the depth of the net, the flexing of the trunk may be increased (20°–90°). Besides that, at any moment there might occur the destabilization of the raftsman's posture due to the movement of the raft in the ocean, or some other intercurrence that demands muscular strength from the human body to maintain the balance.

They sail back to the beach (15–120 min.) It requires the help of other people besides the crew because of the heavy weight of the raft containing the fish which was caught (2 to 200 kg) and because the raft is soaked. Moreover, the moving is done in a sloping area, whose degree varies according to the tide.19.05% of the raftsmen consider this to be the most tiring part of the job. Raftsmen and assistants have postures featuring great flexion of the trunk which may be linked to rotation. The upper limbs are kept flexed. Lower limbs which are kept flexed and in extension move with great strength. Feet are placed in dorsal flexion position.

They make sure the raft is docked on a safe place, clean it and organize the fishing tools (10–40 min.).

Finally they sell the fish, go back home (10–15 min.) and rest until the expedition or some other fishing-related activity.

3.2 Repercussions of the activity: Prevalence of pain/discomfort

The results of applying the pain diagram show that 23.8% of the raftsmen have musculoskeletal pains before the expedition in one (9.52%), two (9.52%) or three (4.75%) parts of the body. One of the most reported areas was the lower back (19.05%) with intensity between 4 to 8 on a 0 to 10 scale. The prevalence of pain/discomfort after the expedition confirms the lower back to be the area of highest incidence with higher occurrence (42.86%) and intensity (between 3 and 10) (Table 2).

Concerning the experience index it is demonstrated that the intermediate group (26 and 50% of the age in the job) did not report pain before the expedition, the experienced group (51 and 75% of the age doing the job) reported some pain (30 to 40%) and the group of experienced raftsmen (51 and 75% of the age in the job) had a significant increase in the report of pain before the activity (60 to 70%).

When questioned about the occurrence of pain in the last seven days and twelve months, 95.24% of the raft-men reported to have felt some pain on the last seven days and 100% in the last twelve months.

50% of the fishermen who reported to have felt pain on the last seven days said to have from 3 to 11 painful parts of the body, being the highest concentration between 3 and 8 parts of the body.

Table 2. Prevalence of pain in the raftsmen before and after the capture expedition.

Part of the body	Before the expedition %	Before the expedition Intensity	After the expedition %	After the expedition Intensity
Lower back	6.35	4	4.76	3
	6.35	6	9.51	8
	6.35	8	28.59	10
Right shoulder	4.76	5	0.00	–
	4.76	10		–
Left shoulder	4.76	10	0.00	–
Right arm	4.76	5	0.00	–
Right hand	4.76	3	0.00	–
Left foot	4.76	2	0.00	–
Left knee	0.00	–	23.8	5
			4.76	8
Right/left leg	0.00	–	4.76	5

In the twelve month period 50% of the raftsmen reported from 5 to 14 painful areas.

The part of the body with the highest incidence of pain in the last seven days is the lumbar vertebral spine (71.43%), preceded by the knees (52.38%), the ankle or feet (33.33%). Similarly, in the last twelve months, the lower back (80.95%), cervical (57.14%) and the knees (57.14%) presented the highest levels of pain in the raftsmen (Table 3).

It has been observed an increase of 31.25% in the report of pain/discomfort in the last 7 days (64 reports) related to the report of pain in the last 12 months (84 reports) also showing an increase of 4.35% in the lower limbs (23 to 24 reports), vertebral spine, 45.83% (24 to 35 reports) and in the upper limbs 47.06 (24 to 35 reports) (Table 3).

As far as the experience index is concerned, the intermediate group (activity time between 26 and 50% of the age) was the one which reported less pains in all parts of the body not only on the last seven days but also in the last twelve months. The experienced group (activity time between 51 and 75% of age) had the highest average level of pain the upper and lower limbs in the last 12 months and in the lower limbs on the last seven days. The most experienced raftsmen group (activity time over 75% of their age) had the most painful areas in the upper limbs on the

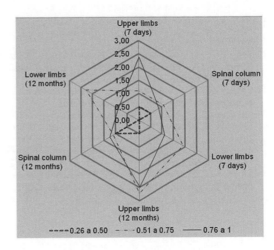

Figure 2. Relation between experience and prevalence of pain.

last seven days and in the vertebral spine in the last twelve months (Fig. 2).

4 CONCLUSIONS

The Artisan Fishing using Rafts represents a high musculoskeletal risk, joining physical effort, adopting forced postures with the rotation of the spine and movement repetition, mainly flexion extension of vertebral spine, aggravated by the demand of strength of the stabilizing muscles of the human body to keep the poise.

It has been highlighted the high percentage of reports of musculoskeletal pain in the spine, and in both lower and upper limbs. The constant painful situations related to the need to fish to support the family and also the lack of orientation on how to correctly do the activity have made the pain evolved to chronic states, and in some cases to the necessity of surgical procedures, mostly herniated lumbar discs.

FOMENTATION

CNPq, CAPES, PROEXT-MEC-SESU, PROPESQ/PROEX-UFRN

REFERENCES

Araújo, N.B.G. (1985). *Jangadas*. Fortaleza: Banco do Nordeste do Brasil S.A, ed. Fortaleza.
Castello, J.P. (2010). O futuro da pesca e da aquicultura marinha no Brasil. *Ciência e Cultura, 62(3),* pp. 32–35.

Table 3. Prevalence of pain in the raftsmen in the last seven days or twelve months.

Part of the body	7 days Freq	7 days %	12 months Freq	12 months %
Upper limbs	17	26.56	25	29.76
Shoulders (R and L)	4	19.05	6	28.57
Right shoulder	2	9.52	3	14.29
Left shoulder	1	4.76	2	9.52
Elbows (R and L)	2	9.52	3	14.29
Right elbow	0	0.0	2	9.52
Left elbow	1	4.76	0	0.00
Hands and wrists (R and L)	4	19.05	5	23.81
Wrist/hand left	1	4.76	1	4.76
Wrist/hand right	2	9.52	3	14.29
Spine	24	37.50	35	41.67
Cervical spine	4	19.05	12	57.14
Dorsal spine	5	23.81	6	28.57
Lumbar spine	15	71.43	17	80.95
Lower limbs	23	35.94	24	28.57
Hip or thighs	5	23.81	7	33.33
Knees	11	52.38	12	57.14
Ankle or feet	7	33.33	5	23.81

Corlett, & Manenica, I. (1980). The efects and meansurement of working postures. *Aplied Ergonomics. 11(1), pp. 7–16.*

Guérin, F; Laville A.; Daniellou, F.; Duraffourg, J. & Kerguelen, A.. (2001). *Compreender o trabalho para transformá-lo.* São Paulo: Edgard Blucher ed.

IBAMA. (2008). *Boletim estatístico da pesca marítima e estuarina do estado do Rio Grande do Norte.*

Jaeschke, A. (2010). Oportunidades de melhorias ergonômicas das exigências físicas da atividade jangadeira em Ponta Negra, Natal-RN, *Dissertação Mestrado Engenharia de Produção,* Universidade Federal do Rio Grande do Norte, Natal-Brasil.

Jaeschke, A.; Saldanha, M.C.W. (2012). Physical demands during the hauling of fishing nets for artisan fishing using rafts in beach of Ponta Negra, Natal-Brasil. *Work (Reading, MA), v. 41, p. 414–421.*

Kourinka, I. (1986). Standardised Nordic questionnaires for the analysis of musculoskeletal symptoms. *Aplied Ergonomics. 18(3), pp. 233–237.*

Saldanha, M.C.W, Carvalho, R.; Oliveira, L.; Celestino, J.; Veloso, I. & Jaeschke, A. (2012). The construction of ergonomic demands: application on artisan fishing using jangada fishing rafts in the beach of Ponta Negra. *Work (Reading, MA), v. 41, p. 628–635.*

Vidal, M.C.R. (2003). Guia para análise ergonômica do trabalho (AET) na empresa. Rio de Janeiro: EVC ed.

Cognitive ergonomics

Occupational Safety and Hygiene – Arezes et al. (eds)
© *2013 Taylor & Francis Group, London, ISBN 978-1-138-00047-6*

A qualitative approach to ergonomic risk existing in pathological anatomy laboratories

R. Rangel
CENCIFOR, Centro de Ciências Forenses, Instituto Nacional de Medicina Legal, I.P., Coimbra, Portugal
INMLCF, I.P., Delegação do Norte do Instituto Nacional de Medicina Legal e Ciências Forenses, I.P.—Serviço de Toxicologia Forense, Porto, Portugal
CITS, Centro de Investigação em Tecnologias da Saúde, IPSN—CESPU, CRL, Gandra PRD, Portugal

A. Dias-Teixeira
CITS, Centro de Investigação em Tecnologias da Saúde, IPSN—CESPU, CRL, Gandra PRD, Portugal

J. Maia
ISLA, Instituto Superior de Línguas e Administração de Vila Nova de Gaia, Vila Nova de Gaia, Portugal
Faculdade de Veterinária, Departamento de Ciências Biomédicas, Universidade de Léon, Léon, Espanhã

E. Maia
ISLA, Instituto Superior de Línguas e Administração de Leiria, Leiria, Portugal

F. D. Baptista
ISLA, Instituto Superior de Línguas e Administração de Vila Nova de Gaia, Vila Nova de Gaia, Portugal
Faculdade de Veterinária, Departamento de Ciências Biomédicas, Universidade de Léon, Léon, Espanhã

M. Dias-Teixeira
REQUIMTE, Instituto Superior de Engenharia, Instituto Politécnico do Porto, Porto, Portugal
CITS, Centro de Investigação em Tecnologias da Saúde, IPSN—CESPU, CRL, Gandra PRD, Portugal
ISLA, Instituto Superior de Línguas e Administração de Vila Nova de Gaia, Vila Nova de Gaia, Portugal
ISLA, Instituto Superior de Línguas e Administração de Santarém, Santarém, Portugal

ABSTRACT: The present study aimed to evaluate qualitatively ergonomic risks in hospitals with pathological anatomy laboratories. A questionnaire was applied to the entire universe, 50 probationers and graduates in pathological anatomy, cytological and thanatological. The data shows that 92.3% of the surveyed who perform their duties in the histology sector, performing repetitive movements, reported symptoms of back pain and upper limb pathologies, indicating that the first signs emerged a few years after the start of their professional activity. It is concluded that these professionals have a high risk of developing musculoskeletal injuries due to the repetitive nature of their tasks.

1 INTRODUCTION

Ergonomics is the scientific discipline concerned with the understanding of interactions between people and other elements of a system to optimize their wellness as well as overall performance of the system itself. This is usually accomplished through the application of ergonomic principles for designing and evaluating manual tasks (tasks that involve lifting, pushing, pulling, carrying, moving, manipulating, holding, hitting or immobilizing a person, animal or object), jobs, products, environments and systems, ensuring that they meet the needs, abilities and limitations of people (Lehto MR & Buck JR, 2008).

When integrated with programs on health and safety, ergonomics can be seen as an asset to reduce the risk of musculoskeletal disorders. Security programs are focused on risks that may result in traumatic injuries. Industrial hygiene is concerned with the dangers that can cause diseases and discusses ergonomics risk factors that can lead to musculoskeletal disorders.

The musculoskeletal disorders related to work (MSDs) are a broad group of conditions referenced in the work environment. MSDs are the subject of great concern by many organizations, including insurers, industry and health services (Amell & Kumar, 2001).

The high rates of injury that affect health professionals are well documented (Adegoke B., Akodu, A., & Oyeyemi, A., 2008; Alexopoulos, E.C., Stathi, I.C., & Charizani. F., 2004; Alnaser, M.Z., 2007; Cromie, J.E., Robertson, V.J., & Best, M.O., 2000; Dawson, A.P., et al., 2007; Fargala, G. & Bailey, L., 2003; Glover, W., McGregor, A., Sullivan, C., & Hague, J., 2005; Holder, N.L. et al., 1999). The technicians who develop activity in pathology anatomy laboratory in the areas of histology and cytology have an increased risk of developing cumulative trauma injuries due to the repetitive nature of pipetting, the use of small instruments (scalpels, forceps and dissecting needles), opening and closing of bottle caps (chemicals and containers of body parts and fluids), the postures forced upon microscopic observation, the uncomfortable position of standing during handling of chemical and/or biological fluids in hoods and biosafety chambers, respectively among other laboratory tasks such as microtomy sector automated processing of histology and cytology in specific equipment in liquid medium. The repetitive strain and cumulative injuries occur(s) when muscles and joints are under stress, with inflamed tendons, nerves tablets, and restricted blood flow (Cotran, R.S. & Robbins, S.L., 1994).

The prevention of such injuries is of paramount importance and the right conduct to adopt may involve the rotation of tasks (Frazer, M.B., Norman, R.W., Wells, R.P., & Neumann, P., 2003), and the ergonomic intervention (Denis, St-Vincent, Imbeau, Jetté, & Nastasia, 2008). Applying ergonomic principles in the workplace is the best method for prevention (Kilbom, S., et al., 1996). The benefits of applying ergonomic principles are not limited to reducing rates of MSDs, but also to improve the productivity and quality of life for workers.

The identification of exposure to risk factors should include consultation to staff, observation of manual tasks and/or review of records at the workplace. Employees should be questioned regarding the tasks they consider to be the most physically demanding and/or with a higher degree of difficulty (Burgess-Limerick. R., 2008).

This study aimed to assess, qualitatively, the ergonomic risk activity for professionals engaged in hospital pathology labor.

2 MATERIALS AND METHOD

In this cross-sectional study (was used) a descriptive methodology based on a structured questionnaire was used, personally delivered (personally) to the students and Graduates in Pathological Anatomy, Cytological and Thanatological, who exercise their professional activity in three major hospitals in the district of Porto, in the year 2010.

The questionnaire, consisting of objective questions closed on facts, knowledge, and attitudes with an approximate duration of 5 minutes, was returned immediately by a sample of fifty students and pathological anatomy technicians, of a total universe of fifty-five, who voluntarily filled in under anonymity.

The issues are as follows:

- How many years (are) have you been in functions?
- In which sector do you perform(s) duties?
- Is there job rotation in the same sector?
- Do you perform movements of the arms above the shoulders during activity in the lab?
- How do you evaluate movements/activities?
- Do you present injuries and/or illnesses caused and/or aggravated by work? What are they?
- Has (been) the laboratory been the target of an ergonomic evaluation?

SPSS was used (SPSS) for statistical analysis, using the χ^2 test and Mann-Whitney tests, to check whether there are significant differences between groups. Data was coded; we used measures of central tendency and descriptive graphs to summarize the data.

3 RESULTS AND DISCUSSION

3.1 *Quality control of air measurements*

In this study, the universe was achieved through 91% of voluntary participation. Of the respondents, 44% are students and 56% are graduates in pathology, cytology and thanatology. Of the graduates, 36% have (between) 1–5 years in office, 18% (between) 16–20 years, 14% (between) 6–10 years, 11% (with) less than 1 year and (between) 11–15 years, 7% (among) 26–30 years and 4% (between) 36–40 years. About 50% of these professionals are working in the field of histology, 18% in cytology (was 18%) and 32% perform (its) their functions in both sectors. About 86% of the students take the stage curriculum in both sectors and only 14% in the area of histology (Table 1).

Of the respondents who perform in the field of histology, 88% have job rotation. On the contrary, in the field of cytology is verified that there is no such rotation. In turn, 86% of interns that perform the functions in the two sectors, and performing work rotation (Figure 1). The Mann-Whitney test reveals that there are statistically significant differences in turnover between sectors trainees and graduates ($p < 0.05$). The professionals who perform their duties in the cytology laboratory should join the rotation task, but since this is an administrative

Table 1. Framework of professional status and sector of activity.

		Sector activity			
		Histology (%)	Cytology (%)	Both (%)	Total (%)
Professional Status	Students	14	0	86	44
	Graduate	50	18	32	56
Total		34	10	56	100

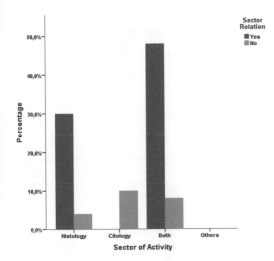

Figure 1. Distribution of sectors according to histology and cytology turnover sectors.

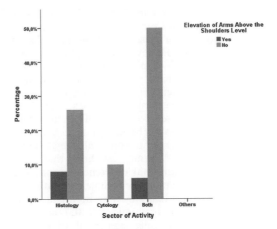

Figure 2. Relationship between lifting the arms above the shoulders level and sector of activity.

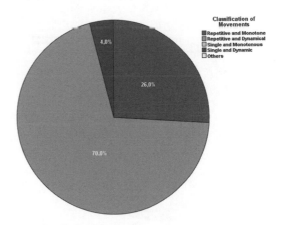

Figure 3. Percentage of classification of movements.

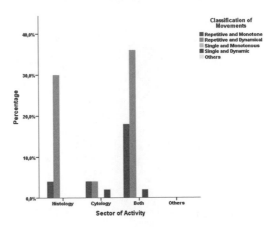

Figure 4. Distribution of sectors of histology and cytology according to the classification of movements.

control measure(s) recommended for reducing the risk of MSDs (Frazer, M.B., et al., 2003). However, this evidence is scarce in the literature.

Only 14% of respondents execute(s) lifting the arms above the shoulders (Figure 2), which allows us to infer that the benches and/ or equipment are set up to these professionals by limiting the duration of tasks in an uncomfortable position. The Mann-Whitney test reveals that there are no statistically significant differences in the performance of that movement (between) among the trainees and graduates ($p = 0.948$, $\alpha = 0.05$).

(According to) The analysis of Figure 3 reveals that 70% of respondents classified the movements they perform as repetitive and dynamic, 26% (of) as repetitive and monotonous, and 4% (of) as single and dynamic. The repetitive movements are mostly dynamic, in the sector of histology, according to respondents (Figure 4). The Mann-Whitney test reveals that there are no statistically significant differences in the classification of movements performed (between) among trainees and graduates ($p = 0.099$, $\alpha = 0.05$).

The analysis of Figure 5 shows that 74.0% of respondents did not present any type of diseases and/or injuries (to) caused (and/) or exacerbated by work activity. However, out of those having pathologies, 92.3% reported neck pain and presented frames of back pain, and back pain and upper limb disorders (Figure 6), affecting mainly respondents who perform their duties in the field of histology (Figure 7), who classify their repetitive movements as... (Figure 8) and begin to appear a few years after the onset of activity (Figure 9). The Mann-Whitney test shows that there are statistically significant differences among the trainees and the graduates technicians with disease and/or injury that may have been caused by the risk factors for ergonomic ($p < .05$), however, there are no statistically significant differences between the symptoms shown by both categories ($p = 0.154$, $\alpha = 0.05$), and no statistically significant differences between the symptoms and years

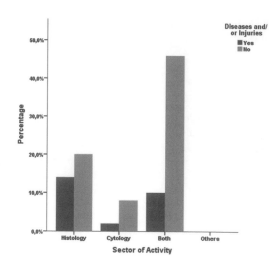

Figure 7. Distribution of sectors which were probably acquired diseases and/or injuries.

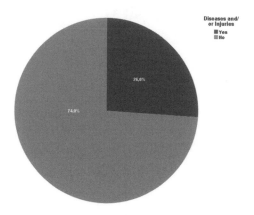

Figure 5. Percentage of diseases and/or injury (with) originated and/or exacerbated by work activity.

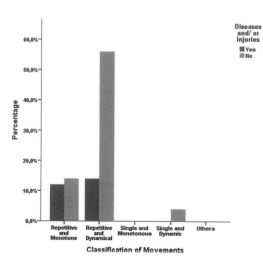

Figure 8. Distribution of the classification of movements by diseases and/or injuries acquired and/or exacerbated by work activity.

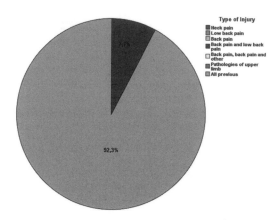

Figure 6. Percentage of types of injuries.

of service ($p = 0.072$, $\alpha = 0.05$). In the field of histology, especially in the stages of inclusion and microtomy, professionals perform those functions with very repetitive nature. Only in the course of a workday, a pathological anatomy graduate may have to include the histological fragments (using the histological forceps) in over 100 cassettes, whose products (paraffin blocks) must be cut and chopped on a microtome manual rotator, by rotating the steering wheel forward at least 2000 times. This is not just a repetitive task, but

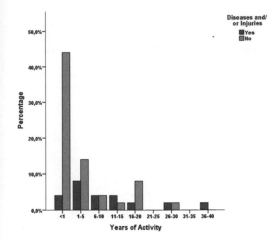

Figure 9. Distribution of years of service by diseases and/or injuries acquired and/or exacerbated by work activity.

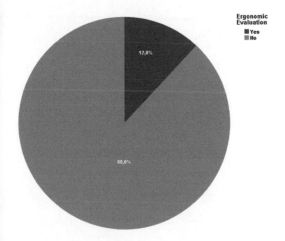

Figure 10. Percentage of responses on the ergonomic evaluation of pathological anatomy laboratories.

also requires force or exertion. In addition, the substitution of paraffin blocks in the holder and the exchange of the cutting blades, in (said) the referred equipment, increases the probability of acquiring MSDs.

All respondents argue that the labs where they perform their duties should have ergonomic equipment [the Mann-Whitney test reveals that there are no statistically significant differences among the categories surveyed (p = 1, α = 0.05)], however, only 12% of respondents say that the facilities of the places where they perform their duties were the target of an ergonomic evaluation (Figure 10), which (does) is not statistically significant [Mann-Whitney test (p = 0.755, α = 0.05)]. The risk assessment in such laboratories should not be neglected, since it is the key factor in a (workplace) safe and healthy workplace.

4 CONCLUSIONS

Among all respondents, professionals performing repetitive tasks in the field of histology have a (at) high risk of developing MSDs due to the nature of these tasks. The neck pain, back pain, back pain and upper limb disorders are the most common painful conditions referenced by this professional class, as a result of injuries exacerbated (resulting and/or exacerbated injuries) by work activity. They evaluate their work habits, take precautions according to their symptoms and take the necessary preventive measures.

REFERENCES

Adegoke, B., Akodu, A., & Oyeyemi, A. (2008). Work-related musculoskeletal disorders among Nigerian Physiotherapists. *BMC Musculoskeletal Disorders,* 9(1), 112.

Alexopoulos, E.C., Stathi, I.C., & Charizani, F. (2004). Prevalence of musculoskeletal disorders in dentists. *BMC Musculoskeletal Disorders 5*, 16.

Alnaser, M.Z. (2007). Occupational musculoskeletal injuries in the health care environment and its impact on occupational therapy practitioners: a systematic review. *Work: A Journal of Prevention, Assessment and Rehabilitation, 29*(2), 89–100.

Amell, T., & Kumar, S. (2001). Work-Related Musculoskeletal Disorders: Design as a Prevention Strategy. A Review. *Journal of Occupational Rehabilitation, 11*(4), 255–265.

Burgess-Limerick, R. (2008). Procedure for Managing Injury Risks Associated with Manual Tasks.

Cotran, R.S., & Robbins, S.L. (1994). *Patologia Estrutural e Funcional* (5ª ed.). Rio de Janeiro: Guanabara Koogan.

Cromie, J.E., Robertson, V.J., & Best, M.O. (2000). Work-related musculoskeletal disorders in physical therapists: prevalence, severity, risks, and responses. *Journal of the American Physical Therapy Association, 76*(8), 827–835.

Dawson, A.P., McLennan, S.N., Schiller, S.D., Jull, G.A., Hodges, P.W., & Stewart S. (2007). Interventions to prevent back pain and back injury in nurses: a systematic review. *Occupational & Environmental Medicine, 64*(10), 642–650.

Denis, D., St-Vincent, M., Imbeau, D., Jetté, C., & Nastasia, I. (2008). Intervention practices in musculoskeletal disorder prevention: A critical literature review. *Applied Ergonomics, 39*(1), 1–14.

Fargala, G., & Bailey, L. (2003). Addressing occupational strains and sprains Musculoskeletal injuries in hospitals.. *Official journal of the American Association of Occupational Health Nurses 51*(6), 252–259.

Frazer, M.B., Norman, R.W., Wells, R.P., & Neumann, P. (2003). The effects of job rotation on the risk of reporting low back pain *Ergonomics 46* (9), 904–919.

Glover, W., McGregor, A., Sullivan, C., & Hague, J. (2005). Work- related musculoskeletal disorders affecting members of the Chartered Society of Physiotherapy. *Physiotherapy 91*(3), 138–147.

Holder, N.L., Clark, H.A., DiBlasio, J.M., Hughes, C.L., Scherpf, J.W., Harding, L., et al. (1999). Cause, prevalence, and response to occupational musculoskeletal injuries reported by physical therapists and physical therapist assistants. *Journal of the American Physical Therapy Association, 70*(7), 642–652.

Kilbom, S., Armstrong, T., Buckle, P., Fine, L., Hagberg, M., Haring-Sweeney, M., et al. (1996). Musculoskeletal Disorders: Work-related Risk Factors and Prevention. *International Journal of Occupational and Environmental Health, 2*(3), 239–246.

Lehto, M.R., & Buck, J.R. (2008). *Introduction to Human Factors and Ergonomics for Engineers* (1 ed.). New York: Taylor & Francis Group.

Occupational Safety and Hygiene – Arezes et al. (eds)
© *2013 Taylor & Francis Group, London, ISBN 978-1-138-00047-6*

Background noise and its influence on the brain waves related to attention

E. Tristán, I. Pavón & J.M. López
Centro de Acústica Aplicada y Evaluación No Destructiva (CAEND),
Universidad Politécnica de Madrid, Madrid, Spain

ABSTRACT: In this work, we present a study whose objective is to prove the influence of background noise produced inside university facilities on the brain waves related to attention processes.

Recordings of background noise were carried out in study areas inside university facilities. Volunteers were asked to perform an attention test without any background noise but also while being exposed to the sound recordings, and their cerebral activity was recorded through electroencephalography (EEG). After the application of the test in both conditions, changes in the frequency bands related to attention processes (beta 13–30 Hz and theta 4–7 Hz) were studied.

The results of this study show that when the students were performing the test while being exposed to background noise, both beta and theta frequency bands decreased statistically significantly. Because attentional improvement is related to increases of the beta and theta waves, we believe that those decreases are directly related to a lack of attention caused by the exposure to background noise. Nevertheless, the results do not allow us to conclude that background noise produced inside university facilities has an influence on the attentional processes.

1 INTRODUCTION

In 2009 the National Institute for Occupational Safety and Health (NIOSH) established the National Services Agenda, which includes the safety and health goals for the educational sector (Nora 2009). Seen from the occupational safety and health point of view, the noise produced inside university facilities and its influence on the quality of life of both university students and teachers is a topic that has received little attention. Students spend a lot of time inside the campus with levels of noise higher than the recommended values (Tristán et al. 2012 a,b). Normally, the students themselves are the main source of noise. The high noise levels have a direct impact on their physical health and may cause hearing disorders, stress, etc., but they also directly impact their psychological health and may cause attention and memory disorders. Especially a lack of attention badly affects the academic and work performance (Baker & Holding 1993) and increases the risk of accidents (Toppila et al. 2009). Few researches have revealed this type of effects of noise on attention, but there is a consensus that noise, of whatever kind and at whatever intensity level, has significant repercussions on attention and memory (Santiesteban et al. 1994). Therefore, the noise produced inside university facilities could certainly be considered as a risk factor for the health and safety of students and educational workers.

The brain and the waves it produces (i.e. Beta and Theta) are the most reliable and complete sources of information there are to understand the different human behaviors. This information is collected using the electroencephalogram (EEG).

Many studies show the relationship between attentional processes and changes of specific cerebral waves. Attentional improvement is associated with increases of the beta band 13–30 Hz and theta band 4–7 Hz (Basar et al. 2001; Barry et al. 2003a, b; Klimesch et al. 1997; Klimesch 1999). All researches present similarities in the sound conditions used: Mainly white and pink noise, and pure tones were used as a sonorous stimulus. The silent and noisy conditions were also used to make comparisons.

2 SCOPE OF STUDY

The objective of this study is to prove the negative effects of background noise produced inside university facilities on attentional processes.

We seek a deeper understanding of the risks and consequences that noise has on the psychological health of university students through the study of the cerebral behavior.

It has been hypothesized that if an individual is exposed to background noise during the performance of a specific task, then beta and theta bands will decrease.

The conclusions can be applied to other occupational areas where attention is an important aspect to be considered.

3 MATERIALS AND METHOD

3.1 Participants

A total of 33 volunteers (17 male, 16 female) between the ages of 19 and 34 years old participated in the experiments. 12 of the participants were college students, 9 were undergoing their master degree and 12 were doctorate researchers/teachers. Before the experiments, the current state of auditory health of the participants was evaluated according to ISO 8253-1:2010—Audiometric test methods, Part 1: Pure-tone air and bone conduction threshold audiometry. The participants did not exceed 20 dB HL in the frequency range 125-8 kHz.

3.2 Recording and playback equipment

A total of 33 volunteers (17 male, 16 female) between the ages of 19 and 34 years old participated in the experiments. 12 of the participants were college students, 9 were undergoing their master degree and 12 were doctorate researchers/teachers. Before the experiments, the current state of auditory health of the participants was evaluated according to ISO 8253-1:2010—Audiometric test methods, Part 1: Pure-tone air and bone conduction threshold audiometry. The participants did not exceed 20 dB HL in the frequency range 125-8 kHz.

3.3 Background noise

The background noise used consisted of 6 stereo recordings carried out inside university facilities; specifically in areas were students usually develop their studying activities such as classrooms during an exam *(CE)*, classrooms during a normal lesson *(CL)*, libraries *(LIB)*, computer rooms *(CR)*, hallways *(HW)* and halls adapted for study activities *(AH)*. Each sound clip was between 1:30 and 2 minutes long. Furthermore, an acoustic analysis

Table 1. Situation of the participants.

Sex	N	Age	Current educational level	N
Male	17	19–34	College students	12
Female	16	21–32	Master students	9
Total	33		Doctorate students	12

Table 2. Sound clips used in the experiment.

Environment	CE	LIB	CR	CL	HW	AH
Time (sec)	90	90	90	90	120	120
$L_{eq,T}$ (dB)	68.1	76.5	76	72.6	81.7	83.5

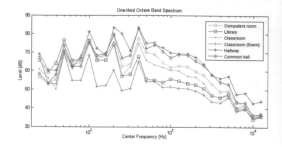

Figure 1. One-third octave band spectrum of each.

of each of the above-mentioned environments was carried out using the platform Pulse Reflex Special Version 16.0.0.500.

The sound pressure levels ($L_{eq,T}$) were in the range 68.1–83.5 dB (Table 2). Figure 1 shows the one-third octave spectrum of all environments.

The acoustics tests and the calibration of the audio equipment were performed in controlled acoustic conditions (in this case an anechoic chamber).

3.4 Attention test

The evaluation of attention was carried out through the application of the Toulouse-Pieròn test. This test is commonly used to evaluate attention capacity and perception. It consists of the identification of figures, which are randomly distributed in 40 rows and 40 columns (10 target figures per row).

During 10 minutes, participants had to identify the largest number possible of targets (400) from a total of 1600 figures. The direct score (DS) is evaluated through three task efficiency indicators: successes (S), mistakes (M) and omissions (O). These variables are integrated in the formula $DS = S - (M + O)$.

3.5 EEG recordings

The experiments were carried out in an audiometric room in order to avoid sonorous and visual distractions. The total time of the experiments was 27 minutes following the timeline shown in Figure 2. The brain activity of the volunteers was recorded in both silent and noisy conditions while they were performing a specific task (the Toulouse-Pieròn attention test), and was registered through EEG.

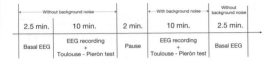

Figure 2.   Experiment development timeline.

The equipment used for the experiments was the Brainquiry PET EEG 2.0 neurofeedback equipment with 5 electrodes. The software Bioexplorer V. 1.5 was used to capture and process the data.

The electrodes were placed following the standard position referred to in the international 10–20 system. The 3 active electrodes were placed on the frontal part of the head on FpZ, Fp1 and Fp2, the reference electrode was placed on the right mastoid A1 and the negative electrode on the left mastoid A2.

Although the alpha band is related to attentional processes, the study of this frequency band was not performed. The applied test consists of a visual task and alpha waves practically disappear when the eyes are open. The beta and theta bands are also perfectly related to attentional processes and it is possible to study them when a person is awake and with his/her eyes open. Thus, in order to identify a possible direct relationship between the exposure to noise and negative influences on the attention, we studied changes in the frequency bands beta (13–30 Hz) and theta (4–7 Hz.).

The statistical analyses were performed using GraphPad Prism version 5.00.

4   RESULTS

4.1   Attention test

The main assumption was that background noise should generate a negative influence on attention.

The results of the attention test show that 55% of the participants increased their direct punctuation (DP) when they were exposed to background noise. The remaining 45% decreased their direct punctuation.

Furthermore, differences between the direct punctuation in silent condition and the direct punctuation with background noise exposure were observed, but the paired differences t-test shows that these differences were not significant for P < .05, the

P value is P = 0.0945. Thus, the results do not allow us to confirm the affirmation that background noise has an important influence on attention (Table 3).

4.2   Changes in beta an theta bands

Figure 3 shows the EEG mean of the 33 participants and its changes in the beta band (4–7 Hz) and Figure 4 shows theta band (13–30) frequency bands. It is possible to observe changes between the silent and noisy condition.

In order to determine if these changes are significant, a statistical analysis was carried out.

In the case of the beta band in silent condition, the lowest amplitude registered was 11.37 µV and the highest 52.43 µV. The mean of all participants was 28.81 µV. On the other hand, when the participants were exposed to background noise the lowest amplitude registered for the beta band was 6.88 µV, the highest 40.3 µV and the mean 22.08 µV. With regard to the theta band, the lowest amplitude registered in silent state was 26.51 µV and the highest 159 µV. The mean of all EEG recordings was 66.9 µV. In the noisy condition, the lowest amplitude registered for the theta band was 19.5 µV, the highest 12.2 µV and the mean 54.2 µV.

Based on the above-mentioned results, we can observe that there are highly significant changes in the beta and theta bands. For P < 0.05 in both the beta and theta band the P values are P < 0.0001. Furthermore, the difference between the silent and noisy condition for the beta band is 6.72 µV and for the theta band 12.78 µV.

A global summary of all these results is shown in Table 4.

4.3   Decreases in beta an theta bands

A decrease tendency in the beta and theta bands was observed when the participants were exposed to background noise. The beta band of 67% of the participants (N = 22) and the theta band of 76% (N = 24) of the participants decreased.

This tendency was analyzed and the paired differences t-test (Table 5) show that in both the beta and theta band there are significant differences for P < 0.005 and a low correlation between the silent and noisy condition. For beta

Table 3.   Paired differences t-test of the application of Toulouse-Pieròn test with and without background noise exposure (P < 0.05).

				95% CI				
	M.D.	S.D.	S.E.	Lower	Upper	t	df	P
Without background noise vs. with background noise	−11.9	48.811	33.852	−25.99	2.174	1.723	32	0.0945

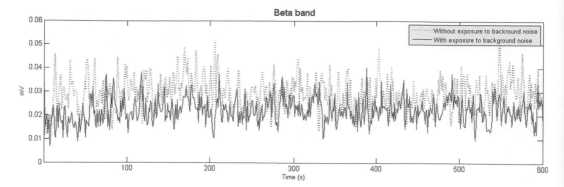

Figure 3.   EEG mean of all participants for beta band without exposure to noise (grey line) and with exposure to noise (black).

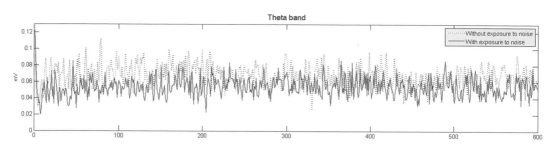

Figure 4.   EEG mean of all participants for theta band without exposure to noise (grey line) and with exposure to noise (black).

Table 4.   Summary of statistical results of the EEG recordings of all the participants.

	Min.	Median	Max.	Mean	S.D.	S.E.	95% CI Lower	Upper
β without background noise	0,01137	0,02835	0,05243	0,02881	0,006833	0,0002790	0,02826	0,02935
β with background noise	0,00688	0,02184	0,04030	0,02208	0,005302	0,0002164	0,02165	0,02251
Θ without background noise	0,02651	0,06643	0,1590	0,06697	0,01368	0,0005584	0,06588	0,06807
Θ with background noise	0,01958	0,05326	0,1220	0,05420	0,01175	0,0004797	0,05325	0,05514

Table 5.   Paired differences t-test of EEG recordings with and without background noise exposure ($P < 0.05$).

	M.D.***	S.D.	S.E.	95% CI Lower	Upper	r	R^2	t	df	P < 0.05
β WO-BN* vs. β W-BN**	0.01356	0.02003	0.01742	0.3292	0.8459	0.6593	0.4348	3.740	21	0.0012
Θ WO-BN vs. Θ W-BN	0.02215	0.03294	0.02353	0.5347	0.8962	0.7717	0.5955	4.691	23	0.0001

* Without background noise; ** With background noise; *** Mean of differences.

band: $P_\beta = 0.0012$, R-squared$_\beta = 0.4348$, Mean of differences$_\beta = 13.56$ µV. For theta band: $P_\Theta = 0.0001$, R-squared$_\Theta = 0.5955$, Mean of differences$_\Theta = 22.15$ µV.

5 CONCLUSIONS

In this work we studied the effects that background noise has on attention, especially amongst university students.

We applied a psychometric test in order to evaluate the students' attention in both silent and noisy conditions. The application of this test did not show any data that allow us to establish a direct relation between the lack of attention and background noise. We therefore believe it is difficult to develop a correct evaluation of the influence of noise on the attention in a psychometric way.

On the other hand, we believe that the study of the brain activity gives us a clearer understanding about the effects that noise could have on the cognitive health of people.

EEG recordings were performed to observe the cerebral behavior of volunteers during the exposure to background noise and while they were developing a specific task (attention test). Literature describes the direct relation between increased of the beta and theta band and the improvement of attention.

Our EEG recordings revealed important information about the cerebral behaviour of our volunteers. When they were exposed to background noise, we were able to observe that the beta and theta waves tend to decrease significantly. This is an important indicator of the relation between the exposure to background noise and the negative effects on attention and in general on psychological health.

Students usually spend almost the same or even more time in the same environments as teachers and other educational workers. They are therefore exposed to the same health and security risks. Nevertheless, the existing standards and recommendations do not consider them as people who are exposed to those risks.

That is why we believe that the study of the sonorous factors which affect the quality of life, security and health of the users of school buildings not only has to focus on educational workers, but also on students. We are thus convinced that our work contributes with important data that could enable future studies and delve deeper into the topic in order to try to proof the negative effects on attention as a consequence of occupational noise. The method used in our experiments can be very useful to perform an objective evaluation not only of attention, but also of psychological health aspects related to noise generated in occupational environments. The study of cerebral behavior with and without noise exposure can be implemented on educational workers and in the future in other occupational areas.

REFERENCES

Baker, M.A. & Holding, D.H. 1993. The effects of noise and speech on cognitive task performance. *Journal Gen Psycholog* 120: 339–355.

Barry, R.J.; Clarke, A.R. and Johnstone, S.J. 2003. A review of electrophysiology in attention-deficit/hyperactivity disorder: I. Qualitative and quantitative electroencephalography. *Clin Neurophysiol*, 114(2): 171–83.

Barry, R.J.; Johnstone, S.J. and Clarke, A.R. 2003. A review of electrophysiology in attention-deficit/hyperactivity disorder: II. Event-related potentials. *Clin Neurophysiol*, 114(2): 184–98.

Basar, E.; Schurmann, M. and Sakowitz, O. 2001. The selectively distributed theta system: functions. *Int. J. Psychophysiol* 39: 197–212.

ISO 8253-1:2010—Audiometric test methods, Part 1: Pure-tone air and bone conduction threshold audiometry.

Jasper, H.H. 1958. The ten-twenty electrode system of the International Federation. Electroencephalogr. *Clin. Neurophysiol* 10: 370–375.

Klimesch, W.; Doppelmayr, M.; Schimke, H. and Ripper, B. 1997. Theta synchronization and alpha desynchronization in a memory task. *Psychophysiol* 34:169–176.

Klimesch, W. 1999. EEG alpha and theta oscillations reflect cognitive and memory performance: a review and analysis. *Brain Res. Rev.* 29:169–195.

Kozou, H.; Kujala, T.; Shrytov, Y.; Toppila, E.; Starck, J.; Alku, P. & Näätänen, R. 2005. The effects of different types of occupational noises on processing of speech and non-speech sounds: an ERP study. *Hearing Research* 199: 31–39.

Lawrence, M.W. 2012. Synchronous neural oscillations and cognitive processes. Trends in Cognitive Sciences 7(12): 553–559.

National Occupational Research Agenda (NORA) Services Sector Council. 2009. National Services Agenda. *National Institute for Occupational Safety and Health (NIOSH): 8–10.*

Santiesteban, C.; Sebastian, E.M. & Santalla, Z. 1994. Efectos de Ruidos Cotidianos Sobre el Recuerdo. *Psicothema.* 6(3): 403–416.

Toppila, E.; Pyykkö, I. & Pääkkönen, R. 2009. Evaluation of the increased accident risk from workplace noise. *Int. J. Occup Saf. Ergon.* 5(2): 155–62.

Tristán, E.; Pavón, G &, López, J.M. 2012. University sound environments. *In P., Euronoise. Prague, 10–13 June 2012.*

Tristán, E.; Pavón, G. & López, J.M. 2012. Evaluation of acoustic quality in university facilities. *In P., ISMA2010-USD2010 Conference, Leuven, 17–19 September 2012.*

Construction safety

Occupational Safety and Hygiene – Arezes et al. (eds)
© *2013 Taylor & Francis Group, London, ISBN 978-1-138-00047-6*

Profile of the construction safety coordinator in Portugal

P. Santos
ISISE, Department of Civil Engineering, University of Coimbra, Coimbra, Portugal

P.A.A. Oliveira
Faculty of Engineering, University of Porto, Porto, Portugal
Higher Institute of Languages and Administration (ISLA), Leiria, Portugal

E.F.P. Almeida
Department of Civil Engineering, University of Coimbra, Coimbra, Portugal

ABSTRACT: Construction workers are exposed to particularly high risks, therefore the accidents levels are often high. In order to fulfill the European Directive 92/57/EEC (*Construction Sites Directive*), the main duties for the coordinator for safety and health matters can be found on the Portuguese decree-law 273/2003, on its 19th article. This law transposes into the Portuguese legislation the above mentioned European directive. However, it is written on article n. 9 (3rd point) that it will be published a special law concerning the requirements for their formation and training. Until today (nine years later), nothing is yet published, creating a legal gap concerning the Safety Coordinator's professional status, weather in project stage or in construction stage! There is no general agreement among the experts regarding this subject. This paper aims to point the different perspectives from the various entities involved in this legislative process, which originates the lack of consensus in establishing the regulation for the construction safety coordination (CSC) activity in Portugal.

1 INTRODUCTION

The construction industry exposes workers to particularly high risks (Almeida 2012). It has several specific characteristics, which make it so distinguish from other economic activities. This industry has a high precarity level, marked also by its rotational work characteristic, often subcontracting (Oliveira 2010).

Regarding the occupational safety (according to the article n. 118-A from the Rome Treaty – 1957), the European Union established a professional preventive politic concerning all the economic activities with the Directive 89/391/EEC from the Council in June 12, named *Framework Directive*. In order to respond to the unique characteristics of the Construction Industry, the European Community adopted a preventive philosophy for this area with the Directive 92/57/EEC from the Council in June 24, also known as *Construction Site Directive*.

In order to fulfill the Construction Site Directive, the obligations related to Safety and Health co-ordination can be found in point n. 19 of the Decree-Law (DL) 273/2003 from October 29. However, it is written in article n. 9 (3rd point) that it will be published a special law concerning this matter. The lack of this regulation, during nine years, created a legal gap concerning the Construction Safety Coordinator (CSC) in project stage and in construction stage, discrediting his activity. It has also generated some controversy among some construction safety experts. Dias (2004) considers that *"..., the CSCs must have a basis formation on the construction area (construction engineering or architecture), regarding the different construction types and its categories according to Portuguese legislation published in 1972, and the additional training regarding the Health and Safety Coordination in construction."*

Vieira (2008), as a specialist of the *"Autoridade para as Condições de Trabalho (ACT)"* (Portuguese authority for health and safety work conditions), mentioned in a paper published by ACT that *"..., the CSCs have a crucial role if they are suitably qualified and if they have technical autonomy regarding the construction owner and to the other stakeholders. In order to be effective, the CSC must have professional experience and know-how on the construction industry, i.e. they must have specific technical skills to fulfill this task."*

However, the working market has responded in an improper way, which allowed a policy of risk tolerance. This enabled professionals, whose education is commonly apart from the construction industry, to act as CSC. Many of these don't have any bond to the construction sector, ignoring its unique characteristics and specificities, which make it one of the most deadly working sectors in Portugal. The ACT (2012) activity report indicates that among the 161 fatal working accidents occurred during 2011, 44 of them were directly related to the construction area (27.3%).

2 MAIN GOALS

Given the delays in the publication of special legislation to define the competencies and skills of the CSC in Portugal, and self-regulation of the market, this study focuses on the comparison of the positions supported by the social partners (professional orders and business associations) for the elaboration of special legislation referred to DL 273/2003. The aim is to identify the different perspectives of the various entities involved in the legislative process that will regulate the safety coordination activity in project stage and construction stage, and has been in the origin of the lack of consensus on this issue.

3 METHODOLOGY

This paper was developed based on partial contents and also as a result of a Master thesis performed at the Department of Civil Engineering of the Faculty of Sciences and Technology of the University of Coimbra, entitled *"Profile of the Construction Safety Coordinator: Action, Requirements and Training"* (Almeida 2012). The applied methodology is based on two main components of literature review: Scientific and legislative.

4 RESULTS

After completing a bibliographical research and contact with the social partners in order to collect existing information, it was designed a table for comparing the positions of the various social partners (Table 1), in order to understand their points of convergence and disagreement about the Decree-Law project (BTE 2009).

The proposal of the working group constituted by: Order of Engineers (OE), Order of Architects (OA), National Association of Technical Engineers (ANET), and the Portuguese Association for Health and Safety at Work (APSET) differ from the project of Decree-Law on three main issues.

The first issue is that the working group defends the autonomy of technical functions of the CSC, not agreeing with the accumulation of functions of inspection and coordination of safety level 3, whenever the construction owner (client) is simultaneously the builder entity (contractor).

Table 1. Comparing the positions about the Decree-Law project to regulate the Construction Safety Coordinator (CSC) activity in Portugal.

Decree-Law project (BTE 2009)	Working group of OE, OA, ANET & APSET (OET 2012)	FEPICOP (Campos 2008)
ARTICLE N.° 2-Arrangements for Health and Safety Coordination		
Two arrangements for Health and Safety Coordination: 1-project preparation stage; 2-construction stage, regarding DL 273/2003 (art. n.3)	Agree.	Agree.
ARTICLE N. 3-Competency levels for health and safety coordination		
Three levels of safety Coordination depending on the construction work value*: Level 1-any construction license class; Level 2-value up to class 6 license; Level 3-value up to class 3 license. Complex engineering works require level 1 license.	Partially agree and propose a change: Level 3-construction work value not exceeding class 2 license.	Partially agree and propose a change: Contemplate level 3 for architecture and engineering technicians.

(Continued)

Table 1. (Continued).

ARTICLE N. 4-Technical autonomy		
CSC technical autonomy only allows to accumulate with inspection functions and in private construction works for Level 3, where the construction owner is the builder.	Defends the technical autonomy and functions of the CSC.	Claim to be outside the scope of paragraph 6 of article 9 of DL 273/2003. Propose a revision for the relationship between Promoter and Builder.
(..)		
ARTICLE N. 6-Minimum guarantee of effective coordination in project stage		
CSC allocation cannot exceed 100% in project stage despite no Previous Notification. Construction works value up to class 9 requires a Level 1 adjunct technician. Each fraction greater than 300% of class 8 value requires an adjunct technician with allocation of 100%.	Agree.	Do not mention it. However, claim discussion for the allocation time.
ARTICLE N. 7-Minimum guarantee of effective coordination in construction stage		
The allocation of CSC in construction stage cannot exceed 100% despite no Prior Notification. Construction works value up to class 9 requires at least one Level 2 adjunct technician. Each fraction greater than 150% of class 8 value requires an adjunct technician with allocation of 100%.	Agree	Do not mention it. However, claim discussion for the allocation time.
(…)		
ARTICLE N. 9-Authorization to engage activities of coordination for health and safety in project stage		
CSC cumulative requirements in Project Preparation Stage: Level 1 Project Coordinator or designer qualification for this level; minimum experience of 5 years in the sector; approved initial specific training. Level 2 Project Coordinator or designer qualification for this level; minimum experience of 3 years in the sector; approved initial specific training. Level 3 Project Coordinator or designer qualification for this level; minimum experience of 3 years in the sector; approved initial specific training.	Disagree and propose the following: CSC cumulative requirements in BUILDINGS Project Preparation Stage: Level 1-Project Coordinator or designer qualification of this level; minimum experience of 5 years in the sector; approved initial specific training. Level 2-Project Coordinator or designer qualification of this level; minimum experience of 3 years in the sector; approved initial specific training. Level 3-Project Coordinator or designer qualification of this level or Level IV Certified Health and Safety Technician (CHST); minimum experience of 3 years in the sector; approved initial specific training.	Disagree and propose the following: Level 1 Architect, civil or other engineering domain, technical engineer near the sector with minimum experience of 5 years in construction site or risk prevention; approved initial specific training. Level 2 Architect, civil or other engineering domain, technical engineer near the sector with minimum experience of 3 years in construction site or risk prevention; approved initial specific training.

(*Continued*)

Table 1. (Continued).

	CSC cumulative requirements in CIVIL ENGINEERING Project Preparation or Execution (Level 1): Civil engineer or technical engineer with minimum experience of 5 years or architect, or other specialist engineer or technical engineer with minimum experience of 10 years in the sector; approved initial specific training.	Level 3 Technician in architecture or engineering; minimum experience of 3 years in sector; approved initial specific training.

ARTICLE N. 10-Authorization to engage activities of coordination for health and safety in construction stage

CSC cumulative requirements in Construction Stage:	Disagree with proposal:	Disagree and propose the following:
Level 1 Construction site or inspection manager; Level IV Certified Health and Safety Technician (CHST); minimum experience of 5 years; approved initial specific training. Level 2 Construction site or inspection manager, Bachelor in occupational safety or Level VI Certified Health and Safety Specialist (CHSS); minimum experience of 3 years in the sector; approved initial specific training. Must be Level IV CHST whenever don´t have Level VI CHSS. Level 3 Construction site or inspection manager and/or Level VI CHSS; minimum experience of 3 years in the sector; approved initial specific training.	CSC cumulative requirements in Construction Stage: Level 1-Construction site or inspection manager of this level, or other specialty near the sector of civil engineering, minimum experience of 5 years; approved initial specific training. Level 2-Architect, civil or other specialty engineer, technical engineer near the sector or Level VI CHSS; minimum experience of 3 years; approved initial specific training. Level 3-Technician in architecture or engineering; Level IV CHST or VI CHSS; minimum experience of 3 years in sector; passing in initial specific training. CSC cumulative requirements in CIVIL ENGINEERING Project preparation or execution (Level 1): Civil engineer or technical engineer with minimum experience of 5 years or architect, or other specialist engineer or technical engineer with minimum experience of 10 years in the sector; approved initial specific training.	CSC cumulative requirements in construction site: Level 1 Architect, engineer, technique engineer; minimum experience of 5 years or Level VI CHSS or bachelor in occupational safety with minimum experience of 6 years; approved initial specific training. Level 2 Architect, engineer, technique engineer; minimum experience of 3 years or Level VI CHSS or bachelor in occupational safety with minimum experience of 4 years; approved initial specific training. Level 3 Technician in architecture or engineering; Level IV CHST; minimum experience of 3 years; Approved initial specific training.

(…)

ARTICLE N. 13-Validity period and revalidation

| The authorization for the activity of CSC in project preparation and execution stages is valid during 5 years from the concession, being renewed for similar periods. To revalidate the authorization as a CSC is needed to be in activity during at least 2 years and to be approved in one upgrade training. If not exercised for 2 years must be approved in an initial specific training again. | Agree. | Agree. |

(…)

(Continued)

Table 1.   (Continued).

ARTICLE N. 16-Competence for training courses

| Initial specific training and specific update courses may only be administered by institutions of higher education or other similar qualified entities. | CSC courses should be administered only by higher education institutions with qualifying training basis and recognized by their professional associations. | Do not mention it. |

ARTICLE N. 17-Initial specific training

| Initial specific training for CSC with a minimum duration of 200 hours. Of those, 120 hours are theoretical (scientific and technologic component) and 80 hours of practice in real work context. | Initial specific training for CSC must have a minimum duration of 250 hours which corresponds to 15 ECTS. | Agree. |

ARTICLE N. 18-Update specific training

| Update specific training must have a duration exceeding 48 hours including laws, regulations and technological developments relevant to the activity on the real work context.
(…) | Do not mention it. | Agree. |

ARTICLE N. 21-Training access

| The access to initial specific training in project preparation stage is restricted to qualified experienced project coordinators or project authors.
The access to initial specific training in construction stage is restricted to qualified experienced construction work manager or inspection director or level VI CHSS. | Access to initial specific training in project stage restricted to: architects, engineers, if recognized by the professional order and architecture and engineering technician with Level IV CHST.
Access to initial specific training in construction stage restricted to: architects, engineers if recognized by professional order, architecture and engineering technician with level IV CHST or level VI CHSS. | Disagree and propose the following requirements: Academic qualifications in architecture, civil engineering or another domain related with construction specialty projects, Bachelor in occupational safety or Level VI CHSS. |

ARTICLE N. 23-Authorization transitional period

| During transitional period have authorization to perform CSC activity in project stage or construction stage, those that have a minimum of 3 years' experience, at legislation publication date and in subsequent 2 years obtain initial specific training or its equivalence. | Agree. | Disagree and propose: The authorization to perform CSC activity in project stage or construction stage, during transitional period, is given to those who are in already in business (evidenced by Prior Notification) until the end of the contract, if requested to the competent authority and those approved in specific initial training in next 2 years. |

(Continued)

Table 1. (Continued).

ARTICLE N. 26-Contraventions		
It is a Very Serious Infraction attributed to the project owner and CSC: – to exercise CSC activity without any authorization; – to exercise CSC activity for unauthorized level and modality. It is a Serious Infraction attributed to project owner and CSC if the general duties of the CSC.	Agree.	Agree.

*According to the Portuguese Legal Regime of Admission and Practice in Construction Industry.

The second is the discordance about the requirements to obtain the authorization for the CSC activity, which proposes the division of safety coordination into three areas: (1) Project of buildings, (2) building construction works, and (3) other civil engineering works.

Finally, is also dispensed the need of any Professional Certificate (PC) for safety coordination activity in civil engineering works.

The amendments proposed by the Portuguese Federation of Construction Industry and Public Works (FEPICOP) are aimed, in general, to extend the requirements for CSC activities, not only to architects and engineers involved in projects specifically related to construction works, but as well to other graduations related within safety at work and to holders of PCs for occupational health and safety activity.

It also proposes to review the promoter-builder relationship in the context of autonomous functions of CSC. Notice that the associations represented by this federation, joined the Technical Committee Tripartite Construction consensus in which a document was drawn up by the Portuguese former Institute for Development and Inspection of Working Conditions (IDICT) with contrary opinion to that now held by the federation they integrate.

5 CONCLUSIONS

It was concluded that the views expressed by each social partner reveal a constant defence of their own interests individually. Such opinions even devaluate policies to combat risks at work and occupational diseases, contributing explicitly to delay the publication of the special legislation referred in article 9 (paragraph 3) of the Portuguese DL 273/2003, supporting the idea of the existence of entrenched corporative interests. Given the importance of the specific legislation to regulate the access and the requirements for the exercise of coordination safety activity in project stage and in construction stage, which would act as a lever to reduce the high number of accidents at work in construction sector, through legislation to strengthen a culture of safety and prevention of greater liability in the sector, ending a reality that complicates the desirable technical quality for the development of this professional activity.

REFERENCES

ACT 2012. *Annual reports 2007–2011*. Autoridade para as Condições de Trabalho (in Portuguese). Consulted in April 2012: http://www.act.gov.pt/(pt-PT)/SobreACT/DocumentosOrientadores/RelatorioActividades/Paginas/default.aspx.

Almeida, E.F.P. 2012. *Perfil do Coordenador de Segurança em Obra: Ação, Requisitos e Formação*. Master of Science thesis (in Portuguese). Department of Civil Engineering, University of Coimbra, Coimbra.

BTE 2009. *Projeto de decreto-lei que regula o exercício da atividade de coordenação em matéria de segurança e saúde na construção* (in Portuguese). Boletim do Trabalho e Emprego (BTE), Separata n. 2/2009. Ministério do Trabalho e da Segurança Social. Gabinete de Estratégia e Planeamento e Centro de Informação e Documentação, Lisbon.

Campos, M.J.R. 2008. *Análise do projeto de decreto-lei sobre o exercício da coordenação de segurança e saúde na actividade da construção*. Carta enviada ao Chefe de gabinete do Senhor Ministro do Trabalho e Segurança Social (in Portuguese), Porto.

Decree-Law n. 273/2003. *Regulamenta as condições de segurança e de saúde no trabalho em estaleiros temporários ou móveis* (in Portuguese). D.R. n. 251. (Série I-A de 2003-10-29). Imprensa Nacional e da Casa da Moeda (in Portuguese), Lisbon.

Dias, L.A. 2004. Repensar a Segurança e Saúde no Trabalho da Construção em Portugal. *2nd Construction National Conference* (in Portuguese). FEUP, Porto.

Directive 89/391/EEC of 12 June 1989 on the *introduction of measures to encourage improvements in the safety and health of workers at work*. The Council of the European Communities, Luxemburg.

Directive 92/57/EEC of 24 June 1992 on the *implementation of minimum safety and health requirements at temporary or mobile construction sites* (eighth individual Directive within the meaning of

Article 16 (1) of Directive 89/391/EEC). The Council of the European Communities, Luxemburg.

OET 2012. *Proposal of the working group: OE, OA, ANET & APSET*. Ordem dos Engenheiros Técnicos (in Portuguese). Consulted in April 2012: http://www.oet.pt/dowloads/Propostas/ParecceresPropostas-Projecto%20Decreto-lei%20273–2003.pdf

Oliveira, P.A.A. 2010. Serão atualmente os Coordenadores de Segurança e Saúde, profissionais imparciais nas ações que assumem? *Tecnologia e Vida*, n. 7: 26–28. Secção Regional do Ordem dos Engenheiros Técnicos—OET (in Portuguese). Consulted in April 2012: http://www.srnorte.oet.pt/docs/revistas/revista_7.pdf

Vieira, L. 2008. Artigo publicado pela Autoridade para as Condições de Trabalho—ACT (in Portuguese). *Suplemento Fórum Empresarial do Jornal de Negócios*. August 2008.

Occupational Safety and Hygiene – Arezes et al. (eds)
© *2013 Taylor & Francis Group, London, ISBN 978-1-138-00047-6*

Characterization of workers suffering serious electrical accidents in the construction industry

M. Suárez-Cebador, J.C. Rubio-Romero & A. López-Arquillos
University of Málaga, Málaga, Spain

J.A. Carrillo
University of Seville, Seville, Spain

ABSTRACT: In occupational accidents, those caused by electrical contact are characterized to produce severe injuries. This study analyzed the severity of 2.776 accidents caused by direct or indirect electrical contact in the construction industry. In order to characterize the groups of workers which suffer higher severity in such accidents, we analyze several own factors of workers such as age, occupation and length of service. According to results obtained, variables related to age, occupation and length of service influenced the seriousness of injuries produced in an electrical accident. This research provides an insight of the likely worker profile than suffer more serious electrical accidents in the construction site, so governments and safety professionals can focus their strategies to provide specific training on electrical hazards and to develop supervising plans addressed to protect the most vulnerable workers groups.

1 INTRODUCTION

Accidents produced in work performance are a great concern and represent an important challenge for companies, administrations, workers and society in general (Haslam *et al.*, 2005). Among different occupational sectors, the construction industry has been presenting a high ratio of accidents (Salminen, 1995; Ore & Stout, 1996; Dufort *et al.*, 1997; Koh & Jeyaratnam, 1998; Chen & Fosbroke, 1998). Currently, although in recent years it has experienced a reduction (Haslam *et al.*, 2005; Xiuwen & Platner, 2004), this sector remains showing high figures in incidence ratios (Cawley & Brenner, 2012) and huge costs associated with them (Waehrer *et al.*, 2007).

Regarding to occupational accidents, those produced by electrical contacts show a great importance due to their severity and damage than they cause (Chi *et al.*, 2012) and we can also find a disproportionate number of fatal accidents (Cawley & Homce, 2003). As example, Janicak (2008) describes an improvement potential in United States indicating that we could prevent 125 deaths by electrical accident in construction sites each year. Other studies such as those conducted by Williamson & Feyer (1998), McCann *et al.* (2003) or Chi *et al.* (2009) prove the importance and significance of electrical accidents and the need to obtain relevant information of its causes in order to control and prevent risks. In analysis of causes,

in addition to working conditions and environment (Cheng *et al.*, 2010), there are several factors that increase the risk of accidents at work as those related to worker's individual characteristics such as age (Chau *et al.*, 2004), occupation (Chen & Fosbroke, 1998) and experience (Salminen, 2004).

Our goal is to obtain information related to personal characteristics of workers who are injured by electrical accidents in the construction industry. This information allows us to achieve a basic profile of those workers who suffer the most serious consequences based on variables such as age, occupation and length of service. The object of this study is to provide more information to help workers, safety technicians and other responsible people in assessment, prevention and protection of this sort of risks in order to achieve reducing these accidents and their serious consequences.

2 MATERIAL AND METHODS

2.1 Data

In this research, Ministry of Employment and Social Security in Spain provided us data of 1,162,598 occupational accidents occurred during 2003-2008 in the construction industry. These data are related to construction activities under National Classification of Economic Activities (CNAE) and whose codes are 451, 452, 453 and 454. These codes refer respectively to following activities: preparation of construction sites; general construction and civil engineering works; building facilities; completion works.

Subsequently, to limit this dataset to accidents by electrical causes, we select only those produced by direct and indirect electrical contacts based on deviation codes associated with accidents. Thus, we finally obtain a total of 2,776 cases which are classified according to their severity and lesion degree in 2,583 slight accidents, 139 serious accidents, 10 very serious accidents and 44 fatal accidents.

2.2 Statistical analysis

Statistical analysis is based on to prove whether there is a dependency relationship between severity of accidents and variables associated to individual characteristics of workers, and later identify a basic profile of those workers who suffer serious electrical accidents in the construction industry.

To do this, we apply methodology proposed by Camino et al. (2008) and we make contingency tables on which we calculate the value of chi-squared statistic. This statistic associated with a significance level <0.05 allows us to verify with a confidence level of 95% a dependency relationship between chosen variable and accident severity.

Table 1. Electric accidents in construction industry comparing age and severity.

Age	TAR N	%	SLAR N	%	FSAR N	%	FAR N	%
<25	656	23.63	613	23.73	43	22.28	12	27.27
25–30	670	24.14	629	24.35	41	21.24	9	20.45
31–35	448	16.14	431	16.30	27	13.99	7	15.91
36–40	332	11.96	305	11.81	27	13.99	7	15.91
41–45	228	8.21	218	8.44	10	5.18	2	4.55
46–55	318	11.46	282	10.92	36	18.65	6	13.64
>55	124	4.47	115	4.45	9	4.66	7	2.27

Table 2. Electric accidents in construction industry comparing occupation and severity.

Occupation	TAR N	%	SLAR N	%	FSAR N	%	FAR N	%
Electrician	849	30.58	799	30.93	50	25.91	16	36.36
Laborers	408	14.70	383	14.83	25	12.95	5	11.36
Mason	350	12.61	333	12.89	17	8.81	4	9.09
Completion	162	5.84	158	6.12	4	2.07	1	2.27
Structure	107	3.85	95	3.68	12	6.22	1	2.27
Concrete	105	3.78	95	3.68	10	5.18	1	2.27
Team leader	46	1.66	40	1.55	6	3.11	2	4.55
Other	749	26.98	680	26.33	69	35.75	14	31.82

Table 3. Electric accidents in construction industry comparing length of service and severity.

Length of service	TAR N	%	SLAR N	%	FSAR N	%	FAR N	%
<1 month	447	16.10	415	16.07	32	16.58	8	18.18
1–2 months	188	6.77	181	3.63	7	3.63	0	0.00
2–3 months	139	5.01	133	3.11	6	3.11	1	2.27
3–4 months	114	4.11	108	3.11	6	3.11	0	0.00
4–5 months	135	4.86	128	3.63	7	3.63	2	4.55
5–6 months	103	3.71	91	6.22	12	6.22	2	4.55
6–7 months	93	3.35	89	2.07	4	2.07	2	4.55
7–8 months	72	2.59	66	3.11	6	3.11	2	4.55
8–9 months	70	2.52	63	3.63	7	3.63	2	4.55
9–10 months	59	2.13	51	4.15	8	4.15	4	9.09
10–11 months	73	2.63	68	2.59	5	2.59	0	0.00
11–12 months	72	2.59	68	2.07	4	2.07	2	4.55
1–2 years	366	13.18	343	11.92	23	11.92	7	15.91
2–3 years	228	8.21	207	10.88	21	10.88	3	6.82
3–4 years	126	4.54	124	1.04	2	1.04	0	0.00
4–5 years	102	3.67	94	4.15	8	4.15	1	2.27
5–10 years	289	10.41	267	11.40	22	11.40	3	6.82
>10 years	100	3.60	87	6.74	13	6.74	5	11.36

In contingency tables we have used different factors associated with each interesting variables and lesion degrees which show us accident severity.

Also, to facilitate sample description we have added information about frequencies ratios as percentage of total. These ratios are represented by: TAR (Total Accident Rate), SLAR (Slight Accident Rate), FSAR (Fatal and Severe Accidents Rate) and FAR (Fatal Accident Rate).

3 RESULTS AND DISCUSSION

3.1 Age

In the analysis of dependency relationship between workers age and accident severity caused by an electrical contact we have obtained positive results (chi-squared = 13.907 d.f. = 6 sig = 0.031). In addition, our results show that increasing age up to 45 years, reduces likelihood of an occurred accident may have serious consequences. Note that this trend is broken in the group of workers 46–55 years and it is also significant that in this age group, slight accidents ratio and total accidents ratio is very similar (SLAR 10.92%, TAR 11.46%) but serious and fatal accidents ratio is increased considerably becoming in the group with higher proportion of accidents with serious consequences (FSAR/TAR = 1.63).

Also, it is interesting to note in accordance with results obtained that younger workers are the next group more exposed to serious consequences (FSAR 22.28%) and they show the highest frequency in fatal accidents (FAR 27.27%). Values of these last ratios have a special relevance when we compare it with other researches which study all kind of accidents in the construction industry. For example, in López et al. (2012) these ratios would show lower values (FSAR 13.64% & FAR 10.79%). This means that according to our research, electrical accidents occurred in construction sites in this age group show incidence ratios increased by 63% and 252% respectively.

3.2 Occupation

The analysis of variable related to occupation, according to results obtained (chi-squared = 21.951 d.f. = 7 sig = 0.003) electrical accident severity is related to worker's occupation. It seems obvious to think that electricians are the ones most exposed to accidents with electrical origin. Indeed so, and they show a total accident ratio over 30% (TAR 30.58%), but we find that there are other professional groups showing high rates in these accidents as construction labourers (TAR 14.70%) or masons (TAR 12.61%). According to results, more than half of electrical accidents occur in these three groups and each of them individually have similar ratios TAR, SLAR and FSAR. With respect to other professional groups we find significant results in managers and team leaders group who even having a low ratio in total accidents (TAR 1.66%), duplicates this value in serious and fatal accidents (FSAR 3.11%).

This situation where serious and fatal accidents are particularly representative compared to slight accidents is also observed in two groups of workers involved in construction structure (TAR 3.85%, FSAR 6.22% & TAR 3.78%, FSAR 5.18%). However, unlike as described in other studies (Cawley & Brenner, 2012), completion workers have a low serious and fatal accidents ratio (FSAR 2.05%) if we compare it to total accidents ratio (TAR 5.84%).

3.3 Length of service

Length of service analysis also shows a dependency relationship with accident severity (chi-squared = 29.278 d.f. = 17 sig = 0.032). In this case, it is noteworthy that more than half of electrical accidents occur among workers with less than a year's service. Within this group, special attention should be paid to workers with a length of service shorter than one month because they show the highest ratios of accidents in all categories (TAR 16.10%, FSAR 16.58%, FAR 18.18%). It also verifies that electrical accidents suffered by workers with a longer length of service (>10 years) show

more probability to have serious consequences (TAR 3.60%, FSAR 6.74%, FAR 11.36%). This situation could be related to phenomenon of misjudgement of hazards (Huang & Hinze, 2003).

4 CONCLUSION

This research shows that severity of electrical accidents is related to variables associated with individual characteristics of workers such as age, occupation and length of service. Our main conclusions according to results obtained are the following.

Younger workers (<25 years) and those with less length of service (<1 year) need a specific training focused on electrical hazards. Also supervision plans in the work place must be considered for these workers groups.

Workers with more length of service (>10 years) and older ones, specifically those between 46 to 55 years old, need retraining plans focused on electrical hazards in order to limit the misjudgement phenomenon.

After electricians, occupations most affected by this type of accident are construction labourers and masons although managers and team leaders have the highest proportion of accidents with serious or fatal consequences. This should lead us to establish training plans adapted to needs of these occupations regarding with electrical hazards.

REFERENCES

Camino, M.A., Ritzel, D.O., Fontaneda, I., González, O.J. (2008). Construction industry accidents in Spain. Journal of Safety Research, 39 (5): 497–507.

Cawley, J.C., Homce G.T., (2003). Occupational electrical injuries in the United States, 1992–1998, and recommendations for safety research. Journal of Safety Research, 34: 241–248.

Cawley, J.C., Brenner, B.C. (2012). Occupational electrical injuries in the US, 2003–2009. IEEE Paper No. ESW-2012-24.

Chau, N., Gauchard, G.C., Siegfried, C., Benamghar, L., Dangelzer, J.L., Francais, M., Jacquin, R., Sourdot, A., Perrin, P.P., Mur, J.M. (2004). Relationships of job, age, and life conditions with the causes and severity of occupational injuries in construction workers, Int. Archives of Occupational and Environmental Health, 77 (1): 60–66.

Chen, G.X., Fosbroke D.E. (1998): Work-Related Fatal-Injury Risk of Construction Workers by Occupation and Cause of Death. Human and Ecological Risk Assessment: An International Journal, 4(6): 1371–1390.

Cheng, C.W., Leu S.S., Lin C.C., Fan. C. (2010). Characteristic analysis of occupational accidents at small construction enterprises. Safety Science, 48: 698–707.

Chi, C.F., Yang, C.C., Chen, Z.L. (2009). In-depth accident analysis of electrical fatalities in the construction industry. *International Journal of Industrial Ergonomics*, 39: 635–644.

Chi, C.F., Lin, Y.Y., Ikhwan, M. (2012). Flow diagram analysis of electrical fatalities in construction industry. *Safety Science*, 50 (5): 1205–1214.

Dufort, V.M., Kotch, J.B., Marshall, S.W., Waller, A.E., Langley, J.D. (1997). Occupational injuries among adolescents in Dunedin, New Zealand, 1990–1993. *Ann Emerg. Med.*, 30:266–273.

Haslam, R.A., Hide, S.A., Gibb, A.G.F., Gyi, D.E., Pavitt, T., Atkinson, S., Duff, A.R. (2005). Contributing factors in construction accidents, *Applied Ergonomics*, Invited paper, special edition on ergonomics in building and construction, 36 (4): 401–416.

Huang, X.Y., Hinze, J. (2003). Analysis of construction worker fall accidents. *Journal of Construction Engineering and Management-ASCE*, 129 (3): 262–271.

Janicak, C.A. (2008). Occupational fatalities due to electrocutions in the construction industry. *Journal of Safety Research*, 39:617–621.

Koh, D., Jeyaratnam, J. (1998). Occupational health in Singapore. *Int Arch Occup Environ Health*, 71:295–301.

López Arquillos, A., Rubio Romero,J.C., Gibb,A., Analysis of construction accidents in Spain, 2003–2008, Journal of Safety Research (2012), http://dx.doi.org/10.1016/j.jsr.2012.07.005

McCann, M., Hunting, L.K., Murawski, J., Chowdhury, R., Welch, L., (2003). Causes of Electrical Deaths and Injuries Among Construction Workers. *American Journal of Industrial Medicine*, 43:398–406.

Ore, T., Stout, N.A. (1996). Traumatic occupational fatalities in the U.S. and Australian construction industries. American *Journal of Industrial Medicine*, 30 (2): 202–206.

Salminen, S. (1995). Serious occupational accidents in the construction industry. *Construction Management and Economics.,13* (4): 299–306.

Salminen, S. (2004). Have young workers more injuries than older ones? An international literature review. *Journal of Safety Research.* 35, 513–521.

Waehrer, G.M., Dong, X.S., Miller, T., Haile, E., Men, Y. (2007). Costs of occupational injuries in construction in the United States. *Accident Analysis and Prevention*, 39: 1258–1266.

Williamson, A., Feyer, A.M. (1998). The causes of electrical fatalities at work. *Journal of Safety Research*, 29: 187–196.

Xiuwen, D., Platner, J.W. (2004). Occupational fatalities of Hispanic construction workers from 1992 to 2000. *American Journal of Industrial Medicine*, 45:45–54.

Occupational Safety and Hygiene – Arezes et al. (eds)
© 2013 Taylor & Francis Group, London, ISBN 978-1-138-00047-6

Safety requirements for machinery in practice

B. Mrugalska
Faculty of Engineering Management, Poznan University of Technology, Poznan, Poland

P.M. Arezes
School of Engineering, University of Minho, Guimarães, Portugal

ABSTRACT: Ensuring machinery safety is a key issue facing modern manufacturing companies. The machines, tools and equipment, in spite of their construction and technological advance, still contribute to many hazards for their operators. With this purpose, new legal acts binding in the Member Countries of the European Union are being introduced. In this paper a concept of machinery safety with the application of Directive 2009/104/EC is presented. The safety design practices are determined on the example of metal machines. These machines are studied for the aspects of ensuring safety by technology and machinery operation, work environment, process and space, and also information, signal and control elements.

1 INTRODUCTION

In recent years the rapid pace of changes in the practice of product design has been remarkable and astonishing. Innovative production technologies have been introduced to create products in a better and faster way (Eppinger, 2011). Machines are now designed to be used in more and more severe operation conditions (Singh & Kazzaz, 2003). In order to achieve this goal various product robust design methods are applied (Phadke, 1989; Mrugalska & Kawecka-Endler, 2012; Mrugalska, 2013). Moreover, the numerous approaches allowing fault detection of the product or its components have been developed (Ding, 2008; Isermann, 2005; Mrugalski & Witczak 2012; Mrugalski et al., 2008). The application of these methods is especially important in the case of complex products because early fault detection of its components may help to limit the range of the fault and resulting economical losses. However, regardless of increasing know-how it is still necessary to remember to ensure the safety of machinery (Gambatese, 2000; Ridley & Pearce, 2006). The applied safety strategies to this aim offer manufacturers a way of improving their productivity and competitiveness in the market. Safety is perceived as an integrated part of machine functionality rather than after-thoughts added to meet regulations (ABB, 2010). It results from the fact that the effectiveness of producers' activities in the scope of safety decreases in the sequential stages of technical production preparation (Smallwood, 1996; Szymberski, 1997; Christensen, 2003). Thus, the activities taken up in the conceptual design or early design stages have the greatest impact on final product safety. Therefore, engineers engaged in safety issues should be motivated to cooperate with designers' teams in all the design stages of machinery design (Korman, 2001; Beohm, 2003; Weinstein et al., 2005). It is advisable that all the aspects of safety machinery usage should be taken into account on drawing board (Szymberski, 1997; Beohm, 2003; Behm, 2005; Weinstein et al., 2005).

In this paper a particular attention is paid to Directive 2009/104/EC, which is mandatory for work equipment such as machine, apparatus, tool or installation used at work produced before 1st January 2003. For the aim of the practical investigation, 23 metal machinery, which should comply this directive, were chosen and its safety aspects were widely investigated.

2 LEGISLATIVE FRAMEWORK FOR MACHINERY SAFETY

In order to facilitate the trading across the European Union boundaries, the European Commission proposed a policy of harmonization. At first, directives, which aimed at identifying a unified approach to production and trade, were launched into law. These Directives appeared to be prescriptive and to counter this problem the New Approach Directives (CE Directives) were introduced (TUV, 2009). Among these directives the major impact on safety imposes the Directive of the European Parliament and of the Council of 17 May 2006 on machinery, and amending Directive 95/16/EC (recast) called Machinery Directive (2006/42/EC). According to this Directive, machine manufacturers (or their authorized representatives within the EU), must ensure that the machine is consistent with its requirements in the scope of health and safety in the given order by:

– elimination or reduction of risks as much as possible by taking into account safety aspects in machine design and construction phases;
– application of required protection systems or measures against hazards that cannot be eliminated;
– informing users about the residual risks that they might be subjected in spite of all feasible protection measures being taken, specify any requirements for training or personal protective equipment (Safety and functional safety, 2010).

Furthermore, such aspects as principles of safety integration, materials and products, lighting, design of machinery to facilitate its handling, ergonomics, operating positions, seating, control systems, protection against mechanical and other hazards, required characteristics of guards and protective devices, maintenance and information are also discussed. In order to confirm that the machine is in accordance with this Directive, a technical file is prepared, CE marking is affixed and a Declaration of Conformity is signed before the machine is placed on the EU market. During exploitation of machines, users have to ensure that they use, inspect and maintain machines in accordance with the manufacturer's instructions. Furthermore, any modification of the machines can be treated as manufacture of a new machine if the risk assessment requires revision. The need to revise a risk assessment arises if the modification concerns a change in the machine's function or its limits (e.g. position, speed, size). Therefore, the company, which modifies a machine, needs to realize that in such situations a need of issuing a Declaration of Conformity and CE marking appears (Safe Machinery Handbook, 2010). The division of responsibilities between two parties engaged in assuring safety is presented on Figure 1.

The requirements of the Machinery Directive in the aspects of safety and occupational hygiene are rather general (Górny, 2006). Thus, in accordance with the so-called "New Approach Directives of the European Union", detailed technical solutions are included in European harmonized standards. The structure of the European standards for the Safety of machinery is shown in Table 1.

As it can be seen, three types of safety machinery standards are differentiated. The categorization into one of these groups depends on the subject of the standard. These EN standards are adopted by the countries members of the European standards bodies as a national one. In a case of conflicting with any national standard, it must be withdrawn to keep a common standard applied

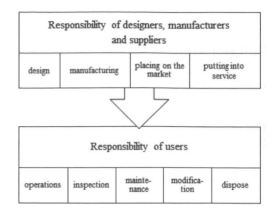

Figure 1. Responsibility of assurance of safety aspects for machinery.

Table 1. Examples of European standards for safety of machinery.

Type of standard	Scope	Examples
Type A	Basic safety standards	EN/ISO 12100-1, EN/ISO 12100-2 EN/ISO 14121-1 EN 614-1, EN 614-12
Type B	Generic safety standards	EN/ISO 13857
Type B1	Safety aspects	EN 349 EN 999 EN 574
Type B2	Safeguards	EN 1088 EN 1760-1, EN 1760-2, EN 1760-3
Type C	Machine safety standards	EN 201 EN 289 EN 692

across Europe. The full conformity to the Directive is only valid if the standard is listed without qualification in the *Official Journal* of the EU. It is also important for manufacturers to check Annex Z of any European standard to be sure which parts of the Essential Health and Safety Requirements (EHSRs) of the Directive are covered. It is important to emphasize that the use of these standards is voluntary, and the manufacturers can design and manufacture their machinery in accordance with these standards only if they wish. However, the EHSRs must still be met, and moreover, information how this was achieved must be explained in detail and attached to the technical file.

The comparison of the revised Machinery Directive 2006/42/EC with the old Machinery Directive 98/37/EC shows that it does not represent any revolutionary change. It combines the achievements of the Machinery Directive as far as free circulation and safety of machinery are concerned. Nevertheless, the scope of the new one is extended (e.g. inclusion of construction-site hoists and cartridge-operated fixing). The most important differences relate to:

– risk assessment;
– risks associated with machinery serving fixed landings extended by construction site hoists and slow-moving lifts;
– noise and vibration emissions;
– application of certain guidelines used to mobile machinery or machinery for lifting to all machinery presenting the risk concerned (Mechanical Engineering…).

The Directive 2006/42/EC is related to other European Directives. This relationship between Directives and the subjects that they may encompass are shown in Table 2.

As it can be noticed, three groups of directives can be distinguished in relation to the Machinery Directive. The machinery, which has to comply with the requirements of this directive, can also be compliant with these three groups of directives. Group 1 concerns general aspects, such as an explosive atmosphere (ATEX), electromagnetic emissions (EMC) etc., group 2 is generally about safety at work and the last one, group 3, tackles special types of machines. All these mentioned regulations implement European Directives and discuss diverse requirements. According to them the machines are identified by the CE mark when the manufacturers declare that they meet all of the requirements relevant to the Directives for that particular product, and it must be shown on the Declaration(s) of Conformity.

In the case of machines produced before 1 January 2003 the mandatory regulation is the Directive 2009/104/EC of the European Parliament and of the Council, from 16 September 2009, concerning the minimum safety and health requirements

Table 2. Relationship between EU directives related to machinery.

Standards in relationship to Machinery Directive 2006/42/EC	
Group 1	Emissions 2010/26/EC
	Low Voltage 2006/95/EC
	EMC 2004/108/EC
	Noise Outdoors 2000/14/EC
	ATEX 94/9/CE
Group 2	Work Safety 89/391/EEC
	Measures 89/654/EEC
	Personal Protective Equipment 89/686/EEC
	Chemical Agents at Work 98/24/EC
Group 3	Waterway vessels 2006/87/EC
	Railway 2004/49/EC

for the use of work equipment by workers (second individual Directive within the meaning of Article 16(1) of Directive 89/391/EEC). This directive replaces Directive 89/655/EEC, from 30 November 1989, concerning the minimum safety and health requirements for the use of work equipment by workers. According to this directive, all the obligations are placed on employers, whereas producers of machinery and their representatives are not even mentioned. The employers are obliged to "take the measures necessary to ensure that the work equipment made available to workers in the undertaking or establishment is suitable for the work to be carried out or properly adapted for that purpose and may be used by workers without impairment to their safety or health" (Directive 2009/104/EC). It is also their responsibility to pay attention to working conditions, characteristics and hazards resulting from the selection of particular equipment by the workers. Thus, for these machines, all practices, mentioned in the directive aiming at ensuring safety, are designed and introduced at the exploitation stage in the industrial setting. This directive concerns work equipment that encompass machines, apparatuses, tools or installations used at work. It provides minimum requirements to them in the range of control, system, safety and protection devices, guards, warnings, markings etc. Additionally, it concerns also to the minimum prerequisites for mobile work equipment, fork-lift trucks, self-propelled work equipment, lifting loads and temporary work at highs. As it can be noticed, in comparison to Machinery Directive, its requirements are more limited in number, what could be already expected based on its title.

3 MATERIALS AND METHOD

The purpose of this study is to determine the most popular industrial design practices to ensure the minimum requirements according to Directive 2009/104/EC. With this aim, three primary activities are undertaken: a review of the directive and regulations enforcing it into the Polish law, an analysis of data from controls done by the Central Labour Inspectorate in industrial companies and a pilot-survey about machines available via Internet. The study was carried out considering the example of metal machines, which are the subject of this directive. The survey of the machines was conducted with five groups of tasks (Mrugalska, Kawecka-Endler, 2011):

– safety of information, signal and control elements,
– safety of technology and machinery exploitation,
– safety of work environment,
– safety of work process,
– safety of work space.

A deeper investigation is also provided with the application of a checklist prepared on the basis of the Regulation of Ministry of Economy of 30 October 2002 with changes in 30 September 2003, concerning minimum requirements for machinery (Dziennik Ustaw No. 191, item 1596 of 18 November 2002 and No. 178, item 1745 of 16 October 2003).

4 RESULTS AND DISCUSSION

As the requirements of the Directive 2009/104/EC came into force in the European countries, many design adjustments have been already introduced. However, there are some machines that still require some adjustments. This phenomenon is particularly visible when statistic data are analyzed. According to data retrieved from the controls, carried out by the Central Labour Inspectorate, it can be noticed that the greatest discrepancy between the real state of machinery and the requirements in force concerns micro and small enterprises (Zając, 2008; 2011). In 2010, the Central Labour Inspectorate checked 2791 metal machinery and equipment due to minimum requirements. It revealed that 787 machines required design changes and 1121 machines had to be adjusted due to some small objections, which corresponds to 68.4% of the controlled machines (Zając, 2011). The details of the investigation are shown in Figure 2.

As can be seen from Figure 2, only 62% of machine guards and protection devices were in accordance to the minimum requirements. The inspectors identified some violations related to the control devices (28%) and energy cut-off devices (16%). Other similar number of features that had to be improved were the emergency stop, lighting, safety signs and colours as only 85% of them were in the required conditions.

The results of the controls carried out by the Central Labour Inspectorate encourage the researchers to investigate what types of adjustments have already been done to fulfil the minimum requirements in the analysed second-hand machinery. With this aim, 23 metal machines were studied with the application of the questionnaire based on the requirements stated in Directive 2009/109/EC. Figure 3 presents the data gathered on this survey.

As can be noticed, the most frequent undertaken activities concerned information, signal and control elements. The second group was the technology and machinery exploitation. Some adjustments were also done in work space, environment and process.

Since the aspect of assuring safety of information, signal and control elements required most of the changes, it was deeply investigated. It was revealed that in this case it was marking what constituted most of the identified problems. It was shown that ca. 20% of signs was not in accordance with the directive and it did not ensure safety for users of machinery. This further investigation is depicted in Figure 4.

It indicates that in most cases marking was unrecognizable because of its filthy (particularly inside machinery). It also happens that it was partially damaged (23%) or missing (9%).

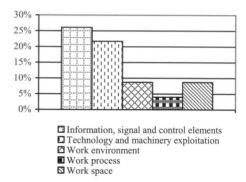

☐ Information, signal and control elements
☐ Technology and machinery exploitation
☒ Work environment
◪ Work process
◪ Work space

Figure 3. Safety requirements for metal machines.

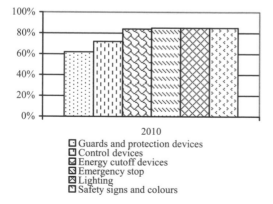

2010

☐ Guards and protection devices
☐ Control devices
☑ Energy cutoff devices
☒ Emergency stop
☒ Lighting
☐ Safety signs and colours

Figure 2. Assurance of chosen safety requirements for metal machines (Adopted from (Zając, 2011)).

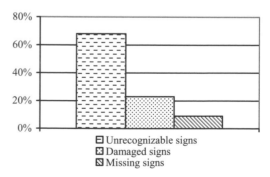

☐ Unrecognizable signs
☒ Damaged signs
☒ Missing signs

Figure 4. Identified marking problems in metal machines.

5 CONCLUSIONS

In recent years, safety is being perceived as a matter of particular importance in machinery design and the number of new regulations regarding machine safety has been enlarged considerably. In spite of this fact, the results of the research based on minimum requirements included in Directive 2009/104/EC revealed discrepancy between the legal requirements and the real condition of the analysed machines. These results also allowed to identify what adjustments in metal machinery are needed to be implemented in order to meet the requested safety requirements. This situation follows from the fact that the rules included in this directive do not concern the machinery manufacturer but, instead, they are focused on the employer who has to ensure safe working conditions. Thus, and according to the obtained results, it seems that the machinery adjustments have to be considered at an early stage and, according to the European Directive, they should be designed and carried out at the exploitation stage of the machinery.

REFERENCES

ABB, 2010. *Safety and Functional Safety. A General Guide.* ABB brochure 1SFC001008B0201. Retrieved November 12, 2012, from www05.abb.com/global/.../-1sfc001008b0201.pdf.

Behm, M. 2005. Linking Construction Fatalities to the Design for Construction Safety Concept. *Safety Science* 43(8): 589–611.

Beohm, R.T. 1998. Designing Safety into Machines. *Professional Safety* 43(9): 20–24.

Christensen, W.C. 2003. Safety through Design. Helping Design Engineers Answer 10 Key Questions. *Professional Safety* March: 32–39.

Ding S. 2008. Model-based Fault Diagnosis Techniques: Design Schemes, Algorithms, and Tools. Berlin/Heidelberg: Springer-Verlag.

Directive 2006/42/EC of the European Parliament and of the Council of 17 May 2006 on Machinery, and Amending Directive 95/16/EC (recast). Official Journal of the European Union. L157/24. Retrieved November 10, 2012 from http://eur-lex.europa.eu /LexUriServ/Lex UriServ.do?uri = OJ:L:2006:157:0024: 0086:EN:PDF.

Directive 2009/104/EC of the European Parliament and of the Council of 16 September 2009 Concerning the Minimum Safety and Health Requirements for the Use of Work Equipment by Workers at Work (second individual Directive within the meaning of Article 16(1) of Directive 89/391/EEC). Official Journal of the European Union. L 260/5. Retrieved September 10, 2012, from http://eurlex.europa.eu- /LexUriServ/ LexUriServ. do?uri = OJ:L:2009:260:0005:- 0019:EN:PDF.

Eppinger, 2011. The Fundamental Challenge of Product Design. *J. Prod. Innov. Manag.* 28: 399–400.

Gambatese, J.A. 2000. Safety in a Designer's Hands. *Civil Engineering* 70(6): 56–59.

Górny, A. 2006. Essential Requirements in Ergonomics Standards Harmonized with the European Parliament and the Council Directive 2006/42/EC, In L.M. Pacholski, J.S. Marcinkowski & W.M. Horst (eds.). *The Role of Education and Researches in Ergonomics and Work-safety in Health Population* (pp. 37–47). Poznan: Publishing House of Poznan University of Technology.

Isermann R. 2005. Fault-Diagnosis Systems: An Introduction from Fault Detection to Fault Tolerance. Berlin/Heidelberg: Springer-Verlag.

Mrugalski M. & Witczak M. 2012. State-Space GMDH Neural Networks for Actuator Robust Fault Diagnosis. *Advances in Electrical and Computer Engineering* 12(3): 65–72.

Mrugalski M., Witczak M. & Korbicz J. 2008. Confidence Estimation of the Multi-layer Perceptron and its Application in Fault Detection Systems. *Engineering Applications of Artificial Intelligence* 21(8): 895–906.

Mrugalska, B. 2013. Design and Quality Control of Products Robust to Model Uncertainty and Disturbances, In: Winth, K., (ed.) *Robust Manufacturing Control, Lecture Notes in Production Engineering* (pp. 495–505). Berlin/Heidelberg: Springer-Verlag.

Mrugalska, B. & Kawecka-Endler, A. 2011. Machinery Design for Construction Safety in Practice, In C. Stephanidis (ed.), *Universal Access in HCI*, Part III, HCII 2011, LNCS 6767 (pp. 388–397). Berlin/Heidelberg: Springer-Verlag.

Mrugalska, B. & Kawecka-Endler, A. 2012. Practical Application of Product Design Method Robust to Disturbances. *Human Factors and Ergonomics in Manufacturing & Service Industries* 22: 121–129.

Phadke S.M. 1989. *Quality Engineering Using Robust Design*, New York: Prentice Hall/Englewood Cliffs.

Ridley, J. & Pearce, D. 2006. *Safety with Machinery* (2nd ed.). Oxford: Elsevier, Butterworth-Heinemann.

Singh, G.K. & Kazzaz, A.S. 2003. Induction Machine Drive Condition Monitoring and Diagnostic Research—a Survey. *Electric Power Systems Research* 64(2): 145–158.

Smallwood, J.J. 1996. The Influence of Designers on Occupational Safety and Health. Proceedings of the First International Conference of CIB Working Commission W99, Implementation of Safety and Health on Construction Sites. September 4–7, Lisbon, Portugal.

Szymberski, R. 1997. Construction Project Safety Planning. *Tappi Journal* 80(11): 69–74.

TUV, 2009. *A practical guide to Machinery Safety*. Edition 4. TÜV SÜD Product Service. Retrieved November 17, 2012, from http://www.hsmsearch.co.uk/stories/articles/ newsletter_ stoies/2009/november_2009/news/ free_ practical_guide_to_machinery_safety/.

Weinstein, M., Gambatese, J. & Hecker, S. 2005. Can Design Improve Construction Safety?: Assessing the Impact of a Collaborative Safety-in-Design Process. *Journal of Construction Engineering and Management* 131(10): 1125–1134.

Zając, T.J. 2008. *Sprawozdanie Głównego Inspektora Pracy z działalności Państwowej Inspekcji Pracy w 2008 r.* Retrieved September 10, 2012, from http://www.pip. gov.pl/html/pl/sprawozd/08/spraw_08.htm.

Zając, T.J. 2011. Sprawozdanie Głównego Inspektora Pracy z działalności Państwowej Inspekcji Pracy w 2010 r. Warszawa.

Occupational Safety and Hygiene – Arezes et al. (eds)
© *2013 Taylor & Francis Group, London, ISBN 978-1-138-00047-6*

Factors that influence the construction safety performance: Overview

F. Rodrigues, A. Coutinho & C. Cardoso

Civil Engineering Department, Geobiotec, University of Aveiro, Portugal

ABSTRACT: The construction sector presents a high diversity of occupational hazards and consequent risks that determine the need for specific intervention different from other sectors. This assumes particular relevance today, because it is still a sector with high number of fatal and serious accidents rates. Therefore, efforts should be made in order to identify the factors that are behind these accidents. So, the aim of this article, is to present a model that correlates the factors contributing to the occurrence of accidents in the construction sector, from the design to the construction phase, taking into account the information provided by several authors in this line of research.

1 INTRODUCTION

The health and safety conditions in the construction workplaces is a social and economic concern due to the high incidence of serious and fatal occupational accidents in this sector (Choi et al., 2012). The construction sector is, in most countries, the one who has the highest rate of fatal accidents at work. To reverse this situation, it should be implemented strong safety policy on construction sites. To achieve this goal the growing interest of companies regarding the minimization of workplace accidents is essential (Kulchartchai & Hadikusumo, 2010). Despite the decreasing of accidents over the last years, at Portuguese and international level, it appears the construction sector stands out systematically from other industries, due to its high number of fatalities, whose main cause is the falling from height (Abdelhamid & Everett, 2000; Suraji et al., 2001; Elbeltagi & Hegazy, 2002; Haslam et al., 2005; Frijters & Swuste, 2008; ACT, 2010; Azevedo, 2010; Kulchartchai & Hadikusumo, 2010; Choi et al., 2012; Koh & Rowlinson, 2012).

Several research papers published in this area identify the essential elements to constitute an efficient health and safety program. Those elements are described by various analytical models showing the relationship of the various factors that contribute to the occurrence of accidents in construction (Hallowell & Gambatese, 2010). The accidents analysis presented in Swuste et al. (2012) refers a considerable percentage of accidents causes related to design. This author also highlights that it is the design phase that offers the greatest potential to positively influence safety. So, as the statistical analysis of the accidents in this sector continues to show high rates comparing to other sectors (Ghosh & Young-Corbett, 2009), it is important a more detailed investigation that seeks to identify and monitor the factors influencing safety performance in the construction. In order to follow up the investigation of several authors focused on the accidents causes in the construction industry this paper proposes a model developed from the analysis of existing models, which correlates all the causal factors, since design to the execution phase. This model is intended to be applied to small and medium enterprises (SMEs), since the construction industry in Portugal, similar to the other countries of the European Union, is majority compounded by SMEs. However, the model developed is applicable to companies of any size. The extreme importance of the accidents in the construction sector justify the identification and understanding of its causes and are essential for the elimination and control of the risk factors that contribute to the occurrence of these accidents (Carvalho, 2005; Roxo, 2004). It thus requires a more detailed analysis at this level to be the basis of a proposal of recommendations for the prevention of such accidents (Hinze et al., 1998; Jeong, 1998; Abdelhamid & Everett, 2000; Haslam et al., 2005).

2 MODEL

2.1 *Proposed model*

The studies that have been analysed refer to the existence of different models and theories, standing as one of the most classic models the domino theory of Heinrich. This theory was developed and updated, highlighting the domino theory of Bird. The emergence of new perspectives responding to the organizations needs, led to more recent theories as the mul-tiple causation models, systems theories and socio-technical approaches (Manu et al., 2012; Roxo, 2004).

Haslam et al. (2005) recommend that should be developed a model that correlates the casual factors that contribute to an occupational hazard in a small or medium construction site. To answer this and in order to follow up the investigation of Haslam et al. (2005) is proposed a model based on bibliographic review directed to small and medium enterprises. However this model can be applicable to any company and correlate the factors at design and at construction level, some of them do not considered by the other authors.

The model is based on the following assumptions:

- The causes of accidents are not restricted to the causes that directly cause the accident, including the immediate causes, but also engage the root causes, which comprise a combination of human, organizational and environmental factors;
- The occurrence of failures giving rise to an accident occurs through a sequential process since the immediate causes until the most basic (Whittington et al., 1992);
- The failures of previous levels increase the likelihood of failures in the next levels, and so on up to the accident (Whittington et al., 1992);
- Associated with each factor exists a set of influences that determine the degree of affectation becoming negative to safety, for example the influence of workers' action due to his health conditions (Haslam et al., 2005);
- Factors in the design phase are considered initializing influencers as the model detailed by Haslam et al. (2005). Although they are not directly related to the accident they led to the appearance of failures that contribute to the occurrence of accidents in the construction sites (Whittington et al., 1992; Suraji et al. 2001; Haslam et al. 2005);
- There are numerous causal factors that contribute to accidents; they establish standard causal relationship of interdependence; it is required that only happens a minimum set of factors to occur the accident (Haslam et al., 2005; Mitropoulos et al., 2005).

2.1.1 *Root causes*

As in the model that was proposed by Whittington et al. (1992) aiming to establish a standard of failures that led to the occupational hazard in construction were considered four principal levels of failures:

- 1st and 2nd level: concerning with failures that occur during the design phase, related with the design options, the selection of the construction processes in the sequence of the client requirements, the financial climate or the deficient health and safety trainee of the design team (Haslam et al., 2005);
- 3rd level: associated with inherent failures of the management system of the small or medium enterprise, responsible for the construction phase,

according to its intrinsic characteristics including financial capacity, health and safety policy and work planning;
- 4th level: related with failures that arise during the execution phase concerning with equipments/materials, workplace, health and safety specifications, unpredictability of tasks, workers and non-human events.

The model level 3 and 4 correspond to the execution phase. It is at this stage that the consequences of failure appear. These four failures levels are considered as the accident root causes (Table 1) that origin the unsafe conditions and/or the unsafe acts. These ones provide the existence of dangerous conditions at work that lead to the occurrence of occupational hazard.

2.1.2 *Failures of level I*
2.1.2.1 Client requirements
The client requirements are one of the initial factors of the model with potential to trigger a sequence of failures, since they influence the project management which can later affect the health and safety performance during the execution phase (Manu et al., 2010; Manu et al., 2012).

The designer's team priority is concerned on satisfying the client needs and requirements included in the preliminary program, in compliance with the legal requirements and taking to a second level the health and safety issues in spite of being also under legal regulations (Cabrito, 2005).

As the designers' choices have a crucial influence on the health and safety during execution, maintenance and demolition work it will be necessary to integrate safety in the overall design objective. So the legal obligations will be complied according to the European Directive 92/57/EEC that has been incorporated into national legislation in all EU countries (Frijters & Swuste, 2008).

Table 1. Root causes.

Failures:	
Level I	– Client requirements; – Financial climate; – Competence of design team.
Level II	– Project management; – Risk management.
Level III	– Financial capacity; – Health and safety policy; – Construction/safety planning.
Level IV	– Materials/Equipments; – Workplace; – Health and safety specifications; – Unpredictability of tasks; – Worker; – Non-human events.

2.1.2.2 Financial climate

From the model perspective the economic climate influences the designer team work leading to a strong intercompany competition for project award to lower prices regarding the scarcity of work and the high number of design companies. This contributes to the development of projects with lower quality. Currently, the lack of investment in the construction sector or the small number of projects that advance occurred in difficult conditions of the market where the competition is a reality that leads to carrying out projects at ridiculous prices, not corresponding to the degree of demand and responsibility for carrying out this work (Ascenso, 2009).

2.1.2.3 Designer team competence

The ability with which the project is carried out is determined by the professional experience and training in both academic and social level of its author. In this way, it becomes important that the teams are made up with elements with experience in the type of project in question and with training in the area of health and safety (Cabrito, 2005). To implement the risks prevention principles it must have a linked cooperation between the health and safety coordinator and the designer team.

2.1.3 *Failures of level II*
2.1.3.1 Project management

The Project management failures can occur in the aim of: project planning and subsequent construction planning, inadequate project schedule, different projects incompatibility, no compliance with the risks prevention principles. These led to the adoption of: hazardous raw materials and dangerous constructive processes, complex architectural solutions with lower safety level, absence of execution details and lack of maintenance, and demolition procedures (Haslam et al. 2005).

2.1.3.2 Risk management

The risk assessment carried out during the design phase as well as the production of documents aiming the risk prevention during the subsequent phases may constitute failures in terms of risk management, when considered as a mere paper exercise or when performed incorrectly corresponding to any specific work (Haslam et al., 2005; Fung et al., 2012). The existence of shortcomings in this area influences later in the workplace the safety and construction planning by the enterprise.

2.1.4 *Failures of level III*
2.1.4.1 Financial capacity

The financial difficulties in SMEs have a notorious impact since have repercussions at the safety investment, also reflecting in a lack of human resources and on the investment to train them (Pinto et al. 2011; Choi et al., 2012). The number and competence of the available enterprise human resources to perform the health and safety functions and activities is a crucial factor to achieve good work conditions.

2.1.4.2 Safety policy

The company health and safety policy determines the importance that is given to these issues, to the safety standards in the Organization reflecting on workers motivation (Mearns et al., 2003; Choi et al., 2012). The shortcomings of the safety policy provides the appearance of faults at the workers level since they consequently ignore the safety practices that avoid the accident, contributing to the violation of the safety rules and to the adoption of unsafe acts and behaviours (Jitwasinkul & Hadikusumo, 2011).

2.1.5 *Failures of level IV*
2.1.5.1 Materials/Equipments

These failures are related with their inadequate conservation, the incorrect storage and the inherent characteristics of the material (weight, shape, size, etc.), no conformity, defective and inadequate tools, and defective collective protective equipments, inexistent of individual protective equipment. According to Abdelhamid and Everett (2000) lead to the occurrence of unsafe conditions at the construction site and consequently to the increase of dangerous conditions.

2.1.5.2 Workplace

The workplace conditions are essential to safety on the construction sites. Several accidents resulting in cuts, trips, falls are causing by lack of organization, slippery floors, materials displaced and accumulation of residues or waste (Fung et al., 2010; Freitas, 2011; Fung et al. 2012).

2.1.5.3 Tasks unpredictability

The unpredictability of tasks compromises the ongoing of the construction process, since it promotes the occurrence of interruptions, by the time that is required to get the new panorama solutions. This fact contributes to the increase of the workload leading to the no compliance with the production objectives and the deadlines initially defined (Mitropoulos et al., 2005).

The reliable planning of tasks and/or operations due to possible changes during the execution phase constitutes an important aspect to reduce hazardous conditions present in the work resulting from the emergence of unplanned work. Thus, it is essential to have a suitable supervision and/or control of the factors that can cause a deviation to the previously defined planning, in order to decrease the likelihood of unforeseen tasks occur (Suraji et al. 2001).

2.1.5.4 Worker

Another important factor considered in the proposed model is the failures associated with individual workers characteristics: physical or psychological. These characteristics contribute to the execution of the work once that compromise the behavior and performance in the tasks execution (Leung, 2012; Yi et al. 2012).

2.1.6 *Immediate causes*

In the proposed model are considered immediate causes of accidents the unsafe acts and unsafe conditions on the construction site which are the result of a sequential process of failures, both at the design and the execution phase level.

Unsafe conditions present in work are associated with the physical space of the construction site being a direct consequence of the equipment, materials, and workplace, unpredictability, and safety specifications failures. The absence of safety procedures involving all work operations contributes to the appearance of these failures.

The lack or inappropriateness of machines and equipment protections, the non-conformity of equipment and tools, the unfavorable working environment conditions such as excessive noise and inadequate lighting, workplace disorganization and lack of cleaning, among others, are examples of unsafe conditions on construction sites.

According to Abdelhamid and Everett (2000) the workers response to unsafe conditions is essential since in its identification they have the ability to stop work and to carry out the correction of the detected situations, or proceed consciously without performing any correction, thereby contributing to the increased likelihood of the accident. However, the employee may not identify the unsafe condition and have no ability to stop the work continuing in a natural way (Abdelhamid & Everett, 2000).

Examples of unsafe acts committed by employees are: the disregard for safety rules, carrying out hazardous work without authorization, the non-use of individual and collective protective equipment, transport vehicles incorrectly, the incorrect handling of loads, staying in forbidden places, the consumption of alcohol or drugs, the use of non-compliant machines and equipment, using equipment improperly or for inappropriate purposes, removal of pro-tective devices.

According to Abdelhamid and Everett (2000) and Liu and Tsai (2011), the existence of conditions and unsafe acts as well as non-human events are factors that originate dangerous conditions at work leading to accidents.

As in the model proposed by Mitropoulos et al. (2005), it is necessary to highlight that the exposure to a risk factor not always inevitably leads to the occurrence of an accident. However, the same inevitably arises from the presence of a dangerous condition, with the potential to cause injury or death, which is triggered by human error, particularly unsafe actions committed by employees and/or by changes in functional conditions, associated with the presence of unsafe non-human conditions and events. It then becomes necessary for accident prevention eliminate unsafe acts and conditions in the workplace. To do this it must be carry out a control of failures that may arise at the design stage and later in the execution phase since all contributing directly or indirectly to the origin of the accident.

3 CONCLUSIONS

The proposed conceptual model lists several factors that may contribute to the occurrence of construction occupational hazards, from the design to the execution phase. As turned out through the literature review they are scarce studies carried out in this framework incorporating a global view of the two phases. The model allows the identification of a simplified sequence of factors, clearly and objectively, and proves the multi causal nature which contributes to the occurrence of accidents in construction activities. Highlight the importance of design phase in safety performance in the later stages, often being neglected by various authors. So, it is necessary that the design team takes into account safety as a preponderant factor in the development of the Project to decrease the percentage of accidents caused by the design unsafe options.

The analysis of some projects and the discussion with the design and site supervision teams, of some SMEs, lead to the conclusion that the health and safety are not really considered during the design phase given no contribution to the accidents prevention. In this construction sites all the safety measures had to be studied and defined during the execution phase without any contribution of the design team. This conceptual model with its set of remarks about accident root and immediate causes can contribute to decrease the construction accident rates and to achieve safest working conditions in any enterprise. To achieve this goal owners and design teams have to understand the real importance of risk prevention during the design phase, the planning phase and the execution phase.

REFERENCES

Abdelhamid, T.S.; Everett, J.G. (2000).Identifying root causes of construction accidents. *Journal of Construction Engineering & Management*, 126 (1), 52.
ACT (2010). Relatório de Actividades. Lisboa, Portugal. Retrieved March 29, 2012, from http://www.act.gov.pt/(pt-PT)/SobreACT/DocumentosOrientadores/RelatorioActividades/Documents/relatorio_actividades_ACT_2010.pdf (in Portuguese).

Ascenso, R. (2009). Projeto: Honorários não acompanham grau de exigência e responsabilidade. Entrevista aos projectistas Odete de Almeida e Paulo Queirós de Faria. Available at: http://www.climatizacao.pt/media/16870/entrevista.pdf (15-03-2012) (in Portuguese).

Azevedo, R.P.L. (2010). Acidentes em operações de movimentação manual de cargas na construção. Tese de Doutoramento, Universidade do Minho, Portugal (in Portuguese).

Cabrito, A.J.R.M. (2005). Construção: A aplicação dos princípios gerais de prevenção na fase de projeto. Instituto para a Segurança higiene e Saúde no Trabalho. Lisboa, Portugal (in Portuguese).

Carvalho, H.I.L. (2005). Higiene e segurança no trabalho e suas implicações na gestão dos recursos humanos: o sector da construção civil. Dissertação de Mestrado em Sociologia, Universidade do Minho. Portugal (in Portuguese).

Choi, T.N.Y.; Chan, D.W.M.; Chan, A.P.C. (2012). Potential difficulties in applying the Pay for Safety Scheme (PFSS) in construction projects. *Accident Analysis and Prevention*, 48 (0), 145–155.

Elbeltagi, E.; Hegazy, T. (2002). Incorporating Safety into Construction Site Management. Proceedings of the First International Conference on Construction in the 21st Century (CITC2002), Challenges and Opportunities in Management and Technology, 25–26 April, Miami, Florida, USA.

Freitas, L.C. (2011). Segurança e Saúde do Trabalho. Edições Silabo. Lisboa (in Portuguese).

Frijters, A.C.P.; Swuste, P.H.J.J. (2008). Safety assessment in design and preparation phase. *Safety Science,* 46, 272–281.

Fung, I.W.H.; Lo, T.Y.; Tung, K.C.F. (2012). Towards a better reliability of risk assessment: Development of a qualitative & quantitative risk evaluation model (Q2REM) for different trades of construction works in Hong Kong. *Accident Analysis and Prevention*. 48 (0), 167–184.

Fung, I.W.H.; Tam, V.W.Y.; Lo, T.Y.; Lu, L.L.H. (2010). Developing a Risk Assessment Model for construction safety. *International Journal of Project Management*, 28 (6), 593–600.

Ghosh, S.; Young-Corbett, D. (2009). Intersection between Lean Construction and Safety Research: A Review of the Literature. Proceedings of the 2009 Industrial Engineering Research Conference, 30 May – 3 June, Miami, USA.

Hallowell, M.R.; Gambatese, J.A. (2010). Population and Initial Validation of a Formal Model for Construction Safety Risk Management. *Journal of Construction Engineering & Management*, 136 (9), 981–990.

Haslam, R.A.; Hide, S.A.; Gibb, A.G.F.; Gyi, D.E.; Pavitt, T.; Atkinson, S.; Duff, A.R. (2005). Contributing factors in construction accidents. *Applied Ergonomics*, 36 (4), 401–415.

Hinze, J.; Pedersen, C.; Fredley, J. (1998). Identifying root causes of construction injuries. *Journal of Construction Engineering & Management*, 124 (1), 67.

Jeong, B.Y. (1998). Occupational deaths and injuries in the construction industry. *Applied Ergonomics*, 29 (5), 355.

Jitwasinkul, B.; Hadikusumo, B.H.W. (2011). Identification of Important Organisational Factors Influencing Safety Work Behaviours in Construction Projects. *Journal of Civil Engineering & Management*, 17 (4), 520–528.

Koh, T.Y.; Rowlinson, S. (2012). Relational approach in managing construction project safety: A social capital perspective. *Accident Analysis and Prevention*, 48 (0), 134–144.

Kulchartchai, O.; Hadikusumo, B.H.W. (2010). Exploratory Study of Obstacles in Safety Culture Development in the Construction Industry: A Grounded Theory Approach. *Journal of Construction in Developing Countries*, 15 (1), 45–66.

Leung, M.-y.; Chan, I.Y.S.; Yu, J. (2012). Preventing construction worker injury incidents through the management of personal stress and organizational stressors. *Accident Analysis and Prevention*, 48 (0), 156–166.

Liu, H.T.; Tsai, Y.l. (2011). A fuzzy risk assessment approach for occupational hazards in the construction industry. *Safety Science*. 50 (4), 1067–1078.

Manu, P.; Ankrah, N.; Proverbs, D.; Suresh, S. (2010). An approach for determining the extent of contribution of construction project features to accident causation. *Safety Science*, 48 (6), 687–692.

Manu, P.A.; Ankrah, N.A.; Proverbs, D.G.; Suresh, S. (2012). Investigating the multi-causal and complex nature of the accident causal influence of construction project features. *Accident Analysis and Prevention*, 48 (0), 126–133.

Mearns, K.; Whitaker, S.M.; Flin, R. (2003). Safety climate, safety management practice and safety performance in offshore environments. *Safety Science*, 41 (8), 641.

Mitropoulos, P.; Abdelhamid, T.S.; Howell, G.A. (2005). Systems Model of Construction Accident Causation. *Journal of Construction Engineering & Management*, 131(7), 816–825.

Pinto, A.; Nunes, I.L.; Ribeiro, R.A. (2011). Occupational risk assessment in construction industry – Overview and reflection. *Safety Science*, 49 (5), 616–624.

Roxo, M.M. (2004). Segurança e Saúde do Trabalho: Avaliação e Controlo de Riscos. Coimbra: Edições Almedina (in Portuguese).

Suraji, A.; Duff, A.R.; Peckitt, S.J. (2001). Development of causal model of construction accident causation. *Journal of Construction Engineering and Management*, 127(4), 337–344.

Swuste, P.; Frijters, A.; Guldenmund, F. (2012). Is it possible to influence safety in the building sector? A literature review extending from 1980 until the present. *Safety Science*, 50, 1333–1343.

Whittington, C.; Livingston, A.; Lucas, D.A. (1992). Research into management, organisational and human factors in the construction industry. Health and Safety Executive. Norwich UK: HMSO Publications Center.

Yi, J.-s.; Kim, Y.-w.; Kim, K.-a.; Koo, B. (2012). A suggested color scheme for reducing perception-related accidents on construction work sites. *Accident Analysis and Prevention*, 48 (0), 185–192.

Occupational Safety and Hygiene – Arezes et al. (eds)
© 2013 Taylor & Francis Group, London, ISBN 978-1-138-00047-6

Use of the recycled plastic roof main beam in the construction of popular houses: Impact on laborer's health

S.F.D. Maia & M.B.F.V. Melo
Federal University of Paraiba, João Pessoa, Paraíba, Brazil

ABSTRACT: Presently made of wood, the main Roof Beam is a structural piece that is placed all over the perimeter of popular houses and has the function to uniformly distribute the roof load. Aiming to prove the disadvantage in the use of the main wood Roof Beam, by emphasizing the impacts on laborer's health, we accomplished a research in the job site of a Housing Program for low income families. The result of the study showed that the use of wood Roof Beam exposes the laborer to wood dust, which causes prejudice to his health and the probability of lesions due to the intense use of the wood hand cutting tool. An innovating product is then presented: The Recycled Plastic Roof Beam, which eliminates the exposition to dust and the use of the wood hand cutting tool, so reducing labor accidents.

1 INTRODUCTION

Popular houses are buildings destined to low income families, whose aim is to somewhat reduce the housing deficit.

According to the "Caixa Popular Houses Standard Project Cahier" (Caixa, 2006), in the constructive process of these buildings, the following stages are accomplished: Foundation or wall basis, Brickwork, Roofing, Wall Plastering, installation of water, sewage, electricity and finally painting. The main Roof Beam setting occurs during the roofing construction phase.

Being a structural piece placed on the walls all over the perimeter of the building, the main Roof Beam has the function to uniformly distribute the roof load as illustrated in Figure 1.

According to the Amapá Center for Higher Education (2009), the material used in the fabrication of this piece must possess physical and mechanical characteristics such as: Resistance to compression (fc) at 15% of moisture equal to of higher than 55.5 MPa and Module of rupture to traction equal or higher than 13.5 MPa.

Presently, wood is used in the fabrication of the main Roof Beam, although there are some disadvantages such as: cost with maintenance, wastes, vulnerability owing to climatic changes, microorganism attacks, further to labor accidents that cause impacts on the laborer's health.

In Brazil, the Law 8,213/1991, the article 19 (Brasil, 1991) establishes that labor accident is what occurs in the exercise of labor while one serves the company, causing body lesion or functional disturbance that leads into death, the loss or reduction

Figure 1. Setting of the main Roof Beam in a popular house.

of the capacity to work, either permanently or temporarily. Labor accidents are classified in three types: Typical Accident (causes immediate lesion), Occupational Disease (alteration of the laborer's health caused by environmental factors associated with labor) and Accident of Transit (occurs on the way to the job).

According to the National Program of Prevention against Labor Accidents (Brazil), there occurred 4,957 Labor Accidents in the State of Paraiba in 2010, out of which 2,166 were typical accidents. In the computation of the afore-mentioned data (2,166 typical accidents), the accidents that resulted in immediate lesions—which took place in the construction of popular houses—are surely included, having been caused by the so-called hand wood-cutting tool, during the use of the main wood-made Roof Beam.

Owing to this evidence, it is detach the importance of the availability of an innovating product in the market, the Recycled Plastic Roof Beam, which aims at the reduction of the impacts on the laborer's health, related with the Typical Accidents, since its adoption eliminates the existence of the referred hand tool. Furthermore, this product aims at reducing the use of wood in the civil construction in exchange for the recycled plastic beam in a responsible and sustainable way.

Aiming to prove the disadvantage of the use of wood Roof Beam in connection with the impacts on the laborer's health, we accomplished a research in a job site of a Housing Project for low income families. During the research, it was assured—throughout the in-loco observation and also the interview with the laborers—that the wood Roof Beam is fabricated in standard dimension and, in most times, this standard does not correspond with the architectonical project, there requiring the use of amendments and cuts throughout the utilization of the wood hand-cutting tool. This tool—a carpenter's and barrel-maker's tool—has a short handle and a cutting steel-sheet, serving to rough-hew the wood and causing labor accidents with immediate lesions in the laborers' lower or upper limbs. Figure 2 illustrates the use of the wood hand-cutting tool and the corresponding impact on the laborer's health.

The result of the study showed that the use of the wood Roof Beam exposes the laborer to dusts that are prejudicial to his health and to the probability of lesions due to the intense use of the manual tool, whose gravity may provoke the removal of this laborer for a long period or permanently.

The innovating proposal to use the recycled plastic in the fabrication of Roof Beam will allow different measures for this structural piece, while its being able to vary in accordance with the project demand on popular housing project, so eliminating the use of the wood hand-cutting tool to make adjustments such as amendments and cuts, thus leading into the reduction of labor accidents and exposition to dusts.

Figure 2.   Use of the wood hand-cutting tool.

2   MATERIALS AND METHODS

The study started with a bibliographical research that allowed to know the proprieties of the recycled plastic and a few concepts about the laborer's health.

Then it was carried out an interview with the workers of a job site in a Housing Project for low income families, which is located in the industrial borough of João Pessoa city—PB/BR.

To do so, it was developed a questionnaire divided into two parts: The first one aimed at collecting data on the main aspects related with the use of wood, its finality, specificities, advantages and disadvantages of its use, required time for reposition, among others.

In a second moment, after the presentation of the recycled plastic Roof Beam prototype, the interviewed workers were asked about the possibility of having the wood pieces replaced by the new product.

The workers submitted the prototype to such tests as: Weight, sandpaper abrasion, fixing and removing nails. After going through these achievements, the interviewed workers could express their opinion about the possible changes that that product would bring.

The interview served to identify the perception of 10 carpenters that were responsible for the construction of the Roofs of those houses in the referred site, as regards the disadvantages of the wood Roof Beam and the possible advantages with the use of the Recycled Plastic Roof Beam.

3   RESULTS AND DISCUSSION

3.1   Technical characteristics of the recycled Roof Beam

Plastic materials can derive from synthetic sources or from natural substances, as oil for instance. According to the Brazilian Association of Composite Materials (Associação Brasileira de Materias Compósitos, 2012), plastic are materials formed by large molecular chains called "polymers". The type of polymer adopted to fabricate the plastic Roof Beam corresponds to High (PEAD) and Low (PEBD) density Polyethylene. The adoption of these polymers occurs by means of Recycling, which—according to the ABNT (Brazilian Association of Technical Norms) NBR 15792:2012 – corresponds to a new process of production as from the residues of materials for a starting use or for other uses.

The prototype (Figure 3) of the innovative recycled Plastic Roof Beam, which is studied in this research, was registered as patent of invention (PI 1001520-5) at the INPI (National Institute of Intellectual Property).

Figure 3. Image of the prototype: Recycled plastic Roof Beam.

Table 1. Prototype tests.

Resistance to:	Compression kg/cm	Flexion kg/cm
Maçaranduba wood	182.3	386.3
Prototype	247.9	288.8

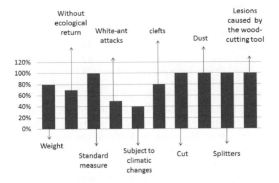

Figure 4. Disadvantages with the use of wood.

Considering that the Roof Beam is a structural resting piece, it must be able to receive a distributed force of 2.70kg/cm^2. Tests were accomplished with the prototype at the SCIENTEC (Association for the Development of Science and Technology) laboratory of the Federal University of Paraíba (UFPB) and some results are shown in Table 1.

The data show that the prototype is ready to stand a distributed force of 386.3kg/cm^2, further to a resistance to flexion equivalent to twice as much that of the wood. These data prove the technical advantages of this new product as regards the required loads for its applicability.

This plastic Roof Beam will be fabricated in accordance with the demanded dimensions as per the architectonical project of each house to be built, so eliminating the possibility of adjustments in the job site.

3.2 Perception of the workers

In a first moment, the interviewed workers were questioned about some relevant aspects in relation with the use of the Roof Beam. It was perceived that there are some diversions as to their understanding of this piece function. Among the answers, a few alternatives as regards the use of the Roof Beam are related with the distribution of the roof load and the improvement of the sustainable capacity of the roof structure of the popular houses.

When asked about something to facilitate the use of the plastic Roof Beam, most of the interviewed workers mentioned the need of pieces developed with varied measures (the more used measure is 3" × 3"). Such procedure would represent a significant reduction in the rates of wastes and residues. Most of the time, these residues do not receive an appropriate treatment as regards its final destination. As far as the final destination is concerned in some job sites, these residues are burned, while they are treated as trash in some others. Some building companies sell the wood residues to bakeries and

the collected money can be destined to the workers or to meet the administrative expenses of the jobs.

Still in this context, some elements related with the use of wood were presented, which were classified as advantages or disadvantages for the workers. Considering that the results are more significant as regards the disadvantages of using wood, some emphasis will be given in the results shown in Figure 4.

The main disadvantages that the interviewed workers identified in the use of wood (Figure 4) are: Standard measure, cut, dust, splitters and lesions caused by the wood hand-cutting tool.

As regards the standard measure, the research identified that the profiles are sized 80mm x 80mm, which are not in accordance with the measure stipulated in the architectonical project of the popular houses.

Another disadvantage resides in cutting the profile to reach the measure required for its fixing. These cuts are accomplished throughout the use of tools such as hand saws, hack-saw and table saw, which tend to cause some lesions on the workers.

This activity of cutting wood also results in the emergence of wood dust that—when inhaled—damages the workers' health.

Furthermore, handling the pieces provokes lesions on the workers' hands owing to their contact with wood splitters.

To fit the dimensions of the wood piece into those that the architectonical project demands, a wood hand-cutting tool is used. That tool exposes the worker to the occurrence of an accident, which may cause lesions to his upper or lower limbs.

Table 2. Perception of the workers.

Recycled plastic advantages	% of concordance
Small weight	90
Deforesting avoided	60
Dimensions as per project	100
No white-ant attack	80
No climatic change interference	70
Open-air stocking	70
Durability	50
No dust produced	80
No splitters	80
No need of the hand-cutting tool	100

In the second moment of the interview, after the presentation of the recycled plastic Roof Beam prototype to the workers, 100% of the interviewed individuals state that the new piece is rather resistant (fact already confirmed by means of the tests accomplished at the SCIENTEC laboratory).

Each worker received a sample of the prototype to be individually tested as for the fixing and removal of nails, weight and sandpaper abrasion. According to the interviewed workers, the piece presented 100% of positive result as per the accomplished tests.

Again the interviewed workers were asked about the possible advantages now related with the recycled plastic Roof Beam, whose results are shown in Table 2.

The advantages of the plastic Roof Beam identified by the workers were: No need of the hand-cutting tool, dimension as per the project, smaller weight, no dust produced, no splitters and no white-ant attack.

As the plastic Roof Beam is produced by the process of extrusion, the pre-fabricated piece will not require the hand-cutting tool to make cuts and amendments, since this project will meet the different measures demanded by the project. The elimination of the use of drill cutting tool (Enxó) means avoiding the accident of work with immediate injury and high gravity, acting in the cause of the problem by practicing the true prevention.

Considering that its main source of raw-material is plastic, its weight can be reduced (specific mass) without reducing its physical and mechanical qualities. For instance, a recycled plastic profile of 3,000mm × 80mm × 80mm weighs 19kg and the wooden one may weigh 22kg, fact that represents the exposure to the risk of lifting and carrying weight and can cause serious health problems for workers related to the spine.

Other advantages of the use of plastic Frechal can be listed: Not release dust that is so harmful to workers' health by causing pneumoconiosis, no use of natural fibers in their production eliminating the existence of barbs in the profiles and considering that this is a synthetic material it does not suffer microorganism attacks.

All in all, the research with the workers showed a strong acceptance of the new product, as they stated that—further to the above-described advantages—this type of material will facilitate the development of the activities related with the job roofing, there still existing the possibility of also cutting off handling inherent risks. Finally, by being a more malleable product, it will allow a reduction of the accidents that occur with workers in the job sites.

4 CONCLUSIONS

The study showed that—being fabricated in the dimensions required by the architectonical project of each house—the recycled plastic Roof Beam will allow its immediate setting-up on the brick-work so as to link the roof with the building, without the need of adjusting the piece as happens with the wood product. We can conclude that the new proposed product will allow a labor process that will reduce the contact of the worker with wood dust and will eliminate the probability of labor accident caused by the use of the hand-cutting tool. The adoption of the studied piece will contribute to the diminishment of the demand on wood pieces in the civil construction, also reducing deforesting. Likewise, having recycled plastic as its main source of raw-material, the production of plastic Roof Beam will contribute to the reduction of plastic trash discarded in dumps, embankments, rivers and sea, turning the world much cleaner.

REFERENCES

Associação Brasileira de Materiais Compósitos. 2012. *Compósitos I: materiais, processos, aplicações, desempenhos e tendências.* In: www.abmaco.org.br

Associação brasileira de normas técnicas. 2012. NBR 15792 - Embalagem Índice de reciclagem: Definições e método de cálculo. São Paulo.

Caixa. 2006. *Cadernos Caixa Projeto Padrão Casa Populares.* In:<http://www.secaplhis.net.br/mdm3/Cadernos_CAIXA_Projeto_padr%C3%a3o_casas_populares.pdf>.

Centro de Ensino Superior do Amapá (2009). *Técnicas de Construção Civil e Construção de Edifícios.* In: http://www.ceap.br/artigos/ART30042009211401.pdf.

Brasil. Lei 8.213. 1991. Presidência da República. In: www.planalto.gov.br/ccivil_03/leis/L8213cons.htm.

Programa Nacional de Prevenção de Acidentes de Trabalho. In: http://www3.tst.jus.br/prevencao/.

Occupational Safety and Hygiene – Arezes et al. (eds)
© *2013 Taylor & Francis Group, London, ISBN 978-1-138-00047-6*

Weighted responsibility in occupational safety management in the construction industry—case study

P.A.A. Oliveira
Faculty of Engineering, University of Porto, Porto, Portugal
Higher Institute of Languages and Administration (ISLA), The Group Lusófona, VN de Gaia, Portugal

Z.A.F. Neves
Faculty of Engineering, University of Porto, Porto, Portugal

ABSTRACT: It is attributed as main cause than infringements of rules Safety legally prescribed and applicable to the sector Construction and Civil Engineering, the lack of culture of the prevention and Safety for workers of this industry. That is aggravated by the "lack" of responsibility for infringements practiced, allied to widespread tendency of improvisation and even the omission of rules and actions legally demanded. Being the purpose of this study, the demand for an instrument that evaluates the degree of responsibility of the various stakeholders in this sector. Based on the assessment of the "degree of compliance" and, in assessing the "weight of responsibility", obtained through the values assigned to legal administrative offenses, provided by law. Of the results obtained, demonstrates that the actors artwork, one of the least penalized are the workers, with the aggravating factor of in general, these are more exposed to the high risks. Because the regulatory laws generally not the contemplate. Being the owner of the work the more penalized intervener, because it is given the responsibility for meeting the other interveners. Given these indicators, it is imperative and indispensable for the "Responsibility ceases to be" dead letter "and that the actors in the sector of Civil Engineering and Construction assume its in full share, from moral responsibility, civic responsibility, professional and ethics.

1 INTRODUCTION

The vast range of the constructive operations in the civil engineering sector is prone to an unfathomable number of risks, which not properly analyzed and accounted for, can give origin to accidents in great numbers, with most of them being of sever gravity (Neves 2011). In 2009, the European Risk Observatory (OER) considered that the accident rate is specifically higher in this sector, where the risk of work accidents is almost the double of the risk of work accidents average of the remaining economical activity sectors (OER 2009). In Portugal, as official statistics from the Strategy and Planning Office (GEP) of the former Labour and Social Security Ministry (MTSS) proves, between 1999 and 2007, the annual work accidents average in the construction sector was, approximately, of 50.000 accidents, evidence that is a value with significance (Oliveira 2011).

Although stating that the non-observance of safety rules legally prescribed results in the lack of a prevention and safety culture/knowledge of our workers, in the construction sector, leading to risks of accident.

In general, in the pragmatic way of live which we live in, the managers' biggest concern is to meet the contract's deadlines; such is to avoid any contractual fine. But this scenario changes in the event of a work accident, or even as the reason to instil a disturbance in everyone's state of mind, thinking of what they could have and should have done but didn't, keeping a constant feel of anguish and uncertainty of what the final condemnation will be, when the investigation is started. In the matter relating to the workers' social responsibility concerning safety and health at work, the legal board couldn't forsake equating the fundamental rights and obligations in a judicial relationship essential to the development of societies. Two legal texts work as an essential reference in this domain, where it can be highlighted some fundamental articles as to understand the employers' responsibilities: i) *the Portuguese Republican Constitution, Constitutional Law n.° 1/2005 of 12 August, which publishes the 7.th and last revision of the fundamental law in Portugal*; ii) *the Penal Code, Law n.°59/2007, of 4 September, which publish the twentieth third alteration to the Penal Code, approved by Law Decree (D.L) n.°400/82, of 23 September*. In this social juridical board extent, matter, by its relevance referring the Deputy Prosecutor's Work, José Albuquerque, in the issues relating the juridical framework and representative of the infringement regarding the work safety rules stating: "*We must therefore proceed with the opening of an investigation whenever the circumstances in which the accident occurs indicating omission of relevant duties by the "entities" responsible, in compliance with the rules of safety, these regulations resulting from statutory or regulatory provisions or rules techniques for the functional performance of the activity where the accident occurred*" (Albuquerque 2006). This functional guideline had consecration in Circular no. ° 19/94 of 9 December 1994, the Attorney General's Office, according to which, for fatal accidents at work, where it is recommended to judges and prosecutors, together with labour jurisdictions that, for such cases and where it is not excluded the existence of criminal responsibility, provide for the immediate opening of an investigation, pursuant to Code of Criminal Procedure. An orientation that was correct and relevant, because in most cases the work accident when occurs gross violation of rules, laws or regulations on safety at work.

2 OBJECTIVES

This research study aims to find a mechanism to facilitate the assumption of the weight of the share of responsibility of the different intervenients, the roles they play in the production process, making it an assertive vehicle as a tool for awareness among workers, weighted by the responsibility for the contribution that the main goal is the minimization of serious and fatal accidents. Wanted also analyze the degree of fulfilment of the parties through the practical confirmation of design, installation and use of the tower crane by its considerable visibility and representativeness productive. Moreover, based on the violations and subsequent offenses provided in the legislation applicable to the construction industry, seeks to formulate a table representing the weight of responsibility that every intervenient necessarily falls within the limits of their duties professionals.

3 METHODOLOGY

This article was developed based on the needs to determine and evaluate the degree of responsibility of the different intervenients involved in the construction industry, according to the share of the responsibility attributed to their professional role. The present study had a research period of one year (2011). To achieve the objectives it was necessary to

collect data and record various information data "*in loco*". To this end we selected and visited 21 temporary or mobile shipyards, from which 24 construction cranes have been analyzed. A questionnaire for collecting data was developed and applied including plug-checking machines, considering the provisions of the articles of Decree-Law (D.L.) n.° 50/2005 of February 25, to determine the degree of legal compliance in the use of equipment (cranes).

In the course of the visits at 21 construction sites for the collection of data, it was found that: 42% of cases the Director of Technical Work was absent, and in 54% of cases the Technical Health and Safety at Work (TSHT) was absent, and in 4% of cases had not responsible for the technical direction, nor TSHT.

Was also used a framework for recording and assessing the degree of legal compliance, with great incision in the legal regime of coordination and planning of Safety and Occupational Health. In particular the DL n° 273/2003 of 29 October which transposes into national law of Directive n° 92/57/EEC of the Council of 24 June concerning the minimum requirements of safety and health at work to apply at temporary or mobile construction sites.

It was also considered that the "weight" of the weighting can be obtained from the "value"of fines provided in the offenses prescribed in the lexicon affect the regulation of activity in construction. Thereby allowing assessing each intervenient in work the weight of the responsibility weighted.

4 RESULTS AND DISCUSSION

Upon completion of the literature search and to determine the desired indicators proceeded as follows:

4.1 *Assessment of the Degree of Legal Compliance (G)*

For this purpose sought to firstly find the "Degree of Legal Compliance," by verification of the "in loco" in the works under way, through a chek-list denominated "Machines Verification Fact Sheet". This is based on the Decree-Law n ° 50/2005 of 25 February, and allows the registration of items NA, NC and C relating to the respects in satisfaction of the legal requirement considered.

Since NA–(No. of items not applicable); NC–(No. of items defaulters) C–(No. of items met-res), and (being n.° questionnaire items = 34), then the degree of breach is given by function (degree of legal compliance) [G] determined by the formula:

$$G = NC/(34 - NA)$$

As mentioned above, this expression is directly or indirectly indexed to a scale of "perception of risk" particularly in the function (Fitness), the variation of n. ° NC's and function (Risk) by the variation of n. ° NC's as a function of n. ° NA's.

Being then the values calculated for the assessment of degree of compliance legal (G), relating to functionality and safety demands of of the cranes installed in construction works, presented in Table 1.

4.2 *Weight Weighted Responsibility (PRP)*

It is based on the product of the "degree of compliance" medium (G) times the "weight of responsibility" (P). This is obtained, as cited by specific legislation that regulates the activity, the number of items covered and the amount of the fines, according to the violations "very serious", "Serious" and "Light."

$$PRP = (G \times P)$$

Then the calculated weight of responsibility to intervene weighted by the study period are presented in Table 2.

The results obtained with the study merited discussion about the validity of technical and scientific theory exposed. Being that towards greater reliability of these should be collected more samples in other diverse areas of applicability and legal requirement.

Should not only cover the equipments, but also as regards application of broadest possible range of legal procedures, relating to other resources.

The deviations from verified legal compliance circumvent the obligations, with application of preventive measures or impromptu influenced by the economic and financial savings. These constitute the highest in risk of default, which generally potentiate accidents and incidents at work. And they also generate losses of every order advertised by different authors, to the "duplicate or tripling the value of investments necessary to preventive action" for protection occupational, concerning to hazards identified and assessed risks, inherent in the productive activities.

From the analysis of the results, there was obtained the graph of Figure 1. That relates to the evolution of non-compliance with the of the degree legal non-compliance.

By the graph of Figure 1 it turns out that exists an increasing influence of the items not applicable in respect of failure.

Table 1. Evolutive analysis of the degree of legal non compliance.

Evolutive relation of items not applied

NC –	1	1	1	1	1	1	1	1	1	1	1	1	1	1	1	1	1	1	1	1	1	1	1	1	1	1	1	1	1	1
NA –	0	1	2	3	4	5	6	7	8	9	10	11	12	13	14	15	16	17	18	19	20	21	22	23	24	25	26	27	28	
G –		2,9	3,0	3,1	3,2	3,3	3,5	3,6	3,7	3,9	4,0	4,2	4,4	4,6	4,8	5,0	5,3	5,6	5,9	6,3	6,7	7,1	7,7	8,3	9,1	10	11,1	12,5	14,3	16,7

Observations.

Table 2. Values of the weight of responsibility weighted by intervening in the construction process.

Weight of responsibility weighted calculus

Intervenients in constructive process	Lexicon responsibilities	Violated articles/ ordinances against*			PRP
The Employer	5	13	42	2	57/5 = 11,40
Draftsman	1	1	2	0	3/1 = 3,00
Coord. Sec. Project	3	2	0	6	8/3 = 2,66
Entity Performer	4	8	6	13	27/4 = 6,75
Coord. Sec. Work	2		0	0	1/2 = 0,50
Supervision	1	0	0	1	1/1 = 1,00
Director of Technical Work	1	0	0	1	1/1 = 1,00
Clerk General	0	0	0	0	0,00
Clerk Front	0	0	0	0	0,00
Safety Technician	2	0	3	3	6/2 = 3,00
Independent worker	3	3	3	4	10/3 = 3,33
Employer	6	5	3	11	19/6 = 3,17
Workers	6	0	0	7	7/6 = 1,17
Occupational Physician	1	0	2	2	4/1 = 4,00

*It means no. of items covered by the intervener and the value of the prescribed offenses and valued according to the degree of severity assigned (Very severe, Severe and Mild).

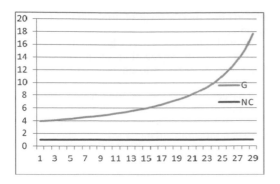

Figure 1. Evolutive analysis of the non compliance degree of legal.

In spite of easy graphical view of this evolution, in a context of decision making, it means the highlight of the importance specific for each prescription in safety and occupational health in the construction sector, in the universe of the legal obligations.

5 CONCLUSIONS

We conclude the present work, Table 1 through whatever the "degree of compliance" or the "burden of responsibility", show that the intervenients work, the least penalized are the main intervenients of the construction workers. Because although prescribed by 6 articles legal, regulatory laws still do not blame them for their actions. The various stakeholders the project owner is the most penalized (11.40), because it is given the responsibility for meeting the other intervenients. The least penalized least are responsible for front and overall (0.00), because legislation hardly gives them any weight of responsibility.

This widely divulged instrument widely will certainly cause a major impact in raising awareness of the responsibilities in the decision to withhold or fulfilment of legal requirements, the material and psychological consequences that afflict all people in general, and workers at all levels in particular.

It is therefore essential that the construction industry stakeholders to take each of themselves, their share from professional responsibility, moral and ethical responsibility to civil, penal and criminal.

REFERENCES

Albuquerque, J.P.R. 2006. *A infração às regras de segurança no trabalho,* (O tipo omissivo do art.º 277, n.º1 da alínea b), 2ª parte do Código Penal, Sesimbra.
Decree-Law n. 50/2005. *Regulamenta as prescrições mínimas de segurança e de saúde para a utilização pelos trabalhadores de equipamentos de trabalho* (in Portuguese). D.R. n. 40. (Série I-A de 2005-02-25). Imprensa Nacional e da Casa da Moeda (in Portuguese), Lisbon.
Decree-Law n. 273/2003. *Regulamenta as condições de segurança e de saúde no trabalho em estaleiros temporários ou móveis* (in Portuguese). D.R. n. 251. (Série I-A de 2003-10-29). Imprensa Nacional e da Casa da Moeda (in Portuguese), Lisbon.
Directive 92/57/EEC of 24 June 1992 on the *implementation of minimum safety and health requirements at temporary or mobile construction sites* (eighth individual Directive within the meaning of Article 16 (1) of Directive 89/391/EEC). The Council of the European Communities, Luxemburg.
Oliveira, P.A.A. 2011. *Modelo de Análise da Sinistralidade Laboral versus Investimento em Prevenção, para o Setor da Construção.* Tese doutoral submetida ao Departamento de Ciências Biomédicas da Universidad de León (UM), para a obtenção do grau de Doutor em Higiene, Saúde e Segurança do Trabalho, pp. 1–219, León.
OER 2009. *Perspectivas 1—Novos Riscos Emergentes para a Segurança e Saúde no Trabalho,* Bruxelas: Serviço das Publicações Oficiais das Comunidades Europeias, pp. 7–21. TE- 81-08-475-PT-N, Bruxelas.
Neves, Z.A.F. (2011). *A Responsabilidade Ponderada na Gestão da Segurança, Higiene e Saúde no Trabalho na Construção Civil.* Dissertação apresentada para a obtenção do grau de Mestre em Engenharia da Segurança e Higiene Ocupacionais, Departamento de Engenharia de Minas da Faculdade de Engenharia da Universidade do Porto (FEUP), Porto.

Occupational Safety and Hygiene – Arezes et al. (eds)
© *2013 Taylor & Francis Group, London, ISBN 978-1-138-00047-6*

Thermal environment as a management tool in high-rise building

H.C. Albuquerque Neto & J. Santos Baptista
PROA/CIGAR/LABIOMEP/Faculdade de Engenharia, Universidade do Porto, Porto, Portugal

B. Barkokébas Jr.
University of Pernambuco, Recife, Brazil

ABSTRACT: The construction industry is a growing sector in countries that are prosperous in economic development, leading to the new studies that drive this sector. Thus, it appears that the buildings have a great relationship with thermal environment, due to their constant exposure to those factors. This can influence the productivity of workers, compromising the execution of construction. This paper aims at investigating the influence of the thermal environment in building at elevated heights, establishing a relationship with productivity. For that, was done a sistematic bibliographic research in this area trying to stablish the stat of the art starting from the papers published in the last ten years. It appears that there is a relationship between the subjects studied, although most of the survey does not include all the issues.

1 INTRODUCTION

The current scientific and technological advances have enabled more efficient procedures in the development of several working activities, together with increasingly efficient equipment. The junction of these features allowed for a greater control of internal and external factors that influence the working activities. On the other hand, some sectors remain oblivious to environmental factors. An example is the construction industry and, more specifically the building branch. In this sense, thermal environment is an external factor for the conditioning of the development of any works and its field of influence embraces directly the way the workers perform their tasks. This line of work proceeds with productivity that may be increased or reduced according to the environment the worker is subjected to. Furthermore, it is necessary to investigate whether the work performed at different height levels affects productivity, since this is one of the environmental variables. Thereafter, there is a need to know the effects that thermal environment can cause in humans, seeking to minimize their reflections on their health and work, thus improving productivity and contributing to more effective methods in the process of building.

Therefore, this paper seeks to reconcile the topic of thermal environment with the construction of buildings in height and its productivity.

2 MATERIALS AND METHODS

Our research, covering at least two of the above mentioned areas, will focus on papers from journals and databases published over the last ten years. At the same time we will also keep in mind the state of the art (during the same period) about these themes. By doing so, we will able to find concepts, models, rules and cases in which the addressed topics—thermal environment, buildings and productivity—are included. For this purpose, more than 30 electronic sites of scientific papers with international scope disclosure were surveyed. The definition of portals to search was based on the list included in the Metalib Exlibris (Metalib, 2012). The portals surveyed are shown in Table 1.

In order to find papers related to the topic under observation eight key words and concepts were chosen, namely: "*thermal environment*", "*thermal comfort*", "*building*", "*construction*", "*model*", "*height*", "*thermal*" and "*skyscrapers*". For the accomplishment of additional screening for the work under way, combinations of keywords using the logical operator "AND" were made. When it was not possible to screen through the abstract, the job title was also used. The combinations performed are shown in Table 2.

Based on these findings, we started reading the publications to confirm which of them might be relevant for the research in progress. The method used for the acceptance/exclusion of the work was:

Table 1. Sites where the surveys were conducted.

Type	Name
Database	CiteSeerX, Compendex, Current Contents, Energy Citations Database, Inspec, PubMED, SCOPUS, Research and Innovate Technology Adminstration, Web of Science e Zentralblatt MATH
Scholarly journals	ACS Journals, Annual Reviews, ASME Digital Library, Cambridge Journals, ASCE, DOAJ, Emerald, Highwire Press, IEEE Xplore, Taylor and Francis, Ingenta, IOPscience, MetaPress, Oxford Journals, SAGE, SciELO, ScienceDirect, Scitation, Springerlink, Wiley Online Library

Table 2. Procedures to the surveys realization.

Survey number	Keyword	Number of publications founds
1	"thermal environment" "building"	1510
2	"thermal environment" "construction"	304
3	"thermal environment" "construction" "model"	135
4	"thermal environment" "height"	199
5	"thermal comfort" "height"	207
6	"thermal" "skyscrapers"	20

the suitability for research purposes, theoretical framework demonstrating scientific reliability; explicit methodological procedures used; criss-crossing and development of the analysis of the results of the keywords used. This research contemplated all these parameters.

3 RESULTS AND DISCUSSION

With the development of economy and society, cities continue to grow both in physical size and population, resulting in an increase in the number of residential neighborhoods (Li et al. 2005), higher density urban planning, urban corridors narrower buildings and structures with increasing hight (Hien et al. 2011). In this sense, the ever growing vertical integration of buildings is a worldwide trend, mainly in big cities due to the gradual increase in concentration of people in the same geographical space.

According to Silva et al. (2009), for the cities to continue to grow, the availability of land is essential, although what happens in big cities is a decrease of free spaces for new housing construction, which drives the trend to vertical integration. Taib et al. (2010) corroborate this idea, by claiming that high-rise buildings are becoming a trend, mainly due to the shortage of land particularly in rapidly developing countries.

According Mollmann & Vollmer (2012), early high-rise buildings emerged in the first half of the 20th century, with rectangular geometries and small windows. Yet according to the authors, in recent decades, two trends have gained strength about skyscraper buildings: they now display more geometric curves, leading to spectacular views at a fascinating distance; most facades involve reflective surfaces, in order to avoid thermal load intensity inside the building during the summer. This latter trend is directly related to the massive use of air-conditioning systems in buildings, especially in seasons that provide thermal stress. Studies of Ghadiri et al. (2011), validate the premise by claiming that previous buildings in hot and humid climates have been traditionally cooled by ventilation. However, amid the current energy and economic crisis, one of the main challenges faced by the construction industry is the need to create solutions to reduce the use of energy in buildings (Wong et al. 2002; Taib et al. 2010). In accordance with this, Berkovic et al. (2012), claim that the architects have studied indoor thermal comfort and its energy needs.

In this context, thermal environment emerges as a relevant factor in defining the housing project because, according to Li et al. (2011), the variation of the thermal environment can significantly affect the health and productivity of people. Thus, it becomes necessary to further study the factors that may influence thermal environment in buildings of great height, and in addition to reduce the use of energy in those buildings.

In compliance with this reality, many papers have been published with the focus on the analysis of thermal environment inside the buildings, such as the study of thermal comfort in dwellings in general (Wang, 2006), the study of buildings without any air conditioning system (Ji et al. 2006), the sensation of comfort in offices (Bluyssen et al. 2011), thermal comfort inside the houses in hot, humid, cold or rainy seasons (Peng, 2010), among others. In addition to these studies, it appears that natural ventilation and the place where the building is constructed is a decisive factor regarding the thermal comfort provided to its occupants.

Wong et al. (2002) found that natural ventilation has proven to be an energy-efficient alternative to reduce the running costs in buildings, achieve thermal comfort as well as maintain a healthy indoor environment. In urban areas, wind conditions have an impact on the potential use of natural ventilation in buildings, on human thermal comfort outdoors and on the dispersion of airborne pollutants (Krüger & Rasia, 2011). Ventilation in a dwelling serves the primary function of providing health, comfort (Prianto & Depecker, 2002) and, better air quality, reduces the thermal load of air-conditioning systems (Zaki et al. 2012).

Therefore, the studies of Deb & Ramachandraiah (2011) concluded that comfortable outdoor locations in a city have multiple advantages ranging from savings in energy to an improved social quality of life. On the other hand, as urbanization progresses, the "urban warming" problem is mainly aggravated because of the reduced density of green vegetation and increased building in the urban environment (Sun, 2011). This intensifies the effect of Urban Heat Islands (UHI), which according to Tan et al. (2010), are characterized by an increase in temperature, which can potentially increase the magnitude and duration of heat waves within cities. In daytime situation, the combination of high temperatures and intense solar radiation create severe problems of heat stress (Emmanuel et al. 2007).

The UHI affects street level thermal comfort, health, environment quality and may contribute for an increase of urban energy demand (Hien et al. 2011). Studies of Hu et al. (2012) demonstrated that the intensity UHI at a height of 80 m is smaller than at a height of 2 m and 10 m. Among the possible explanations for this, is that buildings with high altitude provide shading in the urban environment in which they are located, resulting in a decrease in temperature from solar radiation (Johansson & Emmanuel, 2006; Hien et al., 2011), also concluded that the greater the height of the building, the more it will be exposed to climatic elements such as wind, sun and rain (Taib et al., 2010) the ratio between the height of buildings and the distance between them, influences the amount of incoming and outgoing radiation (Hsieh et al., 2010; Sun, 2011).

On the other hand, there are few studies that focus on the thermal environment during building construction as well as its relation to the workers comfort and productivity, arising from changes in working conditions. It is known that for the construction of buildings, workers are exposed to daily and seasonal variations of the thermal environment where most of the tasks performed in the open air occur, without there being the possibility of cooling from ambient or controlled temperature and moisture. Among these few examples, we can refer to the studies of Mohamed & Srinavin (2002), which relate the productivity of construction workers with thermal environment via thermal comfort, an index which aggregates a set of climatic conditions and parameters of clothing.

The study of these authors is based on PMV (Predicted Mean Vote) indicator comfort, proposed by Fanger, considering all the weather conditions and the nature of the task to be performed by the worker. The results indicate a good fit between the model predictions and actual data. Lopes (2007), in turn, sought to diagnose the conditions of thermal comfort the construction workers were exposed to, over the following stages of construction: construction of foundations and concrete structure, execution of the masonry and installation of technical networks, and other facilities and finishing in general.

From the above mentioned scientific research, papers that measured the influence of thermal environment on the workers' productivity in terms of high-rise building were not found. However, there are evidences that productivity changes with thermal environment and the latter varies with height. In this context, it is considered that the possibility of adding the variable "height" brings measurable benefits to the identification of the relevance of thermal environment and altitude in workers' productivity, allowing in parallel, the improvement of working conditions.

4 CONCLUSIONS

One of the factors with a direct effect on productivity is thermal environment. A possible cause for the decrease in productivity is high temperature, as demonstrated by Seppanen et al. (2004) in many studies, which showed a mean decrease of 2% in the performance of work for each degree Celsius when temperature exceeds 25 °C. However, authors like Eston (2005), point to yield losses that rise to 75% when the temperature reaches 37 °C.

Furthermore, temperature and humidity also vary with altitude. Moreover, studies of Kakon et al. (2010), found that in large buildings, air temperature and humidity are reduced after a certain time, which may favor a greater sense of thermal comfort resulting in higher productivity. When a "high-rise" building is under "construction", the combination of these two factors may be crucial to optimize both working conditions and productivity. According to Wood (2007), open recreational spaces need to be introduced into high buildings, because they improve the quality of the internal environment, which has an impact on the productivity of workers, satisfaction of residents, etc. (which will have indirect financial return). It is verified that workers who are comfortable, tend to work better, increasing their income and subsequently adding the effectiveness of team work, which provides a productivity gain (Niemela et al., 2002; Akimoto et al., 2010).

Consequently, there is a great influence both on the timeline of construction and the profitability of the enterprises, which is essential to any building. By studying the relationship between thermal comfort and productivity at different height levels in buildings, we can provide information that enables a more effective planning of activities in the work. This will lead to a rise of workers productivity, contributing to their comfort and, at the same time, optimizing the construction process. This will help to achieve the stipulated timelines and to avoid additional costs by delays in the completion of the projects. Another important factor regarding high-rise buildings is that according to Wood (2007), the international community is still divided in relation to the sustainability credentials of high buildings as an appropriate typology in our existing and future urban centers. There are those who believe that the concentration of population through high density (therefore reducing transport costs and urban/suburban spread), combined with the economies of scale of building tall, make typology an inherently sustainable option, while others believe that the embodied energies involved in constructing at height, combined with the impact on the urban realm, make them inherently anti-environmental (Wood, 2007). However, it is necessary to emphasize that due to individual needs, the extent of thermal comfort in an environment does not guarantee high productivity and full satisfaction to its occupants (Smith, 2010). For future work, it has been suggested a survey of other factors related to the location where the building is constructed, as they are listed by Desideri et al. (2010): the area's characteristics, the local climatic conditions in different seasons and months of the year, and the typological factors.

REFERENCES

Ahashi, T. 2008. Creating the 'Wind Paths' in the City to Mitigate Urban Heat Island Effects A Case Study in Central District of Tokyo. In: *Construction and Building Research Conference of the Royal Institution of Chartered Sur.* Retrieved November 11, 2012, from http://www.kenken.go.jp/japanese/contents/cib/w101/pdf/04.pdf

Akimoto, T., Tanabe, S., Yanai, T. & Sasaki, M. 2010. Thermal comfort and productivity—Evaluation of workplace environment in a task conditioned office. *Building and Environment*, 45, 45–50.

Berkovic, S., Yezioro, A. & Bitan, A. 2012. Study of thermal comfort in courtyards in a hot arid climate. *Solar Energy*, 86, 1173–1186. doi:10.1016/j.solener.2012.01.010.

Bluyssen, P.M., Aries, M. & Van Dommelen, P. 2011. Comfort of workers in office buildings: The European HOPE project. *Building and Environment*, 46, 280–288.

Deb, C. & Ramachandraiah, A. 2011. A simple technique to classify urban locations with respect to human thermal comfort: Proposing the HXG scale. *Building*

and Environment, 46, 1321–1328. doi: 10.1016/j.buildenv.2011.01.005.

Desideri, U., Proietti, S., Sdringola, P., Taticchi, P., Carbone, P. & Tonelli, F. 2010. Integrated approach to a multifunctional complex: Sustainable design, building solutions and certifications. *Management of Environmental Quality: An International Journal*, 21(5), 659–679. doi: 10.1108/00012531111135646.

Emmanuel, R. & Johansson, E. 2006. Influence of urban morphology and sea breeze on hot humid microclimate: the case of Colombo, Sri Lanka. *Climate research*, 30, 189–200.

Emmanuel, R., Rosenlundb, H. & Johanssonb, E. 2007. Urban shading—a design option for the tropics? A study in Colombo, Sri Lanka. *International Journal of Climatology*, 27, 1995–2004. doi: 10.1002/joc.1609.

Eston, S.M. Problemas de conforto termo-corporal em minas subterrâneas. 2005. *Revista de Higiene Ocupacional*, 4(13), 15–17.

Ghadiri, M.H., Ibrahim, N.L.N. & Dehnavi, M. 2011. The Effect of Tower Height in Square Plan Wind catcher on its Thermal Behavior. *Australian Journal of Basic and Applied Sciences*, 5(9), 381–385.

Hien, W.N., Jusuf, S.K., Samsudin, R., Eliza, A. & Ignatius, M. 2011. A Climatic Responsive Urban Planning Model for High Density City: Singapore's Commercial District. *International Journal of Sustainable Building Technology and Urban Development*, 2(4), 323–330. doi:10.5390/SUSB.2011.2.4.323.

Hsieh, C., Chen, H., Ooka, R., Yoon, J., Kato, S. & Miisho, K. 2010. Simulation analysis of site design and layout planning to mitigate thermal environment of riverside residential development. *Building Simulation*, 3, 51–61. doi: 10.1007/s12273-010-0306-7.

Hu, Z., Yu, B., Chen, Z., Li, T. & Liu, M. 2012. Numerical investigation on the urban heat island in an entire city with an urban porous media model. *Atmospheric Environment*, 47, 509–518. doi:10.1016/j.atmosenv.2011.09.064.

Ji, X.L., Lou, W.Z., Dai, Z.Z., Wang, B.G. & Liu, S.Y. 2006. Predicting thermal comfort in Shanghai's non-air-conditioned buildings. *Building research & information*, 34, (5), 507–514.

Kakon, A.N., Nobuo, M., Kojima, S. & Yoko, T. 2010. Assessment of Thermal Comfort in Respect to Building Height in a High-Density City in the Tropics. *American Journal of Engineering and Applied Sciences*, 3(3), 545–551.

Krüger, E.L., Minella, F.O. & Rasia, F. 2011. Impact of urban geometry on outdoor thermal comfort and air quality from field measurements in Curitiba, Brazil. *Building and Environment*, 46, 621–634. doi:10.1016/j.buildenv.2010.09.006.

Li, J., Wall, J., & Platt, G. 2011. HVAC control strategies for thermal comfort and indoor air quality. *International Journal of Advanced Mechatronic Systems* Vol. 3, No. 1, 24–32.

Li, M., Liu, S., Zhou, H., Li, X., Wang, P. 2005. The Temperature Research of Urban Residential Area with Remote Sensing. In: *International Geoscience and Remote Sensing Symposium* (IGARSS). doi: 10.1109/IGARSS.2005.1526279.

Lopes, R.F. 2007. *Condições de conforto térmico na construção de edifícios*. Dissertation (Master) Porto/PT, University of Porto.

Mohamed, S. & Srinavin, K. 2002. Thermal environment effects on construction workers' productivity. *Work Study*, v. 51, p. 297–302.

Niemela, R., Rautio, S., Hannula, M. & Reijula, K. (2002). Work Environment Effects on Labor Productivity: An Intervention Study in a Storage Building. *American journal of industrial medicine*, 42, 328–335.

Peng, C. 2010. Survey of thermal comfort in residential buildings under natural conditions in hot humid and cold wet seasons in Nanjing. *Frontiers of Architecture and Civil Engineering in China*, year 4, v. 4, 503–511.

Prianto, E, & Depecker, P. 2002. Characteristic of airflow as the effect of balcony, opening design and internal division on indoor velocity A case study of traditional dwelling in urban living quarter in tropical humid region. *Energy and Buildings*, 34, 401–409.

Qiu, L., Wang, C. & Wei, L. 2009. Simulation of Thermal Comfort in Air-Conditioning Room by Aripak. In: *International Conference on Computational Intelligence and Software Engineering* (CiSE) 2009. doi: 10.1109/CISE.2009.5362615.

Qiu, L., Wang, C., Wei, L. & Guo, K. 2010. Research of Indoor Air Quality on an Air-Conditioning Classroom. In: *4th International Conference on Bioinformatics and Biomedical Engineering* (iCBBE). doi: 10.1109/ICBBE.2010.5518152.

Seppanen, O., Fisk, W.J. & Faulkner, D. 2004. Control of temperature for health and productivity in offices. *Lawrence Berkeley National Laboratory*, University of California.

Silva, A.F.S. 2010. Avaliação da qualidade ambiental interior de um edifício climatizado artificialmente, com ênfase na análise do conforto térmico. (Master dissertation). UNIVERSIDADE DE SÃO PAULO, São Carlos.

Silva, L.H., Nascimento, M.L. & Bitoun, J. 2009. A verticalização no Recife: uma análise do bairro do Prado. In: *Anais del Encuentro de Geógrafos de América Latina*. Montevideo.

Sun, C. 2011. A street thermal environment study in summer by the mobile transect technique. *Theoretical and Applied Climatology*, 106, 433–442. doi: 10.1007/s00704-011-0444-6.

Taib, N., Abdullah, A., Fadzil, S.F.S. & Yeok, F.S. 2010. An Assessment of Thermal Comfort and Users' Perceptions of Landscape Gardens in a High-Rise Office Building. *Journal of Sustainable Development*, (3) 4, 153–164. doi: 10.5539/jsd.v3n4p153.

Tan, J., Zheng, Y., Tang, X., Guo, C., Li, L., Song, G., Zhen, X., Yuan, D., Kalkstein, A.J., Li, F., & Chen, H. 2010. The urban heat island and its impact on heat waves and human health in Shanghai. *International journal of biometeorology* (54) 75–84. doi 10.1007/s00484-009-0256-x.

Vollmer, M. & Mollmann, K-P. 2012. Caustic effects due to sunlight reflections from skyscrapers: simulations and experiments. *European Journal of Physics*, 33, 1429–1455. doi:10.1088/0143-0807/33/5/1429.

Wang, Z. 2006. A field study of the thermal comfort in residential buildings in Harbin. *Building and Environment*, 41, 1034–1039.

Wood, A. 2007. Sustainability: a new high-rise vernacular? *The structural design of tall and special buildings*, 16, 401–410. doi: 10.1002/tal.425.

Wong, N.H., Feriadi, H., Lim, P.Y., Tham, K.W., Sekhar, C. & Cheong, K.W. 2002. Thermal comfort evaluation of naturally ventilated public housing in Singapore. *Building and Environment*, 37, 1267–1277.

Zaki, S.A., Hagishima, A. & Tanimoto, J. 2012. Experimental study of wind-induced ventilation in urban building of cube arrays with various layouts. *Journal of Wind Engineering and Industrial Aerodynamics*, 103, 31–40. doi:10.1016/j.jweia.2012.02.008.

Occupational Safety and Hygiene – Arezes et al. (eds)
© *2013 Taylor & Francis Group, London, ISBN 978-1-138-00047-6*

Factors affecting the safety performance in portuguese construction: Summary analysis focused on rehabilitation works

J.P. Couto & C. Gomes
Department of Civil Engineering, University of Minho, Campus of Azurém, Guimarães, Portugal

ABSTRACT: The accidents in the construction are one of the main indicators of the claims recorded at work in Portugal. The identification of the factors and causes related with accidents at work is essential for a correct analysis of the risks and a consequent adoption of procedures and measures to achieve a more effective prevention. The focus on rehabilitation has increased in Portugal and it is in this kind of construction that accidents are more frequent, mainly motivated by the lack of local knowledge to work and skilled manpower. Thus, on the basis of an inquiry implemented to several categories of interveners, this work puts forward the main reasons for noncompliance in safety regarding this type of activity.

1 INTRODUCTION

1.1 Background

The construction sector is considered by many as one of the major pillars of a country's economy, but also one of the most dangerous industries (Enshassi, 1996). The policies adopted to increase productivity in the sector vary from country to country and depend on the strategy of each company. However, occupational safety and health have been widely recognized as two of the most influential aspects in companies' overall performance and have been gradually no longer seen as a luxury and came to be seen as a necessity to avoid losses, injuries or even deaths (Abdul-Rashid et al., 2007). Over time, increasing competencies in project management has enabled greater emphasis on this issue, focusing on workers' safety and health (Wong et al., 2002).

Due to the intense construction over the last decades, there has been a certain neglect of issues concerning construction quality (Araújo, 2009), which has now repercussions at the onset of the appearance of anomalies and premature buildings degradation. Thus, the need to invest more and more in repair and rehabilitation has been growing, which, in the current economic environment, seems an alternative to new construction that is virtually stagnant. In this type of work, the risks with the safety are still high, essentially motivated by the ignorance of the builder to intervene and by the little qualified manpower of the companies who execute them.

Thus, in order to better understand the state of Rehabilitation in Portugal, this study was carried out within the scope of a master thesis in sustainable construction and rehabilitation developed at the University of Minho, where a questionnaire survey was implemented to 57 actors in the field of rehabilitation, with the objective of identifying the key factors that affect safety in rehabilitation. It was concluded that the lack of technical expertise in organization and planning, lack of communication and skilled labor appear as the main factors related to flaws in the coordination and management of the safety of this type of work.

1.2 Safety in international construction

Internationally, there are a substantial number of studies aiming at clarifying the reasons related with safety performance. In Palestine, through an inquiry implemented to 32 construction managers, Enshassi (1996) sought to assess the opinion of interveners on issues related to safety, having concluded that the use of safety methods provides benefits, such as the reduction of costs with accidents, improvements in human relations and consequent increase in the productivity (Enshassi, 1996). For the inquired construction managers, the awareness about safety questions is directly related with worker's age and experience, insofar as, through the obtained results, it may be concluded that the levels of accidents tend to decrease from the age of 30 onwards, showing that the bigger the experience in the accomplishment of the works, the lower the risk of accidents.

Another study on the topic was carried out by Adbdul-Rashid et al. (2007), with the objective of identifying the main factors affecting the safety in great works in Egypt, using an inquiry as boarding methodology, implemented in some companies of the parents. This study was supported on a bibliographical research that made it possible to identify a set of 72 factors, grouped in 12 categories. The factors indicated in table 1 are part, among others, of this set of factors selected for the implementation of the inquiry.

The authors concluded that the factors considered most relevant by companies were the need for awareness of the company's management and project managers for the implementation of the safety management system, and the need for frequent monitoring.

Fang et al. (2002) had also carried out a study to analyze the factors of safety management in construction works in China. The main factors that had been considered to be more influential in safety management were:

– Safety inspection;
– Safety meeting;
– Safety regulation enforcement;
– Safety education;
– Safety communication;
– Safety cooperation;

– Management-worker relationship;
– Safety resources.

In 2008, Aksorn & Hadikusumo carried out a study where they identified the success factors influencing safety in the performance of Thai construction projects. For such, through bibliographical research, they identified a set of 16 critical factors of success (CFS), and implemented a questionnaire to 80 interveners involved in medium or large scale construction projects, with the purpose of perceiving, from the 16 critical factors of success, which were the most influential in safety management. The obtained results are presented in table 2, from the most to the least influential.

On the basis of the presented results, it may be concluded that the most influential factor in works safety of Thai construction is the support given to works management. Moreover, Aksorn & Hadikusumo (2008) consider that these 16 critical success factors may be grouped into 4 groups, namely:

– Worker involvement;
– Safety prevention and control system;
– Safety arrangement;
– Management commitment.

2 SAFETY IN PORTUGUESE CONSTRUCTION

2.1 *Factors of accident in rehabilitation works*

In Portugal, these types of studies are scarce, being only known some exploratory and preliminary approaches. However, it is generally acknowledged

Table 1. Factors affecting safety performance in international literature (Abdul-Rashid et al., 2007).

Literature	Factors affecting safety performance
Ng et al. (2005)	Implementation of safety management system in accordance with legislation; Compliance with occupational safety and health legislation, codes and standards; Definition of safety responsibility; Development of safety policy; Provision of safe working environment; Development of emergency plans and procedures; Development of safety committee;
Fung, et al. (2005)	Effective accident reporting; High line management commitment; Active supervisor's role; Active personal role;
Teo et al. (2005)	Understanding and implementation of safety management system; Understanding and participation in occupational health and safety management system; Quality of subcontractors; Understanding and implementation of safety procedures; Carrying out work in a safe and professional manner; Type and method of construction; Management's attitude towards safety; Monetary incentives; Disciplinary action;

Table 2. Critical factors of success (Aksorn & Hadikusumo, 2008).

Critical factors of success (CFS)	
1	Management support;
2	Appropriate safety education and training;
3	Teamwork;
4	Clear and realistic goals;
5	Effective enforcement scheme;
6	Personal attitude;
7	Program evolution;
8	Personal motivation;
9	Delegation of authority and responsibility;
10	Appropriate supervision;
11	Safety equipment acquisition and maintenance;
12	Positive group norms;
13	Sufficient resource allocation;
14	Continuing participation of employees;
15	Good communication;
16	Personal competency;

that larger works, where load and diversity of resources are very significant and constructive processes are complex, and above all in interventions of buildings retrofitting, usually featured by very specific and differentiated works, are the ones that record more often problems and safety breaches. This results from a set of factors that, in these works, take on particular significance.

Official figures show clearly the preponderance of the construction industry in fatal accidents recorded over the years in Portugal. As documented in table 3 (ACT, 2012), although from 2008 onwards fatal accidents in construction declined significantly, which is probably related to the reduction of economic activity in the sector, the impact of this sector in the overall national industry remained at alarming levels, since the same decreasing trend of fatal accidents occurred in the set of other industries.

In order to organize and systematize the relevant information on this matter, the Authority for Working Conditions (ACT) suggested the organization of the main factors contributing to accidents, and the characterization of the construction industry for economical and organizational terms, as follows:

– The proliferation of small and microenterprises;
– The reinforcement of subcontracting;
– The lack of training of decision makers and workers;
– The economic competitiveness.

The latter term leads to budget cuts and very demanding deadlines, with the intensification of the pace and duration of the work.

Beyond these, the ACT presents other causes of construction accidents, namely the absence of safety structures, the poor work planning and timing, the lack of protection and the use of false protections, among others (Paula, 2008 cited by Araújo, 2009).

In the rehabilitation works, which currently begin to have greater visibility and amplitude in Portugal by virtue of the need for recovery and repopulation of historic centers, the management of this problem has been difficult to prove, due to the unpredictability of work that comes from lack of trustworthy records that indicate how the building was constructed and of projects. This lack of records hinders the works that are performed, mainly by small businesses with limited technical resources and skilled manpower, and whose preparation for the issues of quality, environment and safety and health at work is practically nil. In these works, where various tasks related to demolition, basements and recoveries of facades are carried out, the main causes of accidents are, among others, crushing, falls from height and burial. Against this scenario, it has been widely recognized that, in general, the risks associated with workplace accidents are higher in rehabilitation works than in new construction (Egbu, 1999).

Another of the factors properly considered predominant for accidents in construction, particularly in rehabilitation work, is the lack of prospecting and analysis of the work, i.e., companies start rehabilitation work without analyzing the risk of accidents and adopt prevention procedures thereof.

Through the study conducted in 2007 by Roberts et al, it was concluded that only 35% of the rehabilitation projects consider safety measures in order to prevent risks in carrying out the work on the ground. With this study, conducted through a survey of about 250 of the Rehabilitation professionals in Portugal, it was found that the reasons given by respondents for the lack of preventive measures are:

– Devaluation of safety issues by project owner and designers;
– Lack of risk assessment by designers;
– Lack of safety coordinator during the design phase;
– Poor interaction between team and project coordinator;
– Lack of technical compilation.

As such, it can be concluded that the existence of flaws compromises the existing legislation right from the project phase.

The diversity of participants in this type of work, in representing various enterprises, is also one of the important factors in accidents because many are unaware of the reality of working conditions. Moreover, turnover of skilled labor affects the correct execution of the works, thus leading to the increase in the risk of accidents. Besides this, the pressure on companies to comply with the terms and costs means that the work is carried out under hard conditions, often leading to increased number of people in work, thus increasing the risk of accidents (Couto, 2008).

Table 3. Fatalities in the construction from 2005 to 2009 (ACT, 2012).

	Total construction				
	2005	2006	2007	2008	2009
Total of 12 Months	300 111	253 85	276 103	231 78	217 76
Percentage of accidents in construction	37,00%	33,60%	37,30%	33,80%	35,00%

3 REASONS AFFECTING SAFETY PERFORMANCE IN REHABILITATION

3.1 Background and general characterization

As part of a study carried out at University of Minho, entitled "Optimization of Management Rehabilitation Projects" (Araújo, 2009), a national survey was conducted, with the main objective of analyzing the main reasons for the failures in the management of works rehabilitation.

Based on the literature previously mentioned, it was made a selection of a set of 15 reasons deemed the best fit for this type of project. Through a survey materialized in form of a interview with 57 participants in this type of activity, representing 10 work owners, 20 designers and 27 contractors, with management and leadership positions, it was intended to figure out the reasons for most conditions and failures and poor performance safety verified in that kind of work.

3.2 Analysis of the survey results

Table 4 presents a summary matrix with the main reasons (ranking) for each group of participants, which allows analyzing and comparing the views of different groups on the subject under study (Araújo, 2009).

Through the analysis of table 4, it can be concluded that it is practically consensual that the difficulties associated with the use of labor-intensive,

low-skilled workmanship are the main reason for the no fulfillments in safety. Although with a lower degree of agreement between groups, the lack of a technical phase of work planning and organization also appears as one of the most prevalent reasons. There is equally the importance that contractors attach to the coordination and communication in the work, which certainly results in a more effective knowledge of its importance and consequences. The lack of research and observation of the target area by the contractor was also included among the main reasons, having this reason recorded a relative agreement among the various respondents.

4 CONCLUSIONS

Rehabilitation is, naturally, an added advantage for the prevention and preserving of the country's heritage. However, it seems equally obvious that this should not be developed to the detriment of the safety conditions of those involved. The costs with accidents are considerably higher than the costs involved in measures to foster safety. However, even with these arguments, some construction and rehabilitation companies are still not concerned about this problem or apply preventive measures, thus concurring to the fact that construction is one of the industries that contributes most to accidents in Portugal.

It can be concluded from the survey that the main causes that are associated with defaults and safety flaws in rehabilitation works cover all groups surveyed, as this is a cross-cutting issue for all interveners. The verification of the importance of factors as diverse as the use of skilled labor, shortage of technicians in the phase of work planning and organization, and lack of knowledge of where to work, illustrates well that only a contribution and effort of all sets could lead to a better safety performance.

Table 4. Summary of ranking with the main reasons for the no fulfillments in the safety by group of participants (Araújo, 2009).

Reasons affecting safety performance	Ranking		
	Work owners	Designers	Contractors
Little qualified and specialized workmanship	1	2	1
Shortage of technicians in the phase of work planning and organization	2	6	4
Poor communication and coordination between the various actors in the work		4	2
Lack of research and observation of the target area by the Contractor or inadequate inspection of the workplace	3	3	5

REFERENCES

Abdul-Rashid, I., Bassioni, H. & Bawazeer, F. (2007). Factors affecting safety performance in large construction contractors in Egypt. In: Boyed, D. (Ed), Procs 23rd Annual ARCOM Conference, 3–5, September 2007, Belfast, UK, ARCOM—Association of Researches in Construction Management, 661–670.

ACT (2012). Authority for the conditions of the work, Cabinet of strategy and planning. Accessed October 2012, available at: http://www.gep.msss.gov.pt/estatistica/acidentes/at2009sintese.pdf.

Araújo, J.D. (2009). Optimization of the management of Rehabilitation projects. Dissertation of master's degree in Sustainable Construction (in Portuguese). University of Minho, Guimarães.

Aksorn, T. & Hadikusumu, B.H.W. (2008). "Critical success factores influencing safety program performance in Thai construction projects". *Safety Science 46,* Asian Institute of Institute Technology, Pathumthani, Thailand, 713–715.

Couto, J.P. (2008). Influences in terms of accidents in the Portuguese Construction. *GESCON2008-International Conference on Construction Management, December,* FEUP, Portugal.

Egbu, C. Skills (1999). Knowledge and Competencies for Managing Construction Refurbishment Works. *Construction Management and Economics*, 17, 29–43.

Enshassi, A. (1996). Factors Affecting Safety on Construction Projects. *Department of Civil Engineering, IUG,* Gaza Strip, Palestine, 14–17.

Fang, D.P., Xie, F., Hung, X.Y. & Li, H. (2004). Factor analysis-based studies on construction workplace safety management in China. *Journal of Project Management,* Vol. 22. No 1, 43–49.

Fung, Ivan, W.H., Tam, C.M., Tung, Karen C.F., & Man, Ads S.K. (2005). Safety Cultural Divergences among Management, Supervisory and Worker Groups in Hong Kong Construction Industry. *International Journal of Project Management*, Vol. 23. No 7, 504–512.

Ng, S. Thomas, Cheng, Kam Pong & Skitmore R. Martin (2005). A Framework for Evaluating the Safety Performance of Construction Contractors. *Building and Environment,* Vol. 40. No 10, 1347–1355.

Rodrigues, M.F., Teixeira, J.C. (2009). Rehabilitation of Buildings Operations: Coordinating Health and Safety. *International Seminar on Occupational Safety and Hygiene*, Guimarães.

Wong, F.K.H, Chan, A.P.C., Fox, P., Kenny, T.C. & Easther, F.N. (2002). Identification of Critical Factors Affecting the Communication of Safety-Related Information between Main Contractors and Sub-Contractors in Hong Kong. *Research Project funded by occupational Health and Safety Council,* 1–2.

Occupational Safety and Hygiene – Arezes et al. (eds)
© 2013 Taylor & Francis Group, London, ISBN 978-1-138-00047-6

Specifics of the dams safety coordination

A.M. Márcio & M.R. Cristina
Universidade de Trás-os-Montes e Alto Douro, Vila Real, Portugal

S.O. Carlos
Instituto Politécnico de Viana do Castelo, Escola Superior de Tecnologia e Gestão, Viana do Castelo, Portugal

ABSTRACT: This research it's about the crucial importance of the construction sector that represents a wide range of activities with unique characteristics, thus involving specific risks that must be prevented by eliminating them at the source or minimizing their effects. It was conducted a research study in which aims to adapt and to proceed some corrections of the Safety Risk Management, for Building Security, a software from Tabique—Engineering, Ltd, intended for safety coordinators and technicians.

1 INTRODUCTION

1.1 Legislation

The Decree-Law n. ° 273/2003 of 29 October 2003, revises the regulations relating to safety and health at work at temporary or mobile construction sites. With this directive comes a document about safety planning in the project and construction phases. The health and safety plan should contain all relevant information and advice on safety and health, which are necessary to reduce the occurrence of accidents and to protect the health of workers. Thus, there should be preventive measures aimed at minimizing the risk in the elaboration tasks and measures to protect collective and individual to use. This study also presents the obligations of the safety coordinator in the project and construction phases.

1.2 Dams construction safety

Dams constructions are of great complexity that require coordination of health and safety very specific in order to prevent occupational risks inherent in this type of work. Because of the obligations of the safety coordinator and the characteristics of the type of construction, it's appropriate to conduct a research work that refers to the activity of the safety coordinator in works of dams, in order to ascertain their specificity for many specific constructions, such as dams, and less common in day-to-day. The theme of this research study is based on the safety coordinator activity, that have a major role on the application of safety and health matter at construction. It was decided to keep up with the construction

of the Hydro Plant of Foz Tua (HPFT), promoted by EDP—Gestão da Produção de Energia, SA ("EDP Produção"), along Tabique—Engineering, Ltd, which is responsible of safety coordination on that construction, in order to know the activities of the safety coordinator.

2 METHODOLOGY

2.1 Safety coordinator

Health and safety matter always have been a concern of humanity. It is estimated that 4 million years ago, the humankind already had some concern about safety in the elaboration of the tasks, otherwise, the mankind would have extinguished. There are several marks throughout history that proves the existence of minimum safety, specifically in the Middle Ages, when Georgius Agrícola conducted a study regarding the process of mining and smelting of precious metals, such as gold and silver, which emphasizes work accidents and the diseases that were common to the workers, called "miners asthma". It was proposed the use of protective masks in order to minimize the risk of lung's disease. After World War II, it began to develop a base of safety in enterprises in matters of production, which is currently acclaimed as the birth of integrated prevention. The Portuguese legislation is very exigent in the safety material, ensuring at the project phase (if the project has been developed by a project team), until the construction phase, in which the developer has the duty to nominate the Safety Coordinator in Project (SCP) and the Safety Coordinator in Construction (SCC).

2.2 Safety Coordinator in Project (SCP)

According to the Decree-Law nº. 273/2003 of 29 October, art. 9, paragraph 1, the safety coordinator in project is nominated by the developer, every time the project has been developed by more than one subject, containing such architectural options and complex techniques. It's also required a safety coordinator in project if the work to be performed on the basis of this project has special risks or, if there's a prediction involvement of two or more construction companies. If the conditions are not the mentioned above, the author of the project's role is to ensure coordination in the project phase.

The purpose of the appointment of the safety coordinator in project is to elaborate the Health and Safety Plan (HSP) at the project phase, to ensure the safety and health of all staff involved in the construction site, support the developer for the negotiation of the contract issues and the implementation of preparatory work with regard to safety and health matter. The safety coordinator in project has the duty to organize the technique compilation and to complete the most unfavourable situations, in case there is no safety coordination in construction (SCC). There should be special caution when scheduling tasks, in order not to overlap incompatible tasks in terms of space and time, in order to ensure proper management of simultaneous and successive work.

2.3 Safety Coordinator in Construction (SCC)

The contracting of the Safety Coordinator in Construction (SCC) is made in the event that there is intervention of two or more companies in the execution of the contract, by the developer. If eventually there is only one company in construction phase, it's only applied the normal regime of Health and Safety at Work.

There are a series of obligations relating to the SCC, briefly described:

– Ensuring good organization on the construction site, promoting other obligations of the contractor, subcontractor and independent workers;
– Ensure proper compliance regarding job scheduling;
– Ensure the fulfilment of the work program that invokes special risks;
– Ensure the adaptation of the Health and Safety Plan;
– Ensuring the proper functioning of the hierarchy when it comes to responsibilities;
– Organizing inspections visits to the various works, in order to observe the equipment in operation, materials and work processes;
– Organizing meetings between the coordination and the interveners in the construction site;

– Ensuring the technical compilation updates of possible project changes;
– Ensure the integration of the project owner in terms of relationships with the interveners of the work;
– Conduct inquiries into possible workplace accidents, analyzing the causes of the major accidents;
– Analyze the adequacy of safety procedures sheets, proposing if necessary some changes;
– Promote the fulfilment of HSP.

2.4 Prevention tool

It's on development a program created by Tabique—Engineering Ltd, to assist coordinators and safety technicians when it comes to risks in construction, called Safety Risk Management, for Building Security. Nicknamed by Risk, is a powerful software and an added value to the SCC work, because it makes the interconnection of tasks set forth in the project, to the various equipment and activities. It has as primary objectives, risk prevention and during the monitoring phase of the work, whether in future tasks to perform, as well as preparation of safety procedures sheets. So was carried out the interconnection of all tasks set out in the project, between April 2011 and March 2012. There were two studies: The first where all preventive measures were applied (Flawless Security System) and the second case, the actual conditions in the universe of the Hydro Plant of Foz Tua.

For the first case, for each task, were selected equipment to be used, by consulting the specific safety procedures (SSP) of the task. After the selection of the equipment/activity, was carried out for each of them, the selection of the most appropriate risks to the real situation and therefore, the correspondence and total application of preventative measures. The second case discussed, the procedure was partially identical, but the selection of preventive measures have been conditioned by the work visits inspection (WVI), i.e., no measures were applied to 100%, thereby leading to an approximate reality process.

3 SAFETY RISK MANAGEMENT, FOR BUILDING SECURITY

3.1 Risk classification

The software Risk is equipped with a large risk library, as well as preventive measures. Each risk is classifiable using a scale of 1 to 5, where 1 is the minimum risk, and 5, the maximum risk. To emphasize that no risk assumes the 0 value due to the fact that even with the prevention acting to the utmost, there will always be a residual risk. The degree of risk is obtained by combining the severity of the accident, with the occurrence probability. In the following table, is presented a perception of the risk degree.

After classifying the risk degrees, follows the aggregation of the various risks, with a relationship between them, called family of risk, which aims to gather various types of accidents in a single group, thereby simplifying the process. For instance, a car accident can be grouped along with pedestrian accidents, forming a family risk of type G. In the following table is presented the classification of risk families.

These degrees of risk will have relevance when calculating the risk's graph of the two study cases. The risk graph consists in three series: Maximum risk value (R_{max}), minimum risk value (R_{min}) and risk with applied measures (R_{MA}). For "standards" values, the minimum risk assumes the value of 1, the maximum risk is variable, $1 < R_{max} \leq 5$. The risk with applied measures is calculated by a mathematical formula, having as variables the total number of preventive measures, number of applied measures, minimum risk and maximum risk. To emphasize that in the realization of a work, there's simultaneous activities, as well as the use of different equipment, which results in a risk values sum, i.e., there isn't a proper scale minimum/maximum that can quantify it. The applied measures risk series, is only valid after the application of preventive measures.

3.2 Risks—preventive measures

After the selection of the desired project, but also the task to be performed, it's necessary to choose the equipment/activities adjacent to this task, through consultation of the specific safety procedure (SSP) of the task. After the establishment of groupings (equipment/activities), follows the risk correspondence phase to their preventive measures. Each grouping is characterized by its own risks, thereby varies the type of preventive measures to be implemented. Whatever the equipment/activity, it's possible to choose through the risk available, the risks to consider to the task, in order to fit the task in elaboration.

4 CONCLUSIONS

In the first case, for the Flawless Security System, after the interconnection task process of the equipment/activities, correspondence and full implementation of preventive measures to the risks inherent in these groupings, results in the following line chart:

It can be noted in Figure 1, the complicity of parallel curves—almost coincidental—between the minimum risk series and the measures applied risk series, which indicates that this is a 100% safety effective at work, i.e., all measures proposals are applied by workers and responsible. Note that the graph refers to the period from April 1, 2011 and January 1, 2012.

For the second case, depicts the actual state of construction, in which there's a greater divergence in the risk series.

Table 1. Risk degrees.

G	×	P	=		GR
Low		Low	Minimum		1
Low		Medium	Small		2
Low		High	Medium		3
Medium		Low	Small		2
Medium		Medium	Medium		3
Medium		High	Big		4
High		Low	Medium		3
High		Medium	Big		4
High		High	Maximum		5

Table 2. Risk families.

Types of accident	Risk family
Falls from height	A
Falls to the same level	B
Subsidience	C
Crush/Structures	D
Crush/Machinery	E
Electrocussion	F
Traffic	G
In Itenere	H
Other causes/aggression	I
External lesions	J
Internal lesions	K
Unknown causes	L
Diseases	M

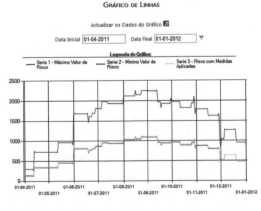

Figure 1. Line chart for the Flawless Security System.

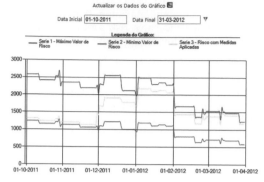

GRÁFICO DE LINHAS

Actualizar os Dados do Gráfico

Data Inicial 01-10-2011   Data Final 31-03-2012   ▼

Legenda do Gráfico:
Serie 1 - Máximo Valor de Risco   Serie 2 - Mínimo Valor de Risco   Serie 3 - Risco com Medidas Aplicadas

Figure 2. Graphical representation of the risks related to the Hydro Plan Foz Tua.

In this case, it's possible to verify a greater applied measures risk line fluency (yellow line), mainly from December. This month, the maximum risk (red line) tended to decreased, however, because of the preventive measures were not applied by the workers or by the responsible of the various work fronts, the level of measures applied risk increased.

4.1 Proposals for improvements

According to the development of the study, it was denoted that the graphic presented with a little fluency, the risk series of measures applied and that same series, never showed values below the minimum risk, which did not translate at all the reality. Faced with these results, it was developed a new calculation formula during the elaboration of this research, in order to correct the aforementioned problems.

$$R_{MA} = \frac{1}{Total_{Med}} \times \left[\frac{R_{max} - R_{min}}{Med_{Apl}} - 2 \times Med_{Apl} \right]$$
$$+ R_{min} + 1$$

Equation 1. Risk with applied measures.

Being:
R_{MA} – Risk with applied measures;
R_{max} – Maximum risk;
R_{min} – Minimum risk;
$Total_{Med}$ – Total number of preventive measures proposed;
Med_{Apl} – Number of preventive measures applied.

This formula was tested theoretically, and it had satisfactory results, showing a higher level of graphical fluidity and also showing for the first time, values below from the minimum risk.

Table 3. R_{MA} tests.

R_{min}	R_{max}	$Total_{Med}$	Med_{Apl}	R_{MA}	
1	1	10	1	1	Original formula
				1,8	Experimental formula
1	2	10	4	1,025	Original formula
				1,225	Experimental formula
1	3	10	7	1,028	Original formula
				0,628	Experimental formula
1	4	20	1	1,150	Original formula
				2,050	Experimental formula
4	5	20	4	4,013	Original formula
				4,613	Experimental formula
1	5	20	20	1,010	Original formula
				0,010	Experimental formula

Note to the last test, in which there is 100% correspondence of the measures implemented with a R_{MA} of 0.010, which is the residual risk value.

Besides the improve proposal for the calculation of the applied measures risk, it was suggested a new input, which is where introduces to a new variable of aggravation risk (with applied measures), during the delay of the application of the preventive measures. As a practical example, the Safety Coordination pays a visit to the building site, developing the report of that visit, which stipulates the immediate placement of guardrails in one particular building site place. Eventually, if the responsible, for a variety of reasons place the guardrails two days after the deadline, it will increase the risk, because the workers will be longer exposed to the risk of falling from height. The following equations were developed in this research, based on learning, in order to make the mechanism to work in practice:

$$Risk_{Aggravated} = N.^{o} Days \times 0,6 \times G.R. - \frac{G.R.}{2}$$

Equation 2. Aggravated risk.

Being:
$N.^{o} Days$ – Number of days of the preventive measure delay;
$0,6$ – Constant;
$G.R.$ – Risk degree, where $1 \leq G.R. \leq 5$
$Risk_{Real} = G.R. + Risk_{Aggravated}$

Equation 3. Real risk.

The final value obtained is the real risk that workers are exposed in the work, according to the number of days of the non-application of the preventive measure. In the Table 4, shows the values of the theoretic tests:

Table 4. Real risk test.

	N.º days	G.R.	Aggravated risk	Real risk
1. Test	1	2	0,2	2,2
2. Test	4	5	9,5	14,5
3. Test	0	4	−2	2
4. Test	0	1	−0,5	0,5

Note to the second test, which is given a risk level of 5 (considered maximum risk), with aggravation risk assuming the high value of 9.5, resulting in a real risk of 14.5. As expected, after 4 days overdue, a task which involves a special risk, for example subsidience, the real risk will assume astronomical values over the time, because the probability of occurring a serious accident increases exponentially, such as death.

To conclude, the Risk is always on development, implementing new preventive measures, as well as equipments/activities. It's in this direction that aims to a better handling for the common user and to a better program operation, so that in the near future, any safety coordinator may hold this powerful tool.

REFERENCES

Ministério da Segurança Social e do Trabalho—Decreto-lei n.º 273/2003, DR—I Série A, n.º 251, de 29 de Outubro de 2003, referentes as condições de segurança no trabalho em estaleiros temporários ou móveis que vem revogar o decreto-lei n.º 155/95 de 1 de Julho de 1995.
A.M. MÁRCIO (2012); Dissertação Mestrado em Engenharia Civil—Especificidades da Coordenação de Segurança em Barragens, Vila Real.
FREITAS, LUÍS (2001); Manual de Segurança e Saúde do Trabalho.
TABIQUE ENGENHARIA, LDA; Safety Risk Management for Building Security, Braga.
M. ALMEIDA SANTOS, FERNANDO; A. DIAS, LUÍS; Análise Económica de Riscos de Segurança na Construção: Um Exemplo Prático.

Occupational Safety and Hygiene – Arezes et al. (eds)
© *2013 Taylor & Francis Group, London, ISBN 978-1-138-00047-6*

Nonlinear analysis of incidents in small construction companies in southern Brazil

A. Falcão & I.G. Guimarães
Federal University of Pampa, Rio Grande do Sul, Brazil

M.P. da Silva
Federal University of Rio Grande do Sul, Rio Grande do Sul, Brazil

L.A.S. Franz
Federal University of Pampa, Rio Grande do Sul, Brazil

ABSTRACT: Over the last years the civil construction sector has shown expressive growth but at the same time presents a significant demand of its Occupational Health and Safety (OHS) comprehension and treatment. This study was conducted in a private construction company located in south of Brazil. The aim was to identify and characterize incidents involving injuries within the last three years. A model of causal accident analysis was applied and the results were extracted through personal interviews with the workers. It was verified that the high turnover rates and the low education of the workers do not seem to be the limiting factors. Evidence suggests that the lack of a proper training and day-to-day dialogue regarding OHS can be determinant factors. The greatest growth potential to advance in OHS can be achieved through better management practices, documentation, evaluation and propositions aiming to isolate the incidence of accidents.

1   INTRODUCTION

Among sectors of the Brazilian economy, the civil construction represents one in the highest growth in recent years (Fochezatto & Ghinis, 2011). At the same time, it has today one of the highest rates of accidents and fatalities according to Egle (2009). More than 54.000 accidents occurred in 2009 on the building industry (AEPS, 2009). Despite these facts, it is possible to see the urgency of several initiatives to improve Occupational Health and Safety (OHS) through government actions, using legislation and inspection. These initiatives include models for evaluating OHS performance, management structures and specific models for the incident analysis.

The incident analysis models that consider human error can provide contributions to OHS. This possibility was initially proposed by Reason (1990) and later adapted to the construction industry by Saurin et al. (2010). Besides incident analysis, the barriers classification model proposed by Rasmussen (1997) could provide great gains in terms of safety maintenance at the work environment. It is considered that human errors are symptoms of problems in a much deeper system and not only the cause of unwanted events (Saurin et al., 2010). These errors

can be identified and classified according to some patterns like Skill-Based Errors (SBE), Rule-Based Errors (RBE), Knowledge–Based Errors (KBE) and simple Violation (VIO) (Reason, 1997). Saurin et al. (2010) also consider the situation where the worker did not made a mistake, calling it a No-Errors (NOE).

Once one understands the paths that led to the incident and its boundary conditions, it is possible to propose and eventually to apply barriers that could prevent the recurrence of such events. Hollnagel (2004) classifies the barriers as Physical or Material (PB), Functional (FB), Symbolic (SB) and Immaterial (IB). Relatively to the Physical Barrier, the human action and the energy or mass transference are prevented by a material or physical limiting element. The Functional Barrier is dynamic and acts by preventing that an undesired action can be completed, using logical or temporal devices. The Symbolic Barrier requires personal interpretation of a conceptual element such as warnings or visual demarcations. In the case of the Immaterial Barriers, the previous knowledge of the worker acts as a barrier that allows the final task to be achieved with safety. As a complement, the conception of a dynamic model was presented by Rasmussen (1997) and divides the routine and environment in three zones: Safe Zone (SZ), Hazard Zone (HZ) and Loss of Control Zone (LZ). One study that incorporates these concepts and the application to the civil construction sector could bring some contribution in terms of understanding and preventing incidents.

This study aims to identify and characterize the injury incidents occurred in two construction companies from different regions from the state of Rio Grande do Sul in Brazil.

2 MATERIALS AND METHODS

This study was performed in two construction companies from different regions from the state of the Rio Grande do Sul in Brazil. The first company (Company A) is located near the central northern part of the state, and the second company (Company B) is located in the central region, and they are distant from each other approximately 220 km. The main activities on Company A involve residential buildings, residential condominiums and commercial buildings. The Company B besides residential and commercial buildings works with reforms and constructions of public companies.

During the data survey, 71 workers were interviewed individually, 35 of the Company A and 36 of the Company B. The data was based on the most important reminders of incidents with or without injury in the three years before the interview period. In the Company A the study was performed in two worksites on the second half of 2011. In the

Company B the visits were made in five worksites (buildings) in the second half of 2010 and first half of 2011. During the investigation each employee was asked to remember in detail the incident where he was involved.

There was no use of structured questionnaire or audio record for the dialogs. A procedure used during the interviews consisted in read for the interviewee the sequence of events taken from your account. The objective of this procedure was to identify if the information taken by interviewer was correct. Besides the interviews, the company's records of the incidents have also been consulted.

The next step consisted in the appliance of the algorithm for the human errors reports classification, in accordance to the model proposed by Saurin *et al.* (2010) and illustrated in Figure 1. This algorithm contains a flowchart with ten questions, leading to five types of possible final answers: slips (SBE-Skill-Based Errors), memory lapses (SBE-Skill-Based Errors), violations (VIO), Mistake (KBE-Knowledge-Based Errors) and not errors (NOE).

Each question of the algorithm, which is sequentially answered, is indicated in Figure 1 by a number. As each question is answered, a path that leads one of the types of errors is established. The numerical sequence obtained is shown in column Pathway, in Table 1.

The research methodology included both types of incidents (a) involving injury (accidents) and (b) not involving injury (near miss). According to Zocchio (2002) this approach is important because even the

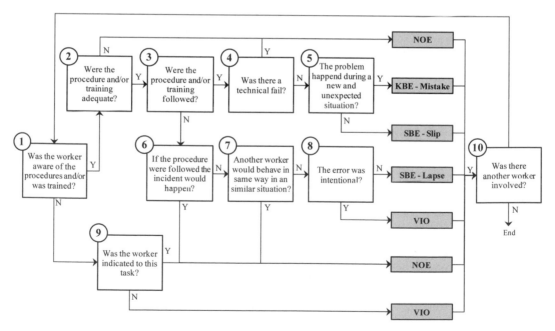

Figure 1. Flowchart proposed for analysis of incidents. Font: Saurin *et al.* (2010).

Table 1. Art of the worksheet where the incidents were compiled and complemented with the proposed barriers.

Type of error	Pathway	Loss control zone (LZ)	Barrier (PB, FB, S B, IB)	Hazard zone (HZ)	Barrier (PB, FB, SB, IB)	Safe zone (SZ)
1 NOE	1 > 2 > 3 > 6 > 7	He released the overloaded wheelbarrow and a piece of metal struck his hand	(PB) fix hand straps on wheelbarrow	Load excessively the wheelbarrow	(IB) training (PB) weight indicator in wheelbarrow (IB) Establish verify procedures	Carring the wheelbarrow at the builder rubble container
2 SBE-Lapse	1 > 2 > 3 > 6 > 7 > 8	Collided his face on scaffolding	(PB) coating to impacts (PB) delimitation of height for passage	Passed under the scaffold	(SB) boundaries indication (IB) standard procedure	Passed beside the scaffold
3 RBE-Mistake	1 > 9	Struck the hand with a power drill	(PB) more resistant gloves (FB) Automatic power off (SB) Orientation/ training	Practical joke in the workplace	(IB) Regras definidas pela empresa (IB) Treinamento e conscientização	Drilling the wooden wall
: :	:	:	:	:	:	:
51 SBE-Lapse	1 > 2 > 3 > 6 > 7 > 8 > 10	Water tank pressed fingers of the worker	(IB) locks during installation	Difficulties in the communication between workers and supervisor	(IB) Preplanning activity (SB) Communication by signals	Align the water tank under the direction of supervisor

near misses have potential to lead to severe injuries. The occupational incidents can affect the worker on different levels of gravity and consequences, as stated by Dalcul (2001).

The information was collected, summarized and analyzed and led to a statistical analysis of the causes of incidents, allowing for a pattern of possible situations to be improved. The proposed improvement suggested was based on the concept of barriers of Hollnagel (2004), and according to this author there are barriers that can help prevent accidents or reduce their consequences. The barriers were studied and applied in the incidents that this research could identify. Different types of incidents use different types of barriers. For example, new projects could use the barriers in a preventive way making the process to be more secure. Even after an incident occurs, the barrier can became a tool capable of preventing similar events and correct future project flaws.

3 RESULTS AND DISCUSSION

The survey was obtained from a total of 51 reports of incidents, including events with injury and no injury. Both Company A and Company B have 100% of male workers. Approximately 40% of

workers were employed in the Company A for less than 1 year. In Company B the number of employees ranged from 5 to 35 individuals during the period of occurrence of the incidents reported. The Table 1 presents part of the worksheet where the incidents were compiled and complemented with the proposed barriers.

Among the incidents investigated, it was found that about 27% of cases happened with no direct influence of the operator or No Error, being caused by organizational weaknesses in the company. In other words, causes of actions or decisions originated from other sources, such as lack of standard procedures, lack of training and security awareness both in Company A and B. This corroborates the perception previously given by Saurin et al. (2010) for a similar application of this work. The authors said that their survey indicated strong evidence that the OHS actions should be primarily directed to management, rather than being focused on the behaviour of workers.

Figure 2 presents synthesis of the percentage for each type of error identified after the analysis of the incidents. In Figure 2 it is possible to notice that most of the incidents (73%) occurred due to the operators human error: 37% of SBE (memory lapse: 25%, slip: 12%), 20% of KBE—Mistake and 16% of Violation occurrences. This demonstrates

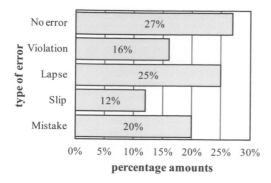

Figure 2. Percentage for each type of error.

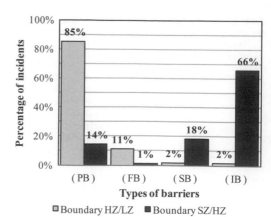

Figure 3. Percentage results obtained regarding the analyses of the types of barriers proposed after incidents.

the necessity of physical barriers and implanting immediate protection often simple and not expensive.

The use of barriers (PB, FB, SB, IB) made it clear from some features on accidents in Companies A and Company B that they could have been avoided through simple but effective measures. Based on the obtained data, it was established that despite companies had provided all required PPE (Personal Protective Equipment) for the safety of their workers, they did not give the necessary attention to other aspects associated with its use, such as appropriate training, supervision and raising awareness for the use. This can be seen from the fact that 25% of the incidents analyzed were caused by problems at the organizational level. Figure 3 summarizes the percentage results obtained regarding to the analysis of types of barriers proposed after incidents.

Another aspect that denotes attention was the large amount of occurred Violations, which demonstrate lack of perception of the importance on the correct execution of the activities for each task and even lack of skills by the senior management. The evidence presented above allows us to infer that the OHS management system is important in companies similar to those studied. Thus, compliance with regulatory standards and the use of protective devices by themselves do not sufficiently reduce the risks on construction sites. A greater commitment is needed in order to apply management models that really consider the risks and that preventively acts upon them.

Regarding to the procedure for incident analysis proposed in this work, some important aspects can be discussed. The flowchart used as a basis for analysis of incidents has two important benefits. Firstly, streamlined format allows for quick comprehension and memory for the analyst, which one gets after a short period of training and use, to develop a large number of analyzes without loss of reliability in the results. As it follows from that,

in a short period (a few months) it is possible for a company to obtain a history with large numbers of incidents analyzed. Another aspect relates to the use of paths in the flowchart which culminates with the possibility of a No Error, which is an important element for identification of weaknesses in terms of organizational management.

An important observation resulting from the applications in this study refers to the combined use of the classification model for errors proposed by Saurin et al. (2010) with the classification model for barriers proposed by Rasmussen (1997). The integrated use of the methods simplifies the surveying and proposition of barriers from the identification of the types of errors and their association with the transposition of boundaries (SZ/HZ/LZ). In addition, the structured worksheet presented in Figure 2 provides a mind map for the analyst that helps to organize the ideas and propose most effective solutions. Finally, after concluding the filling in of the worksheet, it is possible not just to make a quick crossing of information, but also perform an counting down which brings the possibility to perform statistical analysis.

4 CONCLUSIONS

It was observed from the field survey that immediate protective actions may contribute in some degree to reduce incident rates in the studied companies. However, the high occurrence of violations and errors in the level of knowledge indicate the need of greater attention and care with the implementation of more structured management models regarding the OHS. Even with the high turnover of employees in this sector, the use of preventive measures in the management can bring considerable reduction of unsafe acts by employees which even immediate protection devices could not be fully effective.

Regarding the method used, it was identified as benefit the simplification of the analysis without loss of reliability in results, and the possibility of improving proposals more quickly and easily. Furthermore, the proposed method offers the possibility of statistical analysis of the data, especially in cases where there is a large amount of incidents analyzed.

ACKNOWLEDGMENTS

The authors would like to thank Gustavo Schabbach and Ricardo Augusto Maurer by survey and availability of data used in this research.

REFERENCES

AEPS. (2009). Anuário Estatístico da Previdência Social. Base de Dados. Retrieved May 21, 2011, from http://www.previdencia.gov.br.

Dalcul. A.P.C. (2011). Estratégia de prevenção dos acidentes de trabalho na construção cívil: uma abordagem integrada construída a partir das perspectivas de diferentes atores socias. Doctoral Thesis, Business Administration Department. Porto Alegre: Universidade Federal do Rio Grande do Sul, Brazil.

Egle, T. (Dez. 2009). Radiografia da (in)segurança. São Paulo: Revista Techné. (53th ed.).

Fochezatto, A.; Ghinis, C.P. (2011). Determinantes do crescimento da construção civil no Brasil e no Rio Grande do Sul: evidências da análise de dados em painel. Ensaios FEE, vol. 31, Porto Alegre.

Hollnagel, E. (2004). Barriers and accident prevention. London: Ashgate.

Rasmussen, J. (1997). Risk Management in a Dynamic Society: a modeling problem. Amsterdam: Safety Science.

Reason, J. (1990). Human Error. Cambridge: Cambridge University Press.

Reason, J. (1997). Managing the Risks of Organizational Accidents. Burlington: Ashgate.

Saurin, T.A.; Costella, M.G.; Costella, M.F. (2010). Improving an algorithm for classifying error types of front-line workers: Insights from a case study in the construction industry. Safety Science, 48(4), 422–429.

Zocchio, A. (2012). Prática da Prevenção de Acidentes: ABC da segurança do Trabalho. São Paulo: Editora atlas, 7 ed.

Economic aspects of prevention

Occupational Safety and Hygiene – Arezes et al. (eds)
© 2013 Taylor & Francis Group, London, ISBN 978-1-138-00047-6

Organization and use of information to calculate ergonomics financial benefits—a hospital case

M.P. da Silva, V.A.S. Zizemer, M.C. Louzada & F.G. Amaral

Production Engineering and Transportation Department, Federal University of Rio Grande do Sul, Brazil

ABSTRACT: A well-known fact about ergonomics is that the positive effects of its interventions are difficult to be estimated. This represents a barrier to this kind of project in business environment, mainly by the lack of proper financial justification. Although many companies have a good level of information to calculate the cost-benefits of ergonomic projects, usually the necessary information are not complete or organized. In this context, the aim of this paper is to analyse if the organization and the information use regarding ergonomics cost-benefit analysis in a Brazilian hospital is sufficient to justify the required investments. The study was carried out through a systematic model to support the cost-benefit analysis of ergonomics projects. The main outcomes showed that not all the necessary information for a proper quantification of ergonomics benefits was present in the institution. Besides that, the hospital managers seem to be unfamiliar with the possibility of justifying investments in ergonomics projects through administrative and occupational health indicators.

1 INTRODUCTION

A new way for the inclusion of ergonomics in the business community is discussed, and takes into account the need for a financial justification for its implementation. One of the main discussions is that the companies usually do not approve projects that involve some investments without expecting significant results (Kirwan, 2003). Searching for a response to this situation, researches in the area investigate the performance of ergonomic intervention for well-known occupational problems through an economical perspective.

Authors such as Simpson (1990), Anderson (1992), Oxenburgh (1997), Grozdanovic (2001), Beevis & Slade (2003), Hendrick (2003), Morse (2009) and Looze et al. (2010) presented successful cases of ergonomic projects using the cost-benefit approach. These scientific works aimed to supply new information to decision makers through the evidences synthesis founded in studies about financial aspects of ergonomic projects (Goosens; Evers, 1997; Tompa et al., 2009; Nelson; Hughes, 2009; Neumann; Dul, 2010). In general, the focus was the proof that improvements in working conditions may predict the occupational problems reduction that result in cost savings.

Recent researches regarding ergonomics cost-benefit have been undertaken to deal with occupational diseases, absenteeism, turnover, low employee performance and process quality issues (Sommerich, 2003; Collins et al., 2004; Audrey et al., 2006; Driessen et al., 2008). Some researches use governmental worksheets to estimate financial losses due to occupational diseases related absenteeism. Others calculate the cost-benefit relation through historical data regarding quality issues, for example. Nonetheless, an organized form of acquiring and using this kind of information is still unknown for most organizations.

The investment required to start up ergonomics projects can be calculated in a relatively easy way. However, calculating their financial benefits is one of the major difficulties encountered by researchs area. The difficult task to gather relevant information to identify or predict these benefits of ergonomic design has some barriers.

One of the main barriers is the absence of information organized in a logical way for this purpose. Hence, a miscalculation of possible financial benefits of ergonomic intervention caused by the lack of the right set of information is common among companies with different complexity levels. The aim of this paper is to analyse if the organization and information use regarding to ergonomics benefit calculation in a hospital is sufficient to justify the required investments. In a practical way, it is aimed that the firm studied can have conditions to perform financial analysis of projects focused on ergonomics, as well as it already carries out on other matters. Thereby, the improvements in the working conditions may be justified, receiving support from the managers.

2 MATERIALS AND METHOD

To attend the purpose of this paper, a systematic model to support the cost-benefit analysis of ergonomics projects (Silva, 2012) was applied in a Brazilian hospital. The systematic model consists on a framework based on literature review about ergonomic interventions considering cost-benefit analysis in their evaluation. The model objective is to enable the use and application of financial evaluation to the ergonomic projects in enterprises. The main hypothesis is that with the right set of information and projecting possible benefits any ergonomics intervention can be evaluated in the same way as other technology fields. Following the suggestions of the systematic model, ergonomic project can be feasible through cost-benefit analysis and could justify the necessary investments. With that the main barrier that prevents the cost/benefit analysis can be solved, even if partially.

To achieve such goals it is not necessary that the institution have experience in this context because of the model structure and its practical application that involves four steps: (a) introduction of the concepts, considering that those hypotheses are not well known among companies practitioners; (b) definition of the responsible people, for different perspectives between sectors; (c) checklist, to actually compare the possible range of information listed in the model with the indicators that the institution uses in the management; and (d) improvements possibilities and suggestions, for indicate types of information and organization that could facilitate the cost-benefit analysis on the institution ergonomics projects.

In order to start the development of the research into the institution, the main theoretical concepts are presented (a) to a group of managers. The topics that should be discussed are: ergonomics definitions, intervention approach, possible benefits, cost-benefit analysis and published success cases. Questioning and discussion about the topics should be encouraged in order to level the ergonomics understanding.

After that, the definition of the responsible people (b) for each department or sector information takes place. The checklist information is the basis to define the responsible people.

The third step (c) is compounded of individual interviews to feature the information's logics of use by the institution and require the participation of people representing the main sectors such administration, human resources, health and occupational safety. Here, the research investigation is based on the systematic model's checklist.

The Table 1 presents the financial and non-financial information that compose the checklist. Three categories separate the specific data into productive, administrative and occupational health aspects.

Table 1. Checklist of financial and non-financial data (Silva, 2012).

Category of data	Non-financial data	Financial data
Productive aspects	Productivity Work days missed Products with errors, defects or returns Task cycle time	Cost of substandard performance Rework costs
Administrative aspects	Turnover overtime	Compensation cost due to work leave Administrative procedural costs for new hires Judicial proceedings, fines, and compensation costs due to injuries or illnesses Cost of training new workers
Occupational health aspects	Absenteeism due to injury or illness Loss of time due to injury or disease Prevalence and severity of injuries and disease Prevalence of physical pain	Costs of treatment of illnesses or injuries (physicians, tests, prescription drugs, physical therapy, transport)

This was necessary to avoid misunderstandings with the responsible people, but the categories were not exclusionary regarding the interviews. Rework costs data, for example, was questioned even for administrative responsible people. Every type of information on the checklist should be inquired to the respondent in order to understand how the institution manages them and how ergonomics projects can be related to each indicator. The details of the origins of information should also be recorded.

The checklist information that is not used or it is not known on the institution system can indicate the improvements possibilities and suggestions (d). This is a delicate step because the relation of absence of information and suggestion is not direct. It is important to discuss two questions with the responsible people: (1) it is possible to implement the suggested information into an institution's management system? and (2) the information already known on the institution it is considered on the ergonomics cost-benefit analysis context?

The discussion on whether the institution has or not the information conditions to perform cost-benefit analysis on ergonomics projects was based on the results achieved by the research team.

3 RESULTS AND DISCUSSION

The hospital in study is located at Rio Grande do Sul state, south of Brazil. Its facilities consist of general hospital that includes almost every medical specialty, providing services to the community and counting with 2,800 workers. It's a university hospital and besides the medical assistance to the society one of the main objectives is to promote and integration of graduation and post-graduation courses and research groups.

The hospital was undergoing a phase of improvement in its strategic planning and indicators system because of a quality award and an accreditation company. A quality office was created to command these changes that include monthly meetings with managers and leaders from all the areas. The objective of the meetings is introducing the information, carry out training and integrate data from each area.

After initial contact asking for its participation, the researching project was approved by the director of the hospital. The unit was visited by the researchers in the last quarter of 2012 and the method was applied. First, the introduction concepts regarding the financial benefits of ergonomics intervention was presented based on international publications such as Engst et al. (2005) and Chhokar et al. 2005 who also studied hospitals regarding the financial benefits of ergonomic interventions. The definition of which managers were responsible for each type of company data as well as the checklist appliance was conducted at this stage. Two managers were selected to respond to the checklist. The managers are responsible for the administration and occupational health and safety sectors. Open interviews were conducted to gather information that could characterize the management context. After that, suggestions to improve the possibility of cost-benefit analysis of ergonomics projects were explained.

The major problem was identified as the high level of absenteeism, mostly by musculoskeletal disorders (20%). The problem of absenteeism seems to happen like a snow ball effect in the hospital. When a nurse does not go to work, the other nurses in the sector suffer with overload. Then, these nurses who accumulate overwork tend also to absenteeism.

In the hospital practice there is no culture of occupational accidents prevention, and Ergonomics is not viewed by the management as a potential form of improvement. Simultaneously, the hospital has a high turnover rate and that is commonly used as an argument to deny the cause-effect of most of the lesions and occupational diseases of their own workers. This relation seems to characterize some extraordinary case but labor claims while the worker is still on practice is very common in the institution.

Regarding the information checklist, it is important to mention that the hospital areas have peculiar characteristics and that the centralization of data is only used by the high management to administrative aspects. Analysis of improvement projects on any field of knowledge do not seem to count with this centralized data.

The table 2 shows the main results achieved by the research after the interviews guided by the checklist.

Table 2. Main results of the hospital data.

Productive non-financial data
Bed occupancy rate is used as productivity information. Infection rate and health care numbers are also related indicators. However information about loss of work days, task cycle time, errors and rework are not known. This last information are generally more related to the industry field.

Productive financial data
Financial results based on the number of health care and the bed occupancy is the main data registered in this context. As mention before, information regarding rework and its related costs are not accounted.

Administrative non-financial data
Turnover rate, overtime, workers satisfaction level, development level of the managers and many others indicators that are related to management issues and academic courses and research groups. Although this information rarely is considered as an output of poor working conditions.

Administrative financial data
Compensation cost, hiring and training costs, as well as the costs of legal proceedings against the institution. Some of the administrative costs information is just disorganized, but can be calculated through crossing different areas information. Compensation costs, for example, are added to training costs because both are related to overtime.

Occupational health non-financial data
Absenteeism rates and the prevalence of injury and illness. This information is registered by the occupational health and safety department and supplies the quality office. Other information such prevalence of physical pain or loss of time it is not controlled indicators.

Occupational health financial data
Costs of injury or occupational diseases treatment are unknown too. The hospital does not have a health care plan but provides 50% of the cost of a particular health plan to the workers.

It seems to be a cultural restriction when themes like ergonomics evaluation or ergonomics intervention are brought to hospital environment. The hospital that is subject of this paper has similar problems. All the respondents of this research have mentioned that senior managers block almost all the ergonomics projects inside the institution. The root causes for that is not well known but securely relates to occupational safety and health within medical hierarchy and the hospital image in the marketing. Also can be argue that an ergonomics project that presents at least a financial benefits estimate may be a solution to this situation.

The suggestions indicated to the hospital management included not just the creation of important indicators such as costs of treatment of illnesses or injuries, but also a more flexible and integrated data system that could be used and feed by all sectors.

One of the difficulties encountered in the researched hospital was the lack of specificity of the data, in terms of detachment of sectors and working conditions. It is known that a more complete set of information could provide more reliable quantifications and calculations (Grozdanovic, 2001). It should be noted, however, that the barrier on the gathering of information is not a specific problem of Ergonomics, as it is shared with every discipline related to production process management (Rouse; Boff, 1997).

The main outcomes showed that most of the necessary information for a proper quantification of ergonomics benefits was present in the institution. In other words the hospital has the information conditions to perform cost-benefit analysis on ergonomics projects. At the same time the hospital managers seem to be unfamiliar with the possibility of justifying investments in ergonomics projects through administrative and occupational health indicators.

Studies aiming at the benefits of ergonomics intervention on hospitals show positive results even considering only a few types of information. For instance, Franzini et al. (2011) evaluate the costs and cost-effectiveness for the implementation of a telemedicine intensive care unit, concluding that it would decrease hospital mortality without increasing costs significantly. Other examples may be evaluation of an investment in a specific device to reduce healthcare worker injuries (Chhokar et al., 2005) or a musculoskeletal injury prevention program (Collins et al., 2004).

4 CONCLUSIONS

This paper aimed to analyse the organization and use of information regarding ergonomics benefit calculation in a hospital. The institution presented a relevant level of diversity of information types, but a low awareness of the value of this information regarding ergonomics benefits quantification. Some suggestions were made in order to facilitate the cost-benefit analysis and promote the inclusion of ergonomics in the hospital environment.

ACKNOWLEDGMENTS

This research was supported by CNPq, Brazilian National Counsel for Technological and Scientific Development.

REFERENCES

Andersson, E.R. 1992. Economic evaluation of ergonomic solutions: Part II—The scientific basis. *Int. J. of Ind. Ergon.* 10: 173–178.

Audrey, N., Matz, M., Chen, F., Siddharthan, K., Lloyd, J., Fragala, G. 2006. Development and evaluation of a multifaceted ergonomics program to prevent injuries associated with patient handling tasks. *International Journal of Nursing Studies*: 43(6).

Beevis, D., Slade, I.M. 2003. Ergonomics—costs and benefits. *Appl. Ergon.* 34: 413–418.

Collins, J.W., Wolf, L., Bell, J., Evanoff, B. 2004. An evaluation of a 'best practices' musculoskeletal injury prevention program in nursing homes. *Injury Prevention* 10: 206–211.

Chhokar, R., Engst, C., Miller, A., Robinson, D., Tate, R.B., Yassi, A. 2005. The three-year economic benefits of a ceiling lift intervention aimed to reduce health care worker injuries.

Dahlén, P., Wernersson, S. 1995. Human factors in the economic control of industry. *International Journal of Industrial Ergonomics* 15: 215–221.

Driessen, M.T., Anema, J.R., Proper, K.I., Bongers, P.M., van der Beek, A.J. 2008. Stay@work: participatory ergonomics to prevent low back and neck pain among workers: design of a randomized controlled trial to evaluate the (cost-) effectiveness. *Bmc Musculoskeletal Disorders* 9: (145).

Engst, C., Chhokar, R., Miller, A., Tate, R.B., Yassi, A. 2005. Effectiveness of overhead lifting devices in reducing the risk of injury to care sataff in extended care facilities.

Franzini, L., Sail, K.R., Thomas, E.J., Wueste, L. 2011. Costs and cost-effectiveness of a telemedicine intensive care unit program in 6 intensive care units in a large health care system. Journal of Critical Care 26: 329e1–329e6.

Goossens, M.E.J., Evers, S.M.A.A. 1997. Economic evaluation of back pain interventions. *J Occup Rehabil* 7(1): 15–32.

Grozdanovic, M. 2001. A framework for research of economic evaluation of ergonomic interventions. *Economics and Organization*: 1(9): 49–58.

Hendrick, H. 2003. Determining the cost-benefits of ergonomics projects and factors that lead to their success. *Appl. Ergon.* 34: 419–427.

Kirwan, B. 2003. An overview of a nuclear reprocessing plant human factors programme. *Applied Ergonomics* 34: 441–452.

Looze, M.P., Vink, P., Koningsveld, E.A.P., Kuijt-Evers, L., Van Rhijn, G.J.W. 2010. Cost-effectiveness of ergonomic interventions in production. *Human Factors and Ergonomics in Manufacturing & Services* 0: 1–8.

Morse, M., Kros, J.F., Nadler, S.S. 2009. A decision model for the analysis of ergonomic investments. *International Journal of Production Research* 47(21): 6109–6128.

Nelson, N.A., Hughes, R.E. 2009. Quantifying relationships between selected work-related risk factors and back pain: A systematic review of objective biomechanical measures and cost-related health outcomes. *International Journal of Industrial Ergonomics* 39: 202–210.

Neumann, W.P., Dul, J. 2010. Human factors: spanning the gap between OM and HRM. *International Journal of Operations & Production Management* 30(9): 923–950.

Oxenburgh, M.S. 1997. Cost-benefit analysis of ergonomics programs. *American Industrial Hygiene Association Journal* 58: 150–156.

Rouse, W.B., Boff, K.R. 1997. Assessing cost/benefits of human factors. In: Salvendy, G. *Handbook of human factors and ergonomics*: 389–401 John Wiley (ed); Sons, New York.

Silva, M.P. 2012. Proposition of a systematic model to support the cost-benefits analysis of ergonomics projects. (Proposta de sistemática de apoio para análise de custo-benefício de projetos ergonômicos). Doctoral Thesis, Production Engineering and Transportation Department. Porto Alegre: Universidade Federal do Rio Grande do Sul, Brazil.

Simpson, G.C. 1990. Costs and benefits in occupational ergonomics. *Ergonomics* 33(3): 261–268.

Sommerich, C.M. 2003. Economic analysis for ergonomics programs. In: Karwowski, W., Marras, W.S. *Occupational Ergonomics: Design and Management of Work Systems*. CRC Press, New York.

Tompa, E., Dolinschi, R., Oliveira, C., Amick III, B.C., Irvin, E. 2009. A systematic review of workplace ergonomic interventions with economic analyses. *J. of Occup. Rehab.* 20: 220–234.

Occupational Safety and Hygiene – Arezes et al. (eds)
© 2013 Taylor & Francis Group, London, ISBN 978-1-138-00047-6

Costs of safety at work vs. costs of "no" safety at work—Building Sector

Pedro Monteiro & Fernando A. Santos
Tabique Engenharia, Braga, Portugal

Gilberto Santos
Escola Superior de Tecnologia do Instituto, Instituto Politécnico do Cávado e do Ave,
Campus do IPCA, Lugar do Aldão, Barcelos, Portugal

ABSTRACT: The present work focuses on the analysis of data from the construction works of the Modernization of a Secondary School for the beginning of the time period July 2009 to end-September 2010, corresponding to the total of the Work. The investment in prevention of the safety at work, were collected from the data for safety costs, involving human and material resources. From the point of view of the analysis of Risk Assessment, it can be concluded that the Work of Modernization of a Secondary School was associated with a high probability of occurrence of serious accidents or death. As a result, it appears that accidents and incidents were largely caused by human error. It becomes imperative to invest more in prevention and to reduce the economic impact that workplace accidents originate.

1 INTRODUCTION

Heinrich (1931), based on data collected from insurance companies, started the analysis of the costs of accidents and divided them into two classes: direct (if reimbursed by insurance) and indirect (non-reimbursed). Also according to Heinrich (1931) indirect costs would total about 75% of total costs. The iceberg metaphor used by the author to represent the hidden part of the indirect costs, has been since then used by the technical and scientific community. With a new concept, Andreoni (1985), conducted a study in which developed a methodology that analyzes in an integrated way, the costs of accidents and damage, from which are developed additional preventive actions. Introducing a new vision, Pidgeon (1991) suggested that organizational culture is a system of ordinary meaning, that is a system of symbols, ideas, rules and cognitions, with certain observable behaviour and consequences. Moreover, he defined safety culture as "a set of beliefs, norms, attitudes, roles, social practices and techniques that are concerned with minimizing the exposure of workers, managers, customers and public members to conditions considered dangerous or harmful." This meaning is important and legitimate for the group, because this safety culture is created and recreated by the group members repeatedly, so get an idea of natural hazards and safety in the construction industry. In Australia, Holmes et al. (1999) found that the risk of falling from height was

mainly attributed to economic pressure and time, and that such pressure was coupled to prioritize production over safety. Many years after Heinrich cost ratio direct/indirect 1 to 4, Pastore (1999), conducted an analysis of the economic dimension of occupational accidents and diseases in Brazil and obtained a relationship between insured and uninsured costs of 1 to 5. In addition to variables such as lost time, cost of first aid, destruction of equipment and materials, interruption of production, training of skilled manpower, replacement workers, payment of overtime, recovery workers, wages paid to workers apart, administrative expenses, benefits in kind and repair costs, also included the costs incurred by families. From a sectorial manner, Rocha (2000), verified that the building is intended, primarily as a profit for their companies and often, the chosen form for higher profits occurs through the reduction of costs, one of which is the safety. As some professionals do not understand the impact of Work Safety in productivity of the company, often it is left to the background. Corcoran (2002), argues that it is imperative to create a safety culture within companies and organizations, because this will be the most effective way of reducing accidents and consequently, the costs associated with it. According to Lima (2003), the costs of medium and long term—those that are more difficult to identify for the company—may be much higher than the short-term costs. The company cannot remain competitive, for example, with the loss of reputation and motivation of workers, as a result of the history of accidents. Moreover, human resources are the most valuable resource of any company or country, but not always the most valued. Consequently the greatest asset of any organization, any region or any country, are people and their know-how (Santos et al., 2011). On the other hand Waehrer et al. (2007) concluded that in the United States, the total costs of fatal and non-fatal accidents in the building industry was estimated at 11.5 billion dollars in 2002. The average cost per case of fatal and non-fatal accident is $ 27,000 in construction, almost double the cost per case of $ 15,000 for the entire industry in 2002. More recently, Torner (2009), concludes that the complexity of building works require a good and comprehensive safety management, and emphasizes the need for standards to guide staff in many decisions that need to take to do their job, and for the individual commitment of each employee with safety. Safety standards defined by climate of security concerns shared perceptions of policies, procedures and practices in relation to safety in the organization and are therefore "property" of workers, but the antecedents are largely "owned" management. According to data from the International Labour Organisation (ILO) (www.ilo.org/global/) annually lose their lives over 2 million workers, occurs 270 million occupational accidents, more than 1 million workers are disabled and about 160 million contract diseases from causes directly related to work. Within the European Union, it is estimated that there are about 4.6 million accidents per year, which cause the loss of about 146 million work hours, according to data from the European Agency for Health and Safety at Work (EU-OSHA). According to the balance made by GEP—Office of Research and Planning of the Ministry of Labour in Portugal (http://www.poatfse.qren.pt/)—the occupational accidents in Portugal in the last decade, more specifically between the periods 1998 to 2007 found that there were 2773 deaths from 2,269,243 workers in workplace accidents which, led to the loss of 58,230,087 days of work. Despite the concern that these issues are raised, and constant bet on prevention measures, reveals that exposure to elevated risks, the occurrence of accidents and the development of occupational diseases continue to be a reality. Thus, we can conclude that it is possible to do more for workplaces even more safe. According Santos (2013) who carried a study about the certification of Safety Management Systems, in Portuguese SMEs concluded that the main reason for certification mentioned by companies was *"Valuing human capital—Eliminate or minimize risks to workers"*. The aim of this study was to quantify the economic cost that accidents represent in the Building Sector.

2 MATERIALS AND METHODS

The present work focuses on the analysis of data from two actual construction works, and the case study is made on the data of the Work of "Modernization of Secondary School Camilo Castelo Branco-CCB" in Barcelos for the beginning of the time period July 2009 to end-September 2010, corresponding to the total of the Work. It was considered the data from the first nineteen months of the Work of the General Construction Contract, ie, beginning in April 2010 until the end of November 2011 the Work of "Strengthening Power of Venda Nova-Venda Nova III" in Ruivães, municipality of Vieira do Minho, which will be finalized in mid 2015. To identify the investment in prevention of the safety at work, were collected the data for safety costs in the two works under review, involving human and material resources, which allow us to conclude the sum, a percentage of the total cost spent on safety at work against the total cost of the contract, designated Safety Costs (SC). The identification of the Anomalies Cost Correction (ACC), was done by the method of Probabilities and Consequences adapted to Case Studies, to examine all the anomalies of Registers of Inspection Visits (where are described anomalies found at the time in the Work, which were sent to the Contracting Performer for knowledge), Registries Non conformance and registers of accidents and incidents. The procedure was for the suggestion, more economic, of correction of deficiencies found and anomalies that caused accidents and incidents. The total economic cost of these suggestions for improvement will be the Anomalies Cost Correction (ACC). The Safety Costs (SC) added to the Anomalies Costs Correction (ACC) will be the Cost Desired Safety (CDS). CDS = SC + ACC. Thus, Cost Desired Safety (CDS) will be the investment necessary for the Work to be considered Safe. The Costs of "No" Safety at Work (CNS), will be the sum of the Direct Costs (DC), ie, reimbursable by insurance and Indirect Costs A (IC), ie, not reimbursable by insurance. A methodology was followed according Reis (1998). If the Anomalies Costs Correction (ACC) is less than the costs of "No" Safety (CNS), we reach the conclusion it pays to invest in safety against the risk of potential costs of "No" Safety (CNS). Once the accidents have significant impacts on workers, organizations, and society in general, it was intended with the elaboration of this study, to quantify the economic cost they represent of the Building Sector. It was intended to still demonstrate that the lack of investment of safety at work potentiates work accidents and incidents/near accidents and thus pays to invest in the Safety at Work in Building, because the costs of "no" safety can be potentially higher than the costs of safety.

3 RESULTS AND DISCUSSION

3.1 Safety costs

The present work focuses on two case studies mentioned above. Table 1 is presented only one example (remodeling Secondary School CCB) of how the results are presented and discussed.

In table 1 is the ratio of investment in Safety at Work by the Entity Performer and "The owner of Work" on the CCB School with the total value of the Work, verifying that the costs are 240,508.00 € are the Safety Costs (SC) at Work invested by the entities responsible for executing of the Work. These represent an incidence of 2.42% of the total cost of the same.

Having been invested 43,828.00 € in Safety at Work by "The owner of Work", these represent an incidence of 0.44% of the total cost of the Work. The 196,680.00 € invested in Safety at Work by the Entity Performer representing an incidence of 1.98% of the total cost of the Work representing 9,956,356.00 €. Of these 196,680.00 € are invested in Safety at Work by the Entity Performer, it is emphasized that € 130,550.00 are invested in work allocated to Safety, which corresponds to 66% of the investment.

3.2 Analysis of the risks and cost of correcting anomalies

After analysis of 197 total anomalies of the Work of the School Refurbishment CCB, it was obtained the sum of Cost of Anomalies Correction 4193.22 Euros. The sum of Anomalies Cost Correction that caused four accidents correspond to 14.10 Euros, which demonstrates a lack of investment that potentiate the occurrence of accidents at work. In the analysis of 197 anomalies, highlight the incidence of the risk of fall in height, with 87 occurrences, of which 77 occurrences with the Intervention Level I (88.50%). A total of 197 Risks Analyzed, 133 correspond to with the Intervention Level I (67.50%).

Tabel 1. Safety Costs—CCB (Monteiro, 2012).

	Total work	Total spent on safety
Months	Total Value of Work	The sum of Safety Costs of Entity Performer and The Owner of Work
14	9,956,356.00 € (X)	240,508.00 € (Y + Z)

Incidence of safety costs – Entity Performer + The Owner of Work = ((Y + Z) / X) × 100

Incidence of safety costs – Entity Performer + The Owner of Work = 2,42%

3.3 Safety costs desired, degree of safety implementation at work and cost of safety desired in the school CCB: Pessimistic and optimistic

Safety Cost Desired (SCD), SCD = CS + ACC, is the investment required for the Work to be considered safe. Thus, the 244,701.22 € for the Cost Desired, is the cost for the work was considered safe. The implementation level of Safety at Work, called Effective Safety (ES) = (CS × 100)/SCD, in the Work of the School Refurbishment CCB is 98.29%. With regard to the Cost of "no" Safety at Work (CNS) of the School CCB can analyze with a pessimistic view (death and some accidents resulting in a few working days off) and an optimistic (some accidents working days off and accident free days off) for each work. Of the pessimistic point of view, to the Work of the School Refurbishment CCB, is used the ratio of cost of safety for the company 5.60 times higher than the Anomalies Costs Correction (ACC), because it is a work of pure edification. From the optimistic point of view, was used the ratio of non-safety cost 0.40 times higher than the Anomalies Costs Correction (ACC).

4 CONCLUSION

From the point of view of the analysis of Risk Assessment, it can be concluded that the Work of the School Refurbishment CCB is associated with a high probability of occurrence of serious accidents or death. Thus, of the pessimistic point of view of the Anomalies Costs of the "No" Safety at Work should be seen as the likely situation to occur. Thus, it can be concluded that it pays to invest in the Cost of Correction for a 100% safety called "Desired Safety" because the Anomalies Costs of "No" Safety can be 321.20 times more expensive than investing in Anomalies Correction in that refurbishment work of the School CCB. As a result, it appears that accidents and incidents were largely caused by human error, where the preventive measures to correct the anomalies represent an investment of safety of 14.1 Euros. The lack of such investment potentiated workplace accidents and incidents/near accidents. Thus, the decrease in Costs Correction and Costs of "No" Safety is related to the possible causes for the occurrence of anomalies, because if we fight or we eliminate at source the cause of the anomalies, the cost of anomalies correction will be substantially less, as well as, the number of accidents and its associated cost. This study can be used to attract attention of people involved in Building Sector in particular Owners Work, Performer Entity Representatives, Inspection, Coordination of Safety, Managers and

Administrators of Building Companies in general. It is understood that, in a society where it is urgent to produce quality goods and services at the lowest possible cost, it becomes imperative to reduce the economic impact that workplace accidents originate.

REFERENCES

Andreoni, D. (1985). Le Coût des Accidents du Travail et des Maladies Professionnelles, Série Sécurité, Hygiène et Médecine du Travail, n.º 54, Bureau International du Travail.

Corcoran, D.J. (2002). Are accident costs like icebergs: The hidden value of safety, Occupational Health & Safety, June.

GEP—Gabinete de Estudos e Planeamento do Ministério do Trabalho (acesso em Fev. 2012) (www.poatfse.qren.pt/)

Heinrich, H.W. (1931). *Industrial Accident Prevention*. New York-McGraw Hill Mac.

Lima, Francisco. (2003). Os Custos de Acidentes de Trabalho nas Empresas de Construção. Instituto Superior Técnico, Lisboa.

Holmes, N., Lingard, H., Yesilyurt, Z., Munk, F. (1999). An Exploratory Study of Meanings of Risk Control for Long Term and Acute Effect Occupational Health and Safety Risks in Small Business Construction Firms. Journal of Safety Research, Vol. 30, Issue 4, pp. 251–261.

Monteiro, P. (2012). Custos da Segurança no trabalho VS Custos da Não segurança no trabalho—Sector da construção. *Dissertação de Mestrado*. Instituto Politécnico Cavado Ave.

Organização Internacional do Trabalho (OIT) (acesso em Fev.2012) (www.ilo.org/global/lang--en/index.htm)

Pastore, J. (1999). A Dimensão Económica dos Acidentes e Doenças do Trabalho, Comunicação realizada na abertura da Campanha da Indústria da Prevenção de Acidentes do Trabalho, Serviço Social da Indústria, Brasília.

Pidgeon, N. (1991). Safety culture and risk management in organizations. Journal of Cross-Cultural Psychology, v. 22, n. 1, pp. 129–140.

Reis, C.M. (1998). Análise Económica da Segurança na Construção. *Dissertação de Mestrado*. Faculdade de Engenharia da Universidade do Porto.

Rocha et al. (2000). Avaliação da aplicação da NR-18 em Canteiros de Obras. In: Encontro Nacional de Engenharia de Produção, Outubro 2000, Porto Alegre. UFRGS.

Santos, G., Mendes, F., Barbosa, J. (2011). Certification and integration of management systems: the experience of Portuguese small and medium enterprises, *Journal of Cleaner Production*. Vol. 19, pp. 1965–1974.

Santos, G., Barros, S., Mendes, F., Lopes, N. (2013). The main benefits associated with health and safety management systems certification in Portuguese small and medium enterprises post quality management system certification. Safety Science 51. pp. 29–36.

Torner, Marianne. (2009), Safety in construction—a comprehensive description of the characteristics of high safety standards in construction work, from the combined perspective of supervisors and experienced workers. Journal of Safety Research. Volume 40, Issue 6, December 2009, pp. 399–409.

Waehrer, M.G., Dong, X.S., Miller, T., Haile, Y. (2007). Costs of occupational injuries in construction in the United States. *Accident Analysis & Prevention*. Volume 39, Issue 6, November, pp. 1258–1266.

Occupational Safety and Hygiene – Arezes et al. (eds)
© *2013 Taylor & Francis Group, London, ISBN 978-1-138-00047-6*

The role of costs, benefits and social impact on the design of occupational safety programs

D.G. Ramos
Polytechnic Institute of Cávado and Ave, Barcelos, Portugal
University of Minho, Guimarães, Portugal

P.M. Arezes & P. Afonso
CGIT, Department of Production and Systems, University of Minho, Guimarães, Portugal

ABSTRACT: Following an occupational risk assessment, it is necessary to take into account the associated costs and benefits of the preventive measures. Only a cost-benefit analysis can capture all impacts resulting from work accidents and from the measures regarding Occupational Health and Safety. In the present paper a case study related to application of this methodology to a Hospital is presented. An analysis of the work accidents in six selected services has been made. Three of the major types of accidents have been selected: needle stings, falls and excessive efforts. Following the risk assessment, a series of preventive measures have been designed. The Benefit/Cost ratio of these measures has been calculated. Only in some services the ratio is much higher than one, showing very high financial benefit.

1 INTRODUCTION

Safe and healthy workplaces help businesses and organisations to succeed and prosper and also benefit the society as a whole. Work accidents, besides the cost in terms of lost lives and suffering to workers and their families, affect business and also the society. Fewer accidents means less sick leaves, which results in lower costs and less disruption in the production process. It also saves employers from the expense of recruiting and training new staff, and can cut the cost of early retirement and insurance pay-outs (EU-OSHA, 2012).

According to the European Agency for Safety and Health at Work (EU-OSHA, 2012), slips, trips and falls are the largest cause of accidents in all sectors from heavy manufacturing to office work. Other hazards include falling objects, thermal and chemical burns, fires and explosions, dangerous substances and stress. To prevent accidents occurring in the workplace, employers should establish a safety management system that incorporates risk assessment and monitoring procedures.

Several steps are suggested to address risk situation when an organization performs an integrated analysis of risks in evaluating its Occupational Health and Safety (OHS) Management System. The organization should make a detailed analysis of the monetary impact (positive or negative) for the organization of each of the measures considered. However, it is also important to perform an analysis of the impact of each measure on society (externalities). The measures taken by an organization in the prevention of risks can have a positive indirect effect (positive externality) on society, while no action due to costs for the organization, can have a significant negative effect on society (negative externality). Thus, these effects should be duly considered in decision making (Ramos et al., 2012).

Tompa et al. (2006) have made a survey of a large number of studies of workplace-based occupational health and safety interventions. They concluded that very few economic analyses were undertaken amongst such a large number of workplace-based interventions, which shows that economic analysis is rarely regarded as a critical component of an intervention study.

For Varian (1992), the definition of externality is that the action of an agent directly affects the living conditions of another agent. Externalities can also be defined as: "the uncompensated impact of a person's actions on the well-being of a bystander" (Mann and Wüstemann, 2008). The externalities consist of social costs or benefits that come up beyond the scope of the project and influence the well-being of others without monetary compensation (EVALSED, 2009).

Thus, the perspective used shall consider the costs and consequences for the injured worker and his family and also to third parties, in particular public and private players (Silva et al., 1998).

Reniers & Audenaert (2009) have developed a decision-support methodology for safety investments in chemical plants. This methodology includes a cost-benefit analysis that tries to take into account quantifiable and non-quantifiable socio-economic consequences of accidents that can be avoided by preventive measures. Nevertheless, this model only considers costs for the company and for the worker, but not for the society.

Cagno et al., (2013) have made a review of the economic evaluation of occupational safety and health and its way to SMEs, starting with more than 500 studies published since 2000. Despite differences in detail and/or terminology, most authors and institutions adopt the fundamental distinction between direct and indirect for valuing both costs and benefits. Cagno et al., (2013) concluded that this topic needs more multidisciplinary research.

According to Fabela and Sousa (2012), prevention has been encouraged by some European Union countries, including Portugal, through the principle of internalising the costs of workplace accidents. The principle of internalisation of costs is based on the allocation of costs to the employer or the individual that caused the costs.

In this context, cost-benefit analysis should provide answers to the following questions: What investments in OHS should be done? How much should be spent on preventive measures? When should each investment be made? (Ramos et al., 2012).

A robust, properly tested and systematized methodology for economic assessment in the context of risk management will support decision making within the OHS. This represents a natural extension of ISO/IEC 31010:2009 in terms of techniques and tools for economic evaluation in risk management and assessment, using the cost/benefit analysis (CBA).

Table 1 explains the social benefits and costs in terms of their external and internal dimensions.

The authors have recently developed a model for Cost-Benefit Analysis in OHS (Ramos et al., 2012). This model permits to perform economic evaluations of risks and prevention initiatives from both

Table 1. Social benefits and costs.

Benefits and costs	External	Private	Social
Benefits	Agents who benefit from the positive externalities but do not pay for these advantages	Gains earned by agents who pay for	Sum of private and external benefits
Costs	Agents who suffer the negative externalities and who are not compensated	Costs paid by agents that have a direct benefit	Sum of private and external costs

the company and the society perspectives. It is an important tool to support managers and experts on economic analysis and decision making before the beginning of any intervention project related to occupational health and safety. The aim of the current paper is to present the application of this model in the hospital sector.

2 CASE STUDY

The case study presented here regards to a public Portuguese hospital. The hospital is accredited according to CHKS Healthcare Accreditation Standards and has its own internal OHS Services.

This study concentrates on six of the services, which were chosen in collaboration with the OHS services of the hospital, namely: three medicine services, two orthopaedic services and the emergency services.

The occupational accidents in 2011 of these services have been studied using official statistical indexes, which allowed prioritizing the measures to be implemented. Costs corresponding to these occupational accidents have been estimated.

In the present paper, we have used the simple methodology proposed by Heinrich (1959) to calculate the indirect costs of the accidents, as it is the system used by the hospital. According to this methodology, indirect costs can be estimated as being four times the direct costs, so the total costs are five times the direct costs.

The risk assessment process permitted comparing the results of the risk analysis and the criteria to determine the likelihood that the risk and/or the respective magnitude is acceptable or tolerable (DNP ISO Guide 73, 2011). The risk assessment supports the decision about risk treatment. A semi-quantitative method to risk assessment has been applied in the hospital.

As mentioned before, the risk evaluation has been made in six services. Following this risk evaluation, a detailed plan of the preventive measures to be implemented has been designed, with an estimation of the corresponding costs.

An estimation of the benefits of these measures, in terms of the hospital and also for the society, has been made, based on the model developed by the authors (Ramos et al., 2012).

3 RESULTS AND DISCUSSION

Table 2 presents the total costs of accidents occurred in 2011 in the six services that have been studied. Costs with stings are presented autonomously, as these accidents have been studied in more detail.

Table 2. Costs of accidents in 2011 on the six services (in Euros).

Services	Costs with stings	Cost with other accidents	Total costs
Medicine A	0	230	1150
Medicine B	615	0	3075
Medicine C	369	1587	9780
Orthopaedic A	246	3937	20,915
Orthopaedic B	0	2778	13,890
Emergency	984	1510	12,470
Total	2214	10,042	61,280

Total costs include both direct costs (costs presented in the second and third columns) as well as indirect costs. As mentioned before, total costs have been estimated as five times the direct costs, as proposed by Heinrich (1959).

The costs with stings were higher in the emergency department, while the cost of other accidents assumes greater importance in orthopaedic services and particularly in orthopaedic A.

Stings represent about 20% of total accident costs. Other important type of accidents are falls at the same level and situations that lead to musculoskeletal injuries.

The Hospital operates a semi-quantitative method to perform the risk assessment. The semi-quantitative methods determine a numerical value of the magnitude of occupational risk based on the product of the probability of occurrence of the professional risk by the expected severity of injuries:

$$R = P \times S,$$

where R is the level of risk, P the probability and S the severity.

Following the risk assessment evaluation, a series of preventive measures have been defined.

3.1 Needle stings

The main preventive measures identified can be summarized as follows:

– developing training/information actions concerning the use of cut-drilling objects in the workplace;
– place cutting and piercing objects in the appropriate container without exceeding 2/3 of capacity, according to the institutional procedure;
– continuous and annual training accident in the prevention of work accidents—needle stings.

Given the risk of needle stings within the emergency service, the implementation of retractable mechanisms (retractable needle) is specially recommended in this service.

The costs of these preventive measures have been calculated. The costs have taken into account the number of hours needed for information and training actions, including both the working time of the internal trainers and the trainees involved in each service. In case of the use of retractable needles, the excess of cost of these needles respect to traditional needles has also be taken into consideration, based on the average annual consumption of needles in the emergency services.

Table 3 presents the estimated costs in each service.

3.2 Falls

The main preventive measures related to falls that have been identified can be summarized as follows:

– collective Protection: place warning signs that the pavement is wet when it is cleaned and implement corresponding training and information sessions for professionals; clearing of the passageways (especially in emergency service);
– reorganization and restructuring of the jobs: placement and storage of sera media and of the screens (placed in the bathroom of users) in a proper place that does not constitute a risk of falling both of professionals and of users;
– elimination or reduction of risk at source: placement of safety signs, when performing tasks that entail risk of falling; the operational assistants, whenever necessary, should clean the floors that constitute a risk of falling either to health professionals, patients and visitors;
– training and information sessions related to physical agents—falls in the workplace.

The costs of these preventive measures have been calculated. The costs have taken into account the number of hours needed for information and training actions, including the working time of both the internal trainers and the trainees involved in each service.

Table 4 presents the estimated costs in each service, while Table 5 presents the general expected

Table 3. Annual cost of measures related to the prevention of needle stings in the six services (in Euros).

Service	Total costs
Medicine A	1237
Medicine B	1112
Medicine C	808
Orthopaedic A	1323
Orthopaedic B	1159
Emergency	4406
Total	10,045

Table 4. Annual cost of measures related to the prevention of falls in the six services (in Euros).

Service	Total costs
Medicine A	156
Medicine B	194
Medicine C	505
Orthopaedic A	618
Orthopaedic B	599
Emergency	1118
Total	3190

Table 5. Social benefits and costs related top measures to prevent needle stings.

Benefits and costs	Private	Social
Benefits	Reduce sting injuries and the respective costs	Extension of prevention to other entities in the health system Development of new needles with safety mechanism
Costs	Increased cost of retractable needles Training of professionals involved	Increase of public costs

social benefits and costs related to the measures to prevent stings.

3.3 Excessive efforts

The main preventive measures related to excessive efforts that can lead to musculoskeletal injuries can be summarized as follows:

- ergonomic improvements in working equipment and furniture: acquisition of height adjustable beds;
- specific training of staff in handling and service charges and mobilizing users;
- training and awareness about manual handling of loads in the workplace;
- training/information actions about surface cleaning and incorrect postures in the workplace and about mobilization of patients;
- other preventive measures to be adopted: storing heavier materials on shelves at the height of the arms or lower and corresponding training and information of the professionals.

The costs of these preventive measures have been calculated. The costs have taken into account the number of hours needed for information and training actions, including the working time of both the internal trainers and the trainees involved in each service.

Table 6 presents the estimated costs in each service.

3.4 Total costs of preventive measures

The total costs of all the preventive measures mentioned in the previous sections are presented on table 7.

3.5 Cost-Benefit analysis

A simplified cost-benefit analysis has been made for all the preventive measures which can be adopted.

The benefits for the hospital are mainly linked to the reduction of the costs of the accidents. According to an optimistic scenario, supposing that the preventive measures have been well designed and will be successfully implemented, it has been estimated that there will be a reduction of 80% of the accidents after the effective implementation of the preventive measures. The financial benefits for the hospital correspond to the reduction of the costs of accidents.

Table 8 presents the Benefits in the six services, using the above mentioned methodology, while the third column includes the total costs of the preventive measures. The last column presents the B/C (Benefit/Cost) ratio of all the measures in all the six services.

Table 6. Annual cost of measures related to the prevention of excessive efforts in the six services (in Euros).

Service	Total costs
Medicine A	2483
Medicine B	2559
Medicine C	2492
Orthopaedic A	1274
Orthopaedic B	1110
Emergency	4594
Total	14,512

Table 7. Annual cost of all the preventive measures in the six services (in Euros).

Service	Total costs
Medicine A	4919
Medicine B	6379
Medicine C	7753
Orthopaedic A	1756
Orthopaedic B	1919
Emergency	13,499
Total	36,225

Table 8. Calculation of the B/C ratio of all the preventive measures in the six services (in Euros).

Service	Benefits	Costs	B/C ratio
Medicine A	920	4919	0.19
Medicine B	2460	6379	0.39
Medicine C	7824	7753	1.01
Orthopaedic A	16,732	1756	9.53
Orthopaedic B	11,112	1919	5.79
Emergency	9976	13,499	0.74
Total	49,024	36,225	1.35

If the B/C ratio is lower than 1, the preventive measures are not effective in financial terms, as the costs are higher than the benefits. Higher values of B/C ratio represent very effective preventive measures.

It can be concluded that the measures designed for the two orthopaedic services are very effective.

4 CONCLUSIONS

As already mentioned by Pearce (1976), the results of cost–benefit analyses should always be interpreted with care, because estimates of the costs and the benefits of an intervention are never complete and rarely do justice to the complexity of the situation.

The analysis of the financial Benefit/Cost ratio showed that, for the hospital considered in this case study, only part of the preventive measures defined in the risk assessment process are cost effective. The study can be refined by restricting the preventive measures (for example reducing the type and duration of the training actions) in the cases where B/C is lower than 1, trying to improve the cost effectiveness of measures.

The model that has been developed includes not only financial aspects (related to the perspective of the company) but also economic aspects (from the standpoint of view of the society). Data obtained so far is not yet sufficient to calculate the economic B/C ratio. In fact, it is also important to take into account the externalities, namely the benefits for the society, which can also be a tool to refine the preventive measures.

The proposed model is now in a validation phase, which requires adequate planning and a precise definition of responsibility, involving professionals and company structures of the areas concerned. Therefore, taking into account the legal requirements that may exist in the organization, the implementation of this model should be coordinated by the Health and Safety manager, in close collaboration with a wide range of other professionals in the organization, with particular emphasis on the financial, management control and human resources departments (Ramos et al., 2012).

ACKNOWLEDGEMENTS

The authors would like to thank the contribution of the hospital in supplying the data for this study.

This work was partially funded with Portuguese national funds through FCT—Fundação para a Ciência e a Tecnologia, in the scope of the R&D project PEst-OE/EME/UI0252/2011.

REFERENCES

Cagno, E., Micheli, G., Masi, D., Jacinto, C. 2013. Economic evaluation of OSH and its way to SMEs: A constructive review. Safety Science 53, 134–152.

EU-OSHA 2012. European Agency for Safety and Health at Work. Consulted in October 2012, available at http://osha.europa.eu/en/topics/accident_prevention

EVALSED 2009. "A Avaliação do Desenvolvimento Socioeconómico. MANUAL TÉCNICO II: Métodos e Técnicas Instrumentos de Enquadramento das Conclusões da Avaliação: Análise Custo-Benefício". Accessed on February, 2012, at http://www.observatorio.pt/item1.php?lang = 0&id_channel = 16&id_page = 548.

Fabela, S. & Sousa, J. 2012. Os impactes socioeconómicos no âmbito dos acidentes de trabalho. Representações, práticas e desafios à gestão das organizações de trabalho, In H.V. Neto; J. Areosa; P. Arezes (Eds.) – Impacto social dos acidentes de trabalho, Vila do Conde: Civeri Publishing, 99–129.

Heinrich, H.W. 1959. Industrial accident prevention: A scientific approach (4th ed.). New York, NY: McGraw Hill.

ISO Guide 73: 2011. Risk Management. Vocabulary.

ISO/IEC 31010: 2009. Risk management—Risk assessment techniques.

Mann, S. & Wüstemann, H. 2008. Multifunctionality and a new focus on externalities. The Journal of Socio-Economics, 37, 293–307.

Pearce, D. 1976. The limits of cost-benefit analysis as a guide to environmental policy. Kyklos, Vol. 229, Issue 1, pp. 97–112.

Reniers, G.L.L. & Audenaert, A. 2009. Chemical plant innovative safety investments decision-support methodology. Journal of Safety Research 40, 411–419.

Ramos, D., Arezes, P. & Afonso, P. 2012. Ergonomics and occupational health and safety: A cost-benefit analysis model. In Duffy, V. (Edt.), Advances in Human Aspects of Healthcare, Advances in Human Factors and Ergonomics Series. CRC Press. ISBN 9781439870211, Chap. 76, pp. 711–720.

Silva, E., Pinto, C., Sampaio, C., Pereira, J., Drummond, M. & Trindade, R. 1998. Orientações Metodológicas para Estudos de Avaliação Económica de Medicamentos. INFARMED. Accessed on March 2012, at http://www.ispor.org/peguidelines/source/Orien_Metodologicas_EAEM.pdf

Tompa, E., Dolinschi, R., De Oliveira, C. 2006. Practice and potential of economic evaluation of workplace-based interventions for occupational health and safety. Journal of Occupational Rehabilitation 16, 375–400.

Varian, H.R. 1992. Microeconomic Analysis. W.W. Norton & Company, Inc. 3rd Edition.

Education and training in prevention

Occupational Safety and Hygiene – Arezes et al. (eds)
© 2013 Taylor & Francis Group, London, ISBN 978-1-138-00047-6

Evaluation of perceived risk by education professionals in kindergarten

J.M. Tavares & R.P. Azevedo
Instituto Superior da Maia, Maia—ISMAI, Porto, Portugal

M.V. Silva
*Escola Superior de Tecnologia da Saúde do Porto do Instituto Politécnico do Porto,
Vila Nova de Gaia, Porto, Portugal*

ABSTRACT: The present study aims to evaluate the perceived risk by education professionals of 8 Kindergarten of the city of Maia. A survey was applied to a sample of 34 kindergarten teachers and 34 auxiliary teaching staff. The results obtained revealed no statistical difference between risk perception of the two professionals groups, both occupational categories answered in general according to the requirements in the Portuguese legislation. The results also highlight the importance of risk communication and experience sharing as measures to decrease accident occurrence and increase the success of a safety culture.

1 INTRODUCTION

Preschool education is the first stage of the education process of any child which should be accompanied and supported by its related family. This education should complete the development of the child in different dimensions in order to ensure its full insertion as an active member in the society, encouraging values of autonomy, freedom and solidarity (Portuguese Preschool Education Law, Lw 5/97 of 10 February).

In Portugal, preschool education is intended for children between 3 years old and the admission age for elementary school and is yielded in kindergartens. At preschool age the motor expression gets greater definition, as a consequence, the child gains more independence being able to perform different types of locomotion such as walk, run, or jump. According to Erikson (1959) as well as Piaget (1970) referenced by Sprinthall & Sprinthall (1993) this age correspond to the process of discovering the surrounding world. This process increases risk exposure and accident occurrence.

Children safety in preschool age does not depend only in their actions, but mainly what concerns with risk perception of kindergarten education professionals which interact continuously with these children.

The present study aims to evaluate the perceived risk of kindergarten education professionals mainly kindergarten teachers and auxiliary teaching staff, which are the professional groups that are in continuous contact with young children. This evaluation comprised the perception of risk related to different spaces and subjects of the kindergarten (structural conditions, playground, equipment and materials, electrical hazards and emergency organization).

2 MATERIALS AND METHODS

In order to evaluate risk perception a survey was applied to a total sample of 68 professionals, 34 kindergarten teachers and 34 auxiliary teaching staff.

The survey was based in the Portuguese Law of preschool education and school safety as well as good practice documents and international standards. The survey comprised three groups. The first group comprised a characterization of the professional, the second group aimed to collect information about accidents occurred, the third and last group comprised a range of 33 questions which intended to evaluated risk perception of the professional groups about the different spaces and subjects of the kindergarten (structural conditions, playground, equipment and materials, electrical hazards and emergency organization).

Answer options related to the third group of the survey were coded with a Likert scale of 5 values, which 1 correspond to the lowest level of perception and 5 to the highest level of perception.

For statistical analysis, the score obtained by each theme, corresponded to the sum of each score question. A minimum and a maximum score was calculated which correspond to the product of the number of the total questions of each theme by the lowest value of the likert scale or the highest value of the likert scale.

Table 1 present the minimum and maximum scores obtained for each theme in accordance with

Table 1. Minimum and maximum theoretical values for the score of each domain.

	Minimum score (theoretical)	Maximum score (theoretical)
Structural Conditions score	7	35
Playground score	6	30
Equipment and materials score	5	25
Electrical Risk Score	5	25
Emergency Organization Score	10	50

the answers provided by the education professional sample.

The statistical analysis was performed by the software "PredictiveAnalyticsSoftWare" (PASW) Statistics®, version 18.0 for Microsoft Windows®. All the tests were carried out with the significance level $\alpha = 0,05$. Non-parametric tests were used when the assumptions of parametric tests were not met.

Satatistcal diferences of risk perception between kindergarten teachers and auxiliary teaching staff was evaluated using the T-test for independent variables through the comparison. Since this test requires the existence of homoscedasticity, it was applied Levene's test for both groups (Pimenta, 2011).

The analysis of the influence of specific education in the perception of risk was performed trough use of *MANN–WHITNEY test*, since the sample did not fulfill the all criteria for the application of parametric test. The total number of negative answers to the question, *"if already had specific education in safety, hygiene and health"*, was less than 30.

To assess the influence of age on risk perception it was performed the non-parametric *KRUSKAL–WALLIS test*. As p value < α in the two groups, one cannot assume the normality of the variable under study, for this reason the use of parametric tests is not suitable (Pimenta, 2011).

Pearson's coefficient was used to evaluate the influence of professional experience in the perception of risk in different fields.

To check which is the theme kindergarten teachers and auxiliary teaching staff demonstrate less knowlwdege a descriptive analysis was performed for each score evaluated.

3 RESULTS AND DISCUSSION

It was obtained a total of 80 valid answers from eight kindergartens in the city of Maia. However, only 73 answers were from education professionals.

Table 2 shows all the professional categories that answered to the survey. As it can be seen teachers and teaching auxiliary staff were the groups that registered a higher number of responses.

Due to the small number of responses of the remaining occupational professional categories and the fact that teachers and educational auxiliaries are the professional categories that are in close contact with children, the statistical treatment of the data obtained will be treated based on the answers obtained by these two occupational categories.

The answers to the questionnaire were obtained from female professionals. Table 3 shows the distribution of age classes of the sample surveyed, showing that the teachers are aged between 26 to 30 years old and the auxiliaries between 36 to 40 years old.

The Table 4 showed that the majority of respondents have specific education in safety, hygiene and health. Professional reasons were mentioned as main cause for the execution of specific training in safety, hygiene and health. On the other hand, the educative

Table 2. Professional category of respondents.

Professional category	Number of responses
Direction	2
Technical direction	0
Pedagogical coordination	6
Kindergarten teachers	34
Auxiliary teaching staff	34
Other	4

Table 3. Age of respondents according to professional category.

Age group (years)	Kindergarten teachers	Auxiliary teaching staff
16–20	0	1
21–25	3	1
26–30	11	6
31–35	6	6
36–40	5	9
41–45	3	3
46–50	2	6
51–55	3	1

Table 4. Total responses of specific education in safety, hygiene and health.

Specific education	Frequency
Not answered	2
With specific education	50
Without specific education	16
Total	68

auxiliary teaching staff, shown to have more education programs on this area, when compared with kindergarten teachers (Figure 1). This probably happens because the frequency of demands for short courses when inserted in the job market.

Considering the areas assessed in the questionnaire, it was found that both occupational categories answered according to the requirements in the Portuguese legislation (Figure 2). The results obtained indicated no differences on the perception of the risk of kindergarten teachers and auxiliary teaching staff.

When studying the variables age, specific education, area of technical expertise and its influence on the sort of response, the results indicated there is no significant differences in terms of domains studied, as can be seen in Table 5, for the two professional's categories (Figure 3, Figure 4 and Figure 5). It would be expected that younger professionals had a different perception of risk, because they have less experience. However, the training in specific area through targeted courses, may explain the results. It emphasizes the fact that safety, are the youngest

Table 5. P values of the different tests applied in the study.

	T-test	MANN–WHITNEY	KRUSKAL–WALLIS	Pearson's coefficient
Structural conditions score	,916	,506	,690	,317
Playground score	,481	,640	,430	,518
Equipment and materials score	,435	,556	,621	,322
Electrical risk score	,496	,209	,088	,059
Emergency organization score	,771	,489	,780	,492

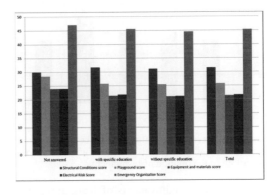

Figure 3. Comparison of risk perception between with and without specific education on safety, hygiene and health at work considering the different domains.

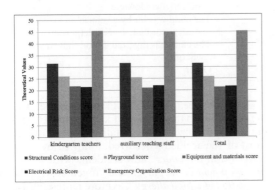

Figure 1. Distribution of responses in function of specific education and professional category.

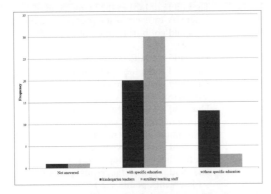

Figure 2. Comparison of risk awareness among educators of children and educational aids for different domains.

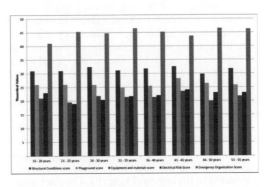

Figure 4. Comparison of risk awareness among the different age groups for different domains.

who follow the rules more easily, generally fulfilling all the tasks that are assigned.

Particular aspects related to the organization of spaces and emergency, were referred by professionals, showing that experience is important and

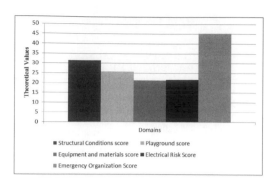

Figure 5. Comparison of risk of awareness between the different domains.

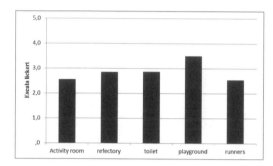

Figure 6. Level of risk attached to the spaces of the kindergarten, where 5 equals much risk and 1 to no risk, according to the opinion of respondents.

that can lead to adjustments of legislation, such as the height of the fire extinguishers that being at the right height from the floor, are excellent as an obstacle to young children.

It is assumed that these results may occur due to the proximity between the kindergarten teachers and the assistants in education, because they share the same living room activities, facilitating the exchange of experiences regarding the safety of children (Silva, 2008). Another presumed reason comes from the sharing of knowledge about the consequences of the occurrence of an accident with a child, increasing awareness about this risk (Bahr, 1997).

The results obtained in the analysis the surveys allow, note that the playground is the place with the greatest potential for danger in kindergartens, not only in the opinion of professionals who responded to the survey (Figure 6). This fact is also corroborated by the results obtained through the analysis of the accidents occurred in the kindergarten, as shown in figure 7.

Table 6. presents the correspondence between the number used in figure 7 and the description of the accident.

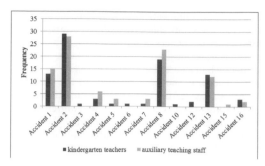

Figure 7. Identification of the typology of accident according to the professional category.

Table 6. Description of the accidents listed in figure 7.

List of accidents	Description of accident
1	The child got hurt while playing with equipment in the playground.
2	The child got hurt when he was playing in the play space
3	The child got hurt in the living room of activity by is not being properly organized
4	The child got hurt because of the deck be wet
5	The child has stifled with the food
6	The child had a food poisoning
7	The child burned himself with the food/ dishes/containers
8	The child wedged his fingers on the door
9	The child has stifled with an object that was in the activity room
10	The child got hurt by being playing with a toy that was not appropriate for her age
11	The child ingested dangerous substances (detergents, paints, etc.).
12	The child had contact with a product that was allergic
13	The child had a fall from a chair/stairs/ balcony/window/etc.
14	The child was hit by a vehicle on arrival or departure of the Kindergarten
15	The child has ingested a food that was allergic
16	Other

As can be seen in figure 7 Playground seems to be, according to the opinion of kindergarten teachers and auxiliary teaching staff, the place where most of the accidents occurred, followed by *"trap fingertips in the doors"* and accidents with *"existing equipment in the playground"* together with the *"ingestion of a food allergy"*, as more representative accidents in the daily routine of a kindergarten.

4 CONCLUSIONS

Risk perception for both groups of educational professionals seems to be quite similar. In this matter, communication, as well as experience sharing seems to have greater influence.

There is a need, according to the responses obtained by professionals surveyed, to increase the attention paid to children in recreational time, this being the place where most accidents occur in the kindergarten. For the same reason, the management of kindergarten, when purchasing equipment for leisure activities, must take into account the recommended requirements for the safety of children and the layout of the playground equipment.

This study highlights the importance of safety culture in school context and provides an analysis of the factors involved in the occurrence of accidents, with the perspective to improve the attitude and behavior of education professionals in kindergarten in order to reduce the occurrence of accidents in these places.

The results obtained also reinforce the importance of communication in this kind of institutions, pointing out the sharing of experiences as one of the measures to be implemented in order to prevent the occurrence of accidents. These findings corroborates the results obtained by Silva (2008) with regard to the implementation of safety and accident prevention.

Children are an important link in the safety culture as they are tomorrow's adults, so, growing up in a safe environment will lead to the learning of safe habits, which later will be incorporated in their daily lives.

REFERENCES

Bahr, N. (1997). System Safety Engineering and Risk Assessment: A Practical Approach. London: Taylor and Francis.

Lw 5/97 of 10 February Portuguese Preschool Education Law.

Marques, R.M. (Outubro de 2009). Os Nossos Alunos e as suas Redes Sociais, Um estudo etnográfico sobre a relação dos alunos com as comunidades virtuais e sua integração na escola. Instituto de Educação e Psicologia—Universidade do Minho.

Pimenta, R. (2011). Estatística Univariada e Multivariada—Mestrado em Ambiente, Higiene e Segurança em Meio Escolar.

Silva, S.C. (Abril de 2008). Culturas de Segurança e Prevenção de Acidentes de Trabalho, numa Abordagem Psicossocial: Valores Organizacionais Declarados e em Uso. Fundação Calouste Gulbenkian, Fundação para a Ciência e a Tecnologia.

Sprinthall, N.A. & Sprinthall, R.C. (1993). Psicologia Educacional. Amadora: McGraw-Hill.

Occupational Safety and Hygiene – Arezes et al. (eds)
© *2013 Taylor & Francis Group, London, ISBN 978-1-138-00047-6*

Continuous training *in loco*: Effects on the symptomatology of WRMD

C. Prüfer, H. Pereira & P.M. Arezes
UMINHO, Guimarães, Portugal

A. Neves, M. Loureiro & P. Soares
Eact-Empresa Activa lda, Porto, Portugal

R. Garganta
FADEUP, Porto, Portugal

F.G. Amaral
UFRGS, Porto Alegre/RS, Brasil

ABSTRACT: Pain and work capacity are not compatible. However the continuous preventive actions, such as training, can reduce work related symptomatology. Investigate the effect of an intervention training program based on ergonomic orientation and preventive physiotherapy, reducing the pain symptomatology of WRMD. Sample consists (233) young adult workers (assembly line). The sensation of pain was assessed with the Nordic Musculoskeletal Questionnaire. The weekly prevention program in the work environment lasted 4 months. The Qui-Square, Kolmogorov-Smirnov and Wilcoxon non-parametritests were used, with 5% positive mean to normality and effect of the program on both moments. A reduction in the prevalence and intensity of the pain symptomatology between (1st and 2nd) time occurred on all body parts. From the obtained results it can be concluded that programs based on ergonomic training and preventive physiotherapy seem to have a positive impact in reducing the pain prevalence and the intensity of the symptomatology of WRMD.

1 INTRODUCTION

The most recent research on labour health has shown the need to act on the prevention of the Work-related musculoskeletal disorders (WRMD), considered the most common problem, – in European Union 24% of the workers report dorsal pain and 22% complaint about muscular pain (Eurofound 2007). Apart from the negative impact on the work capacity, (WRMD) also represent high costs for the companies –43% of EU-27 workers reported absenteeism due to health related problems (Eurofound 2012). On USA the annual costs of compensations due to musculoskeletal diseases are estimated in between 45 to 54 billion dollars (Denis et al., 2008).

In spite of the promotion of the labour health being protected and provided by law and international standards, its practical application still represents a challenge for the health and safety managing workers. The challenge resides on the fact that the WRMD prevention is still incipient and not very assertive; most of these actions are punctual. Most of the times, these actions have the simple purpose of fulfilling the legal requirement, that refers the associated costs to the companies. Inevitably, the results are not very encouraging for both workers and companies alike. The lack of pre-existing indicators on the area of health/ safety/ergonomic, strong enough to validate positive impacts and promote investment stimulus within the companies (Silva et al., 2012).

In order to a change of behaviour and attitude take place, the intervention has to be regular and continued and also suitable to each work reality, i. e., to be held *in loco* and with the active participation of the workers.

The aim of this work is: (1) assess the prevalence of pain sensation/discomfort on different body parts, on a sample of workers from an industrial assemblage; (2) test the effect of the intervention program on the formation and information on working postures and prophylactic physiotherapy on in prevalence of symptoms of pain; and, finally, (3) test the effect of the intervention program on the intensity of pain/discomfort reported.

2 MATERIAL AND METHOD

The sample was formed by 233 workers, male and female, mean age 31,8 from the assembly line defined by 2 types of work: Production/repetitive, present in more than 50% of the jobs and manual Mobilization of load in the rest.

The methodological procedures included a Macro ergonomic Evaluation of the Work of Guimarães (2009) consisting of technical evaluations (ex: RULA and NIOSHI) and participatory evaluation e (open interviews with 30% of the workers and a survey on satisfaction matters, previously included but this time directed to 100% of the workers.

The assessment of the pain caused by (WRMD) was obtained via the *Nordic Musculoskeletal Questionnaire*, adapted to the Portuguese population by Mesquita et al. (2010), presenting a high level of reliability (ICC between 0.7 e 1.0). The severity of the pain was classified by the analogical visual scale (EVA) of 100 mm, Wewers & Lowe (1990), listed from 0 (*no pain*) to 10 (*maximum of pain*), measured at the beginning (first week) and the end (last week) of the program.

Statistic procedures: The exploratory data analysis followed the Kolmogorov-Smirnov normality test. Data was described by mean, standard-deviation and amplitude. During both moments, the Qui-Square and Wilcoxon non-parametric test was carried out in order to evaluate the effect of the intervention program (reduction the prevalence and intensity of the pain symptomatology). The level of significance was kept at 5%. Data was processed in SPSS 19.0.

After identifying the type of work and self-refered pain symptoms, the program begun. The Intervention Program was carried out in the work environment, in a continuous way for the period of 4 months, divided in Ergonomic Orientations—individual training actions on work postures (3 weekly sessions) and Preventive Physiotherapy Orientations—prophylactic training to the workers with painful musculoskeletal symptoms to explain how to perform postural re-education exercises, thermotherapy and cryotherapy at home (1 weekly session). New instructions on more neutral and physiological postures for each kind of task were given every (2) weeks. A total of (8) instructions based on reeducation and global postural strengthening for the spine, upper limb and lower limb (Prufer & Pereira et al., 2012).

Apart from the "*in loco*" actions, during the program were also conducted 3 training sessions in room, (during the 1°, 2° and 4° months of the program). All awareness actions were based on biomechanics observation of the tasks and targeted the understanding of how the fatigue process and the (WRMD) can stimulate the prevention during the daily job of the workers, on the static and repetitive postures and load manoeuvre (Gagnon 2003). During each orientation, new information was added in order to gradually and continuously builds Prevention on the work environment.

Ergonomic modifications (such as resetting platforms, tension and heights for the use of pistols seats to promote posture alterations) were introduced to help the training program.

The multi occupational team was fully skilled to provide orientation on preventive ergonomics and physiotherapy, in order to ensure the maximum prevention of the biomechanics of working positions.

Further to training, periodic meetings with Safety, Production and Work Health Service Departments were conducted to potentiate the mission of the program and develop other ergonomic preventive actions, as boosting up the turnover process, as boosting up the turnover process.

3 RESULTS AND DISCUSSION

The figure 1 ergonomic demand indexes as expected, the critical postures reflected on the Participative Valuation and on the respective identified Ergonomic Demand Indexes. The general result of the AMT in one of the areas identifies, the (1) critical postures of the superior limb, (2) Tiring (3) repetitively and, as the most common discontent questions and the repetition of priority IDEs. Most frequent causes are extreme postures, manual load handling, receptivity and monotony that influence the emergence of (WRMD) signs. As per Silverstein (1985), industrial work show major prevalence of tendinitis in hands and wrists. As well as jobs in manufacturing, presenting prevalence of shoulder disorders, lateral epicondylitis,

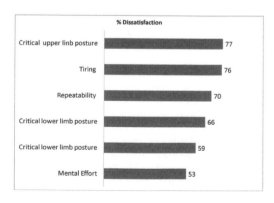

Figura 1. Macro ergonomic work evaluation dissatisfaction Index.

according to (Mccormark et al., 1990). This characterization fosters integration and commitment by the worker to the program, once he feels fundamental part of the program (Hendrick 2001, Grandjean 1981).

The figure 2 presents the prevalence of symptoms of pain in different body regions, before (AV1) and after (AV2) the program.

Initially it is notable a high prevalence of symptoms in the regions most affected by postures adopted for the tasks (Cervical: 42.5%; Shoulders: 55.4%; Fists and Hands: Lumbar and 63.1%: 54.1%). After the 4 months intervention, a significant reduction of the pain symptoms of different body regions was noticed, exception to the thoracic region (Cervical: $\chi^2 = 7,431$, p = 0,006; Shoulders: $\chi^2 = 28,015$, p = 0,001; Thoracic region: $\chi^2 = 1,542$, p = 0,214; Elbows: $\chi^2 = 7,809$, p = 0,005; Fists and Hands: $\chi^2 = 30,157$, p = 0,001; Lumbar: $\chi^2 = 27,718$, p = 0,001; Hips: $\chi^2 = 10,484$, p = 0,001; Knees: $\chi^2 = 7,570$, p = 0,001; Ankles and Feet: $\chi^2 = 17,835$, p = 0,001).

As in the current study, Hsin-Chieh Wu et al. (2009) applied a specific training program on working postures, however, in unique action (1 share 120 minutes) and obtained different results of which are here referred to, noting significant improvements in the prevalence of pain just in the lower limbs. As for the study of Prüfer et al. (2012), reducing the prevalence of pain in the upper limbs and neck was also significant. The study included a group of 322 workers in metal-mechanical industry, which received a program of multidisciplinary interventions for prevention-based training/orientation of preventive physiotherapy, ergonomic and physical labour at work, on a continuous way for 6 months.

In workers who reported symptoms of pain/discomfort, was assessed the respective intensity, considering the last week before the questionnaire.

The figure 3 shows the prevalence of different levels of pain (slight, moderate, intense) in the most affected regions (cervical, lumbar, shoulder, wrists/hands), in which we can ob-serve reduction in both regions. Such data is most evident when comparing the values of pain intensity before and

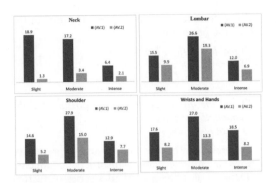

Figure 3. Prevalence of pain intensity.

after the intervention program, finding a significant difference between the two moments (Cervical: Z = –7,495, p = 0,001; Shoulders: Z = –6,312, p = 0,001; Wrists and Hands: Z = –7,592. P = 0,001; Lumbar: Z = –4,557, p = 0,001).

Following the above measured results, we can cite the papers presented by Pereira et al. (2012) & Prüfer et al. (2012) that also foster results in significant reduction in the prevalence of pain, through practices and practical actions and multidisciplinary training for the prevention of (WRMD).

Figure 3 shows some results in reducing the intensity of WRMD, which can also be seen in the practical integration of Preventive Physiotherapy with Ergonomics, occurring simultaneously in the workplace (through prophylactic guidelines, adaptations of work stations) and conventional treatment in office (Mostert—Wentze et al. 2010, Fabrizio 2009) refers another study that also integrated traditional physiotherapy and ergonomic interventions, demonstrating the results of the intervention in reducing painful symptoms; in working postures and on the risk indices Rapid Upper Limb Assessment (RULA), when both actions were held together and not in isolation.

Regarding the intensity of pain, a change in lost days due to WRMD accidents was found by the security department. Therefore, the reduction was found comparing the number of lost days within the respective continuous formation program over the last three years. In 2010 there were 55 lost days; in 2011, 39 lost days; and, in 2012, the year in which the program occurred, 8,5 lost days due to WRMD. Studies such as Guimarães et al. (2012), refer to results in the reduction of symptoms, accidents and lost days due to multifunction directed training. In other words, not only posture but also the capacity to rotate positions, which was also initially encouraged in this study.

This means that, despite the recent and limited scientific evidence in this area, the guidelines of preventive physical therapy seem to have influenced therapeutically to limit the evolution of pain and even fight the symptom already installed.

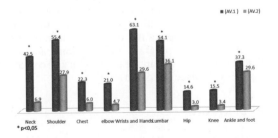

Figure 2. Prevalence of pain/discomfort before and after intervention.

Some comments of employees measured at the end of the program:

"I feel less fatigue and less muscle pain"
"I have less wrist and back pain"
"Before following the guidelines, I had more pain than now"
"I feel the difference in working posture and, by consequence, less pain"
"I have less numbness in hands"

The revision of 47 studies about preventive actions in muscle-skeletal disorders, conducted by Denis et al. (2008), detailed a series of factors about the effectiveness of these studies, which expressed in its majority the actions of formation as one of the most listed with (31studies) followed by the work layout (22 studies) and work organization (21studies). From the Identified problems on the re-search, 28 related to (WRMD), 17 to cervical and upper limb and 12 to lumbar spine. However, in spite of the consistency on the collection and results of the data, a comparison would be impossible, due to the variability of these studies. However, one study verifies that isolated training on manual load manipulation doesn't reflect as well as when carried out along with training on work conditions.

Every worker should have training on risk and risk factors of (WRMD), thus they should be trained on protection as well as participation on an ergonomic program (Guimarães 1999).

However, studies show that the effectiveness of training programs for the reduced pain and prevention of LMERTS reduces with workers over 40 years of age (Brisson et al., 1999). Therefore, any action to prevent (WRMD) must be started as soon as possible.

For this it is important that other actions in addition to training are designed to reduce symptoms of (WRMD) (Cañete 2001). Through a multidisciplinary approach, we include different areas of prevention and intervention to enable the different professionals as (ergonomists, occupational physician, nurse, physical therapist, physical educator, nutritionist, among others) to promote better results. However, the experience shows that the preventive actions within companies normally are sporadic and one time actions (Moraes et al., 1999), limiting the capacity of educational-absorption of these actions.

As per the review of Denis et al. (2008) the most effective results for the preventive actions of (WRMD)occur when integrating various forms of intervention. That is, when we evolve conventional intervention formats and risk new forms of training as such being the case in the study.

However, there are few examples of this type of intervention, i.e., implementation of intervention programs based on training in the work place, in gradual in the work place, in a gradual and continuous/weekly way. Thus the results can be regarded as interesting and promising. A change has been noted in the way the workers perform their tasks and that seems to have made difference in the prevention and/or treatment pain/discomfort.

This change in behaviour, seen as one of the greatest challenges and difficulties in forming Daltroy (1993), reportedly had a positive impact in reducing symptoms of (WRMD), for being a different model of training. Still in testing, allows closer contact with the worker, apparently differentiating this from other actions of conventional format.

Anyway, the area of training is a reality in constant growth and therefore continues to be a source of investment and future in the companies. From 2003 to 2004 the average number of hours applied for training per employee increased from 26 to 32 A.T.S.D. (2005), which encourages actions to have an increasingly solid character. That is to say, the greater the frequency of actions aimed at training in general, the greater the sense of commitment and workers retention of it. This investment invariably returns advantage in human and financial resources for competitive companies (Brum 2007).

4 CONCLUSIONS

We found a high prevalence of symptoms musculoskeletal, in different regions of the body, among the work population in the company's assembly line. The implementation of the prevention, program mainly consisting of training/orientations of correct and safer postures to perform tasks, along with prophylactic of training/orientation on health care regarding the prevention of injury, seems to have fostered a significant reduction in the prevalence of symptoms of pain/discomfort and on its intensity in the majority of body regions analyzed. These results can also be related to the continuous and gradual characteristic training program. This program is still being tested to be boosted with more effective actions in changing positions and exercise program work. In addition to reducing symptoms of (WRMD), programs such as this can assist in the preventive management of the company through the indicators of pain and discomfort to monitor the effects of the program and additional actions that may be necessary in the prevention and correction of workplace sectors in study.

REFERENCES

American Society for Training and Development: (Atsd) State of the Industry Report, Executive Summary http://www.astd.org/astd/research/research_reports

Brisson C, Montreuil S, Punnett L.1999. Effects of an ergonomic training program on workers with video display units Scand J Work Environ Health. Jun;25(3):255–63.

Brum, S. 2007. What Impact Does Training Have On Employee Commitment and Employee Turnover? *Schmidt Labor* Research Center Seminar Paper Series. University of Rhode Island.

Cañete, I. 2001. Humanização: Desafio da empresa moderna a ginástica laboral como um caminho. São Paulo: Ícone.

Daltroy LH, Iversen MD, Larson MG, Zwerling C, Fossel AH, Liang MH. 1993.Teaching and social support: effects on knowledge, attitudes, and behaviors to prevent low back injuries in industry. *Health Educ Q*;20:43–62.

Denis, D., St Vincent M., Imbeau D., Jette C., Nastasia I. 2008. Intervention practices in musculoskeletal disorder prevention. A critical literature review. *Applied Ergonomics*, 39,1–14.

Eurofound 2007. Fourth European Working Conditions Survey, *Office of the European Union*, Luxemb.

Eurofound 2012. Fifth European Working Conditions Survey, *Office of the European Union*, Luxemb.

Fabrizio P. 2009. Ergonomic intervention in the treatment of a patient with upper extremity and neck pain. *Phys.Ther.* 89(4):351–360.

Grandjean, E. 1981. Fitting the Task to the Man: An Ergonomic Approach. *Taylor & Francis*, London.

Guimarães, L.B. 1999. *Abordagem ergonômica: o método macro.* In: Guimarães. Ergonomia de Processo. 1. Porto Alegre:UFRGS/PPGEP.

Guimarães, L.B. 2009. The practice of ergonomics in the south of Brazil from a socio-technical perspective. In: Scott, P.A. (Ed.), Ergonomics in developing regions: needs and applications. *Boca Raton, CRC Press.* 67–87.

Guimarães, L.B., Anzanello, M.J. & Renner, J.S., 2012. A learning curve-based method to implement multifunctional work teams in the Brazilian footwear sector. *Applied ergonomics*, 43(3): 541–547.

Gagnon, M. 2003.The efficacy of training for three manual handling strategies based on the observation of expert and novice workers. *2100 boul Edouard-Montpetit,* Montreal.

Hendrick, H.W. 2003. Determining the cost-benefits of ergonomics projects and factors that lead to their success. *Appl. Ergon.* 34: 419–427.

Hsin-Chieh Wu, Hsieh-Ching Chen, Toly Chen. 2009. Effects of ergonomics-based wafer-handling training on reduction in musculoskeletal disorders among wafer hadlers. *Int Jour Ind Ergonomics.* 39: 127–132.

McCormack RR, Inman RD, Wells A, Berntsen C, Imbus H.R. 1990. Prevalence of tendinitis and related disorders of the upper extremity in a manufacturing workforce.*J Rheumatol.* 17 (7): 958–64.

Mesquita, C., Ribeiro, J., Moreira, P. 2010. Portuguese version of the standardized Nordic musculoskeletal questionnaire: cross cultural and reliability. *Journal of Public Health*, 18: 461–466.

Moraes, M., Alexandre, A, N.M. C, Guiarardello, E.B. 1999. Equipe multiprofissional reduzindo as queixas relacionadas ao sistema músculo-esquelético em costureiras. *Revista de Enfermagem da UERJ*, Rio de Janeiro.7. 1: 19–26.

Mostert-Wentzel, K., Grobler, S., Moore, R., Ferreira, N., Lumley, M., Burelli, K. 2010. Effect of a work-based physiotherapy and ergonomics programme in work-related upper-extremity musculo skeletal disorders in car-seat sea stresses. *Occup Healh S Africa:* 2–7.

Pereira, H., Prüfer, P., Guerreiro, F.M., Soares, P., Garganta, R. 2012. Prevalence of symptoms of low back pain: effects of a preventive intervention program. In: Mondelo, P., Saarela, K., Karwowski, W., Occhipinti, E., Swuste, P., Arezes, P. (Eds.). *Proceedings of the 10th International Occupational Risk Prevention*, ISBN 978–84–615–7900–6, Bilbao, Spain.

Prüfer, C., Pereira H., Guerreiro, F.M., Soares, P., Garganta, R. 2012. Prevalence symptoms of pain in the upper limb: effects of a preventive intervention program. In: Mondelo, P., Saarcla, K., Karwowski, W., Occhipinti, E., Swuste, P., Arezes, P. (Eds.). *Proceedings of the 10th International Occupational Risk Prevention*, ISBN 978–84–615–7900–6, Bilbao, Spain.

Silva, M., Prüfer, C., Amaral F.G. 2012. Is there enough information to calculate the financial benefits of ergonomics projects?. *Work*, 41. 476–483.

Silverstein BA. 1985. The prevalence of upper extremity cumulative trauma disorders in industry, Ph.D. thesis, Ann Arbor: University of Michigan.

Wewers, M.E. & Lowe, N.K. 1990. A critical review of visual analogue scales in the measurement of clinical phenomena. *Research in Nursing and Health*: 13, 227–236.

Emerging risks

Occupational Safety and Hygiene – Arezes et al. (eds)
© 2013 Taylor & Francis Group, London, ISBN 978-1-138-00047-6

Indoor air quality in sports halls

T.S. Filipe, M. Vasconcelos Pinto, J. Almeida, C. Alcobia Gomes,
J.P. Figueiredo & A. Ferreira
College of Health Technology of Coimbra, Coimbra, Portugal

ABSTRACT: Indoor Air Quality (IAQ) is increasingly seen as a widespread environmental problem. This study aims to evaluate IAQ in sports halls during sports activities. The physicochemical parameters [Temperature (T°), Relative Humidity (RH), Carbon Dioxide (CO_2), Carbon Monoxide (CO), Particulate Matter (PM_{10}, $PM_{2.5}$) and Ozone] was measured throughout the duration of the sporting activities and the microbiological parameters [Fungi and Bacteria] at the beginning the middle and the end of those activities. The study sample consisted of a three sports hall in Porto and Lisbon. In this study, the highest recorded values were 24770 CFU/m^3 for bacteria and 4940 CFU/m^3 for fungi, a temperature of 27.7°C, 63.8% relative humidity, 3035 mg/m^3 of CO_2, 2 mg/m^3 of PM_{10}, 0.723 mg/m^3 $PM_{2.5}$, 1.27 mg/m^3 of CO and 0.01 mg/m^3 of ozone. We can observe that some values are above the legal threshold. The sports hall 3 presented the lowest IAQ.

1 INTRODUCTION

Concern regarding indoor air quality (IAQ) emerged following the imperative implementation of energy conservation measures in buildings. These enforced the construction of less spacious facilities with lower ventilation rates, leading to the degradation of IAQ, which also affects the health of their occupants (Cordeiro 2008). People spend on average 80–95% of their time indoors and 6% in enclosed vehicles leading to IAQ becoming a key element in their health and wellbeing (Araújo 2007, Yang *et al.* 2009, Dacarro *et al.* 2003). These studies demonstrate that the levels of pollutants in indoor air may be two to five times, and occasionally one hundred times, greater than the level of pollutants in the outside air (Costa *et al.* 2008). Environments with a low air exchange rate often have high concentrations of carbon dioxide, particles and microorganisms (Freitas *et al.* 2011, Fraga *et al.* 2008). Studies have shown that poor indoor environmental quality in schools can be explained by: (1) insufficient ventilation, especially in winter (Fromme *et al.* 2006, Freitas *et al.* 2011, Cartieauxa *et al.* 2011), (2) scarce and inadequate cleaning of indoor areas (Braniš *et al.* 2011, Freitas *et al.* 2011) and (3) large number of occupants in relation to the area and volume of the rooms with constant lifting of particulate matter off the floor (Freitas *et al.* 2011).

According to Jones (1999), exposure to indoor air may represent harmful effects on human health, since the concentrations of many pollutants are often higher than those found outdoors (Araújo 2007, Yang *et al.* 2009, Demokritou *et al.* 2002).

Nowadays and to a growing extent sports activities have come to be more valued, due to social terms or in terms of professional success [sports activities are] further valued for their potential physical and psychological benefits to the individual (Rocha *et al.* 2007). Although moderate aerobic exercise is recommended for overall good health, adverse health consequences may occur in those who exercise in severely polluted environments (Braniš & Šafránek 2011, Braniš *et al.* 2008, Fromme *et al.* 2006, Braniš *et al.* 2011).

In a sports halls, where many athletes gather, the featured pollutant that is found in large quantities is CO_2, seeing as it is inevitably generated by breathing (Ng *et al.* 2011).

In addition, teenagers are considered more vulnerable than adults are because their health is more susceptible to environmental pollution, as they are still under development (Freitas *et al.* 2011, Yang *et al.* 2009). As result of poor IAQ, student productivity (Braniš & Šafránek 2011, Jones *et al.* 2010), performance (Freitas *et al.* 2011), concentration and memory (Yang *et al.* 2009) can be seriously jeopardized.

2 RESOURCES AND METHODOLOGY

The study sample consisted in three sports hall localized in the urban area of Oporto and Lisbon. The sport performed in these halls is artistic gymnastics, from basic formation to high competition, including Olympic gymnasts.

Table 1. Simple descriptive statistics of the physical-chemical results.

Parameter	Season	Sports Hall 1 Winter	Sports Hall 1 Spring	Sports Hall 2 Winter	Sports Hall 2 Spring	Sports Hall 3 Winter	Sports Hall 3 Spring	Total Winter	Total Spring
Temperature (°C)	Overall Mean	22.2		22.8		20.9			
	Mean	21.5	23.4	22.2	23.8	21.5	23.4	22.2	23.8
	Maximum	23.1	24.8	23.9	25.0	23.1	24.8	23.9	25.0
	Minimum	19.2	22.2	20.8	22.0	19.2	22.2	20.8	22.0
	Standard deviation	1.24	0.93	0.96	1.20	1.24	0.93	0.96	1.20
RH (%)	Overall Mean	42.7		55.7		51.5			
	Mean	44.0	40.8	56.3	54.8	44.0	40.8	56.3	54.8
	Maximum	54.5	45.6	63.8	60.2	54.5	45.6	63.8	60.2
	Minimum	39.5	37.4	57.9	47.4	39.5	37.4	57.9	47.4
	Standard deviation	5.05	3.03	5.00	4.70	5.05	3.03	5.00	4.70
CO_2 (mg/m^3)	Overall Mean	598		1458		514			
	Mean	802	292	1646	1177	802	292	1646	1177
	Maximum	1372	871	3035	1789	1372	871	3035	1789
	Minimum	304	70	325	278	304	70	325	278
	Standard deviation	381	294	913	629	381	294	913	629
PM_{10} (mg/m^3)	Overall Mean	0.201		0.681		514			
	Mean	0.201	0.201	0.658	0.716	0.201	0.201	0.658	0.716
	Maximum	0.426	0.319	1.200	1.270	0.426	0.319	1.200	1.270
	Minimum	0.062	0.088	0.077	0.250	0.062	0.088	0.077	0.250
	Standard deviation	0.120	0.090	0.360	0.380	0.120	0.090	0.360	0.380
$PM_{2.5}$ (mg/m^3)	Overall Mean	0.061		0.301		0.092			
	Mean	0.068	0.050	0.170	0.500	0.068	0.050	0.170	0.500
	Maximum	0.141	0.127	0.405	0.379	0.141	0.127	0.405	0.379
	Minimum	0.014	0.013	0.036	0.018	0.014	0.013	0.036	0.018
	Standard deviation	0.040	0.013	0.120	0.300	0.040	0.013	0.120	0.300
CO (mg/m^3)	Overall Mean	0		0		0			
	Mean	0.4	0	0	0	0.4	0	0	0
	Maximum	2	0	0	0	2	0	0	0
	Minimum	0	0	0	0	0	0	0	0
	Standard deviation	0.7	0	0	0	0.7	0	0	0
Ozone (mg/m^3)	Overall Mean	0		0		0			
	Mean	0.01	0	0	0	0.01	0	0	0
	Maximum	0.01	0.01	0	0	0.01	0.01	0	0
	Minimum	0	0	0	0	0	0	0	0
	Standard deviation	0.01	0	0	0	0.01	0	0	0

The sample collection occurred between January and May of 2012. In order to assess IAQ the measurement of air pollutants (CO_2, CO, PM_{10}, $PM_{2.5}$ and Ozone), weather variables (T° and RH) and microbiological parameters (fungi and bacteria) were undertaken.

It was used the environmental monitor EVM-7 Quest Technologies to assess the concentration of CO, CO_2, Ozone, PM_{10} and $PM_{2.5}$ as well as the weather variables (T° and RH) and the Lighthouse ACTIVE COUNT 90 to assess the microbial contamination (bacteria and fungi).

Measurements were taken on the busiest day of the week during the opening hours. The equipment was placed in a central position and at the level of the respiratory tract of the gymnasts (1.5 ± 0.3 m

Table 2. Simple descriptive statistics of the microbiological results.

Parameter		Sports Hall 1		Sports Hall 2		Sports Hall 3		Total	
		Winter	Spring	Winter	Spring	Winter	Spring	Winter	Spring
Bacteria	Overall Mean	7677.33		10096.67		2842.67			
(UFC/m³)	Mean	10642	3230	11599	7843	2471	3400	8237	4824
	Maximum	>24770	4680	>24770	21160	3790	8360	>24770	21160
	Minimum	2200	1800	270	300	40	280	40	280
	Standard deviation	8476	1129	10443	7999	1388	3034	8583	5170
Fungi	Overall Mean	1274.00		2474.67		1198.67			
(CFU/m³)	Mean	1508	923	2204	2880	924	1610	1546	1804
	Maximum	4640	1280	4260	4940	1620	4940	4640	4940
	Minimum	580	520	440	120	250	340	250	120
	Standard deviation	1296	333	1490	1723	513	1679	1251	1559

from the ground) and at least 3 m from the walls. All the samples were taken during three days in the winter and two days during the spring.

Physicochemical measurements were collected over a period of about four hours. The microbiological collections were performed, in duplicate, at the starting point of the sports activity, during its course and at its conclusion using Trypticase Soy Agar (TSA) for the recovery of bacteria and Sabouraud Dextrose Agar (Sabouraud) for fungi. The air volume collected was 50 L. The incubation temperature of the bacteria was of 35°C for 48 hours and 22°C during 7 days in the case of the fungi.

In order to characterize the capacity of the sports halls, inputs and outputs from 15 to 15 minutes were counted. It wasn't measured the rate of air renovation nor collected reference samples abroad.

3 RESULTS

The study sample consisted of sports hall 1, with an area of 1512 m² and a volume of 10584 m³, sports hall 2, with an area of 895 m² and a volume of 2884 m³ and sports hall 3, with an area of 499 m² and a volume of 3746 m³. Sports hall 1 is located on a ground floor, the second is on a 6th floor and the third is on a 4th floor.

The mean occupancy of sports hall was 68.2 ± 41.5, 33.20 ± 16.62 and 12.60 ± 9.11 people in sports hall 1, 2 and 3, respectively.

It wasn't found any forced ventilation in the main areas, existing only natural ventilation: sports hall 1 has 7 grates with small entrances to the outside at 0.5 m above the ground and sports hall 3 has a disabled duct.

In all the statistical tests, the results were significant to values $p\text{-value} < 0.05$.

After applying the *t-Student* test, no statistically significant differences were verified between winter and spring in relation to the diverse physicochemical parameters. Applying one ANOVA, the parameters regarding RH ($p\text{-value} = 0.004$), CO_2 ($p\text{-value} = 0.007$) and PM_{10} ($p\text{-value} = 0.000$) presented significant differences in each one of the three sports halls. *Tukey's test* demonstrates that these differences are evident between sports hall 1 and 2 and sports hall 2 and 3 in terms of CO_2 and PM_{10}, whereas in terms of RH, the differences stand between sports hall 1 and 2 and, 1 and 3. In order to assess the relation between the average number of people and the mean values of the physicochemical parameters, the *Pearson's Correlation Coefficient* revealed the inexistence of a significant correlation.

When applying ANOVA, both the bacteria ($p\text{-value} = 0.023$) and the fungi ($p\text{-value} = 0.013$) counts presented significant differences between the three sports halls evaluated. *Tukey's* test reveals the existence of statistically significant differences of bacteria between sports halls 2 and 3 ($p\text{-value} = 0.020$). With regard to fungi, the differences were evident between sports halls 1 and 2 ($p\text{-value} = 0.034$) and sports halls 2 and 3 ($p\text{-value} = 0.023$). Through ANOVA, statistically significant differences between the moments of microbiological sampling were noticeable. These differences were observed during the sports activity between the beginning and the middle ($p\text{-value} = 0.000$) and the middle and the end ($p\text{-value} = 0.032$), in the case of bacteria, whereas in the case of fungi, the differences were verified between the beginning and the middle of sports activity ($p\text{-value} = 0.008$).

After applying the *t-Student* test, no statistically significant differences were verified between winter and spring in relation to bacteria and fungi and there is positive correlation between both in each season.

4 DISCUSSION

Using the aforementioned data regarding the dimensions of each sports hall, the amount of fresh air that each possesses at the beginning of any physical activity is different, especially between the Oporto and the Lisbon sports hall. This difference influences the dilution of the various pollutants and it is reflected in the IAQ of the facilities, where those with a greater volume and a lower occupancy rate have better results (Ponsoni & Raddi 2010, Fraga et al. 2008, Ng et al. 2011, Basto 2007, Morais et al. 2010).

After analyzing the results of the parameter temperature, 86.7% of the mean values are higher than the 20°C determined for cooling stations (Decreto-Lei n.°80/2006). However, this is not considered to be a risk to human health (Braniš et al. 2008, Ponsoni & Raddi 2010). In sports facilities, the adjustment of the temperature is very important and should be constant as the needs of each type of setting differ one from another (Ponsoni & Raddi 2010, Xianting et al. 2011).

Regarding the RH parameter, the maximum concentration is the reference range from 25% to 50%, which in this case was exceeded by 53.3%. RH also requires control measures, since it may cause discomfort when practicing sports, subjecting the athlete to greater physical wear caused either by the cooling or heating which is normally associated to sports activities (Cunha 2007).

Regarding CO_2 only sports hall 2 exceeded the average in 20% and in the maximum in 40% of the measurements, since the maximum concentration permitted is 1800 mg/m^3 (Decreto-Lei n.°79/2006), However, sports halls 1 and 3 held acceptable values. Considering that gymnasts undergo intense physical activity, the volume of air inhalated is obligatorily greater, which can lead to adverse health consequences (Basto 2007, Cunha 2007). Apart from this, opening windows during winter causes thermal discomfort to both athletes and other occupants (Yang et al. 2009).

In terms of PM_{10} the maximum concentration reference is 0.15 mg/m^3. The results showed that this parameter was always exceeded (100% of cases). The PM_{10} source is the magnesium carbonate used by athletes to keep hands and other parts of the body dry guaranteeing the desired adherence to equipment essential for this activity (Braniš & Šafránek 2011, Araújo 2007, Bortoleto & Calça 2007, Bortoleto 2007). According to the recommendations of the World Health Organization (WHO), $PM_{2.5}$ concentration should not exceed the limit of 25 μg/m^3 (0.025 mg/m^3) recommended for 24-h in 42% of the days of measurements (Braniš et al. 2008, Braniš & Šafránek 2011). This recommendation was not observed in this study because in all three sports halls the value of 0.025 mg/m^3 was exceeded. Regarding the increasement of $PM_{2.5}$ during the spring, it is possible that this is due to greater training intensity since the measurements was carried out during the week prior to important competitions. Fromme et al. (2007) attribute high levels of particulate matter to the difficulty in maintaining daily cleaning regimes of the facilities (Braniš & Šafránek 2011). These results are worthy of concern because various researches suggests that prolonged exposure to high concentration of particulates can seriously affect human health, causing effects such as infections, allergies, respiratory irritation, etc. (Braniš & Šafránek 2011, Braniš et al. 2011, Cartieauxa et al. 2011, Park & Lee 2003).

We can state that the CO and Ozone concentrations in the pavilions doesn´t represent risk to public health since the values are practically nil, lying below the reference value of 0.2 mg/m^3 and 12.5 mg/m^3

With regard to microbial contamination, the limit value of 500 CFU/m^3 was exceeded in 86.7% of the measurements of bacteria and 88.9% of the measurements of the fungi (Decreto-Lei n.° 79/2006). In the case of bacteria 12.8%, exceeds the maximum measurable value (500 CFU/m^3). The high space occupancy and low income rate of the air contributes to an increased concentration of bacteria in the pavilion, affecting the air quality (Ponsoni & Raddi 2010, Araújo 2007, Fraga et al. 2008, Basto 2007, Morais et al. 2010, Pegas et al. 2010).

Due to the large number of windows in existing, it is possible that the decrease of the count of bacteria in the spring is associated with increased time of exposure to ultraviolet radiation from the sun in the halls (Blanchard 2012). The slight increase in the mold exposure, as explained above, may be explained by the greater intensity of training, which translates into greater use of the sports apparatus and an increase in repeated impacts on the soil/mattress (Dacarro et al. 2003, Pegas et al. 2010).

5 CONCLUSION

Currently, problems have arisen in terms of defining the requirements for sports facilities, the design of air conditioning systems and the operations that ensure both a proper thermal environment and indoor air quality in sports halls.

The sports hall 2 showed a lower air quality. The occupancy factor and volumetric variation between the sports hall contributed to air quality. Even so, the exceedance of the parameters PM_{10}, $PM_{2.5}$ and microorganisms in all the pavilions is of concern, being associated with reduced rate of air exchange. Given the potential health effects that microorganisms can cause, the monitoring of IAQ becomes of boundless prominence. The results highlight the need for air exchange for proper dilution/elimination of pollutants generated during the use of each sport hall.

The equipment maintenance and cleaning regimes of indoor air circulation systems are important practices that can reduce the potential for chemical and microbiological contamination. This maintenance has its significance since several diseases affecting the respiratory system can be transmitted in sports halls, due to deficiencies in air quality.

Thus, it appears necessary and useful future research involving a large number of sports pavilions, as well as in different seasons, in order to compare the variation in time of the pollutants.

REFERENCES

Araújo, A. Teles de. 2007. *Relatório do Observatório Nacional das Doenças Respiratórias.* Lisboa: Observatório Nacional das Doenças Respiratórias.

Basto, José Edson. 2007. *Qualidade do Ar Interno.* Itajaí: CREA.

Blanchard, Laurence. Informações sobre os efeitos do crescimento de bactérias & de luz UV. *Online Directory.* [Online] [Citação: 22 de Maio de 2012.] http://www.fuguitang.com/informacoes-sobre-os-efeitos-do-crescimento-de-bacterias-de-luz-uv.html

Bortoleto, Marco Antonio Coelho. 2007. A ginástica artística masculina (GAM) de alto rendimento: observando a cultura de treinamento desde dentro. *Motricidade*: 323–336.

Bortoleto, Marco Antonio Coelho e Calça, Daniela Helena. 2007. O trapézio circense: estudo das diferentes modalidades. *Revista Digital.*

Braniš, Martin e Šafránek, Jiří. 2011. Characterization of coarse particulate matter in school gyms. *Environmental Research* 111(4): 485–491.

Braniš, Martin, Šafránek, Jiří e Hytychováa, Adéla. 2008. Exposure of children to airborne particulate matter of different size fractions during indoor physical education at school. *Building and Environment* 44(6): 1246–1252.

Braniš, Martin, Šafránek, Jiří e Hytychová, Adéla. 2011. Indoor and outdoor sources of size-resolved mass concentration of particulate matter in a school gym. *Environmental Science Pollution Research* 18: 598–609.

Cartieauxa, E., Rzepkab, M.A. e Cunyc, D. 2011. Qualité de l'air à l'intérieur des écoles. *Archives de Pédiatrie* 18: 789–796.

Cordeiro, Marco Ferreira. 2008. Parâmetros de Qualidade e Conforto Desportivo em Pavilhões Desportivo. *Faculdade de Desporto—Universidade do Porto.* Monograph.

Costa, J.F.B.F.M., *et al.* 2008. Qualidade do Ar em Creches. *Escola Superior de Tecnologia da Saúde de Coimbra* (978-989-8252-02-9) 1: 23–38.

Cunha, Luís Miguel. 2007. Os Espaços do Desporto—Uma gestão para o desenvolvimento Humano. Edições Almedina.

Dacarro, C., *et al.* 2003. Determination of aerial microbiological contamination in scholastic sports environments. Journal of Applied Microbiology 95: 904–912.

Decreto-Lei n.º79/2006, de 4 de Abril. 2006. *Regulamento dos Sistemas Energéticos de Climatização em Edifícios (RSECE).* Ministério das Obras Públicas, Transportes e Comunicações.

Decreto-Lei n.º80/2006, de 4 de Abril. 2006. *Regulamento das Características de Comportamento Térmico dos Edifícios (RCCTE).* Ministério das Obras Públicas, Transportes e Comunicações.

Demokritou, Philip, *et al.* 2002. An experimental method for contaminant dispersal characterization in large industrial buildings for indoor air quality (IAQ) applications. *Building and Environment* 37(3): 305–312.

Fraga, Sílvia, *et al.* 2008. Qualidade do ar interior e sintomas respiratórios em escolas do Porto. *Revista Portuguesa de Pneumologia* 14(4): 487–507.

Freitas, Maria do Carmo e et, al. 2011. Indoor Air Quality in Primary Schools. *Advanced Topics in Environmental Health and Air Pollution Case Studies* (978-953-307-525-9) 361–384.

Fromme, H., *et al.* 2006. Particulate matter in the indoor air of classrooms—exploratory results from Munich and surrounding area. *Atmospheric Environment* 41(4): 854–866.

Jones, Sherry Everett, *et al.* 2010. School Policies and Practices That Improve Indoor Air Quality. *Journal of School Health* 80(6): 280–286.

Morais, Gilsimeire Rodrigues, *et al.* 2010. Qualidade do Ar Interno em uma instituição de ensino superior brasileira. *Uberlândia* 305–310.

Ng, Malcolm Owen, *et al.* 2011. CO_2-based demand controlled ventilation under new ASHRAE standard 62.1–2010: a case study of a gymnasium of an elementary school at west office buildings from the BASE study. *Energy and Buildings* 43 3216–3225.

Park, E. e Lee, K. 2003. Particulate exposure and size distribution from wood burning stoves in Costa Rica. *Indoor Air* 13: 0905–6947.

Pegas, Priscilla Nascimento, *et al.* 2010. Outdoor/Indoor Air Quality in Primary Schools in Lisbon: A Preliminary Study. *Química Nova* 33(5) 1145–1149.

Ponsoni, Karina e Raddi, Maria Stella Gonçalves. 2010. Indoor Air Quality Related to Occupancy at an Air-conditioned Public Building. Brazilian Archives of Biology and Technology 53(1): 1516–8913.

Rocha, Rita Santos, Cardoso, Ana Luísa Alves e Raposo, Pedro Duarte. 2007. Caracterização dos factores de segurança e saúde no trabalho em instalações desportivas—Ginásios. Lisboa: ISHST.

Xianting, LI, *et al.* 2011. *Challenges and Countermeasures for Thermal Environment and Indoor Air Quality in Sports buildings.* Beijing: International Conference on Future Computer Science and Education,

Yang, Wonho, *et al.* 2009. Indoor air quality investigation according to age of the school buildings in Korea. *Journal of Environmental Management* 90 348–354.

Occupational Safety and Hygiene – Arezes et al. (eds)
© 2013 Taylor & Francis Group, London, ISBN 978-1-138-00047-6

Relative risk of accident: Worker collectives in the manufacturing sector

J.A. Carrillo & L. Onieva
University of Seville, Seville, Spain

J.C. Rubio-Romero & M. Suárez-Cebador
University of Malaga, Malaga, Spain

ABSTRACT: Worker collectives in the manufacturing sector are not equally employed and the differences in job assignment can confound the relative risk. In cross-sectional studies, some of those possible confounders can be controlled using a multivariate regression model as long as we have proper information of the exposed workers. The main strength of this paper compared to previous studies is the quality of worker data from the Continuous Sample of Working Lives obtained from Social Security Office. Adjusted relative risk confirms that female workers and young workers have lower injury rates, both for traumatic and musculo-skeletal injuries. Injury rates are also higher for workers of small and medium enterprises in the adjusted model. The methodology presented can be used in the analysis of the risk factors for injury rates in specific type of accidents or activities.

1 INTRODUCTION

1.1 *The Continuous Sample of Working Lives*

The Continuous Sample of Working Lives (hereinafter CSWL) contains a simple random sampling without stratification of an average 4% among all the affiliated workers, working or not, and pensioners. The database is described elsewhere (Jiménez-Martín & Sánchez-Martín 2007) and offers information about the personal characteristics of the worker and also about all their labor history.

This database has been used previously for several purposes but not in the analysis of injury rates. Microdata are available from the Spanish Social Security Agency upon request.

Although other tools such as Working Conditions Surveys (Carrillo et al., 2012a) can provide some insight, the sampling error compared with CSWL is much higher and the quality of data gathered for social security purposes is higher.

The main purpose of this study is to explore the potential use of CSWL for analyzing injury rates of worker categories. At the same time appropriate methodology is proposed to estimate injury rates for the worker categories controlling for possible confounding effects.

Most injury rates are calculated using the number of workers. However the real exposure is related to the time at work. In CSWL the information gathered includes the number of days that each individual worker is active and the number of working hours per day in their contract.

1.2 *Worker characteristics and the risk of accident*

This paper analyzes the influence of worker characteristics in injury rates. Each of them has been analyzed previously in the literature.

It is important to find out what worker collectives have higher relative risk in order to prioritize public enforcement and promotion programs. Estimating injury rates for those categories is difficult as there is no reliable information for the population at risk. Analysis of accident notifications usually lacks information about the worker population characteristics and their exposure.

According to previous studies the individual worker risk factors that should be taken into account are: age, gender, nationality and type of contract. Some specific works and tasks are more likely assigned to male, foreign or non-permanent workers.

The majority of studies have reported that young workers had a higher injury rate, especially if they are men (Salminen 2004). This behavior has also been studied at the company level (Pollack et al., 2007).

There are important differences in terms of job assignment and foreign workers are expected to be employed in more dangerous tasks (Ahonen et al., 2007).

Female workers show lower injury rates as a general trend (Islam et al., 2001) but not for every industry or occupation. In some cases, when analyzing at the company level, female workers seem to have more accidents than their male counterparts

(Taiwo et al., 2008). Some studies conclude that men have higher rates for acute injuries and women for musculo-skeletal ones.

Working in small enterprises is considered a risk factor (Sørensen et al., 2007). Small and micro enterprises have more severe and fatal accidents in comparison with medium and big enterprises. However, some manufacturing activities concentrate more hazardous activities in enterprises with a certain minimum size.

Another important issue is the effect of contract type on safety. Non-permanent contracted workers have more accidents and they are more severe (Benavides et al., 2006).

All of these worker categories need to be studied with multivariate techniques in order to control the confounding phenomena. CSWL provides sampling of real individual workers including data for each worker in all categories of interest.

2 MATERIALS AND METHODS

2.1 Data

The cross-sectional CSWL dataset contains employment variables such as occupational levels, sector of activity, contract type and duration of employment. It also includes relevant variables related to demographics such as location, age and gender and nationality. The duration of the employment comes from the dates of beginning and ending the contract.

Data was gathered from CSWL from 2004 to 2008. However, 2004 was the first year of CSWL, with a lower sampling rate and therefore higher sampling error than in years from 2005 to 2008, as such 2004 was not included in this study. Self-employed worker data are also not considered.

Worker data selected from CSWL are those contracted in the Andalusian Manufacturing Sector. Andalusia is one of the biggest regions in Europe, producing 12% of the Spanish manufacturing sector and employing on average more than two hundred thousand workers.

The manufacturing sector is defined according to the Statistical Classification of Economic Activities in the European Community Council Regulation EEC N°3037/90 (section D, subsections 15 to 37).

2.2 Estimation of the proportion of equivalent work days for each category

The number of equivalent days for each worker is calculated based on the standard work day of eight hours in Spain. The number of effective working days of all workers of the same category in CSWL is added; therefore the equivalent standard days worked for each category can be calculated.

As the number of equivalent work days calculated is a sampling of the real population, the proportion in each of the categories can be calculated.

The data for 51,445 workers and 9,314,179 equivalent work days are available for analysis from CSWL in the Andalusian manufacturing sector from 2005 to 2008.

2.3 Estimation of the relative risk for the worker categories (not adjusted)

The relative risk is calculated as the ratio of the injury rates of two groups of workers, one of them is being considered the reference (Benavides et al., 2006).

Injury rate is estimated as the ratio of two proportions, the proportion of accidents and the proportion of estimated equivalent work days for each category of workers.

Accident notifications in Andalusia are electronically collected in "Official Workplace Incident Notification Forms" (Jacinto & Aspinwall 2004). All accidents that result in an absence from work of one or more days must be reported. In terms of their severity, the accidents can be slight, severe or fatal. Relapses, travelling to work accidents ("in itinere") or accidents of self-employed workers are not included.

We differentiate two main types of accidents in terms of their causation, traumatic accidents (hereinafter TRA) and musculoskeletal disorders (hereinafter MSK). In Spain MSK can be reported both as accidents and as occupational diseases. If the injury is related to movements whereby the injured person's physical exertion exceeded what is normal, it is classified as a MSK accident. If the damage resulted from a long-term influence of working conditions it is reported as a MSK disease and it is not reported as accident but as occupational disease.

Most of the MSK accidents reported are strain and sprain injuries. It is important to note that around of 30% of accidents reported are MSK, although most of them are usually non-slight.

2.4 Worker categories

The categories in each worker variable are defined in Table 1. They are the same used in the National Employment Survey in Spain.

Job and qualification is classified into four groups according to the current version of the International Standard Classification of Occupations, ISCO-88, approved by the International Labour Organization.

ISCO-88 classifies groups on the basis of the similarity of skills required to fulfill the tasks and duties

Table 1. Estimated proportion of effective working days as indicator of population at risk in each category of workers. Data from CSWL of Andalusian manufacturing sector from 2005 to 2008.

Variable	Category (identifier)	%
Contract type	Permanent (1)	67,9%
	Non-permanent (2)	31,8%
Nationality	Spanish (1)	96,9%
of worker	Foreign (2)	3,0%
Sex of	Male (1)	79,4%
worker	Female (2)	20,6%
Age of worker	Less than 30 years old (1)	28,8%
(years)	Between 30 and 44 years old (2)	42,8%
	More than 44 years old (3)	28,5%
Company size	Micro, less than 9 workers (1)	28,4%
(number	Small, from 10 to 49 workers (2)	39,3%
of workers)	Medium, from 50 to 249 workers (3)	21,4%
	Big, more than 249 workers (4)	10,8%
Type of job	High qualification/non-manual (1)	8,3%
and	Low qualification/non-manual (2)	14,9%
qualification	High qualification/manual (3)	53,0%
	Low qualification/manual (4)	23,9%

of the jobs. Two dimensions of the skill concept are used, skill level and skill-specialization. Those groups are the following: High qualification/non-manual (hereinafter HQNM), low qualification/non-manual (hereinafter LQNM), high qualification/manual (hereinafter HQM) and low qualification/manual (hereinafter LQM).

2.5 Panel data set and regression analysis

A panel data set is created with the sum of all working days in CSWL for the workers sampled with the same combination of the categories of the variables (see Table 1 for the identifiers of each of the categories). For example, combination "20431311" consists of the number of working days in CSWL in activity 20 (NACE code for manufacturing of wood and wood products), big enterprises (identifier 4), high qualification and manual (identifier 3), permanent contract (identifier 1), older than 44 years old (identifier 3), male (identifier 1) and Spanish workers (identifier 1).

There are 3,846 possible combinations identified in the panel. There are 1,150 combinations without any affiliation data but with accidents reported (with 15,650 accidents). Another 826 combinations had affiliation data with error in estimation higher than 10% with confidence interval of 95%, and they were also discarded (with 6,869 accidents).

There are 1,869 combinations with enough population to estimate injury rates. They represent the 98.3% of the working days gathered in CSWL and therefore the combinations covered most of the working population. In these combinations, there are 81,920 accidents reported, which represents 78.4% of all accidents reported in the period.

Injury rate is estimated for each of the combinations as the ratio of the real number of accidents reported and the estimated number of workers using the proportion of equivalent working days for the combination of categories in CSWL.

A regression model is adjusted for two dependent variables, natural logarithm of the traumatic injury rate and of musculo-skeletal injury rate. Logarithms are used because most authors considerer that accidents are better modeled with exponential variables (Haviland 2012).

All independent variables are the categories of the qualitative variables available, coded with dummy dichotomous variables. Each qualitative variable with "n" categories is represented by "n − 1" dummy variables. Activities identified with the two first digits of NACE code (from 15 to 37) were also included in the model as controls because injury rates of each activity are very different.

The relative risk of one category to their reference is calculated with the expression e^b because dependent variable is logarithmic. The relative risk is adjusted with the regression model, thus the relative risk calculated considered the effect of the other explanatory variables considered. The relative risks are presented with the confidence interval (95% confidence) for the relative risk. If confidence interval includes 0.0 the relative risk is not significant.

SPPS v.18 was used. Data and models are available upon request to authors.

3 RESULTS

3.1 Estimation of the exposure

The proportion of equivalent work days were calculated for each category of workers (Table 1).

3.2 Relative risk

Results are presented with differentiation of slight and non-slight accidents (Table 2) and separating accidents into TRA and MSK (Table 3).

3.3 Regression model

The regression model has been adjusted for two dependent variables, natural logarithm of the traumatic injury rate and natural logarithm of musculo-skeletal rate. Relative risk estimated with the regression models are presented with confidence intervals (Table 4 and Table 5).

Table 2. Relative risk: Slight vs. non-slight accidents.

Variable	Category	Severity	
		Slight	Non-slight
Contract type	Permanent	0.60	0.66
	Non permanent	Reference	
Nationality of worker	Spanish	0.73	0.46
	Foreign	Reference	
Sex of worker	Male	2.26	3.95
	Female	Reference	
Age of worker (years)	[<30]	1.97	0.91
	[30–44]	1.28	0.90
	[>44]	Reference	
Establishment size (number of workers)	Micro (1–9)	0.72	1.68
	Small (10–49)	1.02	1.78
	Medium (50–249)	1.31	1.52
	Big (>249)	Reference	
Type of job and qualification	HQNM	0.16	0.15
	LQNM	0.17	0.16
	HQM	1.32	1.57
	LQM	Reference	

Table 3. Estimated relative risk: Traumatic vs. musculo-skeletal accidents.

Variable	Category	Type of accident	
		TRA	MSK
Contract type	Permanent	0.36	0.47
	Non permanent	Reference	
Nationality of worker	Spanish	0.64	0.95
	Foreign	Reference	
Sex of worker	Male	2.56	1.91
	Female	Reference	
Age of worker (years)	[>30]	2.38	1.80
	[30–44]	1.43	1.33
	[>44]	Reference	
Establishment size (number of workers)	Micro (1–9)	0.93	0.51
	Small (10–49)	1.22	0.80
	Medium (50–249)	1.45	1.16
	Big (>249)	Reference	
Type of job and qualification	HQNM	0.14	0.22
	LQNM	0.15	0.20
	HQM	1.37	1.37
	LQM	Reference	

4 DISCUSSION

A close examination of the results in Tables 2, 3, 4 and 5 shows that young workers have higher injury rates except for non-slight injuries. The adjusted relative risk for older worker also shows that age is a risk factor in manufacturing (Salminen 2004).

Both Spanish workers and female workers have lower relative risk of injury, especially for traumatic

Table 4. Adjusted relative risk: regression model for natural logarithm of traumatic injury rate. $R^2 = 0.68$, Durbin-Watson 1.36.

Variable	Category	Confidence interval for relative risk	
		Lower limit	Upper limit
Contract type	Permanent	0.68	1.13*
	Non permanent	Reference	
Nationality of worker	Spanish	7.20	16.75
	Foreign	Reference	
Sex of worker	Male	6.14	10.44
	Female	Reference	
Age of worker (years)	[>30]	1.62	3.11
	[30–44]	1.50	2.81
	[>44]	Reference	
Establishment size (number of workers)	Micro (1–9)	1.85	4.48
	Small (10–49)	3.00	7.15
	Medium (50–249)	1.99	4.84
	Big (>249)	Reference	
Type of job and qualification	HQNM	0.02	0.05
	LQNM	0.03	0.06
	HQM	1.67	3.16
	LQM	Reference	

* Not significant p > 0.05: Confidence interval includes 0.0.

Table 5. Adjusted relative risk: Regression model for natural logarithm of musculo-skeletal injury rate. $R^2 = 0.25$, Durbin-Watson 1.61.

Variable	Category	Confidence interval for relative risk	
		Lower limit	Upper limit
Contract type	Permanent	0.84	1.73
	Non permanent	Reference	
Nationality of worker	Spanish	0.60	2.01*
	Foreign	Reference	
Sex of worker	Male	1.06	2.28
	Female	Reference	
Age of worker (years)	[>30]	1.52	3.90
	[30–44]	1.29	3.17
	[>44]	Reference	
Establishment size (number of workers)	Micro (1–9)	0.41	1.46*
	Small (10–49)	0.77	2.68*
	Medium (50–249)	0.60	2.17*
	Big (>249)	Reference	
Type of job and qualification	HQNM	0.21	0.67
	LQNM	0.19	0.51
	HQM	0.47	1.19*
	LQM	Reference	

* Not significant p > 0.05: Confidence interval includes 0.0.

and non-slight injuries. For MSK the situation is not as clear. This is consistent with studies at the company level (Pollack *et al.*, 2006, Taiwo *et al.*, 2009) where with the same assignments the influence of sex and nationality is not as clear.

As other studies have highlighted (Benavides *et al.*, 2006), the relation between non-permanent workers and injury rates is not a straight forward issue, in fact in our adjusted model the relative risk is not significant.

Micro and small enterprises show lower injury rates slight and traumatic accidents. This behavior can be explained with underreporting. In fact, for MSK the relative risk with big enterprises is not significant. MSK accidents are easier to go unreported (Mendeloff *et al.*, 2006).

These results are consistent with other studies in manufacturing. In Carrillo *et al.* (2012a) it was found that being older than 44 years old and being employed as technicians were collectives with higher injury rates. In Carrillo & Onieva (2012) it was found increased severity of accidents in male, foreign and older workers. Similar results have also been found in the construction sector (López-Arquillos *et al.*, in press).

5 CONCLUSIONS

5.1 *Usefulness of CSWL for injury rate estimation*

Sampling error for the worker categories is lower than with other sources of information. The panel data set allows for estimating injury rates for specific combinations of worker categories. Thus, multivariate analysis and control of possible confounders can be performed and adjusted injury rates for each category can be calculated.

The use of the equivalent working days instead of affiliation is a more precise indicator of exposure. Partial-contracts and temporal contracts are included with the real equivalent working days. In those contracts the differences between affiliation and equivalent working days is very important.

One of the big advantages of the methodology presented is that it is possible to estimate the relative risk of accident for different worker categories for each type of accident. Although we have only differentiated traumatic and musculo-skeletal injuries, other specific accident types can be analyzed with this method as long as there are enough accidents to adjust the regression model. Further research in other regions and sectors would provide a more complete insight into relative risk of the worker collectives and the main intervention areas.

5.2 *Implications of the results in safety policies*

The use of CSWL for estimation of the actual exposure of each category of workers for injury rates calculation provides a useful tool in Public Policy design and this paper shows some of the potential uses.

According to the European Safety Framework Directive 89/391/EEC "particularly sensitive risk groups must be protected against the dangers which specifically affect them".

Some of those sensitive risk groups have been identified according to the personal characteristics of workers (European Agency for Safety and Health at Work 2009). In the Andalusian manufacturing sector, specific safety promotion programs are needed for young workers. Also, female workers need specific preventive measures for musculo-skeletal accidents. Smaller enterprises also need attention.

With the economic crisis, some slight changes affects the composition of the working population in the manufacturing sector of Andalusia. Some of those changes can explain, in part, the decrease in injury rates in the last few years, for example the increase in the proportion of female workers and older workers. (Carrillo *et al.*, 2012b).

Prevention programs should be designed considering that intervention in each collective of workers needs to be specific.

REFERENCES

Ahonen, E.Q., Benavides, F.G. & Benach, J. 2007. Immigrant Populations, work and health—a systematic literature review. *Scandinavian Journal of Work, Environmental and Health* 33(2): 96–104.

Benavides, F.G., Benach, J., Muntaner, C., Delclos, G.L., Catot, N. & Amable, M. 2006. Associations between temporary employment and occupational injury: what are the mechanisms?. *Occupational and Environmental Medicine* 63(6): 416–421.

Carrillo, J.A., Gómez, M.A. & Onieva, L. 2012a. Safety at work and worker profile: analysis of the manufacturing sector in Andalusia in 2008. In Arezes, P. et al. *Occupational Safety and Hygiene—SHO2012.*

Carrillo, J.A. & Onieva, L. 2012. Severity Factors of Accidents: Analysis of the Manufacturing Sector in Andalusia. In Arezes, P. et al. *Occupational Safety and Hygiene—SHO2012.*

Carrillo, J.A., Pérez, V. & Onieva, L. 2012b. Modelo de negocio y riesgo de accidente: el caso del sector industrial andaluz. In *"Industrial Engineering: Overcoming the Crisis." Book of Full Papers of the 6th International Conference on Industrial Engineering—CIO2012.*

European Agency for Safety and Health at Work. 2009. *Workforce diversity and risk assessment: ensuring everyone is covered.* DOI: 10.2802/11532.

Haviland, A.M., Burns, R.M., Gray, W.B. & Mendeloff, J. 2012. A new estimate of the impact of OSHA inspections on manufacturing injury rates, 1998–2005. *American Journal of Industrial Medicine* 55(11): 964–975.

Islam, S.S., Velilla, A.M., Doyle, E.J. & Ducatman, A.M. 2001. Gender Differences in Work-Related Injury/ Illness: Analysis of Workers Compensation Claims. *American Journal of Industrial Medicine* 39(1): 84–91.

Jacinto, M.C. & Aspinwall, C. 2004. A survey on occupational accidents' reporting and registration systems in the European Union. *Safety Science* 42(10): 933–960.

Jiménez-Martín, S & Sánchez-Martín, A.R. 2007. An evaluation of the life cycle effects of minimum pensions on retirement behavior. *Journal of Applied Econometrics* 22(5): 923–950.

López-Arquillos A., Rubio-Romero J.C. & Gibb A. (in press). Analysis of construction accidents in Spain, 2003–2008. *Journal of Safety Research* DOI /10.1016/j.jsr.2012.07.005.

Mendeloff, J., Nelson, C., Ko, K, & Haviland, A. 2006. *Small Businesses and Workplace Fatality Risk: An Exploratory Analysis*, USA: RAND Corporation.

Pollack, K.M., Agnew, J., Slade, M.D., Cantley, L., Taiwo, O., Vegso, S., Sircar K., Cullen M.R. 2007. Use of employer administrative databases to identify systematic causes of injury in aluminum manufacturing. *American Journal of Industrial Medicine* 50(9): 676–686.

Salminen, S. 2004. Have young workers more injuries than older ones? An international literature review. *Journal of Safety Research* 35(5): 513–521.

Sørensen, O.H., Hasle, P. & Bach, E. 2007. Working in small enterprises—Is there a special risk? *Safety Science* 45(10): 1044–1059.

Taiwo, O.A., Cantley, L.F., Slade, M.D., Pollack, K.M., Vegso, S. & Fiellin, M.G. 2008. Sex Differences in Injury Patterns Among Workers in Heavy Manufacturing. *American Journal of Epidemiology* 169(2): 161–166.

Occupational Safety and Hygiene – Arezes et al. (eds)
© 2013 Taylor & Francis Group, London, ISBN 978-1-138-00047-6

Risk assessment in processing polymer nanocomposites

S. Sousa & M.C.S. Ribeiro
Institute of Mechanical Engineering and Industrial Management (INEGI), Oporto, Portugal

J. Santos Baptista
PROA/CIGAR/Faculty of Engineering of the University of Oporto (FEUP), Oporto, Portugal

ABSTRACT: The present investigation is aimed at studying the safety assessment in the processing of polymer nanocomposites with focus on control measures for nanoparticles exposure in work environments. The "new" qualitative analyses are discussed and the feasibility of the application of the particles assay standard to nanoparticles is evaluated. For this purpose, a qualitative analysis based on methodologies ISPESL of NanoSafe Nanotool was applied and direct measurements were performed in order to allow the issue charac-terization under the regulatory parameters. A medium risk was obtained in all qualitative methods. As regards to quantitative analysis, obtained results by direct measurements did not exceed or even reach the statutory limit values of safety in Portugal. Obtained results highlight the inefficiency of the existing quantitative meth-odologies for the assessment of safety risks involved in the processing of nano based products. The safety as-sessment criteria must be revised in order to fill the existing gaps.

1 INTRODUCTION

Nanotechnology has been exponentially developed over the last decade, involving the most diverse sectors, bringing promises to create better and new materials with properties never achieved before. It involves the production, processing and application of materials with dimensions 1–100 ηm (Aitken et al., 2006, Amoabediny et al., 2009, Beaulieu, 2009, CE, 2011, Crosera et al., 2009, Gupta, 2011, Iavicoli et al., 2010, NIOSH, 2009).

The new technologies rise is always associated with new risk factors (Beaulieu, 2009, Maynard, 2007). The nanoparticles (NPs) production can expose workers by inhalation, ingestion and absorption through the skin. Exposure can also occur for operators in the downstream processes that use these materials, and for final consumers who use products sold in the market, often without taking great precautions (Aitken et al., 2006, Crosera et al., 2009, Gupta, 2011).

Nanomaterials are here to stay, and polymer nanocomposites are those that are produced and applied in larger quantities (tons/year) (Aitken et al., 2006). It is likely that the nanocomposite production processes using nanopowder or suspension/solution, represent the greatest risk of NPs release. The maintenance of nanocomposites production systems (including cleaning, equipment maintenance and waste disposal) is also likely to result in exposures to NPs (Amoabediny et al., 2009, Tsou and Waddell, 2000).

The present research project aims to study the safety assessment in the processing of polymer nanocomposites, focusing on control measures for nanoparticles exposure in work environments. It is discussed the particles standard assay (ISO 12103-A1 for ultrafine test dust: particles sizes of 0–10 μm) and the "new" qualitative analyses (nanoparticles).

The project has been developed in the Institute of Mechanical Engineering and Industrial Management (INEGI) and is focused in three production line main stages used in the INEGI laboratory, which are usually found in similar industrial processes and are likely to cause exposure.

2 MATERIALS AND METHODS

Nanocomposites were produced mixing an unsaturated polyester polymer matrix with alumina (Al_2O_3) NPs. NPs were handled under the same conditions as usually observed in the industry (general exhaustion). The applied alumina NPs (1344-28-1 CAS number) had a 45,0 ηm average size and a 36,0 m^2/g specific surface area (indicated on the safety data sheet).

It is essential to make measurements before a nanomaterial production or processing to obtain exposure data control. The methodology used in this study was based on qualitative and quantitative analyses. First, was developed a qualitative analysis based on methodologies ISPESL of NanoSafe Nanotool.

The ISPESL (Istituto Superiore per la Prevenzione e la Sicurezza del Lavoro, Italy) developed a structured method to analyse NPs safety. The risk assessment is implemented for each workplace and work activity; it is based on the following 10 factors: A) numerousness of the exposed workers, B) frequency of exposure, C) frequency of direct manipulation, D) dimensions of the nanoparticles, E) nanoparticles behaviour (e.g. dispersion or agglomeration), F) effectiveness of Personal Protection Devices (PPD) used, G) work organization/procedures, H) toxicological characteristics of the substances, I) risk of fire and explosion and J) suitability of workspaces and installations (Giacobbe et al., 2009).

The aforesaid factors are denominated "factors level risk" and one of them may assume three increasing values: 1 (low), 2 (medium) and 3 (high), referred to as "risk levels". Since the use of nanomaterials presents nowadays unknowns about the effective level of danger, the risk assessment takes into consideration these important aspects through the help of an appropriate index denominated "corrective factor". This index assumes a value within the range 0.5 and 2.0 in accordance to the established level of scientific knowledge. This assumes the following values: 0.5—good scientific knowledge; 1.0—enough scientific knowledge; 2.0 – insufficient scientific knowledge (Giacobbe et al., 2009).

The evaluation risk (Eq. 1) is calculate through the factor level risk (flr) sum (from A to J) and multiplied by the corrective factor (cf). The evaluation algorithm covers normal work conditions as well as abnormal and emergency situations. The evaluation result consists in several risk levels subdivided in an increasing way (risk level: "low" 5–15, " medium " 16–35, "high" 36 to 60). The risk level "high" shows that intervention measures are necessary to reduce the immediate outcome of the evaluation, at least to a "medium" risk level. This can be done by intervening in the main factors (low numerousness of the exposed workers; low frequency of direct manipulation; limited frequency of exposure; use of Personal Protection Devices; medium toxicological characteristics of the substances) (Giacobbe et al., 2009).

$$Evaluation\ risk = \sum_{i=A}^{J} (flr)_i \times cf \qquad (1)$$

NanoSafe "Online" research, performed by the Swiss team, got a surprising result: nearly three quarters of two hundred and forty inquired researchers had no basic rules to follow regarding the NPs handling. Based on this study a decision tree was developed to the "nano-laboratories" stabilizing three risk classes, which correspond to similar approaches applied to other hazards types (biological, chemical or radiation) (Groso et al., 2010).

Another safety analysis method to NPs exposure used was the Nanotool matrix. It is a four by four factors matrix that relates severity and probability (Zalk, 2009).

Direct measurements have also been performed to allow the issue characterization under the current regulatory parameters. In order to allow a better exposure interpretation, two different formulations were produced without nanoparticles (control/plane and with microparticles—glass microspheres) to compare the dust exposition with the polymeric nanocomposite production. The equipment used for the measurements was the Dust-Trak monitor aerosol (model 8520). This equipment, measures the fine and coarse particles concentration by weight between 1–10 μm, having been used for this study the 1 μm nozzle.

3 RESULTS AND DISCUSSION

It was observed that NPs pass through different "states" during the nanocomposites production: in the pre-production, NPs are at powder state; in the production, NPs are in a resin/solution; and, at last, in post-production the NPs are inserted into the solidified resin matrix.

Using the ISPESL method the results lead us to a medium risk according to the data contained in Table 1.

The same result was confirmed by the NanoSafe method and that is described in Figure 1.

With the Nanotool method, there was obtained a probable medium severity (Tab. 2, 3).

The application of both methods have convergent and consensual results.

Table 1. ISPESL method results.

Factors	Level*
A-numerousness of the exposed workers	1
B-frequency of exposure	1
C-frequency of direct manipulation	2
D-dimensions of the nanoparticles	2
E-nanoparticles behaviour	1
F-effectiveness of PPD used	2
G-work organization/procedures	2
H-toxicological characteristics of the substances	2
I-risk of fire and explosion	1
J-suitability of workspaces and installations	3
Σ(A – J)	17
Corrective factor	2
Risk assessment	34
Total risk level	Medium

* Values between 1–3.

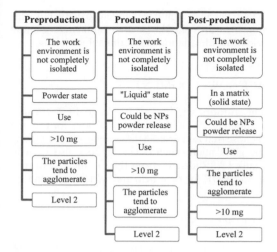

Preproduction	Production	Post-production
The work environment is not completely isolated	The work environment is not completely isolated	The work environment is not completely isolated
Powder state	"Liquid" state	In a matrix (solid state)
Use	Could be NPs powder release	Could be NPs powder release
>10 mg	Use	Use
The particles tend to agglomerate	>10 mg	The particles tend to agglomerate
Level 2	The particles tend to agglomerate	>10 mg
	Level 2	Level 2

Figure 1.   Nanosafe method results.

Table 2.   The Nanotool factors analysis results.

Severity	Probability
Surface chemistry NP: Medium: 5	Estimated amount of NM used during
Particle shape NP: Compact/spherical: 0	operation: >100 mg: 25
Particle diameter NP. <41–100 ηµ. 0	
Solubility NP: Insoluble: 10	Dustiness/mistiness: Medium: 15
Carcinogenicity NP: Unknown: 4.5	
Reproductive toxicity NP: Unknown: 4.5	
Mutagenicity NP: Unknown: 4.5	Number of employees with similar exposure:
Dermal toxicity NP: Unknown: 4.5	6–10: 5
Asthmagen NP: Unknown: 4.5	
Toxicity Parent Material (PM): >1 mg m-3: 0	Frequency of operation: Weekly: 10
Carcinogenicity PM: No: 0	
Reproductive toxicity PM: No: 0	
Mutagenicity PM: No: 0	Duration of operation:
Reproductive toxicity PM: Yes: 4	1–4 h: 10
Asthmagen PM: Yes: 4	

The measurement results obtained with the dust analyser, in the processing of the three different composite formulations, are described in Table 4.

For the comparative analysis, in order to obtain another benchmark, the total dust of the "open plan office" of INEGI (the space for the INEGI employ-

Table 3.   Obtained results by applying the Nanotool matrix.

	Probability			
Severity	Extremely unlikely (0–25	Less likely (26–50)	Likely (51–75)	Probable (76–100)
Very High (76–100)	RL 3	RL 3	RL 4	RL 4
High (51–75)	RL 2	RL 2	RL 3	RL 4
Medium (26–50)	RL 1	RL 1	RL 2	RL 3
Small (0–25)	RL 1	RL 1	RL 1	RL 2

Table 4.   Measurements summary obtained with the dust track during the composites production.

Formulations	Control (Plane)	Microparticles (Glass microspheres)	Nanoparticles (Nano Al_2O_3)
Weighted 8h (mg/m³)	0,014	0,043	0,025
Maximum (mg/m³)	0,060	0,145	0,045
Minimum (mg/m³)	0,003	0,009	0,011

ee's secretarial work) was measured. A maximum value of 0.020 mg/m³ was obtained in the "open plan office", however the maximum reached value, in the laboratory, while was produced the control formulation was 0,060 mg/m³. Although the measurement was not intended to be made only to the nanoparticles, it was found by observation that many tasks involved in the process have different exposure risks to NPs. The statutory limit values of safety in Portugal were not exceeded or even reached (the time weighted average for the glass microspheres and for nano Al_2O_3 is 10 mg/m³). The average value of exposure on the nanocomposite processing was identical to the maximum value measured in the "open plan office". However, one point must be stressed: the equipment applied is standardized and it is only prepared to detect microparticles and not NPs. Therefore, the quantitative results cannot be compared with qualitative results obtained with the different methods.

Analysing the quantitative results, the importance of the exhaust/ventilation in the workplace as well as the isolation of the different places were highlighted. The highest measured values corresponded to the moments when: a) the exhaustion/ventilation was not operating, b) the space was not isolated (door open), or c) tasks like transportation, weighing, mixing and clean-up were in process. It was found that the most critical operation in the nanocomposite entire manufacturing process is handling (weight/mixing) because it deals with NPs in powder, and to which a lack of care procedures, the organizational aspects, or preventive maintenance may lead to typical scenarios of exposure. The most suitable engineering measure is the one that advocates the isolation of all operations involved in the manufacturing process, such as the use of clean rooms, laminar flow chambers or glove box use (suggested by the qualitative methods for a medium risk).

The existing regulation is often based on parameters that may be unsuitable for certain nanotechnology applications. For example, thresholds are often defined in terms of mass or production volume, below which a substance may be exempt from regulation. The thresholds should be reviewed and when appropriate should be revised. In today's global market, economic growth demands innovation, which, in turn, is dependent upon research.

4 CONCLUSIONS

The results obtained by direct measurements were not consistent with the results obtained with the qualitative methods, since the former proceeding follow the legislation established in Portugal. With the actual and impending legislation applicable to nanoparticles, it is possible to implement prevention and protection generic measures. These measures should be taken as soon as possible to ensure a NPs handling responsible development. The adoption process of these measures should always be based on four principles:

- Due to its size and its specific surface, NPs must be regarded as new products and not as mere "smaller" substances/products whose risks and hazards are already known.
- NPs should be systematically assessed. Standard toxicity tests should be performed systematically for all new type of NPs and/or manufacturing/handling processes, according to the tests prescribed on the regulations for the chemical agents.
- Nanomaterials should at least be manufactured/handled according to the regulations for the chemical agents, especially in the powder state, because they can more easily contaminate the work environment and enter into the body through the respiratory system.
- All development programs should be linked to NPs research to assess the safety health and environment.
- It would be also important to ensure the traceability of all the products with nanoparticles capable to being dispersed, by using specific labelling.

More research work is needed to fill existing gaps. Until further information, nanomaterials should be considered hazardous materials. In the future, many NPs types may well have limited toxicity, but caution measures should be used until there is more scientific information about this subject.

ACKNOWLEDGMENTS

The authors wish to acknowledge FCT (Project Nanocrete, PTDC/ECM/110162/2009), for funding the research and also 3M™ company, for donating the glass microspheres. The authors also wish to thank J.A. Rodrigues from INEGI and J.C. Branco from FEUP, for their help and valuable assistance.

REFERENCES

Aitken, R.J., Chaudhry, M.Q., Boxall, A.B.A. & Hull, M. 2006. Manufacture and use of nanomaterials: current status in the UK and global trends. *Occupational Medicine,* 56, 300–306.

Amoabediny, G.H., Naderi, A., Malakootikhah, J., Koohi, M.K., Mortazavi, S.A., Naderi, M. & Rashedi, H. 2009. Guidelines for safe handling, use and disposal of nanoparticles. *Journal of Physics: Conference Series,* 170, 012037.

Beaulieu, R.A. 2009 Engineered Nanomaterials, Sexy New Technology and Potential Hazards *2009 Safety Analysis Workshop.* Las Vegas, NV, United States: Lawrence Livermore National Laboratory (LLNL), Livermore, CA.

CE 2011. Recomendação da Comissão de 18 de Outubro de 2011sobre a definição de nanomaterial. Bruxelas.

Crosera, M., Bovenzi, M., Maina, G., Adami, G., Zanette, C., Florio, C. & Filon Larese, F. 2009. Nanoparticle dermal absorption and toxicity: a review of the literature. *International Archives of Occupational and Environmental Health,* 82, 1043–1055.

Giacobbe, F., Monica, L. & Geraci, D. 2009. Nanotechnologies: Risk assessment model. *Journal of Physics: Conference Series,* 170, 012035.

Groso, A., Petri-Fink, A., Magrez, A., Riediker, M. & Meyer, T. 2010. Management of nanomaterials safety in research environment. *Particle and Fibre Toxicology,* 7, 40.

Gupta, N.S., A. 2011. Issues Associated with Safe Packaging and Transport of Nanoparticles. *ASME Pressure Vessels and Piping Conference.*

Iavicoli, I., Calabrese, E.J. & Nascarella, M.A. 2010. Exposure to Nanoparticles and Hormesis. *Dose-Response,* 8, 501–517.

Maynard, A. 2007. Nanotechnologies: Overview and Issues. *In:* Simeonova, P., Opopol, N. & Luster, M. (eds.) *Nanotechnology—Toxicological Issues and Environmental Safety and Environmental Safety.* Springer Netherlands.

NIOSH 2009. Approaches to Safe Nanotechnology: Managing the Health and Safety Concerns Associated with Engineered Nanomaterials. *DHHS (NIOSH—United States National Institute for Occupational Safety and Health) Publication* no. 2009–125, 104.

Tsou, A.H. & Waddell, W.H. 2000. Fillers. *Kirk-Othmer Encyclopedia of Chemical Technology.* John Wiley & Sons, Inc.

Zalk, D.P., S 2009. Control Banding and Nanotechnology Synergist *The Synergist,* 21, 26–29.

Exposure and perception to electromagnetic radiation of low frequency

A. Rainha, J.P. Figueiredo, N.L. Sá & A. Ferreira
Escola Superior de Tecnologia da Saúde de Coimbra, Portugal

S. Paixão
Escola Superior de Tecnologia da Saúde de Coimbra, Portugal
Centro de Estudos de Geografia e Ordenamento do Território, Portugal

F. Alves
Escola Superior de Tecnologia da Saúde de Coimbra, Portugal
Instituto de Ciências Nucleares Aplicadas à Saúde, Coimbra, Portugal

E. Lankford
Dublin City University, Ireland

ABSTRACT: In residential environment, population is exposed to radiation emitted by several equipments in daily life, for variable periods of time. Thus, it becomes imperative to assess the levels of population exposure in the residential environment, compare them to the legally recommended, as well as to evaluate the level of knowledge on the subject. To fulfill this aim, a descriptive-correlational character (level II) cross-sectional cohort study was performed. The target population were the inhabitants of four blocks of residencies associated with the Polytechnic Institute of Coimbra (IPC), where the sample drawing set was as probabilistic and convenience/sampling techniques. The sample consisted of a total of 180 surveys and 72 measurements of electromagnetic fields. After analysis of the results, we concluded that most of the students have no knowledge regarding electromagnetic pollution. In relation to field intensity measurements, there are some excesses to the reference value set in the Directive 1421/2004 of November 23th, in the frequency ranges of 41 Hz to 58 Hz and 100 Hz to 170 Hz. On frequency ranges of 800 Hz to 1800 Hz and 2000 Hz to 3000 Hz no excesses to the reference value were observed.

1 INTRODUCTION

In the last century there was a progressive increase in exposure to electromagnetic fields (EMF) created by man. This increase was consistent with a greater demand for electricity, with advances in technology and changes in social behavior (DGS 2007).

The radiation from each source can interact with biological systems in different ways, and electromagnetic fields are characterized by the operating frequency and by the intensity of the field. These characteristics determine the interaction with biological systems (WHO 2007). From this interaction it is possible to classify the electromagnetic radiations in two main classes: Ionizing Radiation and Non-Ionizing Radiation (Hardell 2008).

The interaction mechanisms of non-ionizing electromagnetic fields with biological systems can be grouped into two major types: thermal and non-thermal effects, depending on whether they are attributable to the deposition of heat (thermal) or to direct interaction of the field with the substance of the tissue, without significant heating component (non-thermal or heatless) (Dias et al 2002).

Thermal response of a body depends on specific absorption rate (SAR), coverage of the regulatory body, the thermoregulatory system, the physiological condition, the environment, and in the case of irradiation only on a specific body part, the vascularity in the region. Under normal circumstances, blood vessels dilate and heat is removed from the bloodstream. Therefore, the main risk of thermal damage is found in areas of low vascularity, as the eyes and temples (Dias et al 2002).

Many studies have been made in several countries to investigate the alleged connection between certain frequencies of electromagnetic radiation and cancer. Since 1979, epidemiological studies have raised the possibility of a link between the exposure to electromagnetic fields and the onset of child leukemia, cancer of the central nervous system in childhood, breast cancer, lung cancer, colon cancer and other diseases such as amyotrophic lateral sclerosis, Alzheimer's, asthma, allergic diseases, increased incidence of abortion, dermatitis caused by television and computer, cardiac diseases, endocrine disorders, neuro-conductor changes, among others (Freitas et al 2006).

According Poole et al. (1993) and Repacholi (1998), exposure to electromagnetic field can cause dysfunction of central nervous system (CNS), induce nervousness, anxiety, stress, sleep disorders, among others (Ribeiro et al 2007). In 1996, the World Health Organization (WHO) established the International EMF Project to investigate the potential health risks associated with technologies emitting EMF (WHO 2007).

According to the International Agency for Research on Cancer, 2002, a World Health Organization agency, the electromagnetic fields were considered potentially hazardous to health, and were classified as "possibly carcinogenic to humans". This classification is used to designate an agent for which there is limited evidence of carcinogenicity in humans and a less than sufficient evidence of carcinogenicity in experiments on animals (WHO 2007).

In October 2005, WHO formed a working group to assess the health risks associated with exposure to electromagnetic fields in the frequency band from 0 to 100,000 Hz. In this study, the group concluded that there are relevant health issues related to exposure to electromagnetic fields at the levels to which the population in general is usually exposed.

Stacy et al, 2007, of the University of Essex, focused his study on the negative effects of the use of mobile phones on health. In their work, the authors question the relationship between the electromagnetic fields created by everyday objects, as is the case of mobile phones or base stations, and the appearance of idiopathic diseases, during a short term exposure. During this research Stacy et al, sought to get answers about changes to the level of blood volume pulse, heart rate or changes in skin conductance. Through questionnaires distributed concluded that 394 of 429 people reported these symptoms (Alper Çabuk et al., 2009).

The international community has recognized the desirability of establishing protective limits of public health. In this way, in 1998 were defined in a Guide of the ICNIRP (International Non-Ionizing Radiation Committee) values for the limitation of this exposure to variable fields, shed in the Council recommendation faithfully European 519/EC of 1999 and, later, Portuguese legislation, by Directive number 1421/2004 of 23 November (Ulu et al., 2010).

In the residential environment, the population is often in the vicinity of equipment emitting electromagnetic radiation of low frequency, it being necessary to evaluate public exposure to these fields. For the measurement of personal exposure to EMF, specific equipment has been used for this purpose, as is the shoe TAOMA TS/001/UB.

The aim of this study being to assess the levels of EMF exposure of low frequency of population in residential environment in relation to the Directive number 1421/2004 of 23 November, as well as the level of knowledge on the part of the population in the face of electromagnetic pollution.

2 MATERIAL AND METHODS

The study was carried between October 2011 and July 2012, having the data collection period during the month of June. The study applied was level II, descriptive-correlational type and cross-sectional cohort.

The Target Population of the study was living in four blocks of flats belonging to the IPC (A and B Blocks–S. Martinho do Bispo Blocks C and D–Quinta da Nora), in the city of Coimbra, Portugal.

The type of sample used was not probabilistic, and the sampling technique Convenience/accidental.

Before the collection of data was presented a request for authorization to the direction of the institution was obtained.

To collect data was used a survey that was pre-tested to evaluate its reliability. To ensure the confidentiality of data, there was prior information about the aim of the study. The population was informed and comfortable of answering the survey.

The calculation of the sample was taken into account for a finite population (<100,000), and contained a sample of 212 individuals, however only 180 respondents participated in the study. It was divided in 2 parts: Part 1, helped to identify the characteristics of our sample regarding age, gender, education and occupation and Part 2, determined the level of knowledge of citizens about electromagnetic pollution.

After collecting our data, treatment was carried out using the IBD SPSS statistics version 19, and the interpretation of statistical tests was based on a significance level of $\alpha = 0.5$ with 95%. We used simple descriptive statistics: measures of location (mean) and dispersion (standard deviation) and also the Chi-square, T-Student and Rho Spearman's.

For the data collection on levels of exposure to electromagnetic fields, we used a measurement device called TAOMA TS/001/UB. The probe measures in an area between the 15Hz and 100kHz. Data analysis was performed using Microsoft Office Excel 2007.

The areas where the measurements were made, were kitchens and common rooms in 4 blocks (A, B, C, D) belonging to the IPC. And there was a total of 72 measurements, during 3 days. The choice of these places relates to the fact that there are several devices that emit low frequency electromagnetic radiations, and where access is facilitated. Each measurement was carried out at about 1.50 meters in height above the ground and during at times in which the use of the areas is higher. The measurements were carried out in kitchens between 12 to 15 hours and the common rooms between 20 to 24 hours. Each measurements took 6 minutes per area.

For the analyses of the field intensity it was necessary to mind the tridimensional characteristics of the electromagnetic waves during the measurements. The data obtained were grouped in frequency ranges, they, we compared the field intensity values obtained with the reference values in Directive 1421/2004 of 23 November.

3 RESULTS

In order to allow a clear reading of the data, this section was subdivided in two times: the first devoted to the characterization of sample surveys to determine the level of knowledge of the population compared to electromagnetic pollution, having as a goal to answer the questions of the study; in the second time the amounts related to measurements of levels of electromagnetic made in two residences belonging to the Polytechnic Institute of Coimbra.

With regard to the characterization of the sample refers to that of the 180 respondents (the participants) the majority were female (62.8%) and with regard to age, it was found that the majority had 20 years (20%). Regarding the qualifications it was found that the majority of individuals attended a degree (95%), with regard to the area of study we can verify that most participants attended the courses of mechanical engineering, Electromechanical Engineering and computer engineering, with a total of 11 respondents each one (6.1%).

As for the students' knowledge about Electromagnetic Pollution, we found that there is no significant agreement ($p > 0.05$) between knowledge that students claim to have regarding the identification of electromagnetic pollution and their real knowledge about electromagnetic pollution. Only 7.5% of the students who claimed to have some knowledge about electromagnetic pollution, demonstrated those same skills through the identification of electromagnetic fields presented as a suggestion in the survey. In Table 1 it is possible to verify these results.

Table 1. Relationship between the student's knowledge and the identification of sources of electromagnetic fields.

Source of electromagnetic fields					
		No	Yes	Total	p-value
Knowledge of electromagnetic pollution	No	71 (97,3%)	2 (2,7%)	73 (100%)	0,148
	Yes	98 (92,5%)	8 (7,5%)	106 (100%)	
	Total	169 (94,4%)	10 (5,6%)	179 (100%)	

Test Chi-square.

When we tried to verify if that the students study area influences their level of knowledge we found that there was no correlation ($p > 0.05$) between the area of study of knowledge regarding electromagnetic pollution. However, we can see that among students in 52 courses, the Electromechanical Engineering students have a greater knowledge of electromagnetic pollution, with a percentage of 4.5 to 6,1%.

Regarding the level of students' knowledge about electromagnetic pollution and their gender, we found that there was no significant association ($p > 0.05$). However, it was found that the males had a higher level of knowledge.

We also sought to determine whether there was an association between the level of students' knowledge about electromagnetic pollution and the age. There is no association between age and knowledge about electromagnetic pollution ($p > 0.05$). The average age of respondents who had knowledge of electromagnetic pollution (23 years) and respondents who have no knowledge of electromagnetic pollution (21 years) does not allow a differentiation because they are very close.

In relation to students who had knowledge of electromagnetic pollution we sought to determine whether they adopted some measures of protection from electromagnetic radiation. We found no significant association ($p > 0.05$) between these variables. Students who did not adopt or did not know what steps to take to protect themselves from electromagnetic radiations are mostly students who have no knowledge about electromagnetic pollution. For students who claim to have knowledge of electromagnetic pollution, only 12% used protection measures. On the other hand, 88% of the students who claim to adopt protective measure do not have only knowledge about electromagnetic pollution.

We also tried to research whether the students who had knowledge about electromagnetic pollution were concerned about the possible health effects of using equipment emitting electromagnetic waves. We found that there was no significant association (p > 0.05). Most of the students who reported being concerned with possible health effects from the use of equipment emitting electromagnetic radiation, had no knowledge of electromagnetic pollution (96%). Most of the students who said they did not know or expressed no concern about the possible health effects from the use of equipment emitting electromagnetic radiation, had no knowledge of electromagnetic pollution.

When we tried to verify if the students who had knowledge of electromagnetic pollution thought that consider there is sufficient information about the possible health effects of the use of equipment emitting electromagnetic radiation, we found that there was no significant association (p > 0.05) between variables. It was also observed that only 6.3% of the students who had knowledge of electromagnetic pollution, considered information sufficient. Of the students who considered there is sufficient information, 93.8% have no knowledge about electromagnetic pollution.

Through the survey we also sought to determine whether students associated electromagnetic fields to pathologies.

We found that according to students surveyed there was no association between the different diseases (leukemia, cancer, increasing the number of abortions, headache) and the electromagnetic field (p > 0.05). The difference between the numbers of respondents, who associated the emergence of these diseases to electromagnetic fields, when compared to the number of respondents who do not make this association, was small, not allowing a differentiation. However, cancer was the disease which respondents more often associated with electromagnetic field (98.9%).

The existence of Portuguese law establishes limits on human exposure to electromagnetic fields we sought to determine whether students had knowledge of it. We found that of the 179 students who answered the question, 112 said they did not exist or do not know the existence of national legislation that sets limits to human exposure to electromagnetic fields.

When we tried to verify the most important factors to consider in the purchase of electrical equipment, we found that there was a correlation significant (p = 0.000) between equipment efficiency and safety of equipment and people. The respondents believe that the higher the efficiency of the equipment the greater the safety. Regarding the relationship between the variable cost of equipment and the variable aesthetics of the product

there was also a significant correlation (P = 0.000). According to them the higher aesthetics the higher cost.

Note that the reference value established by the law varies according to the frequency range in which the equipment/device operates. The low-frequency electromagnetic fields vary between 3 Hz and 300 KHz. In this particular case ranges of frequencies from 41Hz to 58Hz were considered, which corresponds to the electrical equipment supply, 100 Hz to 170 Hz corresponding to power supplies. We considered a range of frequencies from 800 Hz to 1800 Hz, which operate various telecommunications equipment, such as mobile phones and WIFI, and a range of frequencies above 2000 Hz where they operate telecommunication equipment unconventional. We sought to determine whether the recorded field strength measurements did not exceed the reference values established by the Directive.

Table 2. Important factors in the acquisition of electrical equipment.

		Cost of equip.	Safety of equip.	Aesthetic equip.
Efficiency of the equipament	Correlation coefficient	,115	,328**	−,016
	p-value	,128	,000	,834
	N	177	177	177
Cost of equipament	Correlation coefficent	–	,116	,324**
	p-value	–	,124	,000
	N	–	177	177
People and equipament safety	Correlation coefficient	–	–	−,043
	p-value	–	–	,567
	N	–	–	177

** Correlation is significant at the 0.01 level (2-tailed). Test rho Speraman's

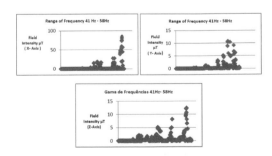

Figure 1. Field strength recorded in the X, Y and Z-axis for the frequency range between 41 Hz to 58 Hz.

As for the characterization of the sample on the intensity measurements of the electromagnetic field, there were two blocks of houses (A and B) located in Bencanta, in the parish of São Martinho do Bispo and two blocks of residences (C and D) located in Quinta da Nora, in the parish of Santo António dos Olivais. The blocks of houses belonging to the IPC.

According to Directive 1421/2004, the reference value for the 25 Hz to 800 Hz frequency range is 5/f, and f is a frequency. The reference value corresponding to this frequency range is between .08 μT to 0.1 μT. We observed that the majority of values of field strength measurement is obtained in close to 0μT. However, there were some higher reference values. These values can be explained by the existence of a damaged equipment (microwave-door cracked), allowing a greater spread of the radiation emitted by the equipment.

3.1 Discussion

Most of the individuals surveyed did not reveal to possess any knowledge on electromagnetic pollution, not knowing the sources of electromagnetic radiation, or the existence of national legislation aimed at protecting the health of the population in general. Most students surveyed consider the information about the possible health effects resulting from exposure to electromagnetic radiation insufficient. When we sought to check whether there was an association between the level of students' knowledge about electromagnetic pollution and their age, gender or level of qualifications, we concluded that none of these factors influenced the levels of knowledge of students about the subject under investigation significantly.

Regarding the adoption of measures for protection against electromagnetic radiation, it was found that most respondents showed a lack of knowledge or awareness of the protection against electromagnetic radiation (only 12% respondents with knowledge about electromagnetic pollution took protective measures). It was also found that the adoption of these measures is not conscious nor is there the intention of using protection against the effects of electromagnetic radiation, since 88% of respondents who adopted those measures had no knowledge about electromagnetic pollution.

When we examined whether students felt concerned about the possible health effects caused by the use of equipment emitting electromagnetic radiation, it was found that most students do not consider its use a source of concern. When we tried to verify the students' perception about pathologies and exposure to electromagnetic fields, it was found that cancer is the most referenced disease.

However, it was not statistically possible to associate these variables.

As for the measurements of the intensity of electromagnetic field, the value established in the Directive 1421/2004, was exceeded. In the frequency range corresponding to the electric supply (41Hz to 58Hz) on which various equipment/devices, that exist in the home environment depend, we observed several values of field strength above the reference value, which may have originated in damaged equipment in the kitchen of Block C of the residences of the IPC. The maintenance of the equipment in good working conditions proves to be extremely important because it allows a more effective safety to the individuals, being referenced by DGS as devices that must be given particular attention by the user (DGS 2007).

The estimate of exposure to electromagnetic fields appears to be of great interest, yet it is difficult to assess due to several factors, such as differences in susceptibility of the population in general. The existence of gaps in current scientific knowledge justifies, the application of the Precautionary Principle, which is reflected in the adoption of precautionary strategies in the management of risk in situations of scientific uncertainty (DGS 2007).

4 CONCLUSION

The subject of electromagnetic pollution, as well as health problems from exposure to this type of radiation, are poorly studied in Portugal. With this research work we intended to fill the little information/studies on the subject. This should be a priority for the government and professionals working in the fields of Environment and Health. Therefore, it is expected that this work gives an added value to the knowledge about the causes and consequences of electromagnetic pollution and an alert to the need for greater information on the subject.

From the results obtained it was possible to conclude that most students do not have any knowledge regarding the subject studied. As for the field intensity in the recorded measurements, the values reference set in the law, there were some excesses in the frequency range 41 Hz to 58 Hz and 100 Hz to 170 Hz. In the frequency ranges of 800 Hz to 1800 Hz and 2000 Hz to 3000 Hz there were no excesses to the reference value.

As a limitation to the study, we stress the characteristics of the population which did not allow us to distinguish the differences in knowledge regarding electromagnetic pollution based on age or qualifications. Initially we had planned to apply 212 surveys, however due to the time of year and since many students go home, we only got 180 responses. Lack of studies and little information about this issue were other difficulties that we found.

We suggest that in the future, this study continues for a broader assessment of people's knowledge and perception of the risks regarding this issue.

REFERENCES

Alper Çabuk, Saye N. K., Hakan Uyguçgil, M. I. GIS and RS Based Location Determination for GSM Transmitters to Minimize the Negative Effects of Electromagnetic Pollution for Improving Quality of Urban Places. Turkey: International Journal of Natural abd Engineering Sciences, 2009, pp. 57–65.

Dias, Maurício H. C. e Siqueira, G. L. *Considerações sobre os Efeitos à Saúde Humana da Irradiação Emitida por Antenas de Estação-base de Sistemas Celulares*. Rio de Janeiro—Brasil. Revista Científica Periódica—Telecomuicações, 2002, Vol. 5, n° 1, pp. 41–54.

Direção Geral de Saúde. *Sistemas de Comunicação Móveis: Efeitos na Saúde Humana.* Lisboa: Polarpess, Lda, 2007. ISBN 978-972-675168-7.

Freitas, Tiago P. M. e Nestor R. M. Monitoramento das radiações eletromagnéticas não ionizantes de baixa frequência em uma creche na cidade de Criciúma—SC. UNIrevista, 2006, Vol. 1, n° 3, pp. 1–10.

Hardell, Lennart e Sage, Cindy. Biological effects from electromagnetic field exposure and public exposure standards. Biomedecine & Phamarcotherapy, 2008, pp. 104–109.

Ribeiro, Edson e Pessoa, Martha. Os efeitos da radiação Electromagnética na Vida do Ser Humano: Uma Análise do Paradigma Ambiental. Revista Tecnologia e Sociedade, 2007, pp. 15–31.

Ulu, Eylem Yilmaz, et al. *Analysis of a Photovoltaic—Fuel Cell Hybrid Energy System in Terms of Electromagnetic Pollution*. International Review of Electrical Engineering, 2010. Vol. 5, n°4, pp. 1600–1608.

World Health Organization. Campos Electromagnéticccos e Saúde Pública—Exposição a campos de frequência extremamente baixa. 2007, n°322.

Occupational Safety and Hygiene – Arezes et al. (eds)
© 2013 Taylor & Francis Group, London, ISBN 978-1-138-00047-6

Risk assessment and control in engineered nanoparticles occupational exposure

F. Silva & P.M. Arezes
Human Engineering Group, Production and Systems Department, University of Minho, Portugal

P. Swuste
Safety Science Group, Delft University of Technology, Delft, The Netherlands

ABSTRACT: The huge research effort and common use of nanomaterials, being an opportunity for economic growth, pose health and safety problems. The research on the nanoparticles health effects performed during the last decade shows the possible harmfulness of several nanoparticles, including those already present in everyday use products. Although the increasing knowledge in the nanotoxicilogy field, and also the occupational hygiene responses in order to develop quantitative methods to evaluate nanoparticles exposure risk, there is a uncertainty climate. The use of qualitative risk assessment methods appears as a suitable way to deal with the uncertainties and to support decisions leading to the risk control. Among these methods, those based in control banding, such as the CB Nanotool and the Stoffenmanager Nano, seems to become applied more frequently. Furthermore, the design approach to safety can be a valuable way to establish the strategy to protect the workers' health focusing in the production process in order to define the most effective measures to control the exposure risk.

1 INTRODUCTION

Nanotechnology is presented as part of a new industrial revolution, creating new opportunities in the areas of energy, materials, health, electronics, information technology and many other areas. Since Richard Feynman gave, in 1959, its conference "There's plenty of room at the bottom," which drew attention to the existing potential in the manipulation of matter at the atomic level (Feynman, 1960), that started the research (first), and then the development and use of hundreds of applications involving nanoscale materials.

According to the Project on Emerging Nanotechnologies (PEN) released information, the number of nanotechnology-based products available to consumers in March 2011 was about 1300 (WWICS, 2011). The main product categories were health and wellness (738), home and garden (209), automobile (126), and Food & Beverage (105) (WWICS, 2011). Massive investments are made worldwide in order to achieve new materials and products with innovative features.

However, this economic and social dawn is undoubtedly overshadowed by questions arising from possible adverse effects, either to human health or to the environment. From the previous experience, with particular emphasis on the issue of widespread use of asbestos and the nuclear technology, lead societies to think about if the scientific and technological development, and hence the economic development, can once again put a serious threat to people's health and well-being or environmental balance. On the other hand, the "precautionary principle" applied to the genetically modified organisms (GMOs), through a moratorium on its widespread use in agriculture, based on the lack of knowledge on harmful long-term effects, raises the doubt about the possible application the same principle to nanotechnology.

In this uncertainty climate, risk management is essential to sustain economic development without jeopardizing the environment and human health, especially in case of the industry and laboratories workers who are exposed to (possibly) dangerous nanomaterials.

In recent years, there has been a great effort in the development of knowledge in this area but the information available is still insufficient to establish whether the parameters for assessing the risk to the health of exposed workers or the exposure limit values that would refer to that same exposure. Both in the field of toxicology and in the industrial hygiene, improvements have been made to better characterize the risk during operations with nanomaterials but the results are still unsatisfactory.

This article goal is to identify the current knowledge on nanoparticles characterization and qualitative occupational risk assessment and control in the nanomaterials field.

2 STATE-OF-THE-ART

In short, nanotechnology can be defined as the nanometer scale matter understanding and control, more specifically material smaller than 100 nm, resulting in size dependent new applications and purposes.

The European Commission published in the Official Journal of the European Union on 20 October 2011 the following definition for nanomaterial:

"«Nanomaterial» means a natural, incidental or manufactured material containing particles, in an unbound state or as an aggregate or as an agglomerate and where, for 50% or more of the particles in the number size distribution, one or more external dimensions is in the size range 1 nm–100 nm."

In Occupational Hygiene the nanoparticle concept is more relevant for the personal exposure assessment. The nanoparticle definition, consistent with the previous concepts, is "particle with a nominal diameter (such as geometric, aerodynamic, mobility, projected-area or otherwise) smaller than about 100 nm" (Technical Committe ISO/TC 146, 2007).

At present, we are witnessing the transition from the first generation of passive nanostructures to the so called second generation nanotechnologies which include active nanostructures (Roco et al., 2011). In a longer term, it is anticipated the third generation nanotechnologies of "Systems of Nanosystems" development and trading and, later the fourth generation, "Molecular Nanosystems" dawn (Renn and Roco, 2006; Bowman and Hodge, 2006). Complexity is increasing as well as the related uncertainties so that, despite the achieved growing acknowledgment on nanomaterials, there are always new conditions that impose new challenges.

2.1 Human health effects

When referring nanomaterials we must consider the variety of materials, both in its composition, shape, size and other characteristics, due to the different behaviors and toxicological effects identified in toxicological tests (Savolainen et al., 2010).

Over the last years, especially in the last decade, toxicological tests have been performed with different types of engineered nanoparticles (NP) (e.g., single-walled carbon nanotubes; ultrafine TiO_2; ultrafine carbon black; silver; etc.), on the attempt to understand their effects on the human body. These are mainly in vitro and in vivo tests performed according to techniques used

for "traditional" materials. The NP tested present different behaviour in the human body when compared with larger particles of the same material. Furthermore, the showed effects in the lungs such as deposit in the alveoli, evade phagocytosis, produce interstitial inflammation, produce fibrosis, produce tumours or induce granulomas, some NP show the ability to pass the body barriers and enter in the circulatory system, penetrate in various organs, (Schulte et al., 2008)

With respect to carcinogenicity, the existing data are inconclusive, although some evidence of possible carcinogenic effect of nanoparticles that do not appear to result from its composition, namely in the carbon nanotubes (Becker et al., 2011).

We are facing a scenario in which there is already a significant amount of information on the health effects of nanoparticles but where the uncertainty is yet large, while the scientific community tries to improve the information quality.

To establish a knowledge base necessary to assess the risk to human health associated with exposure to engineered nanomaterials, staged testing strategies proposals have been made (Savolainen et al., 2010).

A materials physiochemical and toxicity characterization base tests battery, presented in Table 1 was proposed by another author (Warheit et al., 2007).

This tests set does not include all health affection relevant aspects, and can be regarded as a primary diagnosis to the concerned nanomaterial, and, subsequently, must be complemented by other tests to enable a more complete characterization.

The Scientific Committee on Emerging and Newly Identified Health Risks (SCENIHR) sets a wider range of tests for the physicochemical characterization of NM, including the following: Size

Table 1. Nanomaterials base set of hazard tests (Warheit et al., 2007).

Nanomaterial physiochemical characterization	Mammalian hazard tests	Genotoxicity tests	Aquatic screening battery
Size and size distribution	Pulmonary bioassay	Bacterial reverse mutation	Rainbow trout
Crystal structure	Skin irritation Skin sensitization	Chromosomal aberration	Daphnia
Chemical composition	Acute oral toxicity		Green algae
Surface reactivity	Eye irritation		

and size distribution of free particles and fibers/rods/tubes, specific surface area, stability in relevant media (including the ability for aggregating and disaggregating), surface adsorption properties, water solubility, being also recommended the knowledge of chemical reactivity and, depending on the nature of the nanoparticles, photoactivation capabilities and the potential to generate active oxygen (SCENIHR, 2009).

In the same report the SCENIHR, although the information is yet scarce, refer the possibility to infer some effects through the data intercomparison when there are similarities in the characteristics of engineered nanomaterials with other particles already studied and characterized.

Another contribution to obtain reliable results in a more quick and economical way, is the proposed use of *in vitro* tests, specifically designed for nanomaterials and held in co-culture instead of only one type of tissue used for testing (Clift, Gehr and Rothen-Rutishauser, 2011). In another study published in 2008 there has not been found correlation between the results of the *in vitro* and *in vivo* assays made to assess the effects of different types of nanoparticles on the lung tissue, leading to conclude that *in vitro* tests should be more sophisticated in order to better simulate the conditions of the lung (Sayes et al., 2008).

Considering that the basic principles are established to frame the engineered nanomaterials characterization in relation to its harmfulness, it may be considered that the information resulting from it will contribute to workers' risk assessment, considering the necessary precaution whenever information is insufficient or less precise.

2.2 Occupational risks assessment in operations with nanomaterials

The methods used for risk assessment in Occupational Health and Safety can be divided into two groups: Qualitative methods and quantitative methods. With respect to chemical contaminants exposure risk the quantitative methods are preferably used. In general, the methods include the measurement of the concentration of each chemical agent in the air of the worker's breathing zone, and taking into account the duration of the worker exposure, to compare the obtained value with the exposure limit value set for this agent to assess the risk to the exposed worker.

When the agent is a nanomaterial, even well-known and characterized, there are doubts about the best method for concentration measurement (Maynard, 2006) and the occupational exposure limits values are not yet defined, although there are some proposals for a few types of nanoparticles (Schulte et al., 2010).

In a paper on the nanoparticles exposure risk evaluation, an international group of researchers reported that the quantification of risk is full of uncertainties, such as the not yet fully understand contribution that the nanoparticle's physical structure has for its toxicological effects, the differences found among different nanoparticles concerning the behavior in the lung tissue or the absence of consensus on the particles most relevant characteristics to the exposure, i.e., if the specific surface area and/or the size distribution that seem more decisive than the mass (Zalk, Paik and Swuste, 2009).

In this context, several authors refer the Control Banding as an appropriate method for assessing the exposure risk to nanoparticles (Maynard, 2007; Schulte et al., 2010; Beaudrie and Kandlikar, 2011). As examples of Control Banding methods developed for the nanoparticles exposure there are the CB Nanotool (Paik, Zalk and Swuste, 2008) and the Stoffenmanager Nano (Van Duuren-Stuurman et al., 2011).

Other qualitative methods are referred in bibliography, considered as an alternative to the lack of quantitative methodologies for assessing the risk from both occupational and environmental context, in particular, the experts judgment and a more structured variant, the expert elicitation (Murashov and Howard, 2009; Kandlikar et al., 2006) and the multi-criteria decision analysis (Linkov et al., 2007).

At the current state of knowledge regarding the nanomaterials risks, in particular with respect to nanoparticles exposure, the choice to use qualitative risk assessment methodologies seems to be an acceptable option.

2.3 Design analysis approach

Some authors have been defending the need for methodologies that deal with the nanotechnologies risks based on the processes or products design (Fleury et al., 2011; Amyotte, 2011), referring, in particular, the "Design for Safer Nanotechnology" (Morose, 2010).

The importance of the occupational health and safety issues integration in the process design (systems, installations, production lines, machines, tools, etc.) is officially recognized (European Parliament and Council of the European Communities, 2006) but not always considered. Although the occupational safety and hygiene research pays more attention to risk analysis (Swuste, 1996), several authors in this domain have performed some investigation in the safety by design field, specially the Safety Science Group of Delft University of Technology (e.g., Stoop, 1990; Schupp et al., 2006; Hale, Kirwan and Kjellen, 2007). Swuste proposed a systematic approach towards solutions (Swuste, 1996) based on three complementary elements:

- Hazard process model;
- Design analysis;
- Problem-solving cicle.

A simple way to represent the hazard exposure in workplaces is using the model presented in Figure 1.

The term immission is not widely used in occupational hygiene. Instead, it is used the term exposure and the worker is referred as exposed worker. According to this model, it is possible to control de hazard acting on the three phases, eliminating or at least reducing the emission, the transmission and/ or the immission. Both regulatory laws (Council of the European Communities, 1989) and occupational health and safety good practices and standards (IPQ, 2008) set priority on determining or considering the hazard control methods: First of all the hazard elimination or reduction (reducing emission), second acting on the transmission, and finally acting on the exposure. In other words, it is acting from the source to the exposed worker.

More complex models for nanoparticles exposure had been developed such as the conceptual model (Tielemans et al., 2008; Schneider et al., 2011). Although the more elaborated form, the essential aspects are common in both models.

The design analysis methodology allows to study and understanding the workplace conditions. In design analysis the production process is split into three levels of decision (Swuste, 1996), described below:

- Production function: is the highest level and divides the production process into his core activities;
- Production principle: identifies the general process, motive power and operational control methods by which the production function can be achieved;
- Production form: is the lowest level and specifies the detailed design by which the production principle will be accomplished.

If there is a large number of production processes, the type of functions (or unit operations in

rigor), in which each process can be broke down, is relative small. The main unit operations categories are: Material receipt, material storage, transport and feed, processing, packaging, waste disposal.

The processing operations can be subdivided in subcategories that vary from one industry sector to other, and once enumerated will permit to study the more effective and reasonable control measure or set of control measures to apply in each particular situation.

On the occupational safety & health point of view, the focus on the production function will allow to find the less hazardous way to achieve the same production result or to choose the best available technics to control the hazard.

The problem solving cycle has been proposed as a systematic approach to generate solutions in occupational risk management and provides a systematic tool to find solutions to control the existing risks (Hale et al., 1997).

Applying it together with design analysis, it will permit to identify and develop the most suitable risk control measures in each engineered nanomaterial production process even in poor knowledge and high uncertainty situations. Combining different information sources will create synergies and conduct to the best available prevention and protection measures.

The design approach put the focus in the risk control, rather in risk assessment. It provides a tool to eliminate the risk, prevent exposure and/or protect the workers. Adapting the bow-tie model proposed by the safety science group (Ale et al., 2008) to the occupational hygiene field will help to establish the necessary barriers to control the risks arising from different workplace exposure scenarios.

The design analysis approach as described above can be a suitable method to deal with the

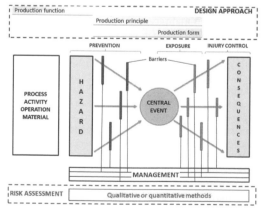

Figure 2. Bow-tie model with arrows representing different exposure scenarios.

Figure 1. Hazard process model (adapted from Swuste, 1996).

nanotechnology occupational risks. The knowledge gap and the related uncertainties can be overcome with a methodology that focuses in solutions (risk control) rather than in the risk evaluation. Moreover, combining together the two focuses will allow to achieve the best practicable preventive actions.

3 CONCLUSIONS

The lack of information and the uncertainty related to NP occupational exposure are an actual problem. The current knowledge is evolving:

- Results from the in-vitro and in-vivo toxicological tests show harmful effects from the nanoparticles;
- Nanoparticles characterization battery tests are already available and will allow to obtain information to exposure risk assessment;
- Quantitative exposure assessment methods are not yet consensual and the same applies to the exposure limit values;
- Qualitative exposure risk assessment methods are in use and gather interest from the experts;
- The design approach to safety is presented as an alternative to develop safer product and processes in the nanotechnologies field.

Thus, there is an opportunity to develop additional research in this area in order to confirm the applicability of the qualitative risk assessment and the design analysis approach in the NP occupational hygiene field. The referred research should include qualitative risk assessment methods in the workplaces where NM are used. Applying the design approach, focusing on the risk control, it is possible to select the production processes that minimize workers' exposure.

REFERENCES

Ale, B.J.M. et al., 2008. Quantifying occupational risk: The development of an occupational risk model. *Safety Science*, 46(2), pp. 176–185.

Amyotte, P.R., 2011. Are classical process safety concepts relevant to nanotechnology applications? *Journal of Physics: Conference Series*, 304 p. 012–071.

Beaudrie, C.E.H. and Kandlikar, M., 2011. Horses for courses: risk information and decision making in the regulation of nanomaterials. *Journal of Nanoparticle Research*, 13(4), pp. 1477–1488.

Becker, H. et al., 2011. The carcinogenic potential of nanomaterials, their release from products and options for regulating them. *International journal of hygiene and environmental health*, 214(3), pp. 231–8.

Bowman, D. and Hodge, G., 2006. Nanotechnology: Mapping the wild regulatory frontier. *Futures*, 38(9), pp. 1060–1073.

Clift, M.J.D., Gehr, P. and Rothen-Rutishauser, B., 2011. Nanotoxicology: a perspective and discussion of whether or not in vitro testing is a valid alternative. *Archives of toxicology*, 85(7), pp. 723–31.

Council of the European Communities, 1989. *Council Directive 89/391/CEE of 12 June 1989 on the introduction of measures to encourage improvements in the safety and health of workers at work.*

European Parliament and Council of the European Communities, 2006. *Council Directive 89/391/CEE of 12 June 1989 on machinery, and amanding Directive 95/16/EC (recast).*

Feynman, R., 1960. There's plenty of room at the bottom. *Engineering and Science*, 23(5), pp. 22–36.

Fleury, D. et al., 2011. Nanoparticle risk management and cost evaluation: a general framework. *Journal of Physics: Conference Series*, 304 p. 012–084.

Hale, A. et al., 1997. MODELLING OF SAFETY MANAGEMENT SYSTEMS. *Safety Science*, 26(1), pp. 121–140.

Hale, A., Kirwan, B. and Kjellen, U., 2007. Safe by design: where are we now? *Safety Science*, 45(1–2), pp. 305–327.

Instituto Português da Qualidade IPQ, 2008. *NP 4397— Sistemas de gestão da segurança e saúde do trabalho. Requisitos.* Caparica: IPQ.

Kandlikar, M. et al., 2006. Health risk assessment for nanoparticles: A case for using expert judgment. *Journal of Nanoparticle Research*, 9(1), pp. 137–156.

Linkov, I. et al., 2007. Multi-criteria decision analysis and environmental risk assessment for nanomaterials. *Journal of Nanoparticle Research*, 9(4), pp. 543–554.

Maynard, A.D., 2006. Nanotechnology: assessing the risks. *Nano Today*, 1(2), pp. 22–33.

Maynard, A.D., 2007. Nanotechnology: the next big thing, or much ado about nothing? *The Annals of occupational hygiene*, 51(1), pp. 1–12.

Morose, G., 2010. The 5 principles of "Design for Safer Nanotechnology". *Journal of Cleaner Production*, 18(3), pp. 285–289.

Murashov, V. and Howard, J., 2009. Essential features for proactive risk management. *Nature nanotechnology*, 4(8), pp. 467–70.

Paik, S.Y., Zalk, D.M. and Swuste, P., 2008. Application of a pilot control banding tool for risk level assessment and control of nanoparticle exposures. *The Annals of occupational hygiene*, 52(6), pp. 419–28.

Renn, O. and Roco, M.C., 2006. Nanotechnology and the need for risk governance. *Journal of Nanoparticle Research*, 8(2), pp. 153–191.

Roco, M.C. et al., 2011. Innovative and responsible governance of nanotechnology for societal development. *Journal of Nanoparticle Research*, 13(9), pp. 3557–3590.

Savolainen, K. et al., 2010. Risk assessment of engineered nanomaterials and nanotechnologies--a review. *Toxicology*, 269(2–3), pp. 92–104.

Sayes, C.M. et al., 2008. Can in vitro assays substitute for in vivo studies in assessing the pulmonary hazards of fine and nanoscale materials? *Journal of Nanoparticle Research*, 11(2), pp. 421–431.

SCENIHR, 2009. *Risk Assessment of Products of Nanotechnologies.* Brussels: European Commission. Available from: <http://ec.europa.eu/health/ph_risk/risk_en.htm.

Schneider, T. et al., 2011. Conceptual model for assessment of inhalation exposure to manufactured nanoparticles. *Journal of exposure science & environmental epidemiology*, pp. 1–14.

Schulte, P.A. et al., 2008. Sharpening the focus on occupational safety and health in nanotechnology. *Scandinavian Journal of Work, Environment & Health*, 34(6), pp. 471–478.

Schulte, P.A. et al., 2010. Occupational exposure limits for nanomaterials: state of the art. *Journal of Nanoparticle Research*, 12(6), pp. 1971–1987.

Schupp, B. et al., 2006. Design support for the systematic integration of risk reduction into early chemical process design. *Safety Science*, 44(1), pp. 37–54.

Stoop, J., 1990. Scenarios in the design process. *Applied Ergonomics*, 21(4), pp. 304–310.

Swuste, P., 1996. *Occupational hazards, risks and solutions*. Technische Universiteit Delft.

Technical Committe ISO/TC 146, 2007. *Technical Report ISO/TR 27628 Workplace atmospheres—Ultrafine, nanoparticle and nano-structured aerosols—Inhalation exposure characterization and assessment*. Geneva.

Tielemans, E. et al., 2008. Conceptual model for assessment of inhalation exposure: defining modifying factors. *The Annals of occupational hygiene*, 52(7), pp. 577–86.

Van Duuren-Stuurman, B. et al., 2011. *Stoffenmanager Nano: Description of the conceptual control banding model. Zeist*.

Warheit, D.B. et al., 2007. Development of a base set of toxicity tests using ultrafine TiO2 particles as a component of nanoparticle risk management. *Toxicology letters*, 171(3), pp. 99–110.

WWICS, 2011. *Project on Emerging Technologies - nanotechnology inventories*. [online] Available from: <http://www.nanotechproject.org/inventories/consumer/analysis_draft/>.

Zalk, D.M., Paik, S.Y. and Swuste, P., 2009. Evaluating the Control Banding Nanotool: a qualitative risk assessment method for controlling nanoparticle exposures.

Environmental ergonomics

Occupational Safety and Hygiene – Arezes et al. (eds)
© *2013 Taylor & Francis Group, London, ISBN 978-1-138-00047-6*

Work Ability Index and thermal and acoustic conditions of municipal schools' teachers

D.A.M. Pereira

Universidade Federal de Campina Grande, Sumé, Paraíba, Brazil

ABSTRACT: This research aimed to analyze the Work Ability Index (WAI) and the thermal and acoustic comfort, of municipal schools teachers in João Pessoa (Brazil). The data that was collected and analyzed were: WAI, age, length of service, PMV, PPD and noise. The verification of reliability of collected data was performed by measuring the BOX-COX. It was observed that 46% had a WAI between low and moderate and 54% among good and great. Teachers reported a PMV from "slightly warm" to "very hot" and a PPD ranged from 61% to 92%. It was found that the noise levels to which teachers are subject are above those recommended by NBR-10.152/87. These results reinforce the idea that teachers are subject to adverse working conditions.

1 INTRODUCTION

The perception that work may have consequences over individual's health is ancient. In Greece, four centuries b.C., Hippocrates' writings already punctuated the diseases caused by work conditions to which workers were submitted. According to Iida, (2005), a great source of tension at work are the unfavourable environmental conditions, such as heat and noise excess, lack of illumination and vibrations. It is shown that the major part of public schools in Brazil does not offer the previous mentioned condition in an adequate way to the development of the activities held there. Such conditions are likely to jeopardize teaching-learning process and teachers' physical-psychological health. This way, schools must have conditions to undergo an adequation corresponding to new teaching-learning activities policy and methodology. Therefore, it is needed that physical installations are adequate to the performance of educational activities, and that the professionals involved in such operations have work conditions that do not cause health harm.

The choice of public schools teachers for a study object was based on social relevance and the evidence, on specialized literature, that these professionals have been suffering with health offense, caused by the effects from both exposal o occupational risks involved in their work activity and the work conditions they are submitted to. Studying the relations between the teaching activity, the conditions under which it is developed, and the possible diseases originated in these professionals are a challenge and a necessity to better comprehend the relation between teachers' health-disease-work,

and consequently pursue ways to preserve either the ability to work or the health of these professionals In this context, the object of the present study is to analyze the work ability as well as thermal and acoustic comfort to which municipal school teachers from João Pessoa (Brazil) are submitted.

2 METHODOLOGY

The present research is classified as quantitative, descriptive type, in which participated 50 teachers from fundamental teaching I, at 16 municipal schools from João Pessoa—PB. Professionals selected to be part of the sample were those who worked two consecutive shifts at the same institution, to guarantee the evaluation of the whole day's work.

Evaluation instrument used to analyze the the work ability was the Work Ability Index (WAI) questionnaire, which estimates the work ability considering physical and mental demands and the workers' resources and health conditions according to their own perception. WAI evaluates seven different dimensions: present work ability compared to the best work from the whole life, work ability related to work demands, number of diseases effectively diagnosed by physicians, estimated work ability loss due to diseases, numbers of days missed at work due to diseases, own prognostic about work ability, and mental resources. Generated score varies from 7 to 49 points, where 7 to 27 corresponds to low work ability, 28 to 36 corresponds to moderate work ability, 37 a 43 mean good work ability, and 44 to 49, great work ability.

Evaluation of heat agent was supported by Rule ISO ISO 7730/2005—*Moderate thermal environments—Determination of the Predicted Mean Vote (PMV) and Percentage of dissatisfied (PPD) index and specification of the conditions for thermal comfort.* There were performed eight measurements a day, with an hour interval between each evaluation, covering, that way, the whole individual's work journey. In order to collect the temperatures, a Heat Stress WBGT Meter-model TGD-300, brand Instrutherm was used. This meter is composed by globe, wet bulb and dry bulb thermometers, WBGT set, and all the thermometers have the accuracy of ±0,5 °C, which indicate, respectively, environmen thermal radiation, air temperature and the influence of humidity on temperature. To determine the PMV and the PPD, the following tasks were completed: application of thermal perception questionnaire; measurement of the temperatures in a point near the professor at the classroom; classification of clothes' resistence (teachers were dressed with clothes equivalent to 0,50 clo) and classification of metabolic energy consumed by teachers according to tables from ISO 7730/2005.

Agent noise evaluation was based on brazilian rules NBR 10.151—Acoustic of noise in inhabited areas, aiming community comfort—Procedures; and NBR 10.152—Noise levels to acoustic comfort. There were performed eight measurements a day, with an hour interval between each evaluation during the whole teacher's work journey, using a sound level meter, model, DEC-470, brand *Instrutherm*, with an accuracy of ±1,5, operated on "A" and slow answer.

The verification of reliability of collected data was performed by measuring the BOX-COX. It is a Bera-Jarque (BJ) normality test, which is a consequence of the study performed by Shenton and Bowman (1997), built with the expressions destinated to asymmetry and skewness. If BJ < 5,91, it represents 2 degrees of liberty and a frequence of 0,95 in the table of χ^2, so the data have normality characteristics.

Studied population data characterization was made by a descriptive analyzis (average, standard deviation, minimum and maximum values).

3 RESULTS AND DISCUSSION

3.1 Sample data analysis

Variables WAI, A (age), LS (lenght of service), PMV, PPD and noise are realiable since the BJ value from each one is lower than 5, 91, according to Table 1. The analysis. The statistic analysis of data has shown that teachers from fundamental teaching I at João Pessoa were aged 47,06 years in

Table 1. Statistics research variables summary.

Variables coefficient	WAI	A	LS	Noise	PMV	PPD
Distortion	−0,160	−0,089	−0,148	−0,590	0,418	0,061
Skewness	−0,284	−0,254	1,164	0,192	−0,913	−1,067
BJ	0,4228	0,426	0,115	0,230	0,707	0,699
Average	36,68	47,06	21,63	81,30	2,040	76,68
Standard deviation	5,407	8,229	8,725	4,313	0,249	10,045
Maximum	25,00	27,00	3,00	72,10	1,690	61,00
Minimum	48,00	63,00	48,00	89,10	2,460	92,00

average, had professional experience of 21,63 years in average, shown an average WAI of 36,68, indicated a PMV = 2,04 and were submitted to noise of 81,3 dB(A) in average.

41 teachers (82%) are aged above 40 years, which shows a reasonable quantity of teachers on middle age. According to Astrand et al., (2006), there is a decline in legs and back muscular strenght of around 60% and a 70% decline at arms musculature from 30 to 80 years. In women, these percentages are still lower, due to the fact that women's muscular mass is lower than men's. These factors may lead teachers to a workload reduction or to a change at the work methods and temporary withdrawal from work, mainly when the time at the same profession is long.

3.2 Work Ability Index

8% of the sample has shown a low WAI, 38% a moderate WAI, 42% a good WAI and 12%, a great WAI. The elevate percentage of 46% of the teachers with work ability between moderate and low stands out. From the inquired 50 teachers, it was verified that, in a scale from 0 (work inability) to 10 (better work ability), the major part has considered that their present work ability is close to their best (Average = 7,72; Standard deviation = 1,67). Regarding the work ability related to physical work demands, it was shown that, in a scale of 1 to 5, the major part of the interviewed teachers considered has themselv with a work ability from moderate to good (Average = 3,56; Standard deviation = 0,704). The evaluation of work ability related to mental work demands, it was verified that, in a scale from 1 to 5, the major part of the teachers has considered themselves with a good work ability (Average = 4; Standard deviation = 0,571).

In the aspect related to number of actual diseases effectively diagnosed by physicians, teachers reported an average of 3,2 health problems, between a list of twenty four diseases. Most frequently diagnosed diseases were, according to those taking place in the study: upper back or neck diseases with frequent pain

(38%), emphysema (30%), allergies or eczema (30%), hearing trouble or reduction (28%) and lower back disease with frequent pain (24%). Teachers' activity was observed and it is suspected that factors related to postures used during work journey, static work and type of furniture used by teachers may explain pain related to reported skeletal problems. Teachers performed repetitive movements as they wrote and erased the board, stood up for too long, bent over students' desks, performed repetitive tasks such as correcting notebooks, tests, students exercises and writing on school transcripts. Direct and habitual contact with chalk dust as well as with dust from the place itself may explain allergies and eczemas problems reported. Teachers who pointed hearing trouble or reduction as their health problems have reasons to have done so. The sound level intensity to which they were submitted was, in average, 81,3 dB (A). Considering that the tolerance limit to industry workers is 85 dB (A), according to what is stabilished by NR-15 (Appendix 1), for eight hours of work journey, and that, according to NBR 10.152, inside classrooms tolerable sound pressure intensity varies from 40 to 50 dB (A), we deduce that the teacher is really exposed to high sound pressure levels and it might cause damage to their hearing system.

Refered diseases complement the item related to estimated work ability loss due to diseases It was verified that, in a scale from 1 to 6 the major part of the teachers reported that although they can actually do their jobs, sometimes they need to diminish their work rhythm or change their work methods (Average = 4,46; Standard deviation = 1,073). This result may indicate that teaching work process is harmful to teacher's health. Regarding numbers of days missed at work due to diseases, it was verified that, in scale from 1 to 5, absenteeism is a maximum of 9 days (Average = 4,36; Standard deviation = 0,693). It means that most part of teachers has took time off work for a relatively short amount of time. However, these data should be analyzed in combination with sick leaving numbers for each teacher along the year.

As far as own prognostic about work ability, and mental resources is concerned, major part of the sample in a scale from 1 to 7 has revealed not to be right about if they will be able to keep performing their teaching activities two years from now (Average = 5,72; Standard deviation = 2,041). Crescent lack of social recognition, low salaries, progressively worse work conditions, function accumulation and consequent increasing work load are factors that may lead the teacher to fear continuing at the same profession. Another factor to be considered is security (or lack of it) at schools. All kinds of aggression performed by students, students' parents, coworkers or school employees and people outside the school were strongly associated to mental and health trouble.

These data confirms what International Labor Organization (ILO) and United Nations Educational, Scientific and Cultural Organization (UNESCO) have affirmed concerning school violence increase thoughout the world. As for mental resources, the three following results were obtained: as for the item "have you been able to enjoy your daily activities recently", major part of the sample has answered, in a scale from 0 to 4, that they often have seized their day-by-day routine (Average = 3,46; Standard deviation = 0,839). The same has happened with the question "have you been feeling active and allert recently. Also a significative number of the sample, inbetween a scale from 0 to 4, has answered that they were often active and allert (Average = 3,48; Standard deviation = 0,614). As for the question "have you been feeling full of hope for the future", teachers have answered in a scale from 0 to 4, that they were often optimistic concerning the future (Average = 3,48; Standard deviation = 0,701). WAI results from the present study are similar to studies taken in another countries (FREUDE et al., 2005; CASTELO-BRANCO, 2006; PEREIRA et al., 2002), which point to a high number of WAI percentages between moderate and low.

3.3 Thermal comfort

Region where the present research was taken is demarcated by a square and is located at Zone 8, as Figure 1 indicates. This zone represents 53,7%—or 99 cities—of those 330 maped by Norma Brasileira Registrada 15220-3—Thermal Performances on Buildings.

Teachers have declared to be thermically dissatisfied with their work environments. According to ASHRAE's seventh scale, 36% of the teachers

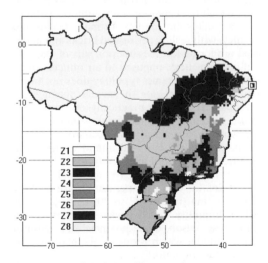

Figure 1.   Bioclimatic brazilian mapping.

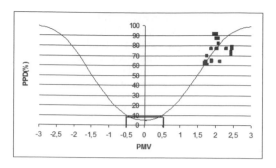

Figure 2.   Teachers PPD in function of PMV.

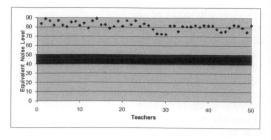

Figure 3.   Equivalent noise levels to which teachers are submitted.

have considered classrroms from "Slightly Hot" to "Hot" while 64% have considered classrooms to be from "Hot" to "Very Hot". PPD has shown a variation between 61 to 92% of thermically dissatisfied teachers in classrooms, according to Figure 2. In order to have an environment between comfort zone stabilised by ISO 7730/2005, the PMV shall be comprehended within –0,5 and +0,5, which corresponds to a maximum of 10% of dissatisfied people, meaning a PPD index ≤ to 10%.

Work done under heat conditions found may cause a reduction in teachers' physical and mental performance. Prolongued thermal stress leads to body fluids loss and damages mental and psychomotor functions (ASTRAND et al., 2006). These teachers' thermal dissatisfaction and discomfort at classrooms findings are similar to some research done in Brazil (NOGUEIRA, 2005; PEREIRA, 2006; RODRIGUES et al., 2006) and in another countries (KWOK, 1998; KWOK, 2003; WONG e KHOO, 2003). These thermal discomfort conditions may come from a great amount of factors, mainly the weather itself, building architecture factors and thermal inertia from the own building. Not all the classrooms are equiped with efficient air inlets and outlets. In many classrooms, air inlets with sealing perforated bricks and undersized are located at the superior part of the walls that contain them, opposed to air outlets. Like it was said before, frames typology does not favour natural ventilation, for the actual openings are too small. It leads to practically no cross-ventilation, turning perforated bricks almost ineffective, what reflects negatively at classroom thermal dissipation ability. Hot and humid weather from the intertropical kind at the city the schools are located is characterized by average anual temperature of 26° C, with a temperature range of 28° C and oscilating from 22° C to 30° C, with average relative humidity of 80%, what potencializes the wet skin sensation, o que potencializa a sensação de pele úmida, which is one of the main reasons for thermal discomfort.

3.4   Acoustic comfort

Noise level inside a classroom depends on four factors: intensity of internal and external sources, quality of material that seal these environments and quality of material that internally cover these environments.

It was verified that noise levels to which these teachers are under in the classrooms are above the acoustic comfort limit recommended by NBR-10.152/87, which varies between 40 and 50 dB (A), according to Figure 3. Noise to which these teachers are exposed have varied between 72,1 e 89,1 dB (A). Noise above 70 dB (A) can increase cardiac ejection fraction, heart rate and blood pressure.

High noise levels can be attributed to the great number of students inside the classroom (an average of 35 students per classroom), the material that internally covers the environment (ceramics) and to acoustically low efficient classrooms' sealing material (simple masonry walls with perforated bricks, plywood walls and Venetian type windows). Even though the ceramic tiles that go up to 1,60 m of height show qualities such as durability and ease of cleaning, it is not the most indicate material to an environment where there is a great noise emission and concentration, because of the fact that ceramics easily reflect acoustic energy, meaning that it is poorly absorbed. Matrial that compose the floor (burnt cement), the buildings' ceiling and walls (painted concrete and ceramics, respectively) are also predominantly reflective, what generates an environment with much reverberation. These noise levels above the limit of acoustic comfort and tolerance limits stabilised by NBR 10.152. Average level of sound pressure inside the classrooms was 81,30 dB(A).

4   CONCLUSIONS

As far as WAI results are concerned, 8% of the sample has shown a low WAI, and 38%, a moderate WAI. Teachers show elevate number of health problems. Os professores apresentaram elevados índices de problemas com a saúde. Mainly reported health diseases were upper back or neck diseases with frequent pain, emphysema, allergies or eczema, hearing trouble or reduction and lower back disease with frequent pain. These results allowed to conclude that there are specific diseased related to the teaching activity. Therefore, additional evaluations and measures in order to improve work ability are recommended, in a way to avoid that the profession perfomance jeopardizes teachers' health and work.

As for thermal comfort, teachers considered the classrooms from "slightly hot" to "very hot". PPD has pointed a variation between 61 to 92% of thermically dissatisfied teachers in classrooms. These thermal discomfort conditions may be attributed to a number of factors, mainly the weather itself.

The only way to reduce temperature and air humidity at the seashore is the usage of mechanical air conditioning systems, which would mean high costs for the city government. The situation can be mitigated by provoking air movement in a speed capable of increasing air loss by natural convection and evaporation. It is possible through ceiling fans, which allow a more homogeneous movement than that from the wall. However, these fans do not reach all classroom points and besides, they elevate noise level.

Natural ventilation shall be pursued with or without mechanical ventilation. Therefore, buildings should have their openings for air inlets as big as possible, turned to the direction of the predominating winds, which not always is possible. Air outlets should be as big as possible as well, but should be located higher than the air inlets, and situated at the opposite wall, in a way for the chimney effect to work efficiently. In these school cases, buildings' architectural style factors (bad orientation and undersizing of the air inlets and outlets, lack of solar protection, inefficient sealings) directly contribute to the thermal discomfort situation reported by teachers.

It was infered that noise levels to which teachers are submitted in classrooms are above those recommended by NBR-10.152/87. The average noise level inside the refered classrooms 81,30 dB(A). This high sound pressure average level leads to teaching with raising voice, in a volume of around 91,30 dB(A). It is shown that an average of 35 students per classroom, material that internally cover these classrooms (ceramics and burnt cement), the ceiling, the walls, and low efficient acoustic isolation regarding material that seal the classrooms contribute for them to be highly reverberant. This way, teachers may not obtain a good performance while they develop their work activity, which may compromise students' learning, due to speaking intelligibility, and also affect their own health, due to high noise levels shown.

Deficiency in ventilation systems inside the classrooms and the consequent thermal discomfort it generates, as well as the noise generated in the classroom are enviromental adverse conditions that influence on teaching activities. These turn the enviroment uncomfortable, bothering and uneasy, both for teachers and students, and require more efforts and more physical and mental demand with negative repercussion on health.

These results reinforce the idea that teachers are submitted to adverse work conditions. When the work is developed under inadequate environmental, organizational and physiological conditions, both damage to the individual's health and decrease of work ability may be accelerated or aggravated due to the work activity.

REFERENCES

ASHRAE. 1997. *Fundamentals Handbook. American Society of Heating, Ventilating and Air-Conditioning Engineers.* Atlanta.

ASSOCIAÇÃO BRASILEIRA DE NORMAS TÉCNICAS (ABNT). 1987. *NBR 10151: Níveis de ruído para conforto acústico—procedimento.* Rio de Janeiro.

ASSOCIAÇÃO BRASILEIRA DE NORMAS TÉCNICAS (ABNT). 1987. *NBR 10152: Avaliação do ruído em áreas habitadas visando o conforto da comunidade.* Rio de Janeiro.

ASSOCIAÇÃO BRASILEIRA DE NORMAS TÉCNICAS. 2005. *NBR 15220-3: desempenho térmico de edificações: parte 3: zoneamento bioclimático brasileiro e diretrizes construtivas para habitações de interesse social.* Rio de Janeiro.

ASTRAND P-O, *et al.* 2006. Tratado de Fisiologia do Trabalho: bases fisiológicas do exercício.. Porto Alegre: Artmed.

CASTELO-BRANCO, M.C. 2006. *Corpo, auto-eficácia e capacidade laboral: na senda do bem-estar docente.* Dissertação de Doutoramento. Aveiro: Universidade de Aveiro.

FREUDE, G. *et al.* 2005. *Assessment of work ability and vitality—a study of teachers of different age groups:* 270–274 International Congress Series.

GASPARINI, M.S, *et al.* (2005). Revista Educação e Pesquisa. O professor, as condições de trabalho e os efeitos sobre sua saúde: 189–199.

IIDA, I. 2005. *Ergonomia:projeto e produção.* São Paulo: Edgard Blucher.

INTERNATIONAL ORGANIZATION FOR STANDARD. 2005. ISO 7730: *moderate thermal environments: determination of the PMV and PPD Indices and Specification of the conditions of Thermal Comfort.* Geneva.

NOGUEIRA, M.C.J.A; NOGUEIRA, J.S. 2005. *Conforto térmico na escola pública em Cuiabá-MT: estudo de caso.* Revista Eletrônica do Mestrado em Educação Ambiental.

PEREIRA, A.M.S.; SILVA, C.F.; CASTELO-BRANCO, M.C.; LATINO, M.L. 2002. *Saúde e a capacidade para o trabalho na docência:* 159–167. In IV Congresso Nacional de Saúde Ocupacional, 29 a 31 de Outubro, Póvoa do Varzim, pp.159–167.

RODRIGUES, R.S. *et al. 2006. Estudo do conforto térmico diurno em Escolas Públicas no município de Bragança-PA.* In: XIV Congresso Brasileiro de Meteorologia. Florianópolis.

SHENTON, L.R; BOWMAN, K.O. 1997. *A bivariate model for the distribuição of (b1)1/2 and b2*: 206–211. Journal of the American Statistical Association.

TUOMI, K. et al. 1997. *Índice de Capacidade para o Trabalho.* São Paulo: FSPUSP.

WONG, N. and S. KHOO. 2003. *Thermal comfort in classrooms in the tropics*: 337–351 Energy and Buildings.

Occupational Safety and Hygiene – Arezes et al. (eds)
© *2013 Taylor & Francis Group, London, ISBN 978-1-138-00047-6*

Thermal suitability in a work environment: Analysis of PMV and PPD

E.E. Broday, A.A.P. Xavier & A.L. Soares
UTFPR, Federal University of Technology of Paraná, Ponta Grossa, Brazil

ABSTRACT: This work analyzes the thermal aspect of the work stations in a metallurgical industry branch. The analysis of thermal comfort is linked to the number of dissatisfied with the environment, so this work sought to check the thermal suitability in a metallurgical environment. The methodology set out for this work is composed of collection, allocation in spreadsheets and statistical data processing. The resources required for the development of this work were the equipment to measure environmental variables and the statistical assistance software. This work environment can be classified in category A, because the PMV calculated is in the interval ($-0,2 <$ PMV $< 0,2$) and the PPD calculated by the equation from ISO 7730 (2005) is smaller than 6% (5,37%). Also, Fanger's interpretation is correct because the curve built between PMV and PPD showed 4,5% of dissatisfied people, with a $R^2 = 0,9928$. Although the PPD values are different, the environment is classified in the same category (A).

1 INTRODUCTION

The employees' quality of work is a reflection of their state of body and mind, and also the welfare of the developed activities generates benefits for everyone. In this context, the importance of the welfare of the worker at the workplace is the main precept of ergonomics. According to Parsons (2000), Ergonomics can be defined as the application of knowledge of human characteristics to the design of systems. People in systems operate within an environment, and environmental ergonomics is concerned with how they interact with the environment from the perspective of ergonomics. From this concern with the environmental work, the present study focuses on the thermal aspect.

An environment can be said thermally comfortable from the appropriate analysis of the various environmental variables that compose it, and also of the occupants' personal variables of that environment. Taking, as a basis, the regulatory standards of thermal comfort, it is possible to determine from pre-established limits whether or not the environment is thermally comfortable for different types of activities (ANDREASI et al., 2010; PINTO, 2012).

According to ASHRAE 55 (2004), Thermal Comfort is "the condition of the mind in which satisfaction is expressed with the thermal environment".

The Thermal environment is a function of four physical variables (air temperature, mean radiant temperature, air velocity and air humidity) and two variables related to people (metabolic rate and clothing insulation) (ASHRAE 55 [2004]).

The physical variables are those that refer to the climatic conditions of the environment in question, while the personal variables are those that related to individuals at the time of the evaluation of the environment.

The air temperature is the temperature of the air around the person. The mean radiant temperature is the uniform surface temperature of an imaginary black enclosure. The air velocity is the rate of air movement at a point. The air humidity is the ratio of the partial pressure of the water vapor in the air to the saturation pressure of water vapor (ASHRAE 55 [2004]).

The clothing insulation is the resistance to sensible heat transfer provided by a clothing ensemble. It is expressed in clo-units.

Metabolism is the energy cost of muscle load associated with the conversion of sugars and fats in thermal and mechanical energies (MALCHAIRE, 2004). In accordance with Havenith et al. (2002), as more intense the activity is performed, the greater the rate of heat is produced. "Met" is the unit used to describe the energy generated inside the body due to metabolic activity, defined as 58.2 W/m^2.

In normal conditions, people maintain their body temperature within a constant level near 37°C. This constant temperature can be explained by the equivalence between the heat production by the organism and heat loss to the surroundings. This can be visualized in equation (1):

$$M - W = Q_{res} + Q_{sk} \qquad (1)$$

where: M = metabolic rate of heat production (W/m^2); W = rate of mechanical efficiency (W/m^2); Q_{res} = total rate of heat loss through breathing (W/m^2); Q_{sk} = total rate of heat loss through the skin (W/m^2).

The mechanisms of heat loss are verified in all stages of life, from newborns to adults. The body loses heat by evaporation, convection, conduction and radiation (ÇINAR and FILIZ, 2006), as shown in figure 1:

In evaporation, heat loss occurs by the evaporation of water from the skin. Heat loss to cooler solid objects in direct contact is conduction and heat loss to cooler surrounding is called convection. When there is no direct contact of the cooler object with the body this is radiation (INCROPERA and DE WITT, 2011).

Thermal equilibrium is reached when the thermal balance is null. In this case, the body temperature does not vary. In order to keep the condition of thermal equilibrium, the human body uses the mechanism of thermoregulation, which keeps the temperature constant, despite external changes of temperature (IVANOV, 2006). So, the balance between the heat produced by the organism and the heat loss from the body must be zero.

According to Fanger (1970), there are three conditions for thermal comfort: thermal neutrality, the skin temperature and the secretion of sweat must be compatible with the activity developed, and the person may not feel any kind of thermal discomfort.

Being in thermal neutrality is a necessary condition to meet thermal comfort, however, is not a sufficient condition. The person may be in thermal neutrality, but may also be subject to some sort of local thermal discomfort.

The skin temperature and sweat secretion rate depend on the activity developed by the person, thus demonstrating that even if being in thermal neutrality can be subject to some form of discomfort.

With respect to local thermal discomfort, there are four distinct ways that someone can find in discomfort (ISO 7730 [2005]):

– Draught: this kind of discomfort can be expressed as the percentage of people bothered by draught;
– Vertical air temperature difference: a high vertical air temperature between ankles and head can produce local discomfort;

– Warm and cool floors: there may be discomfort in the feet, due to the floor being warm or cold;
– Radiant asymmetry: is when the person has part of his/her body hit by radiation while other parts do not. People are more sensitive to asymmetric radiation caused by a warm ceiling than that caused by warm and cold vertical surfaces.

In order to verify the conformity of thermal environment, and to establish requirements for different levels of acceptability, ISO 7730 (2005) proposes the PMV, which represents the average voting index on the seven-point thermal sensation scale: +3 = hot, +2 = warm, +1 = slightly warm, 0 = neutral, −1 = slightly cool, −2 = cool, −3 = cold.

PMV is related to thermal balance, which is connected to the aspects of sensation of body heat. Index is estimated from the PMV, where comes most people's opinion about the Thermal Sensation that an environment offers. PMV equation can be viewed in equation (2):

$$PMV = [0,303 \cdot \exp(-0,036.M) + 0,028] \cdot L \qquad (2)$$

where: PMV = Predicted Mean Vote; M = metabolic rate of heat production (W/m²); L = heat load acting on the body (W/m²).

With the PMV, it is possible to get the percentage of people thermally dissatisfied. The PPD provides the percentage of dissatisfied within a large group. One interpretation is proposed by the Danish researcher Povl Ole Fanger. For him, all people who voted +3, +2, −2 and −3 on the seven-point scale were considered dissatisfied. Fanger also considered that the votes in −1, 0 and +1 demonstrate satisfaction with the environment. The equation that determines the PPD is the equation (3):

$$PPD = 100 - 95 \cdot \exp[-(0,03353. \\ PMV^4 + 0,2179 \cdot PMV^2)] \quad (3)$$

where: PPD = Predicted Percentage Dissatisfied.

As it is already known, there is a relation between PMV and PPD, in which for a certain number of PMV, there is a number of PPD.

The curve representing the relation between PMV and PPD has a minimum value of corresponding PPD in PMV (FANGER, 1970). This means a condition of great comfort, which 5% are dissatisfied and 95% are satisfied. This can be viewed in figure 2:

According to ISO 7730 (2005) there are three categories of thermal environments. In category A, the environment is said comfortable if less than 6% of occupants are dissatisfied ($-0,2 < PMV < 0,2$). In category B, the environment will be comfortable if the percentage of dissatisfied is less than 10%

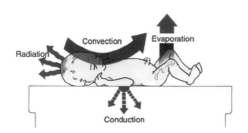

Figure 1. Ways to lose heat; Source: Çinar and Filiz (2006).

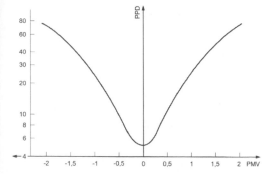

Figure 2.   Relationship between PMV and PPD; Source: ISO 7730 (2005).

(−0,5 < PMV < 0,5), and in category C if it is less than 15% (−0,7 < PMV < 0,7).

The methodology used to analyze the thermal aspect of the work stations in a metallurgical industry branch was based on exploratory quantitative approach developed through research, field research and case studies, interviews and administration of a questionnaire to be completed by the employees.

The survey was conducted in a medium-sized metallurgical company in the south of Brazil. The goal of this work was to verify in which category of thermal environment the company belongs as well as to verify the number of dissatisfied in the work environment.

2   MATERIALS AND METHOD

Figure 3 shows the structure of the methodology of the study for the thermal suitability in metallurgical industry:

Data were collected in October 2011 with the sample selected to represent the target population, according to Pidd (2003). The choice for this company to be analyzed was by "accession", assuming that it has the metallurgical profile and characteristics in the city and surrounding areas.

A total of 32 measurements were carried out, 4 on each spot. When you have a sample number exceeding 30, the sample average distribution can be approximated by a normal distribution. Thus, the Central Limit Theorem allows that approximation and makes it possible to use the normal curve for data evaluation (TRIOLA, 2005).

The data required for this study were personal and environmental variables. In order to obtain environmental data it was necessary to use properly calibrated equipment for a better precision of numeric values. To do that, it was used an equipment called *Confortímetro Sensu®* which provided the following variables: dry Bulb temperature, relative humidity, air

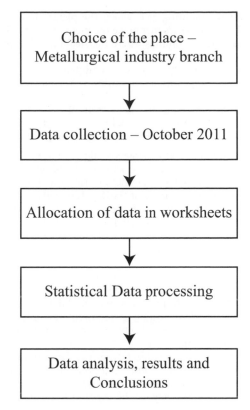

Figure 3.   Flow of study's methodology. Source: The Author.

velocity and globe temperature. Such quantities are measured by sensors that capture the variables and transmit them directly to a computer that makes them visible on the screen and also stores the information.

Thermal preferences and perceptions were obtained through questionnaires issued to employees at the time of their day-to-day activities' completion. The questionnaire was applied in the morning and in the afternoon, during the full development of activities.

Data were collected in 8 distinct points in the company: expedition (A), presses (B), guillotine (C), folders (D), cutters (E), painting (F), pointers (G) and chest of drawers (H).

At the expedition, the furniture is packed and placed in the truck to be transported to its final destination. At the presses, the steel plates are pressed. At the guillotine, the plates are cut so that they get optimal size.

At the folders, plate joints are made for the future shape of the furniture and at the cutters, excesses which may have passed unnoticed by the guillotine or leftovers, depending on the furniture, are frequently removed or will be manufactured.

At the painting, it is where the plates, drawers and all accessories that make the piece are painted. Finally, the H point is where the steel furniture, in the shape of a chest of drawers is manufactured. It was separated this way as the company requested, because they wanted to know the temperature behavior in each sector.

After collecting the data from the 8 stations described earlier, they were allocated in spreadsheets by taking into consideration the following criteria:

a) Environmental variable averages are needed on each workstation: measurements of environmental variables were held during one hour in two distinct periods of the day. It was, therefore, necessary to evaluate these data and get the average of variables for each measurement time.

b) Personal data should be separated according to the workstations: all personal data must be allocated by workstation.

The need for this organization is because the analyses are performed by matching the personal data with the environmental data; so, the correct identification is required not only by workstations, but also by the dates of measurements' completion.

The correct organization of the data obtained from sensation and thermal preference helps to interpret the votes of dissatisfaction with the environment they will be exposed in the course of work.

c) The organization of data is particularly important to facilitate data processing. The software used in this work for processing and treatment of data analysis has similar structure to worksheets; so, if data is well organized from the beginning, the work with the statistical tool is smaller and easier, reducing time and reducing the chance of errors.

3 RESULTS AND DISCUSSION

After data collection, the 32 measurements' average was calculated. PMV index and PPD were found by using specific software. The values used for the calculation of the average of thermal comfort variables are shown in table 1 below:

Table 1. PMV and PPD.

	Average	Standard deviation
AT	24,98	3,95
AV	0,08	0,08
GT	25,38	4,56
RH	59,39	18,28
CI	0,60	0,05
PMV	−0,13	1,5
PPD	5,37	5

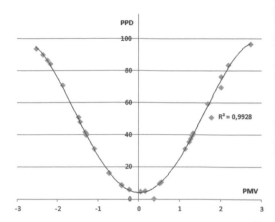

Figure 4. Curve between PMV and PPD; Source: The Author.

where: AT = air temperature (°C); AV = air velocity (m/s); RH = relative humidity (%); GT = globe temperature (°C); CI = clothing insulation (clo); PMV = Predicted Mean Vote; PPD = Predicted Percentage of Dissatisfied.

As showed in table 1, the calculated PMV is −0,13 with a PPD of 5,37. The PPD was calculated by using the equation (3).

This work environment can be classified in category A, because PMV is in the interval (−0,2 < PMV < 0,2) and the PPD is smaller than 6% (5,37%). According to the PPD value, Fanger's interpretation is correct: 5% are dissatisfied and 95% are satisfied.

This data also allow to build a curve representing the relation between PMV and PPD, as showed in figure 4:

Equation (4) represents the curve between PMV and PPD showed above:

$$PPD = -1,4856 \cdot (PMV)^4 + 0,1114 \cdot (PMV)^3$$
$$+ 23,59 \cdot (PMV)^2 - 0,9999 \cdot (PMV)$$
$$+ 4,0206 \qquad (4)$$

where: PMV = Predicted Mean Vote; PPD = Predicted Percentage Dissatisfied.

Replacing the value of PMV = −0,13 in equation (4) above, it is possible to find the PPD, being obtained the value of 4,5%. The coefficient of determination obtained was 0,9928.

4 CONCLUSIONS

In an environmental work you can find satisfied, neutral or dissatisfied people with the thermal aspect. Analysis of thermal comfort is one of the ways to determine if the environment is comfortable or not, with the number of dissatisfied ones.

Comfort occurs when body temperatures are held within narrow ranges, skin moisture is low and the physiological effort of regulation is minimized (DJONGYANG, TCHINDA and NJOMO, 2010). This research has concluded that in this environment of manufacture metal furniture branch, the category A can be applied to analyze the thermal suitability of the environment. In this case study, the environment has no extreme conditions which cannot be reversible.

This research has concluded that in environments of the branch of manufacture metal furniture, the interpretation of Fanger can be applied to analyze the thermal suitability of the environment. Also the environment is classified in category A, because PMV is in the interval (-0,2 < PMV < 0,2) according to ISO 7730 (2005).

The relation made between PMV and PPD gives an equation through is possible to find the number of dissatisfied being 4,5% in this company. Using the equation (3) from ISO 7730 (2005) and the equation (4) made by data, we can find different values to PPD, but this doesn't change the category of the environment.

REFERENCES

American Society of heating, refrigerating and air conditioning engineers, inc. 2004. Thermal environmental conditions for human occupancy. *ASHRAE STANDARD 55*, Atlanta.

Andreasi, W.A. et al. 2010. Thermal acceptability assessment in buildings located in hot and humid regions in Brazil. *Building and Environment*, 45, 1225–1232.

Çinar, N.D.; Filiz, T.M. 2006. Neonatal thermoregulation. *Journal of Neonatal Nursing*, 12, 69–74.

Djongyang, N., Tchinda, R., Njomo, D. 2010. Thermal comfort: a review paper. *Renewable and Sustainable Energy*, 14, 2626–2640.

Fanger P.O. 1970. *Thermal comfort, analysis and application in environmental engineering.* Copenhagen: Danish Technical Press.

Havenith, G., Holmér, I., Parsons, K. (2002). Personal factors in thermal comfort assessment: clothing properties and metabolic heat production. *Energy and Buildings*, 34, 581–591.

Incropera, F., De Witt, D. 2011. *Fundamentos de Transferência de Calor e de Massa* (6th ed.). Rio de Janeiro: LTC.

ISO-International Organization for Standardization. 2005. Ergonomics of the thermal environment-Analytical determination and interpretation of thermal comfort using calculation of the PMV and PPD indices and local thermal comfort criteria. *ISO 7730*, Genebra.

Ivanov, K.P. 2006. The development of the concepts of homeothermy and thermoregulation. *Journal of Thermal Biology*, 31, 24–29.

Malchaire, J. (2004). Travail à la chaleur. *EMC—Toxicologie-Pathologie*, 1, 96–116.

Parsons, K. 2000. Environmental ergonomics: a review of principles, methods and models. *Applied Ergonomics*, 31, 581–594.

Pidd, M. 2003. *Tools for thinking: modelling in management science.* (2nd ed.) John Wiley & Sons Ltd, USA.

Pinto, N.M. 2012. Analysis of the thermal comfort model in an environment of metal mechanical branch. *Work*, 41, 1606–1611.

Triola, M.F. 2005. *Introdução à Estatística* (10th ed.). Rio de Janeiro: LTC.

Occupational Safety and Hygiene – Arezes et al. (eds)
© *2013 Taylor & Francis Group, London, ISBN 978-1-138-00047-6*

Ergonomic analysis based on temperature and thermal stress level

L.F. Monteiro
Universidade Federal de Sergipe, São Cristóvão, Brasil

O.L.S. Alsina
Universidade Tiradentes, Aracaju, Brasil

T.H.S. Starling & V.R. Mendonça
Universidade Federal de Sergipe, São Cristóvão, Brasil

ABSTRACT: This study aimed to compare the climatic parameters obtained *in loco* with values parameterized by the established regulatory agencies followed by analysis of the adequacy of these parameters to the individual characteristics of each worker. The proposed methodology consisted of measuring environmental conditions and compare the obtained data with the parameters advised by the norms and specialized literature. According to climatic characterization of the working places, it was found that it is difficult to maintain optimal temperature in all sectors of the company. It was found that, although the local corrected effective temperature kept on within the normative parameters, it should be considerably smaller in order to achieve levels of adequate thermal stress level for workers. It was necessary to develop technical and economic analyzes to identify ways to minimize and determine the costs and benefits of the identified possible means to avoid this thermal discomfort.

1 INTRODUCTION

Temperature directly influences the productivity of workers. According to Slack (1999), the environment around the worker directly influences the way its activities are carried out, resulting in consequences in its health if the working conditions are relatively hot or cold. Kroemer and Grandjean (2005) assert that the support of a comfortable climate is fundamental to the welfare of workers that perform their tasks with maximum efficiency. The term "climate", specifically, is used to define the physical conditions of the living and working environment. Within the ergonomics science, the climatic analysis in the working place environment approaches aspects such as air temperature, temperature of the bounding surfaces, air humidity, air movement and air quality. (GRANDJEAN and KROEMER, 2005).

Another important variable to be considered in this type of analysis is the thermal stress level. Such variable is directly related to the thermal comfort. Regulatory agencies parameterize that variable to ideal values for work environments. Dul and Weerdmeester (2004), however, assert that the thermal comfort has very personal characteristics for each worker, each one with their individual preferences.

The present work aims to develop an ergonomic analysis related to the climate and the thermal stress level in a sector with five workstations, located on a technology and information management company in Aracaju, Sergipe, Brazil.

This study aimed to compare the climatic parameters obtained *in loco* with values parameterized by established regulatory agencies. Afterwards, a critical analysis of the adequacy of these parameters in relation to the individual characteristics of each worker was carried out.

2 MATERIALS AND METHODS

2.1 Materials

Two instruments were used for measuring environmental conditions in the working places:

– Globe Thermometer/Stress Meter Portable Digital co-terminal – TGD-200 – INSTRUTHERM;
– Termo-higro-decibel-luxim. THDL-400 digital – INSTRUTHERM.

In an auxiliary way, a measuring was used to aid the optimal positioning of the globe thermometer in the working place. A digital timer was also used in the collecting data procedures.

2.2 Methodology

The methodology of this study consisted of three steps:

a) Execution of a bibliographical research in the area of ergonomics, through books, journals, norms, etc., aiming the survey of parameters and

concepts about the influence of climatic aspects in the health and productivity of workers in their workplace;
b) Direct utilization of the measuring instruments of environmental conditions, based on the specifications of their respective technical manuals;
c) Comparison of the data obtained with the parameters specified by the standards and specialized literature.

2.2.1 Measurement

Ina first experiment, the Thermal Stress Globe Thermometer waspositioned near one of the walls, as recommended by the technical manual, near the site with more sunlight incidence.

Temperature was measured using three different probes: globe probe, dry bulb probe and wet bulb probe. Ten measurements for each probe were sampled at intervals of one minute. At the tenth measurement, it was found an oscillation not higher than 0.1 °C between the last three samplings. Thus, for statistical analysis purposes and according to the equipment's technical manual, the eighth (and last) value of the data series was calculated as the arithmetic mean of the eighth, ninth and tenth samplings.

In a second experiment, the temperature probeof the THDL was placedin eight different locations of the analyzed environment. The choice of the measurement points took into account the main sources of heat, localized heat sources and the positions of the workstations.

The methodology consisted in taking five measurements sampled at intervals of 15 seconds, for each previously designated place.

2.2.2 Calculation of the wet bulb index and globe thermometer (WBGT)

The optimal effective temperature will be that one in which the thermal comfort is maximum, then there is no thermal stress or it is minimum (GRANDJEAN and KROEMER, 2005). This condition can be measured by combining the measurements of the dry bulb temperature, wet bulb and globe thermometer. Such temperatures create an index called WBGT Index (Wet Bulb Index and Globe Thermometer), which is used for the evaluation of thermal overload.

According to ISO 7243:1989, the WBGT or wet bulb globe temperature index "is one of the empirical indices representing the heat stress to which an individual is exposed" and the parameters characteristic of the environment used to measure this index are the natural wet bulb temperature and globe temperature and, of course, the measurement of air temperature.

According to annex 3 of the Brazilian norm NR-15/2008,(limits of tolerance to heat exposure), and the norm ISO 7243:1989,equations 1 and 2 express the WBGT index for indoor or outdoor environments without solar load, and for outdoor environments with solar load, respectively.

Thus, for indoors or outdoors without solar load:

$$WBGT = 0.7tbn + 0.3tg \qquad (1)$$

For outdoors with solar load:

$$WBGT = 0.7tbn + 0.1tbs + 0.2tg \qquad (2)$$

where:

- tbn: natural wet bulb temperature;
- tbs: dry bulb temperature;
- tg: globe temperature.

2.3 Climatic characterization of the workstations

The maintenance of the temperature of the analyzed sector is performed by two air conditioners with a capacity of 9,000 and 15,000 BTUs, with temperatureset from 21 °C to 24 °C at the time of the analysis. The devices were positioned near the ceiling, centered in the sector, covering both sides of it, allowing the homogenization of the temperature between the workstations.However, the analyzed sector is located in a panoramic layout open office, beside other sectors of the company. Neighboring sectors are separated by half-height glass partitions, measuring 149 cm. This layout becomes a problem for the maintenance of the optimum temperature in all sectors. When a sector does not use its air conditioners, the temperature of the adjacent sectors becomes uncomfortable for their workers.

The studied sector is in a room with a large glass windows, occupying the entire length of one of the walls. This window is partially covered with white paper. This configuration is responsible for the lowering of the level of thermal comfort in the sector, due to sunlight passing through the glass. Therefore, this sector receives large calorific loads from the solar radiation.

The presence of computers, which remain turned on throughout all the work period in this and in the adjacent sectors, do not contribute to a significant calorific load able to influence the thermal comfort.

Figure 1 ilustratesthe studied sector with its five workstations (yellow arrows) and the globe thermometer used to measure the WBGT index (blue arrow).

Source: Own author.
Figure 1.   Sector where the study was carried out.

2.4   Thermal balance

According to Ruas (2002), thermal balance is essential to human life and it is obtained when the amount of heat produced in the body is equal to the amount of heat transferred to the environment through the skin and breathing. The produced heat is the difference between the rate of metabolism and the mechanical work performed.

ASHRAE (1977) proposed the following equation for expressing this process mathematically:

$$M - W = Q_{sk} + Q_{rs} + S \qquad (3)$$

where:

$$Q_{sk} = C + R + E_{rsw} + E_{dif} \qquad (4)$$

$$Q_{res} = C_{res} + E_{res} \qquad (5)$$

and:

– M: Rate of metabolism, W/m²;
– W: Rate of mechanical work performed, W/m²;
– Q_{sk}: Total rate of heat lost by the skin, W/m²;
– Q_{res}: Total rate of heat lost through respiration, W/m²;
– S: Rate of heat stored in the body, W/m²;
– C + R: Sensible heat loss through the skin by convection and radiation, W/m²;
– E_{rsw}: Rate of heat lost by evaporation of perspiration, W/m²;
– E_{dif}: Rate of heat lost by evaporation of water diffusion, W/m²;
– C_{res}: Rate of convective heat loss in respiration, W/m²;
– E_{res}: Rate of evaporative heat loss in respiration, W/m².

A positive value for the term S represents the body heating, as the same way as a negative value

shows its cooling. When S equals zero, the body is in thermal balance.

The mechanical work performed by the muscles (W) is usually stated in terms of body mechanical efficiency η = W/M. The value of η is at maximum 24% and for the majority of the activities it's near zero. Therefore, the mechanical working is typically neglected and the metabolic rate is equal to the produced heat.

The total heat produced by the organism depends on the type of activity performed and can be calculated according to the equation:

$$Q_p = M \cdot A_{du} \qquad (6)$$

Where:

– Q_p: Is the produced metabolic energy (in W);
– M: Is a value that depends on the activity developed (given in W/m²);
– A_{du}: is the body superficial area (in m²).

The rate of metabolism of a given work is estimated by setting it in one of the activities proposed by ISO 7730:1984.

All the terms of the thermal balance equation are given in the unit of electrical power per area and refer to the superficial area of the bare body. This area is calculated by the equation of Dubois-Poulsem (ASHRAE, 1977):

$$A_{du} = 0.202 \cdot mc^{0.425} \cdot a_c^{0.725} \qquad (7)$$

where:

– A_{du}: Body superficial area, m²;
– m_c: Body mass, kg;
– a_c: Body Height, m.

3   RESULTS AND DISCUSSION

3.1   Assessment of the thermalstress levels

The thermal comfort zone is defined in different ways by the literature. The Brazilian norm NR-17/2007standards allocates this area in the range of effective temperature between 20 °C and 23 °C, with relative air humidity not less than 40% and air velocity not exceeding 0.75 m/s. Annex 3 of the Brazilian NR-15/2008standards says that low metabolic rate activities can be performed continuously at temperatures up to 30 °C, without causing risks to health. Iida (2005) and Grandjean (1998), generally establish the thermal comfort zone between 20 °C and 24 °C of effective temperature for relative humidity of 40% to 80% and air velocity of 0.2 m/s. Table A.1 of Annex A ofISO 7243:1989 presents reference values corresponding to different metabolic rate activities.

Through THDL, an average relative air humidity of 74% was obtained. The corrected effective temperature of the analyzed sector was obtained as 23.64 °C, using the WBGT. The results are not according to the parameters of temperature and relative humidity specified by Brazilian norm NR-17/2007, but they remain under the specifications of Brazilian norm NR-15/2008.

However, the exception already mentioned by Kroemer and Grandjean (2005), says that thermal stress levels have subjective and individual components, fact also verified *in loco*, by non-standardized interviews. Among the five occupants of the workstations in the studied sector, two were satisfied with the local thermal conditions, while the three others reported dissatisfaction with this aspect.

Although the WBGT index remained inside the proper ranges proposed by literature, it should be considerably lower for the achievement of adequate thermalstress levels for the workers. Then, it became clear that the established regulatory standards (NR's and ISO's) does not take into account considerations about the individual preferences. Thus, correctly parameterized environments do not necessarily guarantee correct thermal comfort to the employees.

Figure 2 presents the temperature measurements obtained using THDL,taken from the adjacent working surfaces (for example, table, keyboard, printer etc.).

As mentioned by Iida (2005) and Kroemer and Grandjean (2005), no surface of a workstation must differ no more than ±4 °C from the relative air temperature of the local environment. Through this concept, the upper and lower control limits of the graph were obtained.

It was verified the appropriateness of the obtained values, including theirarithmetic mean

value, with respect to the parameters observed in the literature.

3.2 *Observed thermal sensation*

The subjective thermal sensation was assessed through questionnaires, according to ISO 7730:1984 standard. In order to compare the workers subjective thermal sensation, using data of the relative velocity of the air (Var), relative humidity (RH), thermal resistance of clothing (Iclo), environment temperature (T) and type of activity undertaken, we calculated the estimated average rate of votes (PMV), according to Fanger's model.

Considering the individual differences of workers, there was good agreement between the proposed model and the obtained results.

3.3 *Calculation of the body surface area (A_{du}) and metabolism (Q_p)*

The surface area of the body was estimated by Dubois-Poulsem Equation (7). The metabolic rate was calculated using Equation (6).

It was observed that the mean anthropometric data and the workers metabolism approach the mean pattern of the human body superficial areas, according to ISO 8996:1990.

Considering the tests performed as a light activity, which corresponds to a metabolic rate around 93 W/m², it was found that the found values practically coincide with the standards established by ISO 8996:1990.According to ISO 7243:1989, the level of metabolic rate of the done activities, they were classified as level "1" (low metabolic rate) and, by Annex A, Table A.1, of this norm, thegiven reference value for WBGT is 30 °C or below. According to that, the studied sector's WBGT index is also inside the proper range proposed by that international standard.

4 CONCLUSIONS

The experimental results are consistent with the parameters specified in the standards and concepts observed in the literature. However, it was found that the presence of half-height partitions turns difficult to maintain the thermal comfort due to the influence of air conditioners from the adjacent sectors. Thus, a more precise temperature control only can be achieved in a closed environment, with partitions of type floor-ceiling, which can characterize a trade-off in relation to the flow of information, once the panoramic office shows better levels of information flow with respect to the closed type office (IIDA, 1990 *apud* OGASAWARA, 2004).

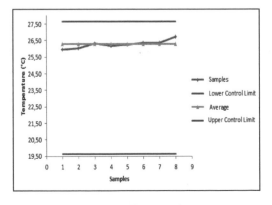

Figure 2.   Graph of the adjacent surfaces temperature.

Likewise, eliminating the problems generated by the presence of the windows could lead to a trade-off regarding to illumination costs.

Contact with workers showed that individual and subjective components related to thermal comfort, together with regulatory standards, should be considered in the design of work environments.

The results showed that the temperature patterns proposed by NR's and ISO's do not guarantee thermal comfort for all workers, because such standards do not take into account individual aspects or any local specificities. Thus, once confirmed the existence of ergonomic discomfort, technical and economic analyzes are necessary in order to identify ways to minimize it, as well as the measurement of theresulting costs and benefits.

REFERENCES

ASHRAE, American Society of Heating Refrigerating and Air Conditioning Engineers. Handbookof Fundamentals. Atlanta, 1977.

BRASIL. Ministério do Trabalho. Normas e Manuais Técnicos: Portaria SIT n.º 13, de 21 de junho de 2007. NR17—Ergonomia. Brasília: Ministério do Trabalho, 2007.

BRASIL. Ministério do Trabalho. Normas e Manuais Técnicos: Portaria SIT n.º 43, de 11 de março de 2008. NR15—Atividades e Operações Insalubres. Brasília: Ministério do Trabalho, 2008.

DUL, Jan; WEERDMEESTER, Bernard. ErgonomiaPrática. 2 ed. rev. e ampl. São Paulo: Edgard Blücher, 2004.

GRANDJEAN, E. Manual de Ergonomia: Adaptando o trabalhoaohomem. Porto Alegre: Bookman, 1998.

GRANDJEAN, E.; KROEMER, K.M.E. Manual de Ergonomia: Adaptando o trabalhoaohomem. 5 ed. São Paulo: Bookman, 2005.

IIDA, Itiro. Ergonomia: projeto e produção. São Paulo: EdgardBlücher, 2005.

INTERNATIONAL ORGANIZATION FOR STANDARDIZATION, Geneva. ISO 7730; moderate thermal environments—determination of the PMV and PPD indices and specification of the conditions for thermal comfort. Geneva, 1984.

INTERNATIONAL ORGANIZATION FOR STANDARDIZATION, Geneva. ISO 7243; Hot environments—Estimation of the heat stress on working man, based on the WBGT-index (wet bulb globe temperature).Geneva, 1989.

INTERNATIONAL ORGANIZATION FOR STANDARDIZATION, Geneva. ISO 8996; ergonomics—determination of metabolic heat production. Geneva, 1990.

OGASAWARA, Érika Lye. Ainfluência da organização do trabalhonacomunicaçãoem um escritório-panorâmico: um estudo de caso. Porto Alegre: Universidade Federal do Rio Grande do Sul-UFRGS, 2004.

RUAS, Álvaro César. Sistematização da avaliação de conforto térmicoemambientesedificadosesuaaplicaçãonum software. 2002. 181p. Tese (Doutoradoem Engenharia Civil, área de concentração de Saneamento). Faculdade de Engenharia Civil, da Universidade Estadual de Campinas, Campinas—SP, 2002.

SLACK, Nigel. et. al., Administração da Produção. São Paulo: Atlas, 1999.

Occupational Safety and Hygiene – Arezes et al. (eds)
© 2013 Taylor & Francis Group, London, ISBN 978-1-138-00047-6

Proposal for a lighting project of a metallurgical company based in the determination of the medium level of illuminance

J.M.N. Silva, A.S.L. Santos Filho, C.M. Souto, W.S. Macêdo, M.B.G. Santos & I.F. Araujo
Universidade Federal de Campina Grande, Campina Grande, Paraíba, Brasil

A.L. Melo
Universidade de Coimbra, Coimbra, Portugal

ABSTRACT: To guarantee the increase in the productivity, many companies have been investing in the improvement of the conditions of labor at labor places. Investments in ergonomics are one of the main focused aspects, for instance, a good condition of illuminance during tasks prevent discomforting glare and some sort of accidents. These added ads to productivity make industries choose the improvement of illuminance conditions at labor places. Aware of some problems related to the field of productions in industries, a study was devised in order to check out the illuminance conditions of a metallurgical industry and so, to propose changes to guarantee better working conditions. In order to determine the level of illuminance site a script based on the method of lumens was used. If the illumination in a particular place is not sufficient, a new design of lighting technique is proposed. The results show that medium illuminance is under what is required by the norms, and that the new lighting design is remarkable relevant.

1 INTRODUCTION

Globalization brings within itself an increasing competitive battle among companies. These companies, therefore constantly attempt to increase their productivity makin use of different methodologies. By using several methodologies, the companies expect to increase their gains as well their competitive advantage on the rivalry. The investiments in ergonomics are among the main changes and breaks of certain paradigms incorporated by many companies in order to improve the performance and results of work, especially to guarantee the comfort during the tasks.

According to Iida (2005) ergonomics supports the improvement, efficiency, confidence and the quality of industrial operations by analyzing, the physical labor, especially temperature, noise and illuminance, among other variations. To Rozenfeld (2006) the discomfort due to environmental factors such as lighting is one main causes of low productivity. According to Silva et al. (2012) there are some physical setbacks related to inadequate illumination (caused by the lack or excess) and the most common symptoms are eyestrain, watery eyes, irritability, and decreased productivity. This discomfort during labor activities might also cause headache, glare, decreased visual efficiency, and sorts of accidents. Faced with the problems mentioned above, companies have been improving the conditions of illuminance at labor places with the aim of increasing the productivity.

Aware of the already metioned problems at labor places, a study was developed in order to check out the illuminance conditions of a metallurgical industry and therefore to propose changes in order to enhance working conditions in the observed industry.

2 MATERIALS AND METHODS

The focused study brings descriptive and exploratory features. To start with, a literature review on the present topic was done, then, through some observation and measurements at labor place was possible to describe the found phenomena. The approach in this academic work is quantitative when it comes to measurement of variables. The present work can also be named qualitative when certain variables are analized and assigning to these variables certain characteristics, such as the cleaning of fixtures which cause the variation of impacts on the environment illuminance level.

In order to determine the level of illuminance in a certain labor place, a methodological script based on the method of lumens was developed. If the illuminace is not enough, so, a new design of illuminance technique is proposed, and for that, it is estimated the number of lamps and fixtures, it is also measured the distance among the lamps and luminaires. The distance among the lamps and the walls are measured in order to find an ideal illuminance for any industry.

In relation to location, the studied metallurgical industry makes use of tippers that help to increase the levels of illuminance. These are positioned at the centers of the sides at twenty six feet from the ground.

2.1 Measurement method

The data were collected throughout five sunny days. The used measurement instrument was THDL-400 by Instrutherm in the luximeter function. The data collections were performed in accordance to the methodology for illuminance measurement of interiors present in NBR 5382 (1985) of the Brazilian Association of Technical Standards (ABNT), in such a way that this method results in average illuminance values with a maximum of 10% (ten percent) of error.

This method is due to make measurements at a factory different sites. For accomplishing the aim of the study, two measurements are made, which were named "p1" and "p2" in two opposal diagonal of the factory. Four measurements named "t1", "t2", "t3" and "t4" the centers of the sides that determine the length of the site. Four measurements named "q1", "q2", "q3" and "q4" at the centers of the sides that determine the breadth of the place. Eight measurements named "r1," "r2", "r3" and "r4", "r5", "r6", "r7" and "r8" made near the center of the place. In figure 1 is possible for us to observe the position of the points on the floor plan.

To go on with the study, the arithmetic average "r", "q", "t" and "p" which respectively give the origin to "R", "Q", "T" and "P". These values are substituted in equation 1 to find out the local luminance (L):

$$L = \frac{[R \cdot (Nlf - 1) \cdot (Mlf - 1) + Q \cdot (Nlf - 1) + T \cdot (Mlf - 1) + P]}{Nlf \cdot Mfl}$$

(1)

when: L = illuminance of the place; R = the average of points "r"; Q = the average of points "q"; T = the average of points "t"; P = the average of the points "p"; Nlf = the number of luminaires per line; Mfl = the number of lines of lamps.

2.2 Ideal illuminance calculus

As expected, the level of illuminance is under the values established by NBR 5413 (1992) of Brazilian Association of Technical Standards (ABNT),

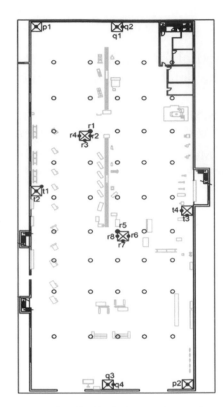

Figure 1. Points in the industry where values were determined with a luximeter device.

so a new method based on the illuminance design lumens is proposed. For that, initially we calculate the local index (K) through equation 2.

$$K = \frac{C \cdot L}{Hm \cdot (C + L)}$$

(2)

where: K = the local index; C = the length of the place; L = the breadth of the place; Hm = the vertical distance between the lamp and the work plan.

In function of the local index (K), and indices of reflection (tpp) from the ceiling, walls and ground, we are able, determine up the utilization factor (U) present in the catalog of the lamp/luminaire used on site. In the pictured below the whole lamp/table lamp and the local index of this set are totally shown.

Then, through site visits, having picked up the light loss factor (Fpl) due to the accumulation of dust and depreciation of lamps and luminaires. It was also noted the number of lamps per fixture (n), the luminous flux (F). The research also claims to the minimum illuminance (E) required by NBR 5413 for such activity. The minimum illuminance value (E) has been chosen takin into consideration

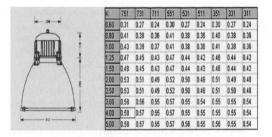

Figure 2. Combination lamp/bulb used in the factory and index table location for such combination.

the age of the developers is less than 40 s, and the job does not require a fast pace and accuracy, and that the reflectance of background task was between 30 and 70%. Then overwrites the values of these variables in equation 3 to obtain the number of lamps required on site (N):

$$N = \frac{(E \cdot C \cdot L)}{(n \cdot F \cdot U \cdot Fpl)} \qquad (3)$$

When N = the number of lamps required on the place; E = the minimum illuminance required by NBR 5413 for such activity; L = the breadth of the site; C = the length of the site; n = the number of lamps per fixture; F = the luminous flux of the lamp; U = the utilization factor; Fpl = the light loss factor.

The value to set the number of fixtures (N) should be rounded and called of Nn. Then it replaces the value of Nn and other variables in the equation aforementioned average illuminance of the room (Em):

$$EM = \frac{Nn \cdot n \cdot F \cdot U \cdot Fpl}{C \cdot L} \qquad (4)$$

When Em = the average illuminance value of the enclosure; Nn = the rounded value of luminaires needed in place; L = the width of the place; C = the length of the place; n = the number of lamps per fixture; F = the luminous flux of the lamp; U = the utilization factor; Fpl = the light loss factor.

Finally, to facilitate the distribution of fixtures at the factory, we calculate the minimum distance among the lamps (eL) using equation 5, and the distance among the fixtures and walls (eLP), using equation 6.

$$eL \leq 1,5 \times Hm \qquad (5)$$

$$eL \leq 1,5 \times Hm \qquad (6)$$

where eL = the minimum distance among luminaries; Hm = the vertical distance between the lamp and the work plan; eLP = the minimum distance among fixtures and walls.

3 RESULTS AND DISCUSSION

3.1 *Iluminance inside the place*

The collected results made throughout five days are listed in the table bellow:

Table 1. Values "r1" to "r8" and "R" for the five days.

Day	Day shift	Points "r1–r8"				Value of "R"
		lux				lux
1°	Morning	60	70	77	73	84,9
		115	110	77	97	
2°	Afternoon	66	72	78	87	84,5
		111	115	82	96	
3°	Morning	72	78	77	80	84,0
		110	112	88	98	
4°	Morning	77	82	75	84	86,0
		106	113	88	98	
5°	Afternoon	66	72	78	87	84,5
		111	115	82	99	

Table 2. Values "q1" to "q4" and "Q" for the five days.

Day	Day shift	Points "q1–q4"				Value of "Q"
		lux				lux
1°	Morning	103	100	125	217	113,8
2°	Afternoon	99	110	133	126	118,0
3°	Morning	113	99	138	130	121,5
4°	Morning	106	98	140	136	121,0
5°	Afternoon	99	110	133	126	118,0

Table 3. Values "t1" to "t4" and "T" for the five days.

Day	Day shift	Points "t1–t4"				Value of "T"
		lux				lux
1°	Morning	125	105	106	67	100,8
2°	Afternoon	129	110	102	77	106,0
3°	Morning	120	119	102	81	110,5
4°	Morning	119	115	101	85	108,0
5°	Afternoon	129	110	102	77	106,0

Table 4. Values "p1" to "p2" and "P" for the five days.

Day	Day shift	Points "p1–p2" lux		Value of "P" lux
1°	Morning	130	60	95
2°	Afternoon	125	77	101,0
3°	Morning	126	80	103,0
4°	Morning	126	62	94,0
5°	Afternoon	130	60	103,0

Table 5. Values of average illuminance (L) following the measurement points on site.

Day	"R" lux	"Q" lux	"T" lux	"P" lux	"L" lux	L average lux
1°	84,9	113,8	100,8	95,0	128,9	130,6
2°	84,5	118,0	106,0	101,0	130,4	
3°	84,0	121,5	110,5	103,0	131,4	
4°	86,0	121,0	108,0	94,0	132,7	
5°	84,5	118,0	106,0	103,0	130,6	

The results show that the industry has medium illuminance below what is required by the rules. For the Brazilian Association of Technical Standards NBR 5413 (1992) states that activities in Bodyshop, presses, shears, and drilling machines, the minimum illuminance value should be 200 lux. Knowing this we calculated the ideal illuminance (Em) to the site.

3.2 Determination of the proper illuminance inside the studied place

Using equation 2 described in the methodology, we calculated the local index (K), reaching a value of 2.21. As the color of the ceiling and floor are dark was determined that the index of reflection of these are 30% and 10% respectively, as the color of the walls is of a light color was attributed to the walls an index of reflection of 50%, thus reaching reflective index (tpp) of 351. Crossing the value of the local index (k) of 2.21 with the index of reflection (tpp) 351, found in Figure 2 a value of the utilization factor (U) of 0.5.

As lamps and fixtures are clean and in good working order is assigned a value of 0.80 for the factor of luminous losses (fpl). Observed during visits that was used by one lamp luminaire, the luminous flux and the same was 12000 lm. I divide to the age of employees is less than 40 years, the work does not require speed and accuracy, and the reflectance of task background is between 30 and 70%, occurred in NBR 5413 that the illuminance required by law is at least 200 lux for the site.

Then replaced with the value of these variables in equation 3 to obtain the number of lamps required on site (N). We found a value of N equal to 93 fixtures. How is the amount of five rows of lamps, can not be divided equally accurate and the 93 fixtures, and being rounded up to 95 the number of luminaires required (Nn).

Then it replaces the value Nn and the other aforementioned variables in equation 4 of the enclosure average illuminance (L). The value found for in is 203.8 lux, this value according to NBR 5413, which requires a minimum of 200 lux for the site. We also calculated the maximum distances between luminaires (eL), and between fixtures and walls (ELP) through equation 5 and equation 6, where I t was found that these values must be below respectively the 14.55 and 7.27 meters. Figure 3 shows a proposed of disposition of luminaire respecting the distances calculated, and the number of luminaries required to have a illuminance of approximately 203.8 lux, and thereby in accordance with the norms.

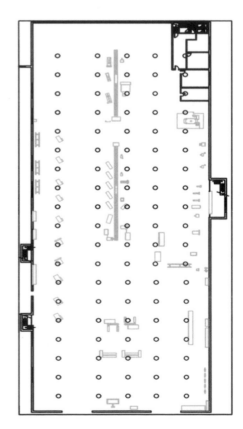

Figure 3. Proposed of disposition of luminaire for the industry.

4 CONCLUSION

It can be concluded that the conditions of illuminance are not enough to guarantee the comfort and safety while performing work activities. From measurements based on the NBR 5382 (1985) found a mean value of illuminance equal to 130.6 lux, a result that is below the limit set by the NBR 5413 (1992), which places as the minimum value of 200 lux for the activities performed in the sector of presses.

Was then calculated the number of lamps required to achieve the set value for the NBR 5413 and supported in equations of the method of the lumens is concluded that the number of ninety-five lamps being distributed in five of nineteen rows, it would be the ideal number to reach a value of 203.8 lux illuminance average throughout the whole industry.

Also it was established that the luminaires are at a distance of less than 14,55 meters from each other, and the walls of 7,27 meters to ensure this illuminance value.

The proposed changes were considered pretty relevant. Besides putting the industry in accordance with auditing standards on lighting, the changes proposed also increase the comfort and safety for employees perform their tasks.

REFERENCES

Brasil. Associação Brasileira de Normas Técnicas (1985). *Norma Brasileira 5382: Verificação de Iluminância de Interiores.* Retirado em 18 de Fevereiro, 2012, de http://pt.scribd.com/doc/46834807/NBR-5382.

Brasil. Associação Brasileira de Normas Técnicas (1992). *Norma Brasileira 5413: Iluminância de Interiores.* Retirado em 03 de Março, 2012, de http://www.labcon.ufsc.br/anexos/13.pdf.

Iida, I. (2005). *Ergonomia: projeto e produção.* (2ª edição revisada e ampliada). São Paulo: Edgard Blücher.

Rozenfeld, Henrique, et al. (2006). Gestão de desenvolvimento de produtos: uma referência para a melhoria do processo. São Paulo: Saraiva.

Silva, L.B., Eulálio, E.J.C., Coutinho, A.S., Soares, E.V.G., Santos, R.L.S. (2012). Analysis on the relationship between the school furniture and the work surface lighting and the body posture of public Middle School students from João Pessoa, Paraíba, Brazil. Work: A Journal of Prevention, Assessment and Rehabilitation. v. 41, pp. 5540–5542.

Occupational Safety and Hygiene – Arezes et al. (eds)
© 2013 Taylor & Francis Group, London, ISBN 978-1-138-00047-6

Environmental disturbances in robust machinery design

B. Mrugalska
Faculty of Engineering Management, Poznan University of Technology, Poznan, Poland

ABSTRACT: In order to increase machinery availability and their performance it is advisable to apply robust design methodology. Therefore, in this paper an attempt was made to specify disturbances, in particularly environmental disturbances, which have an impact on machinery manufacturing and operation. For this aim, the main sources of machines faults such as electrical, mechanical and environmental for electrical motors were identified. Moreover, the investigation of electrical motor requirements concerning operation and storage environment was made. It revealed that most of them concerned contact with chemical substances, cold and hot microclimate, industrial dusts and vibrations. On the other hand, the analysis of industrial working conditions in the European Union countries showed that the working conditions in these aspects were the most hazardous. It allowed to perceive the relationship between motor faults and working conditions and, thus it emphasized the need of application of product design methods robust to environmental disturbances.

1 INTRODUCTION

In today's economy manufacturers, seeking to remain competitive in the market, need to keep up with new technology. They have to rely on their manufacturing engineers to be able to set up manufacturing processes for new products quickly and effectively. In order to achieve this goal it is not enough to typically adjust parameters according to the instructions in the tool catalogues and/or handbooks to achieve the assumed product quality (Saffar et al., 2009). Moreover, it is not sufficient to select parameters for worst-case condition scenarios, in order to avoid machine failures (Qin & Park, 2005). It is obvious that such approaches significantly limit the productivity and accuracy of the manufacturing process. To solve such problems mathematical models and optimization techniques are more and more often applied (Vuchkov & Boyadjieva, 2002). They are based on the choice of a parametric model structure, which aims to the selection of the optimal values of parameters of the model which reflects expected characteristics or product outputs (Mrugalska & Kawecka-Endler, 2012). The obtained model is used as a pattern during manufacturing and therefore, the quality of the manufactured product relies on a model quality. Unfortunately, in most of the product design methods an influence of disturbances affecting manufacturing process and product in its operations is not taken into account (Mrugalska, 2013).

In this paper a particular attention is paid to environmental disturbances. Its correlation to working conditions in industrial companies is investigated.

For this aim, an analysis of sources of machinery faults was widely discussed. On the basis of it, the significance of environmental issues on the performance of machinery was shown. Moreover, the investigation of working conditions in the European Union countries was made. In particular, the environment in Polish industrial companies was widely discussed in the period of last five years. It allowed to noticed that in most cases electrical motor faults were related to the most hazardous issues in working environment. Therefore, the machines should be robust design to environmental disturbances which may appear during their manufacturing and operation.

2 MACHINES FAULTS

Machine fault can be defined as "any change in a machinery part or component which makes it unable to perform its function satisfactorily or it can be defined as the termination of availability of an item to perform its intended function" (Jayaswal et al., 2008). However, before the final fault takes place it is possible to notice incipient fault, distress, deterioration, and damage (Bloch & Geitner, 2005), what makes their part or component unreliable or unsafe for continued use (Mrugalska & Kawecka-Endler, 2011). Thus, it is advisable to do on-line monitoring of the machine variables by taking measurements to diagnose the state of the machine when it enters into the fault mode (Ding, 2008; Isermann, 2005; Mrugalski & Witczak 2012; Mrugalski et al., 2008).

The reliability of many machines and the safety of people using them to a large extent depend on the quality of the produced electrical motors. In Poland, approximately 12 million direct and alternating current (DC and AC) motors and generators, which are important units of many machines and equipment, are produced each year (e.g., 12,118,200 DC motors and generators and 9,556,400 AC motors and generators were produced in 2010, whereas 13,141,000 DC motors and generators and 10,020,900 AC motors and generators were produced in 2011 (Central Statistical Office, 2011; 2012)). In most cases machine fault problems result from failures in rotating electrical motors which derive from their design, manufacturing tolerance, assembly, installation, working environment, nature of load and schedule of maintenance (Singh & Kazzaz, 2003). The electrical motor faults can be categorized into two types: internal and external, as presented in Table 1.

As it can be noticed two types of sources of electrical motor faults can be identified: internal and external. Internal faults refer to components of electrical and/or mechanical part of the motor and thus, they may concern stator and rotor failures etc. (Figure 1). These faults result from the fact that induction motors like other rotating electrical machine are subjected to both electromagnetic and mechanical forces. Their design causes the interaction between these forces in normal operation condition. In such situation this interaction leads to a stable operation with minimum noise and vibrations. However, when the fault occurs, the equilibrium between these forces is lost what results in further enhancement of the fault (Kazzaz & Singh, 2003). On the other hand, the faults, which derive from external sources, effect the final performance of the motor and in the consequence they may also lead to the faults of electrical and/or mechanical part of the motor. They can result from

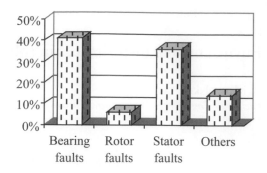

Figure 1. Electrical motor faults (Adopted from (Singh & Kazzaz, 2003)).

being manufactured and/or operated in threatening environmental surrounding which may result in appearance of environmental disturbances.

3 ENVIRONMENTAL DISTURBANCES

Environmental disturbances are regarded as factors negatively affecting manufacturing and operations of products which occur over a period of time and cause undesirable changes. They can result from natural causes or derive from the activities of humans. As they define conditions under which the product must be operated and dictate the performance of the product characteristics (Sutherland et al., 1988) their initial recognition is desirable in the stage of product design as it is depicted on Figure 2.

The disturbances are recognized as "major limiters to next generation manufacturing performance" (Teague et al, 2010) and thus, it is advisable to apply some strategies to improve product robustness to environmental disturbances:

– decouple which means to break all ways which can distribute any disturbance (e.g. protect heat transfer),
– desensitize which is defined as design for minimized response to disturbances (e.g. choose low thermal expansion materials to reduce thermal induced drift),
– control local environment (e.g. temperature controlled rooms (Teague et al., 2010; Winiarski, 2000)).

However, to facilitate the problem not only technical knowledge but ergonomic one is required (Górny, 2012; Moraes et al., 2012). The ergonomic aspects of environmental disturbances can be discussed on the basis of working conditions data in reference to regulations and standards as the products/machines and workers often maintain in the same environment.

Table 1. Sources of electrical motor faults (Adapted from (Singh & Kazzaz, 2003)).

Type of source	Source	Examples
Internal	Electrical	Dielectric failure Rotor bars crack
	Mechanical	Bearing faults Rotor strikes
External	Electrical	Voltage fluctuations Unbalanced voltage
	Environmental	Cleanliness Humidity Temperature
	Mechanical	Pulsating load Over load

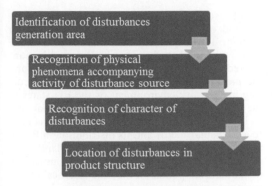

Figure 2. Environmental disturbances in machinery design (Adopted from (Winiarski, 2000)).

Table 2. Operation and storage environmental requirements for chosen electrical motors in industrial setting.

Requirements	Description
Ambient temperature	between $\pm 15^0$C and $\pm 40^0$C
Chemical substances	free
Dust	free
Gas	free
Operation humidity	not exceed 95% (at 30^{0}C)
Storage humidity	not exceed 50%
Storage temperature	between $\pm 5^0$C and $\pm 60^0$C
Vibrations	free

4 MATERIALS AND METHOD

The purpose of this study was to determine the most popular environmental disturbances appearing in the manufacturing and operation of products on the example of electrical motors. This knowledge applied to robust design is very crucial and vital as it provides data about the assumed working conditions of products in advance. In practice the level of disturbances observed in industrial environments and standards concerning industrial environment is often ambiguous. In some environments this issue was ignored for many years—there are many disturbances that do not comply with standard industrial environment requirements (Kałuski et al., 2012). For this aim, three primary activities were undertaken: a review of the literature, an investigation of environmental requirements for a chosen group of machinery such as electric motors and an analysis of data about working conditions in the European Union countries. On the basis of statistical data, the major environmental disturbances were identified and the assessment of correlation between requirements and working conditions was done in Polish industrial sector.

5 RESULTS AND DISCUSSION

For the practical investigation of machinery, electrical motors operated in industrial sector were chosen. The analysis of their operation and maintenance instructions, installation and maintenance manuals allowed to differentiate and precise environmental conditions under which they must be operated and stored, as it is shown in Table 2.

The most important issues concerned temperature and humidity which should be kept in a particular range. Any variations to it must be stated on a nameplate information. There should be no contact with chemical substances as they are corrosive for motor enclosure. The ventilation openings of motors should be clean of dust, dirt or other debris. Moreover, they should be stored indoors, in a clean, dry and gas free location. It is also vital that winding is be protected from excessive moisture absorption.

The investigation of operation and maintenance conditions encouraged to study data about working conditions in the European Union countries. In order to assess the impact of the most environmental hazards occurring in industrial sector data retrieved from the 5th European Working Conditions Survey was analyzed (Figure 3).

As it can be noticed on Figure 2, four main groups of environmental disturbances can be differentiated in the industrial sector. They concern chemical substance, hot and cold microclimate, industrial dust and vibration. The percentage value of chemical substances and vibrations increases within the accession of new EU member states, whereas the other two factors seem to be decreasing.

In order to investigate further working conditions Polish industrial sector was chosen. The gathered data are presented on Figure 4.

In Poland environmental working conditions, in comparison to all European Union countries, were quite different. The analysis of them by indicator „the number of employed working in hazardous conditions per one thousand of the employed in the surveyed population" (Working conditions..., 2011) showed that there were two groups of hazardous factors in manufacturing plants. The first of them concerned industrial dust which values were exceeded about one-third. In the next group the highest percentage participation had: hot microclimate (ca. 6,8%), chemical substances (ca. 6,1%), vibrations (ca. 4,6%) and cold microclimate (ca. 4,7%). The values of particular hazardous factors seem to be decreasing. In spite of this fact, they are above the limits what affirms the need of application of robust design methods where the attention to environmental disturbances occurring in manufacturing and operation of electrical motors should be paid.

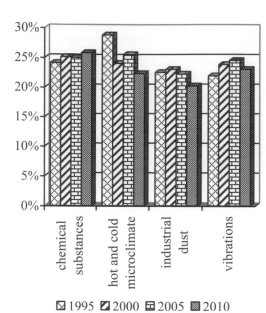

☒ 1995 ▨ 2000 ⊞ 2005 ■ 2010

Figure 3. Environmental conditions in industrial sector (Adapted from (5th European Working…, 2012)).

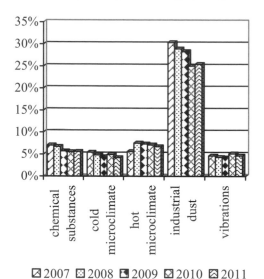

▨ 2007 ⊞ 2008 ◣ 2009 ☒ 2010 ▧ 2011

Figure 4. Environmental conditions in Polish industrial companies (Adapted from (Working conditions…, 2007; 2008; 2009; 2010; 2011)).

6 CONCLUSIONS

Nowadays, in demand for various industries highly accurate components or products are highly required. In order to provide them, not only the application of mathematical models and optimization techniques is often done but also knowledge about disturbances appearing in the manufacturing and operation process should be taken into account. In this paper the attention is paid to the environmental disturbances which should be considered on the stage of product design. Therefore, the most popular environmental hazards, occurring in the work environment of the EU states in the industrial sector, were identified and analysed. In its further investigation the attention was paid to a particular factor and its distribution in Polish industry. It allowed to indicate the overall influence of hazards identified in the work environment of manufacturing plants on the final electrical motor performance.

REFERENCES

5th European Working Conditions Survey. 2012. Luxembourg: Publications Office of the European Union.

Bloch H.P. & Geitner F.K., 2005. *Machinery Component Maintenance and Repair.* Oxford: Elsevier.

Central Statistical Office. 2008. *Working Conditions in 2007.* Warszawa. Retrieved November 12, 2012, from: http://www.stat.gov.pl/cps/rde/xbcr/gus/warunki pracy 2007.pdf.

Central Statistical Office. 2009. *Working Conditions in 2008.* Warszawa. Retrieved November 12, 2012, from: http://www.stat.gov.pl/cps/rde/xbcr/gus/LS working conditions_2008.pdf.

Central Statistical Office. 2010. *Working Conditions in 2009.* Warszawa. Retrieved November 12, 2012, from: http://www.stat.gov.pl/cps/rde/xbcr/gus/warunki pracy 2009.pdf.

Central Statistical Office. 2011. *Working Conditions in 2010* Warszawa. Retrieved November 12, 2012, from: http://www.stat.gov.pl/cps/rde/xbcr/gus/warunki pracy 2010.pdf.

Central Statistical Office. 2012. *Production of Industrial Products in 2007.* Retrieved November 15, 2012, from: http://www.stat.gov.pl/gus/5840_792_PLK_HTML. htm

Central Statistical Office. 2012. *Working Conditions in 2011* Warszawa. Retrieved November 12, 2012, from: http://www.stat.gov.pl/cps/rde/xbcr/gus/PWwarunki pracy_w_2011_r.pdf.

Ding S. 2008, *Model-based Fault Diagnosis Techniques: Design Schemes, Algorithms, and Tools.* Berlin/Heidelberg: Springer-Verlag.

Górny A. 2012. Ergonomics in the Formation of Work Condition Quality. *Work: A Journal of Prevention, Assessment and Rehabilitation* 1, Supp.: 1708–1711.

Isermann R. 2005, *Fault-Diagnosis Systems: An Introduction from Fault Detection to Fault Tolerance.* Berlin/Heidelberg: Springer-Verlag.

Kałuski, M., Michalak, M., Pietranik, M., Skrzypek, K. & Szafrańska M. 2012. Disturbances in Industrial Power Networks. *Przegląd Elektrotechniczny* (Electrical Review) 88(9b): 298–300.

Kazzaz, A.S. & Singh, G.K. 2003. Experimental Investigations on Induction Machine Condition Monitoring and Fault Diagnosis Using Digital Signal Processing Techniques. *Electric Power Systems Research* 65: 197–221.

Jayaswal P., Wadhwani A.K., Mulchandani K.B. 2008. Machine Fault Signature Analysis. *International Journal of Rotating Machinery*. Article ID 583982. Retrieved October 15, 2012 from: http://www.hindawi.com/journals/ijrm/2008/583982/.

Moraes, A.S.P., Arezes, P.M. & Vasconcelos R. 2012. From ergonomics to design specifications: contributions to the design of a processing machine in a tire company. *Work: A Journal of Prevention, Assessment and Rehabilitation* 41, Supp. 1: 552–559.

Mrugalska, B. & Kawecka-Endler, A. 2011. Machinery Design for Construction Safety in Practice. In C. Stephanidis (ed.), *Universal Access in HCI*, Part III, HCII 2011, LNCS 6767 (pp. 388–397). Berlin/Heidelberg: Springer-Verlag.

Mrugalska, B., Kawecka-Endler, A. 2012. Practical Application of Product Design Method Robust to Disturbances. *Human Factors and Ergonomics in Manufacturing & Service Industries* 22: 121–129.

Mrugalska, B. 2013. Design and Quality Control of Products Robust to Model Uncertainty and Disturbances. In: K. Winth (ed.), *Robust Manufacturing Control, Lecture Notes in Production Engineering* (pp. 495–505). Berlin/Heidelberg: Springer-Verlag.

Mrugalski M. & Witczak M. 2012, State-Space GMDH Neural Networks for Actuator Robust Fault Diagnosis. *Advances in Electrical and Computer Engineering* 12(3): 65–72.

Mrugalski M., Witczak M. & Korbicz J. 2008, Confidence Estimation of the Multi-layer Perceptron and its

Application in Fault Detection Systems. *Engineering Applications of Artificial Intelligence* 21(8): 895–906.

Qin, Y. & Park, S.S. 2005. Robust Adaptive Control of Machining Operations. In: J. Gu, P.X. Liu (eds.), *Conference Proceedings of IEEE International Conference on Mechatronics Automation* (pp. 975–979). Niagara Falls.

Saffar, R.J., Razfar, M.R., Salimi, A.H. & Khani, M.M. 2009. Optimization of Machining Parameters to Minimize Tool Deflection in the End Milling Operation Using Genetic Algorithm. *World Applied Sciences Journal* 6(1): 64–69.

Singh, G.K. & Kazzaz, A.S. 2003. Induction Machine Drive Condition Monitoring and Diagnostic Research —a Survey. *Electric Power Systems Research* 64(2): 145–158.

Sutherland, J.W., Ferreira, P.M., DeVor, R.E. & Kapoor, S.G. 1988. An Integrated Approach to Machine Tool System Analysis, Design and Control. *Proceedings of 3rd Int. Conf. on Comp.-Aid. Prod. Engr.* 429–445.

Teague, C., Evans C. & Swyt D. 2010. *Patterns for Precision Instrument Design*. Retrieved October 8, 2012, from: http://www.aspe.net/publications/Newsletters/SeptNewsletter.html.

Vuchkov, I. N., Boyadjieva, L. N. 2002. *Quality Improvement with Design of Experiments: A Response Surface Approach*. Boston: Kluwer Academic Publishers.

Winiarski Z. 2000. Limitation of Thermal Disturbances in Machine Tool Behaviour. *Proceedings of Ecole Centrale Nantes on CD-room, II International Seminar on Improving Machine Tool Performance, Nantes—La Baule*.

Fire safety

Occupational Safety and Hygiene – Arezes et al. (eds)
© 2013 Taylor & Francis Group, London, ISBN 978-1-138-00047-6

Evacuation times sensitivity

Natacha Beleza, J. Santos Baptista & Aura Rua
PROA/CIGAR/LABIOMEP/Faculdade de Engenharia, Universidade do Porto, Porto, Portugal

ABSTRACT: Evacuation training is an essential procedure in order to detect flaws and to train workers exiting buildings in a controlled way. As this involves very high costs as well as production stops, it cannot be performed very frequently It should be planned in order to detect potential failures and correct them in due time to maximize its success and reduce costs. This paper aims at showing the usefulness in applying simple means of prediction, such as to measure the length of the path to go. The simulation of several scenarios was done, and it was possible to identify as critical the frequency whereby workers cross the emergency door. This made possible to identify the training of this matter as essential for the successfulness of the evacuation.

1 INTRODUCTION

The growing concerns in safety conditions, hygiene and health have positively influenced the legislation and, in particular, issues related to Safety Against Fires in Buildings (SAFB).

In Portugal, in spite of the legislation effort in this matter, both the effective execution, and the way it is mentioned in the legislation show serious gaps. In case of fire, that can cause serious human, environmental and economic consequences. In reality, it often happens that the standards do not take into account the practical information supported by computerized models. Nevertheless, these were used in some countries such as the USA, New Zealand and Canada. By structuring the information and facilitating the calculations, these models promote the accomplishment and the adherence of the organizations.

On the other hand, the statistics and other elements of accident analysis, as far as emergency situations in Portugal, are scarce. However, according to the information available on *Segurança Online* (2012) electrical problems, explosions and fires are the most common situations, representing around 0.4% of the total accidents. From these more than 4% are fatal.

In this context, the question about how far is it possible, with the use of simple calculations models, to detect the critical points of an organization in terms of evacuation is an important matter. The existing models of evacuation such as FDS + Evac, Simulex and EvacuatioNZ, are sophisticated and have high liability (Tan 2011). However due to the complexity of its implementation and its high costs, small and medium organizations do not use them. On the other hand, simpler models as proposed by the Coimbra Fire Department, due to their high safety coefficients, result in the determination of excessively high times, which may lead to excessive measures and, as a consequence, too expensive hampering their implementation. Taking into account all these aspects, the best of the two implementations were/was seen as a possible scenario. Thus, the primary objective of this work was to assess the sensitivity to changes in the time required to evacuate a factory plowing through the application of an easy method of calculation and simulation.

2 MATERIAL AND METHOD

This work was took place in an industrial building with the major type of risk category, with 46 workers and 10 workplaces. After analyzing the normative and bibliographic references on SAFB, calculations of the evacuation time have been made using the Coimbra Fire Department equation.

It was compared the time evaluated by the Coimbra Fire Department method with the traveling time between each working place and each building exit door. It was also measured the frequency whereby it was possible for the workers to exiting the building (peoples/second) (Gwynne et al. 2003, Balci 1997, Kuligowski 2009). During these measurements, through an informal inquiry, it was assessed, which exit door the most easily recognized by each worker (Breaux 1976).

Taking these elements into account, a model of evacuation of the building was made from which a sensitivity of the evacuation time analysis was evaluated. Variations of ±5% and ±10% of the traveling time were tested, crossed with .a middle exit frequencies of 1 to 2 workers per second.

3 MODELS DESCRIPTION

3.1 CoFiDe model

The evacuation time through the Coimbra Fire Department proposed equation is calculated through four different parcels.

The variables used in this equation were adjusted for the building under study (table 1).

Simultaneously it was made a small informal inquiry in order obtain reliable information from the workers. During the theme presentation, they were questioned about the number emergency doors there were, and how many they could recall. The *Cais* emergency door, which is the place where workers enter the building was easily noticeable.

3.2 Alternative designed model

The movement times were measured between different workplaces and each available emergency door.

After gathering the results, the occupants were ordered at each exit door by time of arrival and, for each one, was assigned an delay according to the number of people on site.at each exit door. Whenever the time elapsed was equal or under the travelling time of the previous occupant, its delay

time was added (0.5 s or 1.0 s) to the difference between the previous times. This new time became the new evacuation time.

Along with this delay at the exit it was also added the influence of an orderly exit or a messy exit. When the exiting of the occupants was quick and well organized, an exit frequency of 2 people per second was considered. In case of a less organized exit, the frequency was of 1 person per second.

4 RESULTS AND DISCUSSION

The results shown refer to the most unfavorable situation, that is, all employees leaving the building through the same door.

4.1 First results

The time computed with the Coimbra Fire Department algorithm was longer than the real time movement of the people (Figure 1).

The different workplaces are distributed by two floors. The 4 emergency doors to the outside are painted with circles with a continuous line, while the exit/entrance to the available floors are indicated by a dashed line.

The emergency doors on floor 0 are shown in fig. 2.

Figure 3. shows the emergency doors in floor 1 that leads the workers to floor 0.

4.2 Variable influence

Another result obtained shows that there is a great influence in the frequency of the exiting on the evacuation time (Figure 4 and 5) (Nelson & MacLennan 1996, Papinigis 2010). When exit frequency is high (two people per second), which corresponds to a quick and well organized exit, there are no people gathering at the doors and it is possible to evacuate the building within a time similar to the one of the travel from the furthest point to the exit. However,

Table 1. Formula variables definition in the evacuation time calculation.

Var	Variable definition	Function
Te	Evacuation time	$Te = Ts + Tdh + Tde + Tep$
Ts	Evacuation time through the building	$Ts = Et/(Ls \times Ce)$
Tdh	Horizontal lanes motion time	$Tdh = Lh/Vh$
Tde	Stairs motion time	$Tde = Le/Ve$
Tep	Maximum evacuation time of an entire floor	$Tep = Ep/(Lp \times Ce)$
Et	Number of people to evacuate	
Ls	Total width of the exit halls (m)	
Ce	Evacuation coeficiente (1,8 person/m/s)	
Lh	Maximum distance on the horizontal from the furthest point from the exit (m)	
Vh	Circulation speed in horizontal halls (0,6 m/s)	
Le	Maximum distance through stairs from the furthest point from the exit (m)	
Ve	Circulation speed in stairs (0,3 m/s)	
Ep	Number of people to evacuate in the worst floor	
Lp	Total width of the exits in the worst floor	

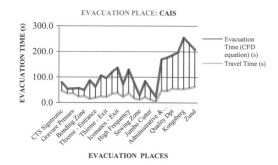

Figure 1. Times calculated with the *CoFiDe* equation and the movement time

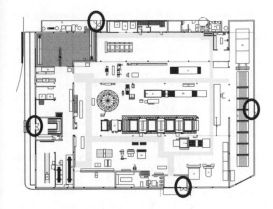

Figure 2.   Floor 0 plant of the industrial building.

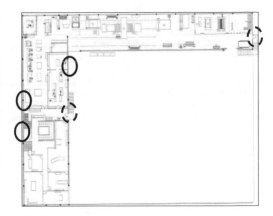

Figure 3.   Plant floor 1 of the industrial building.

even if the movement time is maintained, if the exit is less organized (with trampling, pushing or slowing down in the exit) (1 person per second), the evacuation time rises to around 10 s (~17%) (Figure 4) (Gwynne et al. 2003, Balci 1997, Gwynne et al. 1998, Filippidis et al. 2006, Olsson & Regan 2001, Sargent 1999, Averill et al. 2011).

However, sometimes, there are changes on production layout. That happens, for instance, in the production campaigns of specific marketing materials. In these cases it is shown that if a good organization is maintained with a frequency of 2 people per second, the evacuation time does not suffer any change regarding what was seen in normal conditions of industrial operation. However, reducing the exiting frequency to one person per second, the evacuation time rises to around 20 s (~33%) (Figure 5).

5   CONCLUSIONS

The calculated times through the Coimbra Fire Department algorithm are significantly larger than

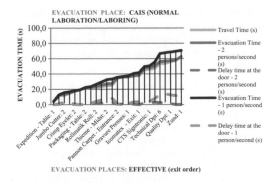

Figure 4.   Evacuation time in normal conditions.

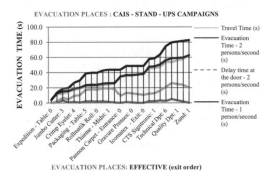

Figure 5. Evacuation time during a *stand-ups* campaign.

the real times, from which it can be seen that its utilization is not adequate to the situation now subjected to analysis.

The informal inquiry showed that at least one of the emergency exits closer to the workers is easily recognized by them. On the same note, it shows that there is a remarkable connection between human behavior and the building's constructive code, highlighted in the moment when the workers fully remember the emergency exit door as the door they used to enter the building (Averill et al. 2011).

On the other hand, for the same number of workers, it can be stated that the evacuation times are influenced by:

– The layout of the production process. It can be assumed that it will also be influenced by the layout of the building itself (Tang & Ren 2008), since the more straight the path to the exit door, the less time the movement will take (Papinigis et al. 2010, Vasudevan & Son 2011);

– Exit frequency. This factor is influenced by the door width and by the organization of the evacuation. With this in mind, there is a need to perform training simulations of this process (Kuligowski 2009, Breaux et al.1976, Gwynne et al. 1998, Bryan 2000).

It was also shown the advantage in simulating the evacuation process. Although the amount of time necessary to elaborate and/or tune the algorithms (Tan 2011, Korhonen & Heliovaara 2011), this process:

- by conducting a sensitive analysis, it is possible to detect potential flaws and weak spots in the evacuation procedure, and to correct them in time (Gwynne et al. 2003, Balci 1997).
- Although it is not possible to substitute the real evacuation training, simulation makes possible to detect and correct different issues, hard to spot and identify with the traditional approach of these processes (Breaux et al. 1976, Zia et al. 2011).

Other aspects should be taken into account for future analysis, such as:

- The relation between human behavior, daily routine and the architecture of the building, since people often leave from where they got in (Averill et al. 2011, Aldewereld 2010, Clark 2010, Zhang et al. 2010).
- The distribution of the highest thermic charges, taking into account the inflammable products, the number and the distribution of the occupants throughout the working area (Filippidis 2006, Yuan, 2011).

In an extensive perspective, it is crucial to start to incorporate the evacuation simulation programs in basic education of safety technicians.

REFERENCES

Aldewereld, H., Álvarez-Napagao, S., Dignum, F., Vásquez-Salceda, J. 2010. Making Norms Concrete. ACM Digital Library. *Proceedings of the 9th International Conference on Autonomous Multiagent Systems*, vol.1, 807–814. ISBN: 978-0-9826571-1-9.

Averill, Jason D., Kuligowski, Erica D., Peacock, Richard D. 2011. After the Alarm Sounds: Historical, Present and Future Perspectives. Historical review and recent trends in egress modeling. Fire Protection Engineering. *Sfpe Magazine.*

Balci, Osman 1997. Verification, Validation and Accreditation of Simulation Models. *Proceedings of the 1997 Winter Simulation Conference.* Blacksburg-Virginia, 135–141.

Breaux, J., Canter, D. e Sime, J. 1976. Psychological Aspects of Behaviour of People in Fire Situations. *5th International Fire Protection Seminar*, Karlsruhe, West Germany, 39–50.

Bryan, J.L. 2000. Human Behaviour in Fire: the Development and Maturity of a Scholarly Study Area. *Fire and Materials*, 249–253.

Clark, Philip. 2010. Contigency Planning and Strategies. Information Security Curriculum Development Conference, 131–140. ISBN: 978-1-4503-0202-9.

Filippidis, L., Galea, E.R., Gwynne, S. e Lawrence, P.J. 2006. Representing the Influence of Signage on Evacuation Behaviour within an Evacuation Model. *Journal of Fire Protection Engineering,* vol.16(1), 37–73.

Gwynne, S., Galea, E.R., Owen, M., Lawrence, P.J., Filippidis, L. 1998. A review of the methodologies used in evacuation modeling. Research article. *Human Behaviour in Fire Proceedings of the 1st International Symposium*. University of Ulster, Northern Ireland.

Gwynne, S., Galea, E.R., Parke, J., Hickson, J. 2003. The Collection and Analysis of Pre-Evacuation Times derived from trials on their applications to Evacuation Modelling. *Fire Technology* vol.39 (2), 173–195. DOI: 10.1023/A:1024212214120.

Korhonen, T., e Heliovaara, S. 2011. FDS + Evac: Herding Behaviour and Exit Selection. *10th Symposium of Fire Safety Science*, 723–734. 10.3801/IAFSS.FSS.10–723.

Kuligowski, Erica D. 2009. The Process of Human Behaviour in Fires. *NIST Technical Note 1632*. s.l.: US Department of Commerce.

Nelson, HE e MacLennan, HA. 1996. Emergency movement. The SFPE Handbook of *Fire Protection Engineering.*vol.2(3), 286–295.

Olsson, P.A., e Regan, M.A. 2001. A Comparison Between Actual and Predicted Evacuation Times. *Safety Science*, vol.38(2), 139–145.

Papinigis, V., Geda, E. e Lukosius, K. 2010. Design of People Evacuation from Rooms and Buildings. *Journal of Civil Engineering and Management*, vol. 16(1), 131–139.

Sargent, Robert G. 1999. Validation and Verification of Simulation Models. *Proceedings of the 1999 Winter Simulation Conference.* Syracuse University. Syracuse, NY, 124–137.

Segurança Online, 2012, *Estatísticas* http://www. segurancaon-line.com/gca/?id=903 (accessed in 03/04/2012) http://sapadoresdecoimbra.no.sapo.pt/Evacuacao.htm (accessed in 17/05/2012) Vasudevan, Karthik e Son, Young-Jun. 2011. Concurrent Consideration of Evacuation Safety and Productivity in Manufacturing Facility Planning using Multi-Paradigm Simulations. *Computers & Industrial Engineering*, vol.61(4), 1135–1148.

Tan, Yong Kiang. 2011. Evacuation Timing Computations using Different Evacuation Models. *University of Canterbury. Department of Civil and Natural Resources Engineering.*

Tang, Fangqin e Ren, Aizhu. 2008. Agen-Based Evacuation Model Incorporating Fire Scene and Building Geometry. *Tsinghua Science and Technology*, vol.13(5), 708–714.

Vorst, Harrie C.M. 2010. Evacuation Models and Disaster Psychology. *Procedia Engineering*, vol.3, 15–21.

Yuan, Weifeng 2011. A Model for Simulation of Crowd Behaviour in the Evacuation from a Smoke-filled compartment. *Physica A: Statistical Mechanics and its Applications,* vol.390(23–24), 4210–4218.

Zhang, X., Qixin, S., Rachel, H., Bin, R. 2010. Network Emergency Evacuation Modelling: A Literature Review. *Optoelectronics and Image Processing (ICOIP) International Conference.* Beijing, China, 30–34.

Zia, K., Riener, A., Ferscha, A., Sharpanskykh, A. 2011. Evacuation Simulation Based on Cognitive Decision Making Model in a Socio-Technical System. *15th International Symposium on Distributed Simulation and Real Time Applications*, Salford/Manchester, UK, ISBN: 978–0-7695–4553–0.

Occupational Safety and Hygiene – Arezes et al. (eds)
© *2013 Taylor & Francis Group, London, ISBN 978-1-138-00047-6*

Analysis and proposal of a role model of the physical arrangement of prevention and fire fighting against panic in a popular shopping mall

W.S. Macêdo, J.M.N. Silva, L.O. Rocha & M.B.G. Santos
Universidade Federal de Campina Grande, Campina Grande, Paraíba, Brasil

A.L Melo
Universidade de Coimbra, Coimbra, Portugal

ABSTRACT: The techniques that lead to workplace safety are emerging and taking a character of science. The research's addressing the security factor has a multidisciplinary profile, and thus can deal with several factors such as, layouts geared to the issue of safety and fire prevention. Within this theme, the present study aims to propose a physical arrangement of prevention and fire fighting and panic, in the floors of the building where they are installed much of the informal traders from the center of Campina Grande. At the end of this paper, was obtained a set of measures that meet the requirements of the security sector and that if implemented will ensure the minimum conditions of security for the occupation, favoring both merchants installed there as well as the large number of consumers who frequent the place.

1 INTRODUCTION

Since ancient times, human beings produce goods, services and knowledge that are consumed throughout society by other human beings. However, many of these activities many cary risks to health, for the worker's life, and generate various kinds of damages. Having this problem exposed, there is a branch of science connected with labor activities, which is named job security, which aims to study and implement measures to protect workers from the risks inherent in their occupational activity, and prevent accidents and reduce their consequent losses. In the reason of its multidisciplinary nature, the labor safety needs various branches of knowledge. Among those branches of knowledge there is the due importance the issue of layout or physical arrangement. So that we can make a good study of safety and prevention in buildings and facilities, it is necessary to use the knowledge and techniques of physical arrangement enabling the correct sizing of environments and allocation of materials in their proper places.

Based on the concepts and knowledge of workplace safety and the physical arrangement, the security professional can initiate the development of a safety project for a company, building or installation of production or service delivery. A project of prevention and fire fighting and panic is a clear example of this situation. To Seito (2008), fire Safety, today is regarded as a science, and has featured on the international scene. According to Barbosa Filho (2010), prevention is the set of measures to prevent accidents arise, but there is no such possibility, they are kept under control by the fighting. According Villar (2004), in developing the physical arrangement of prevention and fire fighting and panic, it is intended to identify the risks of the various facilities, leaving near facilities similar risks, since they will need the same care, which makes design safer and more economical. Also, it is proposed installing of preventives mobile and fixed of the building and the as whole system connected to signage, lighting and emergency exits.

So before all these facts, we build a model of the physical arrangement of prevention and fire fighting and panic in a building has a significant importance as it provides a more professional performance security for the worker and a way to keep them safe capital goods invested in that sector.

Starting from such preliminary information envisioned in the current article, which aims to analyze and propose a physical arrangement of prevention and fire fighting against panic for the floors of the building where they are located the majority of informal traders from downtown of Campina Grande, Shopping mall Edson Diniz, known as the Popular Shopping mall. The specific objectives are:

(1) Conduct a deepening on the theme "Occupational Safety", aiming on prevention and fire fighting and panic;

(2) Scaling the building of Shopping mall with its fixed structures and their boxes installed;

(3) Identify areas of risk classifying them according to their degree of similarity;

(4) Propose layout positioning of mobile and fixed fire extinguishers for building;

(5) Suggest the layout of signs and emergency exits;

(6) Propose installation of emergency lighting.

For Marangoni (1998) several factors are responsible for the spread of fires. This work is justified because it is the physical arrangement of a proposal aimed at security in a place other than his own, and is used widely by people of lower social classes. This study also has a significant importance by the fact that at this location there is a large amount of combustible material, which has a very high destructive potential.

2 MATERIALS AND METHODS

The environment in which this study was developed was the Edson Diniz Shopping Center in the city of Campina Grande-PB, located at the corner comes of AV. Floriano Peixoto and Street Marquês de Herval, opposite the Bandeira Square. This research have in its nature an applied nature and approach qualitatively, also takes a descriptive and bibliographical, being considered a case study, detailing deep aim for a single goal and finally

Figure 1.   Hydrant disabled accumulating the garbage.

having the characteristic of being a research-action to be associated with the action or resolution of the problems.

To obtain a diagnosis of the current situation on-site visits were conducted, in which they used instruments such as structured questionnaire applied to the administration responsible for the activities performed in the building; copies of building plans provided by the Department of City Planning (SEPLAN), and the Award for Technical Inspection conducted by the Center for Technical Activities of the 2nd Battalion of Military Firefighters Campina Grande (CAT/2 BBM), to identify the points of non-compliance with the law accountable. To propose improvements we used basically the guidelines of the Brazilian Regulatory Standards dealing with the issue.

3 RESULTS AND DISCUSSION

After analyzing the place and of reporting conducted by the Center for Technical Activities/2nd Battalion Military Firefighters (CAT/2° BBM) according to the methodological procedures, the study was oriented in two basic lines: diagnosis of the current situation and to proposition an appropriate model.

3.1 *Current conditions of the shopping mall Edson Diniz*

The conditions found in shopping mall Edson Diniz are listed below:

– There are not any copies of structural plans of the building, being necessary to the analysis such, a request joins SEPLAN, which is the only body that has;

– The current situation presents Shopping Mall structural changes compared to plants found in SEPLAN, especially as regards the amount and placement of "boxes or small shops";

– The building has an incomplete system hydrants and disabled, (figure 1), reminiscent of his former occupation, which is devoid of water tank and pumps suitable for the effective workings of the network;

– There is an emergency system of lighting deficient in the building (figure 2);

– There is not a signaling system of emergency;

– Has a protection system per extinguishers acceptable. However, with corrections in positioning of equipment and its installation forms and signaling (figure 3);

– There is an acceptable system of outputs that can be used as emergency exits;

Figure 2. Stairs without emergency lighting.

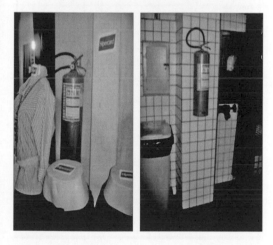

Figure 3. Extinguishers without any identification and obstructed by goods.

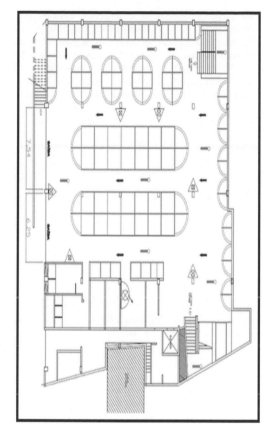

Subtitle	
	Walls
	Boxes or small shops
	Hydrantsshops
	Extinguisher of Water
	Chemical Powder Extinguisher
	CO_2 extinguisher
	Emergency Luminaire
	Emergency Signalling
SAIDA	Emergency Exit

Figure 4. Representation full of the prepositions for the lower floor.

3.2 Proposals for the improvement of safety conditions

Among others, are presented below propositions to improve the conditions of security of the site:

– Provision for copies of flat structural building administration of the shopping mall;
– Creation of plants updated with the correct placement of the boxes and the amount thereof;
– Reactivating hydrants, with the reservoir exclusive creation to fires and the acquisition of centrifugal pump for proper pressurization of the network, while maintaining the placement of exits, as well as guides to the standard of the Brazilian Association of Technical Standards 13714 (BRASIL, 2003);
– Keeping fire extinguishers for protection system, relocating them properly and install them according to the requirements of the standard of the Brazilian Association of Technical Standards 12693 (BRASIL, 2010);

- Creating a system of emergency lighting efficiently and reach all points of the building, giving lighting conditions necessary for a possible evacuation in case of an accident, as well as guides to the standard of the Brazilian Association of Technical Standards 13434-2 (BRASIL, 2004);
- Deployment of the emergency signaling, indicating escape routes, location of stairs and emergency exits, as well as orients the norm of the Brazilian Association of Technical Standards, Brazilian Norm 9077 (BRASIL, 2001);
- Keeping the outputs system indicating the presence of those with appropriate signage, as well as guides to the standard of the Brazilian Association of Technical Standards 13434-2 (Brasil, 2004) (ABNT) and the norm of the Brazilian Association of Technical Standards 9077 (BRASIL, 2001).

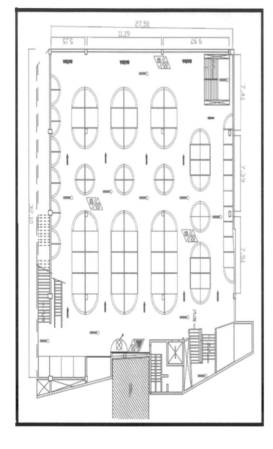

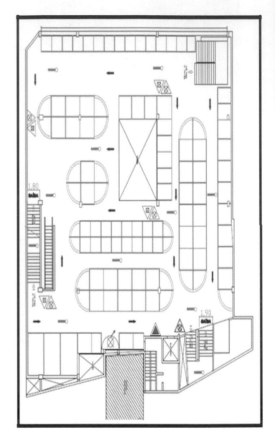

Figure 5. Representation full of the prepositions for the ground floor.

Figure 6. Representation full of the prepositions for the first floor.

3.3 *Proposal of layout for the shopping mall*

The new layout proposes measures that if implemented, will guarantee the minimum safety in the issue of prevention and firefighting and panic in the floors where does work the Shopping Mall.

4 CONCLUSION

Having completion this study through of literature and field research, we could confirm the relevance of the subject to the supervisory agencies, consumers, visitors and especially for merchants installed in the building. With this study, a support was given to reveal the current conditions of the shopping mall center Edson Diniz in Campina Grande—PB as a matter of prevention and fire fighting against panic, as well as proposed actions for implementation and correction of various aspects of non-compliance on the premises to give conditions minimum security to its users.

Elaborated up a new layout with the correct distribution of hydrants, extinguishers, emergency luminaires, emergency signage and emergency exit for the three floors that make up the mall, but this measure does not exclude the need to prepare a project of total prevention and of fighting the fire and panic for the building.

In social vision, this study also intends to includes a class of the society in the which much of their lives living in situations security of inadequates, in an environment that meets the minimum requirements of this sector. Thereby providing a more work worthy for traders of that location, as well as a sense of security to the public that attends.

Still stands as Suggestion the implantation of a permanent fire brigade in this occupation.

REFERENCES

Barbosa Filho, A.N. (2010). Segurança do trabalho e gestão ambiental (3ª Edição). São Paulo: Atlas.

Seito, A.I. (2008). A Segurança contra incêndio no Brasil: fundamentos de fogo e incêndio. São Paulo: Projeto Editora.

Villar, A.M.; Nóbrega Júnior, C.L. (2004). Planejamento das Instalações Industriais (1ª Edição). João Pessoa: Manufatura.

Marangoni, T.T. et al. (1998) Manual do Curso de Formação de Soldados: Técnica e Maneabilidade em Combate à Incêndio (2ª Edição). Rio de Janeiro: Corpo de Bombeiros Militar do Estado do Rio de Janeiro.

Brasil. Associação Brasileira de Normas Técnicas (2004). Norma Brasileira 13434-2: Sinalização de segurança contra incêndio e pânico (Parte 2: Símbolos e suas formas, dimensões e cores). Rio de Janeiro.

Brasil. Associação Brasileira de Normas Técnicas (2001). Norma Brasileira 9077: Saídas de emergência em edifícios. Rio de Janeiro.

Brasil. Associação Brasileira de Normas Técnicas (2003). Norma Brasileira 13714: Sistemas de hidrantes e de mangotinhos para combate a incêndio. Rio de Janeiro.

Brasil. Associação Brasileira de Normas Técnicas (2010). Norma Brasileira 12693: Sistemas de proteção por extintores de incêndio. Rio de Janeiro.

Occupational Safety and Hygiene – Arezes et al. (eds)
© *2013 Taylor & Francis Group, London, ISBN 978-1-138-00047-6*

Fire Prevention and Protection Plan Pathologies

N.L.R. Modro
Universidade do Estado de Santa Catarina—UDESC, Laguna, Santa Catarina, Brasil

N.R. Modro & N.R. Modro
Universidade do Estado de Santa Catarina—UDESC, São Bento do Sul, Santa Catarina, Brasil

Alberto Nascimento Abib
Universidade do Sul de Santa Catarina—UNISUL, Santa Catarina, Brasil

ABSTRACT: This paper presents pathologies detected from inspections at Fire Prevention and Protection Plan (FPPP). Inspections were performed by Fire Department of the municipality of Capão da Canoa—located in southern Brazil—and follow a miscellaneous regulatory collection of 65 itens: Federal and State Laws, Ordinances, Decrees, Regulatory Standards and technical regulations. In recent years have been identified over two hundred and seventy five pathologies. These pathologies were classified in six categories: Liquefied Petroleum Gas (LPG), stairs, extinguishers, hydrant systems, signs and emergency lighting system. It is highlighted that the pathologies were both in the design and execution phase. Identifying these pathologies is critical due a hard measuring value—the human life—in case of an accident.

1 INTRODUCTION

At Rio Grande do Sul State (RS), located in southern Brazil, each owner or responsible for commercial, industrial, entertainment and public residential buildings—with more than one economy and more than one deck—must have a Fire Prevention and Protection Plan (FPPP). The RS Fire Brigade Law N°. 064/EMBM/99 defines FPPP as "A set of documents integrating the Fire Prevention and Protection process" (CGBM, 1999).

The owner building is responsible—voluntarily or after receiving the Notification of Compliance (NC)—for sending FPPP to Military Fire Brigade of Rio Grande do Sul. Military Fire Brigade is the inspector department that verifies FPPP existence, implementation and compliance. About NC, it is fundamental emphasize that it is issued by the Fire Brigade and must observe the 60 days of compliance legal deadlines. Once the owner building has filed FPPP process and has paid examination and inspection fees, Fire Brigade issues the Certificate of Compliance (CC)—a document where is listed and described the Prevention Fire Safety System that shall be inspected. After the owner take notice of which will be inspected and make any changes required, he request the inspection. If any inconsistency is found during the Fire Brigade Inspection will be issue a certificate of correction and the

owner will have 30 days to rectify this discrepancy and place a new inspection. If no inconsistency is found during Inspection shall be granted the Prevention Fire Safety System Fire Brigade Permit. This document has an expiration date, depending on situation, and is required for granting "Habite-se" Permit. "Habite-se" Permit is issued by Production, Industry and Trade Department and it is a concession that enable the building in order to occupancy.

According GNIPPER and MIKALDO JR (2007) already incurs a pathology—a real problem, with symptoms already manifest—or one unconformity—a potential problem or already installed and with no visible symptoms yet—every system or subsystem which does not meet any performance requirement, especially those literally required by specific legislation, regulations or technical standards. Also according to these authors, the importance of the study of constructive pathologies resides in the possibility of preventive action, particularly when they have failures in engineering design.

FREIRE (2009) emphasizes that the study of the fire security process is critical and must not be neglected or pushed into the background because it enables reducing fire hazard or the effects in event of such instance of accident. On the other hand, ONO (2007) cautions that several buildings are particularly vulnerable to fire hazards because they were built at times where fire regulations did not exist.

At RS State many buildings do not have FPPP in disagreement with the current laws. Beyond that some do not have any sort of Fire Safety System. So, it is required develop adaptation projects to correct this situation.

This paper presents the main pathologies detected by FPPP Notifications Correction Inspection issued by the Municipality of Capão da Canoa Fire Brigade. These notifications were objects of actions performed through correction and adaptation projects developed and coordinated by engineer Alberto Nascimento Abib—the Chief Technical Company responsible. As far as possible, due corrections were carried out to eliminate the pathologies found. In some cases, the building had to be modified, taking into consideration currently legislation and the existent infrastructure, in order to fit the Fire Brigade Standards to obtain Prevention Fire Safety System Fire Brigade Permit.

2 LAWS AND REGULATIONS

In order to prepare a FPPP in Rio Grande do Sul State, shall be followed a miscellaneous regulatory collection from Federal and State Laws, Ordinances, Decrees, Regulatory Standards and technical regulations. This collection is quantified in Table 1.

3 RESULTS

Table 2 presents the main pathologies discovered in inspections carried out by the Fire Brigade in order to verify the Prevention Fire Safety System specified in FPPP. These pathologies resulted in Notifications and a new Correction Inspection.

Figures 1 to 10 presents some pathologies of Prevention Fire Safety System specified in FPPP that led to Notifications Correction Inspection.

Table 1. Laws and regulations guidelines to be followed during FPPP elaboration in RS.

Specification	Amount
Federal Law	02
State Law	02
State Decree	03
Ordinances	10
Regulatory Standards	32
Technical Regulations	11
Technical Reports	03
Technical Guidelines	02

Table 2. Main pathologies found that led to Notifications Correction Inspection.

Liquefied Petroleum Gas (LPG)	Amount
LPG cabinet counters inside stair	01
LPG central confined under the ramp to the second floor	01
LPG ventilation and electrical cables at the same pipe	01
Floor drain close to LPG central	04

Stairs	Amount
Improper scaling on the size of the steps	10
Handrail on one side only of the stair, without continuity or improper height	17
Improper width scaling	12
Intermediate landing with tapered steps	12
Improper or missing encapsulation	12

Extinguishers	Amount
Extinguishers with expired inspection term	01
Extinguishers inside enclosed stair	03
Extinguishers placed improperly—in unsuitable places, hard access sites and outside height standard	09
Inexistence of signs at points of extinguishers placement	18
Extinguishers inadequate for fire classes—water type close to electrical controls	05
Extinguishers hidden by flowers	01

Fire hydrant system	Amount
Hydrant inside enclosed stair	04
Hydrants placed behind doors	01
Hydrants hidden by frames (pieces of art)	01
Fire Hose with expired inspection term	20
Absence of nozzles and of key cabinets	18
Hydrant Registry closed, making it impossible to use	03
Hoses lower than the recommended size	04
Hoses bigger than the recommended size	02
Hydrants placed improperly—in unsuitable places, hard access sites and outside height standard	5

Signs	Amount
Inexistence of signs	18
Signs out of technical standards	15
Signs placed improperly	11

Emergency lighting system	Amount
Inexistence of Emergency lighting system	04
Inoperable Emergency lighting system	09
Short batteries lasting	09
Wire Gauge lower than recommended size	09
System lighting level lower than technical standards	18
Poor quality of luminaries	18

Figure 1. LPG ventilation and electrical cables at the same pipe.

Figure 2. Handrail on one side only of the stair. Intermediate landing with tapered steps. Improper width scaling—minimum width less than 1.05 m.

Figure 3. Hydrants placed in unsuitable places, hard access.

Figure 4. Hydrant inside enclosed stair. Fire hose tied up and with expired inspection term.

Figure 5. Hydrant Registry closed—cement embedded and in a condition that prevents the hose coupling, making it impossible to use.

4 FINAL REMARKS

The root causes of pathologies mentioned in this paper might be located both in design or execution phase. Performing a detailed review at design stage enables the detection of possible non-compliance. In this stage, the elimination of non-compliance is simpler and cheaper; being only needed adjust the design to current legislation. Furthermore, pathologies found in buildings are commonly hard to solve technically. It must be taken into account that these "post" corrections could imply at the total cost of implementing the project. Another point to be considered is that either by mistake in the design phase or at the implementation stage,

Figure 6. Extinguishers hidden and obstructed by flowers.

Figure 7. Enclosed stair with tapered steps. Improper encapsulation.

Figure 8. Floor drain close to LPG central.

there is an implicit value difficult to measure: the human life. Such aspect is due to fact that in the event of an accident, the FPPP might not run as required. If this happens, the lives of the users of buildings could be at risk. Thus, designers and those responsible for its implementation,

Figure 9. LPG central confined under the ramp to the second floor, extinguishers removed and placed improperly to install the electric drive motor pump set; storage of flammable chemicals (solvent).

Figure 10. LPG cabinet counters inside stair. Ceiling composed of combustible material. Intermediate landing with tapered steps.

upon learning of recurrent pathologies might take suitable arrangements for prevent further events of these conditions. Finally, it is emphasized that the Military Fire Brigade of Rio Grande do Sul is aware of such liability. Hence, carries out his supervisory role so critically, detecting and aiding at solution of pathologies above described.

ACKNOWLEDGMENTS

The authors gratefully acknowledge Universidade do Estado de Santa Catarina—UDESC, Universidade do Sul de Santa Catarina—UNISUL and Corpo de Bombeiros da Brigada Militar do Rio Grande do Sul.

REFERENCES

Comando Geral da Brigada Militar—CGBM. *POR-TARIA N° 064/EMBM/99*. Regula a aplicação, pelos órgãos de Bombeiros da Brigada Militar, da Lei Estadual n° 10.987 de 11 de agosto de 1997, das normas técnicas de prevenção contra incêndios estabelecidas pela respectiva regulamentação e dá outras providências. Porto Alegre, RS, 18 de novembro de 1999.

Freire, C.D. da R. (2009). *Projeto de Proteção Contra Incendio (PPCI) de um prédio residencial no centro de Porto Alegre*. Monografia de Conclusão de Curso de Especialização. UFRGS, 2009, 49p.

Gnipper, S. F., Mikaldo Jr, J. (2007). *Patologias freqüentes em Sistemas Prediais Hidráulico-Sanitários e de Gás Combustível decorrentes de falhas no processo de produção do projeto*. In: Anais do VII Workshop brasileiro de gestão do processo de projetos na construção de edifícios. Curitiba: Universidade Federal do Paraná.

ONO, R. (2007). *Parâmetros de garantia da qualidade do projeto de segurança contra incêndios em edifícios altos*. Ambiente Construído. Porto Alegre, v.7, n.1. p. 97–113. jan./mar. 2007.

Health monitoring and occupational medicine

Occupational Safety and Hygiene – Arezes et al. (eds)
© *2013 Taylor & Francis Group, London, ISBN 978-1-138-00047-6*

Disbaric disease prevention

Helena Alvim & J. Santos Baptista

PROA/CIGAR/LABIOMEP/Faculdade de Engenharia, Universidade do Porto, Porto, Portugal

ABSTRACT: Hyperbaric medicine has a large field of applicability and is increasingly used as a therapeutic resource. Its field of action is not limited to treat decompression sickness, but was enlarged to all the diseases in which an increase in the amount of dissolved oxygen in the blood has beneficial effects. With this extension has emerged a new class of professionals, the *attendents*, accompanying the sick people within the hyperbaric chamber, which are now, they own, object of study. This work attempts to give an image of decompression sickness, in particular in *attendents* and identify the research developed to describe the effect of O_2 in decompressive phase. Although in decompressive phase the use of O_2 be standard practice, in the research carried out were not found conclusive studies on this subject. It follows the need for further research and development work, methodologically consistent in order to answer questions and troubleshooting.

1 INTRODUCTION

The hyperbaric medicine is practiced in many countries. In recent years, the developments in technology and therapeutic has been significant. In Europe there are 220 Hyperbaric Centers (Fig 1) in 32 countries. Portugal has 4 Hyperbaric Medical Centers (HMC) Two in the Continent, one in Madeira and the other in the Azores Islands (Alvim, 2010). However, their use raises some questions. By one hand, in its current use, should be considered the numerous occupational risk factors, on the other hand, there is not a legislative framework that requires a medical examination plan with appropriate diagnosis for professionals that have as activity the work in a hyperbaric chamber. (Kot, 2003; Kot et al., 2004; Alvim, et al. 2011,b). The preventive measures taken to occupational level, by HMC at international level, are following, in general, the guidelines given in 6th Consensus Conference 2003 for the prevention of disbaric diseases: a) rotation between the *Attendants* within the hyperbaric chamber; b) terminal oxygen (O_2) breathing (Risberg et al., 2004).

One of internationally agreed methods for the prevention of decompression sickness (DCS) in *Attendants*, "100% oxygen pre-breathing", is used empirically. In this work is intended to investigate the research developed at international level in the thematic "Decompression sickness in *Attendants*", in particular in its forms of manifestation, in the measures for its prevention, method(s) employed in their diagnosis, assessment and measurement.

Figure 1. Hyperbaric chamber of hospital pedro hispano.

2 MATERIALS AND METHODS

A systematic review was made in several databases from the browser of Metalib. The search was made with the keywords: *Hyperbaric Medicine*; *Attendants*; *Decompression sickness*; *Venous gas embolism*; *Bubbles*; *Dysbaric Osteonecrosis* and in theses and dissertations with the keywords: *Medicina Hiperbárica*; *Occupational Health*; *Hyperbaric Medicine*; *Decompression sickness*; *Venous Gas Embolism*; *Attendant*; *HSP70*; *Doppler*; *Bubble*; *Magnetic Resonance Imaging*; *Dysbaric Osteonecrosis*; *HBO*; *DCS*. Search in Rubicon Database with the keywords: *Risk*; *Decompression Sickness*; *Hyperbaric stress*; *Doppler*; *Ultrasound*; *Magnetic Resonance Imaging*; *Biomarkers*; *Heat-Shock Proteins*.

3 RESULTS AND DISCUSSION

Was found wide bibliography in thematic DCS referring to its incidence, in particular in hyperbaric medicine (Uzun, 2001). Reports of incidence of DCS in Attendants indicate a variation between 0.0076%–0.076% (Cooper, Broek & Smart, 2009, Uzun, 2011), and 0.03% in diving (Vann et al., 2011).

In the studies of reported cases of DCS was possible to check the effects of bubbles immediately after hyperbaric exposure, in situations such as, for example: a) bubbles by Magnetic Resonance (Fig.2) in a diver (Gempp, 2009); b) gas bubbles in the inferior vena cava and portal veins were found incidentally by computerized tomography through his work-up for abdominal pain, problem reported for the first time in a recreational diver (Bird, 2007); c) arterial gas bubbles in the heart detected by Magnetic Resonance in a tunnel worker (Kutting, 2004).

In spite of its adverse effects, decompression is not fully understood (Brubakk & Mollerlokken, 2009; Gao et al., 2009; Hugon et al., 2009; Møllerløkken et al., 2012). Several authors suggest that the formation of bubbles in various anatomical locations of the body, as for instance formation of venous gas emboli (VGE) or arterial gas emboli (AGE) and its effects on the endothelium may have a role in the mechanism of DCS, including lead-

ing to damage of the central nervous system (CNS) (Gao et al. 2009; Hugon et al., 2009; Møllerløkken et al., 2012).

The bubbles exert their pathophysiological effects by mechanical pressure on the tissues, obstructing blood flow and by biochemical reactions, causing an inflammatory response (Pollock, 2007). There is a relationship between the Bubbles and risk of DCS (Risberg et al., 2004; Cooper et al., 2009; Hugon et al., 2009; Brubakk & Mollerlokken, 2009; Gao et al., 2009; Møllerløkken et al., 2012). The absence of bubbles is a good indicator of safety, but their detection indicates only a 4% probability developing DCS (Møllerløkken et al., 2012). But it is considered that bubbles detection is a good indicator of stress hyperbaric (Pollock, 2007; Møllerløkken et al., 2012).

In several studies, was recognized the correlation between bubbles and Disbaric Osteonecrosis (Cassar-Pullicino, 2008; Sobakin et al., 2008; Gempp, 2009). DON (Disbaric Osteonecrosis) is a type of aseptic necrosis of the bone. The lesions are located mostly in long bones, in the humerus and femur (Cimsit, Ilgezdi & Uzun, 2007). Several researchers have studied DON in workers exposed to high pressure. In hyperbaric medicine in *Attendants* (Zkan, 2008), in divers (Cimsit, Ilgezdi & Uzun, 2007; Gempp, 2009; Kenney & Sonksen, 2010) or in construction workers (Kutting, 2004).

The studies in hyperbaric environment and in animal experiments performed to investigate the detection and diagnosis of disease, have shown the following methods as appropriate: a) with doopler (Fig.3) for the detection of bubbles in the circulatory system (Risberg et al., 2004; Pollock, 2007; Cameron et al., 2009; Blogg & Gennser, 2009; Blogg & Gennser, 2011; Cooper et al., 2009; Gutvik et al., 2010); b) magnetic resonance for detection and diagnosis of DCS (Ors, 2006; Zkan, 2008); c) Biomarkers HSP70 (Rhind, Cameron & Eaton, 2007) to detect cell stress (Cameron et al., 2009).

The use of ultrasound to detect stress after decompression is used for more than 40 years, the

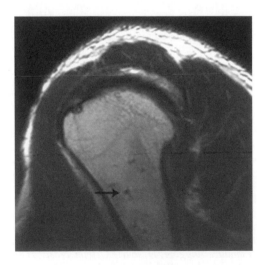

Figure 2. T2 weighed sagittal image revealing multiple unexpected hypo-intense spot in humeral marrow strongly evocative of bubbles (MRI examination 24 hrs following DCS). (Gempp, 2009).

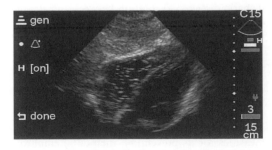

Figure 3. Two dimensional echocardiographic of the chamber of the heart. Use of ultrasound in decompression research (Pollock, 2007).

rating scale of Bubbles of Spencer & Johanson (1974) is still used today, for example the Kisman-Masurel method is based on this scale (Pollock, 2007). The use of ultra-sounds, echography or Doppler is currently considered the only objective method for the quantification of VGE and indicator of stress (Gutvik et al., 2010), having been shown in several studies the statistical correlation of VGE and DCS (Møllerløkken et al., 2012). MRI has a high sensitivity and specificity: a) is more sensitive in identifying changes of bone necrosis than plain radiographs (Zkan, 2008; Kenney & Sonksen, 2010); b) may be used early, soon after exposure to detect brain lesions (Gao et al., 2009), or bubbles in bone (Kenney & Sonksen, 2010) during the treatment and in the staging of the disbaric osteonecrosis. The use of MRI for routine can be justified, in case there is a high risk for bubbles, as in the study performed in recreational divers diagnosed with musculoskeletal decompression sickness (Gempp, 2009).

Analytical studies estimate or probabilistic pathology are used to optimize the tables of decompression and is application in the activity of diving.

The development of estimation studies through computational programs based on mathematical models to predict and evaluate Copernicus VGE (Eftedal, Tjelmeland & Brubakk, 2007; Gutvik & Brubakk, 2009; Gutviket al., 2010; Feng, Gutvik & Johansen, 2010; Mohammadein, 2010; Feng et al., 2011) or on probabilistic models (Hobbs & Gault, 2009; Walkeret al., 2010a) to be applied in user friendly informatic tools in scuba diving activity, are important DCS prevention tools (Møllerløkken et al., 2012).

The biomarkers may also provide important information for stress hyperbaric (Kernagis, Datto & Moon, 2009). Was found correlation between the level of bubbles and the increased level of stress protein (Havnes, 2010), being detected in situations of hyperbaric exposure the formation of protein HSP70 (family of heat shock proteins), (Rhind, Cameron & Eaton, 2007; Møllerløkken et al., 2012).

In the several research studies in animals, these were subjected to high pressures and to different mixtures in respiratory decompressive phase, having been observed the occurrence of DCS and DON by different diagnostic techniques. Use as a preventive method, a 100% oxygen pre-breathing are tested through the induction of the disease. It was done experimentally with positive results in the reduction of bubbles (Blogg et al., 2003) and reduction of DON (Sobakin et al., 2008; Sobakin et al., 2009). The increase in time to breathe oxygen (15 min, 1 hour and 2 hours) has reduced the number of injuries in all long bones submitted to the study (femur, humerus, radius and tibia). The diagnosis was performed by magnetic resonance, pathological anatomy and bone scans, in the last case it was possible to observe the lesions suggestive of DON after hyperbaric exposure (Sobakin et al., 2008), as can be seen in figure 4.

In studies, evaluating the stress in hyperbaric *Attendants* (Table 1) was only found a study done in the context of HMC designed to test the use of oxygen and/or other mixtures in decompressive phase. It was considered a observational study, stress-relief "acceptable", according to the settings of the defense and Civil Institute of Environmental Medicine in Toronto. The Attendants are 100% oxygen pre-breathing for 12 min. The authors consider that the findings of this study are conditioned to percentage (56%) of *Attendants* eligible for the research. The variability inter and intra-individual was considered significantly evident (Cooper, Broek & Smart, 2009). In another prospective study breathing O_2 for 12 min designed to detect VGE, was monitored with doopler and ultrasound. It was then tested a profile O_2 for 20 min and a profile of Nitrox (gas mix 40.5% O_2 and N_2 for 90 min in phase isobaric) to investigate if the gases of breathing during decompression would affect the incidence of VGE. However, the prevalence of VGE tested with the following variables, increasing the time breathing 100% O_2 or fraction

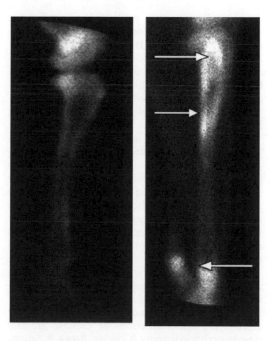

Figure 4.   Right tibia of sheep #174 before (A) and after (B) hyperbaric exposure. (Sobakin et al 2008).

Table 1. Respiratory mixtures HBO (*attendants*).

Type of study	Experimental	Observational
Year of the study	2004	2009
Subjects	19 (9 Attendants + 10 no attendants	28 Attendants
Hyperbaric exposures	240 kPa /115 min	243 KPa (2,4 ata)/90 min
Hyperbaric profiles	−12 min O₂ 100% In the final stage of the decompression; −24 min O₂ 100% In the final stage of the decompression; −2 × 30 min 40.5% O₂, balance N₂, during 3 × 30 min of the isobaric phase	20 min O₂ 100% In the final stage of the decompression
Technique	Doppler	Doppler
Conclusions of the study	Significant decompression stress	'Acceptable' decompression stress

of O_2 gas mixture, in phase of decompression, not brought significant results (Risberg et al., 2004).

4 CONCLUSIONS

Additional multicenter epidemiologic studies are needed if the occupational safety of inside *attendants* is to be enhanced. It was verified that *Attendants* exposed the profile 240 kPa/115 min, are subject to a voltage of decompression significant (Risberg et al., 2004), although it has been considered by some authors an work environment relatively safe (Cooper et al., 2009; Cooper, Broek & Smart, 2009).

Consider that there is the need for epidemiological studies to investigate the risk of Decompression Illness (DCI) and disbaric osteonecrosis (Zkan, 2008) for the *Atendants* that validate their preventive methods used empirically related to: a) the use of O_2 100%, in decompression phase; b) the duration of the use of gases of breathing and the phase of administration; c) the use of other gas mixtures; d) the times of decompression (Risberg et al., 2004).

On the other hand the research performed with a greater number of people in a study, can decide issues such as the ideal age for the cessation of activity as *Attendant* and dwell times appropriate after injury and restrictions of exercise pre and post hyperbaric activity (Cooper, Broek & Smart, 2009).

Prevention should be replaced by the adoption of protocols of occupational health to limit the entry of people with pathology or important individual susceptibility for DCS (Kutting, 2004) and the medical control with specific examinations during the activity of *Attendant* (Alvim et al., 2011a). No studies were found that are evidenced the exposure limit values related to disbaric risks, in particular DCS. But is considered to have an important role in the prevention of disbaric the fulfilment of conditions of work as: a) The rotation of *Attendants* within the chamber (Risberg et al., 2004; Alvim, 2010) the limit of the number and the pressure of exposure; b) working time; c) intervals in working time and between exposure; d) observation time after exposure (Alvim, 2010, Alvim, et al. 2011a). As preventative method, the practice of breathing oxygen during decompression in Hyperbaric Medicine lacks scientific evidence.

REFERENCES

Alvim, Helena 2010. Estudo/Proposta de medidas de prevenção e proteção das doenças disbáricas dos trabalhadores em medicina hiperbárica. *Tese de Mestrado em Engenharia de Segurança e Higiene Ocupacionais, Faculdade de Engenharia*, Porto: Catálogo da Biblioteca da FEUP, 244 pp.

Alvim, Helena, Diogo, Miguel Tato, Leão, Rui Ponce, Camacho, Oscar, Baptista, João, & Nobrega, Júlio 2011. "Fire drills in Hyperbaric Medicine." Ed. R. Mondelo, P., Karwowski, W., Saarela, K., Hale, A.*Proceedings of the 9th International Conference on Occupational Risk Prevention.*Santiago de Chile: Universitat Politècnica de Catalunya.

Alvim, Helena, Diogo, Miguel Tato Leão, Rui Ponce, Camacho, Óscar, & Baptista, João 2011a. "Hyperbaric Medicine Organizational Risks." Ed Pedro Arezes, João Santos Baptista, A.S. Miguel, Gonçalo Perestrelo, Nelson Costa, Mónica Barroso, P. Carneiro, P. Cordeiro, Rui Melo. *Proceedings of the International Symposium on Occupational Safety and Hygiene.*Guimarães: Sociedade Portuguesa de Segurança e HigieneOcupacionais, 89–94.

Alvim, Helena, Diogo, Miguel Tato, Leão, Rui Ponce, Camacho, Óscar & Baptista, João 2011b. "Dispositivos Médicos Sujeitos a Pressão." Ed. J F Silva Gomes, Clito F Afonso, Carlos C António e António Matos. *Proceedings do 6° Congresso Luso-Moçambicano de Engenharia, 3° Congresso Moçambicano de Engenharia.* Maputo: INEGI/FEUP.

Bird, N. 2007."CT finding of VGE in the portal veins and IVC in a diver with abdominal pain: A Case Report." *Undersea & Hyperbaric Medicine*, November-December de 2007: 393–397

Blogg, S, Genncer, M, Loveman, GAM, Seddon, FM, Thacker, JC & White, MG 2003. "The effect of breathing hyperoxic gas during simulated submarine escape on venous gas emboli and decompression illness."*Undersea & Hyperbaric Medicine*, 2003: 163–174.

Blogg, S, & Gennser, M 2009. "Regular Doppler monitoring is necessary to map the progression of Venous Gas Emboli evolution post-dive." *Abstract of the Undersea and Hyperbaric Medical Society. Annual Scientific Meeting.* Las Vegas, Nevada, USA, 2009.

Blogg & Gennser, M 2011. "The need for optimisation of post-dive ultrasound monitoring to properly evaluate the evolution of venous gas emboli. Review." *Diving and Hyperbaric Medicine,* September de 2011: 139–146.

Brubakk, AO, & Mollerlokken, A 2009. "The role of intra-vascular bubbles and the vascular endothelium in decompression sickness. Review." *Diving and Hyperbaric Medicine,* September de 2009: 162–169.

Cameron, B, McLellan, T. Eaton, D & Rhind, S 2009. "The absence of innate inflammatory gene response to acute hyperbaric stress in non-divers following heat acclimation." *Abstract of the Undersea and Hyperbaric Medical Society. Annual Scientific Meetin.* Las Vegas, Nevada, USA., 2009.

Cassar-Pullicino, V. 2008. "Commentary on role of MRI in the detection of marrow bubbles after musculoskeletal decompression sickness predictive of subsequent dysbaric osteonecrosis." *Clinical Radiology,* December de 2008: 1384–1385.

Cimsit, C, Ilgezdi, S, & Uzun, G. 2007. "Dysbaric osteonecrosis in masters and instructors experienced dive." *Aviation Space and Environmental Medicine,* December de 2007: 1150–1154.

Cooper, PD, Van den Broek, C & Smart, David. 2009. "Hyperbaric chamber attendant safety II: 14-Year health review of multiplace chamber attendants." *Diving and Hyperbaric Medicine,* June de 2009: 71–76.

Cooper, PD, Van den Broek, C Smart, DR Nishi, Ron Y & Eastman, D. 2009. "Hyperbaric chamber attendant safety I: Doppler analysis of decompression stress in multiplace chamber attendants." *Diving and Hyperbaric Medicine,* June de 2009: 63–70.

Eftedal, OS, Tjelmeland, H, & Brubakk, AO. 2007. "Validation of decompression procedures based on detection of venous gas bubbles: A Bayesian approach." *Aviation Space and Environmental Medicine,* February de 2007: 94–99.

Feng, L, Gutvik, CR & Johansen, Tor A. 2010. "Optimal decompression through multi-parametric nonlinear programming." *IFAC Proceedings Volumes (IFAC-PapersOnline).* 2010. 1314–1319.

Feng, Le, Gutvik, CR, Johansen, Tor A, Sui, D & Brubakk, AO. 2011. "Approximate Explicit Nonlinear Receding Horizon Control for Decompression of Divers." *IEEE Transactions on Control Systems Technology.* 2011.

Gao, GK, Wu, D, Yang, Y, Yu, T, Xue, J, Wang, X, & Jiang, YP. 2009. "Cerebral magnetic resonance imaging of compressed air divers in diving accidents." *Undersea Hyperb Med,* January-February de 2009: 33–41.

Gempp, E. 2009. "Musculoskeletal decompression sickness and risk of dysbaric osteonecrosis in recreational divers." *Diving and Hyperbaric Medicine,* December de 2009: 200–204.

Gutvik, CR, & Brubakk, AO. 2009. "A dynamic two-phase model for vascular bubble formation during decompression of divers." *IEEE Transactions on Biomedical Engineering,* March de 2009: 884–889.

Gutvik, CR, Dunford, RG, Dujic, Z, & Brubakk, AO. 2010. "Parameter estimation of the Copernicus decompression model with venous gas emboli in human divers." *Medical & Biological Engineering & Computing,* July de 2010: 625–636.

Havnes, MB. 2010. "S100B and its relation to intravascular bubbles following decompression." *Diving and Hyperbaric Medicine,* December de 2010: 210–212.

Hobbs, GW, & Gault, KA. 2009. "Decompression risk evaluation of commercially available desktop decompression algorithms." *Abstract of the Undersea and Hyperbaric Medical Society. Annual Scientific Meeting.* Las Vegas, Nevada, USA., 2009.

Hugon, J, Barthelemy, L, Rostain, JC, & Gardette, B. 2009. "The pathway to drive decompression microbubbles from the tissues to the blood and the lymphatic system as a part of this transfer." *Undersea & Hyperbaric Medicine,* July-August de 2009: 223–236.

Kenney, Ian J, & Sonksen, Camilla. 2010. "Dysbaric osteonecrosis in recreational divers: a study using magnetic resonance imaging." *Undersea & Hyperbaric Medicine,* September-October de 2010: 281–288.

Kernagis, D, Datto, M, & Moon, R. 2009. "Genome-wide expression profiling of decompression stress: a new approach to biomarker discovery and validation." *Abstract of the Undersea and Hyperbaric Medical Society Annual Meeting.* Las Vegas, Nevada, USA., 2009.

Kot, J. 2003. Safety management: Hyperbaric therapy aspects. *6th European Consensus Conference on Hyperbaric Medicine. Prevention of Dysbaric Injuries in Diving and Hyperbaric Work.* Geneva.

Kot, J, Desola, J, Simao, AG, Gough-Allen, R, Houman, R. & Meliet, JL. 2004. *A European Code of Good Practice for Hyperbaric Oxygen Therapy- Prepared by working Group "Safety" of COST Action B14.* (E.C. Medicine, Ed.) available in http://www.echm.org/documents/ECGP%20for%20HBO%20%20May%202004.pdf

Kutting, B. 2004. "Prevention of work-related decompression illness events by detection of a cardiac right-to-left shun." *Scandinavian Journal of Work Environment & Health,* August de 2004: 331–333.

Mohammadein, SA. 2010. "Concentration distribution around a growing gas bubble in tissue." *Math. Biosci,* May de 2010: 11–17.

Møllerløkken, A, Gaustad, SE, Havnes, MB, Gutvik, CR, Hjelde, A, WisløV, U, & Brubakk, AO. 2012. "Venous gas embolism as a predictive tool for improving CNS decompression safety. Review." *European Journal of Applied Physiology* 112, n.° 2 (February 2012): 401–409.

Ors, F. 2006. "Incidence of ischemic brain lesions in hyperbaric chamber inside attendants." *Advances in Therapy,* November-December de 2006: 1009–1015.

Pollock, NW. 2007. "Use of ultrasound in decompression research. Review." *Diving and Hyperbaric Medicine,* June de 2007: 68–72.

Rhind, SG, Cameron, BA, & Eaton, DJ. 2007. "Heat shock protein 70 is upregulated in blood leukocytes from experienced divers in response to repetitive hyperbaric stress." *Abstract of the Undersea and Hyperbaric Medical Society. Annual Scientific Meeting.* Ritz-Carlton Kapalua Maui, Hawaii, 2007.

Risberg, J, Englund, M, Aanderud, L, Eftedal, O Flook, V, & Thorsen, E 2004. "Venous gas embolism in chamber attendants after hyperbaric exposure."*Undersea & Hyperbaric Medicine*, 2004: 417–429.

Sobakin, AS, Wilson, MA, Lehner, CE, Dueland, RT, & Gendron-Fitzpatrick, AP. 2008. "Oxygen pre-breathing decreases dysbaric diseases in UW sheep undergoing hyperbaric exposure." *Undersea & Hyperbaric Medicine*, January-February de 2008: 61–67.

Sobakin, AS, Wilson, MA, Gendron-Fitzpatrick, AP & Eldridge, M. 2009. "Uw sheep dissub rescue profile: accelerated decompression following saturation at 3.7 ATA using oxygen prebreathe." *Abstract of the Undersea and Hyperbaric Medical Society, Annual Scientific Meeting.* Las Vegas, Nevada, USA., 2009.

Uzun, G. 2011."Decompression sickness in hyperbaric nurses: Retrospective analysis of 4500 treatments." *Journal of Clinical Nursing* 20 (January 2011): 1784–1787.

Vann, RD, Butler, FK, Mitchell, SJ, & Moon, RE. 2011. "Decompression illness."*The Lancet* 377, n.º 9760 (January 2011): 153–164.

Walker, JR, Hobbs, GW, Gault, KA, Howle, LE, & Freiberger JJ. 2010. "Decompression risk analysis comparing oxygen and 50% nitrox decompression stops." *Abstract of the Undersea and Hyperbaric Medical Society. Annual Scientific Meeting.* St Pete Beach, Florida, USA., 2010.

Walker, J, Hobbs, GW, Gault, KA, Howle, LE, & Freiberger JJ. 2010a. "The effect of final oxygen decompression stop depth on DCS risk: 20 fsw vs. 10 fsw." *Abstract of the Undersea and Hyperbaric Medical Society. Annual Scientific Meeting.* St Pete Beach, Florida, USA, 2010.

Zkan, H. 2008. "MRI screening of dysbaric osteonecrosis in hyperbaric-chamber inside attendants." *Journal of International Medical Research*, March-April de 2008: 222–226.

Occupational Safety and Hygiene – Arezes et al. (eds)
© 2013 Taylor & Francis Group, London, ISBN 978-1-138-00047-6

Hyperbaric medicine: Occupational medical exams

Helena Alvim & A.S. Miguel

PROA/CIGAR/LABIOMEP/Faculdade de Engenharia, Universidade do Porto, Porto, Portugal

ABSTRACT: The Hyperbaric Medicine is implemented at an international level and in Portugal has been developed since the middle of the last century. The professionals, that accompany patients inside a hyperbaric chamber (attendants), are subject to several significant risks. The control, at an occupational level, plays an important role in the early detection of disbaric diseases. Medical tests, these hyperbaric professionals are subjected to, are defined in Portuguese legislation in two different frameworks: workers in pneumatic caissons and professional diving. The respective review and adjustment to each one of the activities are essential. It is proposed, justifying, a scheme of comprehensive medical examinations and consultations for Hyperbaric Medicine, contemplating the various medical specialties, where there is an evidence of pathological manifestations related with the activity in a hyperbaric environment.

1 INTRODUCTION

Hyperbaric oxygen therapy (HBO) is widespread at international level, existing 213 hyperbaric centers in Europe (Fig. 1).

In Portugal, the hyperbaric medicine has been developed since 1953. There are two hyperbaric chambers in the Hospital of the Navy, in Lisbon, another one in the Local Unit of Health of Matosinhos (ULSM) and two other in the Autonomous Region of Madeira and in the Autonomous Region of Azores (Alvim, 2010). The Pedro Hispano Hospital, belonging to the ULSM gets patients from the whole north and center regions of the country, having performed sessions of hyperbaric oxygen therapy (HBO) to 170 persons in 2011 (Fig.2).

The HBO is a modality of medical treatment, in the field of hyperbaric medicine, through which the patient blows pure oxygen (100%) to an ambient pressure greater than normal atmospheric pressure, aiming at the elimination or control of specific pathological conditions. The attendants who support the patients inside a hyperbaric chamber are subject to various significant risks (Alvim, 2010, Alvim, et al. 2011, a,b). Preventive measures, adopted at the international level, are, in general, the guidelines, given in the 6th Consensus Conference and the European Code for the prevention of disbaric diseases (Kot, 2003, Kot, et al. 2004).

The control, at an occupational level, has an important role in early detection of disbaric diseases. The medical exams, to which the hyperbaric professionals are subject to, are defined in the Portuguese legislation in two different legal frameworks, workers in pneumatic caissons (1982)

and diving professionals (1994), missing a legal framework for the attendants. The appropriate review and adequacy to each one of these activities are essential, on one hand, due to technological advances that the techniques and equipment can provide at the diagnosis level, and on the other hand, to a greater knowledge of disbaric diseases and their prevention.

It is therefore proposed a scheme for medical consultations and comprehensive exams to the hyperbaric medicine attendants, covering the different medical specialties where there is an evidence of manifestation of disbaric diseases.

2 METHOD

The study was developed on the basis of a Master's Degree thesis in Occupational Safety and Hygiene Engineering (MESHO), carried out in 2010 at the Engineering Faculty of University of Porto. A visit to a service of hyperbaric medicine was held "in situ" in 2012, having been collected information on the exposure of professionals in the last three years (2009–2011). In July, it was verified the validity of the Portuguese legislation respecting Safety, Hygiene and Health at Work in hyperbaric environment, in the data base of PCMLEX of the DIGESTO System. Medical examinations/specialties, involved in the occupational control, were subject to discussion, for comparative analysis of workers in hyperbaric environment (diving and caisson workers), in order to establish the most appropriate ones to the attendants. A research was conducted in PubMed' in 2011, with the keywords: hyperbaric chamber, occupational noise,

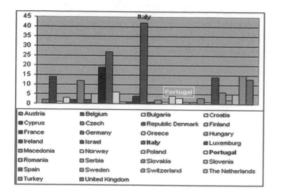

Figure 1. Hyperbaric Centers European. Helena Alvim, study/proposal for preventive and protective measures for disbaric diseases of workers in hyperbaric medicine. Master's Degree thesis in Occupational Safety and Hygiene Engineering FEUP, 2010.

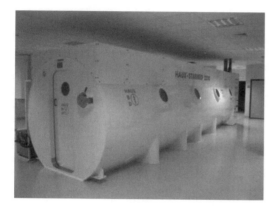

Figure 2. Hyperbaric chamber ULSM.

barotrauma, decompression sickness and diving which allowed to propose more current exams for the prevention of pathology with the highest frequency of occurrence (hearing loss).

3 RESULTS AND DISCUSSION

In Pedro Hispano Hospital, the chamber has an occupation that rounds 100%, having been carried out, in 2011, 5569 routine and 264 emergency sessions to 64 patients (Fig 3).

The Staff of Hyperbaric ULSM consists of five nurses, three technician's hyperbaric chamber operators and two doctors.

The nursery attendants were the most exposed to hyperbaric environment, with 73/71 sessions by watching routine/emergency treatments (maximum values) in 2011 (Fig 4). The number of

HBO urgency sessions (Fig 5) has been increasing, increasing the risk of exposure of attendants in three modalities of exponential conjunction, number and frequency of sessions and exposure to higher HBO pressure/duration levels. In hyperbaric chamber they are subject to pres-

Figure 3. HBO therapy ULSM.

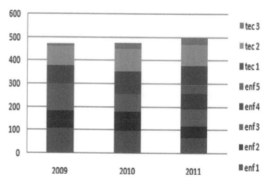

Figure 4. Number of HBO routine by attendants.

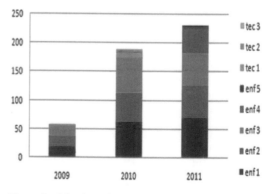

Figure 5. Number of HBO emergency by attendants.

sures of 2.5/3.0 absolute atmospheres (ata) in routine HBO and pressures of 3.0 to 6.0 ata in emergency.

Besides just being defined in the legislation the medical exams to be held by the professionals of diving and caisson workers (Alvim, 2010, Alvim, et al. 2011a), some of the pathologies are only identified in Regulatory Decree no. 6/2001: dysbaric osteonecrosis, vestibular disorders, sub acute media otitis, chronic media otitis and hypoacusis by irreversible cochlear lesion.

Hearing loss is part of a group of consequences where the ear barotrauma has an important role, due to the frequency of its occurrence. Hearing loss is a consequence of three causes that, interconnected, exacerbate its consequences (Smith, 1984, Skogstad, Eriksen & Skare, 2009, Alvim, 2010): pressure/pressure variation, inhalation of gases and exposure to noise

– In the first case, we have the decompression illness (DCI), (Cherry, et al. 2009, Goplen, et al. 2011) and the optic barotrauma (Ledingham & Davidson 1969, Duplessis & Fothergill 2009, Goplen, et al. 2011);
– The inhalation of gases/mixture of different gases of the air can also affect the hearing sensitivity (Skogstad, Eriksen & Skare, 2009, Alvim, 2010). This is the case of mixtures of oxygen and nitrogen (nitrox for use at a depth of approximately 40 m), of mixtures of oxygen and helium (heliox), used for greater depths or mixtures of oxygen, nitrogen and helium (Trimix);
– The noise reaches high intensities, during cycles of compression and decompression in a hyperbaric chamber. The risk of hearing loss is therefore possible. (Nedwell & Needham 1995, Anthony, Wright & Evans, 2009). The noise seems to be the most important cause, in the long term, of hearing loss in professional divers, in the case of absence of manifestation of barotrauma of the inner ear or decompression sickness (Goplen, et al., 2011).

Evidences indicate clearly that the implementation of the pattern of hearing conservation for the normobaric conditions is inadequate for hyperbaric environments (Smith, 1984). The traditional method to detect hearing loss, in the case of workers exposed to hyperbaric environment, is questionable, considering the selection of exams to be held within a framework of medical surveillance subject to optimisation (Duplessis & Fothergill, 2009, Ross, et al., 2010). Differences between the requirements in activities in hyperbaric environment are obvious, regarding occupational control (Table 1).

Table 1. Medical examinations diving/caisson.

Exams	Periodicity	
	Caisson *A	Diving*B*C
ECG Stress/Resting Test	*1	*4/*3
Echocardiography	–	*5
Pulmonary function tests	*1	*4
Electroencephalogram	–	*5
ENT examination	*5	*3
Audiometry testing	*2	*3
Tympanogram Test	–	*3
Vestibular testing form of electronystagmography	–	*3
rx: a) shoulders; b) hip/pelvis; c) knees; d) feet; e) paranasal sinuses; f) chest	*2 a) b) c); *6 f)	*5 a) b) c) d) e) f); *3 f); *4 a) b) c) d)
Bone scan	–	*4
Ophthalmologic examination	–	*3
Clinical Analyzes	*2	*3

*A Decree-law no. 49/82, of 18 February. Regulation of Occupational Hygiene and Safety in pneumatic caissons.

*B Decree-law no. 12/94, of 15 January. Regulation of Professional Diving.

*C Decree n° 876/94, of 30 September. Annex D.

*1 Admission exams and periodic 3/3 months.

*2 Admission exams and periodic 6/6 months.

*3 Admission exams and periodic 1/1 year.

*4 Admission exams and periodic 2/2 year.

*5 Admission exams.

*6 And year after cessation of work.

4 CONCLUSIONS

The professionals, who work in hyperbaric environment, in a generic manner, may suffer from disbaric diseases related to pressure and its variation.

– In the activity of attendants DCI (Sheffield & Piron 1999, Lacerda, et al., 2006, Gerbino, 2008, Fernandes, 2009);
– In the activity of diving, including hearing loss related to the hyperbaric exposure (Skogstad, Eriksen, & Skare, 2009) and DCI (Cherry, et al. 2009);
– In the activity in civil construction (workers in pneumatic caissons or building tunnels), hyperoxic myopia (Onoo, et al., 2002), decompression sickness and aseptic bone necrosis (Ledingham & Davidson 1969).

It is therefore proposed a scheme for comprehensive consultations/medical examinations (Table 2) and the introduction of non-existent specialties in

Table 2. Proposal for medical examinations in hyperbaric medicine.

Speciality	Exams
Cardiology	ECG Stress Test *1
	Echocardiography *2
Pulmonology	Pulmonary function tests *1
Neurology	Electroencephalogram *3
Otorhinolaryngology	ENT examination *2
	Audiometry testing *2
	Tympanogram Test *3
	Vestibular testing *3
Radiology	RM shoulders and hips; rx chest *3
	Densitometry *4
Clinical Pathology	Clinical Analyzes *2
Dentistry	Dental Control *3
Ophthalmology	Medical Control *2
Clinical nutrition	Weight Control Programs *3

*1 Admission exams and periodic 3/3 months.
*2 Admission exams and periodic 6/6 months.
*3 Admission exams and periodic 1/1 year.
*4 For 40 years and periodic 5/5 years.

current legal frameworks, such as: "Nutrition", due to the individual susceptibility/body mass index (BMI) and decompression sickness, "Dentistry", for the prevention of dental barotrauma, "Neurology" for the prevention of neurological changes, caused by intoxication by oxygen. In addition, the completion of more technologically advanced exams in the specialty of Otolaryngology is proposed, as well as, in imaging, exams without the use of ionising radiation (Alvim, 2010).

Although working conditions are distinct (pressure, time of exposure and duration of work) in these activities (diving, workers in pneumatic caissons and hyperbaric medicine) and, in this manner, conditioning the occurrence of disbaric diseases and its repercussions in the long term, it will be necessary to adopt and/or revise a rigorous scheme of evaluation and of medical examination appropriate to each of the activities. However, should be contemplated, for any activity in high pressure environments, all the medical specialties proposed in Table 2 (medical examinations in hyperbaric medicine). There is a need to carry out the measurement and control of the sound level in HBO (Anthony, Wright & Evans 2009), to promote health surveillance of workers, with the support of a physician holding the specialty in Occupational Medicine and with competence in Hyperbaric Medicine, recognised by the Medical Professional Association and to develop organisational measures to include in a normative framework of minimum OSH requirements, for the Hyperbaric Medicine. (Alvim, 2010, Alvim, et al. 2011, a).

ACKNOWLEDGMENTS

We would like to acknowledge the collaboration and information made available under the visit to the Hyperbaric Medicine of Hospital Pedro Hispano ULSM.

REFERENCES

Alvim, H. 2010. Estudo/Proposta de medidas de prevenção e protecção das doenças disbáricas dos trabalhadores em medicina hiperbárica. Faculdade de Engenharia. Porto: Catálogo da Biblioteca da FEUP, 244pp.

Alvim, H., Diogo, M.T., Leão, R.P, Camacho, Ó., Baptista, J. & Nobrega, J. 2011. Fire drills in Hyperbaric Medicine. Ed. R. Mondelo, P., Karwowski, W., Saarela, K., Hale, A. Proceedings of the 9th International Conference on Occupational Risk Prevention. Santiago de Chile: Universitat Politècnica de Catalunya.

Alvim, H., Diogo, M.T., Leão, R.P., Camacho, Ó. & Baptista, J. 2011a. Hyperbaric Medicine Organizational Risks. Ed João Santos Baptista, A.S. Miguel, Gonçalo Perestrelo, Nelson Costa, Mónica Barroso, Pedro Arezes, P. Carneiro, P. Cordeiro, Rui Melo. Proceedings of the International Symposium on Occupational Safety and Hygiene. Guimarães: Sociedade Portuguesa de Segurança e Higiene Ocupacionais, 89–94.

Alvim, H., Diogo, M.T., Leão, R.P., Camacho, Ó. & Baptista, J. 2011b. Dispositivos Médicos Sujeitos a Pressão. Ed. J F Silva Gomes, Clito F Afonso, Carlos C António e António Matos. Proceedings do 6° Congresso Luso-Moçambicano de Engenharia, 3° Congresso Moçambicano de Engenharia. Maputo: INEGI/FEUP.

Anthony, T.G., Wright, N.A. & Evans, M.A. 2009. RR735. Review of diver noise exposure. Health and Safety Executive.

Cherry, A.D., Forkner, I.F., Frederick, H.J., Natoli, M.J., Schinazi, E.A., Longphre, J.P., Conard, J.L., White, W.D., Freiberger, J.J., Stolp, B.W., Pollock, N.W., Doar, P.O., Boso, A.E., Alford,E.L., Walker, A.J., Ma, A.C., Rhodes, M.A. & Moon, R.E. 2009. Predictors of increased PaCO2 during immersed prone exercise at 4.7 ATA. J Appl Physiol, 106, 316–325.

Duplessis, C. & Fothergill, D. 2009. Exploiting otoacoustic emission testing to identify clinical and subclinical inner ear barotrauma in divers: potential risk factor for sensorineural hearing loss. J Otolaryngol Head Neck Surg., 38 (1), 67–76.

Fernandes, T.D. 2009. Medicina Hiperbárica. Acta Med Port., 22 (4), 323–334.

Gerbino, A.J. 2008. Multiplace Hyperbaric Chambers. In T.S. Neuman, S.R. Thom., & D. Meloni (Ed.), Physiology and medicine of hyperbaric oxygen therapy (Vol. Section II Technical Aspects, pp. 36–55). Philadelphia, PA: Saunders.

Goplen, F.K., Aasen, T., Grønning, M., Molvær, O.I. & Nordahl, S.H. 2011. Hearing loss in divers: a six-year prospective study. Eur Arch Otorhinolaryngo, 268 (7), 979–85.

Kot, J. 2003. Safety management: Hyperbaric therapy aspects. *6th European Consensus Conference on Hyperbaric Medicine. Prevention of Dysbaric Injuries in Diving and Hyperbaric Work.* Geneva.

Kot, J., Desola, J., Simao, A.G., Gough-Allen, R., Houman, R. & Meliet, J.L. 2004. *A European Code of Good Practice for Hyperbaric Oxygen Therapy-Prepared by working Group "Safety" of COST Action B14.* (E.C. Medicine, Ed.) available in http://www.echm.org/documents/ECGP%20for%20HBO%20%20May%202004.pdf

Lacerda, E.P., Sitnoveter, E.L., Alcantara, L.M., Leite, J.L., Trevizan, M.A. & Mendes, I.A. 2006. Atuação da enfermagem no tratamento com Oxigeniterapia Hiperbárica. *RevLatino–am Enfermagem, 14* (1), 118–23.

Ledingham, I. & Davidson, J.K. 1969. Hazards in Hyperbaric Medicine. *British Medical Journal, 3*, 324–327.

Nedwell, J. & Needham 1995. 'Noise Hazard in the Diving Environment. *Proceedings of the international conference: SUBTECH '95.* Subsea Challenge.

Onoo, A., Kiyosawa, M., Takase, H. & Mano, Y. 2002. A Development of myopia as a hazard for workers in pneumatic caissons. *Br J Ophthalmol, 86,* 1274–1277.

Ross, J.A., Macdiarmid, J.I., Dick, F.D. & Watt, S.J. 2010. Hearing symptoms and audiometry in professional divers and offshore workers. . *Occup Med (Lond) 60:, 60* (1), 36–42.

Sheffield, P.J. & Pirone, C.J. 1999. Decompression Sickness in Inside Attendants. In W.T. Workman, & W.T. Workman (Ed.), *Hyperbaric Safety: A Practical Guide* (pp. 643–664). Best Publishing Company.

Skogstad, M., Eriksen, T. & Skare, O. 2009. A twelve-year longitudinal study of hearing thresholds among professional divers. *Undersea Hyperb Med., 36* (1), 25–31.

Occupational Safety and Hygiene – Arezes et al. (eds)
© *2013 Taylor & Francis Group, London, ISBN 978-1-138-00047-6*

Thermal environment and cognitive performance: Parameters and equipment

E. Quelhas Costa & J. Santos Baptista
PROA/CIGAR/LABIOMEP/Faculdade de Engenharia, Universidade do Porto, Porto, Portugal

ABSTRACT: The increasing complexity of work activities associated with different environmental conditions, can lead to a situation of thermal stress. The study of the respective impact on workplace has been gaining an increasing importance. It is therefore essential to identify and monitorize variables to evaluate the performance under different environmental conditions. This paper attempts to define the parameters and respective measurement equipment to study mental fatigue in different conditions of temperature and humidity. For such purpose, a literature review was carried out about different topics according to appropriate key words. From the results, it was possible to identify the most relevant parameters and the best suitable equipment for their measurement. It is hoped that the results may contribute to increased research in this field of study.

1 INTRODUCTION

The thermal environments out of the comfort zone can affect the welfare, performance and health of individuals and hence productivity (Costa et al., 2012c) as well as the social equilibrium. The influence of stress on human behavior due to extreme temperatures has been the subject of study since several years, as it allows us to understand the reactions of the individuals in adverse working conditions. As a goal, this study aims to define the respective parameters and measurement equipment to study human cognitive response in different conditions of temperature and humidity in order to determine the conditions of fatigue, in terms of memory, concentration and attention at different levels of tasks in sedentary activity.

2 MATERIAL AND METHOD

The methodology development to define the parameters for studying fatigue in different conditions of temperature and humidity has gone through three stages: a) problem formulation, assumptions, goals, methods and challenges, b) cross-sectional survey from different databases, c) collecting information about the technical requirements to be used.

Three hundred and fifty articles were collected and filtered according to the relevance of the study as well as the goal under survey. The keywords selected were: Thermal Environment, Climatic Chamber, Cognitive Aspects, Thermal Stress,

Workload and Internal Temperature. This step also allowed to identify variables to monitorize (Costa et al., 2012a), for example, skin and core temperature, cerebral electrical activity, heart rate, muscle electrical activity, electrodermal activity; dehydration, as well as subjective variables evaluated by specific questionnaires. Bearing in mind the purpose of the study, the priority was given to the articles approved by the ethics committees of the respective organizations or, at least, with informed consent. Other selection criteria were the calibration/validation of the equipments, usability, reability, comfort under use and the possibility of monitorization in real time.

3 MAIN EQUIPMENTS

These tests requires the use of basic devices such as computers, monitors, precision scales, ruler for height measuring and tape to assure the measurement of all the anthropometric parameters of individuals subjected to the test. Besides all these devices are needed equipment calibrated with a sufficient precision to ensure the quality of the results.

3.1 *Climatic chamber*

The main device is a climatic chamber (FITO-CLIMA 25000EC20)—Figure 1, built according EC standards and directives relating to health and safety requirements. This chamber allows to simulate the exposure to thermal environments of

Figure 1. General appearance of the chamber.

very different workplaces. The temperature inside the chamber can be controlled between –20 °C to +50 °C and the humidity from 30% to 98%. The chamber is also equipped with O_2 and CO_2 sensors, and has been validated for human trials (Guedes et al., 2012).

3.2 Ingestible thermal sensors

In order to recognize the arising fatigue signs from different tasks were sought the respective identifiers. One of the best fatigue identifiers is the core temperature, in particular, when the individuals are exposed to high temperatures. So, the monitorization of this parameter is one of the best methods to minimize the risk of heat damage (Goodman et al., 2009, Byrne & Lim 2007), in addition it also helps to study the cognitive response (Wright et al., 2002).This kind of monitorization has been applied in different studies either in the laboratory or in the real world, especially among athletes, students and military forces.

The internal temperature measurement will be performed with ingestible thermal sensors (ITS) (Costa et al., 2012). This is a ingestible telemetric thermometer with a dimension of 8.7 mm diameter by 23 mm length. This capsule must be swallowed with water at least 10 hours before each test. It travels along the digestive track harmlessly and leaves the body naturally within 24 to 72 hours. The ITS starts transmitting at 15-second intervals, one minute after being activated externally by a specific device. Data are sent by telemetry, with a precision of ± 0.01 °C for the recording system through the Equivital EQ02 Life monitor—Sensor Electronics Module (SEM), in real time, via Bluetooth. The test is stopped if the internal temperature reaches the value of 39 °C (Costa et al., 2012).

3.3 Skin temperature sensors

Skin temperature is also measured at 14 points of the body. This monitorization is made in accordance with standard ISO8996 (Ely et al., 2009). The body points to be monitorized are shown on Figure 2.

Sensors are placed at different locations on the skin with medical adhesive tape. The measurements are made in real time with the help of two bioPLUX synchronized devices.

3.4 EEG for monitoring brain electrical activity

For the cognitive response, the brain activity recording is done through electroencephalogram (EEG) (Berka et al., 2007). The aim of these records is to measure in real time test fatigue results (being the fatigue defined as a transitional state between sleep and wakefulness) (Lal & Craig 2001). Mental fatigue is believed to be a gradual and cumulative process also associated with reduced efficiency,

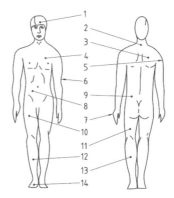

Figure 2. Different sensors location for monitoring the skin temperature. Adapted from standard ISO8996.

Legend:
1. Forehead
2. Neck
3. Right scapula
4. Left upper chest
5. Right arm in upper location
6. Left arm in lower location
7. Left hand
8. Right abdomen
9. Left paravertebral
10. Right anterior thigh
11. Left posterior thigh
12. Right skin
13. Left calf
14. Right instep

alertness and decreased capacity for mental performance (Grandjean 1979).

The brain's electrical activity is classified according to rhythms, which are defined in terms of frequency bands including: *delta*, *theta*, *alpha* and *beta*. The *delta* activity is called slow wave between 0.5 Hz and 4 Hz. These waves appear in the transition between drowsiness and sleep. *Theta* rhythm has a frequency between 4 Hz and 7 Hz. The *theta* rhythms are associated with several of psychological states including low levels of alertness, and consequently are associated with a decrease in information processing. *Alpha* waves have a frequency with a rate in the range of 8 Hz to 13 Hz, which occur during daytime, particularly on the occipital cortex, and can be clearly observed when the eyes are closed, but when they are open this frequency decreases. These rhythms are present both in alertness as on relaxation. The *beta* waves have frequencies between 13 Hz and 30 Hz. His potential is associated with increasing arousal/alertness. These waves are still associated with the reaction time of motor tasks (Lal & Craig 2001). The alpha peak frequency (PFA) is associated with the mental capacities (Raveendran & Ng 2007).

In this study the device Emotive SDK will be used Figure 3a. The electrodes must be installed according to standard positions: parietal and frontal temporal, as shown in Figure 3b. The head set/electrode has to be adjusted so that two reference electrodes are positioned in the mastoid area. The electrodes should be pressed to ensure good contact. In Figure 3c each circle represents a sensor and its color, the contact quality. For best contact quality all sensors must be green.

3.5 *ECG for heart rate monitorisation*

The electrocardiogram (ECG) is another key element to measure the physiological workload (Lal & Craig 2001) and mental effort required for each task (Ryu & Myung 2005) as well as to reflect the degree of thermal comfort (Yao et al., 2008). According Marcora et al., (2009) the increase in HR (Heart Rate) average in a particular task, con-

firms the demanding level of this task in relation to the one of neutral character.

The increase in HR depends on working environment and work load. The HR increasing with the work load and is faster as hotter the environment, as greater the portion of static work and as smaller is the number of the involved muscles (Dinis 2003). The heart rate variability (HRV) refers to the changes of beats per minute. Under resting conditions, the electrocardiogram (ECG) of healthy individuals exhibits periodic variation in RR intervals. These ranges refer to the time between two successive R waves. The effects of the changes in RR intervals of the ECG signal, allows a spectral analysis of HRV that is usually used to study the Autonomic Nervous System (ANS) (Yao et al., 2009).

In testing will be used a 12-lead ECG, with electrodes GE-Healthcare, placed according to Figure 4.

The information will be recorded in real time and all the twelve derivations will be displayed. With this method a complete resting ECG up to 60 minutes can be obtained.

3.6 *EMG for monitoring electrical muscle activity*

The electromyography (EMG) to record the electrical activity of the non-dominant forearm muscle Figure 5, is an important indicator of mental effort. Electromyography (EMG) consists on using surface electrodes to record muscle electrical activity. The monitorization is done with sensors placed on the skin over the muscles whose activity is wished to register. Results are measured at intervals of one millisecond.

3.7 *Electrodermal activity control*

Electrodermal activity (EDA) consists on recording electrical changes that occur in the skin as an indicator of a psychological state. These latest changes are the result of the activity of the Autonomic Nervous System (ANS) (Waard, 1996). EDA is expressed in terms of resistance and electrical conductivity of the skin.

The EDA has been one of the most used methods as response system in the history of psychophysiology. Research involving EDA have been reported in psychology, psychiatry and psychophysiology. The diversity of publications on EDA reflect the fact that its results could be applied to a wide variety of issues, ranging from attention research to the information processing and emotion analysis. The application of EDA for a wide range of issues is largely due to the relative ease of measurement

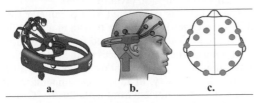

a.　　　　b.　　　　c.

Figure 3. Basic information about Emotiv EEG equipment. (in user manual emotiv software development kit).

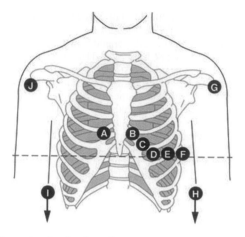

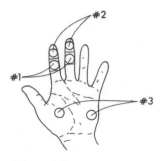

Figure 6.   Example of different placements of electrodes to record electrodermal activity.

Figure 4.   Application of sensors according cardio soft equipment—GE healthcare.

Legend:

A	V1 red	C1 red	Fourth intercostal space at the right edge of the sternum
B	V2 yellow	C2 yellow	Fourth intercostal space on the left edge of the sternum
C	V3 green	C3 green	Intermediate location between locations B and D
D	V4 blue	C4 brown line	Fifth intercostal space
E	V5 orange	C5 black	Anterior axillary line on the same horizontal level with the D
F	V6 purple	C6 purple	Mid axillary line at the same horizontal level with the D
G	LA black	L yellow	Left arm (resting ECG)
J	RA white	R red	Right Arm (resting ECG)
H	LL red	F green	Left foot (resting ECG)
I	RL green	N black	Right foot (resting ECG)

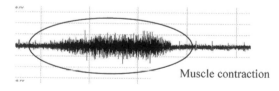

Muscle contraction

Figure 5.   Example of signal acquisition with a monitor emgPlux.

and quantification, combined with its sensitivity to the psychological reaction to states and processes (John Cacioppo, 2007).

Anatomically and physiologically, the skin is a selective barrier which prevents entry of foreign materials into the body. At the same time acts selectively to facilitate the passage of unwanted products from the bloodstream to the exterior. It also helps maintaining the water balance and the core body temperature. These functions are performed primarily through vasoconstriction/dilation and changes in sweat production (John Cacioppo 2007).

Figure 6 shows the different positions of electrodes for recording electrodermal activity. The first (1) involves placing the volar surface of the hand in the medial phalange; the second (2) involves the volar surface of the distal phalanges; the third (3) involves the palm of the hand.

In this study *EDA PLUX* sensors will be used, on "place 3". It is thus possible to accurately detect the events of galvanic skin response. Some of the applications of these sensors include detection of changes in attention, cognitive and emotional states. Within this line of thought, when the level of skin conductivity is greater than the normal, this reflects the excitation/activation levels or response capacity/attention.

3.8   *Other physiological measures*

Additionally breathing and dehydration are also measured and used to support the hypothesis that the cognitive effort coincides with a small increase of energy consumption (Waard, 1996).

The last one is considered as a parameter that affects the cognitive response (Costa et al., 2012b) and is used to calculate the need for fluid replacement. One of the consequences of the dehydration is the increasing of core body temperature during physical activity, in moderate temperature at 23–25 °C, or in hot environments 26–28 °C. For each 1% of dehydration there is an increase of 0.1–0.2 °C in body core temperature. Once dehydrated, the transpiration rate decreases, and the body heat loss by evaporation is significantly reduced (Ely et al., 2009, Kenefick & Sawka 2007).

3.9 Subjective evaluation

While physical parameters are measured, thermal sensation questionnaires are distributed to the voluntaries as well as subjective evaluation of workload, cognitive testing and other specific protocols, properly validated for each purpose. The subjective assessment tools, such as NASA Task Load Index, Staveland & Hart (1988), are used to determine the workload through self-assessment questionnaires.

4 CONCLUSIONS

This study points out the parameters and the equipment needed to perform tests able to identify changes in human behavior by changing the temperature and humidity. The tests are performed within a climate chamber, where a sedentary activity it is simulated at the level of mental fatigue.

The big challenge will be to establish the relationship between the different variables and monitor all parameters continuously doing their real-time control. As a result from this research work, both individual and collective interests, are expected to benefit, although the interests of individuals must always prevail over the interests of science and society.

With regard to individuals subjected to this experiment, they can expect to know their own psychophysiological response to certain environmental conditions which may be of assistance in future professional options. For the general population, greater knowledge leads to improvements in the area of working conditions, welfare, contributing to a safer working environment, especially in tasks where possible cognitive errors can occur and lead to accidents. Knowing and quantifying the physiological response towards different conditions of temperature and humidity and develop the implications of the environment to the level of thermal perception and behavior of individuals, can be a benefit for the society in general in that it results into improvement productivity in a safe manner.

REFERENCES

Berka, C., D.J.L., Michelle N. Lumicao, Alan Yau, Gene Davis, Vladimir T. Zivkovic, Richard E. Olmstead, Patrice D. Tremoulet, & Patrick L. Craven. 2007. EEG Correlates of Task Engagement and Mental Workload in Vigilance, Learning, and Memory Tasks. *Aviat Space Environ Med,* 78(5), 231–244.

Byrne, C., & Lim, C.L. 2007. The ingestible telemetric body core temperature sensor: a review of validity and exercise applications. *Br J Sports Med,* 41(3), 126–133. doi: DOI 10.1136/bjsm.2006.026344.

Costa, E.Q., Guedes, J.C., & Baptista, J.S. 2012. Core Body Temperature Evaluation: Suitability of Measurement Procedures. In XII EAT Congress on Thermology 6–7 September 2012, at FEUP EAT – Faculdade de Engenharia da Universidade do Porto.

Costa, E.Q., Baptista, J.S., & Diogo, M.T. 2012a. Efeitos do Ambiente Térmico em Atividades Sedentárias. Uma breve revisão. *Riscos, Segurança e Sustentabilidade, C.Guedes Soares, A.P. Teixeira, C.Jacinto (Eds),Edições Salamandra, Lisboa,* 2, 1223–1237.

Costa, E.Q., Baptista, J.S., & Diogo, M.T. 2012b. Effects of thermal environment on cognitive response in sedentary activities. A short revision, in International Symposium on Occupational Safety and Hygiene-SHO 2012. *edited by Arezes, P., Baptista, J.S., Barroso, M.P., Carneiro, P., Cordeiro, P., Costa, N., Melo, R., Miguel, A.S., Perestrelo, G.P.ISBN 978–972-99504-9–0. pp 471–477.*

Costa, E.Q., Baptista, J.S., & Diogo, M.T. 2012c. Thermal Environment and Productivity in Sedentary Activities. A Short Review. In International Symposium on Occupational Safety and Hygiene-SHO 2012.*edited by Arezes, P., Baptista, J.S., Barroso, M.P., Carneiro, P., Cordeiro, P., Costa, N., Melo, R., Miguel, A.S., Perestrelo, G.P.ISBN 978-972-99504-9-0.p.p 478–483.*

Diniz, R.L. (2003). Avaliação das Demandas Fisicas e Mental no Trabalho do Cirurgião em Procedimentos Eletivos. Tese de Doutorado.

Ely, B.R., Ely, M.R., Cheuvront, S.N., Kenefick, R.W., DeGroot, D.W., & Montain, S.J. 2009. Evidence against a 40°C core temperature threshold for fatigue in humans. *J Appl Physiol* 107, 1519–1525. doi: 10.1152/japplphysiol. 00577.2009.-Evidence.

Goodman, D.A., Kenefick, R.W., Cadarette, B.S., & Cheuvront, S.N. 2009. Influence of sensor ingestion timing on consistency of temperature measures. *Med Sci Sports Exerc,* 41(3), 597–602. doi: 10.1249/MSS.0b013e31818a0 eef.

Grandjean, E. 1979. Fatigue in industry. *British Journal of Industrial Medicine,* 36(3), 175–186. doi: 10.1136/oem. 36.3.175.

Guedes, J.C., Costa, E.Q., & Baptista, J.S. 2012. Using a Climatic Chamber to Measure The Human Psychophysiological Response Under Different Combinations Of Temperature And Humidity. In XII EAT Congress on Thermology 6–7 September 2012, at FEUP EAT – Faculdade de Engenharia da Universidade do Porto.

Hart, S.F., & Staveland, L.E. 1988. Development of NASA-TLX (Tasl Load Index): Results of Empirical and Theoretical Reserach. *P.A.Hancock & N. Meshkati, Human Mental Workload,* 239–250.

John Cacioppo, L.G.T.G.G.B. 2007. *The Handbook of Psychophysiology.*

Kenefick, R.W., & Sawka, M.N. 2007. Hydration at the Work Site. *Journal of the American College of Nutrition,* Vol. 26, No. 5, 597S–603S (2007), 26.

Lal, S.K.L., & Craig, A. 2001. A critical review of the psychophysiology of driver fatigue. *Biol Psychol,* 55(3), 173–194. doi: 10.1016/s0301-0511(00)00085-5.

Marcora, S.M., Staiano, W., & Manning, V. 2009. Mental fatigue impairs physical performance in humans. *Journal of Applied Physiology,* 106(3), 857–864. doi: 10.1152/japplphysiol.91324.2008.

Parsons, K.C. (2nd ed.). 2003. Human thermal environments: the effects of hot, moderate, and cold environments on human health, comfort, and performance London: Taylor & Francis.

Raveendran, P. & Ng, S.C. 2007. EEG Peak Alpha Frequency as an Indicator for Physical Fatigue. In T. Jarm, P. Kramar & A. Zupanic (Eds.), *11th Mediterranean Conference on Medical and Biomedical Engineering and Computing* 2007 (Vol. 16, pp. 517–520): Springer Berlin Heidelberg.

Ryu, K., & Myung, R. 2005. Evaluation of mental workload with a combined measure based on physiological indices during a dual task of tracking and mental arithmetic. *International Journal of Industrial Ergonomics 35*, 991–1009.

Waard, D. d. 1996. The Measurement of Drivers`s Mental Workload. The Traffic Research Centre VSC. University of Groningen Netherlands.

Wright, K.P., T. Hull, J., & A. Czeisler, C. 2002. Relationship between alertness, performance, and body temperature in humans. *Am J Physiol Regul Integr Comp Physiol*, 283(6), R1370–1377. doi: 10.1152/ajpregu.00205.2002.

Yao, Y., Lian, Z., Liu, W., Jiang, C., Liu, Y., & Lu, H. 2009. Heart rate variation and electroencephalograph - the potential physiological factors for thermal comfort study. Indoor Air, 19(2), 93–101. doi: 10.1111/j.1600-0668.2008.00565.x.

Yao, Y., Lian, Z., Liu, W., & Shen, Q. 2008. Experimental study on physiological responses and thermal comfort under various ambient temperatures.[Comparative StudyResearch Support, Non-U.S. Gov't]. *Physiology Behavior, 93*(1–2), 310–321. doi: 10.1016/j.physbeh.2007.09.012.

Occupational Safety and Hygiene – Arezes et al. (eds)
© *2013 Taylor & Francis Group, London, ISBN 978-1-138-00047-6*

Occupational exposure to radon in thermal spas

A.S. Silva, M.L. Dinis & M.T. Diogo
Faculdade de Engenharia da Universidade do Porto, Porto, Portugal

ABSTRACT: Radon is a natural radioactive gas that may be present in soil air, water, groundwater and indoor as well as outdoor air. Radon is the largest source of natural radiation exposure; it represents almost 50% of the radiation that people are exposed to in a lifetime. Thermal spas are multifaceted projects that have been increasing in the last few years. However, the presence of radon in thermal and mineral waters enhance the risk of professionals being exposed to an additional level of radiation as well as the visitors and patients. Therefore, a growing interest and mostly a concern in protecting people exposed to radiation from radon and short-lived daughters have been drawn much of public attention. The objective of this work is to develop a bibliographical revision about exposure to radon in thermal establishment's spas. The bibliographical revision was carried out by searching key-words in several databases and scientific journals. The majority of the consulted and validated studies concluded that the exposure to high radon concentrations may be a threat to human health; radon is the second most important cause of lung cancer in many countries. As thermal spas are an important economic sector with a significative social and historical components, it is important investigate the magnitude of the exposure in order to take the necessary actions to protect those that in fact may be exposed to radiation in thermal establishments spas.

1 INTRODUCTION

Radon-222 is a colorless and odorless gas, chemically inert, formed by the direct radioactive decay of radium-226. Radon is the only gaseous decay element from the U-238 and Th-232 radioactive decay chains and can be naturally found in the atmosphere, rocks or dissolved in underground waters (Gray et al., 2009). As radon is a gas it can be easily inhaled. With a short half-live of approximately 3.8 days, it decays sequentially to the followings elements of the U-238 decay chain (Po-218, Bi-214, Po-214, Pb-210, Po-210) until the stable form of lead (Pb-206). Radon daughters are solids and easily stick to surfaces, such as dust particles in the air, being able therefore to be inhaled. If radon decays inside the body the energy given off may be highly damaging, increasing the risk of developing lung cancer (Al Zoughool et al., 2009). On the other hand, if radon daughters attached to dust particles are inhaled the effect will be similar although different types of emitting radiation may be present (α and/or β). The aim of this work is to develop a bibliographical revision concerning radon exposure in thermal spas establishments. The relevance of this study is demonstrated in several domains: historical, health, social, economical and environmental.

2 MATERIALS AND METHOD

A systematic review was carried out in several databases with *Metalib* tool. The search was done using a combination of a pair of key-words: "radon" and "spas" refined with "exposure" or "to water" in the subject research field. Others sources were also consulted such as databases from the Portuguese Spas Association (ATP), Turismo de Portugal, Direção Geral da Saúde (DGS), Instituto da Água, IP (INAG), Laboratório Nacional de Energia e Geologia (LNEG) and Direção-Geral de Energia e Geologia (DGEG). Master theses were also surveyed with the same key-words.

The obtained articles were grouped in the following categories: epidemiologist studies (25); radon impact on human health (22); radon concentration in thermal spas indoor air (16); radon concentration in natural mineral waters (16); occupational exposure to radon in thermal spas (13); occupational exposure to radon in workplaces (10); amount of radon in drinking waters (9) and amount of radon in water wells (4).

Twenty-nine articles were selected for review from the following databases: Science & Technology (1), Current Contents (1), Science Citation (8) e Scopus (19).

3 RESULTS AND DISCUSSION

3.1 *Origin of radon*

Indoor radon concentration varies from location to location, depending on the uranium and thorium content present in soil and rocks. In most countries, indoor radon concentration has a magnitude of few tens of becquerel for m³ of air (Bq/m³) (Köteles, 2007). Radon is easily dissolved in water and it is released into air during water usage due to radon outgassing into local air. The amount of indoor radon depends on its concentration in underground waters, volume and aeration as well as the eventual water processing during water circuit and usage points (Ferreira, 2009). Radon generation depends on geologic features (rocks composition and fractures) which are related with local and regional tectonics structures (fractures systems) (Beitollahi et al., 2007), soil porosity and moisture. These fractures systems allow the vertical migration of radon due to the increasing of soil permeability. Also, the common secondary deposits of uranium minerals in fractures systems, due to fluids circulation, contribute to radon generation (Correia, 2010).

Underground water flows in depth through soil pores and rock fractures, generating more or less radon depending on local factors (Ferreira, 2009). Once formed, radon may diffuse through soil pores and fractures systems being dissolved in these waters (Gnoni et al., 2008).

3.2 *Exposure to radon*

Radon exposure in thermal spas has been the target of increasing and recent studies in some countries such as Brazil (Oliveira et al., 1999), Spain (Ródenas et al., 2008), Greece (Nikolopoulos et al., 2007), China (Song et al., 2005) and Turkey (Tarim, et al., 2011), but without any development in Portugal. In particular, a study developed in U.S.A., concluded that exists a relation between internal radon exposure and thyroid cancer. Also, this study associated a high probability for stomach and lung cancers to radon internal exposure, both by ingestion and inhalation (Jalili-Majareshin et al., 2012). Radon inhalation in confined spaces contributes up to 20 000 lung cancer deaths per year in the U.S.A. and it is estimated that radon causes 2000 to 3000 lung cancer fatalities in the United Kingdom every year, (Appleton, 2005 in Ferreira, 2009). And in Great Britain, the risk of developing lung cancer during lifetime is 3 in 1 000 for an average radon concentration of 20 Bq/m³, increasing by 30 in 1 000 for an average concentration of 200 Bq/m³ (Metters, 1992 in Ferreira, 2009). A study carried out on the Norwegian population showed an association between multiple sclerosis and indoor radon exposure (Bolviken et al., 2003). According to Labidi et al., (2006), radon exposure in thermal spas depends mainly on the concentration of radon inhaled, local rate ventilation and exposure time.

3.3 *Supply and demand of thermal spas in Portugal*

The sector of health and well-being is relatively recent but it has been increasing significantly considering user's preferences. In 2004 the total share was about 13%, while in 2011 it increased to 42%. More than 41 thousand users were registered in 2011 representing an increase of 36%, compared with 2010. In 2011 the Spanish market represented about 49% of foreigners who opted for classic Hydrotherapy in Portugal (TP, 2011).

Due to the importance of the sector it is important to make a distinction between thermal balneal establishments and spas. The first ones are defined as units of health-care where the curative properties of a natural mineral water are used as another form of medicine to cure, prevent illnesses and preserve health.

On the other hand, thermal spas are defined as units of health-care where the properties of a natural mineral water are used for cosmetic or beauty treatments as well spiritual or psychological activities (D.L., 2004).

In Portugal, natural mineral waters have aroused a great interest mainly due to the quality, diversity and favorable effects on health. The Portuguese natural waters are one of the most important of the European waters (APIAM, 2010). There are currently 38 thermal spas in Portugal, 14 in the Northern region, 21 in the Center region and 3 in the region of Alentejo and Algarve (figure 1).

Between 2007 and 2011 the demand for these services has increased by 18%; establishments from Lisbon, Alentejo and Algarve regions were the required spas of well-being and leisure, followed by the Center and North regions (figure 2).

In 2011, this sector was composed by 910 professionals being the great majority workers (538) (Rocha, 2011) (figure 3).

3.4 *Impacts*

i) Historical impact: Large part of the thermal establishments has Roman origin, dating from the primordial Christian Age. From a historical point of view, Romans were the pioneers in promoting a ludic context to thermal spas. Thermal establishments developed a pleasure component by creating a thermal service,

Figure 1. Thermal spas in Portugal.

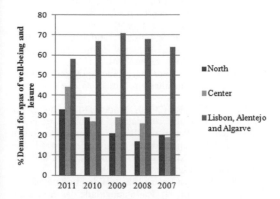

Figure 2. Demand of thermal establishments of well-being and leisure, 2007–2011.

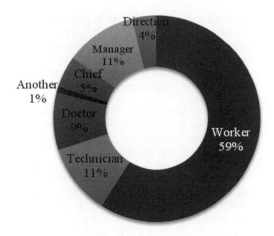

Figure 3. Staff in thermal spas, 2011.

supported by infrastructures of comfort and exclusiveness. The services are offered in buildings with a peculiar beauty and centuries of history. In Europe, it is possible to find several remarkable examples of architectural and constructive innovation buildings preserving the historical tradition (England, Switzerland and Hungary).

ii) Health impact—the exposure to ionizing radiations may cause injury to the living beings. The harmful effects result from a transfer of energy to atoms and molecules in the cellular structure. Ionizing radiation causes atoms and molecules to become ionized or excited, originating biochemists changes. There are several studies showing that the exposure to high radon concentrations is harmful for the human health (Köteles, 2007). According to the World Health Organization (WHO), radon is a sub estimated danger, although recognized as the second leading cause of lung cancer.

iii) Social impact—natural mineral waters are frequently used in thermal spas establishments for diverse purposes. Most of these waters may contain natural radionuclides dissolved, such as radon, and therefore is certain that most of the workers, as well as the public in general, will be exposed to natural radiation. This issue has been followed by an increasing social interest and concern in protecting the workers and the population from the ionizing radiations since the seventies, being the purpose of several studies and discussed in many conferences (Köteles, 2007). In this way, the European Union has established a set of recommendations and regulations to protect the workers, as well as the general public, from the adverse health effects caused by the exposure to natural ionizing radiations.

iv) Economic impact—spas have a high potential economic impact as a multifaceted project and therefore the sector have been trying to eliminate the stigma of thermal establishment's seasonality, to reach a wide range of active individuals, to guarantee the excellence in accommodation, to explore the duality therapeutic/leisure, to offer a gastronomic of excellence, to protect the interface of tourism and the multiple regional valences and to guarantee the sustainability at long-term. In addition, thermal spas contribute to regional development, create employment in its activities and stimulate local population's settings. They require investment either in requalification and modernization but they do generate economic wealth for national, regional and local economy (Frasquilho, 2007).

v) Environmental impact—ionizing radiations are omnipresent in the environment resulting either from natural and man-made sources. All human beings are exposed to natural background radiation from the ground, building materials, water, air, food, and even elements in their own bodies (Ferreira, 2009). Natural radiation varies geographically as it is related with the type of rocks and soils. Much of this variation is due to the differences in radon levels and, consequently, the exposure magnitude will depend mainly on the location, usually higher in granites zones (Sanchez et al., 2012). Thermal spas use underground waters with diversified chemical compositions, containing significant amounts of natural dissolved radionuclides. They offer quality and infrastructures, with sterilized environments, especially designed to appeal to a spirit of high well-being. Therefore, the presence of high radon concentrations in confined spaces air, such as thermal spas, represents a factor of health and environmental risk.

4 CONCLUSIONS

Radon is a radioactive gas present in different quantities in air, water and ground, with higher concentration in granite zones. Thermal spas are places with some peculiarities, where radon can accumulate resulting from the degassing of water often used in the hydrotherapy with high radon concentration dissolved (Carvalho, 2008). High radon concentrations in thermal spas environments have been extensively studied in several countries, for example in Badgestein (Austria), Tuwa (India) and Misasa (Japan) (Oliveira et al., 1999). However, in Portugal, there are no studies on thermal spas. Several studies were carried out to assess exposure to radon in homes, schools and only in one unique thermal spa (central Portugal) (Pereira et al., 2001).

In this way, the social concern in protect population from exposure to natural radiation should be permanent. From the bibliographic revision carried out through this study, we may conclude that exposure to radon may be harmful for the human health, depending on the dose and the duration of exposure. According to the World Health Organization (WHO) radon is an underestimated health hazard although being the second leading cause of lung cancer. Moreover, this pathology has a period of latency from 5 to 50 years; therefore it is necessary to take preventive actions in order to reduce or eliminate, if possible, the exposure.

Several studies revealed that radon exposure is highly influenced by ventilation rate as well as by the duration and frequency of the exposure. In this way, it is necessary to monitor radon concentration in thermal spas workplaces to protect the workers, as well as the common users, with appropriate equipment for radon detection, in the water and air. This should result in the implementation of a radiological protection program, specific for thermal spas.

REFERENCES

Al Zoughool, M., Krewski, D. (2009). Health effects of radon: A review of the literature. International journal of radiation biology, vol. 85, no. 1, p. 57.

APIAM (2010). Águas Minerais Naturais e Águas de Nascente, Livro Branco, APIAM, Associação Portuguesa dos Industriais de Águas Minerais Naturais e de Nascente.

Beitollahi, M., Ghiassi-Nejad, M., Esmaeli, A. and Dunker, R. (2007). Radiological studies in the hot spring region of mahallat, central Iran. Radiation protection dosimetry. Vol. 123, no. 4, p. 505–508.

Bølviken, B., Celius E.G., Nilsen R. and Strand T. (2003). Radon: a possible risk factor in multiple sclerosis. Neuroepidemiology, 22(1):87–94.

Carvalho, F. (2008). Radioactividade de Origem Natural e Radiações Ionizantes em Industrias não-nucleares. Atas do ITN.

Correia, R.M.P.G.R. (2010). Modelação da dispersão da radiação gama correlacionada com a exalação do radão na Península Ibérica, dissertação para a obtenção do grau de Mestre no curso de Mestrado em Engenharia de Minas e Geo-Ambiente, Faculdade de Engenharia da Universidade do Porto.

D.L. (2004), Decreto-Lei n.º 142/2004 de 11 de junho.

Ferreira, A.M.S. (2009). Radioactividade das Águas da Região Subterrâneas do Minho. Tese de Mestrado, Universidade do Minho. Braga, 158p.

Frasquilho, M. (2007). SPA Termal – Oportunidades de investimento e negócio. Espírito Santo Research Sectorial.

Gnoni, G., Czerniczyniec, M., Canoba, A. and Palacios, M. (2008). Natural radionuclide activity concentrations in spas of Argentina. AIP conference proceedings, vol. 1034, p. 242–245.

Gray, A., Read, S., Mcgale, P. and Darby, S. (2009). Lung cancer deaths from indoor radon and the cost effectiveness and potential of policies to reduce them. Bmj, vol. 338, no. jan06 1, p. a3110-a3110.

Jalili-Majareshin, A., Behtash, A. and Rezaei-Ochbelagh, D. (2012). Radon concentration in hot springs of the touristic city of Sarein and methods to reduce radon in water. Radiation Physics and Chemistry. Volume 81, Issue 7, p. 749–757.

Köteles, G.J. (2007). Radon Risk in Spas? vol. 5, no. 41, p. 1.

Labidi, S., Essafi, F. and Mahjoubi, H. (2006). Estimation of the radiological risk related to the presence of radon 222 in a hydrotherapy centre in Tunisia. Journal of Radiological Protection, vol. 26, no. 3, p. 309.

Martín Sánchez, A., de la Torre Pérez, J., Ruano Sánchez, A.B., & Naranjo Correa, F.L. (2012). Radon in workplaces in Extremadura (Spain). Journal of Environmental Radioactivity, 107(0), 86–91. doi: 10.1016/j.jenvrad.2012.01.009.

Nikolopoulos, D. and Vogiannis, E. (2007). Modelling radon progeny concentration variations in thermal spas. Science of the total environment, vol. 373, no. 1, p. 82–93.

Oliveira, J., Mazzilli, B.P., Nisti, M.B. e Sampa, M.H.O. (1999). Radionuclídeos Naturais em Águas Minerais e Lama Sulfurosa Utilizadas em Terapia Termal no Brasil. Instituto de Pesquisas Energéticas e Nucleares – IPEN-CNEN/SP, São Paulo, Brasil.

Pereira, A.J.S.C., Dias, J.M.M., Neves, L.J.P.S. e Godinho, M.M. (2001). O Gás Radão em Águas Minerais Naturais: Avaliação do Risco de Radiação no Balneário das Caldas de Felgueira (Portugal Central). Memórias e Notícias, Publicações do Departamento de Ciências da Terra e do Museu Mineralógico e Geológico da Universidade de Coimbra, n.º 1, Coimbra.

Rocha, A.S.S. (2011). Análise à Oferta Termal Nacional, dissertação apresentada na Faculdade de Economia da Universidade do Porto para obtenção do grau de mestre em Gestão e Economia dos Serviços de Saúde, Porto.

Ródenas, C., Gómez, J., Soto J. and Maraver, F. (2008). Natural radioactivity of spring water used as spas in Spain. Journal of Radioanalytical and Nuclear Chemistry, vol. 277, no. 3, p. 625–630.

Song, G., Zhang, B., Wang, X., Gong, J., Chan, D., Bernett, J. and Lee, S.C. (2005): Indoor radon levels in selected hot spring hotels in Guangdong, China. Science of The Total Environment, p. 63–70.

Tarim, A.U., Gurler, O., Akkaya, G., Kilic, N., Yalcin, S., Kaynak, G. & Gundogdu, O. (2011). Evaluation of radon concentration in well and tap waters in Bursa, Turkey. Radiat Prot Dosimetry, 150(2), 207–212. doi: 10.1093/rpd/ncr394.

TP, (2011). Termas em Portugal, Lisboa.

Occupational Safety and Hygiene – Arezes et al. (eds)
© 2013 Taylor & Francis Group, London, ISBN 978-1-138-00047-6

Wearable monitoring system for locomotion rehabilitation

A. Catarino, A.M. Rocha & M.J. Abreu
Dep. of Textile Engineering, Universidade do Minho, Portugal

J.M. da Silva, J.C. Ferreira, V.G. Tavares & M.V. Correia
FEUP, Rua Roberto Frias, Universidade do Porto, Portugal

A. Zambrano, F. Derogarian & R. Dias
INESC TEC (formely INESC Porto), Portugal

ABSTRACT: Human motion capture systems are used by medical staff for detecting and identifying mobility impairments, early stages of certain pathologies and can also be used for evaluation of the effectiveness of surgical or rehabilitation intervention. Other applications may involve athlete's performance, occupational safety, among others. Presently there is a considerable number of solutions available, however these systems present some drawbacks, as they are often expensive, considerably complex, difficult to wear and use in a daily basis, and very uncomfortable for the patient. With the purpose of solving the above mentioned problems, a new wearable locomotion data capture system for gait analysis is under development. This system will allow the measurement of several locomotion-related parameters in a practical and non-invasive way, comfortable to the user, which will also be reusable that can be used by patients from light to severe impairments or disabilities. The present paper gives an overview of the research that is being developed, regarding the design of the wearable equipment, textile support, and communications.

1 INTRODUCTION

Different techniques have been used in the recent past to capture and analyze locomotion, in order to characterize and improve body postures, and detect early stages of pathologies that cause mobility difficulties (Z. Yuting et al., 2011). Probably the most popular, the technology based on Vision makes use of cameras to capture the spatial location of special identifiers that are attached to the lower limbs. There exist other approaches, such as recording kinematic variables by using accelerometers and gyroscopes fastened to body segments. Electromyography is also used to monitor the muscle activity in the lower limbs. These systems, when combined may give very important information to the professionals. However the present solutions have a major drawback: These systems are expensive and complex, difficult to apply by healthcare staff, difficult to use and uncomfortable for the patient.

2 PROLIMB PROJECT

2.1 General description

With the purpose of simplifying the assembling work and the comfort of the user, a project named *ProLimb* is under development with the main purpose of developing a new proposal of an autonomous, real time monitoring wearable body sensor network for human locomotion data capture. Other objectives involve the development of processing tools for gait analysis in a friendly interface that will be used by health professionals, and the study of the capabilities of transporting digital signals throughout data tracks made with conductive textile yarns.

The proposed system is intended to be dressed by the patient itself or with aid if needed. This is accomplished by means of an instrumented legging, capable of acquiring several human locomotion parameters in a non-invasive way, even for people with strong impairments or disabilities. The system includes the capture of the following parameters:

- Spatial position of the lower limbs, speed and accelerations, by using 3D accelerometers and gyroscopes;
- Electromyographic signals of specific muscles located in the lower limbs.

The system will be conceived to be able to work for long time periods, thus involving typical movement activities under everyday living conditions. This will allow the health professionals to better understand how a patient walks and if the rehabilitation program is being properly followed.

Regarding the inertial sensors, the research team decided to use one inertial module in each segment of the lower limbs, which means that two modules will be placed in each leg.

The surface electromyography will have an important role since it will allow understanding how much and how long the muscles will be called into action. Fatigue will be with no doubt a parameter under study. Since there are several muscles directly involved in locomotion only four muscles were identified at this stage to be followed by sEMG.

The muscles identified as important for the analysis are *quadriceps femoris*, *biceps femoris*, *tibialis anterior* and *gastrocnemius medialis*. The acquisition of several signals simultaneously may reveal activation patterns for different motor actions, such as stepping, walking, climbing stairs or even sitting down, and can be combined and correlated with the kinematic data to more easily expose movement abnormalities, e. g. hemiplegic, Parkinson disease, and cerebral palsy (L. Iezonni et al., 2005). The system can also be used with a preventive purpose, whenever an occupation may require a careful monitoring of the lower limbs.

2.2 Body sensor network

The system uses as support a legging with elastic properties, which allows the correct positioning of the sensors and electronics. This legging is equipped with different sensors that involve the measurement of kinematic quantities, and surface electromyography (sEMG) of several muscles. The sensors communicate by means of sensor nodes (SN) to the central processing module (CPM), which on its turn sends the information by wireless communication, in this case with Bluetooth®. The analogue signals measured with the sensors are immediately converted to digital signals in the sensor nodes (SN), in order to reduce as much as possible the presence of artifacts. The central processing module (CPM) is placed on the waist, while the sensor nodes are distributed on the garment. These sensor nodes share paths between them with the purpose of having alternative paths to send the data collected by the sensors, or fault tolerance capability. Using this approach, the system can select the most favourable path or have alternatives in case of damage on one of the paths. It is important to note that the sensor nodes (SN) do not communicate wirelessly with the central processing module (CPM), rather than communicate through data lines, in this case textile conductive wires that build these paths. The prototype sensor network under development comprises one CPM and eight SN, although capable to be extended to 255 SNs.

It was decided to integrate each sEMG and inertial signals associated with the same limb segment in one SN for a more accurate time alignment. The CPM gathers information from all the nodes, performs some local processing, and sends aggregated data, immediately or later, via a wireless link to a personal computer for further processing.

The communication among SN is performed over a single signal line. Also, since wireless-based systems are prone to be affected by interferences, in order to improve data communication reliability, the CPM module is also equipped with a USB port and a MicroSD card to save and transfer data whenever the wireless communication fails.

Energy efficiency and integration of systems are considered fundamental milestones for the proposed body-area network. The system should work for long periods of time, especially during

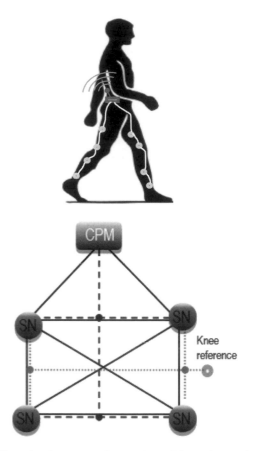

Figure 1.   Interconnection topology of the mesh sensor's network. Solid lines represent data paths, dashed lines ground interconnections, and dotted lines the common-mode reference.

prolonged monitoring. Thus, a reactive, energy-efficient routing protocol, described in (F. Derogarian et al., 2011), was developed and adopted for the network data layer. This protocol does not require each node to possess global information about the network, but still ensures that all data communication uses minimum cost paths. It also handles link and node failures gracefully. Simulation results show that this protocol provides better performance than the standard minimum-cost forwarding protocol samples. Figure 2 illustrates two Sensor Nodes (SN).

2.3 Wearable garment

A wearable garment meant to be comfortable implies an adequate combination of materials, compression effect and preferably with electronic components incorporated in the textile and interconnected with data and power tracks, if possible made with conductive yarns embedded in the fabrics. This solution would allow an easy to dress piece of garment, reusable, and regular methods for cleaning and maintenance. Regarding the sensors to be embedded in the E-legging, the biopotential parameters are the ones that are more successful in terms of reliability. For that reason sEMG electrodes were embedded in the knitted fabric, making use of the technology available on weft knitting. Figure 3 presents the first version of this e-legging.

Although the leggings were conceived for an adult, this solution can easily be adapted to other ages and target publics, by resizing the textile garment. The electrodes would remain with the same size and an anthropometric study should be necessary. It is also important to mention that the e-legging can be used by a broad range of medium sized adults due to the elastic capabilities obtained by using weft knitted fabrics together with a specific structure and elastane.

2.4 Textile sensors

As mentioned above, previous research made by the team has shown that it is possible to measure electric potentials using conductive fibers or yarns instead of conventional electrodes, both as dry as well as wet electrodes (Barros et al., 2011). In order to successfully produce fabrics with conductive yarns, and particularly to build textile sensors, namely sEMG electrodes, two types presenting relatively good electrical properties have been used: A) spun yarns with a mixture of polyester, and stainless steel fibres with linear resistances of 350 Ohm/m; B) and yarns made with twisted filaments, each one a polymeric filament covered with silver with linear resistances of about 30–40 Ohm/m. Tests made with conventional electrodes and textile based electrodes produced according to SENIAM recommendations in simultaneous measurements revealed an excellent correlation between them. Figure 4 illustrates the sEMG

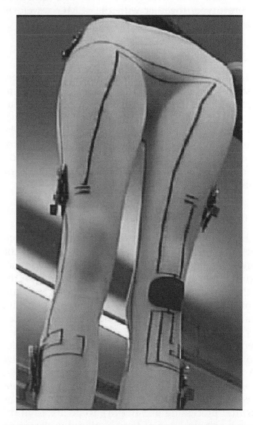

Figure 3. First version of the legging with the sEMG electrodes and inertial modules, corresponding sensor nodes and data/power tracks made in the textile garment.

Figure 2. Two sensor node prototypes connected to each other by regular electric cable.

Figure 4. sEMG electrodes placed on *gastrocnemius medialis* muscle (top image) and sEMG waveform obtained with textile based electrodes (bottom image).

electrodes made with type B yarns and knitted in a prototype, together with the sEMG waveform obtained using these electrodes.

2.5 *Conventional sensors*

Micro accelerometers and micro gyroscopes were used to develop an inertial module. These micro sensors were assembled in the Sensor Nodes due to its complexity regarding the number of connections and electronics needed to properly work. The sensors are directly connected to a 16-bit microcontroller – PIC24. Figure illustrates the accelerometer waveform captured during the proof of concept of the Sensor Nodes.

2.6 *Communications through the legging*

The use of conductive textile yarns in the interconnections raises two issues. On the one hand, the higher impedance (comparing to copper conductors) limits communication frequency and, on the other hand, requires that testing facilities are available to evaluate whether the frequency bandwidth has degraded. The conductive yarn can be modeled as a series RL impedance. This electrical behavior requires using several yarns in parallel in order to obtain a suitable communication frequency. A mesh like network is being used for the interconnected sensor nodes. The worst case interconnection impedance is that of the path between the central processing unit (CPM), to be placed in the user's belt, and one of the SN placed in the shank (Figure 1). Considering a voltage defined signal, it was found that each track should be made with a minimum of 4 yarns to ensure a transmission

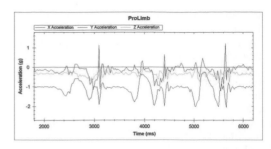

Figure 5. 3D accelerometer waveform during an experiment involving three consecutive steps.

Figure 6. Detail of a sEMG electrode and interconnection line, both embedded in the knitted fabric.

frequency of 10 MHz with rise and falling times shorter than 12 ns. Nevertheless, tracks with six yarns were adopted. Figure 6 illustrates one electrode similar to the eight pairs of electrodes knitted in the legging on figure 3 and its corresponding path of six interlaced lines or courses made with electrically conductive textile yarns.

3 CONCLUSIONS

This paper presented a research that is being developed with the objective of proposing an e-legging for monitoring lower limb movements and help technicians to assess the severity of diseases or accidents on human locomotion. Several issues were presented like the body sensor network, which will be built using conductive yarn, the electrodes embedded in the knitted fabric, as well as the communication and energy path made with textile conductive yarn. Generally, from the results obtained until now, one can say that it is possible to transmit signal at the adequate time rate through textile conductive paths and measure muscle activity with textile based sensors. Several improvements are currently under development in order to achieve the most flexible and reliable system.

ACKNOWLEDGMENTS

The authors would like to thank FCT, which is supporting the project through PTDC/EEA-ELC/103683/2008.

REFERENCES

Barros, L., M.J. Dias, H. Carvalho, A.P. Catarino, 2011. Aquisição de sinais electromiográficos recorrendo a eléctrodos em substratos têxteis, *4° Congresso Nacional de Biomecânica*, Coimbra.

Derogarian, F., J.C. Ferreira and V.M. Grade Tavares, 2011. A Routing Protocol for WSN Based on the Implementation of Source Routing for Minimum Cost Forwarding Method, *Fifth International Conference on Sensor Technologies and Applications*, Jens M. Hovem, et al. (eds), 85–90.

Iezzoni, L.I., B. O'Day, 2005. More than Ramps: *A Guide to Improving Health Care Quality and Access for People With Disabilities*. Oxford University Press, Oxford.

Yuting, Z., S. Markovic, I. Sapir, R.C. Wagenaar, T.D. C. Little, 2011. Continuous functional activity monitoring based on wearable tri-axial accelerometer and gyroscope, *5th International Conference on Pervasive Computing Technologies for Healthcare (Pervasive-Health)*, 370 – 373.

Zambrano, A., J.M. da Silva, 2012. Signal Integrity and Interconnections Test on Technical Fabrics, *18th IEEE Int. Mixed Signal, Sensors and Systems Testing Workshop*, 2012.

Human and organisational factors

Occupational Safety and Hygiene – Arezes et al. (eds)
© *2013 Taylor & Francis Group, London, ISBN 978-1-138-00047-6*

Good practice guide of blind and low vision people integration in workplaces—a proposal

Andreia Costa, Miguel T. Diogo & Aura Rua
Faculdade de Engenharia da Universidade do Porto, Portugal

ABSTRACT: Disabilities, and in this case a visual impairment must not be an impediment to the access to a job and to work. The creation of a Good Practices Guide capable of helping the integration of blind people and with low vision in the workplace revealed to be complementary and fundamental, as the job constitutes preponderant factor for the social and economic independent and autonomy of every citizen and the health and safety at work a fundamental condition for integration.

1 INTRODUCTION

The present article intends to contribute for a larger recognition of the visually impaired when fully integrating them in the workplace.

All around the world there are about a thousand million people with a disability; 785 and 975 million being part of the labor force (ILO, 2012).

Overall approximately 314 million people are visually impaired, of whom about 45 million are blind (WHO, 2010).

Visual information is responsible for a considerable part of apprehending reality. The vision is thus a special sense since it is responsible for transmitting detailed information about objects at great distances. The loss or lack of vision is partially compensated by the action of the other senses, used to explain the use of behavioral alternatives for the visually impaired such as reading Braille (Collignon, et al., 2009).

The classification of visual impairment currently used worldwide is based on the first and second edition of the International Classification of Diseases CID-10, having resulted of a World Health Organization (WHO) study of 1972, on preventing blindness, in order to establish a standardized and comparable definition of visual impairment and blindness.

In Portugal working people with disabilities and over 15 years old, which amounts to 601,583 people, recorded in 2001, a reduced rate of activity (29%, versus almost 50% at the national level) and a high unemployment rate (9.5%, for an unemployment rate of 4.3%), visual impairment representing 9.94% of the total working population with disabilities (INE, 2002). The importance of people with visual impairment, and in particular of the working population carrying this same deficiency

worldwide and at the national level justify therefore the need to develop new instruments that contribute to their integration in the employment context and safeguarding the conditions of safety, hygiene and health in all aspects related to the work.

The disability condition cannot, therefore, in any way, be an excuse for the non-inclusion of people with disabilities, since a workplace accessible and safe for these people is a greater assurance for the other workers (European Agency for Safety and Health at Work, 2004).

This perspective is also an expression of the constitutional principle of equal treatment.

2 MATERIALS AND METHODS

As the employment rates of disabled people in Portugal, particularly the visually impaired, are significantly lower compared to the rest of the population, the need to develop a reference document to support employers arised, in order to integrate blind and low vision people in the workplace— Good Practices Guide on the Integration of the Blind and Low Vision People in the Workplace.

Indeed, it has been proven necessary the creation of observation instruments of observation in order to know the people with visual impairment within the social relations and information at the Occupational Safety and Health (OSH) level as well as the comparison of results with individuals without disabilities. Such instruments are shown in Table 1.

In the actions to raise awareness basic concepts of OSH have been addressed such as hazard and risk, occupational accidents, occupational diseases, prevention management and fire safety, including the definition of fire, the recognition of emergency

Table 1. Instruments used in the study.

Instrument	Goal
Awareness actions (two) to two groups of visually impaired people ($n_t = 20$)	Sensitize groups of visually impaired for fundamental OSH concepts and for procedures and practices of prevention and evacuation
Questionnaire I (6 questions) applied in actions to raise awareness ($n_t = 20$)	Knowing the type of barriers faced by blind and low vision people in society (discriminatory attitudes) and in the access to labor and employment
Questionnaire II (5 questions) applied in actions to raise awareness ($n_t = 20$)	Estimating the degree of knowledge (specifically concepts of risk, preventive attitudes, response to emergencies)
Questionnaire II (11 questions) applied to two groups of sighted ($n_t = 83$)	To compare the responses of the sighted and the visually impaired and assess differences in perception
Simulator of emergency orientation) applied in actions to raise awareness ($n_t = 6$)	To know the capacity of space perception through a model made with prior consultation of the first group of individuals with visual impairment and provide the organization with an emergency plant
Script of orientation of visitors to workplaces applied to individuals with visual impairment in different employers ($n_t = 4$)	To understand on a real life context the integration of blind and visually impaired people on the workplace

situations, as well as strategies and emergency response procedures.

In the course of these actions was also possible to carry out evacuation exercises.

This way procedures and practices were identified with the groups used in evacuation situations including the definition of the Meeting Point and the person in charge of the evacuation process.

The work tool, questionnaire I had as main objective to identify the nature of architectural and social barriers faced by the visually impaired as well as the type of duties performed. Its development was based on the requirements defined on the legal framework of prevention, rehabilitation and participation of people with disabilities and on the legal regime in promoting occupational safety and health (LRPOSH).

The questionnaire II, also developed based on the legal requirements of LRPOSH, was designed with the purpose of estimating the degree of knowledge of the fundamental concepts of OSH

in particular preventive attitudes and response to emergency situations addressed in actions to raise awareness.

Other tool used was the questionnaire III applied to two groups of sighted people. With an approach similar to the questionnaires made to the two groups of the visually impaired aimed, in particular, to know its view regarding the social and architectural barriers to individuals without disabilities as well as perceptions of risk and responses to emergency situations. Obtaining these data was designed to allow comparison with the answers given by blind and low vision people. It will be important to note, that when applying the questionnaire to the groups of the sighted in question it wasn't said that this questionnaire had already been presented to a group of visually impaired. This deliberated omission had the intention of not conditioning the responses of the group.

The emergency orientation simulator that resulted from the first action carried out sensitization was built from a pre-existing model, used in the training center for the spatial orientation of visually impaired people. Its construction was aimed to know the capacity of perception of space of the training center by the blind and low vision people. Apart from the above the organization was also offered an instrument comparable to an emergency plan that meets the needs of individuals with visual impairments. Alongside the alternative symbology general safety instructions in Braille were also prepared (by an individual with a disability) so that people with visual impairment could effectively apprehend those instructions.

In order to provide guidelines for the observation of the workplaces the Guidelines for the orientation of visitors to workplaces were based on legal requirements as well as on the empirical knowledge acquired in the social relations established during the actions to raise awareness. Its structure covers six different dimensions: social, mobility, accessibility, adaptation of the workplace, health and safety and time and work organization.

3 RESULTS AND DISCUSSION

The observation instruments used in the fieldwork enabled the gathering of information relevant to the preparation of a proposal for a Good Practices Guide on the Integration of Low Vision and Blind People in workplaces directed to employers.

Indeed, it is intended that it will contribute to the effective integration of visually impaired workers constituting its operation an aspect of particular importance, as seen in Figure 1.

Thus, the reasons for the construction of the proposed guide are based, initially, on their relevance, given the high unemployment rates of the population visually impaired, 9.5%, in contrast to those without disabilities, 4,3% (INE, 2002).

The election of the employer as a recipiente was a result of several vectors:

- collection of existing legislation, to the level of participation of people with disabilities, of the rights to access to employment, equal treatment in the workplace and promoting safety and health at work, arising for employers responsibilities at various levels.
- be the actor who ultimately makes the decision of hiring also assuming the different legal responsibilities in the integration process, at the level of the public and private sectors.
- benefits of hiring private admissions and integration processes, behavioral aspects, technical aids and finally forms as compliance with legal requirements relating to OHS can contribute to effective integration.
- evidence indicated in the literature review, in particular the numerous types of barriers to the integration of people with visual impairments especially the social ones (Naraine, et al., 2011) and the fears of employers of increased costs resulting from adaptations, insurance premiums, extra time with training and complaints for breach of law, constraints that can be overcome by information and training of the employers (Kaye, et al., 2011).
- be a determining factor in the distribution of income, in safeguarding the health and safety at work and in their quality of life at work, stating that is the understanding of the various authors (Kaye, et al., 2011) (GRACE, 2005) that no action at this level will become effective if there is no firm commitment of top management. The awareness of the employer will allow him to embrace more easily measures to raise awareness for all levels of the organization, that the

integration process is properly monitored and, finally, that the Workers with Visual Impaired (WVI) is privy to all processes related to the promotion of safety and health.

However, this proposal will only produce the desired results if the approach to the target population (the employers) is made according to a process properly weighted and with the interaction of the several actors involved in the integration process, presenting the diffusion project schematically in Figure 2.

Thus, the answer to the question on the mode of dissemination, shown in Image 1, is based on the following:

- article 21. Decree-Law n. 290/2009 of October 12th, when referring that the specific objectives that support the work placing are focused on promoting the labor market integration of people with disabilities, enrolled in employment centers, through a process of mediation between them and the employers, while equating the aspects of accessibility, the adaptation of the workplace, the development of general employability skills, as well as sensitizing those entities to the advantages of hiring this public, and supporting the recipient in active search of employment and in self-employment.
- the assumption that the integration process is initiated by contact between the integration support services and the employer and that those services offer tools that enable the effective integration of WVI, in this case the Good Practices Guide, which despite its overall character, it integrates key issues to consider when integrating any WVI in the workplace.

Figure 2. Interaction of the various actors involved in the process of integration.

What is its usefulness?

Presentation format?

Why addressed to the employer?

Method of disclosure?

Figure 1. Operationalization of the proposal.

- the collaboration of the integration services that can be a facilitator through acceptance of that document by the employer and the enjoyment of their gains, although this process is not, nor has to eventually be the only one.
- consulting the WVI by the employer in order to know their opinion on the adequacy of the contents of the Good Practices Guide to their condition (in the case of written communication) since each individual is unique and of course every job has its specific characteristics (technical aids).

Finally, and in response to the question "Presentation format?" It appears that the dissemination media can be of various kinds: paper or digital format, as desired.

4 CONCLUSIONS

The literature review conducted for this study has revealed that people with disabilities have to overcome many obstacles when trying to acquire or keep a job (Stone, et al., 1996) reflecting unemployment rates that are significantly higher that the remaining population (Crudden, et al., 1999).

It was also noted that the integration services and the employers can play a key role in the insertion and maintenance of these employees to harness their individual skills, as with the necessary adjustments;, physical, technological and social, they can get a performance equal to their sighted peers (Peck, et al., 2001).

Thus, based on legal, scientific and technical research it was designed a set of essential tools for the preparation of a proposal for a good practices guide on the integration of blind and low vision in the workplace within the fundamental scope of safety and health at work.

Applying those instruments one concludes that visual impairment is not a constraint:

- to the understanding of the issues on prevention or emergency when provided the necessary training to individuals in matters of safety and health at work - *Actions of awareness and Questionnaire II*;

- to the level of autonomy required in a situation of evacuation in a familiar space—Practical evacuation exercises conducted in the Actions to raise awareness;
- for perception of spatial reading—design and consultation of the *Emergency orientation Simulator.*

Aimed at the employers, it is intended that the proposal for the Good Practices Guide constitutes a useful tool in the process of professional and social integration, and job retention of the WVI since while top management they are the most accountable in changing existing cultures, instilling diversity values.

Indeed, an individual with a disability should receive equal treatment in all aspects related to work (admission, integration and maintenance of the job) subject to the necessary adaptations.

REFERENCES

Collignon, O., et al. 2009. Cross-modal plasticity for the spatial processing of sounds. Canada: Springer-Verlag, 2009.

European Agency for Safety and Health at Work. 2004. Priority Groups. People with disabilities. ISSN 1681-2166

GRACE-associação portuguesa sem fins lucrativos dedicado à temática da Responsabilidade Social Empresarial (2005). *A Integração de Pessoas com Deficiência nas Empresas: Como Actuar.*

ILO. 2012. Inclusion of persons with disabilities.

INE. 2002. CENSOS 2001. s.l.: INSTITUTO NACIONAL DE ESTATÍSTICA-PORTUGAL, 2002.

Kaye, H. Stephen, Jans, Lita H. e Jones, Erica C. 2011. Why Don't Employers Hire and Retain Workers with Disabilities?.

Naraine, Mala D., et al. 2011. Social inclusion of employees who are blind or low vision. ISSN 1360-0508.

Peck, Bob e Kirkbride, Lynn Trew. 2001. Why businesses don't employ people with disabilities. ISSN 1052-2263.

WHO. 2010. Prevention of Blindness and Visual Impairment. New estimates of visual impairment and blindness: 2010 ISBN 978 92 4 150017 3.

Occupational Safety and Hygiene – Arezes et al. (eds)
© *2013 Taylor & Francis Group, London, ISBN 978-1-138-00047-6*

Human organizational factors and occupational health: A study in a public service entity

M.F.F.M. Catão & M.B.F.V. Melo
Federal University of Paraíba, João Pessoa, Paraíba, Brazil

ABSTRACT: It is in the interlacing of the structure and dynamics of organizations in a certain social context that the human organizational factors dialogue, granting a central place to labor management in the context of the problems of occupational health. This study aims at presenting some outlines of a consulting activity in research on the development of the organization and occupational health in a public entity in northeast Brazil. The accomplished research will include 18 individuals aged between 26 and 45 years, with responsibility in policing the street. We applied two questionnaires: an open one with bio-demographic identification and a closed one aiming at capturing the management system. The study allowed to map the organization as regards the real and the desired management systems and the occupational health in the context of these systems. As initial intervention, we indicated the identification of personal life projects and possible articulations of the life project of the organization.

1 INTRODUCTION

The analysis of the human organizational factors—occupational health, in today's time, requires the knowledge of the predominant labor organization systems in the 20th and 21st centuries (Antunes, 1995; Catão, 1994). It is in the interlacing of the structure and the dynamics of the organization in a certain social environment that these factors sort of dialogue, while granting a central room to the labor organization in the context of the occupational health problems. This research study aims at the analysis of the labor organization system and the configuration occupational health/disease (Dejour, 1986, 1992; Braverman, 1977; Chanlat, 1995; Antunes, 1995).

Labor management system is understood as the set of practices that a company directorship chooses to meet the targets it has established. The management system comprises the conceptions that inspire the establishment with the conditions, the labor relationships and respective practices: the configuration of the hierarchy, the leadership system, communication, motivation, fixing targets, decision making, interaction, evaluation and control of the results. The management system can be a key-factor that is subjacent to the occupational helath/disease (Chanlat, 1995; Davel & Vergara, 2010; Valsiver, 2012, Catão, 1994). A labor organization adapted to the workers at mental and physical levels is determinant to occupational health, mainly is such confirmation is corroborated with studies published in the fields of the industrial

catastrophes, professional stress, labor or ergonomic psychopathology (Chanlat, 1995; Dejour, 1992). Four labor management systems are identified in the 20th and 21st centuries: the Taylorist System, the Humanist System, the Systemic or Open System and the Critical-Analytical System.

The Taylorist System is characterized by a very fragmented labor division and a formal, centralized and authoritarian organization. This system refers to the labor process reorganization wherever it is, though its root and privileged place may have been the internal conflict that existed in the US plants by the end of the 19th century. The same conflict over passed the plants to reach the social institutions in a general way, there included universities, schools, hospitals, associations, public service, unions, by modeling these establishments as well as their professionals in accordance with its foundations and disciplinary principles, warranty of the laborer's time control throughout the notion of "useful time". The formal organization is characterized by the promotion of the separation between labor planning and its execution in the organizations, with one group (patrons, managers) being in charge of planning and the others (laborers) in charge of the execution. According to this system, the human being is conceived as a person that is endowed with physical energy and is solely moved by needs of economic order, the notion of "Homo economicus".

The Humanist System is characterized by the conception of the Homo social. Under this perspective, the human being is seen as someone in need of safety, affection, social approval, prestige and self-accomplishment. This system criticizes the Homo economicus conception as the understanding of the human nature and, to replace it, suggests the Homo social, a being whose behavior cannot be reduced to a simple, *mechanicist* scheme, subject to the social system and the biological order at the same time. Its characteristic is to almost concentrate exclusively on the informal aspects of the organization (informal groups, laborers' social behavior, beliefs, attitudes, expectations etc.), i. e., the set of social relations not foreseen in regulations and organograms, being characterized by its spontaneous and extra-official aspect.

The Systemic or Open System is characterized by the understanding of the organizational totality and by the participation in all levels of the organization. In this system, the relationship laborer/company is not understood in an isolate way. The laborer's personal needs and the organizational needs are interdependent (Catão, 1994; Catão & Sawaia, 2007; Vigostki, 2004). Further to this approach, we incorporate the studies of the Organizational Development theory, the approaches of life quality, participation, strategic planning and recognition of the human capital among the factors that contribute to the organizational efficiency, so confirming the expression referred in organizational literature: "more than an arm, the human being is also a heart and a head".

The Critical-Analytical System is characterized by unveiling the institutions and recuperating the power of the organization and self-management of the labor process. It has the history of the human society and the contradictory reality as its principle, not proposing to consider the models of labor organization, configured through the centuries, as something natural and everlasting. This system promotes the participation in all levels of the establishments, within which the human being grows productively, politically, socially and culturally. This is a way of building up a self-managing society inspired in a mutualism of interests. The self-managing idea is inscribed like the axis of the labor organization of this system. Equality, liberty, autonomy, creativity, solidarity and mutuality, key-expressions of self-management, practically translate the classic principle "a human being, a vote", applied to all measurements of social life, be it in the economic scene, be it in the cultural or the political one. The conception of the human nature of this system gets identified when it places the human being as the agent of his own history.

In the Critical-Analytical System, the individual alienation, in relation to his work and himself as regards the others, is a problem to be overcome in the social relations and the relations of production. Actually, the alienation is a direct consequence of the capitalist division of labor and, to say so, a fundamental evil of the industrial revolution against which participation is the antidote.

Following this theoretical direction and in reply to the request of a public institution of the northeast of Brazil, who asked the Research Nucleus of Intervention in Psychosocial Orientation of the life job project of the Federal University of Paraiba (UFPB) to accomplish the research on growth of the organization and occupational health, we elaborated this manuscript whose target is to present some outlines of the respective carried-out consultation.

2 MATERIALS AND METHODS

Approved by the Committee of Ethics in Research of the Lauro Wanderley University Hospital, protocol 068, the research was accomplished at an institution of public safety in Paraiba State.

The researched institution has 189 servants. Here we will present an outline of the research on 18 policemen who render external public safety services in a community of João Pessoa city/PB, whose age varies from 26 to 45 years, 13 of them being male and 5 female, who have been serving for between 5 and 23 years.

In Brazil, the military police is responsible for the urban patrolling, being in daily contact with the population. So, its action is not only linked with the occurrences of contravention and crimes, but also with the most varied types of mediation and control of social conflicts in an ostensive way, there including manifestations, public protests, strikes and even quarrels among neighbors and family members.

Two instruments were used toward the accomplishment of this research: an open questionnaire and a closed one.

The open questionnaire was applied for biodemographic identification, with questions such as schooling, age, years of service, activity, health, life project. We tried to look for the really experienced situation, as well as for the situation desired by the interviewed individuals.

Adapted to Likert & Likert (1978), the closed questionnaire aimed at the analysis of the labor management system of the studied entity. The referred questionnaire shows a classification of the 04 types of management systems: authoritarian and rigid (typea 1), benevolent authoritarian (type 2), participative (type 3) and analytical participative (type 4). Such systems were built under the theoretical configuration of the four labor management systems: Taylorist, Humanist, Systemic or Open and Critical-Analytical.

The referred questionnaire was applied so as to map the real and the desired profiles of the labor organization, focusing the human and organizational factors: Leadership, motivation, communication, interaction, decision, objective and control.

The procedure used to analyze the collected material was the technique of content analysis proposed by Bardin (1977).

3 RESULTS AND DISCUSSION

With the application of the biodemographic Identification questionnaire as regards the really experienced situation, it was observed: as for the activity accomplished by the interviewed individual—a sort of operational bureaucratic service, with an ostensive, preventive and patrolling control of criminality; as for the occupational health—occurrences of diseases related with the labor activity and the unsatisfactory motivational climate; as for the life project—few actions turned to personal growth and distancing between personal perspectives and the entity organizational planning.

As regards the situation desired by the interviewed individuals, with the application of the biodemographic Identification questionnaire it was verified that: as for the activity, the type of operational and bureaucratic service should be distinguished with the promotion of citizenship with the organization's internal and external preventive actions; as for the occupational health, better conditions and labor relations should be made available in order to reduce the frequency of accidents and occupational diseases; as for the life project, growth and human expansion programs should be implanted as well as organizational planning, in a participative way, with actions to articulate the entity life project and that of the servants.

With the application of the Likert & Likert type of questionnaire it was observed that, in real level, the prevalence of the benevolent authoritarian system of labor organization (type 2) in the following areas: leadership, motivation, communication, interaction, decision and control. On the other hand, the area called establishment of targets showed the authoritarian and rigid system (type 1). The systems 1 and 2, characterized as authoritarians due to the procedures in the conduction of the labor relations, are supported by the conceptions and practices of the Taylorist model that became notable by a very fragmented division of labor. Repetitive tasks: the repetitive aspect of the task, time pressure, the painful physical and mental load, the absence of autonomy, the human reductionism are main aspect responsible for the diseases and the accelerated aging (Chanlat, 1995).

In the desired level, it was observed the absolute choice in all areas for the participative analytical system (type 4), which refers to the labor management systems Open or Systemic and Critical-Analytical, characterized by the participation in all levels of the organization and by the interdependence between the laborers' project of life and that of the institution (Catão, 2001), having the individual needs incorporated by the organizational ones.

In opposition to the authoritarian management system identified in the really experienced situation, which presented negative consequences for those supporting it, the system of participative management identified in the desired situation is the one that gets closer to the condition of human expansion (Vigostki, 2004; Antunes, 2005; Catão, 1994, 2001) and consequently of the occupational health.

Figure 1 shows the researched organizational human profile of the public entity, captured around the analysis of the areas: leadership, motivation, communication, deciding process, determination and control of targets, configuring elements of the management system and occupational health. Figure 1 is presented in 18 items with the following composition: 1 to 3 = leadership; 4 to 6 = motivation; 7 to 9 = communication; 10 to 11 = relationship/cooperation; 12

	SYSTEMS			
	1	2	3	4
	0			20
1. Confidence in the workers		xxx	---	
2. Confidence in the managers		xxx	---	
3. Request and utilization of opinions of the workers/team		xxx	---	
4. Type of motivation used		xxx	---	
5. Responsibility for the attainment of the targets		xxx	---	
6. Cooperation at work				
7. Direction of the flux of the information		xxx	---	
8. Acceptance of descendent communication		xxx	---	
9. Precision of ascending communication		xxx	---	
10. Managers aware of the problems of the workers/team		xxx	---	
11. Workers aware of the organizational policies		xxx	---	
12. Level of decision-making		xxx	---	
13. Involvement of workers in the relative decisions in relation to work		xxx	---	
14. Contribution of the deciding process toward motivation	xxx		---	
15. Way to establish targets	xxx	---		
16. Veiled resistance against the targets	xxx	---		
17. Opposition of informal organization against the formal organization		xxx	---	
18. Distribution of the internal control		xxx	---	
	Real: xxx		Desired: ----	

Figure 1. Real and desired organizational human profile in a public service entity.

to 14 = deciding process; 15 to 16 = determination of targets; and 17 to 18 = organizational controls. To do so, we followed the applied (Likert & Likert, 1978) grouping of questionnaire 2.

Based on the results shown, it can be observed that the configuration of points of tensions and conflict in several areas. In this sense, it is indicated as the initial intervention: to facilitate communication and have the Personal Life Projects articulated with the Organizational Life Project (Catão, 2001) as reference. To create a favorable organizational climate, within which targets and decisions may be discussed and shared, with public safety becoming more and more efficient, as well the quality of life and satisfaction to all those who exercise this labor.

4 CONCLUSIONS

The accomplished study allowed to map the researched organization as for the organization of real and desired labor on the part of the servants and the occupational health within these systems. It was observed the configuration of an authoritarian system, poorly accessed, with centralized responsibility on some hierarchic posts, environment with conflicting relationships, lack of confidence, sensation of discomfort and strangeness in the conduction of labor, stressing environment, depression and alcoholism. As for the organizational system desired by the interviewed individuals, it was identified the need to configure a participative and consulting system, ambience of confidence, interaction and sense of responsibility in all levels of the organization. To conclude, it was pointed out the emergency of organizational policies and practices so as to promote occupational health, about which we point out the need of formation and presence of professional specialists in occupational health in the referred environments, so as to facilitate the work and development of the servants and the organization. Implicit in these principles are found the propositions defended by Vigotski (2004), which sustain the thesis of the determinations of the organizational factors in the behavior of the servants and vice-versa.

REFERENCES

Antunes, R. 1995. Trabalho e estranhamento. In Antunes, R. (ed.) *Adeus ao trabalho? Ensaios sobre as metamorfoses e a centralidade do mundo do trabalho.* (121–134). Campinas: Editora da Universidade Estadual de Campinas.

Bardin, L. 1977. *L'analyse de Contenu.* Paris: PUF.

Braverman, H. 1977 Trabalho e Capital Monopolista. A degradação do trabalho no séc.XX. Rio de Janeiro: Zahar.

Catão, M.F. 1994. Tendência de organização do trabalho: Contexto organizacional e concepção de individuo. In Catão, M.F. *Práticas de recursos humanos em análise.* Dissertação de mestrado: UFPB.

Catão, M.F. 2001. *Projeto de vida em construção na exclusão/ inserção social.* João Pessoa: Editora Universitária UFPB.

Catão, M.F. & Sawaia, B. 2007. Problemas sociais e análise psicológica: questão de método. Relatório de pós-doutorado, São Paulo: Programa de estudos pósgraduados em psicologia social – PUC – SP.

Chanlat, J.F. 1995. Modos de gestão, saúde e segurança no trabalho. In Davel, E & Vasconcelos, J (eds.) *"Recursos" Humanos e Subjetividade.* Petrópolis: Vozes.

Darvel & Vergara. 2010. Gestão com pessoas, subjetividade e objetividade nas organizações. In Darvel & Vergara (eds.) *Gestão com pessoas e subjetividade.* São Paulo: Atlas.

Dejours, C. 1986. *Por um novo conceito de saúde. Revista Brasileira de Saúde Ocupacional.* São Paulo.

Dejour, C. 1992. *A loucura do trabalho.* São Paulo: Cortez

Likert & Likert. 1978. *Administração de Conflitos Novas abordagens.* São Paulo: McGraw-Hill.

Valsiver, J. 2012. *Fundamentos da Psicologia Cultural: mundos da mente, mundos da vida.* Porto Alegre: Artmed.

Vigotski, L.S. 2004. *Teoria e método em psicologia.* São Paulo: Martins Fontes.

Occupational ergonomics

Occupational Safety and Hygiene – Arezes et al. (eds)
© *2013 Taylor & Francis Group, London, ISBN 978-1-138-00047-6*

Occupational low-back pain in hospital nurses

F. Serranheira
Escola Nacional de Saúde Pública, Universidade Nova de Lisboa, Lisboa, Portugal
CMDT, Centro de Investigação em Malária e Doenças Tropicais—Saúde Pública, Lisboa, Portugal
CIESP, Centro de Investigação e Estudos em Saúde Pública, ENSP/UNL, Lisboa, Portugal

M. Sousa-Uva
Escola Nacional de Saúde Pública, Universidade Nova de Lisboa, Lisboa, Portugal

A. Sousa-Uva
Escola Nacional de Saúde Pública, Universidade Nova de Lisboa, Lisboa, Portugal
CMDT, Centro de Investigação em Malária e Doenças Tropicais—Saúde Pública, Lisboa, Portugal
CIESP, Centro de Investigação e Estudos em Saúde Pública, ENSP/UNL, Lisboa, Portugal

ABSTRACT· Symptoms surveys of work-related musculoskeletal disorders (WRMSDs) are performed mainly using short-answer questionnaires. We intended to identify and analyze the prevalence of low-back pain (LBP) linked to tasks and individual characteristics of hospital nurses. Respondents were all hospital nurses (n = 1.396) and show a higher prevalence of LBP in the last 12 months (60.9%) and in the last 7 days (48.8%). The presence of these symptoms appears to be associated with work aspects mainly those with organizational and professional origins such as professional category, type of work, type of service and also other specific work tasks. The patient hygiene in bed had the greater effect in the presence of such symptoms. We conclude that tasks characteristics are the most important factors for the presence of low-back symptoms and therefore we should focus on interventions aimed at preventing LBP and other WRMSD.

Keywords: WRMSD, low-back pain, nurses, hospitals, occupational health, ergonomics

1 INTRODUCTION

Over the past few years there has been an increasing interest on associations between work and musculoskeletal diseases.

Work-related musculoskeletal disorders (WRMSD) are a large group of diseases and other medical conditions that affect the musculoskeletal system.

WRMSDs include, among others, diseases of the muscles, tendons, nerves, and also joint lesions (Fonseca and Serranheira, 2006).

WRMSDs could be divided into two major groups: (i) those related to load mobilization and transport, mainly with low back needs and (ii) those related to the performance of static or repetitive tasks, with or without application of force, frequently using the upper limbs. Several previous studies have already described a high prevalence of low-back pain in nurses [prevalence of 33% of the study participants (Trinkoff et al., 2003); 66.6% (Fonseca and Serranheira, 2006, Trinkoff et al., 2003, Lagerstrom et al., 1995, Smith et al., 2005), and even more than 70% (Alexopoulos et al., 2003)].

Nurses are considered the group of health care professionals more often affected by low-back pain (LBP) (Trinkoff et al., 2003).

Some other studies have investigated aspects related to their work. Aspects like work organization, real characteristics of work activity (Tonges et al., 1998), effects of work on WRMSD, work stressors (Woodcox et al., 1994) (like poor interaction with other health care professionals), and high work demands.

Nursing tasks are often performed in inadequate working situations, often with lack of equipment and working under stress-time conditions, such as shift work or night work.

We found very often at large hospitals a nursing staff shortage with increasing workload due to excessive attributed tasks.

This study aims at the identification and analysis of the prevalence from low back musculoskeletal symptoms, self-reported by nurses working in hospitals in Portugal.

It also intends to identify possible associations between those symptoms and individual aspects, as well as work characteristics and some features related to specific nursing work activities.

2 METHODS

This was a cross-sectional national wide study that targeted all 62.566 Registered Nurses (RN) at the Portuguese Nursing Council (PNC) with 26.920 (43%) of them working in hospitals.

Portuguese nurses were invited to participate in this study through a request made at the Portuguese Nursing Council website.

Those respondents, who accepted our invitation, left their e-mails at the Web "SurveyMonkey Questionnaire Platform", subsequently received a link to be able to answer the questionnaire. The questions were opened online for a period of 8 months (until February 2011). This questionnaire has been adapted from the Nordic questionnaire on musculoskeletal disorders (NMQ—Nordic Musculoskeletal Questionnaire) previously validated and accepted as a survey instrument (Kuorinka et al., 1987, Warming et al., 2009) and already used in Portugal (Serranheira et al., 2003, Serranheira et al., 2008).

It is mainly divided in four sections intended to characterize: socio-demographic aspects; musculoskeletal symptoms; tasks and their symptom relationships and general health status. Statistical analysis was performed using the software Statistical Package for Social Sciences (SPSS) version of PASW Statistics 17. The Chi-square Test of Independence was used to evaluate if the prevalence of low-back symptoms depends on gender, age, Body Mass Index (BMI) and nurses professional category.

To determine the likelihood of increased low-back pain in the past 12 months was estimated the Odds Ratio and the respective 95% confidence interval through the method of Logistic Regression: Forward LR. A level of significance of 5% was set for all statistical tests.

3 RESULTS

A total of 2.140 nurses decided to participate in the study, an amount that represents 3.4% of the total number of registered nurses at the Portuguese Nursing Council, with a total of 1.396 working in hospitals.

The prevalence of low-back musculoskeletal symptoms was 48.8% for the previous 7 days and 60.9% for the previous 12 months, values of equivalent magnitude to other similar studies (Table 1).

In the group of nurses with low back pain in the last 12 months, about 2/3 (64.8%) reported symptoms "6 or more times per day" and 45.8% stated that those symptoms were "intense" or "very intense". The Chi-square Test of Independence revealed that LBP is independent of gender ($\chi2 = 2.37$, p = 0.123), age ($\chi2 = 3.86$, p = 0.27) and BMI ($\chi2 = 1.663$, p = 0.197), but is dependent of job category ($\chi2 = 18.86$, p = 0.001). Individuals with a professional category of "nurses" and "graduate nurses" have a higher prevalence of LBP (23.44% and 22.88% respectively), and professionals with a category of "nurse-specialists", "chief nurse" and "nurse supervisors" have a lower prevalence (9.42%, 4.86% and 0.37% respectively).

Logistic regression allowed us to observe tasks with statistically significant effect on the likelihood of having low-back symptoms of WRMSDs, when performed more than 10 times a day: i) development of invasive procedures; ii) administration of medication; iii) feeding of patients; iv) care of hygiene and positioning with mobilization of patients in bed; v) transfers and transports of patients; and vi) measurements of blood pressure and blood glucose (Table 2).

Patient's hygiene and comfort in bed was the only task that increases the likelihood of low-back pain more than 2 times per day (Table 2).

Logistic regression also allowed us to determine the impact of certain working features in the prevalence of low-back WRMSD symptoms.

Some job variables analyzed here were: i) type of work (fixed or shift work); ii) presence/absence of second jobs; iii) number of breaks per working day; iv) number of hours worked per week; and v) type of service where they work.

Only the type of work and the type of service were found to have a statistically significant effect.

Table 1. LBP previous studies prevalence's.

Study	Symptoms prevalence in the lumbar region (%)
Lagerstrom et al. (1995)	56
Engels et al. (1996)	34
Ando et al. (2000)	54.7
Alexopoulos (2003)	75
Trinkoff et al. (2002)	47
Trinkoff et al. (2003)	32
Smith et al. (2004)	56
Fonseca & Serranheira (2006)	65
This study (2011)	60.9

Table 2. Task impact on LBP symptoms in the lumbar region in hospital nurses.

Tasks	Classes	p-value	Odds Ratio	C.I. (Odds Ratio)	
Invasive procedures	2–5 times per day	0.090	1.30	0.96	1.75
	6–10 times per day	0.190	1.29	0.88	1.88
	More than 10 times per day	**<0.001**	**2.15**	**1.43**	**3.23**
Administration of medication	2–5 times per day	1.880	1.31	0.88	1.96
	6–10 times per day	0.170	1.68	1.10	2.58
	More than 10 times per day	**0.002**	**1.86**	**1.26**	**2.73**
Measurement of blood pressure and blood glucose	2–5 times per day	0.610	1.10	0.75	1.62
	6–10 times per day	0.150	1.35	0.90	2.03
	More than 10 times per day	**0.015**	**1.60**	**1.09**	**2.33**
Care of hygiene of patients in bed	2–5 times per day	**0.037**	**1.36**	**1.02**	**1.82**
	6–10 times per day	**0.011**	**1.63**	**1.12**	**2.37**
	More than 10 times per day	**0.002**	**2.48**	**1.40**	**4.42**
Positioning and mobilization of patients in bed	2–5 times per day	0.520	1.13	0.78	1.62
	6–10 times per day	0.290	1.22	0.85	1.74
	More than 10 times per day	**0.001**	**2.02**	**1.35**	**3.02**
Transfer and transport of patients	2–5 times per day	0.535	1.10	0.82	1.47
	6–10 times per day	0.119	1.32	0.93	1.88
	More than 10 times per day	**0.018**	**1.75**	**1.10**	**2.78**
Feeding of patients	2–5 times per day	**0.040**	**1.36**	**1.01**	**1.81**
	6–10 times per day	0.318	1.22	0.83	1.79
	More than 10 times per day	**0.004**	**2.19**	**1.28**	**3.73**

The likelihood of LBP is increased in shift work compared to fixed work (p = 0.022, OR = 1.32) and also in those services with dominant nursing work compared to non-dominant nursing work (p = 0.024, OR = 1.52).

4 DISCUSSION

These results show a high prevalence of LBP in Portuguese nurses working in hospital settings (48.8% in the last 7 days and 60.9% in the past 12 months) which seem not very different from results described in other similar studies (Fonseca and Serranheira, 2006, Ando et al., 2000, Alexopoulos et al., 2003, Lagerstrom et al., 1998, Smith et al., 2003, Hollingdale and Warin, 1997).

When nurses were asked about WRMSD intensity and frequency of pain (or discomfort) in the lumbar region in the past 12 months, we found a high prevalence of moderate pain (42%) or severe pain (35.2%) accounting frequently more than 10 times a day (42%).

This may be of great importance to the health and well-being of nurses and entails some consequences for the provision of quality health care to patients, as well as absenteeism/sickness and even choosing a job outside of nursing. These results evidence the need for the development and the implementation of effective LBP prevention measures in nursing staff.

There were no statistically significant associations between gender, age and BMI with low-back pain (LBP) in this sample.

Therefore, individual features such as being male or female, being over or underweight, and young or old do not seem to influence the prevalence of lumbar WRMSD symptoms in hospital nurses.

On the other hand, specific professional groups have a statistically significant association with these symptoms ($\chi2 = 18.86$, p = 0.001), with "nurses" and "graduate nurses" groups having a higher prevalence (23.44% and 22.88% respectively).

This may be related to the real working activities performed by professionals in each of these categories, namely the handling of patients.

As a matter of fact, the categories of "nurse" and "graduate nurse" have a greater proximity to the provision of clinical care, involving the handling of patients and hence a greater physical demand associated to the type of tasks performed.

All these tasks were found to have an influence in the presence of LBP clearly increasing the probability of the presence of such symptoms when performed more than 10 times daily, compared to the frequency of 0–1 times daily, which clearly reveals the importance of organizational requirements about the repetitiveness of the work activity in the process of developing those symptoms.

Hygiene and comfort in bed was the task with the highest effect on the probability of occurrence of LBP, and this probability increases when tasks were carried out more than 2 times per day in relation to the 0–1 times per day group and not just to the 10 times a day group (2–5 times per day: OR = 1.36; 6–10 times per day: OR = 1.63; over 10 times per day: OR = 2.48).

This task seems plausible to be most responsible for occupational LBP in hospital nurses, since quite often hospitalized patients get into discomfort positions demanding frequently physical efforts from nursing staff to put them back in bed in a better position and also requiring some other manual handling tasks.

Hygiene care in bed also shows a higher activity impact over those symptoms, requiring long lasting awkward postures of the trunk eventually constituting additional health risks.

Feeding of patients was the only task with an impact on those WRMSD symptoms when performed 2–5 times per day and more than 10 times per day, but with no significant impact when performed 6–10 times per day.

These results could possibly be influenced by other variables that might justify them (confounding variables).

The questionnaire respondents are probably nurses who value them most WRMSD and even those with musculoskeletal symptoms. Thus, there is a high probability the percentage of LBP nurses could be overestimated.

Other limitations of the study are related with the fact of a retrospective study and other possible bias and misclassification.

Additionally nursing staff with higher demanding jobs were probably more interested to participate in the study.

5 CONCLUSIONS

This study has been the first national Portuguese wide study about the prevalence of WRMSD in hospital nurses, showing a high prevalence of symptoms of LBP in the last 12 months (60.9%) and in the last 7 days (48.8%).

Individual characteristics such as gender, age and BMI didn't have a significant effect on LBP prevalence. Aspects related with work organization such as second job, number of hours worked per week, and number of breaks per work day didn't also have a significant effect on prevalence of low-back WRMSD symptoms.

Also the probability of having LBP symptoms and eventually WRMSD, increases when nurses perform shift work as well as work in services with dominant nursing work, relatively to other services.

The type of professional category was statistically significant for the presence of such symptoms.

The majority of nurses with low-back musculoskeletal symptoms were "nursing" or "nursing graduates".

Work characteristics proved to be the most influential factors in the presence of musculoskeletal symptoms in the lower back unlikely those individual characteristics analyzed, evidencing the importance of working conditions and real work activities in the presence of musculoskeletal symptoms and the development of those lesions.

The lack of associations between some individual characteristics such as gender, age and BMI and lower back WRMSD symptoms should not take us to reject this hypothesis in different samples (Serranheira et al., 2012).

The task with the greatest effect on LBP was the patient's hygiene and comfort in the bed.

This shows the need to understand and analyze better this real work activity in order to prevent these disorders.

It is justified, therefore, the importance of the existence of risk management measures (Uva, 2006; 2010) designing LBP prevention programs for nurses in workplace.

These programs should focus on health and safety management and an improved ergonomics system approach.

Job redesign strategies, aspects related to lifting policy and education aimed at LBP prevention like proper use of mechanical devices must be included, for sure, in those LBP prevention measures.

REFERENCES

ALEXOPOULOS, E. C., BURDORF, A. & KALOKERINOU, A. 2003. Risk factors for musculoskeletal disorders among nursing personnel in Greek hospitals. *Int Arch Occup Environ Health,* 76, 289–294.

ANDO, S., ONO, Y., SHIMAOKA, M., HIRUTA, S., HATTORI, Y., HORI, F. & TAKEUCHI, Y. 2000. Associations of self estimated workloads with musculoskeletal symptoms among hospital nurses. *Occup Environ Med,* 57, 211–6.

ENGELS J., VAN DER GULDEN J., SENDEN T. & VAN'T HOF B. 1996. Work related risk factors for musculoskeletal complaints in the nursing profession: results of a questionnaire survey. *Occup Environ Med,* 53(9):636–41.

FONSECA, R. & SERRANHEIRA, F. 2006. Sintomatologia músculo-esquelética auto-referida por enfermeiros em meio hospitalar. *Rev Port Saúde Pública,* Volume Temático, 37–44.

HOLLINGDALE, R. & WARIN, J. 1997. Back pain in nursing and associated factors: a study. *Nurs Stand,* 11, 35–8.

KUORINKA, I., JONSSON, B., KILBOM, A., VINTERBERG, H., BIERING-SØRENSEN, F., ANDERSSON, G. & JØRGENSEN, K. 1987. Standardised Nordic questionnaires for the analysis of musculoskeletal symptoms. *Appl Ergon,* 18, 233–237.

LAGERSTROM, M., HANSSON, T. & HAGBERG, M. 1998. Work-related low-back problems in nursing. *Scand J Work Environ Health,* 24, 449–64.

LAGERSTROM, M., WENEMARK, M., HAGBERG, M. & HJELM, E. W. 1995. Occupational and individual factors related to musculoskeletal symptoms in five body regions among Swedish nursing personnel. *Int Arch Occup Environ Health,* 68, 27–35.

SERRANHEIRA, F., COTRIM, T., RODRIGUES, V., NUNES, C. & SOUSA-UVA, A. 2012. Nurses' working tasks and MSDs back symptoms: results from a national survey. *Work,* 41, 2449–2451.

SERRANHEIRA, F., PEREIRA, M., SANTOS, C. & CABRITA, M. 2003. Auto-referência de sintomas de lesões músculo-esqueléticas ligadas ao trabalho (LMELT) numa grande empresa em Portugal. *Rev Port Saúde Pública,* 2, 37–48.

SERRANHEIRA, F., UVA, A. & LOPES, F. 2008. *Lesões músculo-esqueléticas e trabalho: alguns métodos de avaliação do risco,* Lisboa, Sociedade Portuguesa de Medicina do Trabalho.

SMITH D.R., WEI N., KANG L. & WANG R.S. 2004. Musculoskeletal disorders among professional nurses in mainland China. *J Prof Nurs,* 20(6): 390–5.

SMITH, D.R., CHOE, M.A., JEON, M.Y., CHAE, Y.R., AN, G.J. & JEONG, J.S. 2005. Epidemiology of musculoskeletal symptoms among Korean hospital nurses. *Int J Occup Saf Ergon,* 11, 431–40.

SMITH, D.R., OHMURA, K., YAMAGATA, Z. & MINAI, J. 2003. Musculoskeletal disorders among female nurses in a rural Japanese hospital. *Nurs Health Sci,* 5, 185–8.

TONGES, M.C., ROTHSTEIN, H. & CARTER, H.K. 1998. Sources of satisfaction in hospital nursing practice: a guide to effective job design. *Journal of Nursing Administration,* 28, 47.

TRINKOFF A.M., LIPSCOMB J.A., GEIGER-BROWN J. & BRADY B. 2002. Musculoskeletal problems of the neck, shoulder, and back and functional consequences in nurses. *Am J Ind Med,* 41(3):170–8.

TRINKOFF, A.M., BRADY, B. & NIELSEN, K. 2003. Workplace prevention and musculoskeletal injuries in nurses. *J Nurs Adm,* 33, 153–8.

UVA, A. 2006; 2010. *Diagnóstico e Gestão do Risco em Saúde Ocupacional,* Lisboa, ACT—Autoridade para as Condições de Trabalho.

WARMING, S., PRECHT, D.H., SUADICANI, P. & EBBEHOJ, N. E. 2009. Musculoskeletal complaints among nurses related to patient handling tasks and psychosocial factors--based on logbook registrations. *Appl Ergon,* 40, 569–76.

WOODCOX, V., ISAACS, S., UNDERWOOD, J. & CHAMBERS, L. 1994. Public health nurses' quality of worklife: responses to organizational changes. *Canadian journal of public health. Revue canadienne de sante publique,* 85, 185.

Occupational Safety and Hygiene – Arezes et al. (eds)
© *2013 Taylor & Francis Group, London, ISBN 978-1-138-00047-6*

Occupational ergonomics: Work posture among Brazilian dental students

P.P.N.S. Garcia, D. Wajngarten, A.C.A. Gottardello & J.A.D.B. Campos
School of Dentistry, UNESP, Araraquara, Brazil

ABSTRACT: This study involved observational assessment of work posture in relation to recommended ergonomic posture the requirements necessary for ergonomic posture among students in the final year of a degree program at the School of Dentistry of Araraquara—UNESP/Brazil (n = 73) and investigation of the association of work posture with sex, the type of procedure, four-handed dentistry, and the region of the mouth being treated. The work posture of the students during 250 clinical procedures was observed by means of pictures. Each procedure received a posture classification: Adequate, partially adequate, or inadequate. A descriptive statistical analysis was conducted. The prevalence of final posture classification was calculated using 95% confidence intervals and point estimate. Associations of interest were studied using the chi-square test, with a 5% significance level. It was concluded that the prevalence of procedures performed with partially adequate posture was high, and that the final work posture classification was not associated with the variables of interest.

1 INTRODUCTION

Dentistry is a profession that requires high visual demand, concentration, and accuracy (Rinsing et al., 2005; Lindfors et al., 2006; Gandavadi et al., 2007). Owing to a restricted work area and the need for hand dexterity, dental surgeons eventually adopt inadequate and inflexible work postures (Gandavadi et al., 2007). Furthermore, the long time spent seated without taking breaks and the need for greater hand stability and strength result in static muscular activity that may lead to musculoskeletal disorders (Kerusuo et al., 2000; Valachi & Valachi 2003).

In dentistry, such disorders occur especially in the lower limbs and may compromise the occupational health of dental surgeons, leading to decreased productivity due to the impact on individual performance and sometimes even career termination (Valachi & Valachi 2003).

In this context, in spite of all the evolution that has occurred in dental work systems in recent years, a high risk of occupational ailments related to work posture among individuals in the dental industry has been observed. Lindfors et al. (2006) verified that dental surgeons assessed in Switzerland showed a high risk of developing musculoskeletal disorders. Garcia et al. (2008b) observed that there was a non significant association between the postures adopted by operators while providing care to children and the children's behavior. Thornton et al. (2008) observed that dental surgeons are highly likely to develop work-related bone and muscle

problems. Garcia et al. (2012) verified that the risk of developing musculoskeletal disorders was high among students in the last year of dental school.

In light of the above report, it is clear that such disorders among dental professionals must be managed or mitigated as early as possible, preferably when the individual is still undergoing professional training and education (Chowanadisai et al., 2000), because postural vices can be identified and easily corrected at the learning stage (Rinsing et al., 2005).

Therefore, this work aimed at verifying, through observation, the requirements necessary to obtain an ergonomic work posture in dental students as well as verifying the association of work posture with sex, the type of clinical procedure, four-handed dentistry practice, and the region of the mouth being treated.

2 MATERIALS AND METHODS

2.1 Sampling design

This study was an observational, cross-sectional cohort study. The sample comprised all the students enrolled in the final year of the degree program at the School of Dentistry of Araraquara—UNESP/Brazil (n = 73).

2.2 Recording work posture

Postures adopted by students, at work during 250 clinical procedures conducted in the Integrated Clinical class were assessed by means of pictures.

The pictures were taken from 5 basic angles to allow viewing of the positions of the dental table and tools in relation to the work team, the horizontal and vertical position of the operator's leg posture, the operator's work position, feet support on the ground, lumbar support, the use of the dental stool, spinal inclination, the patient position in the dental chair, the position of the dental light, the distance between the patient's mouth and the team's eyes, and the position of the operator's left and right arms.

2.3 Posture assessment method

The work postures adopted by each student acting as "operator" were classified as adequate, partially adequate, or inadequate (Garcia et al. 2008b) on the basis of the recommended requirements necessary to obtain ergonomic posture taught in the Dental Ergonomics class of the School of Dentistry of Araraquara—UNESP/Brazil (Porto 1994). Each item was scored according to this classification, with 1 point being attributed to adequate, zero to inadequate, and a ½ point to partially adequate. At the end of the assessment, all the items were added to a maximum total of 16 points. Each procedure received a final classification regarding work posture, and the reference adopted was the score reached in relation to the total (Garcia et al. 2008b). The pictures were assessed by the researcher after calibration ($\kappa = 0.87$).

2.4 Statistical analysis

A descriptive statistical analysis was conducted. The data regarding work posture classification were ascertained, and the prevalence was calculated using 95% confidence intervals and point estimate. Associations of interest were studied using the chi-square test. The significance level adopted was 5%.

3 RESULTS AND DISCUSSION

The prevalence of the final posture classification among the students in the study was calculated using 95% confidence intervals and point estimate (Figure 1).

It was found that the prevalence of posture classified as partially adequate was high (70.4%; CI95%: 76.1–64.7%). To understand the reasons that led to such a finding, the behavior of the students in relation to each recommended requirement necessary to adopt an ergonomic posture was observed. This enabled us to determine the presence of some inadequacies.

The first inadequacy observed was linked to the position of the dental chair, and consequently the position of the patient. The patient was positioned

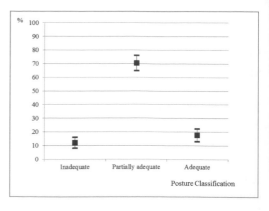

Figure 1. Prevalence of the work posture classification adopted by the "operator" student while performing clinical procedures, as determined using point estimate (p) and 95% confidence intervals (UL: upper limit, LL: lower limit). Araraquara/Brazil, 2012.

adequately in 39.2% of the procedures, that is, with the mouth at knee level (Figlioli 1993; Porto 1994; Valachi & Valachi 2003). In 44.0% of the procedures, the patients were in the semi-recumbent position. Additionally, by analyzing the dental chair, another item that deserves attention was found to be the position of the dental machine, which was put into position according to the region of the mouth being treated, that is, down for maxilla and up for mandible, in only 18.4% of the procedures (Figlioli 1993; Porto 1994). It is interesting to verify that the values for inadequately positioning the patient in the dental chair were similar to those found by Garcia et al. (2008a) and Garcia (2008b), whereas the values regarding the dental machine were similar to those found by Garcia et al. (2008b) and different from those found by Garcia et al. (2008a), who obtained a rate of 77.4% for procedures being performed with the dental machine in an adequate position.

Correct positioning of the patient in the dental chair and the patient's head in relation to the dental machine are the first two requirements to be observed when one wishes to adopt an ergonomic work posture. Fulfillment of these requirements provides an adequate view of the operative field and good access to it without harming the musculoskeletal system (Valachi & Valachi 2003; Garcia et al. 2008a; Garcia et al., 2008b; Grupta, 2011). If these requirements are not met the position of the trunk, neck, arm, and forearm of the operator eventually deviates from a neutral position. This occurs because, to improve visibility and access, the operator needs to increase proximity in relation to the operative field and, consequently, has to keep his/her eyes at a distance less than 30 cm from the patient's mouth. This fact was ascertained in the present study, where 78.8% of the procedures were performed at a distance less than 30 cm, which potentially interfered with lumbar support. It was found that 53.2% of the procedures were performed without lumbar support for the spine, possibly to make it easier to bring the operator's eyes closer to the operative field. According to Garcia et al. (2008a) and Garcia et al. (2008b), a decrease of the recommended distance between the team's eyes and the patient's mouth leads to osteo-muscular-tendinous imbalance and increases the risk of biological contamination. Additionally, according to Thornton et al. (2008) the position of the patient, seated or semi-recumbent in the dental chair, may result in students having more pain and discomfort, as observed in their study.

In addition to the patient's position, some inadequacies in the position of the students were also observed, especially with respect to the location of the legs, both vertically and horizontally. It was found that 48.8% of the procedures were not performed with the legs in an adequate vertical position, that is with a 90° angle between the thighs and lower legs, as observed by Garcia et al. (2008a). An increase or decrease of this angle results in loss of venous return from the lower limbs, consequently leading to changes in the circulatory system, edema, inflammation, and varicose veins (Garcia et al., 2008b; Garbin et al., 2011). With regards to the horizontal positioning of the legs, it was observed that in 55.6% of the procedures the work was performed with the thighs at a 70° angle in relation to each other and on the back rest of the dental chair, and in 8.8% they were parallel to each other and on the back rest. In both situations the distribution of body weight on the ischia is unbalanced, which overloads the spine over time (Valachi & Valachi 2003).

In addition to analyzing the ergonomic posture requirements that interfere with the correct work posture of students, observing the influence of variables such as sex, the type of procedure, four-handed dentistry, and the region of the mouth being treated is also extremely important to design prevention programs.

The distribution of work posture classification, according to the variables of interest, is presented in Table 1.

Non significant association between final work posture classification and sex, the type of procedure, four-handed dentistry, and the region of the mouth being treated was observed. In studies such as those by Finsen et al. (1997), Garcia et al. (2008a), Garcia et al. (2008b), and Teles (2009) a non significant association was also determined, thus suggesting that the postural inadequacies observed in future dental surgeons are the result of the faulty positions acquired during clinical practice, which tend to be aggravated throughout

Table 1. Work posture classification distribution according to the variables of interest. Araraquara/Brazil, 2012.

Variable	Classification of work posture			χ^2	p
	I*	PA	A		
Sex					
– Male	6	37	11	0.380	0.827
– Female	24	139	33		
Procedure					
– Noninvasive preparation	8	49	6	5.935	0.430
– Invasive preparation	4	22	5		
– Restoration	4	32	14		
– Rehabilitation	14	82	14		
Four-handed dentistry					
– Yes	21	123	33	0.456	0.796
– No	9	53	11		
Mouth region					
– Upper region	19	113	28	0.150	0.997
– Lower region	10	57	14		
– Full mouth	1	6	2		

I: inadequate; PA: partially adequate; A: adequate

the years. In contrast, despite the non significant association observed in relation to sex, according to Valachi (2008), female professionals must devote greater attention to their work posture because they are greatly influenced by reproductive and hormonal factors, which put them at a higher risk of developing musculoskeletal problems.

It must be emphasized that the students assessed received training in relation to the principles behind the reasoning and ergonomics applied in dentistry during the second and third years of their degree program, where they were monitored in lab and clinical activities during classes in Ergonomics in Dentistry I and II. Despite this, the current study was able to verify that the application of ergonomic concepts learned by students in their first years in the program was not maintained until the conclusion of the same.

4 CONCLUSIONS

It is concluded that the prevalence of procedures performed with partially adequate posture was high and that the final work posture classification was not associated with sex, the type of clinical procedure, four-handed dentistry practice, and the region of the mouth being treated The findings show the importance of developing future studies in way to identify predictors variables of unhealthy working postures to create strategies to motivate dental students to keep ergonomic work postures throughout their professional lives in order to prevent musculoskeletal disorders.

ACKNOWLEDGMENTS

We thank the São Paulo Research Foundation FAPESP for financial support (case no. 2011/20715-9).

REFERENCES

Chowanadisai, S., Kukiattrakoon, B., Yapong, B., Kedjarune, U. & Leggat, P.A. 2000. Occupational health problems of dentists in southern Thailand. *International Dentistry Journal* 50: 36–40.

Figlioli, M.D. 1993. Postura de trabalho em odontologia. Posições do operador na execução de preparos cavitários em manequim, com dique de borracha previamente colocado. *RGO* 41: 155–60.

Finsen, L., Christensen, H. & Bakke, M. 1997. Musculoskeletal disorders among dentists and variation in dental work. *Applied Ergonomics* 29: 119–25.

Gandavadi, A., Ramsay, J.R.E. & Burke, F.J.T. 2007. Assessment of dental student posture in two seating conditions using RULA methodology—a pilot study. *British Dental Journal* 203: 601–5.

Garbin, A.J.I., Garbin, C.A.S., Diniz, D.G. & Yarid, S.D. 2011. Dental students' knowledge of ergonomic postural requirements and their application during clinical care. *European Journal of Dental Education* 15: 31–5.

Garcia, P.P.N.S., Campos, J.A.D.B. & Zuanon, A.C.C. 2008a. Posturas de trabalho de alunos no atendimento odontológico de bebês. *Revista de Odontologia da UNESP* 37: 253–9.

Garcia, P.P.N.S., Campos, J.A.D.B. & Zuanon, A.C.C. 2008b. Posturas de trabalho de alunos no atendimento odontológico de crianças. *Pesquisa Brasileira de Odontopediatria e Clinica Integrada* 8: 31–37.

Garcia, P.P.N.S., Pinelli, C., Derceli, J.R. & Campos, J.A.D.B. 2012. Musculoskeletal Disorders in Upper Limbs in Dental Students: Exposure level to risk factors. *Brazilian Journal of Oral Science* 11: 148–53.

Grupta, S. 2011. Ergonomic applications to dental practice. *Indian Journal of Dental Research* 22: 816–22.

Kerusuo, E., Kerusuo, H. & Kanerva, L. 2000. Self-reported health complaints among general dental practitioners, orthodontists, and office employees. *Acta Odontologica Scandinavica* 58: 207–12.

Lindfors, P., Thiele, U. & Lundberg, U. 2006. Work characteristics and upper extremity disorders in female dental health workers. *Journal Occupational Health* 48: 192–7.

Porto, F.A. 1994. *The Dental Office*. São Carlos: Scritti.

Rinsing, D.W., Bennett, B.C., Hursh, K. & Plesh, O. 2005. Reports of body pain in a dental student population. *Journal of the American Dental Association* 136: 81–6.

Teles, C.J.C.F. 2009. *Avaliação do grau de conhecimento dos médicos dentistas em relação à aplicação da ergonomia na medicina dentária* [Trabalho de Conclusão de Curso]. Porto: Universidade Fernando Pessoa.

Thornton, L.J., Barr, A.E., Stuart-Buttle, C., Gaughan, J.P., Wilson, E.R., Jackscon, A.D. & Wyszynski, T.C. 2008. Perceived musculoskeletal symptoms among dental students in the clinic work environment. *Ergonomics* 51: 573–83.

Valachi, B. & Valachi, K. 2003. Mechanisms leading to musculoskeletal disorders in dentistry. *Journal of the American Dental Association* 134: 1344–50.

Valachi B. 2008. Musculoskeletal health of the woman dentist: distinctive interventions for a growing population. *Journal of the California Dental Association* 36: 127–32.

Occupational Safety and Hygiene – Arezes et al. (eds)
© 2013 Taylor & Francis Group, London, ISBN 978-1-138-00047-6

Analysis of ergonomics in office work: A case study leading to an intervention in office acoustics

C.S.D. Tavares, T.M. Lima & D.A. Coelho
Universidade da Beira Interior, Covilhã, Portugal

ABSTRACT: Although the current society is characterized by high industrialization and mechanization, which would by itself apparently only bring advantages to human beings, what is seen is that due to the specific features of each profession and due to the current state of the job market, which is demanding and highly competitive, there are increasingly more health problems, not only at physical, but also at psychological level. Therefore, and in view of this scenario, it is increasingly important to provide for the physical and psychological ergonomic needs of workers in their workplace. This study seeks to demonstrate the importance of this issue, by presenting a practical case study (carried out in an office environment). The methodology deployed involved four previously existing instruments which were used to analyze the ergonomic conditions of work, and the related physical and psychological consequences. The sample consisted of 32 workers from the office of the company under focus in the case study, who were subjects in the research process. Subsequently, the results were statistically analyzed. This analysis resulted in statistically significant associations between the several domains of data collected in the study, which also led to the need to modify some of the physical ergonomic conditions, including exposure to noise. Following the analysis of the acoustic reverberation time in the most noisy environment of the office, an improved acoustic surface treatment, albeit of low-cost, in the department most affected by this problem was proposed, leading to a significant decrease in the reverberation time calculated (decreasing from 6.725 to 0.744 seconds).

1 INTRODUCTION

1.1 Background

As time progressed, and industrial developments took place, there were several health problems arising from work, not only of physical but also of psychological origin. In addition to physical health problems, there are increasingly psychological health problems, because increasingly, the job market is more competitive, and therefore more psychologically demanding. Therefore, and in view of this scenario, it is increasingly important to provide for ergonomic physical and psychological needs of workers in their workplace.

1.2 Aims

The manner that should be followed to obtain improved working conditions of workers depends heavily on the characteristics of each specific organization, since each worker has his or her own way of working and each organization has its own way of producing a particular good or service. However, regardless of the form and the type of service they produce, all organizations have to take into account the safety of their workers, and the working conditions they are subjected to. If those same workers do not have the conditions necessary to perform their duties, there is an expected decrease in productivity and increase in diseases and accidents at work, as the organization ceases to be competitive if it fails to offer better quality of working conditions.

In this study, it was sought to demonstrate the importance of ergonomics in organizations, based on theoretical considerations, and by developing a case study, carried out in an office environment.

The first part of the initial goal of the work this paper reports on, aims to study the impact of environmental, physical and psychosocial working conditions on worker well-being through the case study. It was demonstrated through the analysis that there is a fairly strong relationship between working conditions and the environmental, physical and psychosocial aspects of the well-being of workers. The significance of the results confirms the importance of these issues, in relation to the wellbeing of workers and how they are affected physically and psychologically, in the course of their daily work.

For the second part of the overall goal of this study, taking into account the results obtained in the acoustic analysis, it can be concluded that by choosing suitable materials, one can greatly reduce the noise propagated at a site. In the case study reported and through the proposed improvement, significant results in terms of noise reduction could be obtained.

2 STUDY DESIGN AND RESULTS

2.1 Instruments used for data collection

Four instruments were used to analyze the ergonomic conditions of work, and the physical and psychological consequences related to them. Information was collected using DASH (Beaton, Wright & Katz 2005), an ergonomics check-list (Lima & Coelho, 2011), the short version of the COPSOQ questionnaire (Moncada et al. 2005) and a body diagram (Corlett & Manenica 1980) for pain complaints. The sample consisted of 32 workers from the office of the organization the study focused upon. Subsequently the results were statistically analyzed. This analysis resulted in statistically significant associations between the domains of inquiry, which led to the need to alter some of the physical ergonomic conditions, including exposure to noise. Following this, an improved acoustic reverberation setup in the department most affected by this problem was proposed.

2.2 DASH—Disabilities of the Arms, Shoulders and Hands

To evaluate the disabilities of the arms, shoulders and hands there are two highly interrelated questionnaires: DASH and Quick DASH, which are respectively, the short version and the long version of an instrument for the evaluation of the aforementioned disabilities.

In this study the longer version was used—DASH (Portuguese version—DASH Portugal 2005), because it covers more items and provides more information. The optional part was not used because it was not relevant to the study. With this questionnaire it was intended to collect data on the symptoms and the ability of the worker to perform certain activities involving muscular effort and control.

At least 27 of the 30 items must be filled in order to get a DASH score. The values reported in all answers are simply added together giving a mean value. This value is converted into a score by subtracting one and multiplying by 25. A high score indicates greater disability (DASH Portugal 2005).

2.3 Body diagram

Corlett & Manenica's (1980) body diagram was used in the study. This diagram is a qualitative method, and is divided into body regions (right and left) and each one enables the individual respondent to select the zone corresponding to the location where body pain and discomfort is felt at the time of evaluation.

After collecting the data on the affected areas, from the twenty-one body zones identified in the map, these zones were grouped in only three areas (upper limbs, trunk and lower limbs) in order to prepare for the subsequent statistical analysis.

2.4 COPSOQ—Copenhagen Psychosocial Questionnaire

There are various psychosocial risk factors, with stress affecting most individuals worldwide. According to the European Agency for Safety and Health at Work, work-related stress is one of the biggest challenges for health and safety in Europe. Nearly one in four workers is affected by stress, and studies suggest that it is responsible for between 50% to 60% of working days lost. Therefore, this represents a huge cost in terms of human suffering and impaired economic performance.

Thus, stress is a large problem in the world that causes major concern and has led to studies by researchers. The first to address this issue was Hans Selye, when he described the "Adaptation Syndrome" in 1936 (Selye 1976). Since then many other researchers have addressed this issue and investigated it, which led the development of several methods of assessing the factors that cause stress.

In the year 2000 a group of researchers from the National Institute of Labor Health in Denmark, led by Tage S. Kristensen, developed an instrument to assess psychosocial risks. This instrument is called "Copenhagen Psychosocial Questionnaire—CoPsoQ" and was later adapted by the Spanish state by a working group of the Institute of Labor Union, Environment and Health (ISTAS).

There are three versions of the questionnaire:

– Short version: developed for risk assessment of small businesses;
– Medium version: developed for risk assessment in medium to large companies;
– Long version: designed for research.

In this study, we used the ISTAS 21 (CoPsoQ)—short version, since this questionnaire is quickly answered and simple to perform and analyze. This version was developed to identify and measure exposure to six major groups of risk factors for psychosocial health at work:

1. psychological requirements;
2. work and possibility of active development;
3. insecurity;
4. social support and quality of leadership;
5. work/family conflict;
6. self esteem.

2.5 Prevention of Musculoskeletal Disorders (MSDs) in office work checklist

Lima & Coelho's (2011) checklist is based on assessing the postural, seating, equipment and environmental conditions of the office. It is a very comprehensive checklist and easy to use, allowing for results in a very fast and efficient manner. This method is consistent with what is prescribed in Ordinance No. 989/93 of 6 October and Law Decree 243/86 of August 20 (both from the Portuguese Republic) in what regards the conditions necessary for workers to develop their work.

The checklist is composed of eighty-eight domains, divided into four groups of assessment items. The score for the checklist is obtained in terms of the number of mismatches of ergonomic nature found by the analyst for each workstation, considering the worker who labors in it.

Table 1. Demographics of the subjects and checklist and DASH results (checklist results are expressed in terms of non-conformities (NC)).

Work-station	Gender	Age	Total NC	DASH score
1	F	36–45	26	15,0
2	F	26–35	23	14,2
3	M	26–35	23	1,7
4	M	46–55	26	8,3
5	F	36–45	24	30,8
6	F	26–35	24	5,0
7	M	26–35	23	0,0
8	F	26–35	24	14,2
9	M	46–55	26	6,7
10	F	36–45	24	25,8
11	F	36–45	23	25,0
12	F	36–45	23	0,0
13	F	36–45	26	12,1
14	F	26–35	24	5,8
15	F	26–35	25	44,2
16	M	36–45	24	6,9
17	M	46–55	25	1,7
18	M	46–55	22	11,1
19	M	36–45	25	0,8
20	M	36–45	24	0,0
21	F	36–45	24	5,8
22	M	26–35	23	0,0
23	M	36–45	25	2,5
24	F	26–35	22	11,7
25	F	26–35	25	9,2
26	F	26–35	23	5,8
27	M	36–45	26	0,0
28	F	36–45	20	0,0
29	M	26–35	22	0,0
30	F	36–45	24	23,3
31	M	36–45	22	1,7
32	M	36–45	23	4,2

Table 1 presents the results obtained from the DASH and checklist in the company office which was the focus of the case study (due to space limitations more results from CoPsoQ and body diagram can not be depicted).

3 DISCUSSION OF RESULTS

3.1 Ergonomic and psychosocial assessment

Returning to the first part of the initial goal of this study, to study the impact of working conditions (environmental, physical and psychosocial) on physical well-being, through a case study, the results show that there are fairly strong associations among environmental, physical and psychosocial working conditions and physical well-being of workers. The significance of the results (high correlation coefficients with suitable significance, e.g. correlation coefficient of 0.541 with a significance of 0.011 for the pair Work/Family conflict and symptoms of Musculoskeletal Disorders) shows the importance of these issues, in relation to the wellbeing of workers and how they are affected physically and psychologically by their daily work.

Good working conditions that workers need to have to ensure well-being depend heavily on organizations because each one has its own way of working and producing, whether goods or services. However, regardless of the form and the type of production, all organizations should take into account the safety of their workers and their working conditions, because if workers do not have the adequate conditions to perform their functions, there will be a decrease in productivity and an increase in diseases and accidents and, thus, the organization will no longer be competitive.

Taking into account the results obtained in the organization studied in the present case, the organization must be concerned with the physical ergonomic level, particularly for the domain of sitting because several mismatches were observed (thirty workers had more than ten mismatches in this domain), these may be corrected with the use of chairs that comply with current legislation. The domain of devices should also be taken into account (in this context thirty-one workers had more than six mismatches) by providing gel padded mouse support mats, and refitting computer screen monitors in order to offer a more appropriate viewing distance (about 750 mm from the individual) and providing a document-holder for each worker, in order to reduce eyestrain.

Regarding psychosocial conditions, the company should pay particular attention to the psychological climate of great insecurity and psychological demands exerted on workers, as well as to the psychological demands that lead most workers to have adverse psychosocial exposures (twenty one and twenty workers, respectively). In this sense, and taking into account the phase that the organization is going through, talking to the workers and trying to pass confidence and support are recommended actions for the organization studied. In what concerns the low self-esteem encountered in the workforce (eighteen workers with unfavorable psychosocial exposure level), the organization should try to motivate workers and give them more autonomy to carry out their duties. Finally, with regard to the work-family conflict (with eighteen workers reporting an unfavorable psychosocial exposure level), this problem can also be related to the phase of restructuring the company was going through, because with this situation, workers are more anxious, more tense, which makes it necessary to do more to engage both professional tasks as well as household chores, leading, therefore, to more stress and therefore more muscle contractures and pain associated with them, especially in the upper extremities.

3.2 Statistical analysis of the data

The results were analyzed by computing Spearman and Pearson correlation coefficients, according to the nature of the data (discrete or continuous, respectively) (Coelho et al., 2013), as well as by computing ordered logistic regression parameters, using as dependent variables pain in the upper limbs, pain in the lower limbs and pain in the back.

Given the results, we can conclude that the gender most affected by the problems of musculoskeletal complaints is the female gender (having correlation coefficients of −0.583 with appropriate significance of 0.000 for the DASH score, and −0.506 with significance of 0.003 for the pain in the upper limbs), and also that this gender is the one that is most exposed to inadequate psychosocial factors such as social support/leadership quality (coefficient of 0.454 with significance of 0.023), work/family conflict (coefficient of −0.501 with proper significance of 0.021), and self-esteem (coefficient of 0.459 with a proper significance of 0.021). We also conclude that the factors that are most associated with musculoskeletal complaints, particularly those that focus on upper and lower limbs are gender (coefficient of −0,643 with proper significance of 0.037 for pain in the upper limbs and a correlation coefficient of −0.896 with appropriate significance of 0.012 for pain in the lower limbs), age (coefficient of 0.618 with appropriate significance of 0.008 for pain in the lower limbs). Other domains that are associated with MSD complaints include inadequate posture (coefficient of −0.305 with appropriate significance of 0.042 for pain in the upper limbs). It can also be concluded that the most affected area associated

with ergonomic problems and with inadequately set psychosocial factors is the area of the upper limbs (fifteen workers report that this zone is the most problematic).

Finally, it can be concluded that although there are also several problems of psychosocial origin (psychological demands, insecurity, family/work conflict), they are not associated, or directly explained, by physical ergonomic issues. These may be explained primarily by the fact that the company at the time of data collection was going through a very delicate and uncertain moment, undergoing restructuring at national level, which the workers feared might lead to redundancies and mobility of human resources. Given the uncertain future of the company and its employees, the sense of insecurity of the workers may have been higher than it would be otherwise, with workers showing low self-esteem and low commitment in carrying out their functions.

The analysis of environmental conditions carried out led to detecting levels of ambient noise (generated within the office) that, while not posing a threat to auditory health, impaired concentration, and hence productivity and efficiency. An acoustic analysis was carried out within the office room (depicted in Fig. 1) where higher noise readings were obtained, and a proposal for improvement was developed, as described in the following section.

3.3 Acoustic reverberation

Noise is one of the most important ergonomic factors, as excess noise and duration of exposure to which workers are subjected, can cause serious injury. Exposure to excessive noise, may initially cause stress, but as time progresses it can cause more serious illnesses.

For the second part of the initial overall goal of the study, taking into account the results obtained,

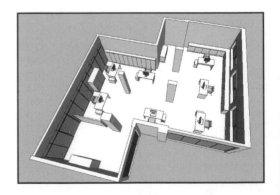

Figure 1. Three dimensional model of the office room where an acoustic analysis and an acoustic intervention proposal (aimed at decreasing the acoustic reverberation time) were developed.

it can be concluded that by choosing suitable materials one can greatly reduce the noise existing at an office. The proposed intervention concerns fitting carpets to the existing linoleum floor, adding noise absorption curtains to the vast window surfaces of the office and fitting compressed wood panels to the previously naked walls of the office. In the case study reported in this study, and through the proposed improvements, quite acceptable results could be obtained both in terms of acoustic reverberation time (which dropped from 6.725 (s) to 0.744 (s), calculated according to Sabine's formula (Beranek 2006)) or the attenuation level of noise (which decreased approximately 10dB, and thus from a mean value of about 65dB of noise to an average value of 55dB). In this case study, the improvement proposal was to amend the provision of materials that had already been implemented, but the ideal is that this improvement be done early in the implementation of the construction project to avoid spending in future interventions later in the building's lifecycle.

4 CONCLUSION

Taking into account the results obtained in the case study focusing on the organization studied, the organization must be concerned with the physical level of the domain of sitting because many ergonomic mismatches were observed (thirty workers had more than ten mismatches in their workplace) in this context, and that can be corrected with the use of chairs that are in accordance with current legislation. It should also take into account the domain-level of work devices (in this context thirty-one workers had more than six ergonomic mismatches identified) and as such provide support with a gel padded mouse mat for wrist support, and increase the suitableness of screen monitors in order to have them offer a more appropriate viewing distance (about 750 mm from the individual) and providing a document holder to each worker in order to reduce eyestrain and inadequate and harmful neck and upper limb postures.

Regarding psychosocial conditions, the company should pay particular attention to the climate of great insecurity and psychological demands exerted on workers, and to the inadequate psychological demands have most workers with adverse psychosocial exposures (twenty and twenty-one workers, respectively). Taking into account the phase that the organization is going through, it is necessary to talk to workers and try to provide confidence and support to them. To cope with the low self-esteem (eighteen workers with unfavorable psychosocial exposure level), the organization should try to motivate workers and give them more autonomy

to carry out their duties, leaving, however, a work structure to follow as a working basis, and providing well identified aims to achieve. Finally, with regard to the work-family conflict (with eighteen workers with unfavorable level of psychosocial exposure), this problem can also be related to the phase of restructuring the company was going through, because with this situation, workers are more concerned, more tense, which makes it necessary to do more to engage both professional tasks as well as household chores, leading, therefore, to more anxiety, more stress and therefore more muscle contractures and pain associated with them, especially in the upper limbs.

For the second part of the overall goal, taking into account the results obtained, it can be concluded that by choosing suitable materials one can greatly reduce the noise propagated at a site. In the case study, and through the proposed improvement, quite acceptable results could be obtained both in terms of acoustic reverberation time and the attenuation level of noise.

Indeed it can be concluded that in this case study the proposed improvement is based on changing the arrangement of materials that have been used, but the ideal is that this improvement is made early in the execution of the construction project, to avoid spending on future interventions.

The results of this case study, a combined study on the domains of ergonomics and acoustics, suggest that much still has to be done to give more importance to ergonomics and psychosocial factors so that these are considered properly in organizations, for the benefit of all parties involved, including society at large. To do so, one must make society more aware of these kinds of problems, and adequate training should be given to the top management of various organizations on aspects such as: What is ergonomics, how it should be applied, the advantages it entails (increasing productivity and competitiveness and reducing accidents and diseases at work) so that it may be put to use for the benefit of society at large.

REFERENCES

Beaton, D.E., Wright J.G. & Katz, J.N. 2005. Upper Extremity Collaborative Group. Development of the QuickDASH: Comparison of three item-reduction approaches. *Journal of Bone & Joint Surgery*. American Volume 2005; 87(5):1038–46.

Beranek, L.L. 2006. Analysis of Sabine and Eyring equations and their application to concert hall audience and chair absorption. *The Journal of the Acoustical Society of America*. 120(3):1399–1410.

Coelho, D.A., Harris-Adamson, C., Lima, T.M., Janowitz, I. & Rempel, D.M. 2013. Correlation between different hand force assessment methods from an ep emiological study. *Human Factors and Ergonomics in Manufacturing & Service Industries*, Volume 23, issue 2, pp. (forthcoming).

Corlett, E.N. &, Manenica, I. 1980. The effects and measurement of working postures. *Applied Ergonomics*, v. 11, n. 1, p. 7–16, march. 1980.

Lima, T.M. & Coelho, D.A. 2011. Prevention of musculoskeletal disorders (MSDs) in office work, *Work—A journal of prevention, assessment and rehabilitation*, Volume 39, issue 4, pp. 397–408.

Moncada, S., Llorens, C., Navarro, A. & Kristensen, T.S. 2005. ISTAS21 COPSOQ: versión en lengua castellana del cuestionario psicosocial de Copenhague [ISTAS21 COPSOQ: Spanish version of the Copenhagen Psychosocial Questionnaire]. *Arch Preven Riesgos Laboral* 2005;8(1):18–29. ISTAS 21 (CoPsoQ)—short version of the Copenhagen Psychosocial Questionnaire. Danish National Working Life Institute, Copenhagen, Denmark.

Selye, H. 1976. *Stress in health and disease*. Reading, MA: Butterworth, 1976.

Occupational Safety and Hygiene – Arezes et al. (eds)
© 2013 Taylor & Francis Group, London, ISBN 978-1-138-00047-6

Worker perception in relation to workplace comfort—an evaluation in metalworking industry

M. Talaia
DFIS, CIDTFF, University of Aveiro, Aveiro, Portugal

B. Meles
DEGEI, University of Aveiro, Aveiro, Portugal

L. Teixeira
DEGEI, GOVCOPP, IEETA, University of Aveiro, Aveiro, Portugal

ABSTRACT: The workplace environment can be evaluated based on thermal comfort, which is defined as the satisfaction expressed when a human subject is exposed to a given thermal environment. This evaluation implies a certain degree of subjectivity, which requires not only the analysis of the physical aspects of the environment, but also the subjective aspects associated with the state of mind of the individual. This study aims to assess the subjective aspects related to thermal comfort of workplaces in an industrial metalworking multinational, based on the perception of their occupants. The data were collected using a questionnaire. The results show that the 'influence of air temperature' is determinative in the responses of the 'sensation of thermal comfort' and the 'workplace occupant satisfaction', demonstrating an association among these three aspects.

1 INTRODUCTION

The intervention area of ergonomics is vast, and can be classified according to the object, the objective or the context of intervention (Rebelo 2004).

In the industrial context, Ergonomics can be evaluated based on several criteria, one of them the thermal comfort of workplaces. The thermal comfort plays an important role in building sustainability (Yao *et al.* 2009) and it is a key factor for a healthy and productive workplace (Akimoto *et al.* 2010, Taylor *et al.* 2008, Wagner *et al.* 2007).

Regarding the concept of thermal comfort, there are different approaches to define it. For example, ISO 7730 (2005) define thermal comfort as "the satisfaction expressed when an subject is subjected to a given thermal environment", which implies a certain degree of subjectivity, requiring not only the analysis of the physical aspects of environment, but also the analysis of the subjective aspects associated with the state of mind of the subject when inserted in a working environment.

Djongyang *et al.* (2010) says that the comfort is not a state condition, but rather a state of mind. Indeed, it is noted that the satisfaction of all subjects within a thermal environment, is an "almost" impossible task, since a thermally comfortable environment for one person may not be to another, i.e. thermal sensation can be different among people even in the same environment.

This study aims to assess the subjective aspects related to thermal comfort of workplaces in the context of an industrial metalworking multinational (Meles 2012), based on the perception of their occupants. For the data collection process, a questionnaire was used, and the results were treated based on descriptive statistical methods.

2 MATERIALS AND METHODS

The field study was performed in an industrial metalworking multinational company located in the centre of Portugal, with geographical coordinates about 40°N e 8°W. In order to accomplish the objectives of this study—to assess the thermal comfort of the workplaces based on the occupants' perception—the most critical workplaces in terms of thermal comfort were selected. The selection criterion was measured based on the *in situ* observation and confirmed by the guidelines of the section managers who, by their experience, were able to identify the most critical workplaces.

After the identification of the workplaces, a questionnaire was applied to a sample of 40 subjects. The data were collected in a single moment in the presence of the researcher. This study, by their nature, can be characterized as a cross-sectional descriptive study, as it focuses on a group of people that will represent the population studied (Ribeiro 1999).

Regarding the process of the construction of the questionnaire, after the identification of the main objectives and respective formulation of the questions that originated the pilot-questionnaire, it underwent a pre-test with a group of 5 individuals (a reduced sample structure homologous to the target population that was intended to study). The pilot-questionnaire was applied in the presence of the researcher, in order to understand any difficulties or other type of reaction by respondents. This pre-test resulted in minor changes, which were included in the final version of the questionnaire. Regarding the structure of the final-questionnaire, this consisted of 9 questions, grouped into three parts: (i) characterization of the sample, (ii) characterization of workplaces and working conditions, and (iii) evaluating the perception of workers regarding the thermal comfort felt in their workplace.

The first part included a set of questions in order to gather data to characterize the sample, particularly in terms of age, gender and jobs.

The second part included a set of questions in order to ascertain the working conditions, including those related to working positions, protections used, and seniority in workplace.

The third part aimed to question employees about some aspects in order to assess the perception of the workplace occupants about (i) the influence of the temperature, (ii) the sensation of comfort; and (iii) the satisfaction in the workplace in summer and winter. On this third part, the answers were given on 5-point *Likert* scale, where the number "1" corresponds to the statements "nothing warm", "nothing comfortable" and "nothing satisfied" and the number '5' to the statements "very warm" "very comfortable" and "very satisfied". The 5-point *Likert* scale was chosen because it is an appropriate scale to measure attitudes and perceptions and, additionally, it is considered as easy to understand by the public studied (Likert 1932). It should be noted that the results obtained from the 5-point *Likert* scale were analyzed and were consistent with the seventh ASHRAE scale (ANSI/ASHRAE55 2004). In practice, the use of a 5-point *Likert* scale suggests the same considerations or strategies to the extreme value of maximum 5 or 7.

Regarding the treatment of the data, it was analyzed based on a set of descriptive statistical techniques.

3 RESULTS OF THE SURVEYS AND DISCUSSION

As already mentioned, the sample used in this study consisted of 40 individuals, 75% males and 25% females.

The age range of the respondents was 23–56 years old; the mean was 40.7 years, and the median value was 42.0 years. 50% of the respondents had ages between 34 and 43 years, followed by 33% of the respondents with 44 to 53 years.

3.1 *Characterization of workplaces and working conditions*

This project surveyed workers who operated in 21 workplaces, spread across 6 factory sections. Of these, the Section D with eight workplaces is considered the most critical by having a very warm environment due to the presence of furnaces.

Regarding the mandatory protection equipment, the occupants should wear certain additional equipment for individual protection, such as gloves, earphones, goggles, mask, apron, bata and sleeves. The results of this study reveal that there are certain jobs where additional personal protective equipment has a representative rate of 100% with respect to its use. There are other stations where is not mandatory certain protective equipment, but the workplaces occupant insists to use it.

Concerning the type of clothing used by workers in the workplace, the results revealed some variation between hottest periods (summer) and the coldest periods (winter), as expected.

Regarding to the seniority of operators on the workplace, the results show that 40.0% are in the same workplace between 1 and 5 years, revealing also that a significant number of workers (17.5%) remain in the same place for over 20 years, Figure 1.

Regarding the rotation of workers, the results show that 42.5% of the workers remain at the same job during the 8-hour shift. It should be noted that

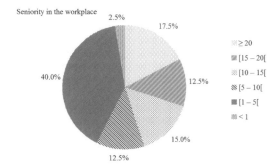

Figure 1. Seniority in the workplace.

314

the Section D presents the most critical situations in terms of rotation.

The occupants were questioned about symptoms that usually felt in the workplace and the results showed that most respondents not felt any symptoms, both in the morning shift (52.4%) and in the afternoon shift (47.4%). One possible explanation for this result could be the habituation of the worker to the workplace. Possibly, the number of years on the job (seniority) influences the answer to some questions. In the case of respondents who said they feel some symptoms, the most frequently reported was fatigue and dry mouth, verifying some relationship between these two symptoms. These symptoms have a higher incidence in warm workplaces, since the subject is exposed to relatively high temperatures.

3.2 Assessment of the perceptions of the workers in relation to the sensation of thermal comfort

Regarding the perception of operators to the thermal comfort they feel in the workplace, the answers were obtained based on a 5-point *Likert* scale, where the number '1' corresponds to a statement "nothing comfortable" and the number '5' to the statement "very comfortable".

Overall, the results revealed that the occupants consider their workplace uncomfortable, particularly in summer, with 57% of the responses in values '1' and '2', as shown in Table 1 and Table 2.

3.3 Influence of temperature in the workplace

Regarding the influence of temperature in the workplace, there are 79% of individuals who consider their workplace very warm in the summer, classified this issue with values '4' and '5' of the scale. While in winter there are 25 individuals (62.5%) that consider the temperature of the workplace 'neutral'.

Table 1. Influence of comfort sensation.

Issue	Values of the Likert scale				
	1	2	3	4	5
Summer	5	17	13	4	1
Winter	0	10	21	9	0

Table 2. Measure of location of the 'comfort sensation'.

Issue	n	Min	Q_1	Median	Q_3	Max
Summer	40	1	2	2	3	5
Winter	40	2	2	3	3	4

Legend: Min-minimum; Q_1-1st Quartile; Q_3-3tr Quartile; Max-maximum.

The "climate" indoor of the factory where the study was conducted, is directly influenced by atmospheric conditions that are registered outdoor. It should be noted that the factory location has the geographical coordinates about 40°N e 8°W.

Moreover, there is evidence to suggest that women are more sensitive to temperature, because in a total of 10 female workers, 70% answered that the workplace is very warm. This situation is according to the study published by Talaia *et al.* (2011).

Overall, the results revealed that the operators consider the workplace too warm, particularly in summer, as shown by the frequencies presented in Table 3 and descriptive statistics using measurements of location presented in Table 4.

3.4 Workplace occupant satisfaction

Regarding the workplace occupant satisfaction, the results revealed a position of low or medium satisfaction, Table 5 and Table 6.

Table 3. Influence of air temperature.

Issue	Values of the Likert scale				
	1	2	3	4	5
Summer	0	0	8	13	19
Winter	0	7	25	8	0

Table 4. Measure of location of the influence of temperature.

Issue	n	Min	Q_1	Median	Q_3	Max
Summer	40	3	4	4	5	5
Winter	40	2	3	3	3	4

Legend: Min-minimum; Q_1-1st Quartile; Q_3-3tr Quartile; Max-maximum.

Table 5. Workplace occupant satisfaction.

Values	Valores da Escala de Likert				
	1	2	3	4	5
Summer	1	13	21	1	4
Winter	0	4	27	6	3

Table 6. Measure of location of the 'occupant satisfaction'.

Issue	n	Min	Q_1	Median	Q_3	Max
Summer	40	1	2	3	3	5
Winter	40	2	3	3	3	5

Legend: Min-minimum; Q_1-1st Quartile; Q_3-3tr Quartile; Max-maximum.

It should be noted that the occupant satisfaction slightly increases during the winter. This situation is in agreement with the effect of the diurnal cycle of solar radiation, i.e. the energy balance is strongly conditioned by free thermal sources, e.g. furnaces.

Overall, the results show that the triangulation of the 'influence of air temperature' is determinative in the responses of the 'sensation of thermal comfort' and of the 'workplace occupant satisfaction'.

3.5 Section D—section considered hotter

As mentioned above, this study surveyed workers who operate in 21 workplaces, spread across 6 factory sections, being the 'Section D' considered the most critical by having a very hot environment due to the presence of furnaces. This section is composed by eight workplaces, and 16 occupants that work in two shifts were inquired. The main conditions of each operating position are: 45 is manual welding booth. This compartment was installed in this section about one year ago, and has an adequate exhaust system; 78 is a place of exit of products from the furnace where the quality inspection is made; 89/3 is a place of entry of products (on the furnace), having a protective mechanism before the entrance; 97 is a station with automatic welding booth. In this space there is a cooling system of products installed; 99 is a manual welding booth. This station receives the product coming from the workplace 97, but it only works when it is necessary; 178 is a manual welding booth. This station has an older system and consequently has an inappropriate exhaust system; 187 is a place of exit of products (from the furnace). The products from this station register a high temperature, being removed and placed in appropriate transport carriages; 191 represent a manual welding booth. Like workplace 178, it has an antique system and consequently an inappropriate exhaust system. Additionally, it is located close to the workplace number 78 and near to the entrance gate.

Figure 2 shows the layout of the section D, since the arrangement of the workplaces can be influenced by the thermal conditions and comfort felt by their occupants.

The results obtained in this section, in relation to the summer season, show that nearly all subjects who work in this section consider the environment in workplace very warm, very uncomfortable and therefore are dissatisfied with the conditions in which they work.

Figure 3 presents the results obtained based on an evaluation scale of 5 points (Likert), showing how the air temperature influences the workplace occupant. Particularly in summer, the values of median in '4' and third quartile in '5' reveal a situation very unfavorable. In the winter the responses are more favorable, with a median value in '3' and

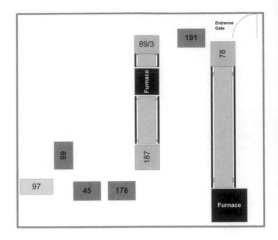

Figure 2.   Layout of workplace in section.

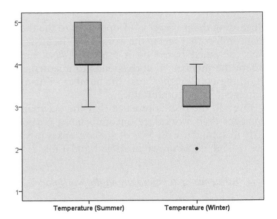

Figure 3.   Influence of air temperature in the workplace.

a maximum value in '4'. The results clearly reveal that the respondents consider the air temperature quite high at the workplace.

Regarding to the sensation of thermal comfort in the workplace, the results obtained in summer season reveal that 50% of respondents evaluated it, using values lower than '2' in scale, indicating that the workplace occupant felt uncomfortable. In the winter, the values were slightly better, Figure 4.

Figure 5 shows the results obtained for the workplace occupant satisfaction, revealing insufficient satisfaction, especially in summer. In the winter season, although some answers are considered outliers, the large concentration of responses has fallen in value '3' of the Likert scale, occupying a neutral position between 'neither satisfied nor dissatisfied'.

Nevertheless, the results also show that the acclimatization operator influences some of answers. For example, the occupant of workplace 178 that

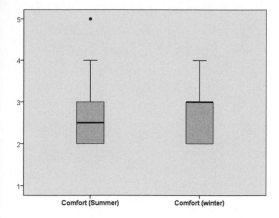

Figure 4. Sensation of thermal comfort in the workplace.

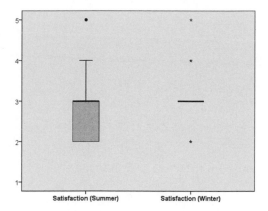

Figure 5. Workplace occupant satisfaction.

work in the morning shift, consider the station very hot, very comfortable and is satisfied with the thermal conditions. This contradiction can be explained by the fact that workers are afraid of being identified. Although the questionnaire is anonymous, the workplace and the section are identified.

4 CONCLUSION

The results showed that it is possible to get the workers perception in terms of the influence of air temperature, the feeling of thermal comfort and the satisfaction in the workplace. A more detailed study was carried out in a critical section (Section D), being considered a very hot environment due to the presence of furnaces.

The acclimatization of the operator has an influence in the improvement of their thermal sensation and the study showed that females seem to be more vulnerable to thermal sensations. This is in agreement with some studies by the specialized literature (Talaia & Alves 2011).

The results unequivocally showed that Section D has a very hot environment for some workplaces (Talaia et al. in press) and that intervention strategies should be taken by the Department of Health and Safety, whether of organizational nature or as individual worker protection.

REFERENCES

Akimoto, T., Tanabe, S.I., Yanai, T. & Sasaki, M. 2010. Thermal comfort and productivity-Evaluation of workplace environment in a task conditioned office. *Building and Environment* 45(1):45–50.

ANSI/ASHRAE55. 2004. Thermal environmental conditions for human occupancy. American Society of Heating, Refrigerating and Air—Conditioning Engineers. *AINSI/ASHRAE Standard* 55:1–26.

Djongyang, N., Tchinda, R. & Njomo, D. 2010. Thermal comfort: A review paper. *Renewable and Sustainable Energy Reviews* 14(9):2626–2640.

ISO 7730. 2005. *Ergonomie des ambiances thermiques—Détermination analytique et interprétation du confort thermique par le calcul des indices PMV et PPD et par des critères de confort thermique local.* Switzerland, International Standardisation Organisation, Geneva, Suisse.

Likert, R. 1932. A Technique for the Measurement of Attitudes. *Archives of Psychology* 140:1–55.

Meles, B. 2012. *Ergonomia Industrial e Conforto Térmico em postos de trabalho.* Dissertação apresentada à Universidade de Aveiro (publicada). Aveiro (*in Portuguese*).

Rebelo, F. 2004. *Ergonomia no dia a dia.* Lisboa: Edições Sílabo, Lda (*in Portuguese*).

Ribeiro, J.L.P. 1999. *Investigação e Avaliação em Psicologia e Saúde.* Lisboa: Climepsi Editores (*in Portuguese*).

Talaia M.A.R, Meles, B. & Teixeira, T. (in press). Guide lines to improve environment conditions in jobs—metalworking industry. *International Symposium on Occupational safety and hygiene, Sho13.* Guimarães, Feb 14–15 (*in Portuguese*).

Talaia M.A.R & Alves, J. 2011. A Condução e o Conforto Térmico na Segurança Rodoviária—Estudo de Percepção. *Proceedings 6° Cong° Luso—Moçamb° de Eng^a e 3° Cong° de Eng^a de Moçambique* (artigo CLME'2011_2806 A, 12 páginas). Maputo, 29 Agosto–2Setembro (*in Portuguese*).

Taylor P., Fuller, R.J. & Luther, M.B. 2008. Energy and thermal comfort in a rammed earth office building. *Energy and Buildings* 40:793–800.

Wagner, A., Gossauer, E., Moosmann, C., Gropp, T. & Leonhart, R. 2007. Thermal comfort and workplace occupant satisfaction—Results of field studies in German low energy office buildings. *Energy and Buildings* 39(7):758–769.

Yao R., Li B. & Liu J. 2009. A theoretical adaptive model of thermal comfort adaptive predicted mean vote (aPMV). *Building and Environment*; 44:2089–96.

Occupational Safety and Hygiene – Arezes et al. (eds)
© *2013 Taylor & Francis Group, London, ISBN 978-1-138-00047-6*

Implications of obesity on occupational health

A. Colim & P.M. Arezes
CGIT, University of Minho, Guimarães, Portugal

P. Flores
CT2M, University of Minho, Guimarães, Portugal

ABSTRACT: Obesity is an emerging public health problem. Therefore, obese subjects are a growing fraction of the workforce. Although different negative effects of obesity on work performance have been demonstrated before, additional studies are required to provide a more complete understanding of those effects on work performance. This study aimed to analyse the perceptions of occupational health practitioners about the obesity implications on some work tasks, especially on manual materials handling tasks. With this purpose, qualitative data were collected from eight semi-structured interviews. The interviewees argued that obese workers present different limitations, such as lower physical resistance, locomotor difficulties and postural restrictions. These individual limitations have negative implications on work, namely on absenteeism and on the decrease of productivity. To prevent these consequences, companies should adopt educational programs, among other initiatives, in order to establish a culture of health promotion at the workplaces.

1 INTRODUCTION

The improvement of social conditions among the population led to an increase of the number of individuals with overweight and obesity. From the worldwide statistics it is possible to see that obesity has more than doubled since 1980 and, actually, more than 1.4 billion adults have overweight (WHO 2012). In Portugal, statistical data show that more than 50% of all population has, at least, an unhealthy weight (INE 2010). Therefore, it is likely that obese people represent already a significant fraction of the workforce.

Obesity is being associated with social, psychological and physical problems, including musculoskeletal disorders (MSD), which can negatively affect the productivity (Lidstone et al., 2006, Morris 2007). Additionally, employees with overweight are absent from work due to illness more often and for longer periods than employees with normal weight (Tsai et al., 2008). Lier et al. (2009) demonstrated that the absenteeism, in obese workers, is frequently related to musculoskeletal complaints.

In the specific occupational environment, other implications of obesity were described in several studies. For example, Renna & Thakur (2010) reported that, in the US, obesity beyond increases the incidence of work disability, it also increases the probability of taking an early retirement. In European countries, individual weight may also affect employment opportunities, probably due to

employers' discrimination toward obese individuals (Atella et al., 2008). Furthermore, the disabilities related to obesity appear also to originate a negative impact on the salaries in some European countries (Brunello & D'Hombres 2007).

Focused on obese people' characteristics, specific studies analysed different activities of daily living. Considering the walking, biomechanical studies demonstrated that body adiposity produces changes through this activity (Wearing et al., 2006). For example, comparing obese and non-obese, the former tend to present higher ground forces and pressures (Sousa et al., 2010), and also a decrease on the kinematic parameters of the motion, as well as a delay of the ankle joint, reducing the range of motion and increasing the muscle overload (Silva 2009).

The individual body conditions, such as the adiposity, may affect the postural balance. However, obesity effects on posture maintenance are rarely studied. In addition, postural analysis tools, frequently used in ergonomic studies, seem to consider only people with normal weight. Based on this assumption, Park et al. (2009) performed a psychophysical research with obese and non-obese participants. These participants performed static box-holding tasks in different working postures. The obese participants reported a higher perceived overload in all postures considered, demonstrating that obesity increases postural stress. In this context, Gilleard & Smith (2007) revealed that the obese subjects showed a more flexed trunk posture and increased hip joint moment and hip-to-bench distance for a simulated standing work task.

Additionally, it was also demonstrated that obese subjects have more problems with work-restricting musculoskeletal pain when compared with normal weight subjects (Peltonen et al., 2008). From the biomechanical point of view, excessive body weight can negatively affect the muscles and spine behaviour during occupational tasks. Different studies have found positive relations between obesity and back pain (Kostova & Koleva 2001), as well as the decrease in the trunk muscle strength (Bayramoglu et al., 2001, Hulens 2001). However, the literature on epidemiologic research does not demonstrate a clear link between obesity and low back pain (Xu et al., 2008).

The back pain is often related to manual materials handling (MMH) tasks. Several findings have shown that these tasks are very common in a wide variety of workplaces and represent an important MSD risk factor, mainly for the low back (Yeung et al., 2002). In this context, some findings suggested that obesity does not seem to reduce the maximum acceptable weight (according to a psychophysical approach) (Singh et al., 2009). However, the referred researchers pointed out that this area requires further studies, which should try to find other type of data, such as biomechanical data. Consequently, a biomechanical research by Xu et al. (2008) evaluated the lifting kinematics and kinetics of individuals with different body constitutions. At the beginning of this study, the authors had the expectation that the heaviest people might lift more slowly, in order to minimize the loading on the muscles and structures of their low back. However, in the obese group greater values for the kinematics trunk variables were registered, comparing with the normal weight participants, which seem to be due to the generation of greater inertial forces.

In the majority of the previous works, the body mass index (BMI) was the only indicator of obesity that was used. However, BMI does not reflect accurately the individual physical constitution. This indicator is based exclusively on individual weight and height and does not distinguish the fat-free mass and fat mass rates (Rezende et al., 2007), foreseeing the need to use alternative methods/indicators for obesity assessment that can be more appropriate and comprehensive (Xu et al., 2008).

Although obesity has been intensively investigated over the past years, the findings still involve some controversy. For example, the effect of adiposity on the function of the locomotor system is not well understood (Wearing et al., 2006). In addition, ergonomic studies are required to provide a more complete understanding of obesity effects on work performance (Williams & Forde 2009). For these reasons, the present work aims at analysing

the obesity effects on work individual capacity for MMH, taking into consideration the opinions of occupational health practitioners. It is expected that this study can contribute to the identification of relevant indicators that can be used to design and develop a biomechanical study focused on the obesity, as a risk factor for the development of MSD when doing MMH tasks.

2 METHODOLOGY

The current paper includes the first stage of a wider research project that is focused on the biomechanical effects of obesity in MMH tasks, in particular in the lifting and lowering tasks. This stage included a search and a critical review of the published and available technical and scientific literature concerning the project subject. A particular attention was given to the impact of obesity on the ability to work, namely in what concerns to the abilities and limitations of obese people to perform MMH tasks.

At this exploratory stage, qualitative data were also collected from eight semi-structured interviews to occupational health practitioners (one nurse, four physicians and three physiotherapists). This convenience non-probabilistic sampling technique was adopted with the aim to obtain detailed information about the research problem (Saunders et al., 2007). All interviews were audio-recorded, transcribed and the opinions obtained were classified into meaningful categories, as it is exemplified in Figure 1. This example shows the construction of the categories, according to the

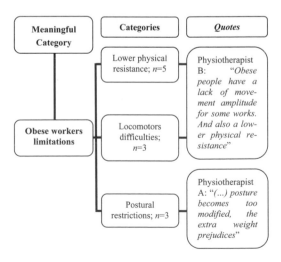

Figure 1. Example of a meaningful category construction (*n* represents the frequencies included in the category).

obtained answers. It is important to clarify that some interviewees' answers fall into more than one category (like the case of the answer of the Physiotherapist B, transcribed in Figure 1).

The topics of the interviews derived from the literature review and discussion within the research team. These are related to MMH tasks, concerning the followings topics: (i) Techniques for body constitution assessment used by the occupational health professionals; (ii) Most frequent limitations, complaints and MSD in obese workers; (iii) Possible occupational improvements recommended to obese workers.

3 RESULTS AND DISCUSSION

The interviewees have a professional experience in occupational health ranging from 10 to 25 years. The workers assisted by them (between 200 to 4200 workers by each practitioner) belong to different companies (such as hospitals, textile and footwear industries, administrative services, storage and retail trade). As mentioned before, the collected answers were classified into meaningful categories, analysed and compared with theoretical framework. Table 1 summarizes some of the main results.

The first part of the interview was related to body composition assessment. This task seems to be undervalued in the occupational health practice. Consequently, most of interviewees reported that they only make a register of weight and height with the sole purpose of calculating the BMI, but typically these data are not analysed and considered in the individual ability evaluation to specific occupational tasks. As mentioned before, obesity is associated to several negative implications at work, therefore it seems that this should be face as a relevant factor for the evaluation and control of occupational health. In this intervention field, it can be also important to apply body constitution assessment techniques with more sensitivity and accuracy than the BMI, such as the waist circumference.

Considering the effects of obesity on work ability, in particular on MMH tasks, the perception of the interviewees are aligned with data obtained in other previous empirical studies (Table 1).

Back, knees, arms and shoulders are pointed out by the interviewees as being the most affected areas by pain and MSD. In addition, these practitioners claimed that the type of MSD that affects obese workers who perform MMH tasks is more dependent on the occupational characteristics, rather than on the body constitution. However, they highlighted that these disorders are more frequent and their treatment more difficult in obese than in normal weight workers. Consequently, and as argued

Table 1. Some of the main results obtained in interviews and the relation to other studies.

Interview themes	Principal results		Previous empirical work
Obesity assessment techniques used	BMI	$n = 6$	BMI does not distinguish the fat-free mass and fat mass rates; other anthropometric data have shown a greater sensitivity in the obesity assessment, comparing with BMI (Rezende et al., 2007).
	Waist circumference	$n = 2$	
	None	$n = 1$	
Most frequent limitations on obese workers	Lower physical resistance	$n = 5$	Positive correlations between obesity and MSD, as well as respiratory disturbs (Lidstone et al., 2006); biomechanical restricted movement in obese (Wearing et al., 2006); greater postural overload perceived by obese during MMH (Park et al., 2009); obese present postural alterations during standing work (Gilleard & Smith 2007).
	Locomotors difficulties	$n = 3$	
	Postural restrictions	$n = 3$	
Implications on work	Increase absenteeism	$n = 4$	The obesity is a debilitating health condition that has a negative effect on productivity (Lidstone et al. 2006, Morris 2007); health complaints reported by obese workers were significantly related to work absence (Tsai et al., 2008, Lier et al., 2009).
	Decrease productivity	$n = 3$	
	None	$n = 2$	

by the practitioners interviewed, the referred limitations have social and economic negative implications on occupational contexts, frequently reflected in a decreased productivity and an increased absenteeism. If no preventive measures are taken, this negative impact will continue to increase.

Therefore, possible occupational improvements were also discussed in the interviews. The development of postural education programs, workplace adaptations, practices and educational programs to prevent obesity constitute the main recommended actions to prevent MSD associated to MMH tasks. These recommendations were mainly oriented to obese workers but can include all workers.

Finally, in the occupational health domain more emphasis should be allocated to the assessment and the prevention of obesity. The tasks allocation and the workplace adaptations must consider this individual factor. Therefore, with this type of measures it is expected to reduce the negative implications of obesity on work. As argued by the interviewees, and also by other authors, such as Tsai et al. (2008), the companies have to adopt educational programs, among other actions, in order to implement a culture of health promotion.

4 CONCLUSIONS

Obesity is a risk factor for the appearance of chronic pathologies, such as MSD, and it has been shown that it negatively affects work. It is important to adapt work according individuals' physical condition, considering obesity as an important factor in the evaluation and control of the workers' health.

Therefore, companies should implement educational programs, among other initiatives, in order to establish a culture of health promotion at the workplaces. With this, it is expected to reduce the negative impact of obesity on work.

The collected qualitative data are also relevant to help in the next stages of the research project and to justify its relevance and opportunity. As mentioned before, this work is part of a biomechanical research project regarding the effects of obesity when performing MMH tasks. Hopefully, this research will result in an important output to increase the knowledge about the effects of obesity at work and to help in the construction of workplaces more suitable to obese workers.

ACKNOWLEDGMENTS

This work was financially supported by the Strategic Plan of the Engineering School of the University of Minho: Agenda 2020, and it was carried out under the scope of the Research Centre for Industrial and Technology Management (CGIT), and was partially funded with Portuguese national funds through FCT—Fundação para a Ciência e a Tecnologia, in the scope of the R&D project PEst-OE/EME/UI0252/2011. This research stage was also directly dependent of the willingness shown by the occupational health practitioners interviewed, who volunteered to contribute to this study.

REFERENCES

Atella, V., Pace, N. & Vuri, D. 2008. Are employers discriminating with respect to weight? European evidence using quantile regression. *Econ Hum Biol* 6(3): 305–329.

Bayramoglu, M., Akman, M.N., Kilinc, S., Cetin, N., Yavuz, N. & Ozker, R. 2001. Isokinetic Measurement of Trunk Muscle Strength in Women with Chronic Low-Back Pain. *American Journal of Physical Medicine & Rehabilitation* 80(9): 650–655.

Brunello, G. & D'Hombres, B. 2007. Does body weight affect wages? Evidence from Europe. *Economics and Human Biology* 5: 1–19.

Gilleard, W. & Smith, T. 2007. Effect of obesity on posture and hip joint moments during a standing task, and trunk forward flexion motion. *International Journal of Obesity* 31(2): 267–271.

Hulens, M., Vansant, G., Lysens, R., Claessens A., Muls, E. & Brumagne, S. 2001. Study of differences in peripheral muscle strength of lean versus obese women: an allometric approach. *International Journal Obes. Relat. Metab. Disord.* 25: 676–681.

INE—Instituto Nacional de Estatística (2010). *Anuário estatístico de Portugal 2009*. Lisboa: INE.

Kostova, V. & Koleva, M. 2001. Back disorders (low back pain, cervicobrachial and lumbosacral radicular syndromes) and some related risk factors. *J Neurol. Sci.* 192: 17–25.

Lidstone, J., Ells, L., Finn, P., Whittaker, V., Wilkinson, J. & Summerbell, C. 2006. Independent associations between weight status and disability in adults: Results from the health survey for England. *Public Health* 120: 412–417.

Lier, H., Biringer, E., Eriksen, H. & Tangen, T. 2009. Subjective Health Complaints in a Sample with Morbid Obesity and the Complaints' Relation with Work Ability. *European Psychiatry* 24(S1): S750.

Morris, S. 2007. The impact of obesity on employment. *Labour Economics.* 14, 413–433.

Park, W., Singh, D., Levy, M. & Jung, E. 2009. Obesity effect on perceived postural stress during static posture maintenance tasks. *Ergonomics* iFirstarticle: 1–14.

Peltonen, M., Lindroos, A. & Torgerson, J. 2008. Musculoskeletal pain in the obese: a comparison with a general population and long-term changes after conventional and surgical obesity treatment. *Pain* 104: 549–557.

Renna, F. & Thakur, N. 2010. Direct and indirect effects of obesity on U.S. labor market outcomes of older working age adults. *Social Science & Medicine* 71: 405–413.

Rezende, F., Rosado, L., Franceschinni, S., Rosado, G., Ribeiro, R. & Marins, J. 2007. Revisão crítica dos métodos disponíveis para avaliar a composição corporal em grandes estudos populacionais e clínicos. *Archivos Latinoamericanos de nutricion* 57(4): 327–334.

Saunders, M., Lewis, P. & Thornhill, A. 2007. *Research Methods for Business Students—4th Ed.* Harlow: Financial Times Prentice-Hall.

Silva, T. 2009. *Análise da Marcha em mulheres obesas e sua relação com índice de massa corporal.* MsD Dissertation, Brasília: University of Brasília.

Singh, D., Park, W. & Levy, M. 2009. Obesity does not reduce maximum acceptable weights of lift. *Applied Ergonomics* 40(1): 1–7.

Sousa, H., Peduzzi, M., Abreu, S., Machado, L., Santos, R. & Vilas Boas, J. 2010. Caracterização Cinemática e Dinamométrica de Indivíduos Obesos. In Roseiro, L. & Neto, A. (ed.). *4º Congresso Nacional de Biomecânica, Coimbra, Portugal, 4–5 February 2010,* 203–208.

Tsai, S., Ahmed, F., Wendt, J., Bhojani, F. & Donnelly, R. 2008. The Impact of Obesity on Illness Absence and Productivity in an Industrial Population of Petrochemical Workers. *AEP* 18(1): 8–14.

Wearing, S., Henning, E., Byrne, N., Steele, J. & Hills, A. 2006. The biomechanics of restricted movement in adult obesity. *Obesity Reviews* 7(1): 13–24.

Williams, N. & Forde, M. 2009. Ergonomics and obesity. *Applied Ergonomics* 40: 148–149.

WHO—World Health Organization 2012. *Obesity and overweight.* Fact sheet N°311. Accessed on: www.who.int/mediacentre/factsheets/fs311/en/index.html.

Xu, X., Mirka, G. & Hsiang, S. 2008. The effects of obesity on lifting performance. *Applied Ergonomics* 39: 93–98.

Yeung, S., Genaidy, A., Huston, R. & Karwowski, W. 2002. An expert cognitive approach to evaluate physical effort and injury risk in manual lifting—A brief report of a pilot study. *Human Factors and Ergonomics in Manufacturing* 12(2): 227–234.

Occupational Safety and Hygiene – Arezes et al. (eds)
© *2013 Taylor & Francis Group, London, ISBN 978-1-138-00047-6*

The effects of local illumination and work-rest schedule on light-on test inspection in a TFT-LCD plant

Li-Jen Twu
Department of Industrial Engineering and Engineering Management,
National Tsing Hua University, Hsinchu, Taiwan, R.O.C.

Chih-Long Lin
Department of Crafts and Design, National Taiwan University of Arts, Taipei, Taiwan, R.O.C.

M.J.J. Wang
Department of Industrial Engineering and Engineering Management,
National Tsing Hua University, Hsinchu, Taiwan, R.O.C.

ABSTRACT: TFT-LCD (thin film transistor liquid crystal display) industry is one of the most important industries in Taiwan. Light-on test is one of the visual inspection tasks in TFT-LCD panels manufacturing. The operators of the light-on test inspection are working in a high luminance ratio (low ambient illumination and highly local illumination) environment. The purpose of this study was to examine the effects of different local illumination and work-rest schedule on inspectors' visual fatigue and task performance. A total of 9 well-trained female operators participated in this experiment. The experiment involved four levels of local illumination: (1) 2600 lx (with original lamp); (2) 6000 lx (with new lamp); (3) 2600 lx (with new lamp plus polarizer panel), and (4) 1900 lx (with new lamp plus polarizer panel and lamp half-covered). The work-rest schedule had two different conditions: (1) working 29 minutes and resting 1 minute, then repeat once, and (2) working 58 minutes and resting 2 minutes. At the beginning and the end of each experiment combination, the subjects took critical fusion flicker (CFF) tests to measure their visual fatigue. The results showed that the local illumination effect was significant on task performance. Using new lamp (6000 lx) and new lamp plus polarizer panel (2600 lx) had lower miss rate than the original lamp (2600 lx). It seems that increasing local illumination to 6000 lx, and using polarizer panel to reduce high contract can enhance the inspection performance. The results also showed that the local illumination and work-rest schedule did not show significant influence on visual fatigue. This might due to the operators were adapted to the work environment. A further study is necessary to involve more environmental illumination factors in evaluating the light-on test inspection.

1 INTRODUCTION

The manufacturing processes of TFT-LCD (thin film transistor liquid crystal display) include array process (Array), cell process (Cell) and assembly process (Module) sequentially. The array process is similar to that of being used to fabricate semiconductor devices, which includes cleaning, photolithography, deposition of thin film and etching. The difference between these two processes is the material being used onto which is a glass substrate in TFT-LCD array process, while onto a silicon wafer in semiconductor process. After finishing array process, the glass substrate is combined with a color filter then cut into different size panels as originally specified, and this is the cell process. The cell process also includes liquid injection, polarizer panel combination and panel inspection. The assembly process is followed to assemble driver IC, backlight unit and PCB with the panel. Then the TFT-LCD panels will become finished product after completing final inspection in assembly process.

In order to assure the high quality TFT-LCD panels output, different inspection tasks are included such as optical inspection, electrical inspection and human eyes inspection in the manufacturing process (Su et al., 2000). In the cell process, the TFT-LCD panels need to be inspected in a high luminance contrast ratio (low ambient illumination and highly local illumination) environment, and this task is called the light-on test inspection for examining the panel defects. Operators of light-on test inspection are required to examine defects on panels using a 15-time magnifying glass under the high luminance contrast ratio environment. They complained about the visual discomfort problem (Wang and Wu, 1999).

The purpose of this study was to identify the important factors that influenced light-on test operators' visual fatigue and task performance, and to propose countermeasures to improve the problems. First of all, a Fishbone diagram analysis was conducted in the workplace and the light-on test operators were requested to fill out a questionnaire to find out the key factors. The factors that had major influence on light-on test inspection were identified as local illumination and work-rest schedule. Different local illumination levels and work-rest schedule conditions were designed as the experiment variables in this study.

2 MATERIALS AND METHOD

2.1 Field investigation

The light-on test inspection is performed by human eyes to check panel's quality under high local illumination (2600 lx) condition first, and then to place the panel on a test machine turning on light from panel's backlight for subsequent defect inspection. This task is performed in a low ambient illumination (98 lx in average) condition. The work environment with a high luminance contrast ratio is shown in Figure 1.

In the first phase of this study, we performed a Fishbone diagram analysis for the light-on test inspection, and the results are shown in Figure 2.

Four critical factors are identified to have important influence on this task, including the experience of operators, the inspection equipment condition, operator's visual fatigue and the defects to be detected. According to operator's questionnaire survey and site surveillance of work environment, we found that the operator's visual fatigue was the major factor to affect task performance. Thus, the

Figure 1. The work environment of light-on test inspection.

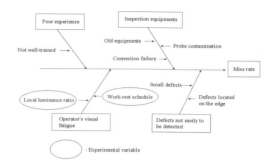

Figure 2. The fishbone diagram of the light-on test inspection.

local illumination ratio and work-rest schedule were considered for experiment evaluation.

2.2 Participants and experimental design

A total of nine well-trained female operators who worked in a TFT-LCD plant participated in this experiment. Their working experience was 1.8 ± 0.5 years. All subjects had good eyesight and had a clear understanding about the experiment procedure.

Four levels of local illumination were specified, including (1) 2600 lx (with original lamp); (2) 6000 lx (with new lamp); (3) 2600 lx (with new lamp plus polarizer panel), and (4) 1900 lx (with new lamp plus polarizer panel and lamp half-covered). The purpose of the polarizer was to reduce high contrast effect to inspectors' eyes. The lamp was a circular fluorescent tube, and the difference between original and the new lamps was the illumination level. The color temperature of the lamp was 6700 K. When the operators inspected the panel at 20 cm under the lamp, they felt the disturbance of the glare caused by the lamp. The four different illumination conditions are shown in Figure 3 to Figure 6.

The two work-rest schedules designed were: (1) working 29 minutes and resting 1 minute, then repeat once, and (2) working 58 minutes and resting 2 minutes. Critical flicker fusion (CFF) tester (Model No. 502, made by Takei Kiki Kogyo Co., Japan) was used to evaluate inspectors' visual fatigue. There were 70 experimental panels being used in this study, including 40 good panels and 30 no-good panels. The inspectors' task performance were classified into four categories: (1) judging good panel as good (OK/OK); (2) judging no-good panel as no-good (NG/NG); (3) judging good panes as no-good (OK/NG), and (4) judging no-good panel as good (NG/OK).

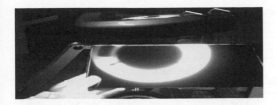

Figure 3.   2600 lx (with original lamp).

Figure 4.   6000 lx (with new lamp).

Figure 5.   2600 lx (with new lamp plus polarizer panel).

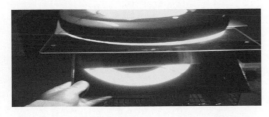

Figure 6.   1900 lx (with new lamp plus polarizer panel and lamp half-covered).

2.3   Experiment procedure

At the beginning of the experiment, all subjects took CFF test four times including starting from CFF 10 Hz upward and from CFF 60 Hz downward with two times each, then calculating the average of the four trials as the pre-CFF value. After the experiment task, the same CFF test procedure was repeated, and the average value was taken as the post-CFF value. The difference value between the post-CFF and pre-CFF value was taken as the indicator of visual fatigue.

Each subject had to finish a total of 8 experiment combinations (four illumination levels by two work-rest schedules). The panels were randomized each time to make sure that the subjects would not be able to predict the defect conditions of the panels. Subjects should keep the same inspection pace as they usually did at work. If the subject did not inspect more than 40 panels in an experiment run, she should continue the experiment until reaching the specified number of panels. After each experiment run, the task performance in terms of OK/OK, NG/NG, OK/NG and NG/OK percentage were obtained.

2.4   Statistical analysis

One way analysis of variance (ANOVA) was performed to evaluate the effects of different illumination level and work-rest schedule, on the visual fatigue and task performance. Then, Duncan's multiple range test was conducted as a post-hoc testing.

3   RESULTS AND DISCUSSION

Table 1 shows that mean values of CFF test and the inspection performance under different experiment conditions. Except for inspecting under 2600 lx (with original lamp), the inspectors' post-CFF value was lower than the pre-CFF value.

In TFT-LCD industry, the ratio of judging no-good panels as good panels is called leakage rate. In here, the leakage rate was expressed as NG/OK. The effect of different illumination level was significant on leakage rate (NG/OK) performance. The results of Duncan's post-hoc testing showed that inspecting under illumination level 2 (6000 lx, with new lamp) and level 3 (2600 lx, with new lamp plus polarizer panel) had significantly lower leakage rate (5%) than that of under illumination level 1 (2600 lx, with original lamp) (8%). But there was no significant difference in leakage rate between illumination level 2 and level 3. The results showed that increasing local illumination to 6000 lx, and using polarizer panel to reduce high luminance contrast can enhance the inspection performance. Except for using the original lamp had a positive difference in CFF value (0.32 Hz), the CFF difference values of this experiment were all negative (from -0.28 Hz to -1.02 Hz). But the results showed that the local illumination and work-rest schedule did not show any significant influence on visual fatigue. This might be due to that operators had been adapted to the work environment.

Table 1. Mean value of CFF difference and task performance.

		Illumination level				Work-rest schedule	
		2600 lx (with original lamp)	6000 lx	2600 lx (with new lamp plus polarizer panel)	1900 lx	W29R1	W58R2
CFF difference value (Hz)		0.32	−0.28	−1.02	−0.28	−0.33	−0.3
Task performance (%)	OK/OK	59	60	60	58	60	59
	NG/NG	32	33	34	34	33	34
	OK/NG	1	2	1	1	1	1
	NG/OK	8*	5*	5*	7*	6	6

*$p < 0.05$

4 CONCLUSION

The finding of this study indicates that increasing local illumination and using polarizer panel to reduce high contrast can enhance the inspection performance. This information is useful for the TFT-LCD panel manufactures for improving the quality of the products. Although there is no significant difference on visual fatigue under different local illumination and work-rest schedule, it still requires further studies with longer period of experiment time to confirm the relationships. In the next stage of the in-depth study, additional illumination designs and task factors can be considered to find the best working condition that can facilitate both the effectiveness of the TFT-LCD production as well as the quality of operators' working life.

REFERENCES

Bullimore M.A., Fulton E.J. and Howarth P.A. (1990). Assessment of visual performance. In: Wilson J.R. and Corelett E.N. (Eds.), Evaluation of human work: A practical ergonomics methodology. Taylor and Francis, London.

Ferguson D.A., Major G. and Keldoulis T. (1974). Vision at work, visual defect and the visual demand of tasks, Applied Ergonomics, 5, 84–93.

Horie Y. (1987). A study of optimum term of working hours with rest intervals for VDT workers. Japanese Journal of Ergonomics, 23(6), 373–383.

Lu, C., Sheen, J., Su, S., Kuo, S., Yang, Y., and Kuo, C. (2007). Work environment and health effects of operators at light-on test process in TFT-LCD plants. Ergonomics and Health Aspects, HCII 2007, LNCS 4566, 113–117.

Su, C., and Huang, C. (2000). LCD light-on test instrument. J Mechanical Ind. 207, 146–151.

Wang, M.J., and Wu, S.C. (1999). Application of ergonomics in semiconductor industry. Industrial Safety Technology, 30, 33–41.

Occupational Safety and Hygiene – Arezes et al. (eds)
© 2013 Taylor & Francis Group, London, ISBN 978-1-138-00047-6

Workplace accommodation for people with disabilities: A literature review

B.M. de Guimarães & L.B. Martins
Federal University of Pernambuco, Brazil

B. Barkokébas Jr.
University of Pernambuco, Brazil

A.M. das Neves
Federal University of Pernambuco, Brazil

ABSTRACT: The study objective was to conduct a literature review on the topics that are covered within the sphere of the accommodations of workstations for People with Disabilities (PD) and verify the truths and uncertainties found in this context. The literature review was performed from research in PubMed, Scopus and Web of Science electronic databases. From the research in databases, it was obtained 40 papers and the main topics discussed were: Factors that can facilitate accommodation of the workplace for the PD, evaluation methods and workstation adapting for the PD and the types of accommodations of the workplace for PD. It appears that, in the various issues related to work inclusion of PD, there is some consensus and various uncertainties, since in certain approaches there is little research and few experimental studies that describe and compare different types of accommodations of workplaces for PD and the effectiveness between the same types and between different types of disabilities.

1 INTRODUCTION

People with disabilities represent about 15% of the world population, or one billion people (WHO, 2011). In Brazil, according to the 2010 Census, 23.9% of the population, that is 45.6 million, is PD (IBGE, 2011).

The inclusion of this population in the workforce has been discussed and encouraged in many countries through various laws, such as with stipulated quotas for the employment of people with disabilities. Germany has a quota of 5% for the employment of people with severe disabilities in companies with more than 20 workers. In South Africa, government departments must be formed by at least 2% of disabled workers. Turkey has a quota of 3% for disabled workers in companies with more than 50 employees (WHO, 2011). While in Brazil, the laws establish a quota of 20% for public sector enterprises and 2% to 5% for private companies with more than 100 employees.

Despite attempts to include people with disabilities in employment, the number of individuals seeking employment and those receiving job opportunities remains low (Westmorland & Williams, 2002). A recent study showed that in 27 countries, people with disabilities at a working age experience significant disadvantages and worse labor market outcomes than people without disabilities (OECD, 2010). Meanwhile, in Brazil among the 46.3 million people employed in 2011, only 325,000 were declared as people with disabilities, representing 0.7% of the total amount of workers (Brazil, 2011). Unfortunately, workers with disabilities are often seen as a problem to be dealt with, instead of an opportunity that can be leveraged (Morton, Foster & Sedlar, 2005).

Workstation accommodations are individualized solutions that enable people with disabilities to perform the work activities and be more productive (Zolna et al. 2007). Examples of adaptations include ensuring that the procedures for employment selection are accessible to all, there are adaptations of the work environment, and there are Assistive Technologies, there is a workload modification and that there is a redistribution of non-core tasks to other workers. Thus, the implementation of workstation accommodations is a vital tool to increase the employment of individuals with disabilities.

Therefore, it is important to understand the main topics discussed in the literature in the workstation accommodations context for the PD. Therefore, the objective of this study was to conduct a literature review on the topics that are covered within the workstation accommodations sphere for persons with disabilities and verify the truths and uncertainties found in this context.

2 MATERIALS AND METHOD

The literature review was performed from electronic researches in the Portal of Electronic Journals of the Coordination for the Improvement of Higher Education Personnel (CAPES), which is a virtual library that collects and provides to the education and research institutions in Brazil the best of the national and international scientific production (Portal of Electronic Journals CAPES, 2012). The research was performed in the databases PubMed, Scopus and Web of Science during the months of March and April 2012 and papers published between the years 1992 and 2011 were found.

In order to achieve the collection of the papers in the databases searched, we used the following combinations of keywords: Job accommodation disabled person, Job accommodation people with disabilities, workplace accommodation disabled person, Workplace accommodation people with disabilities, workplace adaptation disabled person, Workplace adaptation people with disabilities. Only journal articles were included in the sample, therefore books and congress articles were excluded.

From the research with the keywords we found 255 papers in the Pubmed base, 457 in Scopus and 138 in the Web of science. After reading the summary of all papers, articles dealing with workstation accommodations and the PD, and which were available for free, were downloaded. Thus, it was possible to obtain 50 papers in the Pubmed base, 80 in Scopus and 21 in the Web of science, but some of these papers were repeated, since they are indexed in more than one of these bases. Finally, after reading all the articles obtained, 63 in total, 23 were excluded because they had no connection with the subject matter.

3 RESULTS

From the research in databases, we obtained 40 papers on the topic workplace accommodations for the PD. After analyzing each one, it was found that 31 articles have been developed in the United States, 4 in Canada, 1 in Sweden, 1 in Japan, 1 in Taiwan, 1 in Croatia and 1 in Spain. Research centers that had more publications on the subject were: International Center for Disability Information (ICDI) from West Virginia University, with 4 papers, and Center for Assistive Technology and Environmental Access at the Georgia Institute of Technology, with 4.

Regarding the main authors on the subject, it was found that Denetta L. Dowler was the researcher who participated in more papers, five in all, Richard T. Walls and Tatiana I. Solovieva participated in 4, and Deborah J. Hendricks and Dory Sabata participated in 3 published articles. As for the year of publication, it was found that the year 2006 had the most publications, with 7 in total, followed by the year 2011, with 6, and 2004 with 5 publications. Finally, the journals with the highest number of articles published were: Work, with 9 papers, International Journal of Industrial Ergonomics, with 4, and Disability and Health Journal, with 3 articles. From the research in databases, we found that the main issues related to the workstation adaptations for people with disabilities are: Factors that can facilitate accommodations of the PD workplace, methods of evaluation and accommodations of PD workplace, and the types of accommodations of workplace for PD. Within this latter issue, the following subtopics were found: generic accommodations, assistive technologies, personal assistance service and work outside of the workplace.

The theme found to have more published articles was about the types of accommodations of workstation for people with disabilities, with 21 papers. As it was stated before, this theme is subdivided into four sub-themes, and the one that had more publications was the sub-theme of generic accommodations, with 10 articles, followed by the sub-theme personal assistance service, with 5, assistive technologies, with 4, and work outside the workplace, with 2. The second theme with more publications was the topic factors that can facilitate accommodations of the PD workstations, with 13 publications, followed by the theme evaluation methods and workstation accommodations for the PD, with six papers.

4 DISCUSSION

From the results of this literature review, it was found that on the subject in matter there are no authors or research centers with a significant number of published papers in relation to others. This is because the author that has more publications was the researcher Denetta L. Dowler with 5 articles, or 12.5% of the sample, while the International Center for Disability Information at West Virginia University and Assistive Technology and Environmental Access at the Georgia Institute of Technology were the research centers that had more publications, four in all, or 10% of all publications. The same is not true for the country of origin of the research. In the United States a significant percentage of 77.5% of the total sample papers were on the subject.

Another factor that draws attention was that articles on the topic accommodations of workplace for PD, developed in Brazil, were not found in this study. Thus, this result confirms what Guimarães, Barkokébas & Martins (2012) claim that there is little scientific literature on the labor inclusion of PD in Brazil.

In the Work journal 22.5% of the sample papers were published, however, it is important to note that 5 of the 9 articles published in this journal, resulted from a special issue on accommodations of workstations for the PD, in 2006. This is also why it was the year with the most publications on the subject, seven in all.

In the general accommodations sub-theme it was found that among the 6 papers that conducted experiment, two were case studies, being that one of the articles described 2 cases, one of a subject visually impaired and another of a subject with physical disabilities. In the other paper, the case study described was of individuals with motor disabilities. In the other four articles where experiments were performed, various types of disabilities were studied. Thus, these findings are not in agreement with the results of Butterfield & Ramseur (2004) that performed a literature review on adaptations of workplaces and found that 19 of the 30 papers were case studies of a single individual. According to Sanford & Milchus (2006), few studies describe and compare different types of accommodations of workstations for PD and its effectiveness between the same types and between different types of disabilities. This can be confirmed in this research, given that none of the papers found makes this type of comparison. Thus, it appears that there is a lack of research in the literature on what accommodations have better outcomes for different types of PD in the execution of their job duties.

It is observed that there are few papers in the sample on the sub-theme personal assistance service, as we found only five publications. According to Stoddard (2006), some individuals with more severe disabilities may require a helpdesk staff to adequately perform some labor or complementary activities. The helpdesk personnel can include help at work, such as readers, interpreters, help to load or reach objects, as well as help in activities of daily living, such as using the bathroom, eating and dressing (Dowler, Solovieva & Walls, 2011). In the articles found, there seems to be a consensus on the fact that the personal assistance service can facilitate the tasks of work and consequently the job inclusion of individuals with disabilities. However, they still lack studies on the costs, consequences in the workplace and on the methodology applied in this type of accommodation for PD.

The sub-theme work outside the workplace is a kind of adaptation of the PD workspace that allow individuals to accomplish their tasks outside the work environment, such as at home. According to Kaplan et al. (2006), working outside of the workplace is a useful accommodation of the PD and should be evaluated case by case. Meanwhile, for Baker, Moon & Ward (2006) work outside the workplace is a viable adaptation, which can increase the inclusion of PD in the workplace, but some policy research, outreach and interventions should be taken to reduce the isolation of these workers. Thus, it appears that this issue is rarely addressed in the literature, considering that only two items were found in the databases searched and both papers conclude that this is a useful type of accommodation, which can facilitate the job inclusion of the PD. However, they still have a lack of studies on this topic focused on cost, methodology to be applied and the effects on the workers themselves and on the coworkers.

Surprisingly, in the sub-theme of assistive technologies, only 4 papers were found in this literature review. We hoped to find more articles on this topic due to its importance and wide use in society. After reviewing the papers, it was found that the surveys come to the conclusion that the application of assistive technology devices can help in job accommodation of individuals with disabilities. However, it is found that 50% of searches are case studies, focusing on a particular device assistive technology such as, for example, in the study by Schuyler & Mahoney (2000) in which was evaluated the efficacy of manipulation for individuals with motor disabilities via a robotic device. Thus, literature has a lack of studies of various types of devices and their adaptation to different kinds of people with disabilities.

Meanwhile, there were six papers on the topic of evaluation methods and accommodations of workstations for PD. From the analysis of these publications, it appears that there is a consensus in which it is necessary to compare the demands of work with the functional capabilities of individuals with disabilities to carry out the adaptations of the workplaces. Moreover, it was observed that in most cases, five in all, new methodologies were based on case studies with a small sample size and with few types of workers with disabilities.

After analyzing the thirteen papers of the topic factors that can facilitate accommodations of workstations for the PD, it was found that the use of the principles of ergonomics and of universal design in the workplace could facilitate the adaptation of workstations for the PD. Furthermore, it was found that the main benefits of accommodations of the workplace for workers with disabilities are: minimal cost or no cost in adapting the workplace; retention of skilled workers; increased productivity; avoiding the cost of hiring and training new employees. Thus, knowledge of these benefits can also encourage the implementation of accommodations of workplaces for workers with disabilities by entrepreneurs. However, it is important to note that in this theme only five papers performed experiments. More practical studies are required to better understand the importance of each of these factors.

5 FINAL REMARKS

From the survey results, it was found that the vast majority of published papers on accommodations of workplace for the PD were developed in the United States. In addition, the journal Work presented most of the publications on this theme and 2006 was the year in which there was the highest number of papers published in the databases searched. Regarding the issues addressed in these articles, the one with more publications was about the different types of accommodations of workplaces for persons with disabilities.

Furthermore, it is observed that in the databases surveyed 40 papers were found about the accommodations of the workplace for PD, considering that it is a very broad topic and with different approaches, ranging from factors that can facilitate the accommodations of workplaces for people with disabilities, methods of assessment and adapting of workplaces for the PD, and types of accommodations of workplaces for the people with disabilities. Beyond that, it was found that there is consensus and uncertainties on different issues, since there is little research and few experimental studies that describe and compare types of accommodations of workplaces for the PD and the effectiveness of these accommodations among the types of disabilities found in the documentary research.

Therefore, it is important that further studies be conducted on adapting workplaces for people with disabilities that include a larger number of individuals with different characteristics and different accommodations of the workplace. Finally, there is also a need for longitudinal research to verify or control the determination, the settings and the consequences of these adaptations. Thus, aiming for a better understanding of the subject matter in order to facilitate and encourage the accommodations of the workplaces for workers with disabilities.

REFERENCES

Baker, P.M.A., Moon, N.W. & Ward, A.C. (2006). Virtual exclusion and telework: Barriers and opportunities of technocentric workplace accommodation policy. Work 27: 421–430.

Brasil. Ministério do Trabalho e Emprego. (2011). RAIS—Relação Anual de Informações Sociais, 2011.

Butterfield, T. & Ramseur, H. (2004). Research and case study findings in the area of workplace accommodations including provisions for assistive technology: A literature review. Technology and Disability 16: 201–210.

Dowler, D.L., Solovieva, T.I. & Walls, R.T. (2001). Personal assistance services in the workplace: A literature review. Disability and Health Journal 4,: 201–208.

Guimarães, B.M., Martins, L.B., Barkokébas Junior, B. (2012). Issues Concerning Scientific Production of Including People with Disabilities at Work. Work 41; Suppl 1: 4722–4728.

IBGE-Instituto Brasileiro de Geografia e Estatística. Censo Demográfico 2010. Rio de Janeiro: IBGE, 2011.

Kaplan, S., Weiss, S., Moon, N.W. & Baker, P. (2006). A framework for providing telecommuting as a reasonable accommodation: Some considerations on a comparative case study. Work 27: 431–440.

Morton, L., Foster, L. & Sedlar, J. (2005). Managing the mature workforce. New York: NY: The Conference Bord.

OECD-Organization for Economic Co-operation and Development. (2010). Sickness, disability and work: breaking the barriers. A synthesis of findings across OECD countries. Paris, Organisation for Economic Co-operation and Development.

Portal de Periódicos da CAPES [Retrieved March, 1, 2012, from http://www2.periodicos.capes.gov.br/portugues/index.jsp?urlorigem = true.

Sanford, J.A. & Milchus, K. (2006). Evidence-Based Practice in Workplace Accommodations. Work; 27: 329–332.

Stoddard, S. (2006). Personal assistance services as a workplace accommodation. Work 27: 363–369.

Westmorland, M.G. & Williams, R. (2002). Employers and policy makers can make a difference to the employment of persons with disabilities. Disability and Rehabilitation, 24 (15): 802–809.

WHO—World Health Organization. World report on disability 2011 (2011). Genebra: World Health Organization.

Zolna, J.S., Sanford, J., Sabata, D. & Goldthwaite, J. (2007). Review of accommodation strategies in the workplace for persons with mobility and dexterity impairments: Application to criteria for universal design. Technology and Disability 19: 189–198.

Occupational Safety and Hygiene – Arezes et al. (eds)
© 2013 Taylor & Francis Group, London, ISBN 978-1-138-00047-6

Ergonomic analysis applied to chemical laboratories on an oil and gas industry

C.P. Guimarães
National Institute of Technology, Brazil
Anhanguera University, Brazil

G.L. Cid, M.C. Zamberlan, V. Santos, F.C. Pastura, J. Oliveira, G. Franca & A.G. Paranhos
National Institute of Technology, Brazil

ABSTRACT: The aim of this paper is to present an ergonomic study applied to chemical laboratories of an oil and gas research center. The ergonomic work analysis was the main methodological approach. The ergonomic study was conducted in thirty laboratories of an oil and gas company in order to analyze current work conditions, redesign and simulate new work environments proposals. The EWA methodology was adapted to cover three stages: 1) reference situation diagnosis and recommendations; 2) ergonomic design concept establishment and 3) evaluation of the new work condition. The results showed that even considering each chemical laboratory specificity, some ergonomic problems were common, such as lack of space between workstations and on the workstation itself; repetitive and static awkward postures as squatting, trunk and neck forward bending, shoulder flexion and abduction over 90 degrees; manual material handling activities, as lifting, carrying, pushing and pulling heavy loads. The recommendations based on the diagnosis were presented in two different tools: a) Reference Human Activities Database tool—applied to work-related situation diagnosis for storing and organizing the reference situations collected data; b) 3D virtual simulation tool.

1 INTRODUCTION

The aim of this paper is to present an ergonomic study applied to chemical laboratories of an oil and gas research center. The ergonomic work analysis was the main methodological approach applied to this study (Wisner, 1995). The Ergonomic Work Analysis (EWA) is a methodology in which, as the result of studying behaviors in the work situation, provides an understanding of how the operator builds the problem, indicates any obstacles in the path of this activity, and enables the obstacles to be removed through ergonomic action (Wisner, 1995). The central point of this methodology is the analysis of real work activities performed by workers. Ergonomic work analysis can also be defined as a method which enables, through observation and work activity analysis, the determinants and constraints recognition of work situations and their consequences for the workers health and the production system. This method can be accomplished by following steps: Demand identification, modeling activities: Use of methods and techniques for search, collection and analyses of data and information to produce the positive transformations in work situation and results: Specification that ensures the positive transformation of labor reality.

Taking that on account, some ergonomic risk interactions have to be included in the study:

- Risk aspects inherent to the worker—involve physical, psychological and non-work-related activities that may present unique risk factors;
- Risk aspects inherent to the job—concern work procedures, equipment, workstation design that may introduce risk factors;
- Risk aspects inherent to the environment—concern physical and psychosocial "climate" that may introduce risk factors.

The ergonomic design concept is a set of references that should be established in the very first beginning of a new work condition design that defines future work. These ergonomic references give input to work and learning process evolution, work environment design—architectural and layout—and informational and interface devices design, to mention some (Santos and Zamberlan, 1992). These references can be presented as architectural drawings (blueprints), 3D digital *maquettes* and ergonomic standards that should be adopted by the design team. The ergonomic design helps to define the future work condition concept with respect to human intervention. That guarantees that the design concept match users variability and work performance variability in different use contexts.

In that sense laboratories must be prepared to adapt to changing technologies. Incorporating new technologies can substantially alter demands placed on the laboratory environment because physical alterations in the lab, including utilities, often follow the inclusion of new technologies (Mortland, 1997).

It is believed that ergonomic deficiencies in industry are a root cause of workplace health hazards, low levels of safety, and reduced worker productivity and quality. Although ergonomics applications have gained significant momentum in developed countries, awareness remains low in developing regions.

Based on our experience in ergonomic studies, we have developed a methodology for work condition ergonomic design that involves three stages: Reference situation diagnosis and recommendations; ergonomic design concept establishment and evaluation of the new work condition (Santos and Zamberlan, 2008).

This methodology incorporates 3D work simulation as a tool to improve the work environment proposed concept. By using that tool workers directly involved in the process work with designers and manufacturing engineers on design stage to improve the ultimate product and work environment layout.

2 MATERIALS AND METHOD

The ergonomic study was conducted in thirty laboratories of an oil and gas company in order to analyze current work conditions, redesign and simulate new work environments proposals.

The EWA methodology was adapted to cover three stages: 1) reference situation diagnosis and recommendations; 2) ergonomic design concept establishment and 3) evaluation of the new work condition. The reference situation diagnosis is concerned with the analysis of distinct work activities performed by laboratories' teams. These work activities were previous selected by the laboratories technicians and managers. For data collection a questionnaire, previously tested by the ergonomics team, and video capture were applied.

Data analysis was conducted at the Ergonomics Laboratory of National Institute of Technology from Brazil (INT).

The ergonomics team visited each one of the thirty laboratories three times a week for one year in a scheduled time. Firstly, the laboratory supervisor was interviewed. Then the research team followed the laboratory technicians' team to their specific lab where previous selected laboratory test activities were followed and registered by video (see Figure 1).

The laboratory technician's team was also interviewed and the activities were videotaped. The ergonomics team made activities registration and took measurements of the workplace, tools and equipments for anthropometric and layout analysis. All data was presented to workers for validation. The circulation study shown at Figure 2 gave information about which laboratory area needed to be changed in order to accommodate equipments and minimized circulation problems that could interfere on laboratory safety.

3 RESULTS AND DISCUSSION

The results showed that even considering each chemical laboratory specificity, some ergonomic problems were common, such as lack of space between workstations and on the workstation itself (see Figure 3); repetitive and static awkward postures as squatting, trunk and neck forward bending, shoulder flexion and abduction over 90 degrees; manual material handling activities, as lifting, carrying, pushing and pulling heavy loads (see Figure 4 and 5).

All these physical problems had pointed out the need for ergonomics improvements in order to minimize the potential work related musculoskeletal problems.

Figure 1. Laboratory test activity registration.

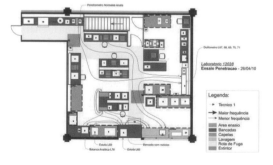

Figure 2. Example of a circulation study.

Figure 3. Lack of space to reach the utilities equipment.

Figure 4. Forward static trunk bending and MMH.

Figure 5. Forward static trunk bending and MMH—preparing sample.

Other risk and safety problems were also found such as: workstations and trashcans located on escape routes; oil samples handling outside chemical chapels and intermediate oil sample storage in chapels and cabinets (see Figure 6 and 7). Some environ-

Figure 6. Trashcan near the emergency door and escape route.

Figure 7. Lavatory at emergency area.

ment conditions were also pointed out as common problems in special with reference to heat and noisy.

4 CONCLUSIONS

The recommendations based on the diagnosis were presented in two different tools: a) Reference Human Activities Database tool—applied to work-related situation diagnosis for storing and organizing the reference situations collected data; b) 3D virtual simulation tool.

The Reference Human Activities Database tool optimized the organization and storage of the collected data. It enabled workers visualization of real work problems and it facilitates company's view of direct costs (errors, re-work, occupational diseases etc) and indirect costs (company image damage, loss of earnings etc). The Reference Human Activities Database tool can also be used as a training tool for new workers (see Figures 8 and 9). The use of computerized systems for collecting, storing

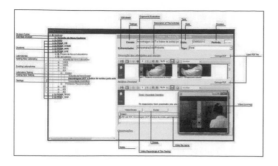

Figure 8. Human reference database—ergonomic diagnosis information.

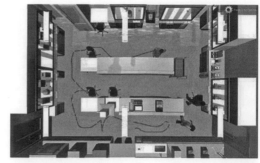

Figure 10. Example of lab simulator—circulation study.

Figure 9. Human reference data base—simulators window.

and processing the data has been of great help in the ergonomic work analysis. (Jastrzebska-fraczek et al, 2006). These systems facilitate the application of data for later study and allow comparisons with ergonomic actions. These also help map the existing health problems in the implementation of activities at work, so you can better prepare the strategies for correction on workstations, for training and qualification of workers. Entering data over the months provides control for the reduction of symptoms and health problems related to work in the long term.

The 3D virtual simulation tool (see Figure 10) helps on design validation process, minimizing design errors and conflicts between users and designers. By accurately reproducing the proposed work environment—furniture, operators and work activities performed, it was possible to observe how these elements interacted and interfered at the tasks performed. The use of virtual environments gives the possibility of negotiation, changes and creation of better results due to more graphic and visual interface for non-architects and designers professionals (Reed, 2001). That tool helps on understanding and on discussing of new workspace layout leading to a better design alternative choice.

These tools facilitate an effective application of ergonomics in work system design and can achieve a balance between worker characteristics and task demands. This can enhance worker productivity, provide worker safety and physical and mental well being, and job satisfaction (Shikdar and Sawaqed, 2004).

REFERENCES

Guimarães, C.P.; Cid, G.L.; Santos, V.S; Zamberlan, M.C.P; Pastura, F.C.H; Abud, G.M.D; Lessa, C.; Batista, D.S.; Fraga, M. M. (2012). Human Activity Reference Database. 18th World Congress on Ergonomics. IEA 2012, Recife, Brazil.
Guimarães, C.P.; Cid, G; Paranhos, A.G.; Pastura, F.; Santos, V.; Zamberlan, M.; Streit, P.; Oliveira, J. (2012).Ergonomic Work Analysis applied to Chemical Laboratories. 18th World Congress on Ergonomics. IEA 2012, Recife, Brazil.
Jastrzebska-fraczek I., Bubb H. (2003). Software Design and Evaluation by Ergonomics Knowledge and Intelligent Design System (EKIDES). PsychNology Journal, 1(4), 378–390. Retrieved from www.psychology.org
Mortland, K.K. (1997). Laboratory Design for Today's Technologies. Med TechNet Presentations. pp. 1–14.
Reed, M.P. (2001). Creating Human Figure Models for Ergonomics Analysis from Whole-body Scan Data. Proceedings of the Human Factors and Ergonomics Society 45th Annual Meeting, pp. 1040–1043.
Santos, V; Guimarães C. P.; Cid, G. L. (2008). Simulação Virtual e Ergonomia. In: xV Congresso brasileiro de Ergonomia,VI Fórum Brasileiro de Ergonomia, 2008, Porto Seguro. V Congresso brasileiro de Ergonomia, VI Fórum Brasileiro de Ergonomia.
Santos, V.; Zamberlan, M.C.P.L.; Pavard, B. (2009). Confiabilidade Humana e Projeto Ergonômico de Centros de Controle de Processos de Alto Risco. Rio de Janeiro: Synergia,. v. 1. 316 p.
Shikdar, A.A and Sawaqed, N.M. (2004). Computers & Industrial Engineering 47: 223–232.
Wisner, A. (1995). A Inteligência no Trabalho. Textos selecionados em ergonomia. São Paulo: Fundacentro.

Occupational Safety and Hygiene – Arezes et al. (eds)
© *2013 Taylor & Francis Group, London, ISBN 978-1-138-00047-6*

Ergonomic analysis applied to work activities at a pilot plant of oil and gas industry

M.C. Zamberlan, C.P. Guimarães, G.L. Cid, A.G. Paranhos, J. Oliveira & F.C. Pastura
National Institute of Technology, Brazil

ABSTRACT: The aim of this paper is to present an ergonomic study of maintenance and support work activities at a pilot plant of an oil and gas company. The Ergonomic Work Analysis (EWA) was the main methodological approach. The study concerns the analysis of different maintenance and support activities previous selected by workers and managers. The maintenance activities were classified as insulation, instrumentation, electric and mechanics. The support activities can be performed inside or outside the unit plant as well as inside chemical lab. The EWA methodology involved three stages: Reference situation diagnosis and recommendations; establishing an ergonomic design concept; evaluation of the new work condition and/or improvement of the maintenance tools. Based on activity analysis, some problems of physical ergonomic arose. When comparing body discomfort analysis between maintenance and support work activities the low back pain, neck and upper arm shown the highest scores. The results also showed that even considering plant unity specificities some ergonomic problems were common in maintenance and in support activities. The results and recommendations based on the ergonomic diagnosis were presented to managers at periodic meetings where the ergonomists team, designers and managers discussed issues to establish the ergonomic design concepts that would be applied on the maintenance and support work conditions.

1   INTRODUCTION

The aim of this paper is to present an ergonomic study of maintenance and support work activities at a pilot plant of an oil and gas company. The Ergonomic Work Analysis (EWA) was the main methodological approach. EWA is a methodology in which, as the result of studying behaviors in the work situation, provides an understanding of how the operator (worker) builds the problem, indicates any obstacles in the path of this activity, and enables the obstacles to be removed through ergonomic action (Wisner, 1995). The central point of this methodology is the analysis of real work activities performed by workers. Taking that in account, some ergonomic risk interactions have to be included in the study:

- Risk aspects inherent to the worker—involve physical, psychological and non-work-related activities that may present unique risk factors;
- Risk aspects inherent to the job—concern work procedures, equipment, workstation design that may introduce risk factors;
- Risk aspects inherent to the environment—concern physical and psychosocial "climate" that may introduce risk factors.

At a pilot plant of oil and gas industry, the maintenance work activities as well as the support work activities can be characterized as a work involving different tasks. This kind of work is distinguished by a wide variety of tasks that are part of the operator's expertise and know-how. It also involves a set of tasks that underlies a large number of operations that are not always organized in a specific work cycle. These tasks can be performed at places that vary considerably from one to another. For example, a mechanic is required to perform different tasks, such as maintaining and repairing equipment throughout different chemical processing units at the pilot plant. On the support activities the operators can perform work activities also inside the chemical lab as well as inside or outside the unit plant.

Despite the great variability in work places and job tasks, the human body can be considered as a constant. In order to make design decisions taking into account human factors, it is important to understand how the body responds to and moves about in its environment. Work in the oil industry involves diverse activities including work in rigs, workshops and offices. Heat stress as a potential safety and health hazard has been recognized in the literature and guidelines for exposure have been formulated (Hancock and Vasmatzidis, 1998). Ergonomic deficiencies in industry are a root cause of workplace health hazards, low levels of safety, and reduced worker productivity and quality.

Manual material handling as lifting, pushing and pulling and carrying have been usual activities that maintenance and support workers has been performing in their working day. These working activities have been pointing out as cause of back and upper limb problems.

2 MATERIALS AND METHOD

The study concerns the analysis of different maintenance and support activities previous selected by workers and managers. The maintenance activities were classified as insulation, instrumentation, electric and mechanics. The support activities can be performed inside or outside the unit plant as well as inside chemical lab. The EWA methodology involved three stages: Reference situation diagnosis and recommendations; establishing an ergonomic design concept; evaluation of the new work condition and/or improvement of the maintenance tools.

The reference situation diagnosis were conducted using a questionnaire, previous tested by the ergonomists team and video capture Figure 1 and 2.

Figure 1. Maintenance activities registration.

Figure 2. Support laboratory activities registration.

Video recording section took time because the operator performed many operations and the task involved many variable factors. The operator did not remain in one place at his workstation, but may have to move from one machine to another. The ergonomist must follow the operator so that nothing is missed concerning the operator and his work activity (operations performed, risk factors, safety problems). If a given production condition cannot be filmed, you will have to ask the participants to remember the difficulties encountered under this condition and the strategies adopted to avoid them. In all cases, the ergonomist tried to record all the operations relevant to the analysis, such as unexpected situations, incidents or special conditions. Extra information was provided, such as work zones, traffic zones, circulation of material and storage areas when information from video recording and questionnaire answering were combined.

The study of body areas discomfort was conducted using the discomfort/pain diagram (Corlett and Bishop, 1976). The data analysis was conducted at the Ergonomics Laboratory of National Institute of Technology. The ergonomists team went to the pilot plant three times a week for six months at a scheduled time. First the supervisors were interviewed and then the ergonomists followed the maintenance and support workers group to the unit of the plant where the maintenance activities as well as support activities were performed. These groups were also interviewed and activities were videotaped, the ergonomists team conducted activities registration and took measurements of workplaces, tools and equipment for anthropometric and layout analysis.

3 RESULTS AND DISCUSSION

The main problems associated with performance of work operations, and their causes were brought out from the diagnosis stage. The operators were encouraged to engage in a free and open discussion of the problems encountered in the different operations condition. For long-cycle tasks, the purpose of the analysis technique is to get the operators to verbalize based on the video recordings. Based on activity analysis, some problems of physical ergonomic arose. When comparing body discomfort analysis between maintenance and support work activities the low back pain, neck and upper arm shown the highest.

The results also showed that even considering plant unity specificities some ergonomic problems were common in maintenance and in support activities, such as lack of work space at the unity Figure 3 and 4; repetitive, static and forceful activities in awkward postures as squatting, trunk and neck forward bending, shoulder flexion and abduction over 90 degrees, wrist flexion and extension combined with abduction, Figure 5 and 6 manual material handling activities as lifting, carrying, pushing and pulling heavy loads. Figure 7 and 8.

Some environment conditions Figure 9 and 10 were also pointed out as common problems, in special with reference to heat, lighting and vibration,

Figure 3. Lack of space to take oil barrel from kiln (support activity).

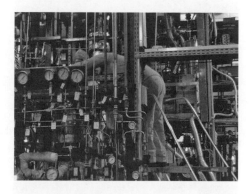

Figure 6. Awkward posture—disassembly a unit (maintenance activity).

Figure 4. Lack of space to reach part of equipment (maintenance activity).

Figure 7. MMH pulling barrel to Laboratory (support activity).

Figure 5. Awkward posture—cleaning unit (support activity).

Figure 8. MMH—pulling coke barrel—(maintenance activity).

Figure 9. Vibratory tool—support activity cleaning the coke barrel.

Figure 12. Using navy ladder.

Figure 10. Vibratory tool disassembly the coke barrel (maintenance activity).

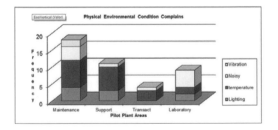

Figure 11. Physical environmental condition complains at different areas of pilot plant.

shown at Figure 11, this last one mainly to maintenance activities

Some safety and risk problems were also frequent such as high inclination ladders or navy ladders used to get to work places carrying loads and tools that can predispose to slipping and falls in

special to maintenance work activities (Figure 12) (Zamberlan et al, 2012).

4 CONCLUSIONS

The results and recommendations based on the ergonomic diagnosis were presented to managers at periodic meetings where the ergonomists team, designers and managers discussed issues to establish the ergonomic design concepts that would be applied on the maintenance and support work conditions. As previously explained, the description of the problem may refer to different dimensions of the work situation, ranging from its design/planning and its organization to its consequences for the workers (injuries, pains) or the production system (equipment failure, production shutdown, quality impairment). Some of the strategies to implement the recommendations were: Managers must be well informed of the benefits of ergonomics and prevention of injuries through ergonomics implementation, tools remodeling and improvements on maintenance and support workstations in order to minimize manual material handling activities; employees need to be trained systematically in ergonomics in order to improve ergonomic conditions and safety; the work and workplace design should be carried out using ergonomic guidelines and recommendations should consider user population with special emphasis to maintenance workers. It should also be given adequate consideration to environment, mainly with regards to temperature (heat) and vibration of equipment's that can be remodeled or changed to new ergonomics ones.

The ergonomic analysis gave access to information to recommending changes at equipments of the unit plant. As shown at Figures 8, 9 and 10, different work activities were performed at the coke barrel in order to maintain and clean the barrel. Some changing at the support cart was proposed to improve the work activities, to facilitate pushing and pulling the cart and the manipulation of the barrel over the cart. These recommendations were explained below: The original height of the cart with the barrel over it was 177 cm (Figure 13). The new design propose a height of 112 cm (Figure 14) and changing of the area of barrel support in order to facilitate manipulation, changing awkward posture and minimize physical loads to workers. See Figures 15, 16 and 17.

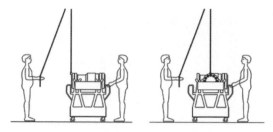

Figure 16 and 17. New cart design—to improve manipulation.

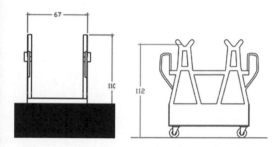

Figure 13 and 14. Original e propose cart height (design changing).

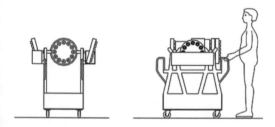

Figure 15. New cart design.

REFERENCES

Corlett, E.N. & Bishop, R.P. (1976). A technique for assessing postural discomfort. Ergonomics, v. 19, p. 175–182.

Hancock, P.A. & Vasmatzidis, I. (1998). Human occupational and performance limits under stress: The thermal environment as a prototype example. Ergonomics, 41, 1169–1191.

Shikdar, A.A. (2004). Identification of Ergonomic Issues That Affect Workers in Oilrigs in Desert Environments International Journal of Occupational Safety and Ergonomics (JOSE) 2004, Vol. 10, No. 2, 169–177.

Shikdar, A.A. & Sawaqed, N.M. (2004). Ergonomics, and occupational health and safety in the oil industry: a managers' response. Computers & Industrial Engineering 47, 223–232.

Wisner, A. (1995). A Inteligência no Trabalho. Textos selecionados em ergonomia. São Paulo: Fundacentro.

Zamberlan, M.C.P.L., Guimarães, C.P., Cid, G.L., Paranhos, A.G. & Oliveira, J.L. (2012). Ergonomic Work Analysis applied to maintenance activities at a pilot plant of oil and gas industry In: 4th International Conference on Applied Human Factors and Ergonomics (AHFE). London: USA Publishing, 2012. p. 7749–7754.

Occupational Safety and Hygiene – Arezes et al. (eds)
© 2013 Taylor & Francis Group, London, ISBN 978-1-138-00047-6

3D Digital Human Models and collaborative virtual environments: A case study in oil and gas laboratories

Venétia Santos
Pontifícia Universidade Católica do Rio de Janeiro—PUC-Rio, Brasil

Maria Cristina Palmer Lima Zamberlan
Instituto Nacional de Tecnologia—INT/MCTI, Brasil

Priscilla Streit
Escola Superior de Desenho Industrial—ESDI/UERJ, Brasil

José Luis Oliveira, Carla Patrícia Guimarães, Fernando Cardoso Ribeiro &
Flávia Hofstetter Pastura
Instituto Nacional de Tecnologia—INT/MCTI, Brasil

ABSTRACT: This paper will present the workflow developed for the application of serious games in the design of complex cooperative work settings. The project was based on ergonomic studies and development of a control room among participative design process. Our main concerns were the 3D human virtual representation acquired from 3D scanning and motion capture technologies, human interaction, workspace layout and equipment designed considering ergonomics standards. Using a game engine platform to design the virtual environment, the virtual human model can be controlled by users on dynamic scenario in order to evaluate the new work settings and simulate work activities. The results obtained showed that this virtual technology can change the design process by improving the level of interaction between final users and, managers and human factors team.

1 INTRODUCTION

Serious games concepts have been discussed in collaborative virtual environment as a tool for knowledge transfer and experience gaining through simulation and non-physical interactions through life-like experiences using various techniques to embody human-artifacts interactions. There is a clear need for considering new frameworks, theories, methods and design strategies for making serious games applications and virtual environment technologies more effective and useful as part of education, health and training. Virtual simulation has been used in Ergonomics for the design of control centers, transport design and product evaluation. (Santos, et al., 2009, Santos, et al., 2008, Guimarães, et al., 2010).

This paper presents a project in which virtual reality and game engines were used to improve the workflow and interaction between teams in the design process of a series of oil and gas laboratories. These interactive environments provided the possibility of realistic scenario based drills among the use of Digital Human Models (DHM) built from 3D anthropometric data and Motion Capture (MOCAP),

aiming to evaluate layout proposals and training new personnel based on the virtual simulation of activities performed by the workers themselves. The benefits of using 3D DHM with each technician's own features is the possibility of recognition not only by himself, but also by the team. One who is acquainted can recognize others movements and sometimes predict actions without a need for verbal communication. By mapping these interactions in a visual manner, it is possible to use the information acquired to train new personnel. (Guimarães, et al., 2012).

The project was based in the research center of one of the biggest oil company in the world. Among the stages of the project were considered: (1) gathering information of thirty laboratories and personnel data among its diagnosis based on Ergonomic Work Analysis, which provided the multidisciplinary team with data and knowledge on the procedures, in order to build new work environment layouts, (2) 3D scanning of the workers in several postures aiming the design of workstations, proposals of new working conditions and the development of their 3D DHM for the virtual interactive environment and (3) MOCAP of the same workers performing daily activities in order to have their 3D DHM with their own movements.

2 MATERIALS AND METHODS

The development of the simulators were segmented on the following stages: (1) building the 3D environment based on 2D CAD representation of the new proposal, (2) creation of a furniture and workstations database according to ergonomic recommendations and the application of both, Brazilian and International standards, (3) creation of an equipment database, which was based on real equipments located in the laboratories—dimensions, utilities, textures, hierarchy, etc., (4) development of 3D DHM based on 3D laser scanning and inertial MOCAP technology and (5) implementation and setup in the game engine.

All the equipments that compose the equipment database (Fig. 1) were meant to be used with high quality pictures intending to increase visual recognition of its particularities at first sight, also increasing the familiarity for the users in the proposed environment.

3D DHM were developed using each technician's own features given by the laser scanning process, and its movement identity acquired with the motion caption system (Figs. 2–3). All these features are used to complement this complex DHM, that has the capability of representing the technicians with visual accuracy, and still give them a human like movement, much more trustable than a robotic or parametric movement created by algorithms.

In addition to the interactive DHM and to ensure Brazilian and International standards were applied in its best way, some functional poses were acquired with the laser scanners and used to have a most efficient evaluation of internal area and occupancy, avoiding conflicts between the adopted postures of the DHM animated by motion capture and the exact space required when assuming certain postures represented by the functional poses.

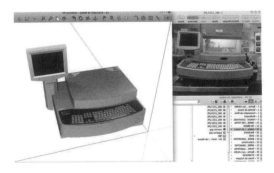

Figure 1.   Equipments database.

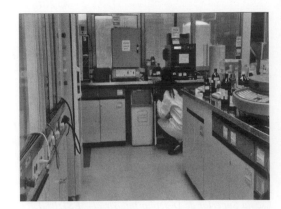

Figure 2.   Technician performing daily activities and wearing inertial MOCAP system.

Figure 3.   Preliminary DHM studies based on technician inertial MOCAP.

3 RESULTS AND DISCUSSION

One of the simulator's goals was to allow multiple users to interact with each other using their own avatars, participating in a pervasive method of interaction design, described by Kaptelinin & Nardi (2006) as something that comprises all efforts to understand human engagement with digital technology and all efforts to use that knowledge to design more useful and pleasing artifacts. So far, it has allowed all stakeholders, from laboratory personnel to managers and design team to engineers, to discuss and evaluate solutions, avoiding mistakes and misunderstandings between both parts when concerning about new solutions. Being able to visualize 3D environment and populate it with 3D DHM based on the users themselves and equipments from the current laboratories has shown to be more productive than a 2D floor plan traditional approach, once it leads to knowledge and awareness democratization. It has also made possible to study occupancy and analyze the workflow in the laboratories (Figs. 4–5), where all activities were simulated at once, in this virtual environment, with controlled

Figure 4. Occupancy studies in one of the proposed laboratories.

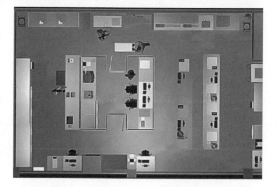

Figure 5. Top view of the simulator.

situations and multiple points of view for the same scene, even been able to rewind and see everything. The technologies used in this project, regarding the development of the 3D DHM are still being improved in order to decrease production time and enhance both movement and visual accuracy of the models. The MOCAP system used in the project is based on inertial sensors, which have the disadvantage of losing reference when near metal surfaces. Although the data acquired from this system can easily be edited, new systems are being studied to complement MOCAP data.

4 CONCLUSIONS

Nowadays, industrial projects are developed with the use of 3D software engines instead of 2D tools, allowing the main focus to be human labor and not only the project itself. Allowing it to the study of social interactions at work helps to project better environments. (Santos, et al., 2011).

The use of virtual environments gives the possibility to discuss, change, create and deliver a better result as it is more graphic and visual for non architects and designers professionals, to understand and discuss the new layout of the work space where can be chosen a better design alternative, optimize the interfaces, integrate countless projects and a great number of professionals involved.

Transparency of the future project allows adjustments and error recovery throughout the design process. Projects become more robust, since the scenarios and future activities may be simulated and also the risks involved studied.

These simulations may be used to evaluate technology, industrial safety and/or human performance. One may map process risks, ergonomic and architectural problems, escape routes, displacement of people in crisis situations, assembly and maintenance problems.

Therefore the conclusion is that virtual simulators of social interactions contribute towards: the activity of designers in the occupation of three-dimensional space; evaluation of possible alternatives; detailing the environment; validation of the future project by users, managers, and others; safety, health and environment evaluation; training of human resources.

As observed in this work, engineers can use this simulation platform in the design of future laboratories, thus minimizing the time required for similar projects and increasing the compliance of these environments to standards.

Even if character interaction cannot always be done in the virtual space, stakeholders can easily project their knowledge of the working situation in order to assess part of the new working space characteristics improving the participative dimension of the project.

Organizational decisions were taken around these tools; they help people to project themselves in their future working spaces and furthermore it was a great tool to improve the feeling of participation.

REFERENCES

Guimarães, C.P., Cid, G.L., Paranhos, A.G., Pastura, G.C.H., Franca, G.A.N., Santos, V., Zamberlan, M.C.P.L., Streit, P., Oliveira, J.L., Correa, T.G.V. (2012). Ergonomic Work Analysis Applied to Chemical Laboratories on an Oil and Gas Research Center. In: Vincent G. Duffy. (Org.) *Advances in Applied Human Modeling and Simulation*. 1ed.Boca Raton: CRC Press, 2012, v. 1, p. 471–477.

Guimarães, C.P., Pastura, F.C. H., Pavard, B., Pallamin, N., Cid, G., Santos, V., Zamberlan, M.C.P.L. (2010). Ergonomics Design Tools Based on Human Work Activities, *3D Human Models and Social Interaction Simulation IHX Congress*, Miami.

Kaptelinin, V., Nardi, B. A. (2006). Acting with Technology: Activity Theory and Interaction Design, The MIT Press, Cambridge.

Santos, V., Guimarães, C.P., Cid, G. (2008). Simulação Virtual e Ergonomia, XV *Congresso Brasileiro de Ergonomia, VI Fórum Brasileiro de Ergonomia,* Porto Seguro.

Santos, V., Zamberlan, M.C.P.L., Pavard, B. (2009). *Confiabilidade Humana e Projeto Ergonômico de Centros de Controle de Processos de Alto Risco,* ed., Synergia, Rio de Janeiro.

Santos, V., Zamberlan, M.C.P.L., Pavard, B., Streit, P., Oliveira, J.L., Guimarães, C.P., Pastura, F.C.H. (2011). Social Interaction Simulators: Serious games for the design of complex socio-technical systems. *DHM 2011—First International Symposium on Digital Human Modeling,* UCBL-Université Claude-Bernard Lyon.

Occupational Safety and Hygiene – Arezes et al. (eds)
© 2013 Taylor & Francis Group, London, ISBN 978-1-138-00047-6

Evaluation of the ergonomic changes made in the presses sector in an metallurgical industry

J.M.N. Silva, A.S.L. Santos Filho, J.C.M. Santos, A.M.D. Nunes, C.M Souto & I.F. Araujo
Universidade Federal de Campina Grande, Campina Grande, Paraíba, Brasil

A.L. Melo
Universidade de Coimbra, Coimbra, Portugal

ABSTRACT: Ergonomics is gaining space within industries, because improvements related to quality, efficiency, safety and comfort they are perceived when ergonomic adaptations are made. These improvements may be on posture, seating, environmental conditions, in machinery, tools, and even the method adopted by the employee to perform his duties. After changes ergonomic is necessary to verify whether these actually will solve the problems of the job. Thus, this study aims to changes ergonomic analyze that occurred in the industry presses with actuation per pedal electro-hydraulic comparing the current situation after the ergonomic improvements, and conditions previously encountered. Methodologically we used a set of interviews with employees, in which they indicated what changes perceived and how such changes have improved their labor activity. The results show that ergonomic problems encountered previously have been resolved, and that work is now performed more comfortably.

1 INTRODUCTION

According to Iida (2003) the adaptation of the labor place of employees' needs is the main goal of ergonomics. This multidisciplinary science had gotten even more credibility in the industries, the improvements related to quality, efficiency, safety and comfort are easily noticed when adjustments ergonomics are executed. Filus *et al.* (2012), Wisner (1994), and Vidal (1992) affirm that this same ergonomics purpose is now widely observed in new manufacturing processes, as predicted in 1967.

To start with, making improvements into the labor place with the aim of conducting a survey of the ergonomic demands, also called ergonomic assessment. In the survey on ergonomic demands there are several points there are mapped and suggested to improve and safety, in particular the quality of life and health of employees. These improvements may be about posture, seating, environmental conditions, machinery, tools, and even in the adopted method by the employee to perform his tasks. According to Vidal (2003), the ergonomic demand may have several sources such as internal demands of the organization related to leadership, management, employees and external demands as union demands, society comprehensively, among others. In most cases the demands are related to the workers who are preferably carried out in Brazilian companies, mainly due to a slow process of innovation.

The second step is to make ergonomic improvements to diagnosis where they are ergonomic in-depth studies on the problems with higher priority, which are those that cause increased wear to man (GIOVANNI DI, S/D). Generally this step is carried macro ergonomic analysis of the task. The third step is where the projection ergonomic features ergonomic design and all information related to this.

The fourth step is to assess or validate the changes made in the workplace. Thus, even after the changes are made ergonomic it is necessary to verify whether these actually solve the problems of the job, and if indeed bring improvements in comfort to perform work activities. These changes are intended to eliminate not only accidents, but also occupational diseases, which according to Reis *et al.* (2000) "usually involves the removal of the employees in a fully productive age, causing exit the market where workers could be contributing to the development of the economy and passing, instead, to rely on Social Security benefits." It is expected mainly to avoid the appearance of such modifications illnesses such as back pain and neck pain.

The present article aims to analyze the ergonomic changes occurring in the industry presses electro-hydraulic pedal comparing the current situation after the ergonomic improvements, and conditions previously encountered, noting that parallel what benefits these changes have brought to the development of labor activity in this sector.

2 MATERIALS AND METHODS

The study is exploratory, it performs site visits, where through videos and photographs are analyzed situations encountered, previously ergonomic changes. Also through videos and photographs these situations are confronted and compared with the current state, so that one can easily notice the changes. The approach is characterized as qualitative as it indicates and not quantifies the improvements not found. Methodologically we used a set of interviews with employees, in which they indicated what changes perceived and how such changes have improved their labor activity. All employees press operators were interviewed. Finally, it was found through the diagram proposed by Corlett and Manenica (1980) that the parts of the body that interviewed employees, have more pain during or end of the Labor Day, to compare to the diagram collected in previous researches. Expected to check enhances at the end of a labor day, and the workers are supposed not to complain about body pain.

3 RESULTS AND DISCUSSION

3.1 *Physical arrangement*

The figure 1 and figure 2 shows images of the job before the ergonomic adaptations, and these are listed and flagged irregularities that require changes. The irregularities were noticed and listed as bellow:

1. Part of the power transmission machine unprotected which can cause several sorts of accidents. This irregularity is in total disagreement with the Norm 12, the Ministry of Labor and Employment (BRASIL, 2010).
2. Uncomfortable seats and inappropriate to perform any kind of task. The Norm schedules 17 (seventeen) points of requirements regarding the characteristics of the seats, which among others may cite shaped backrest slightly adapted to the body to protect the lower back, rounded front edge, features little or no basis in conformation, and base upholstered with material density of 40 to 50 kg/m^3 (BRASIL, 2012).
3. The pedal are at high position and not parallel to the floor contributing to poor body posture. this irregularity is at odds with the regulatory norm 17, which states that for work requiring the use of the feet also pedals and other commands to drive the feet must have positioning and dimensions that enable easy reach, as well as proper angles between the various parts of the body of the worker, depending on the characteristics and peculiarities of the work to be performed (BRASIL, 2012).

Figure 1. Posture adopted and showing the points of ergonomic inadequacy in performing the work.

Figure 2. Posture adopted back view.

4. Worker performs its function container with a between the legs which also contributes to wrong posture, when it gets filled container is loaded manually to the next job. This working method is in disagreement with the Regulatory Standard 17, which requires that the jobs where the work is performed in a sitting position arms and legs stay in comfortable positions. The same rule also directs that in order to limit or facilitate manual transport of these cargoes should be made by appropriate technical means. (BRASIL, 2012).
5. Where the metal plate is pressed without protection or any device that prevents the hands, fingers,

or arm is pressed. The regulatory standard 12 states that "the danger zones of machinery and equipment must have safety systems, characterized by fixed protection" (BRASIL, 2010)

6. Seat too low, no height adjustment. Norm 17 states that the seats height need to be adjustable upper surface, from the floor, among 37 and 50 centimeters (BRASIL, 2012).

7. The employee in the shadow of the work machine, and without illumination required pressed in place, something perceived by window position in the image. such a procedure is at odds with the regulatory standard 17, which directs that the lighting must be designed and installed to avoid shadows on jobs where tasks are performed (BRAZIL, 2012).

Figure 3.   Diagram where are highlighted areas in which employees feel pains.

3.2   *Workstation ergonomic after changes*

The Figure 3 and figure 4 shows the corrections made in order to resolve the points detected before. The changes indicated are as follows.

1. Part of power transmission now cloistered put adequate shields, as well as guides the Norm 12. This rule states that where hazardous parts can occur grasping, such as belts, steering wheels, gears, pulleys, must have protection designed (BRASIL, 2010). A number of accidents can be prevented such as amputations, scalping and even death;

2. Transportation of pressed parts no longer done manually through container, but through appropriate cart, within the requirements for cargo transportation in Standard Regulated 17. Back pain and neck pain can be developed due to wrong shipping charges;

3. Comfortable seat with height adjustment and appropriate for work as well with the demands Norm 17. It can be argued improves posture by changing the seat. Good posture may be regarded as one in which the body is in a position favorable to perform the task (OLIVER 1998);

4. Lighting directed, facilitating the visualization of the location pressed, avoiding shadows, glare and directs Norm 17. As stated by Fiedler *et al.* (2010) should always favor natural lighting, but in some cases the lighting directed in machines solve problems illumination;

5. Pedal in the proper position, which is parallel to the floor in order to avoid a poor body posture, according to norm number 17;

6. Location suitable for depositing the dough pieces into the machine, which eliminates the need to work with receptacle between the legs, thereby to require the Norm 17. According to Barbosa Filho (2010) among the factors causing occupational ills is the posture adopted (and maintained)

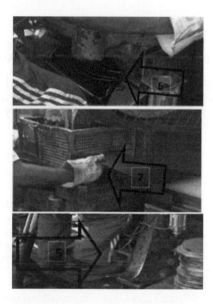

Figure 4.   Detailed image of parts modified.

to perform work. Therefore actions to maintain good posture should always be implemented;

7. Placing the pressings happens properly protected to avoid accidents, in total agreement with what we walked into Norm 12. This rule directs that pinch points must have protection (BRASIL, 2010).

3.3   *Parts painful of the body*

Coury (2009) affirms that musculoskeletal disorders have been associated with individual risk factors present in the environment and occupational biomechanics. Although many symptoms are associated with work-related musculoskeletal

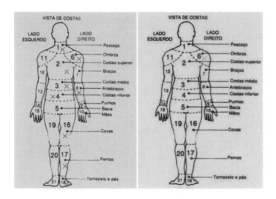

Figure 5. Body painful before and after the ergonomic changes.

disorders, one of the most notable is the pain (STRAZDINS AND BAMMER, 2004).

The figure 5 below shows two diagrams based proposed by Corlett and Manenica (1980), where the left image show up, the body parts where employees feel pain during and after the workday before ergonomic changes in the workplace, and the right image, also marked, which body parts they feel pain during and after the workday after the changes made in such a job.

4 CONCLUSION

Having faced the evidences of the importance of ergonomics to improve employment and working conditions, mainly in Brazilian factories it was observed that the points that required ergonomic changes undergone the necessary changes, and that in order to ensure comfort and safety to perform work activities was accomplished.

The figure 3 and 4 shows improvements in posture of the employee, due to changes in relation to pedal position, the type of seat, the deposit and expenditure pressed in an appropriate place, and not in a container placed between the legs of the employee. These changes make the job less stressful, and prevents diseases like back pain and neck pain. Such figure 3 and 4 also shows that improvements were carried out efficiently with respect to lighting, transportation pressings, and protection of power points.

Thus all the points raised before suffered the necessary changes and are in accordance with Brazilian law, especially the Standard Regulator 12 and Norm 17.

Through the diagram proposed by Corlett and Manenica (1980) can be observed that the number of parts of the body sore was drastically decreased.

But the same diagram shows that improvements still need to be implemented because employees still bothered by pain in his left shoulder, and thus studies are needed to propose changes in order to eliminate the pain felt by all employees.

REFERENCES

Barbosa Filho, A.N. (2010). Segurança do trabalho e gestão ambiental (3ª Edição). São Paulo: Atlas.
Brasil. Ministério do Trabalho e Emprego (2010). Norma Regulamentadora 12: Segurança no Trabalho em máquinas e equipamentos. Retirado em 28 de janeiro, 2012, de http://portal.mte.gov.br/data/files/8 A7C812D350 AC6F801357BCD39D2456 A/NR12%20(atualizada%202011)%20II.pdf.
Brasil. Ministério do Trabalho e Emprego (2012). Norma Regulamentadora 17: Ergonomia (70ª Edição). Manual de Legislação. São Paulo: Atlas.
Corlett, E.N., Manenica, I. (1980). The effects and measurement of working postures. Applied Ergonomics, 11(1), 7–16.
Coury, H.J.C.G., Moreira, R.F.C., Dias, N.B. (2009). Efetividade do exercício físico em ambiente ocupacional para controle da dor cervical, lombar e do ombro: uma revisão sistemática. Revista Brasileira de Fisioterapia. São Carlos, v. 13, n.6, pp. 461–479.
Di Giovanni, J.R.M., Silveira, C.S. (s/d). Intervenção Ergonômica de Posto de Trabalho: Um estudo de Caso da Indústria de Toldos. Retirado em 28 de janeiro, 2012, de http://www.ergonomianotrabalho.com.br/analise-ergonomica-toldos.pdf.
Fiedler, N.C., Guimarães, P.P., Alves, R.T., Wanderley, F.B. (2010). Avaliação ergonômica do ambiente de trabalho em marcenarias no sul do espírito santo. Revista Árvore, Viçosa, Minas Gerais, v.34, n.5, pp. 907–915.
Filus, R., Wruca, R., Charleaux, V., Ortega, A., Ferreira, C., Jesus, L., Stramari, A., Neufel, M., Maia, U. (2012). Ergonomics at Volkswagen Brasil, Multidisciplinary work to equalize health, Productivity and Quality. Work: A Journal of Prevention, Assessment and Rehabilitation. ISSN 1051–9815, v. 41, Supplement 1, pp. 4418–4421.
Iida, I. (2003). Ergonomia: projeto e produção. (9ª edição). São Paulo: Edgard Blücher.
Oliver, J., Middleditch, A. (1998). Anatomia funcional da Coluna vertebral. Rio de Janeiro, RJ: Revinter.
Reis, R.J., Pinheiro, T.M.M., Navarro, A., Martin, M. (2000). Perfil da demanda atendida em ambulatório de doenças profissionais e a presença de lesões por esforços repetitivos. Rev. Saúde Pública, v. 34, n. 3, pp. 292–298.
Strazdins, L., Bammer, G. (2004). Women, work and musculoskeletal health. Soc Sci Med.58(6):997–1005.
Vidal, M.C. (1992). Textos selecionados los Ergonomia Contemporânea. GENTE/COPPE/UFRJ, Rio de Janeiro.
Vidal, M.C. (2003). Analise Ergonomica do Trabalho I—Demanda Gerencial. Apostila do Curso de Especialização Superior em Ergonomia. Fundação COPPETEC. COPPE. UFRJ.
Wisner, A. (1994). A Inteligência no Trabalho: Textos selecionados de Ergonomia. São Paulo. Fundacentro.

Occupational Safety and Hygiene – Arezes et al. (eds)
© 2013 Taylor & Francis Group, London, ISBN 978-1-138-00047-6

Development of a job rotation scheme to reduce musculoskeletal disorders: A case study

H. Fonseca, I.F. Loureiro & P.M. Arezes
Universidade do Minho, Guimarães, Portugal

ABSTRACT: In addition to the interest on productivity, organizations are focused on the employee's wellbeing to achieve their economic purpose. In a production process, activities involving ergonomic risk factors may contribute to the development of musculoskeletal disorders (MSDs), which seems to be largely associated to an increase in absenteeism. The implementation of a job rotation system is often performed to reduce the risk of MSDs. This case study aims to develop of a job rotation scheme and identify the factors that can have influence on the job rotation system. It was necessary to identify and characterize the activities and then applying risk assessment. Individual and organizational factors were also analyzed. As a result it is considered that the implementation of the job rotation system should take into account the joint analysis of the risk of MSDs and the identification of factors associated with the workplace.

1 INTRODUCTION

1.1 *The strategic value of ergonomics*

Ergonomics contributes to the optimization of the human being (a social goal) and total system performance (a strategic goal) (Dul & Neumann 2009). Ergonomics can add value to a company business strategy of reaching the business goal profit, or intermediate business goals such as cost minimization, productivity, quality, delivery reliability, responsiveness to customer demands, or flexibility. By considering ergonomics in the design of the production system, including the job and workplace designs, or human work elimination by mechanization or automation of inefficient, unhealthy or hazardous tasks, the costs per unit can be reduced and work productivity increased. Nowadays, increasing productivity and reducing costs seems to be the path to the success of organizations (Dul & Neumann 2006).

Managers usually associate ergonomics with occupational health and safety, not with business performance. By contributing to the goals of business performance, ergonomics will also be able to reach its long-established health and safety objectives. Ergonomics can contribute to the many strategies and business outcomes and is a way to carry out a sustainable growth without high social costs, such as the work-related illnesses (Dul & Neumann 2009).

Ergonomics knowledge and applications have evolved over the time, as work organization has progressed. The emergence of several ergonomic contexts with a certain level of complexity may, in some way, affect human activities and individual performances. As changes arise in the work organizations, it is necessary to understand the role of ergonomics in the design of systems organizations, jobs, machines, software, interfaces and products. The deeper the influence of the external environment on the equilibrium of the organization is, a more holistic approach to the problems of the organization is required. Manufacturing industries are under increasing pressure to improve productivity, quality, to reduce delivery times and costs. This pressure is likely to be transferred to workers and can have negative effects on health (Van Rhijn et al., 2005).

1.2 *Absenteeism and MSDs*

Absenteeism is an important indicator and that can have impact on several levels of the system: economic, social, health promotion, family and employee satisfaction.

Several studies indicate that absenteeism and Musculoskeletal Disorders (MSDs) are in some way related (Widanarko et al., 2011). Indeed, MSDs are the more common occupational health problem in European Union. Numerous studies show that musculoskeletal disorders continue to be a tremendous burden in industry, being low back and shoulder disorders among the most common disorders (Ferguson et al., 2011). Activities with identified risk factors, such as the adoption of awkward postures, repetition and excessive force, contribute to the development of MSDs and, consequently, to an increase in absenteeism.

This situation has a negative impact on productivity and satisfaction amongst workers (Widanarko et al., 2011). Therefore, it is important to develop actions that can reduce the risk of MSDs.

1.3 Job rotation system

The implementation of a job/task rotation system is usually performed on the organizations contributing to achieve economic (productivity point of view) and social (wellbeing point of view) benefits (Dul & Neumann, 2009).

MSDs assessment can lead to strategies for redesigning activities involving manual materials handling activities in order to reduce the associated risk factors. Ergonomic redesign strategies have sometimes indicated the need to change from lifting, lowering, and carrying tasks to pushing and pulling tasks (Ciriello et al., 2007). Job rotation must not be considered as an alternative to redesign workplaces with relevant risk factors. Job rotation does not reduce the risk instead it redistributes the risk among the workers. The idea underlying job rotation is that, by alternating tasks between muscle groups, it will provide rest periods and reduce the overall muscle activity, thus reducing muscular overload (Mathiassen, 2006).

Several studies related to jobs' monotony have found that there is a significant correlation between repetitive task and boredom. Boredom is a serious problem and can be associated to reduction of performance, dissatisfaction and accidents (Azizi et al., 2010). In these cases, job rotation can have a positive effect.

Rotation between different workers means that they have sufficient skills and knowledge to carry out the different assigned tasks. If workers do not have suitable training to develop the activities, the organization must provide the adequate training (Diego-Mas et al., 2008). A rotation system to be considered as an effective MSDs prevention measure is required regular assessments of the results based on valid and reliable parameters. (Aptel et al., 2008).

1.4 Parameters that can have influence on the job rotation system

Some parameters must be taken into consideration during the implementation of a job rotation system, such as: detailed description of the rotation system, identification of the characteristics of the workers population that might have influence on the process, professional experience, knowledge and risk assessment, in this case taking into consideration the posture, the repetitive movements and manual handling of loads (Takala 2007). Uva et al., (2008), identified three major groups that can have influence on the risk of MSDs: (1) ergonomic factors

(e.g. repetitive movements, frequent manual handling, postures, vibration, extreme temperatures), (2) organizational factors (e.g. excessive work rates, insecurity or job dissatisfaction, shift work) and (3) individual factors (e.g. smoking, alcohol intake, obesity, among others). These factors are likely to affect job rotation implementation. According to Widanarko et al., (2011), it is important to characterize and identify, among organizations, which factors can contribute to the success of the rotation implementation reducing the occurrence of MSDs.

1.5 Objectives

It is expected that factors which can have influence on the implementation of the rotation scheme will be characterized and identified in order to develop of a job rotation between tasks that contributes to reducing the risk of work-related MSDs. The current study was carried out within a textile industry were several complaints related to MSD were previously identified and were the cause of a high percentage of absenteeism among the workers. The main purpose of this work was to identify the factors that can have influence on the performance of the worker activity in order to develop a rotation scheme to be implemented.

2 MATERIALS AND METHOD

This case study was carried out in a textile industry located in Portugal and belonging to a German company that produces tires and other car systems with many other factories all over the world. In this organization, dedication and outstanding performance from the employees are required. Working conditions are created to enhance and foster quality performance among workers. The organization is committed to provide safe and healthy workplaces. In this type of industry, mainly due to the high manual materials handling and repetitive activities, workers are reporting a high number of MSDs.

A methodology to develop a job rotation scheme suitable to this type of industry was developed according to the following steps:

– Identification and characterization of the activities presenting high number of diagnosed MSDs. The different activities developed by workers were observed, directly and indirectly, and the measurements relevant to the study were done;
– Selection of methodologies for risk assessment taking into consideration the factors posture, repetitive movements and manual handling of loads—Rapid Upper Limb Assessment (RULA) (McAtamney and Corlett, 1993) and Manual Materials Handling guide (Mital et al.,1997).

The results were presented using a 3-color scale: red (R), representing a critical situation, orange (O), representing a medium-term intervention, and green (G), identifying a non-critical situation;
- Application of risk assessment for the identified activities;
- Identification and assessment of individual risk factors such as age, height, gender, seniority and some organizational factors;
- Identification of the workers' skills (skills from a lower level of responsibility (G1) to high level of responsibility (G4)-designation defined by the organization);
- Development of a matrix database able to relate each worker's skills to the identified activity risk;
- Analysis of the obtained results taking into consideration the relation between risk, identified relevant factors (age, seniority, height, gender) and workers' skills;
- Establishment of a job rotation system based on previous results.

In order to identify the activities that presented a high number of MSDs, it was necessary to study the associated textile process, which is divided in three main steps: twisting, weaving and dipping. In the twisting process several activities can be identified. These activities are related to manual materials handling when workers are loading or unloading the twisters. In these activities, a higher number of complaints by workers have been reported to the organization's health and safety department.

In the weaving process, repetitive activities can be observed, in particular when workers are tying the cords. The dipping process is divided in two steps (twisting and dipping) and several activities can be identified. These activities are related to manual materials handling when workers are loading or unloading the twisters and the dipping machine.

In order to carry out a risk assessment of the identified activities; posture, repetitive movements and manual handling of loads were studied. For each activity, heights, distances, times, frequencies, among other were measured.

In the proposed job rotation scheme, it was also important to identify the qualifications required for each activity development, since rotation is dependent on the skills and knowledge of the workers. In this industry, workers are divided into groups according to their skills. These groups were designated as G1, G2, G3 and G4. G1 is the less qualified group. With the increasing of the standard of the skills, workers are qualified to perform a greater number of tasks and tasks are per se more demanding. Subsequently, the risk assessment related to the possibility of

MSD occurrence and the identification of the skills, compelled the development of a matrix database that was able to relate these two variables. In a first approach, an empirical methodology of colours in a 3-point scale was used. The main purpose of the colours was to identify the risk level of MSD development related to each analysed task. Three principal colours were used: red (R), orange (O) and green (G). The red colour represents a demanding task regarding the MSDs possibility occurrence. Orange represents a task with a certain possibility of MSDs development and the green category identifies a non-critical situation. The rotation must be done, whenever possible, from red to green. The rotation must take into consideration that a group (G) does not always have the required skills to perform the tasks of the group immediately above. An experimental matrix database is presented in Table 1. In this matrix, it is possible to observe that task 1 is developed by workers included on the three identified worker skills' groups (G1. G2 and G3), The same can be observed on task 2. Regarding task 3 it is possible to observe that this task is performed by workers with a level of skills related to G1 and G2 groups. This factor is very important as it shoes that the level of skills should be taken into consideration on the rotation scheme, Indeed it should be guaranteed that workers can participate on the rotation scheme if they have competences to perform a given task. Individual factors such as gender, age, seniority, height, skills, and some organizational factors were also analysed since they can have influence on the development of a given task.

A job rotation scheme was developed taking into consideration the influence of the analysed factors on the development of a task. This was based on the risk assessment (identified as green, orange and red), individual and organizational factors identified and analysed (gender, age, seniority, stature and level of skills). Figure 1a, b represents an example of a job rotation scheme. The main idea is to adapt a rotation in order to alternate workers between different tasks taking in consideration their skills (figure 1a). In figure 1b the idea is to rotate workers between different tasks taking in consideration the risk of each task.

Table 1. Matrix database concerning risk factors for each activity.

Task	Risk	Age	Seniority	Stature	Level of skills		
					G1	G2	G3
1	Green	x	x	x	x	x	x
2	Red	x	x	x	x	x	x
3	Orange	x	x	x	x	x	

x applicable.

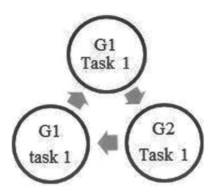

Figure 1a. Example of a job rotation scheme based on workers skills (adapted from Fonseca et al., 2012).

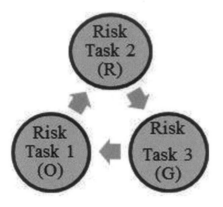

Figure 1b. Example of a job rotation scheme based on risk assessment (adapted from Fonseca et al., 2012).

3 RESULTS AND DISCUSSION

The company sections selected for the study were related to the twisting, weaving and dipping process. A sample of 70 workers was considered in this study. Results of the risk assessment for manual materials handling and repetitive tasks performed in these sections identified the tasks with higher risk of musculoskeletal disorders.

Regarding the individual factors, about 93% of the workers are male and, on average, have 35 years (SD = 11.491; interval range 19–61 years old). The majority of men have ages in the cluster [19–44] years old (Fig. 2), and are allocated to the productions sectors (twisting and dipping sections) (Fig. 3). This seems to indicate that these sections are more demanding in terms of physical effort. Women are predominantly allocated to the weaving section. These results were somehow expected considering the physical demanding of the others sections.

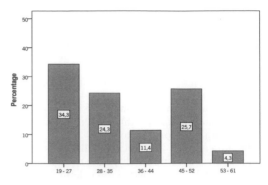

Figure 2. Workers distribution concerning age factor.

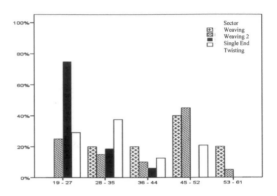

Figure 3. Workers age distribution by sector.

Results showed that most of the workers presented a low level of skills, G1 (63%), 1% is in the G4 level (Fig. 4). 34,4% of the workers has ages in the cluster [19–27) years old. The older workers are in the G1 level. Results also show that the younger workers (cluster [19–27]) perform the most demanding activities while the older (cluster [53–61] are assigned the less demanding tasks.

The taller workers (cluster [173–183] cm) perform their activities in the productive sections (twisting and dipping process) (Fig. 5). These results were also expected as, in these sections, being tall is a request feature to carry out some of the tasks. Regarding the weight, results showed that this factor is independent of the workers' distribution.

The taller workers (cluster [173–183] cm) perform their activities in the productive sections (twisting and dipping process) (Fig. 5). These results were also expected as, in these sections, being tall is a request feature to carry out some of the tasks. Regarding the weight, results showed that this factor is independent of the workers' distribution.

Based on these results, a rotation scheme was proposed. As an example, Table 2 presents factors

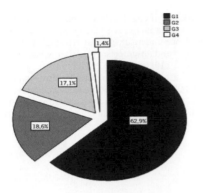

Figure 4. Workers distribution by level of skills (G1 to G4).

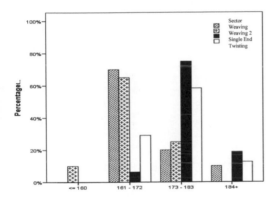

Figure 5. Workers stature distribution by sectors.

Table 2. Factors that influence the job rotation scheme on the twisting section.

			Level of skills		
Task	Risk	Stature	G1	G2	G3
1	Green	x	x	x	x
2	Red	x	x	x	x
3	Orange	x	x	x	x

x applicable.

that have influence on the rotation scheme for the twisting section (tasks 1, 2 and 3).

In this case, the rotation scheme was based on the workers' skills and stature, as well as on the results of the material handling risk assessments. The main idea was to adapt a scheme in order to alternate between high-risk tasks (red) and other tasks at a lower risk (green).

4 CONCLUSIONS

The development of a job rotation scheme aims at reducing the risk of musculoskeletal disorders and consequently the absenteeism, which is strongly associated to this subject. The identification of the factors that can have influence on the scheme implementation is a very important achievement. The current study identified some of the individual and organizational factors that could be responsible for the success of this process. The analysis of the results showed that individual factors, such as gender, age, seniority, height, skills and other associated factors, are important to the job rotation scheme implementation, but the rotation should only be implemented between tasks of the same sector.

It is expected that the implementation of this job rotation scheme will contribute to reduce the risk of MSDs associated to each job, thus contributing also to the economic and social goals of the organization. Future work should focus the improvement of the tasks, implementation and re-evaluation of the rotation system proposed.

REFERENCES

Aptel, M., Cail, F., Gerling, A., Louis, O. (2008). Proposal of parameters to implement a workstation rotation system to propetec agaist MSDs, *International Journal of Industrial Ergonomics* 38, 900–909.

Azizi, N., Zolfaghari, S., Liang, M. (2010). Modeling Job rotation in manufacturing systems: The study of employee's boredom and skill variation. *International Journal Production Economics*, 123 (69–85).

Ciriello, V., Dempsey, P., Maikala, R., O'Brien, N. (2007). Revisited: Comparison of two techniques to establish maximum acceptable forces of dynamic pushing for male industrial workers. *International Journal of Industrial Ergonomics*, 37 (11–12), 877–882.

Diego-Mas, J.A., Asensio-Cuesta, S., Sanchez-Romero, M.A., Artacho-Ramirez, M.A. (2008). A multicriteria generic algorithm for the generation of Job rotation schedules. *International Journal of Industrial Ergonomics, doi:10.1016/j.ergon.2008.07.009.*

Dul, J., Neumann, W.P. (2006). The strategic business value of ergonomics. *The International Ergonomics Association's 16th World Congress on Ergonomics*, Maastricht, NL, 2006.

Dul, J. and Neumann, W.P. (2009). Ergonomics contributions to company strategies. *Applied Ergonomics*, 40, 745–752.

Ferguson, S., Marras, W., Allread, W., Knapik, G., Splittstoesser, R. (2011). Musculoskeletal disorder risk during automotive assembly: current vs. seated. *doi:10.1016/j.apergo.2011.10.001.*

Mathiassen, S.E., 2006. Diversity and variation in biomechanical exposure: what is it, and why would we like to know? *Applied Ergonomics* 37, 419–427.

McAtamney, L., Corlett, E. (1993)- RULA: Rapid upper limb assessment - A survey method for the investigation of work-related upper limb disorders. *Applied Ergonomics*, 24(2), 91–99.

Mital, A., Nicholson, A. & Ayoub, M. (1997). *A Guide to Manual Materials Handling*, 2nd Edition. London: Taylor & Francis.

Takala, J. (2007). Lighten the Load—Foreword. *Magazine of the European Agency for Safety and Health at Work*, 10 (1).

Van Rhijn, J.W., de Looze, M.P., Tuinzaad, G.H. Groenesteijn, L., de Groot, M.P., Vink, P. (2005). Changing from batch to flow assembly in production of emergency lighting devices. *International Journal of Production Research*, 43, 3787–3701.

Widanarko, B., Legg, S., Stevenson, M., J. Devereux, J., Eng, A., Mannetje, A., Cheng, S., Pearce, N. (2011). Prevalence and work-related risk factors for reduced activities and absenteeism due to low back symptoms. doi.org/10.1016/j.apergo.2011.11.004.

Uva, A., Carnide, F., Serranheira, F. Miranda, L., Lopes, M. (2008). Guia de orientação para a prevenção das lesões músculo-esqueléticas e relacionadas com o trabalho: programa nacional contra as doenças reumáticas. *DGS*.

Occupational Safety and Hygiene – Arezes et al. (eds)
© *2013 Taylor & Francis Group, London, ISBN 978-1-138-00047-6*

Design of personal protective equipment: To promote of Brazilian artefacts with ergonomic attributes

C.R. Leite da Silva
Federal University of Pernambuco, Recife, Brazil

J.E.G. dos Santos
Univ. Estadual Paulista (UNESP), Bauru, Brazil

ABSTRACT: This research aims to investigate a topic sometimes overlooked among the research of ergonomics. Particularly, we treat the garments as personal protective equipment. In this study we abode on clothing produced and used in Brazil, making notes about the management of occupational safety for the use of such equipment. In turn, identify Relevant requirements for the design of personal protective equipment used in Brazil, in line with what is being studied in global forums on the subject.

1 INTRODUCTION

In Brazil the number of accidents has decreased year after year, due to a culture implemented, which encourages appreciation of safety at work. Despite the effort by organizations and government in preventing accidents, about 700 thousand cases of occupational accidents are recorded on average each year in Brazil, excluding cases not officially reported (Portal Brazil, 2012).

This research aims to investigate the legal requirements of projects of clothing as personal protective equipment (PPE), produced by Brazilian manufacturers, in accordance with current legislation. And to meet these goals, methodologically, we did a literature review of theses, dissertations and journals on the design in your protective clothing decades. And confront some data collected with the present there visiting some of these garments production centers in Brazil.

In many societies, clothing is meant to express wealth, status, occupation, age, occasion, gender, among others (Coates, 2005). However, we realize that in certain professional activities, such as surgeons, military, agricultural workers, fire-fighters, among others, the function of clothing means more.

In turn, it was essential thematic works (LY, 2001; NIELSEN, 1991; GIBSON, 2008), is addressing the variables that delineate comfort in occupational clothing or even reporting on questions of safety, strength, durability and other functionality of textiles.

In Brazil, the problem of inadequacy of PPE ergonomic and environmental conditions is also no stranger.

Brazilian agriculture, especially in small rural communities, it is common to come across rural workers without required PPE during handling and application of pesticides.

For Coutinho (1994), one of the main reasons for not using PPE lies in the fact that many of PPE used in agriculture, due to its inadequacy, can cause thermal discomfort, making them quite uncomfortable to use and can lead, in extreme cases, heat stress of rural worker.

2 THE DESIGN OF PERSONAL PROTECTIVE EQUIPMENT BRAZILIAN

The use of PPE, while not required, should be considered as protection technology available within an integrated and systemic approach to occupational problems.

The efficiency of whole of health and safety system (HSS) at is closely related to how it is conducted and balanced decision-making process, the selection of alternatives for prevention, protection and control (ILO, 2001).

However, an efficient design of HSS should contemplate a systemic approach, the integration of all elements relevant to establish policies and strategies appropriate to each situational reality.

Effective management of a program must aspire HSS also increases productivity in work processes with reductions in risk.

The personal protective equipment, whether clothing or devices, used for individual use by the employee, for the protection of risks likely to threaten the health and safety at work (MTE, 2011).

In Brazil, they are normalized of brazilian regulation number 6, which requires the government to monitor the effectiveness of these. However, besides the registration of the organization, it is for the manufacturer to prove their efficiency from the technology employed. That is, legislation determines the product design, but does not define the minimum requirements for the same.

This gap that problems are perceived inadequacy of PPE to their users. The attributes of comfort in the design of these garments (Figure 1), in Brazil, have evolved as safety practices are established, short-demands of the workers.

According Veiga (2007), several problems can cause the inadequacy of PPE to certain conditions. Some of the desirable characteristics for IPE and were designed to provide greater security may be introducing operational difficulties in many work situations. For example, increased resistance to permeability of a fabric, a greater resistance to shock, increased heat resistance can be associated with weight gain, low thermal comfort and less portability of PPE.

Another important aspect is the difficulty of the adequacy of PPE to anthropometric characteristics and environmental conditions of each locality.

2 CONTRIBUTIONS OF RESEARCH IN ERGONOMICS

In some research: as those conducted by Sarraf (2004), which sought to map the ergonomic aspects of work uniforms; those of Kagiyama (2011), on variables of comfort in apparel; research

and Oliveira (2011), about design of PPE for pesticide application to cropland, recognize their forest clearance, merit and relevance to the area.

Still, it is possible to note that Brazilian studies about the guidelines of comfort and effectiveness of clothing still fall short, falling short of significant industrial growth in Brazil, highlighted in recent years.

In countries like Brazil, with continental characteristics, as stated Sayad (2000), adds itself, many "Brazils", where their regional differences may correspond to differences between nations, we must strive to build a reality where the artefacts surrounding the occupational safety are there to the worker in accordance with their individual characteristics to their effective use.

Legislation in many countries in Europe already requires re-examining the risks associated with the use of Personal Protective Equipment (PPE). between these risks would be the harm to health caused by hyperthermia. In surveys of Crockford (1999), we find that beyond the problems to human health, the lack of thermal comfort in the workplace, caused by the use of PPE, turns out to have effects also economical, since direct influence on productivity and quality of work performed.

Therefore it would be correct to say that the room temperature affect job performance. Body temperatures beyond optimal limits (36.5° to 37.5° C) affect the physical and mental performance, and may lead to physiological and pathological impairments.

For Ahasan (2002) and Rodahl (2003), the human body, failing to maintain thermal equilibrium, the rate of increase heat retention by influencing the efficiency and productivity of workers.

Figure 1.   PPE clothing used as the application of pesticides.

However, the limits in the legislation for thermal comfort and temperature extremes are based on acute reactions of workers exposed to heat and not in their chronic effects. Therefore, one can say that the literature on prolonged and continuous exposure of workers to the heat still needs further studies (Wood, 2004).

This is an interesting concern of ergonomics. Consistent efforts to secure a job without impacts on productionand productivity of organizations.

3 RESULTS AND DISCUSSION

Of protective clothing that we analyze, to design this study, we evidenced two types. Those that are necessary to protect workers from thermal hazards and ensuring protection of chemical risks.

The protective clothing against thermal hazards are those against convection flames, heat contact, against sparks, molten metal droplets, against the intense cold and frost. They are a prime requirement of protective clothing is able to work, leisure or sports.

Already, advances in technology have brought with them an increase in the types of chemical hazards to which the worker is exposed, as liquid (spray), gases, occurring in industries such as chemical, pharmaceutical, petrochemical, electroplating industry and agriculture (fertilizers). They require the use of clothing that is waterproof and resistant to chemicals, provides a tight seal against toxic gases, or dust filters dangerous.

We realize that manufacturers of industrial garments with protective characteristics, tend to use parameters and productive foreign references, confirming the findings of Sarraf (2004).

These patterns sometimes clash with the particularities of Brazilians. Are caused by problems related to modelling clothes designed for a worker with diverse biotype. Or by cultural issues, where the habits and organizational culture construct their own strategies.

We identified some important stages for defining the design requirements of industrial garments. They are: (i) market research, (ii) studies of wear, (iii) objective evaluation of the characteristics of clothing and (iv) objective evaluation of tissue characteristics employee. This categorization is consistent with research by Das; Alagirusamy (2010). And, in turn, helps to structure a rationale for research and construction of the project requirements of garments surveyed.

The ISO (1986) provides that security is freedom from unacceptable risks and damage. The concepts proposed here define the use of safety clothing is clearly not the first choice among the security measures. However, it remains essential in all types of protective clothing.

We emphasize the understanding of comfort in apparel extrapolates issues only and usability should be thought ahead multidisciplinary, whether by doctors, psychologists, physiotherapists, ergonomists, designers, and others, for health and safety.

4 CONCLUSIONS

Discuss the past, present and future trends of studies on safety clothing is required. Protective clothing has a long history. Since trellised vines of Adam and Eve, the armors of medieval knights and warriors.

We stopped in this study, the research of last decades, when we realize at the same time, a development of modern protective clothing. All in accordance with the excellence of production of smart textiles, which provide greater protection for its users.

The development of protective clothing, as observed in this research presents a considerable improvement in recent decades. Since the clothes from normal use, with some protective properties to design of complex systems and multifunction protection, using sophisticated and modern materials and manufacturing techniques.

Studies on PPE, generally no problems emphasize design of these products. Are studied mainly contamination suffered by workers and the environment. Although there are reports on difficulties in using the equipment, especially for recovery by the employee, the immediate perception of discomfort versus the perception of health risks in the long term.

Brazilian law is naive compared to PPE when universally accepted that the use of these products should eliminate or neutralize the unhealthy, assuming that the protection of the worker to use the EPI is efficient.

In some of the situations studied, the partial protection resulted in contamination of the worker. However, the penalty would depend on the responsible oversight. Unfortunately, in Brazil such supervision is almost always conditioned by the number of cases.

With the passage of time, other sectors, not least important, has used the principles of protective clothing, such as sport and leisure sector. These, in turn, is raising funds for research and development. What makes everything more complex technical point of view, because the designer is replaced by the elements that combine protection with other factors, such as fashion or performance in sports.

At the same time, when we see the impacts, to improve quality of life, health and safety in the work environment, we feel comforted and blessed with the challenge.

REFERENCES

Ahasan, R. (2002). Strenuous tasks in a hot climate: a case study. Work Study, v. 51, n. 4/5, p. 175–181.

Coates, J.F. (2005). From my perspective: the future of clothing. Techn. Forecasting & Social Change, 72, 2005. 101–110.

Coutinho, J.A.G. (1994). uso de agrotóxicos no município de Pati do Alferes: um estudo de caso. Caderno de Geociências, n. 10, p. 23–31.

Crockford, C.W. (1999). Protective clothing and heat stress: introduction. Ann. occup. Hyg., v. 43, n. 5. p. 287–288.

Das, A., Alagirusamy, R. (2010). Science in clothing comfort. New delhi, Cambridge, Oxford: WPI.

Gibson, P. (2008). Water-repellent treatment on military uniform fabrics: physiological and comfort implications. Journal of Industrial Textiles, July, 38 (1). 43–54.

ILO-International Labour Office (2001). Guidelines on occupational safety and health management system. Geneva: ILO-OSH.

Kagiyama, W. (2011). Design de vestuário íntimo: o sutiã sob abordagem de conforto. Porto Alegre: UFRGS.

Li, Y. (2001). The science of clothing comfort. Textiles Progress, 31 (1/2). 1–138.

MTE-Ministério do Trabalho e Emprego. (2011). NR6—Equipamento de Proteção Individual.

Nielsen, Ruth. (1991) Work Clothing. International Journal of Industrial Ergonomics, 7, 77–85.

Oliveira, V.P.Z.T. (2011). Condições de segurança operacional e proposta de uso de novo tecido para confecção de equipamentos de proteção individual: estudo de caso. Bauru: UNESP.

Portal Brasil. Acidentes de Trabalho. (2012). Retrieved from http://www.brasil.gov.br/sobre/saude/saude-do-trabalhador/acidentes-de-trabalho.

Rodahl, K. (2003). Occupational Health Conditions in Extreme Environments. Ann. occup. Hyg., v. 47, n. 3. pp. 241–252.

Sarraf, R.A. (2004). Aspectos ergonômicos em uniformes de trabalho. Porto Alegre: UFRGS.

Sayad, A. (2000). O retorno: elemento constitutivo da condição do imigrante. Travessia: Revista do Migrante. São Paulo, XIII.

Veiga, M.M. (2007). Contamination by pesticides and Personal Protective Equipment (PPE). Rev. bras. Saúde ocup., São Paulo, v. 32, n. 116. pp. 57–68.

Wood, L. (2004). Heat Resistant. Occupational Health. v. 56, n. 7, p. 25–29.

Occupational Safety and Hygiene – Arezes et al. (eds)
© *2013 Taylor & Francis Group, London, ISBN 978-1-138-00047-6*

A comparison between two participatory approaches in a manufacturing company

A.S.P. Moraes & P.M. Arezes
University of Minho, Guimarães, Portugal

R. Vasconcelos
University of Porto, Porto, Portugal

ABSTRACT: Participation has been established as a new trend for managers, designers, and social scientists. Occupational Health and Safety and Ergonomics moved to the same direction. However, participation cannot be considered a single notion. Different participatory approaches can be supported by different types of participation. The objective of this paper is to show how an existing participatory tool, which could have been used as a facilitator, actually hampered the introduction of a second participatory approach. The already established tool was part of the Total Productive Maintenance program, whilst the new one was part of an ergonomic intervention, with the main goal of contributing to the design of a processing machine. From the difficulties that emerged during the intervention, three issues were identified as having a direct impact over its success. As a conclusion, it is showed that different types of participation, based on different frameworks, do not necessarily facilitate each other.

1 INTRODUCTION

Participatory approaches have become a trend among organizations through the last decades: users and workers are asked to give their contribution, express their points of view and help on the search of solutions for the production problems and the improvement of products and processes.

Different approaches involving workers and users participation have been established in management, design and social sciences, and concepts like autonomous workgroups, human or user-centred design, empowerment movement, and job enrichment became part of the organizational jargon. According to Fischer (2011), the development of new technologies, combined with specific sociotechnical environments, have facilitated a shift throughout society, from the consumer cultures, specialized in producing finished artifacts that are going to be consumed passively, to cultures of participation, in which all people are provided with the means to participate and to contribute actively in personally meaningful problems.

Following this tendency, some initiatives were developed in the field of Occupational Health and Safety (OHS). The importance of workers participation in preventing and managing risks became officially accepted. An example is the Guidelines on Occupational Safety and Health Management Systems published by the International Labour Organization, which state that the "worker participation is an essential element of the OSH management system in the organization" (ILO, 2001).

Ergonomics drifted in the same direction, in particular in a specific domain of ergonomics that is called Participatory Ergonomics (PE). This concept founded its basis in the early 1980's, and since then, followed the remarkable studies of Noro and Imada (1991), Nagamachi (1995) and Wilson (1995a). As stated by Imada (1991), participatory ergonomics should involve the end-users in planning, developing and implementing workplace changes. The subjacent idea is that "workers know very well what kind of ergonomic problems there are in the workplace" (Jong, 1995), and so they can be useful in the search for solutions.

Some PE studies show the positive impacts of workers' participation in preventing musculoskeletal disorders and reducing physical workload (de Jong and Vink, 2002; Hignett, Wilson, & Morris, 2005). Other studies focus the results of workers' participation in the design of hand-tools (Vink & van Eijk, 2007), final products (Sundin, Christmansson, & Larsson, 2004), workspaces (Seim & Broberg, 2010), equipment (dos Santos, Farias, Monteiro, Falcão, & Marcelino, 2011), among others. Thus, Participatory Ergonomics Design (PED) emerged as a branch of PE, in the cases "where participatory ergonomics efforts and achievements are made in the design phase" (Sundin et al., 2004).

Furthermore, authors like Haines and Wilson (1998) discuss the needed framework to promote participation, and Bødker (1996) emphasizes the needed conditions to implement participatory tools.

However, if there is not a "single, unifying model of participatory ergonomics" (Vink et al., 1992), the same cannot be expected from the diverse participatory approaches. The focus here is not to discuss or compare specific participatory approaches and their theoretical basis, but what type of participation is practiced. So, the issue is identifying the similarities and differences among types of participation, suggested by different participatory approaches.

The study context in which this paper is based on was conducted in a large manufacturing factory, where an ergonomic intervention was developed as part of a cooperation agreement between the company and the research party. Its prime objective was to contribute to the design of a second processing machine, which was going to be installed in the factory, brought from another production plant, located in a different country but of the same group. In the moving process, the machine was going to be disassembled, and, despite the design's main concept being to replicate the machine that was already in operation, it would undergo some technical improvements, especially regarding its electronic and automation systems. Its installation would represent to the plant an increase in production rates and less dependence on a single machine, which was already a bottleneck and was working over the recommended rate, by the maintenance and safety managers. When the cooperation started, the new machine design was already being developed, and some decisions were already taken. However, according to the project manager engineer, some modifications in the design could still be done, if they would represent a benefit to the project. The conditions and framework to seek for improvements and solve design gaps through an ergonomic intervention were settled. The idea was to avoid future problems regarding bad design choices and bring ergonomic criteria to the design. However, the intervention did not achieve the planned goals due to some issues: among them, the impact of a previously existing participatory tool in the company, which was under the responsibility of an organizational techno-structure.

The objective of this paper is to show how this existing participatory tool, which could have been used as a facilitator, actually hampered the introduction of the participatory ergonomic approach.

2 MATERIALS AND METHOD

The ergonomic intervention was based on an action-research strategy and was planned to happen in longitudinal time horizon (Saunders et al., 2009). Following the "future work activity" approach (Daniellou, 2002), it consisted of two phases and workers' involvement was a fundamental key to ensure the success of both of them. The idea was the same defended by Wilson (1995b): workers being given the opportunity and power to use their knowledge to address ergonomic problems related to their own working activities.

The first intervention phase applied the Ergonomic Work Analysis method (Guérin, Laville, Daniellou, Duraffourg, & Kerguelen, 2001). It lasted for five months, and took around 35 data-collecting days. Machine workers' involvement was granted through the observation of their work activities, and open and semi-structured interviewing. Spontaneous and concurrent verbalizations were also gathered. Additional interviews were conducted on higher hierarchical levels according to their relevance to the study. The goal of this phase was to get closer to the machine workers and have a better understanding of the machine's production process.

In the second intervention phase, the main idea was to contribute to the design of the new machine. Starting with the organization of data collected by the researcher during the first phase, which was done mainly individually with the workers, the second phase was focused on a collective construction of ergonomic demands and the development of design criteria. It would also involve possible changes in the initial design and monitoring the machine's installation and start-up. Participatory Ergonomics and Participatory Design methods and tools, such as participatory meetings and on site work simulations, were proposed. It lasted two months, in a total of seven data-collecting days. Furthermore, three participatory meetings were held and their planning and procedures are described below.

2.1 The design participatory meetings planning

In the first round, the design participatory meetings were planned to be a place for the workers to express their opinions about their work routines, difficulties, and share some individual perceptions with their colleagues. Its main goal was to collectively formulate new design criteria, based on the point of view of the activity, and to search for solutions for problems that might exist in the second machine.

Subsequently, a second round of meetings was planned, including representatives from diverse fields of knowledge and with different technical expertise, such as managers, Total Productive Maintenance (TPM) team members, and project and safety engineers. It would widen the possible design solutions and bring different points of view to the issues in debate.

2.2 The design participatory meetings procedures

The design participatory meetings should be held outside the shop floor, where workers could be fully focused. Besides the researcher, only the machine workers should take part in the first round. As there are only thirty-five machine workers, full attendance would be possible, at least in the first round of meetings. This would ensure that all workers had, at least, an opportunity to express their points of view. The meetings were negotiated and approved by the department head and by the OHS department director. Workers were personally invited by the researcher to attend the meetings. In addition, the machine supervisors were aware of the meetings.

The first difficulty regarding the participatory meetings was arranging time for them to be held, due to the availability of workers, apart from their work time. The solution was to promote the meetings during the programed stop of the machine for basic maintenance, which happens once a week. According to the shift system of the company, this meant that it would take at least three weeks to promote the meetings with three of the five shift working groups. In the case of the other two, the weekend shifts, another solution was necessary, as there is no programed maintenance stop in their working hours.

In order to conduct the meetings, a power point presentation, consisting of three parts, was prepared by the research party: firstly presenting the study, its main objectives and basic Ergonomics concepts; secondly, workers should state what most bothered them in the machine and later give opinions on their "ideal machine"; and thirdly, presenting the new machine design project, after that discussing their first impressions and ideas about possible future difficulties.

Three meetings were held, and lasted, in average, 90 minutes each. The meetings were not recorded, but their ideas and verbalizations were annotated.

3 RESULTS AND DISCUSSION

As soon as the second intervention phase started, some difficulties in promoting the involvement of workers emerged. After three meetings were held and the field simulations started, the intervention had to be interrupted.

Some difficulties were related to technical aspects: (i) scarce information in the presented blueprint, such as specifications on materials, production flow and human-machine interfaces, which has an impact in the capability of representation of the future work situation; (ii) the research party's inability to answer some of the questions posed by the workers; and (iii) the time constraints and design change possibilities. Alternative solutions, such as field simulations, would help to solve the first difficulty, and the presence of a design team member to answer the workers doubts, as it was planned to happen in the second round of participatory meetings, would solve the second and third difficulties.

But also, there was an obvious impact on the workers involvement by the social aspects. There was a visible weakening of the workers collectiveness, which hampered their involvement: (i) very less suggestions were given by the workers, when compared to the number of suggestions in the first intervention phase, when data collection was done individually; (ii) deeper discussions between workers were avoided, to not stimulate new nor old conflicts; (iii) there was a common and uncomfortable feeling by the workers for being put in a position that they were not accustomed with.

Also in the social aspects, a remarkable difficulty was the recognition, by the design team members, of a possible contribution from the workers on solving design gaps. This had a direct impact over the implementation of the second round of meetings, as there was an evident lack of willingness to attend them.

Other impacts on workers involvement were directly related to the already existing participatory tool in the company. Next, it will be explained how this tool is organized, how it was considered to be used by the ergonomic intervention, and the main issues that impacted the success of the ergonomic intervention.

3.1 The possible usage of the TPM data in the ergonomic intervention

When the ergonomic intervention started, many suggestions, made over the existing machine, were already compiled by another participatory approach, under the responsibility of the TPM group. Although they were not initially planned to be used by the research party, they were a good source of data, thus should not be ignored or discarded.

However, not only the use of this data became a conflict point among the different stakeholders, which discouraged its use, but also the TPM participatory tool impaired workers' involvement in the new participatory approach. According to it, three issues regarding these influences will be explained in the following items, along with a brief explanation on how the TPM participatory tool works.

3.1.1 The TPM participatory tool

The already implemented participatory tool was structured in the company around 2003, and underwent many changes throughout the years. Through this tool, workers can make suggestions on any aspect or department of the company. It functions as follows: using a pre-defined form, workers fill it in with suggestions and deposit it in a ballot box. Later, the TPM personnel collect the suggestions and proceed with its analysis. Further analysis is taken by a committee formed by representatives from the engineering department and safety department. They assess its relevance, technical feasibility and cost-benefit ratio. The profit obtained through such suggestions is the main criteria for adopting or not the proposed idea. If the suggestion is implemented, the worker responsible for it receives a monetary reward, proportional to the profit generated. It was pointed out by the TPM coordinator that, apart from suggestions, further worker involvement only exists if some uncertainty over the given suggestion is observed.

3.1.2 The reward issue

The first issue raised by this study was related to the workers' willingness to take part in the design, because a suggestion made and implemented by the TPM, after the new machine installment, could represent future monetary gain for the workers. When the possibility of using the TPM suggestions in the design of the new machine was raised by the research party, the issue of monetary reward emerged. The main question was if their implementation in the design of the new machine should be rewarded or not by the TPM system. No agreement was achieved in this issue.

3.1.3 The trust and transparency issue

The workers' opinion on the TPM suggestion tool drifted apart: some of them were very proud of the amount of suggestions they made and were implemented. On the other hand, some were not so convinced about the efficiency of the suggestions or did not agree with the modifications proposed by their colleagues. Other workers mentioned episodes in which their suggestions were discarded by the TPM group, but were put into operation a few months later, as being a solution proposed by the engineering department. In such cases, workers felt "betrayed".

There was a visible lack of confidence in the TPM participatory tool by a significant number of workers. The impact of this feeling over the ergonomic intervention tool could not be ignored as it discouraged some of the workers' involvement, mainly those who expressed no interest in contributing to the design of the machine.

3.1.4 The collective issue

The TPM tool is based on individual suggestion and prizes. This feature leads to individualism and stimulates rivalry: even discussing an idea with a colleague could mean losing it to him. The need for secrecy was one reason claimed by the TPM members not to discuss the suggestions with other workers. Due to this, workers' suggestions are discussed away from them, by the TPM committees.

Discussing ideas collectively revealed itself as being a difficult task, because the group was fragmented, there were conflicting interests, and the workers' power to influence decision making inside the TPM group was limited.

3.2 Comparing participation

A recent concept refers to participation as a mechanism for the involvement of employees in management decision-making by means other than information and consultation, and that can be done in a 'direct' way, when employee participation is practiced face-to-face or individually between employees and managers, or 'indirect' way, when it occurs through employee representation (European Industrial Relations Dictionary, 2009). In this sense, it is questionable if the TPM participatory tool should be considered as a "true" participation. Without considering this aspect, a comparison between the 2 participatory approaches can still be done.

The TPM system was organized in a way that does not require overtime from workers, or a need to stop the production. However, promoting a collective participation as it was suggested by the ergonomic intervention, face this practical difficulty: there was no established place and time for this kind of workers' involvement.

According to Heckscher (1995), participation can lead to different results depending on the way that it is organized, i.e. it can reinforce bureaucracy or overcome it. The type of participation suggested by the TPM tool is based on a bureaucratic way and lead to a reinforcement of it. The workers participation happens in a very specific and controlled framework, where workers can give suggestions, but the power to accept them and take decisions are still in the hands of the technical staff and higher hierarchical levels. In the other hand, the participation suggested by the ergonomic intervention could help on "breaking the walls" of bureaucracy, as it would depend on a social process, where different points of view would be put together, and the decision would be accorded collectively.

Also, keeping the process individually and confidential, allow a better control over the suggestions and the workers idea. Here, it is possible to note what Cattani (2011) affirms as being a possible answer to why participatory management is not adopted in a more systemic and permanent way in organizations: the possible deployments that can be resulted from the gaps in the hierarchical power hold by the managers. The level of control in the participation proposed by the ergonomic intervention, by the other side, was very reduced. As the discussion departure points were not necessarily a suggestion, but problems and difficulties experienced by the workers, any kind of idea could emerge, and maybe there was no easy answer for some of them.

4 CONCLUSIONS

Nowadays, participation is one of those words that can mean almost anything (Heckscher, 1995) or be considered as a panacea by organizations (Wilson, 1995b). Participatory tools can be useful to reach similar objectives—obtain the workers contribution, get their points of view, search for solutions and improve processes—but the way to achieve them may not be the same, that is to say, participation can assume different forms, need different frameworks, and lead to different final results.

The case presented was a unique opportunity to promote a participatory ergonomics design approach. This happened because it was conducted simultaneously with the installment project, which was mainly developed by an internal engineering group, and would have the participation of the same workers from the existing machine. However, the intervention did not accomplish its intended results.

Among other difficulties, the most prominent were the difficult in promoting a collective construction of design solutions and the impact of the existing participatory tool over workers' involvement.

In conclusion, even pursuing similar objectives, different means of participation can be employed, and if they are not based on similar frameworks, they do not necessarily facilitate one another.

ACKNOWLEDGMENTS

This work was partially funded with Portuguese national funds through FCT—Fundação para a Ciência e a Tecnologia, in the scope of the R&D project PEst-OE/EME/UI0252/2011.

REFERENCES

Bødker, S. (1996). Creating conditions for participation: Conflicts and resources in systems design. *Human Computer Interaction*, 11(3), 215–236.

Cattani, A.D. (2011). Gestão Participativa. In: Antonio David Cattani & Lorena Holzmann (eds.), *Dicionário de Trabalho e Tecnologia*. Porto Alegre, RS: Zouk.

Daniellou, F. (2002). Métodos em ergonomia de concepção: A análise de situações de referência e a simulação do trabalho. In: Francisco Duarte (ed), *Ergonomia e projeto na indústria de processo contínuo*. Rio de Janeiro: Lucerna.

de Jong, A.M., & Vink, P. (2002). Participatory ergonomics applied in installation work. *Applied Ergonomics*, 33(5): 439–448.

dos Santos, I.J.A.L., Farias, M.S., Monteiro, B.G., Falcão, M.A., & Marcelino, F.D. (2011). Using participatory ergonomics to improve nuclear equipment design. *Journal of Loss Prevention in the Process Industries*, 24(5): 594–600.

Eurofound. (2009). Participation. In Eurofound, *European Industrial Relations dictionary*. [Online] Available from: http://www.eurofound.europa.eu/areas/industrialrelations/dictionary.

Fischer, G. (2011). Understanding, fostering and supporting cultures of Participation. *Interactions*, 18(3): 42–53.

Guérin, F., Laville, A., Daniellou, F., Duraffourg, J., & Kerguelen, A. (2001). *Compreender o trabalho para transformá-lo: a prática da ergonomia*. São Paulo: Edgard Blucher.

Haines, H.M., & Wilson, J.R. (1998). *Development of a framework for participatory ergonomics*. Contract Research Report 174/1998. Norwich: Health and Safety Executive.

Heckscher, C. (1995). The failure of Participatory Management. *Across the Board*, 54(Nov/Dec): 16–21.

Hignett, S., Wilson, J.R., & Morris, W. (2005). Finding ergonomic solutions – participatory approaches. *Occupational Medicine*, 55(3): 200–207.

Imada, A.S. (1991). The rationale and tools of participatory ergonomics. In Kageyu Noro & Andrew S. Imada (eds.), *Participatory Ergonomics*. London: Taylor & Francis. pp. 30–51.

International Labour Organization (2001) ILO-OSH 2001. *Guidelines on occupational safety and health management systems*. Geneva: International Labour Office.

Nagamachi, M. (1995). Requisites and practices of participatory ergonomics. *International Journal of Industrial Ergonomics*, 15(5): 371–377.

Noro, K., & Imada, A.S. (1991). *Participatory Ergonomics*. London: Taylor & Francis.

Saunders, M., Lewis, P., Thornhill, A. (5th ed) (2009). *Research methods for business students*. Harlow: Prentice Hall.

Seim, R., & Broberg, O. (2010). Participatory workspace design: A new approach for ergonomists? *International Journal of Industrial Ergonomics*, 40(1): 25–33.

Sundin, A., Christmansson, M., & Larsson, M. (2004). A different perspective in participatory ergonomics in product development improves assembly work in the automotive industry. *International Journal of Industrial Ergonomics*, 33(1): 1–14.

Vink, P., & van Eijk, D.J. (2007). The effect of a participative product design process on user performance. *Safety Science*, 45(5): 567–577.

Wilson, J.R. (1995a). Ergonomics and Participation. In John R. Wilson and E. Nigel Corlett (eds.), *Evaluation of Human Work: a practical ergonomics methodology*. 2nd ed. London: Taylor & Francis. pp.1071–1096.

Wilson, J.R. (1995b). Solution ownership in participative work redesign: The case of a crane control room. *International Journal of Industrial Ergonomics*, 15(5): 329–344.

Occupational hygiene

Occupational Safety and Hygiene – Arezes et al. (eds)
© 2013 Taylor & Francis Group, London, ISBN 978-1-138-00047-6

Chemical mixtures—is a risk assessment actually necessary?

P.E. Laranjeira
CIICESI, ESTGF, Instituto Politécnico do Porto, Felgueiras, Portugal

ABSTRACT: Although workers are continually exposed to mixtures of chemicals, it is clearly neither feasible nor scientifically appropriate to consider every possible combination of chemicals to which the population or the environment might be exposed. The circumstances that will define the need to conduct a risk assessment for a mixture will therefore depend on the context of exposure. In the present paper, author aims to recall the difficulty behind the development of new/usage of existing chemical mixtures risk assessment methodologies, delimiting therefore the situations where the need to conduct a mixture risk assessment is considered mandatory. Furthermore a list of most relevant risk assessment methods is presented.

1 INTRODUCTION

Current risk assessment practices are largely based on evaluating the toxicity of single chemical entities. However, in reality workers are exposed to a large number of chemicals simultaneously (the so-called "chemical cocktail").

There are regular expressions of concern that exposure to this "chemical cocktail" could result in adverse health effects unforeseen by current risk assessment practices, but EU legislation does not currently contain a systematic mechanism for the comprehensive and integrated assessment of the effects of chemical mixtures that takes account of different exposure routes and different types of products.

Given the extreme complexity of such assessment, this issue cannot be solved rapidly.

In the present paper, author aims to recall the difficulty behind the development of new chemical mixtures risk assessment methodologies and/or the usage of existing methodologies, delimiting therefore the situations where the need to conduct a mixture risk assessment is considered mandatory.

2 THE CHALLENGE ON DEVELOPING A NEW RISK EVALUATING METHODOLOGY

When evaluating a mixture of compounds one of the main points to consider is whether there will be either no interaction or interaction in the form of either synergism or antagonism.

These basic principles of combined actions of chemical mixtures are purely theoretical and one often has to deal with more than one of the concepts at the same time when mixtures consist of more than two compounds and when the toxicity targets are more complex.

Even though methodologies for assessing the risks of priority mixtures exist, many data and knowledge gaps persist.

The issue is further complicated by the fact that much EU legislation targets specific groups, covering plant protection products, biocides, cosmetics, pharmaceuticals, veterinary medicines, and so on. This can be an obstacle to coordinated, integrated assessments of mixtures containing substances that fall under different pieces of legislation.

As the number of potential chemical combinations is very large, the first challenge under any new approach will be to identify priority mixtures, so that resources can be focused on the most potentially harmful combinations.

3 NEED TO CONDUCT A RISK ASSESSMENT?

Although workers are continually exposed to mixtures of chemicals, it is clearly neither feasible nor scientifically appropriate to consider every possible combination of chemicals to which the population or the environment might be exposed. The circumstances that will trigger the need to conduct a risk assessment for a mixture will depend on the context in which exposure to the mixture occurs.

Certain mixtures, usually those that are commercially supplied, fall within the scope of regulatory risk management schemes. These tend to be mixtures that are commercially supplied. For these mixtures, the data reporting schemes generally dictate the framework in which the risk assessment is conducted.

In other situations, for mixtures that are not commercially supplied, there are no clear guidelines to indicate the circumstances when a mixture risk assessment is required.

Traditionally substances that are present in such mixtures have been regulated based on single substances evaluations such as occupational exposure limits (OELs), acceptable daily intakes (ADIs) and maximum residue levels (MRLs).

These evaluations generally do not include an assessment of the effects of co-exposure to other chemicals.

However, in some cases, where groups of similarly acting chemicals are known to occur together, group evaluations are performed and group ADIs assigned.

But where mixtures fall outside the scope of reporting schemes and there is no existing requirement to conduct a mixture risk assessment, is a risk assessment actually needed? The need for mixture assessments should be, in author's opinion, mandatorily considered in any situation where:

– There is the potential for significant human exposure to occur and there is direct evidence for toxicity of the mixture;

or

– There is evidence for a synergistic interaction between substances that are known to occur together;

or

– For individual components in the mixture, the margins between measured/predicted levels of exposure and thresholds of toxicological effect are narrow or there is a concern that exposures may exceed thresholds of effect;

or

– There is the likely presence of other similarly acting substances;

or

– Chemicals are present together that share aspects of their absorption, distribution, metabolism and elimination and there is reason to believe that this may affect the levels of a toxicant at its target site.

4 EXISTING RISK ASSESSMENT METHODS

Several risk assessment methods are available, particularly for complex circumstances. Table 1 presents a list of most relevant risk assessment

Table 1. Risk assessment methods.

Organization/Region/Source
ECB
http://ecb.jrc.ec.europa.eu/home.php
IPCS
www.inchem.org/pages/about.html
Japan
www.env.go.jp/chemi/communication/senmon.html
www.mhlw.go.jp/bunya/roudoukijun/anzeneisei14/index.html
www.safe.nite.go.jp/english/index.html
www.safe.nite.go.jp/english/ghs/pdf/guidance_e.pdf
http://unit.aist.go.jp/riss/crm/index_e.html
OECD
www.oecd.org/department/0,3355,en_2649_34373_1_1_1_1_1,00.html
REACH
http://guidance.echa.europa.eu/docs/guidance_document/information_requirements_en.htm
US EPA
www.epa.gov/risk
WHO
www.who.int/ipcs/methods/en

methods. Despite the selected method, the following basic steps are part of any risk assessment process:

– Hazard characterization: dose-response determination, determining the relationship between the magnitude of exposure to a hazard and the probability and severity of adverse effects;
– Exposure assessment: identifying the extent to which exposure actually occurs;
– Risk characterization: combining the information from the hazard characterization and the exposure assessment in order to form a conclusion about the nature and magnitude of risk, and, if indicated, implement additional risk management measures. Risk characterization is an iterative process. There might be several circles of assessment necessary before concluding that the substance can be handled safely.

5 HAZARD INFORMATION SOURCES

Hazard information used in an assessment must be reliable and current. For commercially available substances, the principal sources are:

– Product labels;
– Chemical Safety Data Sheets;
– Information from governmental entities;
– Internet sites (see Table 2 to Table 5);
– Specialty handbooks.

Table 2. Hazard information internet sites: International organizations.

Organization web site
Environmental Chemicals Data and Information Network (ECDIN, European Union) http://ecdin.etomep.net/
The European Agency for Safety and Health at Work http://europe.osha.eu.int/
International Program on Chemical Safety (IPCS), a program from WHO, ILO, and UNEP http://www.who.ch/programmes/pcs/index
International Chemical Safety Cards http://www.cdc.gov/niosh/ipcs/icstart.html
International Labor Organization (ILO) http://turva.me.tut.fi/cis/home.html
United Nations Environmental Program (UNEP) http://irptc.unep.ch
World Health Organization (WHO) http://www.who.int/home-page/http://ecb.jrc. ec.europa.eu/home.php

Table 3. Hazard information internet sites: National organizations.

Organization web site
Agency for Toxic Substances and Disease Registry (ATSDR) Country: USA http://www.atsdr.cdc.gov/
Cancer Information Service, National Cancer Institute Country: USA http://cis.nci.nih.gov
Cancer Research Campaign Country: UK http://www.crc.org.uk
Health and Executive Country: UK http://www.open.gov.uk/hse/
National Institute of Environmental Health Sciences (NIEHS) Country: USA http://www.niehs.nih.gov/
Occupational Safety and Health Administration (OSHA), U.S. Department of Labor Country: USA http://www.osha.gov

Table 4. Hazard information internet sites: Nongovernmental organizations.

Organization web site
American Conference of Governmental Industrial Hygienists, Inc. Country: USA http://www.acgih.org
British Occupational Hygiene Society Country: UK http://www.bohs.org
Chemical Abstracts Service (CAS) http://info.cas.org

Table 5. Hazard information internet sites: Universities and institutes.

Entity web site
Canadian Centre for Occupational Health and Safety Country: Canada http://www.ccohs.com/
Cornell University, Material Safety Data Sheets Country: USA http://msds.pdc.cornell.edu/msdssrch.asp
Finnish Institute of Occupational Health Country: Finland http://www.occuphealth.fi/e/
Institute of Occupational Safety and Health (IOSH) Country: UK http://www.iosh.co.uk/home.html
International Agency for Research on Cancer (IARC) Country: France http://www.iarc.fr
IARC Monographs http://193.51.164.11/default.html
Karolinska Institute Library, Occupational Diseases Country: Sweden http://www.kib.ki.se/index_en.html
University of Lund, Occupational and Environmental Medicine Country: Sweden http://www.ymed.lu.se
McGill University, Occupational Health Services Country: Canada http://www.mcgill.ca/occh/
National Institute for Occupational Safety and Health (NIOSH) Country: USA http://www.cdc.gov/niosh/homepage.html
NIOSH Pocket Guide to Chemical Hazards http://www.cdc.gov/niosh/npg/pgdstart.html
Swedish Institute for Working Life Country: Sweden http://www.niwl.se
University of Occupational and Environmental Health Country: Japan Web site: http://www.uoeh-u.ac.jp
University of Vermont, Vermont SIRI Material Safety Data Sheet Collection Country: USA http://hazard.com/msds2/
University of Uppsala, Department of Occupational and Environmental Medicine Country: Sweden http://www.occmed.uu.se/english/index.html

Usually, when dealing with well-known substances, the Safety Data Sheets produced by manufacturers should permit assessment of the hazard. Unfortunately, information on these data sheets is not always reliable.

For recently introduced commercial substances, similar information will be available as a result of the requirement in many countries for notification of a "base set" dossier of toxicological and other data.

However, it should be noted that for most substances not recently introduced, the toxicological data are inadequate and intelligent deduction is the only substitute. Also, when dealing with processes, it will sometimes be completely unclear which substances are formed during the process. In this case, only measurements will give an insight into the hazards.

6 CONSEQUENCES OF RISK ASSESSMENT FAILURE

The output from risk assessment is used in decisions about control of risks—either to avoid or replace the chemical of concern or to impose controls which reduce the exposure to levels where the risk is insignificant or acceptable, given the use of the chemical. For this reason, risk assessment should be based on logical, scientific principles. It should also strive to be accurate and to state the basis of the assumptions made in the assessment.

When risk assessment has failed, occupational diseases may appear. A common division of occupational diseases is based on the cause of the disease:

– Chemicals;
– Physical processes, such as radiation, noise, heat, and pressure;
– Mechanical stress, such as heavy load and dynamic burden;
– Psychological burdens, such as stress.

As there are about 120 000 chemicals with a Chemical Abstracts Service (CAS) Registry Number used in industry, one may expect that occupational diseases are described for a fraction of these chemicals only. A possible way to simplify prediction of chemically induced diseases is to base it on classes of substance (e.g., solvents), which produce characteristic effects such as, for present example, neurological disturbances. A second possibility is to look at homologous series (e.g., formic acid is corrosive to eyes, lungs, and skin). For the related homologous series of short-chain fatty acids [acetic acid (C2), propionic acid (C3), butyric acid (C4), isobutyric acid (C4), isovaleric acid (C5), hexanoic acid (C6)], medium-chain fatty acids [octanoic acid (C8), capric acid (C10), lauric acid (12)], and long-chain fatty acids [myristic acid (C14), palmitic acid (C16)], similar effects are expected, even though reducing intensity as the chain increases. A third possibility is to look at the properties of functional groups such as alcohols, organic acids, and esters.

A quite different approach is to look first at the disease (e.g., occupational asthma, cancer, effects on the reproduction system, and dermatology), since there are a number of chemicals that may cause these pathological conditions.

Once a disease is diagnosed, its cause has to be established from the case history of the patient, including workplace description, and, if possible, biological monitoring or environmental monitoring data. After establishment of a possible occupational cause, an assessment has to be made to find out if the exposure levels were high enough to cause the disease, considering that a minimum exposure time existed.

7 CONCLUSIONS

The number of chemical combinations that workers are exposed to is enormous. Assessing thus every conceivable combination is not therefore realistic, and improved predictive approaches must still be implemented in risk assessment.

Despites the difficulty to establish guidelines on how to evaluate the "chemical cocktail" risk, there are several situations where the need for a mixture risk assessment must be considered, despite the uncertainty of currently available methodologies. A list of existing risk assessment methods, particularly for complex circumstances, is presented.

Occupational Safety and Hygiene – Arezes et al. (eds)
© *2013 Taylor & Francis Group, London, ISBN 978-1-138-00047-6*

Occupational exposure to Aflatoxin B_1 in Portuguese swine farms

S. Viegas, L. Veiga, P. Figueredo, A. Almeida, E. Carolino & C. Viegas
ESTeSL, IPL, Lisbon, Portugal

ABSTRACT: In 1987, the International Agency for Research on Cancer concluded that there was sufficient evidence for carcinogenicity of naturally occurring aflatoxins in humans. Regarding occupational exposure to this chemical agent, farmers and other agricultural workers present a higher risk due to airborne aflatoxin via inhalation of dust. This study was carried out in 7 swine farms located at the district of Lisbon, Portugal. Blood samples were collected from a total of 11 workers. In addition, a control group (n = 25) was included that conducted administrative tasks in an educational institution without any type of agricultural activity. Results obtained suggest that occupational exposure to AFB_1 by inhalation occurs and represents an additional risk in this occupational setting that need to be recognized, assessed and, most important, prevented.

1 INTRODUCTION

Aflatoxins were first isolated about 40 years ago after outbreaks of disease and death in turkeys (Williams et al., 2004) and cancer in rainbow trout (Rucker et al., 2002; Williams et al., 2004) fed with rations formulated from peanut and cottonseed meals. These toxins are secondary metabolites produced under certain conditions of temperature, pH and humidity predominantly by *Aspergillus flavus* and *Aspergillus parasiticus* fungi species (Bhatnagar et al., 2006).

Among 18 different types of aflatoxins identified, major members are aflatoxin B_1, B_2, G_1 and G_2. Aflatoxin B_1 (AFB_1) is normally predominant in cultures as well as in food products. AFB_1 was shown to be genotoxic and a potent hepatocarcinogen (IARC, 1993; Dash et al., 2007). This mycotoxin is metabolized by the mixed function oxidase system to a number of hydroxylated metabolites including the 8,9-epoxide. The latter is considered to be the ultimate carcinogen that reacts with cellular deoxyribonucleic acid (DNA) and proteins to form covalent adducts (Autrup et al., 1991; Richard, 1998; Brera et al., 2002; Dash et al., 2007).

In 1987, the International Agency for Research on Cancer concluded that there was sufficient evidence for carcinogenicity of naturally occurring aflatoxins in humans (IARC, 1987). This conclusion was reaffirmed in two subsequent re-evaluations (IARC 1993, 2002), based upon results from several cohort studies in China and Taiwan that reported associations between biomarkers for aflatoxin exposure and primary liver-cell cancer.

Occupational exposure to this mycotoxin may occur by inhalation of dust generated during the handling and processing of contaminated crops and feeds. Therefore, farmers and other agricultural workers present a higher risk for occupational exposure due to airborne aflatoxin via inhalation of dust (Flannigan and Gillian, 1996; Ghosh et al., 1997; Brera et al., 2002).

To confirm exposure, mycotoxins and/or mycotoxin metabolites may be detected in biological samples using biomarkers (Hooper et al., 2008).

Swine production is known to be an occupational setting that involves high occupational exposure to particulate matter and fungi (Donham et al., 1989; Vogelzang et al., 2000; Duchaine et al., 2000; Kim et al., 2007; Kim et al., 2008). Thus it is conceivable that swine production workers are exposed via inhalation to aflatoxins. The aim of this study was to determine whether swine workers in Portugal were exposed to aflatoxin (AFB_1).

2 MATERIALS AND METHODS

This study was carried out in 7 swine farms located at the district of Lisbon, Portugal, between January and May 2011. The pig buildings investigated in this research were all classified as the manure removal system where manure can be removed from the pig building completely several times a day. In some swine's places, such as in the maternity and where the males were confine, the floor is cover with straw or journal paper.

The ventilation modes of the pig buildings were mechanical ventilation by wall exhaust fans and natural ventilation by operation of a winch-curtain.

Blood samples were collected from a total of 11 workers. In addition, a control group (n = 25) was included that conducted administrative tasks in an educational institution without any type of agricultural activity. All subjects were provided with the protocol and signed a consent form.

For quantification of AFB_1 the RIDAS-CREEN®AflatoxinB130/15ELISA(R®Biopharm) was used. Before ELISA determination five hundred microliters of serum was incubatedfor 18h at 37°C with pronase (Calbiochem, 50U per 5 mg protein). The supernatant was washed in a pre-wet C18 column (RIDAC18 column, RBiopharm) to remove small peptides and amino acids. Subsequently AFB_1 was purified using an immunoaffinity aflatoxin column (easi-extract aflatoxin; (R@Biopharm).

The basis of the test is the antigen-antibody reaction. The assay is calibrated with aflatoxin standards ranging from 1 to 50 ng/ml. Values below 1 ng/ml are not detectable. For testing, samples or standards were pipetted into the wells, which are coated with capture antibodies directed against anti-aflatoxin. Prior to addition of AFB_1-antibody solution, AFB-enzyme conjugate was added. After 30 min incubation, the wells were washed thrice. Color development, was obtained by addition of substrate/chromogen solution to each well and reaction stopped after 15 min with a stop solution. Absorbance was measured at 450 nm and results assessed with Ridasolf Win software version 1.73 (R®Biopharm).

Statistical analysis was performed with SPSS for Windows statistical package, version 19.0.

3 RESULTS

Six workers (54.5%) had detectable levels of AFB_1 (values ranging between >1 ng/ml and 8.94 ng/ml, with a mean value of 1.61 ng/ml). In the control group, the AFB_1 values were all below 1 ng/ml and these findings corroborate the hypothesis that occupational exposure to AFB_1 by inhalation occurs in swine production. However, significant differences were not found between workers and the control group (Mann-Whitney test; $p = 0.723$).

4 DISCUSSION

Among all aflatoxins, the AFB_1 is the most potent hepatocarcinogenic substance known, recently after a thorough risk evaluation, it has been proven to be also genotoxic (Van Egmond and Jonker, 2004; Zain, 2011; Ferrante et al., 2012).

In the present study a biomarker of internal dose was used providing information regarding recent exposure to AFB_1 and its intensity. Therefore, the results obtained based on AFB_1 quantification are related to intensity of environmental contamination and absorption rates (Zhang et al., 2003).

Mycotoxins are not volatile but when found in respirable air are associated with mold spores or particulates (Robbins et al., 2004; Bush et al., 2006) and therefore, in occupational settings the preferential route of exposure to AFB_1 is through inhalation. Therefore, the differences in AFB_1 results found between workers can be related with tasks developed by some of them that are associated with higher exposure to dust, such as manual feeding, manure procedures, cutting piglets' tails, vaccination, among others, that increase expending time in the confinement pig house, and consequently exposure to bioaressols. Moreover, workers were probably exposed to airborne particulates because of the lack of prevention and protection devices, such as masks and ventilating systems.

Although, in human liver is where AFB_1 is essentially metabolized, the human lung have also metabolic activity that is in general low (Wheeler et al., 1990). Taking this in consideration is possible to observe some AFB_1 biotransformation in the lung (Donelly et al., 1996). This aspect is significant when analyzing biomarker results since only AFB_1 concentration and not their metabolites was determined. As a result, the amount quantified by the biomarker used is only the AFB_1 that has not been metabolized in lung and this may result in an underestimation of exposure.

Is also important to consider that there are sufficient experimental and epidemiological data to suggest that the lung is, in addition to the liver, a target for AFB_1 (Dvorackova and Pichova, 1986; Donnelly et al., 1996; Oyelami et al., 1997; Massey et al., 2000).

Additionally, is necessary to take in account that, in this occupational setting, a potential exposure to more than one mycotoxin, since in addition to *A. flavus*, other fungal species recognize as mycotoxin producers were found (published elsewhere). Therefore, the effects of possible interactions need to be considered in the risk assessment process (Sexton and Hattis, 2007; Ferrante et al., 2012).

For instance, the interaction between AFB_1 and ochratoxin A—a mycotoxin produce by *A. ochraceus*—was demonstrated in an animal model in a study developed in 2001 by Sedmíková and colleagues (Sedmíková et al., 2001). Those results showed that the ability of ochratoxin A to increase the mutagenic effect of AFB_1 may be due to the relation of these two toxins to proteosynthesis (Creppy et al., 1995; Sedmíková et al., 2001).

5 CONCLUSIONS

Results obtained suggest that occupational exposure to AFB_1 by inhalation occurs and represents an additional risk in this occupational setting that need to be recognized, assessed and, most important, prevented.

ACKNOWLEDGMENTS

This study was funded by the Portuguese Authority for Working Conditions and would not have been possible without the assistance of the Portuguese Ministry of Agriculture, Portuguese Ministry of Health and swine farmers.

REFERENCES

Autrup, J., Schmidt, J., Seremet, T. & Autrup, H. 1991. Determination of exposure to aflatoxins among Danish workers in animal-feed production through the analysis of aflatoxin 81 adducts to serum albumin. *Scand J Work Environ Health*, 17, 436–440.

Bhatnagar, D., Cary, J.W., Ehrlich, K., Yu, J. & Cleveland, T.E. 2006. Understanding the genetics of regulation of aflatoxin production and Aspergillus flavus development. *Mycopathologia*, 162, 155–66.

Brera, C., Caputi, R., Miraglia, M., Iavicoli, I., Salerno, A. & Carelli, G. 2002. Exposure assessment to mycotoxins in workplaces: aflatoxins and ochratoxin A occurrence in airborne dusts and human sera. *Microchem J., 73,* 167–173.

Bush, R.K., Portnoy, J., M., Saxon, A., Terr, A. & Wood, R. 2006. The medical effects of mold exposure. *J Allergy Clin Immunol., 17,* 326–333.

Creppy, E, Baudrimont, I, Betbeder A, 1995. Prevention of nephrotoxicity of Ochratoxin A, a food contaminant. *Toxicology Letters.* 82 3869–3877.

Dash, B, Afriyie-Gyawu, E., Huebner, H.J., Porter, W., Wang, J.S., Jolly, P.E. & Phillips, T.D. 2007. Determinants of the variability of aflatoxin-albumin adduct levels in Ghanians. *J. Toxicol. Environ. Health A, 70,* 58–66.

Donham, K., Haglind, P., Peterson, Y., Rylander, R. & Belin, L. 1989. Environmental and health studies of farm workers in Swedish swine confinement buildings. *British Journal of Industrial Medicine*, 46, 31–37.

Donnelly, P.J., Stewart, R.K., Ali, S.L., Conlan, A.A., Reid, K.R., Petsikas, D. & Massey, T.E. 1996. Biotransformation of aflatoxin B1 in human lung. *Carcinogenesis.* 17, 2487–2494.

Duchaine, C., Grimard, Y. & Cormier, Y. 2000. Influence of Building Maintenance, Environmental Factors, and Seasons on Airborne Contaminants of Swine Confinement Buildings, *AIHAJ – American Industrial Hygiene Association*, 61, 1, 56–63.

Dvorackova, I. & Pichova, V. 1986. Pulmonary interstitialfibrosis with eviodence of aflatoxin B₁ in lung tissue. *J. Toxicol. Envioron. Health* 18: 153–157.

Ferrante, M., Sciacca, S. & Conti, G.A. 2012. Carcinogen Role of Food by Mycotoxins and Knowledge Gap, Carcinogen, Margarita Pesheva, Martin Dimitrov and Teodora Stefkova Stoycheva (Ed.), ISBN: 978–953–51–0658–6, InTech, Available from: http://www.intechopen.com/books/carcinogen/carcinogen-role-of-food-by-mycotoxins-and-knowledge-gap

Flannigan, B. & Gillian, E.W. 1996. Moulds, mycotoxins, and indoor air, Proceedings of the 10th International Biodeterioration and Biodegradation Symposium, Hamburg, 15–18 September.

Ghosh, S.K., Desai, M.R., Pandya, G.L. & Venkaiah, K. 1997. Airborne aflatoxin in the grain processing industries in India. *American Industrial Hygiene Association Journal.* 58: 583–586.

Hooper, D., Bolton, V. & Gray, M.R. 2008. Fungal mycotoxins can be detected in tissue and body fluids of patients with a history of exposure to toxin producing molds, 2006. Availablebfrom:http://www.realtimelab.com/documents/MycotoxinPosterMay1620 7.pdf.

IARC. 1987. Overall evaluations of carcinogenicity: an updating of IARC Monographs volumes 1 to 42. IARC Monogr Eval Carcinogen Risks Human Suppl. 7: 1–440. PMID:3482203.

IARC. 1993. Some naturally occurring substances: food items and constituents, heterocyclic aromatic amines and mycotoxins. IARC Monogr Eval Carcinogen Risks Human 56: 1–599.

IARC. 2002. Some traditional herbal medicines, some mycotoxins, naphthalene and styrene. IARC Monogr. Eval Carcinogen Risks Human 82: 1–556. PMID:12687954.

Kim, K.Y., Ko, H.J, Kim, H.T, Kim, Y.S., Roh, Y.M, Lee, C.M. & Kim, C.N. 2007. Influence of Extreme Seasons on Airborne Pollutant Levels in a Pig-Confinement Building. *Archives of Environmental & Occupational Health*, 62, 1, 27–32.

Kim, K.M., Ko, H.J., Kim, Y.S. & Kim, C.N. 2008. Assessment of Korean farmer´s exposure level to dust in pig buildings. *Ann Agric Environ Med., 15,* 51–58.

Massey, T.E., Smith, G.B.J., & Tam, A.S. 2000. Mechanisms of aflatoxin B₁ lung tumorigenesis. *Exp Lung Res.* 26, 673 683.

Oyelami, O.A., Maxwell, S.M., Adelusola, K.A., Aladekoma, T.A. & Oyelese, A.O. 1997. aflatoxins in the lungs of children with Kwashiorkor and children with miscellaneous diseases in Nigeria. *J. Toxicol. Environ. Health A,* 51: 623–628.

Richard, J.L. 1998. Mycotoxins, toxicity and metabolism in animals—a system approach overview, in: M. Miraglia, H.P. van Egmond, C. Brera, J. Gilbert (Eds.), Mycotoxins and Phycotoxins—Developments in Chemistry, Toxicology and Food Safety, Alaken, pp. 363–397.

Robbins, C.A., Swenson, L.J. & Hardin, B.D. 2004. Risk from inhaled mycotoxins in indoor office and residential environments. *Int J Toxicol.* 23, 3–10

Rucker, R.R., Yasutake, W.T. & Wolf, H. 2002. Trout hepatoma—a preliminary report. *Prog Fish Cult,.* 23, 3–7.

Sedmíková, M. Reisnerová, H. Dufková, Z. Bartá, I. Jílek, F. 2001 Potential hazard of simultaneous occurrence of aflatoxin B1 and ochratoxin A. *Vet. Med.* 46, 169–174.

Sexton, K. & Hattis, D. 2007. Assessing cumulative health risks from exposure to environmental mixtures—Three fundamental questions. *Environ. Health Pers,.* 115, 825–832.

Van Egmond, H.P. & Jonker, M.A. 2004. *Worlwide regulations for mycotoxins in food and feed in 2003.* The Food and Agriculture Organization of the United Nations (FAO).

Vogelzang, P.F.J., van der Gulden, J.W.J., Folgering, H., Heederik, D., Tielen, M.J.M. & van Schayck, C.P. 2000. Longitudinal Changes in Bronchial Responsiveness

Associated With Swine Confinement Dust Exposure. *Chest*, 117, 1488–1495.

Wheeler, C.W., Park, S., S. & Guenthner, T.M. 1990. Immunochemical analysis of a cytochromc P-45OLA1 homologue in human a lung microsomes. *Mol. Pharmacol.* 38, 634–643.

Williams, J.H., Phillips, T.D. Jolly, P.E., Stiles, J.K., Jolly, C., M. & Aggarwal, D. 2004. Human aflatoxicosis in developing countries: a review of toxicology, exposure, potential health consequences, and interventions. *Am J Clin Nutr,*. 80, 1106–1122.

Zain, M.E., 2011. Impact of mycotoxins on humans and animals. *Journal of Saudi Chemical Society*, 15, 129–144.

Zhang, J., Ichiba, M., Hanaoka, T., Pan, G., Yamano, Y. & Hara, K. 2003. Leukocyte 8-hydroxydeoxy-guanosine and aromatic DNA adduct in coke-oven workers with polycyclic aromatic hydrocarbon exposure. *Int. Arch. Occup. Environ. Health* 76, 499–5404.

Occupational Safety and Hygiene – Arezes et al. (eds)
© 2013 Taylor & Francis Group, London, ISBN 978-1-138-00047-6

Exposure to particles and fungi in Portuguese swine production

C. Viegas, S. Viegas & E. Carolino
ESTeSL, IPL, Lisbon, Portugal

R. Sabino & C. Veríssimo
Mycology Laboratory, INSA, Lisbon, Portugal

ABSTRACT: A number of studies have shown that exposure to airborne dust and microorganisms can cause respiratory diseases in humans. Agricultural workers, such as pig farmers, have been found to be at high risk of exposure to airborne particles. The aim of this study was to detect contamination caused by particles and fungi in 7 swine farms located in Lisbon district, Portugal. Environment evaluations were performed during the winter season of 2011 with a portable direct-reading equipment (Lighthouse, model 3016 IAQ) and it was possible to obtain data concerning contamination caused by particles with 5 different sizes (PM0.5; PM1; PM2.5; PM5; PM10). To assess air contamination caused by fungi, air samples of 50 liters were collected using a Millipore Air Tester (Millipore) by impaction method at a velocity of 140 L/minute and at one meter height, using malt extract agar supplemented with chloramphenicol (0.5%). Air sampling was also performed outside premises, since this is the place regarded as reference. All the collected samples were incubated at 27 °C for 5 to 7 days. Results from particles' contamination showed that higher values were associated with PM5 and PM10 sizes and that smaller particles exhibit lower contamination values. Concerning the fungal load of the analyzed swine, the highest obtained value was 4100 CFU/m^3 and the lowest was 120 CFU/m^3. Forty six different fungal species were detected in the air, being *Aspergillus versicolor* the most frequent species found (20.9%), followed by *Scopulariopsis brevicaulis* (17.0%) and *Penicillium* sp. (14.1%). Data gathered from this study corroborate the need of monitoring the contamination by particulate matter, fungi and their metabolites in Portuguese swine.

1 INTRODUCTION

Several studies have shown that human exposure to airborne dust and microorganisms, such as bacteria and fungi, can cause respiratory diseases. Agricultural workers have been found to be at high risk of exposure to airborne particles (Radon et al., 2003; Predicala and Maghirang, 2003; Baur et al., 2003; Rautiala et al., 2003; Dosman et al., 2005).

From a human health perspective, dust exposure in pig farming is the most important because of the large number of workers needed in pig production and the increasing number of working hours inside enclosed buildings (Iversen et al., 2000). In pig buildings, particulate matters like dust play a role in not only deteriorating indoor air quality but also in causing adverse health effects on workers (Donham et al., 1990; Pearson and Sharples, 1995; Mackiewicz, 1998; Kim et al., 2008). Generally, dust is recognized to adsorb and transport odorous compounds (Carpenter, 1986) and biological agents (Robertson et al., 1984; Kim et al., 2005) such as fungi (HSE, 2008). All these bioaerosoles

in high concentrations together with fungal metabolites pose agricultural workers, and especially pig farmers, at increased risk of occupational respiratory diseases.

Animal confinement, such as pig farming, tends to increase the overall microbial load in the production environment caused by high amounts of feed and organic residuals (manure and wastewater) present in those environments. The number of animals and the handling and management required to work in these settings also contribute to enhance that microbial load (Clark et al., 1983; Cole et al., 2000; Douwes et al., 2003; Zejda et al., 1994). Exposure to bioaerosols in swines may vary depending upon the stage of the animals' growth, density, manure management procedures, used floor coverage, among others (Mc Donnell et al., 2008). Gathering temporal information about occupational exposure to particles and fungi is necessary to better understand eventual adverse health symptoms of workers.

The aim of this study was to determine contamination due to particles and fungi in 7 swine farms located in Lisbon district, Portugal.

2 MATERIALS AND METHODS

Environment evaluations regarding particle matter contamination were performed during the winter season of 2011 with a portable direct-reading equipment (Lighthouse, model 3016 IAQ) and it was possible to obtain data concerning contamination caused by particles discriminated in 5 different sizes (PM0.5; PM1; PM2.5; PM5; PM10). This differentiation between particle size fractions is important because it allows the estimation of the penetration and deposition of dust within the respiratory system and, consequentely, the possible health effect related. Vincent and Mark (1981) demonstrated that the respirable dust is the fraction of airborne dust that reaches regions of the lung where the gas exchanges occur. It is composed with particles with less than 7 μm aerodynamic diameter (size from PM0.5 to PM5).

Measurements were conducted in the vicinity of nasal area of the workers and during the performance of different tasks. In the swine farms, 3 to 11 measurements were undertaken and the mean value obtained for each particle size was the one considered. All measurements were done continuously and during 5 min. In all the studied swine farms, workers did not use respiratory protection devices.

To assess air contamination caused by fungi, air samples of 50 liters were collected using a Millipore Air Tester (Millipore) by impaction method at a velocity of 140 L/minute and at one meter height, using malt extract agar supplemented with chloramphenicol (0.5%). Air sampling was also performed outside premises, since this is the place regarded as reference. All the collected samples were incubated at 27 °C for 5 to 7 days.

After laboratory processing and incubation of the collected samples, quantitative (colony forming units/m³—cfu/m³) and qualitative results were obtained, with identification of the isolated fungal species (Hoog et al., 2000).

To ascertain the existence of statistically significant differences between contamination results of different types of particles, it was used the Friedman test. Statistical analysis was performed with SPSS for Windows statistical statistical package, version 19.0.

3 RESULTS

3.1 Particles

Friedman's test showed the existence of statistically significant differences between the five sizes of particles ($\chi_4^2 = 228$, $p = 0,000$). In an exploratory data analysis, it appears that smaller particles exhibit lower contamination values.

The distribution of particles size showed the same tendency in all swine farms (higher concentration values in PM5 and PM 10 sizes). Farms B and D, however, presented higher levels of contamination, particularly in PM5 and PM10 (Table 1). These two farms were the ones having only natural ventilation as ventilation resource. The others swine farms have a combination between natural and mechanical (exhaust) ventilation.

3.2 Fungi

Concerning the fungal load of the analysed swines, the highest obtained value was 4100 CFU/m³ and the lowest was 120 CFU/m³. Forty six different fungal species were detected in air, being *Aspergillus versicolor* the most frequent species found (20.9%), followed by *Scopulariopsis brevicaulis* (17.0%) and *Penicillium* spp. (14.1%) (Table 2).

All the swine presented fungal species indoor different from the ones isolated outdoor, and more than two places with fungal load higher than outdoor.

4 DISCUSSION

The majority of the previous studies estimated particles' exposure by measuring the total mass concentration; very few studies on agricultural farms investigated the exposure regarding to particles' size. The size of the particles, however, affects their deposition in the respiratory system, resulting in different

Table 1. Particles measurements obtained in each swine farm (mean value/mg.m⁻³).

Farms	N° *	PM 0.5	1.0	2.5	5.0	10.0
A	11	9.1×10^{-4}	1.4×10^{-3}	5.1×10^{-3}	4.9×10^{-2}	2.4
B	7	1.9×10^{-3}	5.4×10^{-3}	1.6×10^{-2}	1.1	4.9
C	6	2.5×10^{-4}	7.4×10^{-4}	4.6×10^{-3}	4.5×10^{-2}	2.0
D	5	2.8×10^{-4}	9.3×10^{-4}	8.8×10^{-3}	1.1	5.8
E	3	2.3×10^{-3}	4.3×10^{-3}	1.2×10^{-2}	6.0×10^{-2}	1.9
F	7	1.4×10^{-4}	8.6×10^{-4}	7.6×10^{-3}	7.9×10^{-2}	3.5
G	11	3.9×10^{-4}	7.6×10^{-4}	3.8×10^{-3}	4.5×10^{-2}	2.3

* Measurements done in each farm.

Table 2. Fungal species most frequently found in the air from the analyzed swine farms.

Fungal species	Frequency (CFU/m³) (N; %)
Aspergillus versicolor	3210; 20.9
Socpulariopsis brevicaulis	2620; 17.0
Penicillium spp.	2160; 14.1
Others	7380; 48.0

types of health effects (Lee et al., 2006). Our study gives information concerning 5 different sizes and this information permits the achievement of more detailed information concerning contamination with particles and their possible health effects.

Our data showed higher values in PM5 size and, predominantly in PM10, indicating that swine dust can penetrate into the gas exchange region of the lung (PM5) and may also produce disease by impacting in the upper and larger airways below the vocal cords (PM10) (Vincent and Mark, 1981).

Wathes and colleagues (1998) found that the inhalable dust emissions from pig buildings were 40% higher in the summer than in the winter, while respirable dust emissions were not affected greatly by the season. Considering this aspect, we can point out that there is a possibility that PM10 values can be even higher in the summer time.

In a European project developed in England, the Netherlands, Germany, and Denmark, stationary measurements in 256 animal buildings were performed and the mean value for inhalable dust in pig buildings was 2.19 mg.m^{-3} (Seedorf, 1998; Takai and Pederson, 2000; Iversen et al., 2000). In three of our seven studied farms (B, D and F) the obtained mean values were higher.

In a study developed by Donnell and colleagues (2008) in five Irish swine farms the same tendency was found on respect to the distribution of the particles' size, namely a median value of 2.99 mg.m^{-3} for inhalable and 0.19 mg.m^{-3} for respirable dust (Donnell et al., 2008).

The amount of dust in the air of livestock buildings is correlated to environmental factors such as ventilation, feeding practices, bedding materials, dung and slurry handling, and animal activity (Takai and Pedersen, 2000). A well designed and managed ventilation system will control the levels of gases, dusts and vapours, and it is an important factor in controlling odours from swine confinement buildings (Chastain, 2000). The absence of a ventilation system in B and D farms can contribute to explain the higher results obtained, particularly in PM5 and PM10 sizes.

Regarding the fungal load, different fungal counts were obtained when comparing with a study performed by Duchaine et al., (547 CFU/m^3–2862 CFU/m^3 versus 120 CFU/m^3–4100 CFU/m^3). This difference maybe due to different procedures of building maintenance (Duchaine et al., 2010). In a study published by Jo and Kang (Jo and Kang, 2005) Aspergillus spp. and Penicillium spp. were also the most frequent fungi found in swine.

All the swine had one or more spaces with fungal load higher than outdoor and fungal species that differed from the ones isolated outdoor, suggesting fungal contamination from within (Kemp et al., 2003). Aspergillus versicolor, the most

frequent species isolated, is known as being the major producer of the hepatotoxic and carcinogenic mycotoxin sterigmatocystin. The toxicity of this mycotoxin is manifested primarily in liver and kidney (Engelhart et al., 2002). Due to their easier detection, fungi are often used as an indirect indicator of mycotoxins presence both in agricultural and occupational settings. Because of that, we must consider the eventual exposure not only to fungal particles, but also to mycotoxins (Thrane et al., 2004). The mycotoxin sterigmatocystin is closely related to the mycotoxin aflatoxin, as a precursor of aflatoxin biosynthesis (Barnes et al., 1994) and it is classified by the International Agency for Research on Cancer as a class 2B carcinogen (i.e., as possibly carcinogenic to humans) (International Agency for Research on Cancer, 1987).

5 CONCLUSIONS

Data gathered from this study corroborate the need of monitoring the contamination by particulate matter, fungi and their metabolites in Portuguese swine. Results demonstrate high levels of particulate matter in the swine farms studied, particularly regarding PM5 and PM10 sizes. This study also raises the concern of occupational treat due not only to the detected fungal load, but also to the toxigenic potential of Aspergillus versicolor. Therefore, in this setting, inhalation should be considered as a route of exposure to the mycotoxin sterigmatocystin.

The evidence of respiratory disease in this occupational setting documented in many studies supports the need for the development of health protection programmes within the workplace.

ACKNOWLEDGMENTS

This study was funded by the Portuguese Authority for Working Conditions and would not have been possible without the assistance of the Portuguese Ministry of Agriculture, Portuguese Ministry of Health and swine farmers.

REFERENCES

Barnes, S., Dola, T., Bennett, J. & Bhatnagar, D. 1994. Synthesis of sterigmatocystin on a chemically defined medium by species of Aspergillus and Chaetomium. Mycopathologia, 125, 173–178.
Baur, X., Preisser, A. & Wegner R. 2003. Asthma due to grain dust. Pneumologie, 57, 335–339.
Carpenter, G.A. 1986. Dust in livestock buildings – review of some aspects. J Agric Eng Res, 33, 227–241.
Chang, C., Chung, H., Huang, C. & Su, H. 2001. Exposure of Workers to Airborne Microorganisms

379

in Open-Air Swine Houses. *Appl Environ Microbiol*, 67(1), 155–161.

Chastain, J. 2000. Air Quality and Odor Control From Swine Production Facilities [Online] http://www.clemson.edu/extension/livestock/livestock/camm/camm_files/swine/sch9_03.pdf [24th September 2012].

Clark, S., Rylander, R. & Larsson, L. 1983. Airborne bacteria, endotoxin and fungi in dust in poultry and swine confinement buildings. *Am. Ind. Hyg. Assoc. J.*, 44, 537–541.

Cole, D., Todd, L. & Wing, S. 2000. Concentrated swine feeding operations and public health: a review of occupational and community health effects. *Environ. Health Perspec.*, 108, 685–699.

Donham, K.J., Merchant, J.A., Lassie, D., Popendorf, W.J. & Burmeister, LF. 1990. Preventing respiratory disease in swine confinement workers: intervention through applied epidemiology, education and consultation. *Am J Ind Med*, 18, 241–261.

Donnell, P.E., Coggins, M.A., Hogan, V.J. & Fleming, G.T. 2008. Exposure assessment of airborne contaminants in the indoor environment of Irish swine farms. *Ann Agric Environ Med*, 15, 323–326.

Dosman, J.A., Lawson, J.A., Kirychuk, S.P., Cormier, Y., Biem, J. & Koehnce, N. 2005. Occupational asthma in newly employed workers in intensive swine confinement facilities. *Eur. Respir. J.*, 24, 698–702.

Douwes, J., Thorne, P., Pearce, N. & Heederik, D. 2003. Bioaerosol health effects and exposure assessment: progress and prospects. *Ann. Occup. Hyg.*, 47, 187–200.

Duchaine, C., Grimard, Y. & Cormier, Y. 2000. Influence of Building Maintenance, Environmental Factors, and Seasons on Airborne Contaminants of Swine Confinement Buildings. *AIHAJ – American Industrial Hygiene Association*, 61, 1, 56–63.

Engelhart, S., Loock, A., Skutlarek, D., Sagunski, H., Lommel, A., Harald, F. & Exner, M. 2002. Occurrence of Toxigenic *Aspergillus versicolor* Isolates and Sterigmatocystin in Carpet Dust from Damp Indoor Environments. *Applied and Environmental Microbiology*, 68 (8), 3886–3890.

Hoog, C., Guarro, J., Gené, G. & Figueiras, M. 2000 *Atlas of Clinical Fungi*. (2th ed). Centraalbureau voor Schimmelcultures.

HSE. "Statement of evidence: Respiratory hazards of poultry dust Health and Safety" Executive 03/09 14 pages.

Iversen, M., Kirychuk S., Drost, H. & Jacobson, L. 2000. Human Health Effects of Dust Exposure in Animal Confinement Buildings. *Journal of Agricultural Safety and Health*, 6(4), 283–288.

International Agency for Research on Cancer. 1987. *Some naturally occurring substances*. Monographs, vol. 10, Suppl. 7, p. 72. International Agency for Research on Cancer, Lyon, France.

Jo, W. & Kang, J. 2005. Exposure Levels of Airborne Bacteria and Fungi in Korean Swine and Poultry Sheds. *Archives of Environmental & Occupational Health*. 60, 3, 140–146.

Kemp, P., Neumeister-Kemp, H., Esposito, B., Lysek, G. & Murray, F., Changes in airborne fungi from the outdoors to indoor air; Large HVAC systems in non-problem buildings in two different climates. *American Industrial Hygiene Association*, 64, pp. 269–275, 2003.

Kim, K., Ko, H., Kim, Y. & Kim, C. 2008. Assessment of Korean farmer´s exposure level to dust in pig buildings. *Ann Agric Environ Med*, 15, 51–58.

Kim, K.Y., Ko, H.J., Lee, K.J., Park, J.B. & Kim, C.N. 2005. Temporal and spatial distributions of aerial contaminants in an enclosed pig building in winter. *Environ Res*, 99, 150–157.

Lee, S., Adhikari, A., Grinshpun, S.A., McKay, R., Shukla, R. & Reponen, T. 2006. Personal Exposure to Airborne Dust and Microorganisms in Agricultural Environments. *Journal of Occupational and Environmental Hygiene*, 3, 118–130.

Mc Donnell1, P., Coggins, M., Hogan, V. & Fleming, G. 2008. Exposure assessment of airborne contaminants in the indoor environment of irish swine farms. *Ann. Agric. Environ. Med.*, 15, 323–326.

Mackiewicz, B. 1998. Study on exposure of pig farm workers to bioaerosols, immunologic reactivity and health effects. *Ann Agric Environ Med*, 5, 169–175.

Pearson, C.C. & Sharples, T.J. 1995. Airborne dust concentrations in livestock buildings and the effect of feed. *J Agric Eng Res*, 60, 145–154.

Pedersen, S., Nonnenmann, M., Rautiainen, R., Demmers, T.G.M., Banhazi, T. & Lyngbye, M. 2000. Dust in Pig Buildings. *Journal of Agricultural Safety and Health*. 6(4), 261–274.

Predicala, B.Z. & Maghirang R.G. 2003. Field comparison of inhalable and total dust samplers for assessing airborne dust in swine confinement barns. *Appl. Occup. Environ. Hyg.*18, 694–701.

Radon, K., Garz, S., Riess, A. et al. 2003. Respiratory diseases in European farmers—II. Part of the European Farmers' Project. *Pneumologie, 57*, 510–517.

Rautiala, S.J., Kangas, K., Louhelainen & Reiman, M. 2003. Farmers' exposure to airborne microorganisms in composting swine confinement buildings. *Am. Ind. Hyg. Assoc. J.* 64, 673–677.

Rimac, D., Macan, J., Varnai, V., Vuĉemilo, M., Matkovic', K., Prester, L., Orct, T., Trošic',I. & Pavičic, I. 2010. Exposure to poultry dust and health effects in poultry workers: impact of mould and mite allergens. *Int Arch Occup Environ Health*, 83, 9–19.

Robertson, J.H. & Frieben, W.R. 1984. Microbial validation of ven filters. *Biotechnol Bioeng*, 26, 828–835.

Seedorf, J., Hartung,M.J., Schröder, K., Linkert, H., Pedersen, P., Takai, H., Johnsen, J.O., Metz, J.H.M., Grook Koerkamp, P.W.G, Uenk, G.H., Philips, V.R., Holden, M.R., Sneath, J.L., Short, R.P., White, C.M. & Wathes, C.M. 1998. Concentration and emissions of airborne endotoxins and microorganisms in livestock buildings in Northern Europe. *J Agric Eng Res* 70, 97–109.

Senthilselvan, A., Dosman, J., Kirychuk, S., Barber, E., Rhodes, C., Zhang, Y. *et al.* 1997. Accelerated lung function decline in swine confinement workers. *Chest*, 111, 1733–1741.

Takai, H. & Pederson, S. 2000. A Comparison Study of Different Dust Control Methods in Pig Buildings. *App. Eng. in Agri.*, 16(3), 269–277.

Thrane, U., Adler, A., Clasen, P., E., Galvano, F., Langseth, W., Lew, H., Logrieco, A., Nielsen, K.F. & Ritieni, A. 2004. Diversity in metabolite production by Fusarium langsethiae, Fusarium poae, and Fusarium sporotrichioides. *Int. J. Food Microbiol*, 95,257–266.

Vincent J & Mark D: The basis of dust sampling in occupational hygiene: a critical review. *Annals of Occupational Hygiene*, 24: pp. 375–390, 1981.

Zejda, J.E., Barber, E.M., Dosman, J.A., Olenchock, S.A., McDuffie, H.H., Rhodes, C.S. & Hurst, T.S. 1994. Respiratory health status in swine producers relates to endotoxin exposure in the presence of low dust levels. J. *Occup. Med.*, 36, 49–56.

Wathes, C., Phillips, V., Holden, M., Sneath, R., Short, J., White, R., Hartung, J., Seedorf, J., Schro, M., Linkert, K., Pederson, S., Takai, H., Johnsem, O., G root, K., Uenk, G., Metz. J., Hinz, T., Caspary, V. & Linke, S. 1998. Emissions of Aerial Pollutants in Livestock Buildings in Northern Europe: Overview of a Multinational Project. *J. Agri. Eng. Research,* 70, 3–9.

Occupational Safety and Hygiene – Arezes et al. (eds)
© 2013 Taylor & Francis Group, London, ISBN 978-1-138-00047-6

The wrist vibrations measured with anti-vibration gloves in a simulated work task

José Miguel Cabeças, Rui Messias, & Bernardo Roque
Faculdade de Ciências e Tecnologia da Universidade Nova de Lisboa, Caparica, Portugal

Susana Batista
Cesnova, Faculdade de Ciências Sociais e Humanas da Universidade Nova de Lisboa, Lisboa, Portugal

ABSTRACT: The main objective of the research was to analyze the effectiveness of anti-vibration gloves in terms of the vibration level that reaches the wrist, during a simulated work task with a reciprocating saw. Operations were repeated with bare hand and with four different types of anti-vibration gloves. ISO 5349–1:2001 (r.m.s.) accelerations were measured with an accelerometer fixed in the wrist of the operators (n = 40) and fixed in the handle of the tool. The main finding of the research was that anti-vibration gloves did not attenuate the total value of frequency un-weighted r.m.s. vibration components (one third octave band), measured in the wrist of the operators, in the dominant range in the wrist ~31.5–200 Hz, neither in the whole one third octave band range. Also, non-significant differences in ISO-weighted total r.m.s values of a_{hv} measured in the wrist, were found between bare hand and gloved operations, during blade cutting operations.

1 INTRODUCTION

A reciprocating saw (sabre saw) is a type of saw in which the cutting action is achieved through a push and pulls reciprocating motion of the blade. Considering professional electrical sabre saw tools, typical working values vary between ~1,000 to 1,300 watts, ~3.7 to 4.2 kg, vibration total values (a_h, triax vector sum) determined according to EN 60745 ~14.6 to 50.3 m.s^{-2} (mean value of 30.6 m.s^{-2}; n = 23 tools, right hand, according to C.D.C.).

1.1 The transmissibility of vibrations from the handle of the tool to the arm of the operator

Adewusi et al. (2011) considered that all of the reported data in their research showed rapid decrease in the vibration transmissibility of the hand–arm segments with increasing frequency and distance from the source of vibration. The vibration at frequencies below 100 Hz was transmitted to the forearm, and below 40 Hz was transmitted to the upper arm; vibration above 200 Hz was confined to the hand. Considering the bent-arm posture, an increase in the transmissibility (un-weighted acceleration values) was observed in the wrist to frequencies below ~60–70 Hz and a decrease in the transmissibility between ~60–70 Hz and ~200 Hz.

The vibration transmission to the finger-hand-arm system has been studied by different investigators

(Dong et al. 2004). These studies found that vibration at frequencies below 40 Hz could be transmitted to the arms, shoulders and head, vibration at frequencies above 100 Hz are limited to the hand and less than 10% of vibration at frequencies above 250 Hz may be transmitted to the wrist and beyond. A study developed by Xu et al. (2009), aimed to investigate the characteristics of the wrist and elbow vibrations transmitted from the handles of impact wrenches and their association with the ISO-weighted acceleration measured at the tools. The predominant components of the vibration measured at the wrist were between 20 and 200 Hz. To frequencies between ~10 Hz and ~40–50 Hz it was observed wrist transmissibility > 1. The authors stated that vibration exposure duration can be detected accurately and reliably using the on-the-wrist method with its advantages of posing the least interference with working tasks and avoiding the dc-shift problem.

1.2 The effect of anti-vibration gloves in the transmissibility of vibrations

According to Rakheja et al. (2002), the vibration isolation performance of a glove may depend upon the nature of tool vibration (magnitude and frequency range), visco-elastic properties of the glove material, arm posture, and magnitudes of hand-grip and feed forces. However, the term glove isolation effectiveness is used to indicate the extent to which a glove attenuates the effective vibration on a handle. It depends on the spectrum of vibration on the handle, the transmissibility of the glove (expressed as a function of frequency) and the frequency weighting (Griffin 1998). It is clear that the vibration isolation effectiveness of the glove depends on not only the dynamic properties of a glove but also the biodynamic properties of the hand-arm system (Welcome et al. 2012). A study by Dong et al. (2005) reveals that there is a strong linear correlation between the isolation effectiveness of a typical anti-vibration glove and the biodynamic characteristics of the human hand–arm system in a broad frequency range (40–200 Hz).

Field measurements were carried out by Pinto et al. (2001) in different hand tools, during simulated work procedures in five samples of anti-vibration gloves. The authors concluded that laboratory tests on glove performance are valid only in work situations where the feed force and/or the shape of the spectrum is similar to that used in the laboratory tests. Griffin (1998) also stated that the test method ISO 10819:1996, cannot predict the vibration isolation performance of anti-vibration gloves when used with a specific tool, since the vibration spectra of various hand-held power tools differ considerably from those of the idealized M- and H-spectra. Welcome et al. (2012) also stated that the transmissibility measured with the standardized method may not represent the vibration isolation effectiveness of AV gloves used in some workplace environments, and laboratory-measured transmissibility should not be directly used to account for vibration reduction in risk assessments. Rakheja et al. (2002) concluded that both the grip and feed forces affect the frequency response characteristics of the gloves, specifically at frequencies above 250 Hz. However, the contributions due to variations in grip and feed forces may vanish when the frequency weighting (ISO-5349: 1986) is applied.

Alternative methods to evaluate the vibration isolation effectiveness of anti-vibration gloves are reported in the literature. Rakheja et al. (2002), reports a methodology to estimate vibration isolation effectiveness of anti-vibration gloves. Griffin (1998) proposes that the glove isolation effectiveness may be calculated from the measured transmissibility of a glove, the vibration spectrum on the handle of a specific tool and the frequency weighting indicating the degree to which different frequencies of vibration cause injury. Hewitt (2010) implemented a triaxial investigation of the performance of a glove, by measuring the transmissibility of the glove in three axes, consecutively, using a shaker and an instrumented handle. Reynolds and Wolfe (2005) found that it is possible to replace the M and H spectra currently defined as inputs in ISO 10819 with a single F spectra. Welcome et al.

(2012) proposed a revised version of the standard to improve the reliability of the anti-vibration (AV) glove test defined in the current standard.

In the present study, a reciprocating saw is operated in a simulated work task, with bare hand and four different anti-vibration gloves. The main purpose of the study is to evaluate the effectiveness of anti-vibration gloves in the attenuation of total r.m.s. values of a_{hv} measured in the wrist of the operators.

2 MATERIALS AND METHODS

2.1 The equipment

The hand tools used in the research was a Bosch reciprocating saw GSA 1200 E professional sabre saw 220V, equipped with metal saw blade S1122BF (225 mm/9") , with a no-load stroke rate between 0–2,800 spm, 3.7 kg weight without cable, vibration total values (triax vector sum) determined according to EN 60745: $a_h = 19$ m.s^{-2}, uncertainty K = 5.5 m.s^{-2} (cutting wood values). The tool was operated at maximum rate during the operations (2,800 spm). The sabre saw performed transversal cuts at a 20 mm diameter Mild Steel Round Bars.

Four types of anti-vibration gloves (meets ISO 10819: 1996) were used by the subjects in the research: (a)Type 1 glove, molded Gfom padding on the palm, fingers and thumb, 100% grain cowhide leather, sewn with Kevlar; (b) Type 2 glove, air glove, utilizing patented air technology in the palm, fingers and thumb, driver style cuff; (c) Type 3 glove, with polymer in the palm, the back of the glove covered by a tough, micro-injected dorsal shell, a woven elastic cuff features a rugged; (d) Type 4 glove, neoprene/leather/elastic cuff, incorporates patented polymer, pigskin leather palm and fingers, closure with woven elastic cuff.

2.2 The subjects

Forty volunteer adult subjects participated in the study (Table 1). The maximum handgrip strength tests were performed with the participants seated, the forearm supported on a table with a 90° flexed

Table 1. Mean (SD) and range of age, anthropometrics and muscular strength for the subjects (n = 40).

Demographic data	Mean (SD)	Range
Male subjects (n)	40	–
Age (years)	24.6 (2.5)	19–33
Height (cm)	176.5 (5.4)	160–186
Weight (kg)	73.0 (7.4)	58–88
Right Handgrip strength (kg)	41.7 (8.9)	21.3–66.7

elbow, exerting a power grip in a Hand Dynamometer (Lafayette Hand Dynamometer, model 78010, Lafayette Instrument Company, USA).

2.3 The vibrations measurement

The vibration transmitted to the hand–arm system was measured using a triaxial accelerometer (Model 3023 A2, 10 mV/g, 4 grams, Dytran Instruments, Inc.) and the signal recorded with a VI-410 PRO vibration analyzer, real-time frequency analysis (Quest Technologies), which complies with the specifications of the standards ISO 8041:2005, ISO 5349-1,2:2001, ISO 2631-1,2:2003, ISO 10816 and IEC 61672-1. Two groups of measurements were performed with the vibration meter: (a) with the accelerometer attached to the tools according to recommendations of ISO 5349-2:2001 and (b) with the accelerometer attached to the wrist of the subject's dominant arm by means of a special device fixed to the wrist.

The device was attached to the operator's wrist by means of a Velcro strip. Special care was taken in the pressure exerted by the strip in the wrist of the operator. The strip was tight enough to avoid undesirable movements of the accelerometer, allowing however for free flow of blood and assuring subjects comfort.

The acceleration frequency-weighted r.m.s. values of a_{hwx}, a_{hwy}, a_{hwz} and a_{hv} (m.s^{-2}) were recorded in the vibration meter during operations time. The log rate time was set to 1 s in order to analyze the variability of the results. Full octave band (1/1) frequency vibration spectrum (non-weighted) was recorded. The vibrations were measured in the dominant arm of the operator, the one operating the trigger of the tools. During operations time, the progress of the acceleration frequency-weighted r.m.s. values of a_{hwx}, a_{hwy} and a_{hwz} (m.s^{-2}) were carefully observed in the vibration meter. Stable values of r.m.s. acceleration were observed in the end of each measurement time.

2.4 Statistical analysis

The research design included the same subjects operating with the reciprocating sabre-saw in different conditions (bare hand and the four types of gloves). In order to compare the variability of a_{hv} measured in the handle of the hand tools or in the wrist of the operators, repeated measures ANOVA—which uses an F statistic to determine significance—were conducted.

In addition to the usual assumptions of these analysis (independence of observations and normality), repeated measures ANOVA implies sphericity. If this assumption was not met by Mauchly test, Greenhouse-Geisser (if epsilon <0.75) or Huynh-Feldt (>0.75) correction were considered to determine a more accurate p-value. Whenever significant differences were found, we used post-hoc comparisons using the Bonferri correction to compare different pairs of variables.

When comparing globally handle and wrist measurements, two-sample independent t-test were conducted, since the subjects tested were not the same. All tests were conducted in SPPS—version 18.0 for Windows.

3 RESULTS

3.1 The variability of vibration total value of frequency-weighted r.m.s. acceleration

Forty operators performed operations with the sabre saw in simulated work tasks over a period of 20 s. The log rate time of the vibrations measurements was set to 1 s in order to analyze the variability of ISO-weighted total r.m.s. a_{hv} values (CV (%) = SD/ a_{hv} × 100) during the operations. The variability was measured in the handle of the tool and in the wrist of the operators (Table 2).

A repeated measures ANOVA was conducted to test if the six variables had the same mean. The test corrected with Greenhouse-Geisser delivered no significant difference between the six variables in the handle measurement (F(1.478; 10.348) = 4.123, p = 0.057) and between the six variables in the wrist measurement (F(3.208; 99.460) = 1.734, p = 0.161).

The CV of the ISO-weighted total value r.m.s. acceleration measured in the handle of the tool and in the wrist of the operators, with the blade cutting the material are not significantly different from the values with the blade held off the material, which means that individual cutting operations didn´t introduce a significant variability in the total value

Table 2. The variability of ISO-weighted total r.m.s. acceleration values measured in the handle of the sabre saw (n = 8 subjects) and in the wrist of the operators (n = 32 subjects) during simulated operations (log rate time = 1 s; 20 readings per subject).

	Handle measurement CV (%) Mean (Range)	Wrist measurement CV (%) Mean (Range)
Blade held off the material		
Bare hand, tool on	2.4 (0.7–4.6)	9.3 (2.1–20.0)
Blade cutting the material		
Bare hand	5.2 (1.2–10.8)	9.9 (3.5–25.7)
Type 1 glove	2.4 (1.2–4.6)	8.2 (3.3–17.1)
Type 2 glove	2.5 (0.7–6.0)	7.8 (3.8–14.8)
Type 3 glove	1.9 (0.8–3.2)	7.6 (3.4–16.8)
Type 4 glove	2.2 (0.8–5.2)	8.5 (1.9–16.3)

of frequency-weighted r.m.s. acceleration. However, comparing globally the variability of a_{hv} measured in the handle of the hand tools (n = 48 readings) and in the wrist of the operators (n = 192 readings) according to two-sample independent t-test, the variability of accelerations measured in the wrist is higher than in the handle of the tool (t(145.686) = 14.102, p = 0.000). The transmission of vibration from the handle of the tool to the wrist of the operator is affected by different variables as the grip and push forces and the biodynamic response in the hand-arm system (Dong et al. 2004, 2004a, 2005, 2005a, Hartung et al. 1993, Griffin 1998, Griffin 2011, Marcotte et al. 2005, Adewusi et al. 2011); most probably the variability increase of a_{hv} measurements in the wrist is a result of the variability in the different variables that affect the transmissibility to the wrist.

3.2 The vibration total value of frequency-weighted r.m.s. acceleration measured in the handle of the hand tool and in the wrist of the operators

Acceleration values a_{hv} were evaluated in the tool handle and in the wrist of the operators. Measures in the tool handle (Table 3) and in the wrist (Table 4) were taken with a log rate of 1 s, during operations time of 20 s. Mean and range values were calculated based on 20 readings per operation.

Comparing globally the ISO-weighted total r.m.s values of a_{hv} measured in the handle of tool (n = 48 readings) and in the wrist of the operators (n = 192 readings), according to two-sample independent t-test, the acceleration measured in the handle of the tool is ~7.5% higher than in the wrist of the operator (t(145,686) = 14,102, p = 0.000). To each operating conditions, according to two-sample independent t-test, non-significant differences (p > 0.05) were found between ISO-weighted total r.m.s values of a_{hv} in the handle of the tool and in the wrist of the oper-

Table 3. The ISO-weighted total r.m.s. values of a_{hv} measured in the handle of the tools during simulated operations with the sabre saw (n = 8 subjects; 20 s operation time).

	Handle measurement a_{hv} (m.s⁻²) Mean (Range)
Blade held off the material	
Bare hand, tool on	17.94 (16.00–20.90)
Blade cutting the material	
Bare hand	19.91 (16.80–26.20)
Type 1 glove	18.29 (15.60–20.80)
Type 2 glove	17.05 (15.20–19.10)
Type 3 glove	18.06 (15.20–20.60)
Type 4 glove	17.78 (15.40–20.90)

Table 4. The ISO-weighted total r.m.s. values of a_{hv} measured in the wrist of the operators during simulated operations with the sabre saw (n = 32 subjects; 20 s operation time).

	Wrist measurements a_{hv} (m.s⁻²) Mean (Range)
Blade held off the material	
Bare hand, tool on	14.12 (7.71–23.00)
Blade cutting the material	
Bare hand	17.34 (6.94–33.10)
Type 1 glove	17.57 (7.05–29.00)
Type 2 glove	17.10 (6.19–27.00)
Type 3 glove	16.32 (8.69–25.70)
Type 4 glove	19.02 (7.12–31.70)

ators (exception to the operating condition "bare hand, tool on" in which the accelerations measured in the handle of the tool was higher than in the wrist of the operator (t(31,745) = 4,138, p = 0.000)).

4 DISCUSSION

The accelerations measured at the tool handle and at the wrist of the operators were not simultaneous. Hewitt (2010) considers important to measure the vibration simultaneously at the surface of the handle or surface imparting the vibration and between the hand and glove i.e. inside the glove, to assess the transmissibility of an anti-vibration glove. Regarding measurements in the hand-arm system, simultaneous measurements were applied in their research by Adewusi et al. (2011) and Xu et al. (2009). In this research, the CV of the ISO-weighted total value r.m.s. acceleration measured in the handle of the tool and in the wrist of the operators, with the blade cutting the material, were not significantly different from the values with the blade held off the material, which means that individual cutting operations didn't introduce a significant variability in the total value of frequency-weighted r.m.s. acceleration.

The variability of ISO-weighted total r.m.s values of a_{hv} measured in the wrist of the operators was higher than in the handle of the tool (t(145.686) = 14.102, p = 0.000). The transmission of vibration from the handle of the tool to the wrist of the operator is affected by different variables as the grip and push forces and the biodynamic response in the hand-arm system (Dong et al. 2004, 2004a, 2005, 2005a, Hartung et al. 1993, Griffin 1998, Griffin 2011, Marcotte et al. 2005, Adewusi et al. 2011); most probably the variability increase of a_{hv} measurements in the wrist is a result of the variability in the different variables that affect the transmissibility to the wrist.

Comparing globally the ISO-weighted total r.m.s values of a_{hv}, measured in the handle of the tool ($n = 48$ readings) and in the wrist of the operators ($n = 192$ readings), the acceleration measured in the handle of the tool was ~7.5% higher than in the wrist of the operator ($p = 0.001$). However, when comparing specific operating condition, non-significant differences were found between ISO-weighted total r.m.s a_{hv} values in the handle of the tool and in the wrist of the operators.

For the Sabre saw, and for the type of operation performed with the tool in this research, the dominant vibration frequency was in the 250–315 Hz range and in the 40–50 Hz range (one third octave band). With bare hand, a substantial reduction in the vibration total value of frequency-un-weighted r.m.s. components (one third octave band) above 160–200 Hz was observed in the wrist measurements; a similar reduction was not observed in the peak acceleration values at 40–50 Hz measured in the handle of the tool and in the wrist. The rapid decrease in the vibration transmissibility of the hand–arm segments above 200 Hz may be related to the increasing distance from the source of vibration. As stated by different authors, the frequency attenuation in the wrist is particularly visible in the vibration components above 200 Hz (Adewusi et al. 2011, Xu et al. 2009, Dong et al. 2004).

Non-significant differences ($p > 0.05$) in ISO-weighted total r.m.s values of a_{hv} measured in the wrist, were found between bare hand and gloved operations, during blade cutting operations. Globally, ~50% of the operators, revealed gloved a_{hv} values in the wrist higher than bare hand a_{hv} values; only in 25% of the operators ($n = 32$) bare hand a_{hv} values were consistently higher than gloved values.

5 CONCLUSION

The anti-vibration gloves did not attenuate the total value of frequency un-weighted r.m.s. vibration components (one third octave band), measured in the wrist of the operators. The wrist measurements with bare hand and with anti-vibration gloves did not reveal significant differences in the vibration components in the dominant range in the wrist ~31.5–200 Hz, neither in the whole one third octave band range. Also, non-significant differences in ISO-weighted total r.m.s values of a_{hv} measured in the wrist, were found between bare hand and gloved operations, during blade cutting operations.

REFERENCES

Adewusi, S.A., Rakheja, S., Marcotte, P. & Boutin, J. 2011. Vibration transmissibility characteristics of the human hand–arm system under different postures, hand forces and excitation levels. *Journal of Sound and Vibration* 329: 2953–2971.

C.D.C.-Centers for Disease Control and Prevention. *PowerTools Database: Reciprocating Saw*. Available at http://wwwn.cdc.gov/niosh-sound-vibration/ (April 4, 2012).

Dong, R.G., Schopper, A.W., McDowell, T.W., Welcome, D.E., Wu, J.Z., Smutz, W.P., Warren, C. & Rakheja, S. 2004. Vibration energy absorption (VEA) in human fingers-hand-arm system. *Medical Engineering & Physics* 26: 483–492.

Dong, R.G., McDowell, T.W., Welcome, D., Barkley, J., Warren, C. & Washington B. 2004a. Effects of Hand-Tool Coupling Conditions on the Isolation Effectiveness of Air Bladder Anti-Vibrations Gloves. *Low Frequency Noise, Vibration and Active Control* 23(4): 231–248.

Dong, R.G., Wu, J.Z. & Welcome, D.E. 2005. Recent advances in biodynamics of human hand-arm system. *Industrial Health* 43(3): 449–71.

Dong, R.G., McDowel, T.W., Welcome, D.E. & Smutz, W.P. 2005a. Correlations between biodynamic characteristics of human-arm system and the isolation effectiveness of anti-vibration gloves. *International Journal of Industrial Ergonomics* 35: 205–216.

Griffin, M.J. 1998. Evaluating the effectiveness of gloves in reducing the hazards of hand-transmitted vibration. *Occupational and Environmental Medicine* 55: 340–348.

Hartung, E., Dupuis, H. & Scheffer, M. 1993. Effects of grip and push forces on the acute response of the hand-arm system under vibrating conditions. *Occupational and Environmental Health* 64(6): 463–467.

Hewitt, S. 2010. RR795-Triaxial measurements of the performance of anti-vibration gloves. *Health and Safety Executive (HSE)*.

Marcotte, P., Aldien, Y., Boileau, P.-É., Rakheja, S. & Boutin, J. 2005. Effect of handle size and hand–handle contact force on the biodynamic response of the hand–arm system under z_h-axis vibration. *Journal of Sound and Vibration* 283 (3–5): 1071–1091.

Pinto, I., Stacchini, N., Bovenzi, M., Paddan, G.S. & Griffin, M.J. 2001. Protection effectiveness of anti-vibration gloves: field evaluation and laboratory performance assessment, in: *9th International Conference on Hand-Arm Vibration, Nancy, France, 05–08 Jun 2001*.

Rakheja, S., Dong, R., Welcome, D. & Schopper, A.W. 2002. Estimation of tool-specific isolation performance of antivibration gloves. *International Journal of Industrial Ergonomics* 30: 71–87.

Reynolds, D.D & Wolfe, E. 2005. Evaluation of Antivibration Glove Test Protocols Associated with the Revision of ISO 10819. *Industrial Health* 43: 556–565.

Welcome, D.E, Dong, R.G., Xu, X.S., Warren, C. & McDowell, T.W. 2012. An evaluation of the proposed revision of the anti-vibration glove test method defined in ISO 10819. *International Journal of Industrial Ergonomics* 42: 143–155.

Xu, X.S., Welcome, D.E., McDowell, T.W., Warren, C. & Dong, R.G. 2009. An investigation on characteristics of the vibration transmitted to wrist and elbow in the operation of impact wrenches. *International Journal of Industrial Ergonomics* 39: 174–184.

Occupational Safety and Hygiene – Arezes et al. (eds)
© *2013 Taylor & Francis Group, London, ISBN 978-1-138-00047-6*

Risk assessment of chronic exposure to magnetic fields near electrical apparatus

F.O. Nunes
Instituto Superior de Engenharia de Lisboa/Centro de Eletrotecnia e Eletrónica Industrial, Lisbon, Portugal

E.F. Margato
Instituto Superior de Engenharia de Lisboa/Centro de Eletrotecnia e Eletrónica Industrial, Lisbon, Portugal
Center for Innovation on Electrical and Energy Engineering, Lisbon, Portugal

ABSTRACT: The paper presents a numerical model to calculate ELF magnetic fields for typical conductor arrangements of electrical power lines. After an introduction to the problems arising from the exposure of human beings to low frequency magnetic fields, a description of the numerical model for the magnetic field evaluation is presented, discussing its application and presenting some examples. The attention is then devoted to the analysis of possible solutions for magnetic field reduction.

1 INTRODUCTION

The electromagnetic field results from the joint presence of the magnetic field and electric field and can be caused by natural or artificial phenomena. However, when there is a significant prevalence of one of its components over the other, it is reasonable to consider and study separately the magnetic field or the electric field. With regard to effects on the human body, concern is focused on the magnetic fields, which by their nature are able to, unopposed, penetrate biological tissue. As for the electric field, when varying and intense enough, its effect on the human body is due almost exclusively to the movement of electric charges surface (inside the body the electric field is reduced by 5–6 orders of magnitude). These fields may cause the development of high intensities of current and become harmful, causing burns by action of the Joule effect.

Terrestrial magnetic field and the magnetic field originated in the atmospheric discharges (lightning) are identified primarily regarding the magnetic field originating from natural sources. As for artificial sources, they are intrinsically linked to technological development seen in modern societies, supported by widespread and intensive use of electricity.

The artificial magnetic fields generated by electrical currents existing in electrical installations of production, transmission and use of electricity, appear in the vicinity of such facilities, either with industrial frequency (50 Hz) or with multiple frequencies (harmonics) due to equipment operation for conversion of electrical energy. These fields thus occur in a frequency range classified as extremely low frequencies (ELF), between 3 and 300 Hz and fall within the scope of non-ionizing radiation.

Since many physiological functions are electrochemical in nature and some biological activities, such as brain activity and cardiac conduction, involve measurable electrical current, it is plausible that the magnetic field inside the human body can interfere with his "electrical system".

In fact, Faraday's law of induction states that a time varying magnetic field induces in a considered loop an electromotive force that is proportional to the rate of change of the magnetic flux through the loop. Consequently those generated electromotive forces may be responsible for electric currents in biological tissues that are, to a greater or lesser degree, conductors.

Continued exposure to ELF magnetic fields has been reported in the scientific literature identifying potential adverse effects on individuals and on public health (Hardell & Sage, 2008, Fazzo & Comba, 2009). However, many adverse effects attributable to prolonged exposure to ELF magnetic fields may occur after many years of exposure, making the causeeffect relation very difficult to establish. To date scientific studies have failed to show clear evidence of the relationship between exposure to ELF magnetic fields and problems to human health.

Still, several epidemiological studies have reported numerous biological effects and behavioral disorders. Among them are mentioned various types of cancer (Loomis et al., 1994, Beniashvili et al., 2005), cardiovascular diseases (Hakansson et al., 2003, Savitz et al., 1999), reproductive disorders (Robert, 1993), DNA damage (Winker et al., 2005) and even depressive symptoms (Vercasalo et al., 1997).

The results of numerous available studies have been compiled by different organizations in order to establish internationally harmonized values that can serve as a benchmark in assessing the risks of exposure to magnetic fields.

The measurements of the magnetic field present in the vicinity of electrical apparatus, and the results obtained with calculation models based on the classical electromagnetic theory can be used to assess the risks of prolonged exposure to these magnetic fields compared to reference levels available.

2 METHODOLOGY

2.1 Reference levels

Despite all the scientific inconsistencies, the International Commission on Non-Ionizing Radiation Protection (ICNRP, 1998) established reference levels for occupational exposure and for the general public to electromagnetic fields of frequencies up to 300 GHz. These reference levels include in the range of ELF (3–300 Hz) values ranging, to occupational exposure, from 83.3 µT to 22.22 mT, passing for 500 µT at power frequency (50 Hz). For the general public reference levels range from 16.7 µT to 4.44 mT, passing for 100 µT at power frequency (50 Hz). These guidelines were later followed by the European Union (Council of the European Union, 1999). Based on established evidence regarding acute effects (short term exposure) those reference levels for power frequency were changed, doubling to 1,000 µT and 200 µT, respectively, for occupational exposure and for the general public (ICNRP, 2010). The Institute of Electrical and Electronics Engineers Standards Association (IEEE, 2002) also defined exposure levels to magnetic field at power frequency (50 Hz): 904 µT (head and torso) and 75,800 µT (arms and legs).

However, the World Health Organization (WHO, 2007), have focused on epidemiological studies on childhood leukaemia in residential exposures to ELF magnetic fields in the range of values from 0.3 to 0.4 µT, which refers to prolonged exposure to this type of non-ionizing radiation and with the values above referred as a risk factor for cancer. Calculations based on case studies of controlled exposure to ELF magnetic fields and childhood leukaemia also results in similar values that, with a precautionary approach, can thus be assumed to be alert level for chronic exposures (long term exposure).

In the same way the International Electromagnetic Fields Alliance (IEMFA, 2011) adopted new exposure guidelines to protect public health and the health of future generations. Seletun Scientific Panel based on the available evidence, settled at 0.1 µT the exposure limit for extremely low frequency (fields from electrical power) for all new installations, such as power lines, indoor electric appliances, household items, TVs, radios, computers, and telecommunication devices, based on findings of risk for leukaemia, brain tumors, Alzheimer's, ALS, sperm damage and DNA strand breaks.

2.2 Numerical modeling of magnetic field

The magnetic induction field at a point (x,y) due to a section of rectilinear conductor traversed by a current intensity I is obtained from the Biot and Savart law:

$$d\vec{B} = \frac{\mu_o}{4\pi} \frac{I\, d\vec{s} \times \hat{r}}{|\vec{r}|^2} \tag{1}$$

Making the cross product between the current element $I\,d\vec{s}$ and the unit vector $\hat{r}$ of the vector $\vec{r}$ that links to the point $(x, y, 0)$ where you want to calculate the magnetic induction field (figure 1):

$$|d\vec{s} \times \hat{r}| = ds \cdot sen\,\beta = ds \cdot \cos\alpha \tag{2}$$

and given that:

$$tg\,\alpha = \frac{s}{R} \Rightarrow ds = \frac{R}{\cos^2\alpha}d\alpha$$

$$|\vec{r}|^2 = \frac{R^2}{\cos^2\alpha}$$

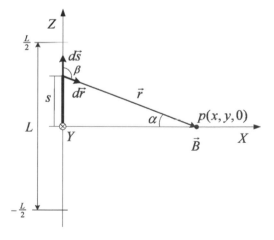

Figure 1. Magnetic induction field at a point $(x, y, 0)$ due to a section of rectilinear conductor with current intensity I.

equation (1) may take the form:

$$dB = \frac{\mu_o I}{4\pi} \frac{\frac{R}{\cos\alpha}}{\frac{R^2}{\cos^2\alpha}} d\alpha = \frac{\mu_o I}{4\pi R}\cos\alpha d\alpha \qquad (3)$$

So the amplitude of the magnetic induction field at the analyzed point is given by the following equation:

$$|\vec{B}| = \frac{\mu_o I}{4\pi R} \int\limits_{-\frac{L}{2}}^{\frac{L}{2}} \cos\alpha d\alpha \qquad (4)$$

To consider the entire length L of the conductor, it is necessary to integrate equation (4) according to the variable α for the entire length, $-L/2$ to $L/2$.

Thus, the amplitude of the magnetic induction field due to the entire conductor, of length L, located in a plane which intersects the conductor, can be calculated by:

$$|\vec{B}| = \frac{\mu_o I}{4\pi R} \int\limits_{-tg^{-1}\frac{L}{2R}}^{tg^{-1}\frac{L}{2R}} \cos\alpha d\alpha$$

$$= \frac{\mu_o I}{4\pi R}\left[sen\left(tg^{-1}\frac{L}{2R}\right) - sen\left(-tg^{-1}\frac{L}{2R}\right)\right] \qquad (5)$$

For a conductor of infinite length the value of the magnetic induction field intensity B is directly proportional to the current intensity I and inversely proportional to the distance to the conductor R and can be calculated simply by:

$$B = \frac{\mu_o I}{2\pi R} \qquad (6)$$

The error introduced by assuming that a conductor of length L can be considered of infinite length is illustrated in figure 2 as a function of the ratio L/R.

With $L > 5R$ the error rate is less than 2%, and less than 0.5% for $L > 10R$.

For purposes of further generalization of the sum of magnetic induction fields generated at a point $(x, y, 0)$ by different conductors carrying different currents, may one consider a conductor of infinite length intersecting the $X - Y$ plane at the point $(dx, dy, 0)$ as illustrated in figure 3.

If the current which traverses the conductor is alternating (sinusoidal) this can be represented at power frequency (50 Hz) by a complex vector,

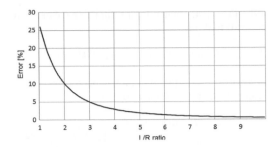

Figure 2.   Error introduced by assuming that a conductor of length L can be considered infinite length, as a function of the ratio L/R.

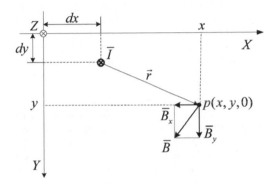

Figure 3.   Magnetic induction fields generated at a point $(x, y, 0)$ by infinite conductor perpendicular intersecting the $X - Y$ plane at the point $(dx, dy, 0)$ and carrying current I.

respectively, with intensity I and phase θ. So magnetic induction field can be calculated by:

$$\vec{B}(x,y,0) = \frac{\mu_o}{2\pi} \frac{\vec{I} \times \vec{r}}{r^2}$$
$$= \frac{\mu_o}{2\pi} I \frac{\vec{e}_z \times [(x-dx)\vec{e}_x + (y-dy)\vec{e}_x]}{(x-dx)^2 + (y-dy)^2} e^{j\theta} \qquad (7)$$

The components of the magnetic induction field at a given point $(x, y, 0)$ exist only in the $X - Y$ plane so that the respective components of the magnetic induction field equation (8) due to the sum of the N currents with intensity I_i e phase θ_i existing in N parallel conductors, can be calculated by the contribution of the different complex vectors involved.

$$\vec{B}_x = -\frac{\mu_o}{2\pi} \sum_{i=1}^{N} \frac{(y-dy_i)\cos\theta_i + j(y-dy_i)sen\theta_i}{(x-dx_i)^2 + (y-dy_i)^2} I_i$$

$$\vec{B}_y = \frac{\mu_o}{2\pi} \sum_{i=1}^{N} \frac{(x-dx_i)\cos\theta_i + j(x-dx_i)sen\theta_i}{(x-dx_i)^2 + (y-dy_i)^2} I_i$$
(8)

Some particular cases can be tested with equation (8), such as, the calculation of the magnetic induction field created by a bifilar line (phase and neutral) making

$$I_1 = I_2 = I \quad \text{and} \quad \theta_1 = \theta_0; \theta_2 = \theta_0 \pm 180°$$

or groups of conductors in a balanced three-phase system that can be simulated with

$$I_1 = I_2 = I_3 = I \quad \text{and} \quad \theta_1 = \theta_0;$$
$$\theta_2 = \theta_0 \pm 120°; \theta_3 = \theta_0 \pm 240°$$

Some typical arrangements of conductors used in transmission lines are shown on figure 4.

In order to exemplify the application of the model we present in figure 5 the magnetic induction field intensity calculated at different vertical distances H from the bottom of the group of conductors (figure 4). All the values were obtained with a current intensity I of 1 kA for a constant distance D of 1 m between conductors.

For practical situations ($H > 1$ m) one can see the influence of the double-circuit line with proper phase sequences in the reduction of magnetic induction field intensity. A diminution for less than 1/3, with double power transmission, can be obtained for a vertical distance H of 6 m. In this case, even for a short vertical distance H of 6 m, reducing intensity current I to 100 A, one can obtain magnetic induction field intensities lower than the alert level of 0.3–0.4 µT, previously referred.

It is also noticeable that vertical arrangement of the conductors is always better than the horizontal.

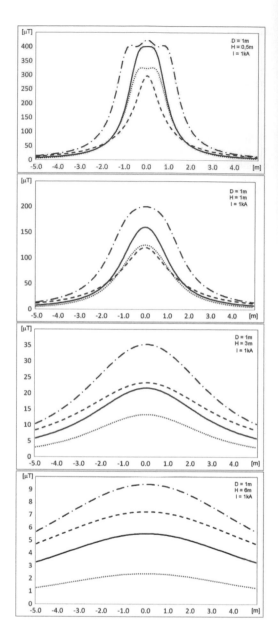

Figure 5. Magnetic induction field intensity B in µTesla as a function of horizontal distance d in meters from the centre of the conductors group in meters, considering different vertical distances from the bottom of conductors groups.

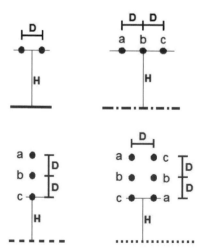

Figure 4. Four common conductors arrangements.

2.3 Possible solutions for magnetic field reduction

The reduction of ELF magnetic field can be obtained through various approaches. Available techniques include design reconfiguration of wiring schemes and shielding. For each magnetic field source, one must find design options that reduce magnetic field exposure without altering the function for which the system was intended.

The magnetic field generated by transmission or distribution lines is substantially conditioned, besides the phase currents, by the distance from the conductors, the distance between phases and the line geometry. Therefore, by acting on the conductor arrangements a significant reduction of the magnetic field can be obtained.

The transformation of a single circuit line into a double-circuit line with the suitable phase arrangement, besides the substantial reduction of magnetic field as seen in figure 5, the advantage to allow the use of masts whose design is already available.

As in multi-circuit lines (such as double-circuit lines), the geometrical arrangements of the single phases in each circuit can play an important role. This concept is the basis of the idea which is applied in the split-phase line. A split-phase line is a three-phase line where one or more of the phases are split into two or more conductors to achieve considerably field reduction.

Magnetic field exposure may also be reduced if the volumes occupied by people are shielded. Shielding methods, suitable for cable systems, can be divided into: shielding obtained by induced currents in high conductivity materials; shielding obtained by modification of magnetic flux patterns using high permeability materials and; active shielding obtained by superposing another magnetic field to the primary one, obtaining a global reduction.

Shielding mechanism of high conductivity screening elements arranged near the source is based on the currents induced in the screening elements themselves. An additional field source is then superposed to the primary one, producing a compression of the field lines in the region between conductors and shield.

High magnetic permeability shields using ferromagnetic materials provide a low reluctance path for the magnetic field, thus deflecting the field lines near the shield. Their presence strongly reduces the magnetic field intensity behind the sheet.

However, in some cases, numerical models and experimental investigations showed that screens made of conductive material can be more effective than analogous ferromagnetic screens in the reduction of the magnetic fields.

Finally, active shielding includes the rearrangement of the conductors and the disposal of passive circuits to flow induced currents by the presence of the external field to be reduced.

3 RESULTS

The availability of a computational tool is a practical alternative to performing measurements, particularly in tracing the profiles of the magnetic field for the establishment of "safe zones".

It is also an essential tool in the prediction of the profiles of the magnetic field of electrical installations in the design phase, enabling timely the necessary changes/corrections.

Results are presented of the spatial distribution of the magnetic field obtained for different geometries of conductors used in electrical installations, including typical arrangements in three phase installations, calculated with the developed tool.

4 CONCLUSIONS

Classical electromagnetic theory has proved sufficiently precise calculation of the unperturbed magnetic fields and thus to develop suitable tools for the simulation of the magnetic field created by certain arrangements of conductors.

It was possible to confirm that vertical arrangement of the conductors is preferable than the horizontal one and double-circuit line with proper phase sequences reduces magnetic induction field intensity in the underneath or proximity volumes occupied by people.

The results obtained are consistent with published power transmission lines studies (Mamishev et al., 1996, Conti et al., 2003; Turri & Moro, 2012), thus validating the model.

REFERENCES

Beniashvili, D., Avioach'm I., Basov D. & Zusman I. 2005. The role household electromagnetic fields in the development of mammary tumors in women: clinical case-recorded observations. Medical Science Monitor, International Medical Journal for Experimental and Clinical Research, Vol. 11, N.° 1, CR10–13.

Comba P. & Fazzo L. 2009. Health effects of magnetic fields generated from power lines: new clues for an old puzzle. Ann. Inst. Super. Sanitá, Rome, Vol. 45, N.° 3, pp. 233–237.

Conti, R., Giorgi, A., Rendina, R., Sartore, L. & Sena, E.A. 2003. Technical Solutions to Reduce 50 Hz Magnetic Fields from Power Lines. IEEE Bologna Power Tech Conference, June 23th–26th.

Council of the European Union. 1999. Council Recommendation on the limitation of exposure of the general public to electromagnetic fields (0 Hz–300 GHz). 12 July [1999/519/CE].

Hakansson, N., Gustavson, P., Sastre A. & Floderus B. 2003. Occupational Exposure to Extremely Low Frequency Magnetic Fields and Mortality from Cardiovascular Disease. American Journal of Epidemiology, Vol. 158, N.° 6, pp. 534–542.

Hardell, L. & Sage, C. 2008. Biological effects from electromagnetic field exposure and public exposure standards. Biomedicine & Pharmacology, Elsevier Masson, Vol. 62, pp. 104–109.

ICNIRP—International Commission on Non-Ionizing Radiation Protection. 1998. Guidelines for Limiting Exposure to Time-Varying Electric, Magnetic and Electromagnetic Fields (Up To 300 Ghz). Health Physics, Vol. 74, N.° 4, pp 818–836.

ICNIRP—International Commission on Non-Ionizing Radiation Protection. 2010. ICNIRP Guidelines for Limiting Exposure to Time–varying Electric and Magnetic Fields (1 Hz–100 kHz). Health Physics, Vol. 99, N.° 6, pp. 494–522.

IEEE—Institute of Electrical and Electronics Engineers. 2002. IEEE Standard for Safety Levels with Respect to Human Exposure to Electromagnetic Fields, 0–3 kHz. Std C95.6.

IEMFA—The International Electromagnetic Fields Alliance. 2011. Seletun Statement. Oslo, Norway, February3rd.[http://iemfa.org/index.php/publications/seletun-resolution].

Loomis, D.P., Savitz, D.A. & Ananth, C.V. 1994. Breast cancer mortality among female electrical workers in the United States. Journal of the National Cancer Institute, Vol. 86, N.° 12, pp. 921–925.

Mamishev, A.V., Nevels, R.D. & Russell, B.D. 1996. Effects of conductor sag on spatial distribution of power line magnetic field. IEEE Trans. on Power Delivery, Vol. 11, N.° 3, pp. 1571–1576.

Moro, F. & Turri, R. 2012. Accurate Calculation of the Right-of-way Width for Power Line Magnetic Field Impact Assessment. Progress In Electromagnetics Research B, Vol. 37, pp. 343–364.

Robert, E. 1993. Birth defects and high voltage power lines: an exploratory study based on registry data. Reproductive Toxicology, Elsevier, Vol. 7, N.° 3, pp. 283–287.

Savitz, D.A., Liao, D., Sastre, A., Klekner, R.C. & Kavet, R. 1999. Magnetic Field Exposure and Cardiovascular Disease Mortality among Electric Utility Workers. American Journal of Epidemiology, Vol. 149, N.° 2, pp. 135–142.

Vercasalo, P., Kaprio, J., Varjonen, J., Romanov, K., Heikkila, K. & Kuskenvuo, M. 1997. Magnetic Fields of Transmission Lines and Depression. American Journal of Epidemiology, Vol. 146, N.° 12, pp 1037–1045.

WHO—World Health Organization. 2007. Extremely Low Frequency Fields. Environmental Health Criteria 238, WHO Press.

Winker, R., Ivancsits, S., Pilger, A., Adolkofer, F. & Rudiger, H.W. 2005. Chromosomal damage in human diploid fibroblast by intermittent exposure to extremely low-frequency electromagnetic fields. Mutation Research/Genetic Toxicology and Environmental Mutagenesis, Elsevier, Vol. 585, N.° 1–2, pp. 43–49.

Occupational Safety and Hygiene – Arezes et al. (eds)
© 2013 Taylor & Francis Group, London, ISBN 978-1-138-00047-6

Indoor air quality: Analytic assessment and employees' perception

A. Ferreira, L. Jesus & J.P. Figueiredo
College of Health Technology of Coimbra, Portugal

A. Carvalho
Technological Center for Ceramic and Glas, Coimbra, Portugal

ABSTRACT: Air quality and its influence in the human health is one of the biggest challenges of the modern society. Nowadays air pollution levels inside limited spaces often exceed the ones that are registered outside these spaces and can be the cause of several health problems. This study has evaluated the IAQ in a company and the perception of its employees regarding their exposure in the workspace. The main conclusion is that the most common symptoms indicated by the employees at their workspace (regarding air quality) are "rhinitis" and "headache". Most employees considered that their workspace has a "reasonable" or "good" air quality. Some buildings reveal highest levels of CO and fungi concentration despite of not showing quantities that evince risk to the employees.

1 INTRODUCTION

Air quality is the term that it is used, usually, to reflect the level of pollution in the air we breathe (Agencia, Portuguesa et al.,). The concerns associated with the effects of air quality on public health, have generally been associated to air pollution in the buildings. But according to several studies, the level of air pollution inside buildings can reach values of two to five times, occasionally one hundred times higher than the outdoor air pollution (Déoux 2001; European, Environmental et al., 2005). The modern man lives, works and passes most of the time in closed spaces where the lack of air flow causes an accumulation of pollutants, often harmful to human health (Sodré, Corrêa et al., 2008). Consequently, contamination of indoor air presents itself a major concern since most individuals are most of the time in the interior of buildings, from 65% to 90%, inhaling an average of about 10 m^3 air daily (Déoux 2001; European, Environmental et al., 2005).

The indoor air contamination in a building results from the interaction of its location, outdoor pollutants, the ventilation system, microorganisms, cleaning products, furniture, work processes and activities, materials and equipment, and the number of building occupants (Matos, Brantes et al., 2010).

Therefore, the main pollutants that contribute to the degradation IAQ, being the most common pollutants in the atmosphere, are the chemical, physical and biological parameters (Gomes 2001). Most health problems result from occupational diseases that constitute a direct consequence of more or less continued exposure to a physical, biological or chemical threat that exists during the exercise of profession or as result of the conditions in which professional activity is exercised (Freitas 2008). These problems can be feelings of discomfort or, in extreme cases, can cause serious diseases or even death (Déoux 2001).

In addition to direct effects on public health, we should also consider the impact on the productivity of the buildings' occupants. Indeed, several studies show the relevance of the high level of absenteeism, behavior changes, dissatisfaction and low work performance in closed spaces due to deficient air quality (Ministério das Obras Públicas 2006; Carbon 2008). When about 20% of the occupants of a building have headaches, eye irritation, nose and throat, dry cough, dry skin, dizziness and nausea, difficulty concentrating, fatigue and sensitivity to odors and these symptoms decrease or disappear when occupants leave this kind of environment, we are dealing with indicators of "Sick Building Syndrome" (SBS) (Universidade.; Silva 2005). The IAQ is also related to the ventilation conditions. It is important that all the ventilation and air conditioning systems work properly in order to have good quality environmental conditions (Garcia 2006; Piteira 2007). The promotion of a good IAQ in the buildings must be a constant concern for all that has political, technical and scientific responsibilities, according to the existing professional relationship between the involved areas (Santos 2008).

The present study aims to evaluate the IAQ, taking into account the pollutants defined under the National Energy Certification and Buildings' Indoor Air Quality (RSECE-IAQ) in a company as well as the perception of its employees regarding the air quality.

2 MATERIAL AND METHODS

The study was developed in 2011 and 2012. The data collection was performed on the 6th of February 2012. The study was conducted in CTCV, a company that provides public services—privately held and headquartered in Coimbra, and created especially for technical and technological promotion of national industries of ceramics, glass and related/complementary sectors.

A proper assessment of air quality must take into account, not only the concentration of chemical contaminants, biological and physical parameters, but also study the characteristics of the building, type of operation and average occupancy (Carmo and Prado 1999).

Data collection was organized into two stages of research, being the first stage the evaluation of the IAQ in the three buildings (A, B and C) that constitute the CTCV, measuring chemical contaminants such as CO_2, CO, O_3, HCHO, VOC's and PM_{10}, physical parameters like temperature, relative humidity, air velocity and biological pollutants such as bacteria and fungi, in the winter; and the second stage, the perception of CTCV employees regarding the IAQ, via a survey that was sent in an electronic format. It is a study of Level II—an observational and transversal study. The type of sampling was not probabilistic and the technique for convenience. The universe of the study was composed by the three buildings of the company and all its workers. Initially a previous visit was made to know the physical conditions of the different areas that would be evaluated.

We used the following formula contained in the Technical Note NT-SCE-02 (RSECE 2009): $N = 0,15 \times \sqrt{Ai}$. Where "Ai" is the area of the "i" zone in m^2, in this case, each building. The building A has an approximate area of 4578 m^2, the building B has 327 m^2 and the building C has 688 m^2. So, the sample size included 21 locations within the company distributed on different floors: 14 in Building A, (reception desk, mechanical testing laboratory, administrative services, meeting rooms, room 1, room 2, environmental monitoring lab, cafeteria, technological proficiency testing, room 3, an auditorium, SGA customers, measurement and energy management and room 4), three in the building B (Industrial Property Office Support's room—GAPI, training room and group work room) and four at the Building C (room 5, auditorium, room 6 and room 7). On the outdoor space it was made the parameters' measuring at a point that is near the entry of all the three buildings. Regarding the response to the survey, 44 from the 71 CTCV's employees have answered to the questions.

The measurements of the chemical, physical and biological parameters were performed during a period of 5 minutes, being the result of each measurement an average of the two measurements that were made at the same location. IAQ measurements took place during the normal activity of the company, between 9 am and 6 pm, being the equipment positioned at the most possible central position of each place on a desk, and about the height of the respiratory tract of the employees in the sitting position. Having NT-SCE-02 as a reference, the collection of samples was performed at a height of 1 m from the ground and at least 3 m from the walls (RSECE 2009). The measurements of Outdoor Air Quality (OAQ) were carried out at a point near the entrance of the three buildings at the same height of the measurements that were taken for the IAQ, but at least 1 m away from exterior walls of the buildings. As for organic pollutants, the sample was reduced to 10 places. Prior to each collection of organic pollutants, the sampler has been disinfected with 70% ethyl alcohol. Samples of 360 liters were collected at each place. These collections were performed in duplicate for each place in Petri plates with culture medium Plate Count Agar (PCA) for bacteria and Malt Extract Agar (MEA) for fungi. The collection method was made by impact on semi-solid medium. A blank sample was also collected in each place of measurement, as a control. It should be noticed that measurements were made with the closed windows, as mentioned in the Technical Note NT-SCE-02 (RSECE 2009).

The equipment used for the analytical assessment IAQ were: Brand: SKA, Model: Haz-Dustepam 5000, Serial Number: 8072152, Parameters: PM50; Brand: Photovac, Inc., Model: 2020 ppb Pro, Serial Number: PBYN0001, Parameters: COV'S; Brand: PPM, Model: Formaldemeter Htv-m, Serial Number: F266, Parameters: HCHO; Brand: Aeroqual, Model: Serie 500, Sensor Ozono, Parameter: O_3; Brand: TSI, Model: 7545, Serial Number: T75450741005, Parameter: CO_2; Brand: TSI, Model: 7545, Serial Number: T75450741005, Parameter: CO; Brand: PBI, Model: Duo-SAS-Super 360, Serial Number: 07-DC-06268, Parameters: fungi and bacteria; Brand: Testo, Model: 435–2, Serial Number: 1406135/719, Parameters: Temperature, Relative Humidity and Air velocity.

We considered as the maximum concentration for CO_2 1800 mg/m^3 (984 ppm), 12.5 for CO mg/m^3 (10.7 ppm) for O_3 0.2 mg/m^3 (0, 10 ppm), HCHO to 0.1 mg/m^3 (0.08 ppm) for VOC 0.6 mg/m^3 (0.26 ppm—isobutylene) and PM_{10} 0.15 mg/m^3 as referred to in Decree-Law No. 79/2006 of April 4th (Ministério das Obras Públicas 2006). According to Decree-Law no. 80/2006, dated April 4th, the environmental comfort conditions of reference are situated between the 20°C and the 24°C in order to create a temperature suitable for the winter season (Ministério das Obras Públicas 2006). The rule ISO 7730 discloses a range of relative humidity from 30% to 70% and the values established for the air velocity should be from 0 to 0,2 m/s (7730 2005). Regarding the biological parameters, according to Decree-Law No. 79/2006 of April 4th, the reference maximum concentration is 500 CFU/m³ is for bacteria and fungi (Ministério das Obras Públicas 2006).

Regarding the second stage of the investigation, the questionnaire applied to the employees.

The statistical analysis of the data was performed by the software IBM SPSS Statistics version 19.0 for Windows. The tests used were: Chi-squared tests, Kruskal-Wallis test, t-test for a sample. The interpretation of the statistical tests were performed based on a significance level of p-value = 0.05 with Confidence Interval (CI) of 95%. We have also used simple descriptive statistics: measures of location (mean and median), dispersion (variance and standard deviation) and frequency (absolute and relative).

3 RESULTS

3.1 Sample description

Forty-four employees were analyzed and most of them were women (54.5%). This profile is also present at the ages class <30 years and ≥ 40 years. Regarding the education of the employees, most of them (77.3%) had a University Degree (Table 1-Appendix). In the same table we can also confirm that 64.7% of the employees with a University Degree (n = 34) were women. Nonetheless, most of the 8 employees with "High school studies" were men.

3.2 Analysis of the results

We want to analyze if the estimated average values at the different places were different from the reference value that is established by Law. Please analyze the tables 1 and 2.

We registered average differences significantly below the values established by Law (1800 mg/m³) in the buildings A and B. Other statistical meaningfully average differences were also registered between the CO analytic values collected in the 3 buildings (A, B and C—Tables 1 and 2), taking into consideration the reference value (12.5 mg/m³). The values were significantly below the established limits.

Regarding the average concentration values of HCHO, none of the buildings exceeded the reference value (0.1 mg/m³), despite of a residual value of this pollutant material in the Building C.

Table 1. Average values of the chemical, physical and biological parameters (Building A and B).

| | Measurement places | | | |
| | A (n = 14) | | B (n = 3) | |
	Av.	S.D.	Av.	S.D.
CO_2	1365,6*	270,4	1211,7*	13,3
CO	1,19*	0,25	1,03*	,21
HCHO	0,00	0,01	0,00	,00
PM10	0,04	0,03	0,03	,01
T°	18,14	1,8	14, 7	1,5
Hr	48,37	5,4	52,2	2,8
Air Velocity	0,05	0,05	0,01	,01
Bacteria	51,8	28,6	57,3	34,9
Fungi	101,9	98,2	15,7	9,5

Av. – Average; S.D. – Standard Deviation; t-Student test for 1 sample; * p < 0,001.

Table 2. Average values of the chemical, physical and biological parameters (Building C and Outdoor).

| | Measurement places | | | |
| | C (n = 4) | | Exterior (n = 1) | |
	Av.	S.D.	Av.	S.D.
CO_2	1404,0	333,3	749,00	–
CO	0,85*	0,1	6,00	–
HCHO	0,01	0,01	0,01	–
PM10	0,03	0,03	,03	–
T°	17,75	1,0	15,50	–
Hr	53,72	2,7	60,10	–
Air Velocity	0,01	0,00	–	–
Bacteria	30,00	18,1	3,00	–
Fungi	54,50	12,8	103,00	–

Av. – Average; S.D. – Standard Deviation; t-Student test for 1 sample; * p < 0,001.

The concentration of O_3 and COV's inside the three buildings was null.

In relation to the average concentrations of PM10 (Tables 1 and 2), we can detect considerable differences (p < 0,01) considering the reference value (0,15 mg/m³), being the average value of all the buildings below the values established by Law. Regarding the physical parameters—there were important statistical differences in the building A considering the reference value (20°C to 24°C), presenting average values, during the winter season, below to the value proposed by Law. As for the two other buildings, this difference was not so substantial. The Building B presented lower values. Nonetheless, the humidity had values in the reference range (30% to 70%). Last but not least, the biological parameters—bacteria and fungi—the estimated values were significantly lower (p < 0.001) having the reference value (500 UFC/m³) into account.

We made the evaluation of the different chemical, physical and biological parameters according to the place of data collection (building A, B and C). Using the Kruskal-Wallis Test we verified that there were big differences between the three places regarding the CO parameters, Temperature and Fungi (p < 0.05).

The Table 3 presents us the differences between medians by place of employees' exposure, namely the parameters that presented the main differences.

We verified that, despite of the low risk values of CO, the Median value of CO at Building A (Table 3) was substantially higher (1.3 mg/m^3), having the other evaluated buildings into consideration (B – $P_{50\%}$ = 1.1 mg/m^3; C – $P_{50\%}$ = 0.9 mg/m^3).

As we can see on Table 3, the building A identified a set of spaces where the extreme values of Fungi concentration (289 UFC/m^3) were higher than the maximum limits of Fungi concentration in the Buildings B (25 UFC/m^3) and C (67 UFC/m^3). The values of the Medians of Fungi Concentration were: A – 70.5 UFC/m^3; B – 16.0 UFC/m^3; C – 54.0 UFC/m^3. The Temperature estimated values were higher in the Buildings A ($P_{50\%}$ = 18°C) and C ($P_{50\%}$ = 17.5°C). The Building B had a Temperature median value of 15°C. Regarding the characterization of the occupational exposure, most of the employees (80,95%) considered that the IAQ of the buildings impacts their health (p < 0,001).

We verified the non-presence of a significant statistical association between perception of air quality in the workplace in what concerns to air renewal (or not) in the same workplace (p > 0.05). However, 39.5% of the employees have consid-

Table 4. Perception of the air quality in the workplace and the presence of air renewal in the same place.

Air quality in the workplace		Air renewal in the workplace		
		Yes	No	Total
Bad	n	1	0	1
	% row	100,0	0,0	100,0
Reasonable	n	12	3	15
	% row	80,0	20,0	100,0
Good	n	16	1	17
	% row	94,1	5,9	100,0
Very Good	n	10	0	10
	% row	100,0	0,0	100,0
Total	n	39	4	43
	% row	90,7	9,3	100,0

Independence Chi-squared Test.

ered the air quality as "Good" and 34.9% "Reasonable". We have also verified that 9.3% of the employees say that there is no air renewal in their workplace. Last but not least, we applied the Chi-squared Test. We verified with this analysis that the more common "simptoms/diseases" (p ≤ 0,05) of the employees at their workplace were "headache" (32%) and "allergy and rhinitis" (20%).

4 DISCUSSION

Modern societies are more and more looking for a better quality of life. It is essential to have an adequate air quality, not only because of aspects that may cause diseases but also the aspects that affect comfort and the productivity of the people that are lots of hours in closed spaces (Massa 2010).

After the analysis of the obtained results we can verify that the values of the chemical, physical and biological parameters of the IAQ (in the three buildings) did not exceed the reference values, excluding one of the physical parameters, the temperature, with lowest values during the winter season (20°C). The satisfaction with the thermal environment results from a thermal balance between the human body and the environment – and it is important to take into account individual vulnerability. The comfort temperatures can be optimized with the use of mechanical means of acclimatization, taking into account the type of activity, the air velocity, and dress typical for each season (Pinto, Freitas et al., 2007; Matos, Brantes et al., 2010). Therefore, although the value found, it is noted that in all locations exist HVAC equipment that can be connected and which are temperature control by de user, so it is hot considered that this is not in accordance.

Table 3. Quantiles of the chemical, physical and biological parameteres in diferente places.

Parameters		Measurement places		
		A	B	C
CO	P - $_{25\%}$	1,0	0,8	0,8
(p 0,05)	P - $_{50\%}$	1,3	1,1	0,9
	P - $_{75\%}$	1,3	1,2	1,0
	Minimum	0,8	0,8	0,7
	Maximum	2,0	1,0	1,0
T (°C)	P - $_{25\%}$	17,0	13,0	17,0
(p 0,05)	P - $_{50\%}$	18,0	15,0	17,5
	P - $_{75\%}$	19,0	16,0	18,5
	Minimum	14,0	13,0	17,0
	Maximum	21,0	16,0	19,0
Fungi	P - $_{25\%}$	34,0	6,0	43,5
(p 0,05)	P - $_{50\%}$	70,5	16,0	54,0
	P - $_{75\%}$	111,0	25,0	65,5
	Minimum	24,0	6,0	43,0
	Maximum	289,0	25,0	67,0

Efficient ventilation allows people not only to maintain an appropriate level of comfort but also to improve health conditions in the buildings. The ventilation has two main objectives: to evacuate the air contaminated by various pollutants, renew the air to dilute contaminants, keep the humidity values in the range of reference, the condensation and bad odors, and introduce new air and "clean" from the outside to renew the air we breathe inside the building (Déoux 2001; Matos, Brantes et al., 2010).

None of the buildings presented a CO_2 value above the Law proposed levels. However, this was the most significant pollutant. We should create conditions to maintain the level of CO_2 below the legal value. This pollutant comes from our own metabolism and the more people there is in a closed space, the more concentration of CO_2 will be in that location. Given the results with regard to the measured parameters, we can conclude that there is no health risk for CTCV's employees.

According to the collected information via survey, most of the employees admitted that air quality can affect their health and they also considered that there is an adequate air quality inside the buildings. For the two individuals who suffer from asthma, the conditions are good in the buildings and, according to their opinion, the air quality in the workplace do not aggravate their disease. In this study, the symptom/disease with the highest prevalence was headache, followed by allergy (rhinitis).

5 CONCLUSIONS

The evaluation of the buildings has identified some issues that, despite of their legal compliance, can be improved. Thus, we present some recommendations.

In order to avoid the presence of high concentrations of pollutants a good ventilation of spaces must be promoted, like a daily process to renew the air by opening windows and doors. In the interior rooms people should not close the doors or alternatively place ventilation equipment to allow air circulation.

In the laboratories where there are HCHO, COV's e CO people should turn the exhaust system on, namely the hottes. Moreover, people should have the routine of making the correct cleaning and maintenance of the acclimatization systems' filters in all buildings. It is also necessary to ensure that no permanent signs of humidity and if it happens, the situation should be resolved immediately. The cleaning process of fixed surfaces and furniture must be made by means of vacuum to decrease dust. Books, magazines and papers must be maintained whenever possible in closed rooms to prevent the accumulation particles that are released

during its subsequent handling. We recommended as a preventive measure to remove the carpets in some rooms (including rooms of the 1st floor and 3rd floor of building A), because it is a source of particles, bacteria and fungi. We also recommend registering all the cleaning activities in all the buildings, as well as the chemical specifications of the used cleaning products. Spaces such as sanitary and food facilities should have a hygiene with bactericidal disinfectant environment friendly. Materials with a biological origin should be avoided, like food and plants in internal spaces. Meals should be performed only in areas specifically designated for this purpose (i.e. cafeteria).

REFERENCES

7730, I. (2005). Ergonomics of the thermal environment — Analytical determination and interpretation of thermal comfort using calculation of the PMV and PPD indices and local thermal comfort criteria. ISO. Suíça, ISO: 52.

Agencia, Portuguesa, et al. Qualar - Base de Dados online sobre Qualidade do Ar.

Carbon, G. (2008). "Ar interior afecta milhares de pessoas em Portugal."from http://www.carbonoverde.pt/news_detail.php?lang=0&id_channel=2&id_page=2&id=68.

Carmo, A.T. and R.T.A. Prado (1999). *Qualidade do Ar Interno*. São Paulo, Escola Politécnica da USP/Departamento de Engenharia de Construção Civil.

Déoux, S., Ed. (2001). *Ecologia é a Saúde*. Portugal, Instituto Piaget.

European, Environmental, et al. (2005). Environment and Health. E.E. Agency. Report n.º 10.

Freitas, L.C. (2008). *Manual de Segurança e Saúde do Trabalho*. Lisboa, Edições Sílabo.

Garcia, R. (2006). Sobre a Terra - Um guia para quem lê e escreve sobre Ambiente. Lisboa, Público.

Gomes, J.F.P. (2001). *Poluição Atmosférica - Um manual universitário*. Porto, Publindústria.

Massa, A.C.F.A.A. (2010). Auditoria à Qualidade do Ar Interior nos Edifícios da Universidade do Minho em Azurém. *Escola de Engenharia*. Minho, Universidade do Minho. Master: 128.

Matos, J., J. Brantes, et al. (2010). *Qualidade do Ar em Espaços Interiores - Um Guia Técnico*. Amadora, Agência Portuguesa do Ambiente.

Ministério das Obras Públicas, T. e. C. (2006). Certificação Energética e Qualidade do Ar nos Edifícios. Lisboa, Dário da República.

Ministério das Obras Públicas, T. e. C. (2006). Regulamento dos Sistemas de Climatização em Edifícios. Lisboa, Diário da República. 67: 2416.

Ministério das Obras Públicas, T. e. C. (2006). Regulamento dos Sistemas Energétios de Climatização em Edifícios. Lisboa, Dário da República. I Série A: 2416.

Pinto, M., V. Freitas, et al. (2007). "Eficiência Energética nos Edifícios. Qualidade do Ambiente Interior em edifícios de Habitação." *Engenharia e Vida*(38): 43.

Piteira, C. (2007). *A qualidade do Ar Interior em Instalações Hospitalares*. Lisboa, LIDEL - Edições Técnicas.

RSECE (2009). Metodologias para auditorias periódicas da QAI em edifícios de serviços existentes no âmbito do RSECE. *Nota Técnica NT-SCE−02*. RSECE.

Santos, J.P.C.M. (2008). Avaliação Experimental dos Níveis de Qualidade do Ar Interior em Quartos de Dormir Lisboa, Faculdade de Ciência e Tecnologia da Universidade Nova de Lisboa. Phd.

Silva, F.C. (2005). "Síndrome dos Edifícios Doentes." Retrieved 30 April 2012, 2012, from http://saudepublica.web.pt/05-PromocaoSaude/054-SOcupacional/SED.htm.

Sodré, E.D., S.M. Corrêa, et al. (2008). "Principais carbonilas no ar de locais públicos no Rio de Janeiro " *Quim. Nova* 31(2): 249–253.

Universidade., U.R.d. "Síndroma do Edifício Doente." Retrieved 27 Maio de 2012, 2012, from http://biblioteca.universia.net/html_bura/ficha/params/id/37237011.html.

Occupational Safety and Hygiene – Arezes et al. (eds)
© *2013 Taylor & Francis Group, London, ISBN 978-1-138-00047-6*

Bioaerosols exposure in hospital context

Cláudia Vieira
Centro Hospitalar São João, E.P.E, Alameda Prof. Hernâni Monteiro, Porto, Portugal

J. Santos Baptista
PROA/CIGAR/LABIOMEP/Faculdade de Engenharia, Universidade do Porto, Porto, Portugal

ABSTRACT: Bioaerosols are important sources of infection in hospital settings. In this paper was performed a short bibliographic review about the importance of this kind of contamination source for health care personnel. It was carried out a search in several scientific data bases. Were selected 39 papers with peers review with special interest for this problem. It was not found an extensive knowledge on the subject which leaves open an enormous research potential and may explain the lack of proper regulation and the difficulty of implementing effective mitigating measures.

1 INTRODUCTION

Biological agents are important sources of infection transmission in the hospital settings (Rezayee et al., 2011). They represent a significant role in the definition of indoor air quality (Camacho, 2010) with implications in different areas.

Aerosols have direct effects, not only on the occupational healthiness of the health professionals, but also on public health and ecology. Through its eventual global transmission, it is possible that humans face increasing threats of infection caused by aerosols of biological agents. The recent episodes of infections caused by H1N1 virus, anthrax attacks after the disaster at the World Trade Center and the potential danger of avian flu pandemic, triggered a remarkable progress achieved in this area (Xu et al., 2011). In the current economic context, the associated costs to occupational diseases should be considered by organizations, particularly in the health sector. Furthermore, a poor indoor air quality can be associated with loss of productivity or absence of professionals. The existing bioaerosols in hospital settings come from several processes, including through breathing, coughing, sneezing, intubation, noninvasive ventilation, medication administration by nebulization and surgeries (Fang et al., 2008; Quadros, 2008) and through HVAC systems (heating, ventilation and air conditioning) (Silva, 2008). Some pathogenic biological agents can be spread through the air as, for instance, *Legionella spp. Mycobacterium tuberculosis, Staphylococcusspp, allergens Penicillium spp Alternaria.spp. Bacillus subtilis, Bacillus cereus*, and *Actinomyces spp*. and *Aspergillus fumigatus* fungal agents (Camacho, 2010; Chen et al., 2012). Injuries resulting from inhalation of biological agents have direct repercussions on health and can lead to disability or reduced work capacity and are associated with a range of adverse health effects, including infectious diseases, acute toxic effects, allergies, asthma, inflammatory lung disease and cancer (Ji et al., 2007). The presence of aerosols can also compromise the health and comfort of patients and may lead to the emergence of nosocomial infections and interfere with recovery time. Therefore, it becomes essential to have multidisciplinary teams with different experiences, but with common interests in aerosol science, such as environmental engineers, biomedical engineers, epidemiologists, microbiologists, chemists, physicists and others to link each adverse health disease and eliminate and prevent disease outbreaks (Xu et al., 2011; Srikanth et al., 2008).

The risk assessment for health professionals inherent to aerosols inhalation, although significant (Singh, 2009), is difficult to relate by the few data available on the real workers exposure. A real knowledge about this reality is only possible by measuring the exposure and knowing dose-response relationships (Schlosser et al., 2008), whose is scarce (Gehanno et al., 2009). Traditionally the prevention and control of aerosols exposure have three levels of intervention: the elimination of the source or, when that is not possible, it should be necessary to monitor and control the exposure and the source (Xu et al., 2011). Therefore and to minimize the presence of bioaerosols it is recommended to implement the following measures: humidity control, equipment maintenance, promotion of natural ventilation, placement of exhaust air filters (HEPA), use of disinfectants to decontaminate the air and surfaces (Vaquero et al., 2003; Srikanth et al., 2008), keep the doors closed and restricting access (Vaquero et al., 2003). Another measure is the development of contingency plans and monitoring programs of health professionals (Vaquero et al., 2003). The use of personal protective equipment available is critical, such as airway protection through protective face masks or surgical masks for respiratory protection (Eisenkraft et al., 2002; Vaquero et al., 2003).

In this context, this paper aims to conduct a literature review on the identification and characterization of exposure of health workers to biological agents transmitted by aerosols in the hospital, in order to identify knowledge gaps on this subject.

2 MATERIALS AND METHOD

The survey was developed through a systematic review using keywords in search browser "MetaLib" of "Exlibris", Google scholar, as well as in scientific journals, theses and dissertations. Were used several combinations of keywords, including "Biological Agents"; "Aerosols", "Hospitals", "Bioaerosols". The emphasis has been given to articles with publication with date greater than or equal to 2000. The exclusion criteria used were the absence of reference biological agents transmitted by aerosols. The application of the above keywords on mentioned conditions, yielded a total of 352 papers. From all of them, 185 papers resulted from databases and scientific journals 167. From all these 352 papers, 313 were excluded, 58 for publication before 2000, 7 were repeated and 248 articles did not address the topic under study. The reasons for exclusion included the fact that the papers address the issue of exposure to bioaerosols in other sectors. The same combinations of keywords were applied in theses and dissertations not have being obtained any study.

A large proportion of papers studied the exposure to bioaerosols associated with bioterror attacks and potential consequences for public health.

According to the inclusion criteria were selected 39 articles for characterizing exposure to bioaerosols in hospital settings. On these remaining selected papers, were analysed the bibliographic references trying to found complementary relevant information and the most cited authors in order to proceed with supplementary search.

3 RESULTS AND DISCUSSION

From the literature review conducted on the identification and characterization of health workers exposure to biological agents transmitted by aerosols in the hospitals, it was found that there are many published papers addressing the topic. However, from the 39 paper, only 6 were carried out in Portugal is scarce. Most articles searched effected the national evaluation and characterization of microbiological air, four hospitals in context, including two operating rooms and a study in the school environment. It was also found an article in assessing the risk of exposure to biological agents in healthcare institutions. The number of studies carried out in Portugal is scarce, particularly regarding the study of atmospheric concentration and diversity of microorganisms in the air in hospital setting (Smith, 2009). The review highlights the fact that the health care services represent a unique set of microflora inside as bioaerosols, which can be a source of infection for healthcare professionals and patients, may cause nosocomial infections and occupational risks (Srikanth et al., 2008). The dispersion and transport of aerosols (droplets or solid particles) is determined by their physical properties (Camacho, 2010) and by many and complex physical processes, including atmospheric conditions (Blanty et al., 2011). These exposures can have consequences for health care professionals in hospitals. These agents can cause allergies and simple to severe respiratory diseases (Ji et al., 2007). The understanding of the main sources of aerosols, their influence on indoor air and potential threat to health professionals is relevant in determining strategies for infection control (Singh, 2009).

The authors emphasize the different methodologies available and used for bioaerosol monitoring by the control services/departments (Srikanth et al., 2008; Tables, 2008, Xu et al., 2011). The application of particular methods depends on different factors, including greater or lesser concentration of bacteria or fungi. For example in places with higher concentrations of bacteria and fungi may be used active sampling techniques, such as filter methods and impinger (Srikanth et al., 2008). As sample active methods can be used the aerosol impaction (Quadros, 2008) and filtration (Srikanth et al., 2008; Xu et al., 2011). Several papers have mentioned the use of ultraviolet laser-induced-fluorescence to detect bioaerosols, in order to measure in real time the distribution of bioaerosol size particles (Foot et al. 2004; Cabalo et al., 2004; Sickenberger, 2005; Poldmae et al., 2006; Cabalo et al., 2008).

However, the absence of regulations specifying exposure limit values, criteria and sampling media may hinder the selection of appropriate interpretation methodologies of obtained data.

However, the absence of regulations specifying exposure limit values, criteria and sampling media may hinder the selection of appropriate interpretation methodologies of obtained data (Srikanth et al., 2008). The papers address the need to adopt preventive measures and adequate protection in order to minimize exposure of health care workers to bioaerosols (Eisenkraft et al., 2002; Vaquero et al., 2003; Srikanth et al., 2008; Xu et al., 2011). The absence of specific risk assessment methodologies is also mentioned as a gap hindering the perception of risk and restricting the activities of organizations (Schlosser et al., 2008; Gehanno et al., 2009).

The authors are unanimous on the need for developing strategies to mitigate exposure to bioaerosols both in terms of implementing engineering measures and in the implementation of organizational measures. As an example it is mentioned the ventilation of the workplaces and the definition of restricted areas. Is also mentioned the need of implementation of prevention programs and staff health surveillance as well as the use of appropriate personal protective equipment are also mentioned (Eisenkraft et al., 2002; Vaquero et al., 2003; Srikanth et al., 2008).

4 CONCLUSIONS

The undertaken literature review had shown the importance of the knowledge about biological agents as risk factors for people health, with the ability to cause serious illnesses and epidemics (Miller, 2011). Despite the difficulty in establishing limits for the concentration of aerosols in hospital settings, it is essential to monitor environment and people (Su et al., 2012) and suitable programs development for maintaining a low microbial concentration in these environments and the absence of microorganisms with potential to cause damage to health care professionals. The air monitorization should be recommended in order to define acceptable amounts of bacterial contaminants (Pereira et al., 2005; Sue et al., 2012). In Portugal, specific legislation should be implemented for hospitals and health units (Quadros et al., 2009), as the one that already exist in other countries. The current Portuguese legislation is generic for all kind of buildings, regardless to the activity inside and does not specify standards for air monitoring (Smith, 2009). Despite there are several recommendations, no sampling or analysis methods have been implemented and the relationship between the levels of microorganisms in the atmosphere and infection rates are not clearly defined (Smith, 2009). The maximum reference concentration of microorganisms in the air is too permissive to certain locations inside hospitals, especially for surgical rooms and places for hospitalization of patients with an weakened immunitary system (Camacho, 2010).

REFERENCES

Arruda, V.L. (2009). Estudo da Qualidade Microbiológica do Ar em Ambiente Hospitalar Climatizado e sua Relação como Elemento de Risco para o aumento de Infecções: estudo do caso do Hospital Regional de Araranguá, SC. Dissertação Pós Graduação em Ciências Ambientais. Universidade do Extremo Sul Catarinense. Brasil.

Blatny, J.M.; Fykse, E.M.; Reif, B.A.P.; Andreassen, O.; Skogan, G.; Olsen, J.S.; Waagen, V. (2011). Tranckin pathogenic Biological Agents in Air – A Case Study of the Outbreak of Legionellosis in Norway. Acedido a 09 de Julho de 2012 em http://www.nbcsec.fi/nbc/nbc2009/proceedings/BLATNY.pdf.

Cabalo, J.; Sickenberger, R.; Underwood, W.; Sckenberger, D. (2004). Micro UV Detector. The Internacional Society for Optical Engineering; 73–81.

Cabalo, J.; DeLucia, M.; Goad, A.; Lacis, J.; Narayanan, F.; Sickenberger, D. (2008). Overview of the TAC-BIO detector (Conference Paper). Proceedings of SPIE— The International Society for Optical Engineering.

Camacho, R.A.P. (2010). Detecção de Bactérias no Ar em Ambiente Hospitalar com Recurso a Técnicas Moleculares. Dissertação de Mestrado em Biodiversidade e Conservação. Universidade da Madeira.

Chen, P.S.; Li, C.S. (2005). Sampling Performance for Bioaerosols by Flow Cytometry with Fluorochrome, Aerosol Science and Technology, 39:3, 231–237.

Eisenkraft, A.; Cohen, A.; Krasner, E.; Hourviz (2002). Personal Protection Against Biological Warfare Agents (Review). Harefuah; 105–110.

Fang, M.; Lau, A.PS.; Chan, C.K.; Hung, C.T.; Lee, T.W. (2008). Aerodynamic Properties of Biohazardous aerosols in hospitals. Hong Kong Medicine Journal; 14: 26–28.

Foot, V.J.; Clark, J.M.; Baxter, K.L; Close (2004). Characterising Single Airborne Particles by Fluorescence Emission and Spatial Analysis of Elasttic Scattered Ligth. The Internacional Society for Optical Engineering: 46; 262–299.

Gehanno, J.F.; Louvel, A.; Rysanek, E.; Pestel-Caron, M.; Nouvellon, M.; Kornabis, N.; Touche, S.; Ripault, B.; Buisson-Valles, I.; Sobaszek, A. (2009). Biological risk assessment among healthcare workers. Journal Archives des Maladies Professionnelles et de l'Environnement:70; 36–42.

Ji, J.H.; Bae,G.N.; Yun, S.H.; Jung, J.H.; Noh, H.S.; Kim, S.S. (2007). Evaluation of a Silver Nanoparticle Generator Using a Small Ceramic Heater for Inactivation of S. epidermidis Bioaerosols, Aerosol Science and Technology, 41:8, 786–793.

Monteiro, L.M.R. (2011). Contributos para o Estudo da Exposição dos Trabalhadores a Agentes Biológicos nos Agrupamentos de Centros de Saúde. Tese de Mestrado em Engenharia de Segurança e Higiene Ocupacionais. Faculdade de Engenharia. Universidade do Porto.

Poldmae, A.; Cabalo, J.; Lucia, M.; Narayanan, F.; Strauch, L.; Sickenberger, D. (2006). Biological Aerosol Detection with the Tactical Biologicaç (TAC-BIO) Detector. The Internacional Society for Optical Engineering. Article 63980E.

Quadros, M.E. (2008). Qualidade do Ar em Ambientes Internos Hospitalares: Parâmetros Físico-Químicos e Microbiológicos. Dissertação de Pós Graduação em Engenharia Ambiental. Universidade Federal de Santa Catarina. Brasil.

Quadros, M.E.; Lisboa, H.M.; Oliveira, V.L.; Schirmer, W.N. (2009). Qualidade do Ar em Ambientes Internos Hospitalares: estudo de caso e análise crítica dos padrões actuais. Engenharia Sanitária Ambiental; 14(3): 431–438.

Rezayee, A.; Ramin M.; Ghanizadeh Gh.; Valipour F. (2011). Designing of bioaerosol production system for removing Escherichia coli from contaminated air using bone char. Journal of Military Medicine: 13(2); 89–95.

Schlosser, O.; Huyard, A. (2008). Bioaerosols in composting plants: Occupational exposure and health (Review). Environnement, Risques et Sante; 37–45.

Sickenberger, D. (2005). Trend Towards Low Cost, Low Power, Ultra Violet (UV) based Biological Agent Detectors. The Internacional Society for Optical Engineering. Article 59940I.

Silva, E.R.S.S. (2008). Avaliação microbiológica do ar em ambiente hospitalar. Dissertação de Mestrado em Microbiologia. Universidade de Aveiro.

Singh, T.S.; Mabe, O.D.(2009). Occupational exposure to endotoxin from contaminated dental unit waterlines (Review). Journal of the South African Dental Association; 8, 10–12, 14.

Srinkanth, P.; Sudharsanam, S.; Steinberg, R. (2008). Bio-aerosols in indoor environment: composition, health effects and analysis. Indian Journal of Medical Microbiology. 302–312.

Soares, I.C.M. (2009). Aeromicologia Hospitalar. Dissertação de Mestrado em Biologia Molecular e Celular. Universidade de Aveiro.

Su, W.C.; Tolchinsky, A.D.; Sigaev, V.; Cheng, Y.S. (2012). A wind tunnel test of newly developed personal bioaerosol samplers. Journal of the Air & Waste Management Association: 32(7); 828–837.

Vaquero, M; Gómez, P.; Romero, M.; Casal, M.J. (2003). Investigation of Biological Risk in Mycobacteriology Laboratories: a multicentre study. Internacional Journal Tuberculosis Lung Disorders: 7(9); 879–885.

Xu, Z.; Wu,Y.; Shen, F.; Chen, Q.; Tan M.; Yao M. (2011). Bioaerosol Science,Technology, and Engineering: Past, Present, and Future, Aerosol Science and Technology: 45:11; 1337–1349.

Occupational Safety and Hygiene – Arezes et al. (eds)
© *2013 Taylor & Francis Group, London, ISBN 978-1-138-00047-6*

Electromagnetic radiation: Risk perception among workers of telecommunications companies

A. Ribeiro, J. Almeida, M. Pinto, J. Figueiredo & A. Ferreira
College of Health Technology of Coimbra, Portugal

ABSTRACT: In our daily life we are exposed to various radiation sources. The workers who maintain telecommunication base stations are an example of occupational exposure to radio frequency. Therefore, the aim of this study is to assess risk perception on Telecommunication Workers. The study was constituted by two stages: the first one was the administration of a questionnaire, and the second consisted on processing and analyzing workers exposure values to electromagnetic radiation. The statistical analysis was performed using the following tests: Kruskal-Wallis test, Mann-Whitney U test and Student's t-test. From this study it can be concluded that the risk perception varies with the working sector and the qualifications of the respondents. The training may also influence this perception, yet it was not found any statistically significant relation. To fill the gaps in knowledge about electromagnetic fields it's necessary to invest in training and information.

1 INTRODUCTION

Since the beginning of mankind, Man has been surrounded by electromagnetic fields (EMF) (DGS, 2007). Initially those were natural, caused by the sun (for example), however with advancing technology and the changes of social and working habits, there has been a substantial increase of exposure to various Man-made radiations sources (Sebastião et al., 2009).

The electromagnetic spectrum can be divided into ionizing radiation and non-ionizing radiation. The difference between them is the quantum of transported energy: non-ionizing radiation by having a lower frequency wave has not the required energy to break atoms bounds (Vargas, 2004). Each part of the electromagnetic spectrum has different applications on our life. It ranges from high-voltage lines up to X and Gamma rays, passing through microwave, infrared and ultraviolet radiation, visible light and radio frequencies (RF) (DGS, 2007).

RF emits non-ionizing radiation (NIR) and occupies frequencies between 3 kHz to 300 GHz in the electromagnetic spectrum. One of its main applications is the studied sector, i.e., telecommunications (monIT, 2002). These have great impact on today's society as they allow users to communicate without location restrictions. The geographic distribution of base station, forms a relatively regular net. This happens because mobile communication systems work in a bidirectional way, and cell phones have a limited range (monIT, 2002). So, sometimes it's necessary to increase the number of antennas to have a better cell signal reception area (APRITEL, 2007). This number increase, generally leads to a decrease of the power of each antenna (Vargas, 2004).

In telecommunications industry, technicians and maintenance workers, often have to work close to base stations where EMFs are strongest, becoming necessary to assess the occupational exposure to EMF (Kos et al., 2011). For the measurement of personal exposure to EMF, several methods have been used including: digital simulation, on-site measurements or using dosimeters. The last one increasingly used since it provides continuous data about exposure (Chauvin et al., 2009).

The parameter used to characterize the radiation absorbed by the body is Specific Absorption Rate (SAR) which represents the measure of absorbed power per mass of tissue. However SAR cannot be measured in the human body non-invasively, so the most common practice is to use a model, as realistic as possible, of the human body (Kos et al., 2011).

After gathering exposure data, it shall be compared with the limits set by several international organizations. The safety limits are the maximum permissible levels for radiation absorbed by the human body (monIt, 2010).

For radiofrequencies the safety limits are established for SAR which cannot be measured non-invasively. Due to this constraint, limits for some easily measurable electromagnetic quantities outside the body, were also set and designated by reference levels (Sebastião et al., 2009).

In 1988, the International Committee on Non-Ionizing Radiation published exposure limits to electromagnetic radiation for global application which provided the basis for the current International Commission on Non-Ionizing Radiation Protection (ICNIRP) guidelines (Chauvin et al., 2009). At European level there are two documents regarding EMF exposure: Recommendation No. 1999/519/CE and Directive 2004/40/CE.

The Recommendation No. 1999/519/CE was transposed into national law by the Ordinance No. 1421/2004 from November 23rd, which applies to population's EMFs exposure. The occupational reference levels are established by Directive 2004/40/CE, and should have been be transposed until 30 April 2008. However, this period was twice amended by Directive 2008/46/CE (which established the deadline until April 30, 2012) and by Directive 2012/11/UE which established the current deadline (October 31, 2013).

Currently, the possible effects of EMR on workers' health are still being discussed. There are various symptoms associated with exposure to NIR including, feelings of dizziness, nausea, headache (Sebastião et al., 2009) and tissue heating, being this last one, the only proven biological effect (monIT, 2002). Although these symptoms are temporary, they can have major impacts on the workers' safety leading to work accidents (Sebastião et al., 2009). This study focused on the evaluation of risk perception and occupational exposure to EMFs.

2 MATERIAL AND METHODS

The study was conducted between October 2011 and September 2012, and data collection took place during the month of May, 2012. The universe was constituted by telecommunications companies, and the target population was the employees of these companies. The sample consisted of two companies: Company A and Company B. The type of sampling was not probabilistic, and the sampling technique was by convenience.

The study had two stages: the first one was the administration of a self-administered questionnaire; the second consisted on processing and analyzing workers' exposure values to electromagnetic radiation.

At the first stage of this study the sample consisted of 25 employees from Company A. At second stage the sample was constituted by 14 exposure measurements of both companies (A and B).

The questionnaire applied to Company A, aimed to assess workers' risk perception. It was composed by 12 questions and was divided into three main parts: sociobiographical data, exposure characterization and knowledge/safety culture evaluation. Sociobiographical data allowed identification of workers' age average, gender, qualifications and their work sectors. Regarding exposure characterization, it was intended to assess workers' perception against various risk situations. Each question owned six answer choices: 0 (no opinion) 1 (no risk) 2 (low risk) 3 (moderate risk), 4 (high risk), 5 (very high risk).

To evaluate knowledge and safety culture, each question was classified as: yes, no, no opinion. It has been attributed to each question, one point. The variable of the total points was subsequently adjusted to the percentage.

For the analysis of electromagnetic radiation exposure values, permission to access to data on workers' occupational exposure was requested. Both companies provided some measurements, of which 14 were selected since they contemplated the time of climb and descent of workers. According to the information given in the selected measurements databases, companies used the equipment EME Guard brand SATIMO. The reference value used was based on Directive 2004/40/CE.

Data was statistically treated with IBM SPSS Statistics 19.0. Simple descriptive statistics were used: mean and standard deviation; non parametric tests for independent samples: Kruskal-Wallis test, Mann-Whitney U test, and Student's t-test. Statistical tests interpretation was based on a significance level of $\alpha = 0.05$ with a confidence interval (CI) of 95%. Towards a significant α ($\alpha \leq 0.05$), differences or associations between groups were observed, however for $\alpha > 0.05$, these differences or associations were not considered statistically significant.

3 RESULTS

The sample was composed by 25 respondents (11 females and 14 males), with an age average of 37 years. Regarding workers' qualifications, it was found that a large number of respondents (10 respondents) had higher education. The sector which presented the highest number of workers was the production sector (10 respondents). In relation to employment contracts, 15 respondents claimed to have a permanent contract (Tab. 1).

Regarding exposure characterization, statistically significant differences were observed between the risk of "Exposure to NIR" and the risk of "Fall height". The workers from the telecommunications sector were the ones who considered to be exposed to higher risk levels. As for the remaining listed risks ("Exposure to cold/heat" and "Exposure to adverse weather conditions"), there were not found

Table 1. Respondents sociobiographical characterization.

		Gender (N)		
		Female	Male	Total
Qualifications	4th grade	0	4	4
	6th grade	0	1	1
	9th grade	0	2	2
	12th grade	2	6	8
	Higher education	9	1	10
	Total	11	14	25
Work sector	Telecommunications	3	3	6
	Production	2	8	10
	Health and safety	1	1	2
	Administrative	4	2	6
	Total	10	14	24
Employment contract	Permanent	9	6	15
	Fixed-term contract	0	8	8
	Others	2	0	2
	Total	11	14	25
Work schedule	Flexible	2	1	3
	Rigid	6	12	18
	Schedule exemption	3	1	4
	Total	11	14	25

Table 2. Knowledge by work sector and qualifications.

		Total of knowledge %		
		N	Mean	Std. Dev
Work sector p-value = 0,036*	Telecommunications	6	77,78	16,15
	Production	10	50,00	29,69
	Health and safety	2	90,00	4,71
	Administrative	6	42,22	22,57
	Total	24	58,33	28,23
Qualifications p-value = 0,041*	4th grade	4	25,00	3,33
	6th grade	1	6,67	–
	9th grade	2	60,00	9,43
	12th grade	8	64,17	25,93
	Higher education	10	70,00	23,78
	Total	25	57,60	27,88

Kruskal-Wallis test | * p-value ≤ 0,05.

statistically significant differences. It was verified, however, that the former situation is maintained, i.e., workers from the telecommunication and production sectors considered themselves to be exposed to higher levels of risk compared to workers from the remaining sectors.

The level of awareness of respondents was evaluated through a series of questions about EMF and safety culture. The answers provided by respondents showed that, in general, the population has presented a reasonable level of knowledge since: 19 respondents knew that in their work they are exposed to RF, 14 said they can recognize an exclusion zone, 14 stated not to work towards the directive gain of an antenna and 14 knew that EMF levels vary with the distance and the traffic of the antenna.

For a more detailed assessment of respondents' knowledge level, tests to check for significant differences between the level knowledge (%) and other variables (work sector, qualifications, gender, employment contract, work schedule and training) were performed. Statistically significant differences (p-value < 0.05) were only observed between work sector and qualifications (Tab. 2).

The workers who showed a higher knowledge level were those that work at the sectors of Health and Safety (90%) and Telecommunications (77.78%). Regarding qualifications, respondents with higher levels of education were the ones that had higher means of knowledge, particularly

those with higher education (70%) and 12th grade (64.17%).

The remaining variables (gender, employment contract, work schedule and training), did not present any statistically significant differences (p-value > 0.05). However, it must be noted that respondents who reported having attended to some kind of training, showed a higher knowledge level (62.22%) compared to the other. Despite of not having been verified any statistically significant differences, the comparison between the knowledge level and gender revealed that female respondents were more enlightened in matters of safety culture (61.82%) compared to male respondents (54.29%).

Regarding self-reported symptoms, 28% of respondents reported having felt symptoms, related to their activity. Headaches and nauseas were the most reported symptoms.

The assessment of workers' individual exposure to RF consisted of data provided by two telecommunications companies (company A and company B). It was observed that the mean exposure values to electric field and power density of both companies showed no significant differences (Tab. 3). However, Company B presents a slightly higher mean value of exposure.

The reference value recommended by Directive No. 2004/40/CE is 50 W/m². As can be seen, the values did not exceed the limit. The exposure values average of both companies were very similar, there was however, large values dispersion.

4 DISCUSSION

The concept of risk has different meanings for different people (Slovic, 2010), so the risk perception

407

Table 3. Comparison between the mean exposure values of workers from Company A and Company B.

	Company	N	Mean	Std. Dev
Electric field (V/m)	A	14	5,5262	9,78337
p-value = 0,922	B	14	6,8037	9,43524
Power density (W/m²)	A	14	0,3351	0,82504
p-value = 0,378	B	14	0,3572	0,77489

Studen's t-test | p-value ≤ 0,05.

also varies from person to person. This perception is the apprehension or opinion of the likelihood of risks associated with performing a certain activity, and of the consequences that may result (Nance, 2005; NIOSH, 2008).

Some of the workers surveyed, in particular those who work in the sectors of Telecommunications (who perform the maintenance of the antennas) and Production (performing activities in construction), considered to be exposed to a very high risk of falling from height. Indeed, given the developed activity, this is a situation which is likely to occur (since these workers often perform work at height) and can have very serious implications.

There are several factors that influence risk perception, namely personal experience, risk awareness (Nance, 2005) and psychological determinants (Antonucci et al., 2010; Kleef et al., 2010). Another factor that is also very important face to risk perception is knowledge. If there is not a sufficient knowledge level that allows understanding the risk, it may be over or underestimated (Nance, 2005; Antonucci et al., 2010). Studies show that men generally tend to assess the risks in a less problematic way than women (Slovic, 2010), which proves that the gender is also an influential factor in risk perception. Indeed, this situation was observed in this study, since it were women that showed a higher level of safety culture, proving to be more careful when it comes to risky situations.

Over the past years it has been recognized that the perception of the existing risks in the workplace influences safety behaviors (NIOHS, 2008; Stewart-Taylor et al., 1998). One of the matters that needs attention in this activity is workers' health face to NIR. Even though telecommunications technicians are aware that they should never work in the exclusion zone (zone where radiation levels exceed the limits), it's not always possible for them to remain totally apart from these areas.

As noted, there were no overexposure cases during measurements analysis, however, the high values dispersion may be due to the fact that the electromagnetic field varies with distance from the antenna. Regarding the issue of exposure, it must be taken into account that the used dosimeter did only measured instantaneous exposure, not evaluating the levels accumulation throughout a working day.

Prolonged exposure is worrying as it may lead to adverse effects on workers' health. This is a controversial issue since there are several studies claiming that there's a relationship between exposure to radio frequencies and tumors (Baldi et al., 2010; Forssén et al., 2004) yet, in none of them are found significant associations or consistent relationships with exposure doses or times (Forssén et al., 2004). One of the biological proven effects of NIR is tissues heating so, what needs to be understood is to what extent this slight warming (less than 1 °C) may affect the balance of human body's different systems, at long exposures (monIT, 2002).

To protect workers from radiation there are several actions that can be taken such as: workers' training; minimize the sites power; or shut down sites in maintenance (which is not always possible). Workers information, particularly with regard to work practices or repair on the back of the antennas, is also important. Usually antennas are directional, so the exposure levels will be lower in their back (Sebastião et al., 2009).

Training presents itself also as an important topic, because workers can learn to recognize potential risk situations and respective preventive measures (Mild et al., 2009). In this study, it was observed that some workers answered questions with "no opinion", which may represent unawareness. The question which stated "I understand the information on health and safety at work" had 21 respondents answering "No". This situation reveals itself as worrying because if workers don't understand the information related to their safety, they can put themselves into potentially harmful situations. It is presented however, a curious question: it is stated that health and safety information are not understood, but the same number of respondents said they always work safely.

Therefore, it is important that all workers, without exception, know and understand the information about health and safety. The employer and the responsible for health and safety need to ensure the non-existence of doubts about this issue. It is also necessary that workers understand the importance of communicating accidents or perceived hazards during the development of their work. This knowledge can be transmitted through training.

5 CONCLUSIONS

Gaps in knowledge about EMF highlight the need for more and better training of workers. From this study several conclusions can be drawn, namely that the work sector and the qualifications of the respondents influence risk perception. Training is a factor that may also influence the risk perception, although there were not found any statistically significant results. Regarding workers' exposure measurements to NIR, it can be concluded that the levels of radiation to which workers of companies A and B are exposed, did not exceed the established levels. However it's important to note that the presented values are instantaneous and do not represent the cumulative exposure at the end of a workday.

The study had some limitations as: the impossibility of applying the questionnaire to both companies; the small size of the sample; the lack of a defined methodology for conducting measurements of individual exposure; the impossibility to go on field to perform individual exposure measurements having to use existing measurements. The used dosimeter, also presents the limitation of collecting only instantaneous data. It does not measure an exposure accumulation that would allow to state whether radiation levels to which workers are exposed exceeds, or not, the established reference values at the end of a workday .

In future research it would be interesting to continue the study, involving a larger number of companies in this business, to investigate whether there are or not significant differences between them. It could also be developed a study on accidents and their causes in this area, or even study the influence that psychological determinants have on these workers risk exposure.

REFERENCES

Antonucci, A., Giampaolo, L., Zhang, Q., Sicilianoc, E., D'abruzzo, C., Niu, Q. & Boscolo, P. 2010. Safety in construction yards: perception of occupational risk by italian building workers. *European Journal of Inflammation* 2(8): 107–115.

Associação dos Operadores de Telecomunicações (APRITEL). 2007. *Campos Electromagnéticos de Comunicações Móveis—Posição da APRITEL*. Lisboa.

Baldi, I., Coureau, G., Jaffre, A., Gruber, A., Ducamp, S., Provost, D., Lebailly, P., Vital, A., Loiseau, H. & Salamon, R. 2010. Occupational and residential exposure to electromagnetic fields and risk of brain tumors in adults: a case–control study in Gironde, France. *International Journal of Cancer*, 129: 1477–1484.

Chauvin, S., Gibergues, L., Wuthrich, G., Picard, D., Desreumaux, J. & Bouillet, J. 2009. Occupational exposure to ambient electromagnetic fields of technical operational personnel working for a mobile telephone operator. *Radiation Protection Dosimetry* 136: 185–195.

Direcção Geral da Saúde (DGS). 2007. *Sistemas de Comunicações Móveis—Efeitos na Saúde Humana*. Polarpress, Lda. Lisboa.

Forssén, U., Rutqvist, L., Ahlbom, A. & Feychting, M. 2004. Occupational Magnetic Fields and Female Breast Cancer: A Case-Control Study using Swedish Population Registers and New Exposure Data. *American Journal of Epidemiology*, 161(3): 250–259.

Kleef, E., Fischer, A., Khan, M. & Frewer, L. 2010. Risk and Benefit Perceptions of Mobile Phone and Base Station Technology in Bangladesh. *Risk Analysis* 30(6).

Kos, B., Valic, B., Kotnik, T. & Gajsek, P. 2011. Exposure Assesment in Front of a Multi-Band Base Estation Antenna. *Bioelectromagnetics*, 32: 234–242.

monIT. ABC das OEM: noções básicas. [Online] 2002. Instituto de Telecomunicações, Instituto Superior Técnico. [Citation: 20, April, 2012] http://193.136.221.5/item/info_bas _oem2.htm.

monIt. Glossário. [Online] 2010. [Citation: 25, July, 2012] http://monit.it.pt/index.php?id=31&letter=76.

Mild, K., Alanko, T., Decat, G., Falsaperla, R., Gryz, K., Hietanen, M., Karpowicz, J., Rossi, P. & Sandström, M. 2009. Exposure of Workers to Electromagnetic Fields. A Review of Open Questions on Exposure Assessment Techniques. *International Journal of Occupational Safety and Ergonomics*, 15(1): 3–33.

Nance, P. 2005. Risk Perception. *Encyclopedia of Toxicology*, 2: 743–744.

National Institute for Occupational Health and Safety (NIOSH). 2008. Critical Elements for Contract Worker Risk: A Contractor Safety Initiative. Division of Safety Research, Analysis and Field Evaluations Branch.

Sebastião, D., Ladeira, D., Antunes, M. & Correia, L. 2009. Exposição Ocupacional a Campos Electromagnéticos na banda das Radiofrequências. monIt. Lisboa.

Slovic, P. The Psychology of Risk. 2010, *Revista Saúde e Sociedade*, 19(4): 731–747.

Stewart-Taylor, A. & Cherrie, J. 1998. Does Risk Perception Affect Behaviour and Exposure? A Pilot Study Amongst Asbestos Workers. *The Annals of Occupational Hygiene*, 42(8): 565–569.

Vargas-Marcos, F. 2004. La protección sanitaria frente a los campos electromagnéticos. *Gaceta Sanitaria,* 18: 239–244.

Occupational Safety and Hygiene – Arezes et al. (eds)
© *2013 Taylor & Francis Group, London, ISBN 978-1-138-00047-6*

Thermal comfort evaluation of an operating room through CFD methodology

N.J.O. Rodrigues, S.F.C.F. Teixeira & A.S. Miguel
CGIT, School of Engineering, University of Minho, Guimarães, Portugal

R.F. Oliveira & J.C.F. Teixeira
CT2M, School of Engineering, University of Minho, Guimarães, Portugal

J. Santos Baptista
PROA, CIGAR, LABIOMEP, Faculty of Engineering, University of Porto, Porto, Portugal

ABSTRACT: Although the heating, ventilation and air conditioning systems consume a large amount of energy, they are necessary to provide thermal comfort conditions and good air quality. Specifically in operating rooms, the environmental conditions should be suitable for medical staff performance and patients' safety. This study focuses on the evaluation of the thermal comfort sensation felt by surgeons and nurses, in an orthopaedic surgical room of a Portuguese hospital. Computational fluid dynamic tools were applied for evaluating the Predicted Mean Vote (PMV) index. This enlightened us to the fact that using average ventilation values to calculate the index does not provide a correct and enough descriptive evaluation of the surgical room thermal environment. As reported, surgeons feel slightly hotter than nurses. The nurses feel a slightly cold sensation under the air supply diffuser and their neutral comfort zone is located in the air stagnation zones close to the walls.

1 INTRODUCTION

Operating rooms are exigent places where it is important to set the appropriate air quality patterns as well as the aseptic conditions to ensure the success of the operations. These conditions are obtained using Heating, Ventilation and Air Conditioning systems (HVAC), which despite consuming a large amount of the buildings energy (Macario et al., 1995), they are necessary to prevent the risk of infection to the patient during the surgical procedures, what ensures safety and an appropriate comfort level for both staff and patient. To ensure good settings for all the operating rooms, the parameters are standardised in the number of air changes per hour, temperature range, relative humidity and pressurisation (Balaras et al., 2007, ASHRAE Standard 170P 2006). In a study elaborated by Zwolińska and Bogdan, it was concluded that the surgeon, working with a high metabolic rate, feels uncomfortable after a time (Zwolińska & Bogdan 2012). While the human body tries to metabolically adapt to the environment conditions, the necessary effort will result in thermal discomfort sensation and, therefore, in lower work performance with fatigue and irritability (Zwolińska & Bogdan 2012, Parsons 2002). This means that studies concerning indoor thermal conditions are very important in defining, for instance, the satisfactory comfort temperatures range in health care facilities. Thermal comfort is often assessed by Fanger's model, which is considered a good method for most of the cases. In fact, several authors comprehensively studied the thermal comfort in hospital facilities using this index (Pourshaghaghy & Omidvari 2012, Ho et al., 2009, Memarzadeh & Manning 2000). The Computational Fluid Dynamic (CFD) models are a practical, fast and cost effective way to predict fluid behaviour in complex situations (Versteeg & Malalasekera 1995). These techniques have suffered major developments in recent years and became a low-cost state-of-art tool for the design of efficient HVAC systems. In a study elaborated by Paul Roelofsen (Roelofsen 2011), it was demonstrated that the CFD analyses are a great asset to the ISO 7730 standard, allowing better predicting of the draughts and local gradients that can cause discomfort. In this study, thermal comfort of an operating room is assessed, specifically in an orthopaedic ward of a Portuguese hospital.

A CFD simulation was carried out accounting for surgical lamps' radiation heating, air humidity, airflow patterns and temperature distribution. The post processing of data has allowed the calculation of PMV index values for the entire domain.

2 MATERIAL AND METHODS

2.1 CFD Model

The CFD simulation was elaborated using FLU-ENT™ 14.0, from ANSYS™. In addition to the mass and momentum balance, the simulation includes the energy model which accounts for thermal exchanges between boundaries and fluid, the radiation model, the species model for humidity matters and the turbulence model.

2.1.1 Solver configuration

For the turbulence modelling, a low-Reynolds number model proposed by Menter in 1993 (Menter 1994), the k-ω Shear Stress Transport (SST), was used. It was selected due to its precision and stability for predicting the fluid behaviour close to boundaries and heat transfer surfaces (Zhai et al. 2007). It combines the k-ε and the k-ω models using a blending function and applies them according to their performance. Thus, in the near wall regions, the SST model activates the k-ω model and, for the rest of the flow domain, the k-ε model is selected (Zhai et al. 2007, Stamou & Katsiris 2006). This model was previously used for indoor airflow simulations, obtaining highly correlated results (Stamou & Katsiris 2006).

Regarding the radiation exchange between the surfaces in the operating room, the surface-to-surface (S2S) radiation model was used. To calculate the energy transfer between the surfaces, a geometric function, "view factor", was used. It depends on the surface's area, distance between them, and reflective angle of orientation. This model is described thoroughly elsewhere (ANSYS 2011).

The mixing and transport of chemical species was included in the simulation, by a model that solves the conservation equations which describe the convection and diffusion (ANSYS 2011). The fluid in the domain was considered as a typical mixture of H_2O, O_2 and N_2, being introduced in the room through the supply grille in the ceiling, with different mass fractions of each chemical component, according to their corresponding concentration.

In all simulations the ventilation is considered independent of the thermal conditions. This approximation is acceptable because the inertial forces are greater than the buoyancy forces. By calculating the Archimedes number, the validity of this approach can be accessed (Stavrakakis et al. 2008, Carrilho da Graça et al. 2002).

The simulation was carried out in a steady state, and the boundary conditions were assumed to be constant. The SIMPLE algorithm scheme for the pressure-velocity coupling was used. For spatial discretization of the pressure the Standard scheme was used, while for the momentum, k, ω, species and energy, the Second Order Upwind scheme was selected. Solutions were obtained iteratively and the convergence was accepted when the residuals for pressure, momentum, k, ω were below 1E-4, and for the energy and species were below 1E-6.

2.1.2 Geometry and computational grid

In Figure 1b, the surgical room is represented, which is approximated to a rectangular parallelepiped shape, with L: 6.8 m × W: 5.6 m × H: 2.7 m. Two surgical lamps and an operating table in the centre of the room were also included. The air supply grille in the ceiling (simplified to forty eight rectangular diffusers) and the six exhaust grilles (two different types) dimensions were measured at the hospital. In the model, two rectangular ceiling lamps and an X-Ray board in the wall were also included.

For engineering purposes, some assumptions were implemented in the original geometry. One concerned the shape of the air supply diffuser, which was assumed to be rectangular although the actual air supply in the room had a round type grille mesh. The same simplification was introduced in the exhaust zones. For the model simplification, a smaller number of air supply diffuser entries were also considered, although keeping the same area, for air flux and inertial sake.

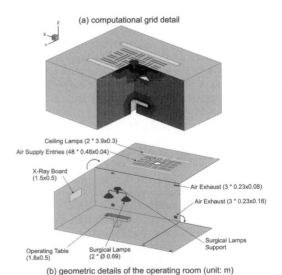

(a) computational grid detail

(b) geometric details of the operating room (unit: m)

Figure 1. Representation of the surgical room computational grid and geometry specifics.

To obtain the computational grid of the domain, the ANSYS™ software was used (Fig. 1a). The domain is fully discretised by tetrahedral elements, using refinements in the most relevant boundary zones for accurate heat transfer calculation. An advanced size function that accounts for the proximity and curvature of the geometry was implemented.

The grid accuracy was tested using three sets of grids with 1.9, 4.2 and 7.4 millions of elements, respectively. Taking the average PMV, calculated as described in section 2.2, as a reference criterion, the finer grid has shown a good convergence ($< 1.2\%$ difference). The apparent order of the Grid Convergence Method (GCI), described elsewhere (Celik et al. 2008), ranging from 0.5 to 40.0, with an average value of 10.8. Oscillatory convergence occurs at 24% of the 40 points selected.

2.1.3 Boundary conditions

The data for the boundary conditions assumed in the computational simulation, was measured using a Brüel & Kjær indoor climate analyzer Type 1213. This is a device with several sensors that allows for the measurement of surface and air temperature, air velocity, relative humidity and radiative temperature in a confined environment.

The air supply was defined as a velocity inlet condition, supplying the fluid at a constant velocity (0.65 m/s) and temperature (21.4 °C), which was considered a mixture of species for humidity (47.0%) assessment purposes. From the experimental data, some ventilation parameters were calculated. The Air Changes per Hour (ACH) was found to be around 20 (ASHRAE Standard 170P 2006), the airflow rate was 2084 m³/h corresponding to 55 m³/m²/h (Balaras et al. 2007). The turbulence at the air supply, defined as turbulence intensity (5.81%) was given as a function of the hydraulic diameter (73.85 mm) of the air supply grille. A pressure outlet condition on the air exhaust boundary guaranteed a positive pressure (5 Pa) in the room, assuring the ASHRAE standard (ASHRAE Standard 170P 2006). It was also considered that the room walls do not perform a significant role in the heat transfer, being defined as adiabatic although the concrete material properties were applied for radiative calculation purpose. For the lamps in the room (ceiling, X-Ray board and surgical lamps), it was assumed that they had the thermal properties of the glass to simulate more precisely the thermal flux and radiative transfer. With a heat flux value of 260 W/m² for the surgical lamps, and 50 W/m² for the ceiling and X-Ray board lamps. The material from the surgical lamps support was defined as aluminium and considered adiabatic. For the surface of the operating table, the properties of cloth (cotton) material were assumed once that surface is covered by bed sheets, and treated as adiabatic. All

the used material thermophysical properties were taken from the literature.

2.2 PMV-PPD index

As previous referred, the PMV-PPD index was used for the thermal comfort evaluation, because it is a well validated, simple and comprehensive method (ISO 7730: 2005). This index is based on an empirical investigation of how people react to differing environments. It characterises the thermal comfort sensation in a subjective scale and is influenced by several variables as the temperature and velocity of the air, relative humidity, metabolism, clothing, and radiative temperature. The scale used for the thermal evaluation is a scale of 7 points, where the zero value corresponds to thermal neutrality/thermal comfort (ISO 7730: 2005).

For the present calculation, the PMV model was implemented through a script coded in Python programming language as a subroutine for post-processing the CFD results. In this way, a field of PMV values for all the domain nodes is obtained. It provided useful information related to the different zones of thermal comfort inside the operating room. In this study, two different cases (surgeon and nurse) were assumed, each with its own values of metabolic rate (MET) and clothing insulation (CLO). For the surgeon was considered a metabolic rate of 1.93 MET and a clothing insulation value of 0.95 CLO, while the nurse had a 1.32 MET and 0.6 CLO. These values where obtained from a survey made on the personnel working on this operating room, subsequently validated by direct comparison with some more accurate data from the literature (Konarska et al. 2007). The surgeon was dressed using a long sleeve shirt and a surgical apron F-1, while the nurse was using a simple eagle scrub, corresponding, respectively, to C and A cloth configuration presented in the work of Konarska and collaborators.

3 RESULTS AND DISCUSSION

3.1 Thermal comfort assessment

As described in section 2.2, the PMV index is a function of metabolic rate and clothing insulation once the environment conditions (i.e. velocity, temperature and relative humidity) are defined as constant (ISO 7730: 2005). From the present CFD data the average value for these three environment variables was calculated for the entire computational domain, with the following values: velocity 0.124 m/s; temperature 21.873 °C and humidity 45.656%.

The PMV model is implicit and a Python code was implemented combining these average values.

In total 72 combinations of metabolic rate and clothing insulation were made and an iso-contour plot was calculated, as shown in Figure 2. The iso-contour plot for the PMV values (Fig. 2) was obtained by polynomial regression where the grey zone corresponds to neutral comfort PMV values ranging from -0.5 to 0.5. In Figure 2 the results for the combination values of metabolic rate and clothing insulation used for the surgeon and the nurse are also represented.

The surgeon PMV value predicted for the average CFD ventilation conditions of the surgical room is around 1, which means he feels slightly warm. For the nurse, the value is almost 0, which represents a neutral thermal sensation. These results show that there is a clear difference between the thermal sensation experienced by surgeons and nurses inside the same surgical room. The results are in agreement with previous studies in the literature, supporting the idea that surgeons have a hot thermal sensation (Zwolińska & Bogdan 2012).

The metabolic rate value considered for the surgeons in this study may be underestimated because the survey responses corresponded to the orthopaedic surgery service, which is more metabolically exhaustive than any other. There is a lack of good experimental data in the literature for the metabolic rate of orthopaedic surgeon during a procedure, because it is difficult for them to execute their job with a measuring device (i.e. a device that measures the amount of consumed O_2).

3.2 CFD vs Experimental results

Experimental measurements at two locations of the room were made for CFD model validation purposes. Point A is located above and centre of the operating table, and point B close to the floor and the opposite wall from the X-Ray board. Experimental and CFD values for point A and point B are shown in Table 1. The CFD results appear to be in good agreement with the experimental measurements.

3.3 Local PMV results

Using the Python post-processing routine for calculating the PMV value for each grid node, as previously discussed in 2.2, the values are represented in contours at the three partial orthogonal planes, as shown in Figure 3.

Table 1. CFD and experimental data for point A and point B.

		Velocity (m/s)	Temperature (°C)
Point A	Experimental	0.05	21.4
	CFD	0.0877	21.609
Point B	Experimental	0.02	21.5
	CFD	0.0196	21.850

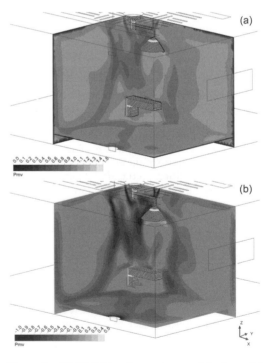

Figure 3. PMV field values represented in 3 partial orthogonal planes for two different metabolic rate and clothing insulation combinations: (a) surgeon configuration, (b) nurse configuration.

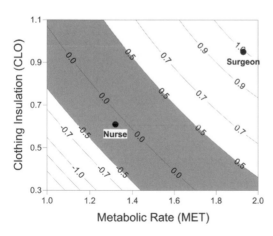

Figure 2. PMV evaluation of the surgical room thermal comfort for several metabolic rates vs clothing insulation.

Using the configurations for the metabolic rate and clothing insulation, described in 2.2, two different PMV fields were obtained, one for the surgeon and other for the nurse, as shown in Figure 3a, b respectively.

The surgeon PMV field (Fig. 3a) goes from neutral thermal sensation of 0 to a slightly warm sensation of 1.5, while the nurse PMV thermal sensation (Fig. 3b) ranges from a-1 (slightly cold) to a maximum of 0.5 (warm/neutral). It is noticeable that for the same domain locations each one has distinct thermal sensations. It is also important to mention that, although the average PMV values for the room indicated 1 for the surgeon and 0 for the nurse, the PMV local value inside the surgical room varies with the location where the person is positioned.

The PMV value for the surgeon and nurse is higher close to the heated surfaces of the lamps (due to the high temperature and low humidity of the air) and lower at the air exhaust locations (caused by high velocity of the air). Although the surgeon seems to only have a neutral thermal sensation close to the room walls (Fig. 3a), the nurse has a cold sensation zone located at room's centre (Fig. 3b), where a draught of air flows from the air supply onto the operating room floor. The nurse neutral thermal sensation zone is located away from the centre of the room, at the stagnation zones, previously referred in 3.2.

4 CONCLUSIONS

The main objective of the present work was the evaluation of the thermal comfort sensation by surgeons and nurses inside a Portuguese orthopaedic surgical room, by analysing the PMV index. This was achieved using CFD tools, saving time and resources. Calculating the PMV for each domain node, enlightened us to the fact that using average ventilation values to calculate the PMV does not provide a correct and enough descriptive evaluation of the surgical room thermal environment.

It is noticeable that surgeons and nurses feel different thermal sensations for the same surgical room, which bring the, already reported, "battle" for the ventilation control between both classes of professionals. For the studied case, the surgeon feels the room environment hotter than the nurse. The nurse feels a slightly cold sensation under the air supply diffuser and a neutral zone located in the air stagnation zones close to the walls. As expected for both cases, the surfaces of the lamps provide an uncomfortable sensation.

For future work, the iso-contour plot for PMV can be used to adequate the clothing insulation of surgeons and nurses alike to fit them in the neutral thermal sensation zone, according to their metabolic rate values. More experimental measurements shall be made to evaluate the correct metabolic rates and ventilation characteristics inside the room. A simulation of the surgical room shall also be made, including representation manikins for surgeons and nurses, to evaluate how thermal comfort changes when more individuals are added to the operating room.

REFERENCES

ANSYS 2011. *ANSYS FLUENT Theory Guide*. Canonsburg, PA, USA: ANSYS Inc.

ASHRAE Standard 170P 2006. *Ventilation of Health Care Facilities*. Atlanta, GA: American Society for Heating, Refrigerating and Air-Conditioning Engineers Inc.

Balaras, C.A. Dascalaki, E. & Gaglia, A. 2007. HVAC and indoor thermal conditions in hospital operating rooms. *Energy and Buildings* 39(4): 454–470.

Carrilho da Graça, G. Chen, Q. Glicksman, L. & Norford, L. 2002. Simulation of wind-driven ventilative cooling systems for an apartment building in Beijing and Shanghai. *Energy and Buildings* 34(1): 1–11.

Celik, I.B. Ghia, U. Roache, P.J. Freitas, C.J. Coleman, H. & Raad, P.E. 2008. Procedure for Estimation and Reporting of Uncertainty Due to Discretization in CFD Applications. *Journal of Fluids Engineering* 130(7): 078001.

Ho, S.H. Rosario, L. & Rahman, M.M. 2009. Three-dimensional analysis for hospital operating room thermal comfort and contaminant removal. *Applied Thermal Engineering* 29(10): 2080–2092.

ISO 7730: 2005. Ergonomics of the thermal environment—Analytical determination and interpretation of thermal comfort using calculation of the PMV and PPD indices and local thermal comfort criteria. Geneva: International Organization for Standardization.

Konarska, M. Sołtynski, K. & Sudoł-szopińska, I. 2007. Comparative Evaluation of Clothing Thermal Insulation Measured on a Thermal Manikin and on Volunteers. *Fibres & Textiles in Eastern Europe* 15(2): 73–79.

Macario, A. Vitez, T.S. Dunn, B.B.A. & McDonald, T. 1995. Where are the costs in perioperative care? Analysis of hospital costs and charges for inpatient surgical care. *Anesthesiology* 83(6): 1138–1144.

Memarzadeh, F. & Manning, A. 2000. Thermal Comfort, Uniformity, and Ventilation Effectiveness in Patient Rooms : Performance Assessment Using Ventilation Indices. *ASHRAE Transactions* 106(2): 748–761.

Menter, F.R. 1994. Two-equation eddy-viscosity turbulence models for engineering applications. *AIAA Journal* 32(8): 1598–1605.

Parsons, K.C. 2002. Human Thermal Environments: The Effects of Hot, Moderate, and Cold Environments on Human Health, Comfort, and Performance. 2nd ed. Taylor & Francis.

Pourshaghaghy, A. & Omidvari, M. 2012. Examination of thermal comfort in a hospital using PMV-PPD model. *Applied ergonomics* 43(6): 1089–95.

Roelofsen, P. 2011. Evaluation of draught in surgical operating theatres: proposed revision to (NEN)-EN-ISO-7730. *Journal of Facilities Management* 9(1): 64–70.

Stamou, A. & Katsiris, I. 2006. Verification of a CFD model for indoor airflow and heat transfer. *Building and Environment* 41(9): 1171–1181.

Stavrakakis, G.M. Koukou, M.K. Vrachopoulos, M.G. & Markatos, N.C. 2008. Natural cross-ventilation in buildings: Building-scale experiments, numerical simulation and thermal comfort evaluation. *Energy and Buildings* 40(9): 1666–1681.

Versteeg, H.K. & Malalasekera, W. 1995. An introduction to computational fluid dynamics: the finite volume method. Harlow, England: Longman.

Zhai, Z.J. Zhang, Z. Zhang, W. & Chen, Q.Y. 2007. Evaluation of Various Turbulence Models in Predicting Airflow and Turbulence in Enclosed Environments by CFD: Part 1—Summary of Prevalent Turbulence Models. *HVAC&R Research* 13(6): 853–870.

Zwolińska, M. & Bogdan, A. 2012. Impact of the medical clothing on the thermal stress of surgeons. *Applied Ergonomics* 43: 1096–1104.

Occupational Safety and Hygiene – Arezes et al. (eds)
© *2013 Taylor & Francis Group, London, ISBN 978-1-138-00047-6*

Vibration exposure in mechanical olive harvesting: Workers' perception

N. Costa, P.M. Arezes & C. Quintas
University of Minho, Guimarães, Portugal

R.B. Melo
CIPER, Human Kinetics Faculty, Technical University of Lisbon, Lisbon, Portugal

ABSTRACT: In manufacturing, forestry and agricultural work, mining and construction, public works, among others, portable tools are used, and expose the hands of the workers to excessive levels of vibration. Currently it is recognized that exposure to harmful Hand-Arm Vibration (HAV) on a regular basis can induce different health problems, especially in the upper limbs. This study aims to contribute to the knowledge of the HAV exposure risks of workers performing seasonal olives picking for oil, especially regarding vibrations arising from the mechanic olive harvester, and how these risks are perceived by the exposed workers. To this end we interviewed 75 workers (33 exposed to HAV) and 11 different models of mechanic olive harvester were assessed in terms of HAV. Twenty two (66%) of the 33 workers exposed to HAV reported symptoms related to exposure and 21% of the evaluated harvesters presented vibration exposure values above the legally prescribed exposure action value. However, only a fraction of the exposed operators (12.1%) perceived a whitening of the fingers during the workday.

1 INTRODUCTION

Portable tools that expose the hands of the workers to vibrations are used in the manufacturing industry, forestry and agricultural work, mining and construction, public works, among others. This type of vibration exposure is called Hand-Arm Vibration (HAV) and it is estimated that about 1.7% to 3.6% of workers in European countries and the United States are exposed to potentially harmful HAV (Buckle & Devereux 1999).

The vascular changes are the most frequent and the most widely studied effects of exposure to HAV. In general, workers exposed to HAV may experience episodes of white or pale fingers. This disorder is a vascular circulatory alteration due to a temporary interruption of blood circulation in the fingers (Griffin 1990). With regard to the possibility of musculoskeletal abnormalities, workers exhibit local pain, inflammation and stiffness in different areas of the upper limbs that may be associated with degeneration of bones and joints. There are also cases of tendinitis, tenosynovitis and carpal tunnel syndrome (Rao 1995).

The olive harvesting has hardly changed over the past centuries. It is a craft hard work that takes place on the outdoor, between December and February, in very low temperatures, as well as on rough terrain and with the adoption of discomforting postures from the ergonomics' point of view. For the collection of olives one can used two different

ways: manually (strongly shaking the branches with sticks) or mechanically (with the aid of machines). The mechanized harvesting of olives covers two distinct types of equipment: the mechanic olive harvester and the telescopic vibrators attached to tractors. The mechanic olive harvester is operated manually and aims to vibrate the olive tree branches so as to facilitate the fall of olives (fig. 1). The mechanic olive harvester has a much lower cost than telescopic vibrators and for that reason is widely used in the district of Bragança.

Figure 1. Example of a mechanic olive harvester.

Considering the potential role of workers' risk perception in their protective behaviour (Arezes and Miguel, 2005 and 2006), this study aimed to contribute to the knowledge of the HAV exposure risk of workers performing the seasonal task of harvesting olives for oil, especially with regard to the perception that these operators have of the effects of such exposure. Thus, it was necessary to collect information on this type of exposure, on the main risk factors associated with it and check the main symptoms reported by operators exposed to high vibrational levels. A characterization of the exposure of these operators, in order to propose measures to minimize the risk of exposure to HAV, was also performed.

2 MATERIALS AND METHOD

This study took place in the district of Bragança, which occupies the extreme Northeast of Portugal and has an area of approximately 6599 km². The primary sector is the second most important sector of activity in the district including nearly 37% of the employed population in this region. Within this sector of activity olive growing is a culture of relevant economic and social importance, which, according to the Agency for the Control of European Aid to olive oil, in the campaign of 1998/99 had enrolled about 8 million trees in this district (INE 2001).

The territorial scope of development of this research comprised six counties of the Bragança District with great representation regarding the production of olive oil. The sample comprised 22 workgroups distributed among the counties of Bragança, Macedo de Cavaleiros, Mirandela, Mogadouro, Vila Flor and Vimioso. The application of this survey to all workers (exposed and unexposed to HAV) was performed in order to allow the differentiation between the symptoms experienced by both groups of workers.

The construction of the questionnaire observed the remarks cited by Foddy (1996), in particular it were considered short and simple questions, avoiding the double negative.

For the measurement of HAV transmitted from the mechanic olive harvester to the operator it was used the technical criteria established by ISO 5349:2002, Parts 1 and 2. In particular, the measurements were made at the entry point of the vibration in both hands, respecting the frequency range of interest, expressed in one-third octave bands. The vibrations transmitted to the hands were measured in the appropriate directions (X_h, Y_h, Z_h) of an orthogonal coordinate system and the acceleration was expressed as frequency weighted equivalent acceleration (A_{hv}, q, T, in m/s²).

Due to the specific nature of the task, five measurements of 16 seconds were performed for each hand of the operators, with the exception of the equipment that was operated with only one hand.

The qualitative and quantitative evaluation of the questionnaire data and the treatment of the vibrational data were performed using the program PASW Statistics (version 19 2010).

3 RESULTS AND DISCUSSION

Of the total of the 75 valid questionnaires, 51 (68%) were answered by men. It was observed that only 33 (44%) of the respondents performed tasks which involved the use of mechanic olive harvester and hence the consequent exposure to HAV. Most respondents (61.3%) had completed the 6th Grade and had no other activity other than agriculture (66.7%).

Twenty six mechanic olive harvesters of 11 different models were evaluated. Models characteristics are specified in table 1. Emphasis should be given to the weight sustained by the operators, while performing the harvesting task, which is relatively high and ranges between 10.8 kg (Model VM60-S6 Active) and 16.4 kg (Cifarelli model SC700). The exception to this situation refers to the model of Pellenc Olivium that was the only electric powered equipment with batteries being transported in a backpack carried by the operator and for that not included on the weight referred.

Unfortunately, it was not possible to determine the power of some of the evaluated models, not even in the equipment manual nor in the manufacturer web page. However, it can be noticed by examination of the values in Table 1 that the power of these machines normally exceeds 2 kW.

Table 1. Resume of the main characteristics of the mechanical olive harvesters.

Brand	Model	kg	Engine (cc)	Power (kW)
Active	VM60-S6	13.8	51.3	2.2
Cifarelli	SC700	16.4	52.0	nd
Cifarelli	SC800	14.9	52.0	nd
Cifarelli	SC105	10.8	52.0	nd
Efco	SO5300ER	14.5	52.5	2.1
Kawasaki	TH48	nd	nd	nd
Pellenc	OLIVIUM	2.7	na	nd
Stihl	SP480	13.9	48.7	2.2
Stihl	SP450	14.5	44.3	2.1
Tekna	TK650	14.4	52.0	nd
Vibroli	M.POWER	15.0	52.0	2.0

na: not applicable.
nd: not determined.

It was verified that 44% of the respondents referenced prior exposure to HAV resulting from the operation of mechanic olive harvesters on previous campaigns. Of the 33 exposed workers 69.7% reported a daily personal exposure up to 4 hours, which is the most referenced exposure time (42.4%). These 4 hours daily exposure values were used to determine the personal daily exposure to vibrations, A(8) column in table 2.

Table 2 also shows that the mean A_{hv} values obtained for the different evaluated machines ranged from 1.3 m/s² (±1.53 sd) for the Vibroli Maxi-power to the 4.0 m/s² (±1.64 sd) for the Stihl SP480.

During the campaign for the evaluation of the vibrational levels of the mechanic olive harvesters some models were found to have greater representation in relation to others. This fact is probably related to the promotional activity or dissemination of equipment accomplished by the brands representatives, resulting in a difference in the diversity of work situations in which the equipment was assessed. It can be seen in Table 2 that the most popular models are the Cifarelli's SC700 and SC800, followed by the Sthil SP450.

It was also found that some machines were operated mainly with only one hand. Therefore, machines like the Kawasaki TH48, and Pellenc Olivium Vibroli Maxi-Power were evaluated for this particular hand only.

From the analysis of Table 2 it is also possible to verify that there is a relatively large set of machines (36%) transmitting vibration values (A_{hv}, q, T, in m/s²) higher than the legal vibration exposure action level

Table 2. Resume of the A_{hv} values found during the survey (mean ± sd).

Brand	Model	n	Right-H A_{hv} (m/s²)	Left-H	A(8)* m/s²
Active	VM60-S6	3	2.8 ± 1.40	2.9 ± 1.36	2.9
Cifarelli	SC700	6	1.5 ± 0.39	2.3 ± 1.01	2.3
Cifarelli	SC800	5	1.7 ± 0.45	2.5 ± 0.27	2.5
Cifarelli	SC105	1	2.0 ± 0.44	2.4 ± 0.71	2.4
Efco	SO5300ER	2	1.0 ± 0.23	2.4 ± 0.14	2.4
Kawasaki	TH48	1	n.d.	3.7 ± 0.62	3.7**
Pellenc	OLIVIUM	1	n.d.	2.3 ± 0.24	2.3
Stihl	SP480	1	4.0 ± 1.64	2.8 ± 0.59	4.0
Stihl	SP450	4	2.3 ± 0.20	3.0 ± 0.96	3.0
Tekna	TK650	1	2.2 ± 1.04	2.5 ± 0.61	2.5
Vibroli	M.POWER	1	n.d.	1.3 ± 1.53	1.3

*considering the maximum exposure time reported by the operators.
**A(8) value significantly different from the others (p < 0.05).
n.d. not determined.

(2.5 m/s²), considering the time of operation of these devices an eight hours shift.

A_{hv} acceleration values found for the evaluated machines showed some variability, namely for the right and left hands and also within the same machine. In the first case, the asymmetry of the handles on the machines seems to justify the differences.

As for the differences within the same machine, represented in Table 2 by the values of standard deviation (sd), these appear to result from two distinct factors.

One factor is related to the difference in the operation of the machine, verified whenever similar equipment's were operated by different workers. Costa & Arezes (2009) also found that individual characteristics can play an important role concerning the risk of vibration exposure, namely, more experienced and faster workers are exposed to a higher risk of vibration exposure than those with minor experience and less skills.

The second factor results from the task itself, namely from the difference in operating the equipment regarding the height of the branch intended to be harvested, its robustness, and the amount of olives it presents. Joshi et al. (2001) found similar results when evaluating mechanical clamping. In particular these authors found variations in vibration values depending on the posture assumed during the task of clamping. The lower values of vibration were found to hold the position involving tightening the mechanical equipment slightly above the elbow height of the operator, while the highest vibration values were found during the position which involved holding the device with arms raised to shoulder level (Joshi et al., 2001).

In Table 2 it is also possible to observe an apparent difference in the values of A(8) found for the different machines evaluated. However, the t-test performed for different pairs of samples revealed only significant differences (p < 0.05) for the machine Kawasaki TH48 with the A(8) value of 3.7 m/s², when compared with any other machine evaluated. Similar statistical test was undertaken for the Sthill SP480 model because of the 4.0 m/s² A(8) vibration level. However, the p value of 0.722 led to the rejection of the null hypotheses that A(8) values found for this machine were significantly different from the values found for the other machines.

The characterization of the thermal environment performed while evaluating the vibrational levels of the machines shown in Table 2 revealed that the mechanical olive harvesting tasks were performed with air temperature ranging from a minimum of 0.1 °C to a maximum of 12.9 °C (average 6.7 °C) and with globe temperature between 2.2 °C and 21.7 °C (average of 12.4 °C). Low temperatures and the fact that the task is performed outdoors are, according to several authors, risk factors that potentiate the adverse effects of occupational exposure to HAV (Rao 1995, Griffin 1990, Necking 2004, Palmer et al. 2001, Tomida et al. 2000). This is mainly due to the vasoconstriction that occurs at the level of the hands, as a physiologic response to low temperatures (Necking 2004).

The vibrational levels of the evaluated machines and the specific exposure conditions are in accordance with the reported percentage (81.8%) of hands numbness, which is usually associated with the Raynaud syndrome. However, only a fraction of the exposed operators (12.1%) perceived a whitening of the fingers during the workday. Burström and Neely (2006) found that although there are no differences in thresholds of perception between men and women, there are differences in sensitivity to the intensity of the vibrations. It is therefore possible that the reported perceived symptoms associated with occupational exposure to HAV have been influenced by gender differences among the questioned individuals (68% of respondents were male).

An additional question posed to these operators revealed that the white finger disease happened in both hands (75%) and occurred only at the extremity of the fingers (75%). This seems to configure a less advanced stage in the evolution of Raynaud's phenomenon, namely the second stage, characterized by discoloration of one or more fingers, usually confined to numbness at the winter season (Tomida et al. 2000).

4 CONCLUSIONS

It was noticed that during the campaigns of olive harvesting, workers who use mechanic olive harvesters are exposed to HAV levels that exceed, in some cases, the action limit of exposure imposed by law. Confounding factors associated with the task itself, namely low temperature and extreme positions of the arms and torso potentiate the evolution of Raynaud's syndrome.

It was also noticed that the exposed workers have misperceptions of the effects of exposure to HAV. Namely, only a small fraction (12.1%) perceived a whitening of the fingers during the workday but a much larger fraction (81.8%) reported feeling numbness at the hands, during the same period of time.

This misperception can pose a health problem because it means that no remediation or preventive measures are undertaken by these workers during these episodes, like halting the task, warm the hands with massages or protecting them with warmer gloves.

ACKNOWLEDGMENTS

This work was partially funded with Portuguese national funds through FCT—Fundação para a Ciência e a Tecnologia, in the scope of the R&D project PEst-OE/EME/UI0252/2011.

REFERENCES

Arezes, P.M., Miguel, A.S. 2005. Individual Perception of Noise Exposure and Hearing Protection in Industry. *Human Factors*, 47 (4): 683–692.

Arezes, P.M., Miguel, A.S. 2006. Does risk recognition affect workers' hearing protection utilisation rate? *International Journal of Industrial Ergonomics*, 36:1037–1043.

Buckle, P. & Devereux, J. 1999. *Work-related neck and upper limb musculoskeletal disorders*, European Agency for Safety and Health at Work, Luxembourg.

Cifarelli 2012. *News from Cifarelli Spa.* Retrieved June, 29, 2012, from http://www.cifarelli.it/

Costa, N., Arezes, P.M. (2009). The influence of operator driving characteristics in whole-body vibration exposure from electrical fork-lift trucks, International Journal of Industrial Ergonomics, 39(1), 34–38.

Foddy, W. 2006. Como perguntar: Teoria e prática da construção de perguntas em entrevistas e questionários. Oeiras: Celta Editora.

Griffin, M.J. 1990. *Handbook of human vibration*, Academic Press, London.

INE 2001. *Censos 2001, resultados definitivos, Norte.* Instituto Nacional de Estatística, Lisboa.

ISO 5349–1:2001. Mechanical vibration—Measurement and evaluation of human exposure to hand-transmitted vibration—Part 1: General requirements.

ISO 5349–2:2002. Mechanical vibration—Measurement and evaluation of human exposure to hand-transmitted vibration—Part 2: Practical guidance for measurement at the workplace.

Joshi, A., Leu, M. & Murray, S. 2001. *Ergonomic analysis of fastening vibration based on ISO Standard 5349*, Applied Ergonomics, 43, 2012, 1051–1057.

Neely, G. & Burström, L. 2006. *Gender differences in subjective responses to hand-arm vibration*, International Journal of Industrial Ergonomics, 36, 135–140.

Necking, L.E., Lundborg, G., Lundström, R., Thornell, L.E. & Fridén., J. 2004. *Hand muscle pathology after long-term vibration exposure, Journal of Hand Surgery*, 29(5): 431–437.

Palmer, K.T., Griffin, M.J., Syddall, H.E., Pannett, B., Cooper, C. & Coggon, D. 2001. *Exposure to hand-transmitted vibration and pain in the neck and upper limbs*, Occupational Medicine, vol. 51(7): 464–467.

Rao, S.S. 1995. *Mechanical Vibrations, Third Edition*, Addison-Wesley Publishing Company, Reading, Massachusetts.

Tomida, K., Miyai, N., Yamamoto, H., Mirbod, S.M., Wang, T.-K., Sakaguchi, S., Morioka, I. & Miyashita, K. 2000. *A cohort study on Raynaud's phenomenon in workers exposed to low level hand-arm vibration*, Journal of Occupational Health, 42, 292–296.

Occupational Safety and Hygiene – Arezes et al. (eds)
© 2013 Taylor & Francis Group, London, ISBN 978-1-138-00047-6

Risk of exposure to xylene in a pathologic anatomy laboratory

M. Dias-Teixeira
REQUIMTE, Instituto Superior de Engenharia, Instituto Politécnico do Porto, Porto, Portugal
CITS, Centro de Investigação em Tecnologias da Saúde, IPSN—CESPU, CRL, Gandra PRD, Portugal
ISLA, Instituto Superior de Línguas e Administração de Vila Nova de Gaia, Vila Nova de Gaia, Portugal
ISLA, Instituto Superior de Línguas e Administração de Santarém, Santarém, Portugal

R. Rangel
CENCIFOR, Centro de Ciências Forenses, Instituto Nacional de Medicina Legal, I.P., Coimbra, Portugal
INMLCF, I.P., Delegação do Norte do Instituto Nacional de Medicina Legal e Ciências Forenses,
I.P.—Serviço de Toxicologia Forense, Porto, Portugal
CITS, Centro de Investigação em Tecnologias da Saúde, IPSN—CESPU, CRL, Gandra PRD, Portugal

A. Dias-Teixeira
CITS, Centro de Investigação em Tecnologias da Saúde, IPSN—CESPU, CRL, Gandra PRD, Portugal

V. Domingues
REQUIMTE, Instituto Superior de Engenharia, Instituto Politécnico do Porto, Porto, Portugal

S. Abajo Olea
Faculdade de Veterinária, Departamento de Ciências Biomédicas, Universidade de Léon, Léon, Espanha

ABSTRACT: This study intends to assess risk of occupational exposure to xylene in a pathologic anatomy laboratory, of a higher education school. In this study, we estimated xylene in the workplace air using a GC-FID system. Xylene concentrations in laboratory environments were found at the range of 63–174 ppm and varied along the week. The whole average concentration obtained exceeded the Threshold Limit Values—Time Weighted Average (TLV-TWA), established in Decree-Law 24/2012 (50 ppm), and 43.2% of the average concentration was higher to 100 ppm (established in standard NP 1796:2007).

1 INTRODUCTION

Laboratories are inherently potentially dangerous environments and there will be always a level of risk associated with the work there undertaken. In laboratories (where a variety of hazards exist) the workers must be closely supervised at all times. People who work in a histology laboratory and related disciplines are at risk from exposure to risk agents. These risks can be traumatic for individuals, as well as extremely toxic (Buesa, 2007; Vecchio, Sasco and Cann, 2003). A hazardous chemical by the Occupational Safety and Health Administration (OSHA), is a substance that may cause health effects in short—or long-term exposed employees, based on statistically significant evidence from at least one study conducted using established scientific principles (OSHA, 1994). It is certainly a broad definition that applies to all, or almost all of the chemicals typically used in laboratories.

In pathology and histology laboratories, the chemicals more used are xylene, formaldehyde, (the) acids and ethanol, among other toxic substances that easily contaminate the air (Roy, D. R., 1999), as well generate hazardous waste (xylene) (Environmental Protection Agency, 2000).

Commercial xylene is a colourless liquid at normal room temperature and produced from petroleum or coal tar (IPCS, 1997; ATSDR, 2007). Has a sweet odour that exists in three isomers; meta-xylene, ortho-xylene and para-xylene. The odour threshold for xylene is between 0.2–2 ppm. It is an important component of the routine, almost indispensable in laboratories (ATSDR, 2007).

Many industries rely on the use of xylene, including the paint and painting industry, automobile garages, the metal industry and furniture refinishers, as a process chemical in the rubber and leather industries, in the formulation of pesticides, as an interme-diate in the manufacture of certain polymers, in petroleum distillation, and in histology laboratories. So, the occupationally exposed populations to xylene are: paint, varnish, ink, glue manufacturers and applica-tors; pesticide formulators and sprayers; histology laboratory workers; Rubber, leather, petroleum and some other chemical industry workers (IPCS, 1997).

Three men were involved in one situation of a severe exposure, painting a double-bottomed storage tank, with the retrospective estimation of probable, xylene airborne levels was 10.000 ppm (Morley, et al., 1970). The Time-weighted values of the exposure to xylene obtained in the breathing zone air of coke plant work-ers were 0.002–0.18 ppm (Bieniek, G. & Lusiak, A., (2012)). The personal exposure air samples of thirty workers from ten gasoline stations revealed xylene concentrations between 0.06 to 0.57 ppm (Chen, M.L., et al., 2002). The average work-shift exposure of tech-nicians working in the air at the chemical laboratory to xylene was between 0.12 to 0.39 ppm (Panev, T., Pavlova, M, & Tzoneva, M. 1998).

Exposure to xylene can occur via inhalation, ingestion, eye or skin contact, and, to a small extent, by absorption through the skin. However, information on dermal toxicity and skin absorp-tion is limited. Dermal absorption is supported by available literature information, but occurs to a lesser extent than either oral or pulmonary absorp-tion (Environment Agency, 2009).

The mechanism by which xylene produces toxicity is not know. Information on the toxicity of xylenes to humans is almost exclusively limited to case reports of acute exposures and studies of occupational exposures in which persons often inhaled a mixture of hydrocar-bon solvents 8 hours per day, 5–6 days per week. These studies often have incomplete information on the air-borne concentrations of xylene and other hydrocar-bons. Health effects of xylene exposure include eye, nose, and throat irritation, difficulty in breathing, impaired lung function, and nervous system impair-ment. In addition, inhalation of xylenes can adversely affect the nervous system (ATSDR, 2007).

After exposure to about 700 ppm xylene (calcu-lated) for up to one hour, headache, nausea, irrita-tion of the eyes, nose and throat, dizziness, vertigo and vomiting have been reported (Klaucke et al., 1982). The percent of 50 healthy subjects exposed to 100, 200, or 400 ppm mixed xylenes for 30 minutes reporting eye irritation was 56% for controls (clean air), 60% at 100 ppm, 70% at 200 ppm, and 90% at 400 ppm (Hastings, et al.,1984). Ten healthy human volunteers exposed for periods of 3 to 5 minutes to estimated concentrations of 100 or 200 ppm technical grade xylene, reported eye, nose, and throat irritation at 200 ppm but not at 100 ppm (Nelson, et al., 1943).

These data are consistent with a human no observed adverse effect level for eye irritation of about 100 ppm for at least a 30-minute exposure (ATSDR, 2007). The irritation has been chosen as the critical end point because it occurs at low levels after short exposures. Ten male volunteers who were exposed 4 hours to 100 ppm xylene (purity not specified) or 100 ppm toluene or a mixture of 50 ppm of each, during four exposure sessions, revealed that xylene had the most adverse effect on simple reaction time and choice reaction time, while the combined exposure gave weaker effects than xylene alone but stronger than toluene alone (Dudek, et al., 1990). The threshold for adverse effects is 100 ppm, without effects on reaction times seen in 15 vol-unteers exposed to xylenes at 100 and 299 ppm for 70 minutes. However when subjects exercised for the first 30 minutes of the 70 minute exposure to 299 ppm xylenes, reaction times were increased and short term memory was impaired (Gamberale, et al., 1978).

Persons exposed only to xylene vapor do not pose substantial risks of secondary contamination. Persons whose clothing or skin is contaminated with liquid xylene can cause secondary contami-nation by direct contact or through off-gassing vapor. Hospital personnel can be secondarily con-taminated (ATSDR, 2007).

The exposure to xylene is often perceived by the probationers in pathological anatomy, cytological and thanatological as a source of problems for the health (Teixeira, M., et al., 2009).

So, the exposure to xylene in pathological anat-omy labs and the opportunities for effective con-trol should be monitored and the results should be compare with reference values. The workplace exposure limits, they are guidelines designed for use by industrial hygienists in making decisions regard-ing safe levels of exposure to various chemical sub-stances found in the workplace (ACGIH, 2012).

The Portuguese Economy and Employment Ministries (Ministério da Economia e do Emprego of Portugal), through the Decree-Law 24/2012 recommended exposure limits (RELs) for xylene (are) to be 50 ppm (221 mg/m³) as a Time Weighted Average (TWA) for a normal 8-hour workday and a 40-hour workweek and 100 ppm (442 mg/m³) for periods not to exceed 15 minutes as a short-term limit (STEL) (Ministério da Economia e do Emprego, 2012). The NP 1796/2007 has assigned xylene a threshold limit value (TLV) of 100 ppm (435 mg/m³) as a TWA for a normal 8-hour workday and a 40-hour workweek and a STEL of 150 ppm (655 mg/m³) for periods not to exceed 15 minutes. The NP 1796/2007 limit(s) is based on (xylene's) the potential of xylene to cause central nervous system depression and respiratory and ocular irritation (Instituto Português da Qualidade (IPQ), 2007).

The above exposure limits are for air lev-els only. When skin contact also occurs, may be overexposed, even though air levels are less than the limits listed above.

This study intends to assess risk of occupational exposure to xylene in a pathologic anatomy laboratory, of a higher education school.

2 MATERIALS AND METHOD

2.1 Air sampling of pathological anatomy laboratory

Air sampling was performed from October 2008 to January 2009. Air samples (111) were collected using activated charcoal cartridges, between 9:00 am and 5:00 pm.

Personal sampling was performed by passive diffusive samplers, placed in the breathing zone of the worker (a graduate in Pathological Anatomy). Three samplings were held per day (15 minutes, each), considering sufficiently representative samples, to allow the calculation of 8-h TWA exposures. The sampling strategies general were: sample directly exposed workers; take sufficient samples to allow the 8-h TWA exposures to xylene to be estimated; and use active sampling.

Environmental sampling was carried out by active sampling tubes (SKC—Anasorb® CSC: coconut charcoal, 20/40 mesh, 50/100 mg), with the aspiration flow fixed at 0.13 liters/minute, according to a NIOSH analytical method 1501 for aromatic hydrocarbons. All activities (histological staining, slide mounting and chemical waste disposal) with manipulation of xylene, were developed with a ventilation system.

2.2 Sample preparation

At the end of the sampling period, tubes were stored at 4 °C and analyzed within 24 h. Each sample was eluted using ultrapure carbon disulfide (CS_2 99.9% low benzene content, Aldrich 34.227–0) (1 mL) and subsequent 1 µL of extraction solvent injected into a gas chromatograph (GC).

2.3 Chromatographic analysis

The samples were analyzed using a GC Chrompack CP-9000 Series equipped with a flame ionization detector (FID) and capillary column VF-5 ms, 30 m × 0.225 mm ID, film 0.25 µm. The carrier gas was nitrogen (purity ≥ 99,999%) and the flow rate through column was 0.5 mL/min.

The calibration curve was prepared from known concentrations of xylenes. Separate standard of, xylene (m), xylene (p), xylene (o), and mixed standards were prepared as described in standard NIOSH 1501 and were injected directly into the gas

chromatography device and retention time of each analyte was obtained and peaks of mixed standards were identified according to their retention times. Xylenes (ortho, para and meta with 99%) were purchased from Sigma-Aldrich (Spain). GC thermal program was constant at 39 °C for 10 minutes.

3 RESULTS AND DISCUSSION

3.1 Quality control of air measurements

The detection limit of p-Xylene, m-Xylene and o-Xylene were 4.327, 5.850 and 5.743 ppm respectively, calculated for the 3 aromatics with a signal-to-noise ratio ranging from 5 to 1. The precision of the analysis was determined by analyzing 1 sample 20 times.

The mean of the 21 injections into GC/FID recorded xylene as 160 ppm (SD = 0.5) with a coefficient of variation of 3%. Recovery (p-Xylene: 92%; m-Xylene: 80%; o-Xylene: 81%) was determined using the previously described in NIOSH method n.° 1501. When the signal-to-noise ratio ranged from 5 to 1, there was no apparent benzene contamination of CS_2 bottles.

The personal monitor air measurements were compared to the values of atmospheric air at the three histological stages (histological staining, slide mounting and chemical waste disposal.

Figure 1 shows the results obtained for the ambient air contaminants. Each bar represents the TLV-TWA by day, during 12 weeks. The values of average concentration (TWA) obtained exceeded the Threshold Limit Values—Time Weighted Average (TLV-TWA), established in Decree-Law 24/2012 (50 ppm) (Ministério da Economia e do Emprego, 2012), and 43.2% of the TWA exceeded the limit of 100 ppm (established in standard NP 1796:2007) (Instituto Português da Qualidade (IPQ), 2007). The mean of STEL values was 160 ppm. The values obtained in 15 minutes exceeded the STEL limits established in the Decree-Law 24/2012 and NP 1796:2007, 100 ppm and 150 ppm, respectively. In the first day

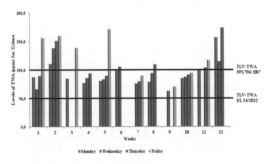

Figure 1. Xylene levels (TWA) obtained from air monitoring in twelve weeks.

of the week, the workers are exposed to xylene at lower exposure levels and on Friday to higher levels of exposure, due to the renewal of xylene of each container and chemical waste disposal.

The results of this study show higher concentrations of xylene to does found identical studies (Panev, T., Pavlova, M, & Tzoneva, M., 1998).

4 CONCLUSIONS

The environmental concentrations found exceed the TLV-TWA, established in Decree-Law 24/2012. Concentrations above 100 ppm may endanger the professionals in the development of their activity in this laboratory.

From a general point of view, the need to implement strategies of prevention in this higher school laboratory is evident:

– monitoring activities should include both air and urinary analyses;
– there is a need for a better control on exposure to xylene, to be realized by means of improvement and development of ventilation systems and enforcing personal respiratory protection for high intensity exposure tasks;
– workers in these environments should be made aware of the potential risks by means of information and training;
– workers should be forced to use, at least during critical working phases, respiratory protection devices.

ACKNOWLEDGMENTS

The authors acknowledge their thanks to CESPU, CRL—Cooperativa de Ensino Superior Politécnico e Universitário, the financier of this study.

REFERENCES

ACGIH. (2012). TLVs. Threshold limit values and biological exposure indices. Cincinnati, OH: American Conference of Governmental Industrial Hygienists.
ATSDR. (2007). Toxicological Profile for Xylenes. Atlanta, GA: Agency for Toxic Substances and Disease Registry, US Deparment of Health and Human Services.
Bieniek, G., & Lusiak, A. (2012). Occupational exposure to aromatic hydrocarbons and polycyclic aromatic hydrocarbons at a coke plant. *Annals of Occupational Hygiene.* 56(7): 796–807.
Buesa, R.J. (2007). Histology safety: now and then. *Annals of Diagnostic Pathology.* 11(5): 334–339.
Chen, M.L., Chen, S.H., Guo, B.R., & Mao, I.F. (2002). Relationship between environmental exposure to toluene, xylene and ethylbenzene and the expired breath concentrations for gasoline service workers. *Journal of Environmental Monitoring.* 4(4): 562–6.

Dudek, B., Gralewicz, K., Jakubowski, M., Kostrzewski, P., & Sokal, J. (1990) Neurobehavioral effects of experimental exposure to toluene, xylene and their mixture. *International Journal of Occupational Medicine and Environmental Health.* 3: 109–116.
Environmental Protection Agency. (2000). Healthy Hospitals: Environmental Improvements Through Environmental Accounting Washington. Retrieved February. 2008. From http://www.epa.gov/oppt/library/pubs/archive/acctarchive/pubs/hospitalreport.pdf
Environment Agency. (2009). Human health toxicological assessment of contaminants in soil. Science Report SC050021/SR2. Bristol: Environment Agency. From http://www.environmentagency.gov.uk/static/documents/Research/SCHO0309BPQL-e-e.pdf
Gamberale, F., Annwall, G., &Hultengren, M. (1978). Exposure to xylene and ethylbenzene: III. Effects on central nervous functions. *Scandinavian Journal of Work, Environment & Health.* 4: 204–211.
Hastings, L., Cooper, G.P., & Burg, W. (1986). Human sensory response to selected petroleum hydrocarbons. In: MacFarland HN, ed. *Advances in modern environmental toxicology. Vol. 6. Applied toxicology of petroleum hydrocarbons.* Princeton, NJ: Princeton Scientific Publishers, 255–270.
Instituto Português da Qualidade (IPQ). Segurança e saúde do trabalho. Valores limite de exposição profissional a agentes químicos. Instituto Português da Qualidade, 2007.
Klaucke, D.N., Johansen, M., & Vogt R.L. (1982). An outbreak of xylene intoxication in a hospital. *American Journal of Industrial Medicine.* 3: 173–178.
Ministério da Economia e do Emprego. *Decree-Law n.º 24/2012.* Diário da República, 1ª série, 2012.
Morley, R., Eccleston, D.W., Douglas, C.P., Greville W.E.J., Scott, D.J., & Anderson, J. (1970). Xylene poisoning: a report on one fatal case and two cases of recovery after prolonged unconsciousness. *British Medical Journal* 3: 442–443.
Nelson, K.W., Ege, J.F. Jr, Ross, M., Woodman, L.E., & Silverman, L. (1943). Sensory response to certain industrial solvent vapors. *Journal of industrial hygiene and toxicology.* 25: 282–285.
OSHA. (1994). Hazard Communication. *Toxic and Hazardous Substances.* Retrieved 04 of November. 2008. from http://www.osha.gov/pls/oshaweb/owadisp.show_document?p_table=STANDARDS&p_id=10099
Panev, T., Pavlova M., & Tzoneva M. (1998). An assessment of the exposure of technicians working in a chemical laboratory for aromatic hydrocarbons at Neftochim, Burgas. *International Archives of Occupational and Environmental Health.* 71: S60–3.
Roy, D.R. (1999). Histology and pathology laboratories. Chemical hazard prevention and medical/health surveillance. *Official Journal of the American Association of Occupational Health Nurses.* 47(5): 199–205.
Teixeira, M., Rangel, R., Teixeira, A., & Domingues, V. (2009). Perception and Risk of Exposure to Xylene by Pathologic Anatomy Students. *Acta Medicinae Legalis et Socialis.*
Vecchio, D., Sasco, A.J., & Cann. C.I. (2003). Occupational risk in health care and research. *American Journal of Industrial Medicine.* 43(4): 369–397.

Occupational Safety and Hygiene – Arezes et al. (eds)
© *2013 Taylor & Francis Group, London, ISBN 978-1-138-00047-6*

Determination of noise levels in the intensive care unit of the Coimbra Hospital and University Centre

D. Castro, H. Simões, J.P. Figueiredo, Ó. Tavares, H. Braga, M. Negrão,
R. Gonçalves, V. Bizarro, D. Ferreira & A. Ferreira
College of Health Technology of Coimbra, Portugal

H. Braga, M. Negrão, R. Gonçalves, V. Bizarro & D. Ferreira
University Hospital of Coimbra, Portugal

ABSTRACT: Occupational noise is present in greater or lesser degree in all workplaces. It is known as an important risk factor for workers, affecting their physical and psychological health and safety, while decreasing the quality of work and productivity. This study aims to explore noise exposure and its impact on the health of nurses in one intensive care unit of the Coimbra Hospital and University Centre, by trying to understand the data collected according to the guidelines provided by the World health Organization. We found that the noise in this area is not a great danger to medical personnel in the ICU; we obtained an average value estimated for Leq of 64.78 dB (A), that is approximately 15.20 dB (A) lower in average than the value established by Decree-Law 182/2006. The estimated mean value of Lmax originated a value of 84.01 dB (C), which is approximately 50.97 dB(C), less than the average value set by legal document mentioned earlier, we may also observe that Monday and Tuesday are the days of greater noise exposure and weekend days present lowest values.

1 INTRODUCTION

In the last years, several studies have been developed in the context of assessing the level of exposure to noise in patients, medical staff and nurses in hospitals (Anjali, J et al, 2007). Hearing is an essential condition to acquire an adequate equilibrium and concentration, both in terms of concentration and equilibrium level, as well as at the level of physical well-being and to allow for proper psychological services, as well as the level of personal health. Nowadays, the population acknowledges the importance that hearing holds for health and quality of life. However, as the Max-Plank Institute for Physiology of labor observed in 1980, "the noise is presently the largest disturbance of the environment," it is estimated that about 20% of the population of industrialized countries is steeped in very intense sound levels (Simões, Hélder, 2009).

In a study on the effects of noise in intensive care patients, it was found that the Intensive care units (ICUs) are areas where noise is excessive, due to ventilators and alarms. Some studies show that hospital noise is a reason for the potential cause of stress in patients. However some authors argue that the most common sources of noise are in 65% of the cases caused by conversations amongst the medical staff, 54% of patients, 42% due to alarms,

39% intercom and 38% due to mobile phones and pagers (Yoder, JC, et al, 2012).

We can consider that high sound pressure levels in hospitals are common throughout the world, as reported in several studies: average between 60 and 65 dB (A) at a hospital in Austria, 55 dB (A) in a hospital of the University of Valencia in Spain and 68 dB (A) at ICU in Manitoba, Canada (Macedo, I, et al, 2009).

The number of medical staff and equipment existing in an ICU, makes the environment very fulfilled and can create states of sensory overload, over stimulation and makes it difficult for the adaptation and subsequent recovery of patients (Susan, E. 2010).

Plus, the noise level in hospitals and the later effect on the health and the recovery period has been common in many studies, since we are facing a high risk for occupational exposure to chronic noise. The continuous and excessive noise exceeding 85 dB (A) can cause physiological and psychological effects on the healthcare team (Macedo, I, et al, 2009).

This study intends to take into account both the occupational health and public health, i.e. not only evaluating the exposure of the body provider of health care, but also evaluating the exposure of patients.

The World Health Organization recommends 45 dB (A) for the daily schedule and 35 dB (A) for night-time in ICU's (Akansel, N, 2007) (Pediatrics, A.A., 1997) (WHO, 1999).

In Portugal there is no specific legal framework for noise in these health units. The studies on this topic are scarce, so we felt the need to investigate the acoustic environment that is found in these spaces.

With this, the intention is to determine the noise level in an Intensive Care Unit, to test the disturbance produced at the level of the patients as well as at the body care provider.

2 MATERIAL AND METHODS

This study was conducted in the intensive care unit of the University Hospital of Coimbra.

The study was developed in the academic year 2011/2012 and the period of data collection took place during the month of June 2012. The target population was composed by the medical and paramedical personnel, present in this unit, as well as by patients who were here hospitalized.

This study was classified as Level II, and type descriptive correlation of Transversal nature.

The equipment to collect noise level was installed in the study area. Therefore, a sound level meter was installed (CESVA SC310), for measuring the level of noise present in the unit. There were also measurements to evaluate the level of personal exposure, resorting to a dosimeter (CESVA DC112). The equipment was calibrated before and after each measurement.

In order to get a representative value of noise exposure, the measurements were made during the working week for sampling of 8 hours (taking into account the work shift of 8 hours day and then divided into periods of 15 minutes for a better interpretation of the results) in a total of 416 measurements, among which 355 using the sound level meter to evaluate the exposure of patients and 61 dosimeters to evaluate the exposure of the working personnel.

As the sampling was 8 hours daily, the parameter of Leq assumes equal values regarding to the parameter of Lex,8h.

For the analysis of the values obtained, we took as the standard set out in the Law-Decree 182/2006, which establishes the legal framework for the protection of workers againts risks due to the exposure of noise at work.

In addition, we also took into consideration the guidelines of both the World Health Organization and the American Institute of Standards, recommending to patients rooms no more than 45 dB (A) during the day and 35 dB (A) overnight, stating that above these values, capacity to resting and recovering of patient is disturbed. (WHO, 1999).

PASW Statistics 18 software was used for statistical analysis, evaluating assumptions the type of statistic to apply (parametric and nonparametric), resorting to the T-STUDENT test for a sample, applied independently for each sampling and ANOVA test for independent samples.

The interpretation of statistical tests were based on a significance level of p-value = 0.05 with a confidence interval (CI) of 95%. For a value of $\alpha \leq 0.05$ there are differences observed and/or significant associations between the groups.

This investigation has exclusively academic interest, excluding any financial or economic interest.

3 RESULTS

In this study, 416 measurements were made, thereby obtaining an estimated mean value of Leq equal to 64.78 dB (A), i.e. 15.20 dB (A) below the value set by Decree 182/2006. The estimate mean of Lmax (84.01 dB (C)) was significantly lower compared to the value established by the legal document ($d\bar{x} = 50.97$ dB (C)). Therefore, we can say that to a p-value <0.05, there are significant differences between the estimated mean values of Leq and Lmax and to the legal values, respectively 80dB (A) and 135dB (C) (lower action values). (Table in Appendix I).

From the 416 measurements regarding the parameter, Leq, 5.5% of these (23 measurements) had levels higher than the legally established. Taking into account the parameter Lmax, only one measurement showed higher value than the legally established (0.2%). (Table in Appendix II).

In terms of measurements in the ICU room, the estimated average value was 62.58 dB (A), significantly higher compared to the reference value established by WHO ($d\bar{x} = 17.59$ dB (A)). We also verified that all measurements were above the recommended value. Regarding the measurements of exposure of the staff in the ICU, there was 100% risk absence for the estimated average values of Lmax (Table in Appendix III).

Regarding the measurements to verify the level of exposure of the medical staff in the ICU, we proceeded to analyze 61 measurements using a dosimeter, and there was 100% risk absence in relation to the estimated average values of Lmax. However, compared to the estimated average values of Lex, 8h we found that 22 measurements (36.1%) had higher values compared to the lower action value, 80 dB (A) (Table in Annex IV).

Comparing the estimated average of Leq obtained in the ICU and the estimated average of Lex8h obtained in the measurements of personal exposure, we can say that only 5.5% of the measurements were in the presence of risk (Table in Appendix V).

For a better understanding of the results obtained, further analyses were made, according to the weekdays. Therefore observing the following graphs:

Considering the graph 1, which corresponds to estimated global average values, (joining both the values recorded in the ICU room, and the measured values on the personnel) we can verify that the estimated average value of Leq decreases along the week, a decrease of about 12.56 dB (A). However, there is an increase of approximately 7.26 dB (A) to the minimum value registered, on Wednesday with the estimated average value of 59.59 dB (A). Therefore, we see an increase tendency from Wednesday to Thursday. We observed that the maximum value was registered on Monday and we can see a decrease of 5, 72 dB (A) to the value registered on Thursday. Therefore we can say that there are statistically significant differences to a p-value <0.05 compared to values obtained from day to day throughout the week.

In relation to the estimated average values of Lmax, the results are not statistically significant, however we saw a tendency to decrease throughout the week with variation between the maximum (Tuesday) and the minimum Lmax (week-end) of 12.52 dB (C). The same happens in the chart which refers to the estimated average values of Leq. From Wednesday to Thursday, we observe an increase of approximately 5 dB (C). (Table in Annex VII).

Analyzing the graphics, we can conclude that on Monday and Tuesday higher values occurred, we found that these values still increased on Thursday. Falling back on the remaining days, we can also say that although they refer to different parameters, regarding their variation over the week, they have similarities.

Taking into account the estimated average values measured regarding to patients, i.e. all values recorded in ICU room, except the values measured on the nurses, regarding the parameters Lmax and Leq measured in the ICU room, once again we observe a tendency to decrease throughout the week, with the maximum at Tuesday, respectively an average of 66.75 dB (A) and 100.6 dB (C).

In this case, the difference between the beginning and end the of the week, was approximately 6dB (A) and there is an increase of 7dB (A), on Thursday, observing again a tendency to decrease in the following days.

In relation to the estimated average values of Lmax, we find that the results obtained had a tendency to decrease throughout the week, with a variation between the maximum (Tuesday) and the minimum Lmax (week-end) of approximately 21dB (C), such as it happens in the graphic referring to estimated average values of Leq furthermore from Wednesday to Thursday we can see an increase of about 6 dB (C), falling back on the remaining days. (Table in Annex VII).

Analyzing the graphics in conjunction, we conclude that on Monday and Tuesday occur higher values, although the graphics deal with different parameters, they feature similarity regarding their variation over the week, although the variation of Lmax is not as significant as previously verified.

Looking now at Figure 2, which represents the average values of Leq with standard deviation per hour of measurement.

Verifying the measurements we can observe that the period of hours during which the average values of Leq approach the risk value 80 dB (A), occur within the 10th until the 18th hour of measurement (10h until 18 hours of afternoon), in other words in within 8 hours of a total of 24 hours of measurement, corresponding about 34% of the measurements, in other words regarding the period between 10 and 18h, being the highest values registered on the 10th and 12th hour.

We can see that in the first 10 hours of measurement, there is a growing tendency, and in the following 8 hours they tend to approximate the risk values, returning to decrease in the next hours.

We can also compare that only 2 in 24 measurements, representing 8.3%, exceed the risk value.

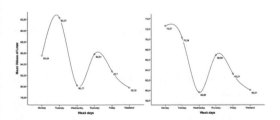

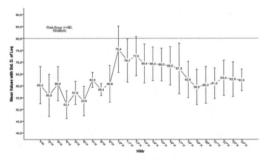

Figure 1. Average values of estimated Leq and Lmax by day of week.

Figure 2. Average values of Leq with standard deviation for time measurement.

In the graphic below we can see the average values of Lmax with standard deviation of measurement, per hour.

In relation to the estimated average values of Lmax, we find that the results obtained had a tendency to decrease throughout the week, with a variation between the maximum (Tuesday) and the minimum Lmax (week-end) of approximately 21dB (C), such as it happens in the graphic referring to estimated average values of Leq furthermore from Wednesday to Thursday we can see an increase of about 6 dB (C), falling back on the remaining days.

Verifying the measurements we can observe that the period of hours during which the average values of Leq approach the risk value 80 dB (A), occur within the 10th until the 18th hour of measurement (10h until 18 hours of afternoon), in other words within 8 hours of a total of 24 hours of measurement, corresponding about 34% of the measurements, in other words regarding the period between 10 and 18h, being the highest values registered on the 10th and 12th hour.

We can see that in the first 10 hours of measurement, there is a growing tendency, and in the following 8 hours they tend to approximate the risk values, returning to decrease in the next hours.

We can also compare that only two in 24 measurements, being 8.3%, exceed the risk value.

We can also verify that the average values of Lmax shows a consistency of results, and that most values are below the threshold 135 dB (C), and it was found that at the 21st hour measurements, (21H00), a maximum average value of 112.63, has been registered, exceeding the risk value at certain times.

The graphic below represents the average of global Lex,8h, recorded in the ICU, analyzing measurements per day and hour.

Comparing the global average of Leq with the hours and days of the labor week, we can certify that these average values do not exceed the lower action value.

Between the 9th and the 18th hour of measurement (9 to 18 pm), the values are higher than the average, especially on Monday and Tuesday.

The following graphic shows the average values of Leq, recorded in the measurement performed in the ICU, taking into account only the patients.

Taking into account only the measurements performed in the ICU, we can determine that the values do not exceed the lower action value, 80 dB (A). We can also certify that the days when the average values of Leq are closer to the lower action value are on Monday, Tuesday and Thursday, between the 10th and the 18th hour of measurement (10 to 18 pm), as well as on Thursday and Friday during the 21st and 23rd hour measurement (21 to 23 pm).

With the analyzed results obtained on the nurse's measurements we proceeded with the preparation

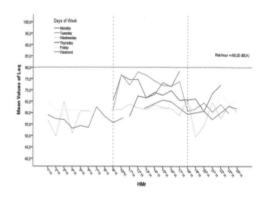

Figure 3. Average values with standard deviation of hourly Leq measurement regards weekdays.

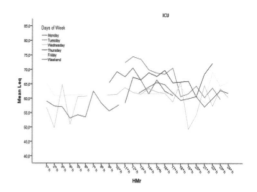

Figure 4. Average Values with global standard deviation of hourly Leq measuring regards weekdays at ICU.

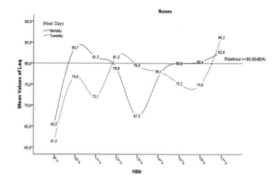

Figure 5. Average values of Leq for time measurement for the medical staff.

of the following chart, taking into account the values of Leq, days and hours of measurement.

Concerning the measurements performed regarding the personal exposure to noise levels, we can say that the limit, 80 dB (A) is exceeded in average, between the 10th and the 12th hour on

428

Mondays, and between the 16th and 17th hour of measurement, both on Monday and Tuesday.

The variation recorded at the 10th hour of measurement, was about minus 7dB (A) to the value obtained on Tuesday regarding the value obtained on Monday, this variation was also denoted in the 11th hour, verifying that on Tuesday was registered an approximate value of 10dB (a) lower than the value recorded on Monday.

In addition, on Monday, on the 13th hour, the measurement was 12dB (A) lower compared with the same measurement period on Tuesday. After this time, denote a tendency of growing on Monday, until the end of measurements.

On the contrary, the measurement of Tuesday denotes a certain tendency to decrease since the 12th hour of measurement up to the 16th hour of measurement, culminating in an average maximum of 86.2 dB (A) during the 17th hour of measurement.

4 DISCUSSION

The interpretation of results is essential to achieve and infer various conclusions.

The WHO recommends a maximum noise level for the hospital environment in the range of 45 dB. The tolerance of the human ear is less than 80 dB, less noise than produced in a noisy street (100 dB), for a horn-type vuvuzela (120 dB) or by a jet engine (140 dB) (Concha-Barrientos M, et al, 2004).

Studies show that the increase of the noise level is associated with the reduced quality of sleep, through its fragmentation and reduction of periods of sleep-REM (rapid eyes Movement), which may contribute to vital function changes, including production of certain hormones and slowing down clinical recovery process (Stanchina ML et al., 2005).

After analyzing the results, we can certify that the noise to which patients are exposed in UCI is harmful for them, since the noise levels exceed the values recommended by the WHO (45dB (A)).

Results can be expected due to the high levels of sound pressure in hospital environments, which are common throughout the world, as reported by several studies: average between 60 and 65 dB (A) in hospital in Austria, 55 dB (A) in hospital in Spain (Macedo, I, et al., 2009) obtaining in this study estimated average value Leq of 64.78 dB (A).

On the positive side, we note that in relation with the issue on occupational health, with the target population of ICU medical staff, we can certify that the noise they are exposed to is not harmful, because only 36% of measurements register values above the lower value of action 80 dB (a), and that less than 1% of the measurements Lmax register values above the lower action value, 135 dB (C). We can still verify that globally the evaluation results obtained in measurements performed either at patients level, as the personal exposure level, these risk values decrease to 5.5% for Leq and 0.2% for Lmax.

According to some authors, noise reduction on the ICU may undergo reduction in the intensity of audible monitoring alarms, increased sound absorption, for example, by altering the characteristics of floor coverings and ceiling used; avoid conversations at ICU, absorbs impacts caused by opening and closing doors (Anjali, J et al., 2007).

Pereira et al, found an overall average noise at ICU of 65.36 dB, verifying that the night time noises were less intense than during the day, in this study, according to the different days it was found that average values of Leq and Lmax were higher on Monday and Tuesday, especially between the 10th and the 18th hour of measurement, between 10 and 18 o'clock. The average value of Leq and Lmax obtained showed a decrease in the weekend. Because during the weekend there are less people in the unit, due to the fact that there are fewer professionals in the service, and because they made fewer exams (x-rays, ultrasounds, etc.), leading to lower dragging equipment, and movement of people, which leads us to believe that the ICU, on the weekend presents a higher level of tranquility.

5 CONCLUSIONS

Noise levels measured during this study are not in accordance with the values recommended by WHOM for hospitals at any time, referring to the target population of patients.

However, according to occupational health, we can say that the global values obtained do not exceed the lower limits of action, i.e. the average Leq obtained was 64.78 dB (A), about 15.20 dB (A) less than the average value established by Law-Decree 182/2006. Regarding the estimated average value of Lmax obtained a value of 84.01 dB (C), which is about 50.97 dB (C) less than the average value set by the legal document mentioned above.

Despite not having obtained alarming noise values, it is still important trying to reduce them. Researchers suggest that environmental interventions are effective and efficient in reducing the noise level in hospitals, choosing to increase the acoustic conditions of the building (Anjali, J et al, 2007).

We believe that this article will be the first step towards the development of new studies where these can review, check and mitigate all the limitations that are met, to certify the results obtained.

REFERENCES

Akansel, N. & Kaymakçi, S. Effects of intensive care unit noise on patients: a study on coronary artery bypass graft surgery patients. Jounal of Clinical Nursing. 2008 The Authors. Journal compilation, 2007.

Anjali, J. & Roger, U. Sound Control for Improved Outcomes in Healthcare Settings Healthdesign. [Online] 2007.http://www.healthdesign.org/chd/research/sound-control-improved-outcomes-healthcare-settings.

Concha-Barrientos M; Campbell-Lendrum D & Steenland K. Occupational noise: assessing the burden of disease from work-related hearing impairment at national and local levels. Geneva: World Health Organization; 2004. (WHO Environmental Burden of Disease Series, No. 9).

Macedo, I, et al. Avaliação do ruído em Unidades de Terapia Intensiva. Brazilian Journal of Otorhinolarygology. 75, 2009, Vol. 6.

Medeiros, L.B. Ruído: Efeitos Extra Auditivos. Ruído: Efeitos Extra Auditivos. Porto Alegre, 1999.

Pediatrics, A.A. American Academy of Pediatrics. [Online] Outubro 4, 1997. http://HYPERLINK "http://www.pediatrics.org/cgi/content/full/100/4/724" www.pediatrics.org/cgi/content/full/100/4/724.

Pereira RP; Toledo RN; Amaral JLG & Guilherme A. Qualificação e quantificação da exposição sonora ambiental em uma unidade de terapia intensiva geral. Rev Bras Otorrinolaringol. 2003; 69(6): 766–71.

Saúde, Organização Mundial de. WHO. Guidelines for Community Noise [Online] 1999. http://www.whqlibdoc.who.int/hq/1999/a68672.pdf.

Simões, Hélder. Ruído nos Locais e Postos de Trabalho. Coimbra: ESTeSC, 2009. Higiene do Trabalho, Manual.

Stanchina ML; Abu-Hijleh M; Chaudhry BK; Carlisle CC & Millman RP. The influence of white noise on sleep in subjects exposed to ICU noise. Sleep Med. 2005; 6(5): 423–8.

Susan, E. Hospital Noise and the Patient Experience. HYPERLINK "http://www.healingheath.com" www.healingheath.com[Online]2010. http://www.healinghealth.com/images/uploads/files/hhs_hospital_noise_whitepaper.pdf.

Yoder, JC, et al. Hospital Noise Puts Patients at Risk. Arch Intern Med. 4, 2012, Vol. 112.

Occupational Safety and Hygiene – Arezes et al. (eds)
© *2013 Taylor & Francis Group, London, ISBN 978-1-138-00047-6*

Musicians noise exposure in a Portuguese orchestra—a case study

M.A. Rodrigues, P. Alves, L. Ferreira & M.V. Silva
*Research Centre on Environment and Health, Allied Health Sciences School of Polytechnic of Porto,
Vila Nova de Gaia, Portugal*

M.P. Neves & L. Aguiar
Environmental Health Department, National Health Institute, Ricardo Jorge, Porto, Portugal

ABSTRACT: Orchestra musicians can be exposed to high sound levels, which can result in hearing damages. This issue is not yet well characterized and there are no studies for Portuguese orchestras. Therefore, this study aims to analyse the noise exposures of musicians from a Portuguese orchestra. The musicians' noise levels were monitored along group and general rehearsals of 2 different repertoires. Test subjects were selected in accordance with their position in orchestra. Participants were requested to wear noise dosimeters during the entire rehearsals. A sound meter was used to analyse the conductor exposition. Results show that musicians of the orchestra analysed are exposed to high noise levels, which can lead to hearing damages. The results also showed that the sound levels vary with the instrument, the repertoire and the position. This study suggest that musicians are at risk of hearing loss, being essential pay more attention to these professionals.

1 INTRODUCTION

Exposure to high noise levels is broadly recognized as being one of the most significant and frequent risk factors in occupational environments, particularly at industrial settings (Arezes *et al.*, 2012). Such exposure can result in several effects for the workers' health, mainly in the development of noise-induced hearing loss (NIHL), which is the most frequent occupational disease in Europe (EU-OSHA, 2002). However, noise can be a problem in other jobs and workplaces, and, therefore, other professionals groups are also considered as important. The professional orchestras musicians are one of these groups.

Previous studies show that orchestral musicians are exposed to loud music (Lee *et al.*, 2005; MacDonald *et al.*, 2008; O'Brien *et al.*, 2008), which can have a great impact on their health. Hearing loss, tinnitus, hyperacousis and diplacusis, are the more frequently health effects referred in previous studies (Juman *et al.*, 2004; Laitinen, 2005; Morais *et al.*, 2007; Laitinen & Poulsen, 2008; Jansen *et al.*, 2009). It is also important to highlight that, for musicians, these effects are particularly critical, considering that they hearing ability is a relevant part and a crucial tool of their professional activity and performance (Jansen *et al.*, 2009; Kähäri *et al.*, 2004). Musicians need to identify different tones and perceive the sound level that their instrument is producing, as well as interacting in the meantime with other instruments that surround them. If musicians cannot do that, their performance will be diminished and its permanence in the orchestra called into question.

However, despite of the relevance of this issue for orchestral musicians, it seems that the problematic of noise exposure is still not well characterized, and the strategy for the noise exposure assessment are not well established. Furthermore, the Portuguese legislation, in particular the Decreto-Lei nº 182/2006, does not considers the specific case of musicians, and there is no code of conduct provided in accordance with Directive 2003/10/CE that establishes the guidelines about how musicians should be protected from noise exposure. Therefore, in Portugal this professional group has remained out of the noise exposure concerns. However, in the last years, some studies and technical reports have being developed, clearly showing the noise exposure of these particularly professionals, promoting the awareness about this problem.

As previously referred, the strategy for the noise exposure assessment is still not well established. Different methodologies are being used to assess the noise exposure, making difficult the comparison of the results obtained in different studies (see e.g. Lee *et al.*, 2005; Morais *et al.*, 2007). In some cases the procedures applied were inaccurate and unclear, which can put in question the validity of some previous results. Without a reliable risk assessment of musicians' exposure, it is not possible to compare sound pressure levels with the current guidelines and it is also very complex to define and implement an effective strategy to reduce the risk of NIHL (Arezes *et al.*, 2012). Besides, it is difficult to define the periods of musicians' noise exposure, due to its variability, situation that difficult the noise exposure level determination. It is also essential for characterization of the musicians' noise exposure to consider other factors as the number of rehearsals and repertoires, different assessment points and venues, because the noise levels vary among them (Lee *et al.*, 2005; O'Brien *et al.*, 2008; Schmidt *et al.*, 2011). In this context, only a few number of studies tried to characterize the noise exposure levels (see e.g. Lee *et al.*, 2005; MacDonald *et al.*, 2008). Most of previous studies only determined the equivalent continuous sound pressure level ($L_{p,A,eqT}$) and the peak sound pressure level ($L_{p,Cpeak}$) in the course of some rehearsals and performances, which can give an image of the noise levels that musicians are exposed in the course of these activities, but not a value that allows to infer if musicians are at risk of NIHL or not.

Faced with this problematic, and the lack of any characterization of noise exposure level for Portuguese orchestral musicians, this study aims to be a first attempt to characterize the noise level of exposition of Portuguese orchestral musicians, through a case study.

2 MATERIALS AND METHOD

Sound level measurements were developed on musicians of one Portuguese orchestra. Two different repertoires during group rehearsals and general rehearsals were assessed.

2.1 Instrument

Measurements were performed using dosimeters and a sound level meter. Two Quest Noise-Pro dosimeters were used to measure $L_{p,A,eqT}$ and $L_{p,Cpeak}$. Three CESVA DC112 dosimeters and one CESVA SC-310 sound level meter provided $L_{p,A,eqT}$, $L_{p,Cpeak}$ and octaves frequency data.

After the field measurements, the data were transferred to the Capture Studio Editor Software and QuestSuite™ Professional Software, which processes all the data collected.

2.2 Measurement procedure

Before starting the study, its aim and procedures were explained to all musicians and all their questions clarified. Based on information delivered by the orchestra manager, test subjects were selected in accordance with their instrument and position in the orchestra structure. In each rehearsal 10 musi-

cians were evaluated, as well as, the conductor at the same time. 7 of the assessed musicians were the same. The others were different in order to include as many situations as possible. Between both repertoires 6 musicians were the same.

Participants were requested to wear noise dosimeters during the entire rehearsals. The microphone was located on the left or right shoulder of the test subject, close to the most exposed ear, without restricting movement. In the case of string instruments, the microphone was positioned on the opposite shoulder of the instrument. A sound meter was used to analyse the conductor exposure, and the equipment was fixed on the support at the height of the conductor ear.

2.3 Results treatment

According to the Portuguese legislation, Decreto-Lei nº 182/2006, it is only possible to determine the level of daily exposure or weekly average exposure. Consequently, the average sound level for a year, as presented in previous works, is beyond the scope of Portuguese legislation. Therefore, exposure level normalized to a nominal week of five 8h working days ($L_{EX,8h}$) was determined, considering all time that musicians' spent with the orchestra. In accordance with the ISO 9612:2009, a $L_{p,A,eqT} = 70$ dB was used to the remaining periods as a conservative estimate.

3 RESULTS AND DISCUSSION

The orchestra was divided into six groups: strings, woodwinds, brass, percussion and timpani, piano and conductor. Strings includes violin I and II, viola, cello and contrabass. Woodwinds comprise bassoon, saxophone, flute, clarinet, oboe, recorder and piccolo. Brass includes trombone, tuba, trumpet and French horn.

For each repertoire, values of $L_{p,A,eqT}$ and $L_{p,Cpeak}$ were measure and the range of its results are presented in Table 1 and Table 2. According the results, the Portuguese orchestral musicians are exposed to high sound levels in the course of rehearsals, which are in approbation with the results presented on previous studies (Lee *et al.*, 2005; MacDonald *et al.*, 2008; O'Brien *et al.*, 2008). Values of $L_{p,A,eqT}$ vary in accordance with the instruments type. Values of $L_{p,A,eqT}$ vary, in general in both rehearsals, between 81.8–91.9 dB(A) for strings, 87.6–94.4 dB(A) for woodwinds, 87.7–95.1 dB(A) for brass, 88.6–95.4 dB(A) for percussion and timpani, and 80.6–84.4 dB(A) for conductors. Piano was only included in repertoire B an the $L_{p,A,eqT}$ vary between 84.8–86.2 dB(A). Noise levels were found, in general, higher for percussion and timpani, brass and woodwinds and lower for piano and conductor.

Table 1. Summary of $L_{p,A,eqT}$ and $L_{p,Cpeak}$ data by group of instrument type in repertoire A.

	Group rehearsal 1		Group rehearsal 2		General rehearsal	
	$L_{p,A,eqT}$ dB (A)	$L_{p,Cpeak}$ dB (C)	$L_{p,A,eqT}$ dB (A)	$L_{p,Cpeak}$ dB (C)	$L_{p,A,eqT}$ dB (A)	$L_{p,Cpeak}$ dB (C)
Strings	81.8–88.7	114.7–124.0	81.9–89.7	115.9–128.9	83.1–91.9	115.1–117.3
Woodwinds	89.3–92.8	119.3–122.2	87.6–96.8	115.4–123.2	–	–
Brass	90.0–92.6	128.2–126.5	90.4–92.7	122.6	87.7–91.9	118.6
Timpani	88.6	128.8	90.8	130.4	92.8	130.5
Conductor	80.6	110.7	81.9	107.7	82.9	111.8

Table 2. Summary of $L_{p,A,eqT}$ and $L_{p,Cpeak}$ data by group of instrument type in repertoire B.

	Group rehearsal 1		Group rehearsal 2		General rehearsal	
	$L_{p,A,eqT}$ dB (A)	$L_{p,Cpeak}$ dB (C)	$L_{p,A,eqT}$ dB (A)	$L_{p,Cpeak}$ dB (C)	$L_{p,A,eqT}$ dB (A)	$L_{p,Cpeak}$ dB (C)
Strings	88.7–89.4	117.8–123.9	88.7–89.3	120.6–120.8	88.3–89.4	117.8–120.7
Woodwinds	91.5–92.8	133.9–134.0	93.4	125.6	92.0	121.6
Brass	92.6–95.1	123.2–132.3	93.2	134.4	94,2	133.7
Percussion	92.2	133.9	95.4	136.5	–	–
Piano	86.2	132.7	–	–	84.8	116.4
Conductor	84.4	115.1	84.2	116.4	84.3	115.6

The results also show that $L_{p,A,eqT}$ values vary with the repertoire. The sound levels were higher in repertoire B. These results can be related to differences in the program, in the number of musicians and in the type of instruments included (O'Brien et al., 2008). It is also possible to analyze a great variation in $L_{p,A,eqT}$ values in the same instrument group. These differences are resulting from differences in the instruments in analysis and the impact of the musician stage position (O'Brien et al., 2008), due to the influence of the musicians that surround them (Behar et al., 2006). For example, a musician that play a violin and that was positioned closely to woodwinds is more exposed to loud music than a musician that play the same instrument, but was positioned at the outer limits of the orchestra structure (Figure 1 presents the orchestra structure in repertoire B).

The $L_{p,Cpeak}$ exceeded the lower exposure action level presented in Decreto-Lei n° 182/2006 [135 dB(C)] for percussion in repertoire B. For the other instruments group, this value was never achieved, being these results similar with the obtained by O'Brien et al., (2008). This suggest that it is common for these musicians group be exposed to high $L_{p,Cpeak}$ values, being essential implement adequate risk reduction measures, in particularly the use of hearing protectors (Lee et al., 2005).

As the rehearsals of each repertoire were carried out over one week, it was determined the $L_{EX,8h}$ for each repertoire, being the results by each analysed instrument presented on Table 3.

It is possible to observed that the obtained $L_{EX,8h}$ values exceeded the lower exposure action

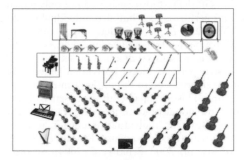

Figure 1. Orchestra structure in repertoire B.

Table 3. $L_{EX,8h}$ by analysed instrument.

	$L_{EX,8h}$ [dB(A)]	
	Repertoire A	Repertoire B
Violin I	82.4	82.3
Violin Ia	77.6	–
Violin II	84.1	81.9
Violoncello a	78.4	–
Violoncello b	77.1	81.8
Flute	83.1	85.3
Clarinet	89.2	85.8
Oboe	83.3	–
French horn	86.5	87.7
Trumpet	84.0	86.0
Trombone	–	87.5
Timpani	84.8	–
Percussion	–	87.0
Piano	–	78.9
Conductor	76.1	77.9

Table 4. Summary of octaves frequencies analysis for repertoire A and B.

Instrument	Octave frequencies							
	63 Hz	125 Hz	250 Hz	500 Hz	1 KHz	2 KHz	4 KHz	8 KHz
Strings	44.6–59.0	58.6–70.2	70.7–77.6	77.9–84.6	78.5–86.6	75.3–83.5	65.2–79.6	47.9–64.1
Woodwinds	43.0–55,2	60.7–74.3	71.5–78.7	82.7–89.2	84.0–94.3	79.8–92.8	68.1–76.6	49.7–63.3
Brass	47.4–59.8	65.2–77.4	72.5–83.1	83.2–90.1	85.5–91.9	80.2–88.4	66.8–81.7	48.9–74.2
Percussion and timpani	54.9–57.9	67.7–84.5	80.7–86.9	85.8–88.2	83.1–88.2	77.3–84.5	64.8–80.4	51.3–75.9
Conductor	70.2–75.7	76.6–82.8	77.5–81.7	77.5–83.1	75.2–80.9	72.5–77.7	67.6–71.7	53.2–60.6

level presented in Decreto-Lei n° 182/2006, i.e., 80 dB(A) for all instrument groups of the orchestra, except for the conductor, piano and the strings musicians positioned at the outer limits of the orchestra (Violin a and Violoncello a). The higher exposure action level, 85 dB(A), was exceeded by flute, clarinet, French horn, trumpet, trombone and percussion. These results show that musicians are at risk of NIHL. These results are in accordance with Laitinen et al., (2003). However, the authors in their study considered not only rehearsals and performances, but also individual practice. In this study $L_{EX,8h}$ were determined considering only group rehearsals and general rehearsals, where the last one was also seen as a representation of the orchestra performances. Individual practice was not considered due to the difficult in characterize the same. This is related with the absence of musicians' individual practice in orchestras' rehearsals rooms. However, individual practice is an important source of noise exposure (Royster et al., 1991; Laitinen et al., 2003; Behar et al., 2004). According previous studies the noise levels in the course of individual practices can be higher than in the course of rehearsals or performances (Royster et al., 1991; Laitinen et al., 2003). Therefore, this suggests that the musicians' noise exposure in this orchestra may be higher than the presented in this study. But, it is important to highlight that it were considered in for the study two noisy repertoires. Moreover, it was considered a conservative value of 70 dB(A) for the remaining periods. These situations are responsible for the high levels of exposition obtained in this study.

The octaves frequencies were also analysed in some musicians and the results are summarized in Table 4. According Portuguese legislation (Decreto-Lei n° 182/2006), this analysis is essential to determine a suitable hearing protection. Results show differences in octaves frequency among the instruments types. In general strings, brass and woodwinds are more affected between 500 Hz to 2000 Hz frequencies, percussion and timpani between 250 Hz to 1000 Hz frequencies, and finally

conductor in 125 to 500 Hz frequencies. Therefore, the characteristics of noise exposure are different among instruments type. In this context, it is important to consider the frequencies analysis for the implementation of noise exposition reduction measures, in particularly the assignment of the appropriate hearing protection. The use of hearing protectors was the most referred measure (Juman et al., 2004; Lee et al., 2005; Morais et al., 2007; MacDonald et al., 2008). However, if they are not suited to the noise characteristics, musicians will be reluctant to its use.

4 CONCLUSIONS

The results obtained in this study showed that the orchestral musicians are exposed to high sound levels and it was dependent of instrument type, repertoire and position, suggesting that orchestra musicians may be at high risk of hearing loss. Percussion are exposed also to high Peack sound levels. In this context, more attention needs to be provided to this professional group, and the specific case of musicians are urgent to include in the law, in order to protect them. Also the average sound level for a year must be considering in legislation in order to include all variability related with repertoire.

REFERENCES

Arezes, P.M., Bernardo, C.A. & Mateus O.A. 2012. Measurement strategies for occupational noise exposure assessment: A comparison study in different industrial environments. International Journal of Industrial Ergonomics 42 (1): 172–177.

Behar, A., Wong, W., & Kunov, H. 2006. Risk of hearing loss in orchestra musicians: Review of the literature. Medical Problems of Performing Artists: 164–168.

Decreto-Lei n.° 182/2006, de 6 de Setembro. Diário da República n°172—1ª Série. Lisboa.

Directive 2003/10/EC, of the European Parliament and of the Council, of 6 February 2003.

EU-OSHA. (2002). *Data to describe the link between OSH and employability*. European Agency for Safety and Health at Work. ISBN 92-95007-66-2.

ISO 9612:2009. Acoustics—*Determination of occupational noise exposure—Engineering method*. International Organization for Standard.

Jansen, E.J.M., Helleman, H.W., Dreschler, W.A. & de Laat, J.A.P.M. 2009. Noise induced hearing loss and other hearing complaints among musicians of symphony orchestras. *International Archives of Occupational and Environmental Health* 82: 153–164.

Juman, J., Karmody, C. & Simeon, D. 2004. Hearing loss in steelband musicians. *Otolaryngology– Head and Neck Surgery* 131 (4): 461–465.

Kähäri, K., Zachau, G., Eklöf, M. & Möllerc, C. 2004. The influence of music and stress on musicians' hearing. *Journal of Sound and Vibration* 277: 627–631.

Laitinen, H. 2005. Factors affecting the use of hearing protectors among classical music players. *Noise and Health* 7 (26): 21–29.

Laitinen, H. & Poulsen, T. 2008. Questionnaire investigation of musicians' use of hearing protectors, self reported hearing disorders, and their experience of their working environment. *International Journal of Audiology* 47 (4): 160–168.

Lee, J., Behar, A., Kunov, H. & Wong W. 2005. Musicians noise exposure in orchestra pit. *Applied Acoustics* 66: 919–93.

MacDonald, E.N., Behar, A., Wong, W. & Kunov, H. (2008). Noise exposure of opera musicians. *Canadian Acoustics* 36 (4): 11–16.

Morais, D., Benito, J.I. & Almaraz, A. 2007. Acoustic Trauma in Classical Music Players. *Acta Otorrinolaringol Esp.* 58 (9): 401–7.

O'Brien, I., Wilson, W. & Bradley, A. 2008. Nature of orchestral noise. *Acoustical Society of America* 124(2): 926–939.

Schmidt, J.H., Pedersen, E.R., Juhl, P.M., Christensen-Dalsgaard, J., Andersen, T.D., Poulsen, T. & Bælum, J. 2011. Sound exposure of symphony orchestra musicians. *Annals of Occupational Hygiene* 55(8): 893–905.

Royster, J.D., Royster, L.H. & Killion, M.C. 1991. Sound exposures and hearing thresholds of symphony orchestra musicians. *Journal of the Acoustical Society of America* 89(6): 2793–2809.

Occupational Safety and Hygiene – Arezes et al. (eds)
© 2013 Taylor & Francis Group, London, ISBN 978-1-138-00047-6

Evaluation of noise generated by propagation equipment beat stakes construction site

Felipe Mendes da Cruz, Eliane Maria Gorga Lago & Béda Barkokébas Jr.
Universidade de Pernambuco, Recife, Brasil

ABSTRACT: This paper aims to develop a methodology for mapping of noise by equipment slam types: Metallic, Propeller and Continuous Pre-molded, within the limits of the construction site using geostatistics. This method aims to model the noise helping to identify areas that are located in the bed under the influence of noise. The work was structured into 4 distinct phases, the first being the literature review, the second the sampling plan, the third and fourth field surveys geoprocessing. The extent of propagation of cuttings preformed metallic and showed little difference between them, however in relation to the continuous helical gear rays presented to 5.7 times higher. With the development of this type of mapping by future studies, it will be possible to improve the quality of appraisal reports relating to noise emitted by construction sites at both occupational level as in environmental and community comfort.

1 INTRODUCTION

Noise pollution is according to the World Health Organization (WHO), the type of pollution that reaches the largest number of people in the world, after the air pollution from exhaust emissions, and water pollution (WHO, 2010).

The term "noise" has several definitions. According Bistafa (2006), noise can be defined as a sound without harmony and generally has a negative connotation. However, according to Lida (2005), more subjective in its definition, noise is nothing more than an unwanted sound. However, it is important that a sound can be undesirable for an individual, but not for another.

Noise, more formally, can be characterized as a physical phenomenon vibratory characteristics of indefinite pressure variations (in this case air) as a function of frequency, i.e. for a given frequency can exist at random over time, Variations of different pressures. It can also be defined as an auditory stimulus that does not contain useful information for running task (DO RIO, 2001).

Workers in the construction industry, most of the activities are not adequate protection to their health and physical integrity and among the main problems reported in the industry shows the effects caused by excessive noise from equipment that are routinely used in construction sites. It may be mentioned hearing loss, difficulty in communication, stress, lack of concentration and even physical and psychic disorders. And the damage is not adequately assessed and there are economic reasons that hinder social and technical evaluation in this sector of the economy (MAIA, 2001).

Workers exposed to high levels of sound pressure can have, over the years, an irreversible sensorineural hearing loss (hearing loss due to exposure to high sound pressure levels). Initially, you may experience temporary changes in hearing threshold (TTS—Temporary Threshold Shift), ie a short-term effect of the reduction in hearing sensitivity, which gradually returns to normal after cessation of exposure. The change in hearing threshold depends on the exposure time, the sound level of acoustic emission, the frequency of the sound and individual sensitivity. Through continued exposure may occur permanent changes in the threshold of hearing (WU AND DING, 1998).

According to Gerges (2011), noise is one of the most harmful physical agents to worker health in their work environment. Exposure to high levels of sound pressure can cause irreversible hearing loss for workers, which is indispensable in effective measures for its reduction and control.

The same author also states that the physical understanding of noise sources and dynamics of each machine, with the main control techniques, is the best tool for specification, design and problem solving noise.

According Sirvinskas (2007), the great challenge of urban centers is the control of noise pollution (commercial, industrial, traffic, etc.). A survey conducted by the World Health Organization (WHO) found that Brazil is the country of the deaf, in view of the intensity of the noise produced, especially in large urban centers.

The same author shows that noise pollution is the emission of unpleasant noise levels that exceeded legal way and over time may cause in a given period of time, damage to human health and the welfare of the community.

According to the Ministry of Social Safety—MPS were noted between January and September 2011, 33 occurrences of diseases classified under Code Indicator Diseases-10, H-83, according to the national classification of diseases are classified as "Other Disorders Inner Ear," which features one of the contributors to hearing loss and noise-induced acoustic trauma.

Recent data show that the MPS construction industry in 2009 was responsible for 17 incidents linked to the damages hearing.

Through empirical knowledge acquired in field activities can be seen that the noise levels emitted by equipment works have values around 100 dB which under Brazilian law provides risk the health and safety of exposed individuals, when they do not have adequate protection. Medeiros (1999) shows that disturbances attributed to exposure depend on factors such as the frequency of noise, intensity, duration, rhythm, exposure time, individual susceptibility, and the attitude of each individual with the sound.

Gerges (1995), adds: nervousness, mental fatigue, frustration, irritability, poor adjustment in new situations, and social conflicts among workers exposed to noise.

Hence the need to study the noise more precise with the application of geostatistics and geoprocessing tools for better assessment of its propagation in the medium, as well as analyze the catchment areas and the workers are exposed to this influence what motivate this work with the results used to set up measures that minimize impacts to human health and the environment, helping to reduce the spread and the preservation of the health of workers on the site.

The geoprocessing, understood as "a set of techniques for collecting, viewing and processing information spatialized" (RODRIGUES, 1990), allows the joint analysis of a range of socio-environmental.

Considering the environment as a system composed of variables distributed in space and time, it must be studied from models that represent the territoriality and inspection of possible relationships between these variables. Therefore, the digital representation of the environment has been extremely helpful (Christofoletti, 1999).

The spatial statistics is the area of statistics that deals with understanding the spatial distribution of data from geographic phenomena occurring in space. That is, scientific studies methods for the collection, description, analysis and visualization of data that have geographic coordinates (CAUMO, 2006).

According to this author the visual perception of the spatial distribution of the data is very effective in translating existing standards with objective considerations, as well as the perceived association with possible causes, directing and supporting decision making.

The overall objective of this research was to develop a methodology to define by mapping the area covered by the noise emitted by equipment types beating stake: Metallic, Propeller and Continuous Pre-molded in vertical construction sites using geostatistics.

2 MATERIALS AND METHODS

The research was conducted in construction sites during these foundations being in step pile driving where we selected three types of equipment, namely continuous helix, precast and metal. Was monitored six (6) being two sites for each type of stakes.

To develop this work was initially performed a literature search through a review of national and international literature. Then the plants were acquired lease the construction sites being prepared a sampling plan by creating an imaginary net points obeying a spacing of ten (10) meters, the amount of points being dependent on the extent of the construction sites. The field survey consisted of a traversal of all points determined by the sampling plan for each point collected were established Cartesian coordinates (x, y). To obtain the Sound Pressure Level (SPL) at each grid point, we used a sonometer noise Quest Technologies Serial No: QIE070075.

After collecting field for the next phase of geoprocessing were extracted images generated by NOAA through the software Google Earth, followed in the geographic coordinates extracted with the same program and then performed the treatments of data through Software Surfer 8 of Golden software, after the generation of isolines propagation was done compiling the information through the software Autodesk Auto CAD 2010, generating thus the propagation of noise maps.

3 RESULTS AND DISCUSSION

One can verify that equipment type propeller continuous plot situated in an area A showed the spread to the tolerance limit corresponding to 170.10 m^2 which means that the radius of coverage to the limit of 85 dB (A) is 7, 36 m from the machine, as to the level of action has an area of 622.30 m^2 corresponding to a radius of coverage for the value of 82 dB (A) 14.07 m which is approximately twice the distance regarding the limits of tolerance (Fig. 1).

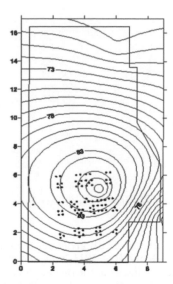

Figure 1. Propagation beat stake—type propeller continuous—site A.

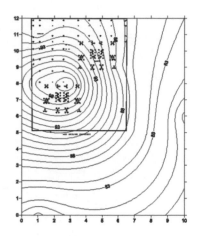

Figure 2. Propagation beat stake—type precast—site B.

Referring to satake precast construction site located at B, one can see that it presented a coverage area of 5372.74 m² for the tolerance limit of 85 dB (A) and 5477.75 m² to the level of action which is radius equal to 41.36 m and 41.76 m, respectively. (Fig. 2).

Regarding stake metal situated in the median C, it can be seen that the area of influence corresponding to the tolerance limit of 85 dB (A) is 1392.00 m² and the area corresponding to the action level is 1607.00 m² which corresponds to a radius of 21.05 m equivalent to the value of tolerance and 22.62 m for the action level (Fig. 3).

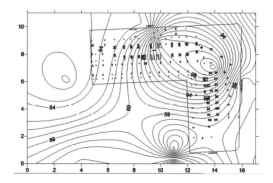

Figure 3. Propagation beat stake—type metal—site C.

You can check that the stake kind of continuous propeller bed D showed an area of 406.11 m² to the level of action around the equipment since the tolerance showed an area of 230.23 m², which represents a radius of action of 8.56 m from the emission source, as to the level of action has been of a range of 11.37 m. (Fig. 4).

Referring to square metal lying in bed and can verify that the catchment area for the tolerance limit issued by crimping equipment stakes plot C was 4433.9 m² which corresponds to a radius of 37.57 m propagation, for now action level was not possible to define your area, because of the noise corresponding to 82 dB is situated outside the observation area defined for this study, their analysis can be explored in future studies. (Fig. 5).

From the evaluation of stake precast piles of construction sites E, you can see that it presented a catchment area for the tolerance limit for noise of 85 dB corresponds to 6938.94 m² and the action level showed an area 9545.34 m² of which means

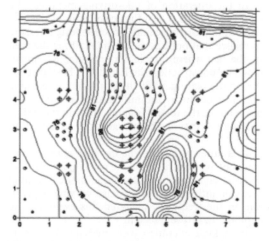

Figure 4. Propagation beat stake—type propeller continuous—site D.

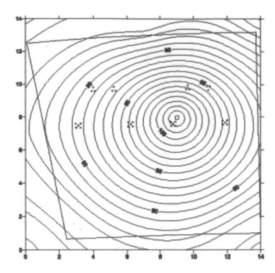

Figure 5.　Propagation beat stake—type metal—site E.

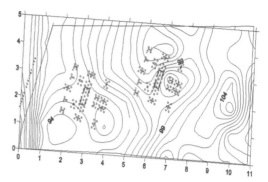

Figure 6.　Propagation beat stake—type precast—site F.

that the rays of sonic performance are 47 m and 55.13 m for the tolerance and action level respectively. (Fig. 6).

Analyzing the noise emitted by the beat stake type propeller continuous kind you can check the coverage areas of the limits of tolerance and action level, if both A and D are lower than if the areas of their beds in order that 2942.9 m² even presented to the construction site and 1239.4 m² to constriction site B.

In the case of piling with beat stake type propeller continuous should be noted that its noise operation receives the influence of the concrete pump is seen that the activity is essential for the use of the same besides being characterized as a continuous noise and no impact.

In the operation of pile driving metal can be seen that the areas of influence of noise is the corresponding tolerance limit or action level referring to realize that both exceeded the limit for the area of the bed where they had a area of 3.823 m² to 6.446 m² construction site E and C. Thus, depending on the placement of the equipment, the whole plot will be under the influence of occupational noise above the tolerance limits making the use of hearing protection since entering the construction site.

It is noteworthy that the metallic sound shows traces of resonance where the assessment through filters frequency band can help in determining areas of influence with coverage varying depending on the frequency analyzed this event can be explored in future research.

In staking operation with precast components is possible evidence that the area affected by the noise as much for the tolerance limit of 85 dB as

the action level of 82 dB are larger than the areas of the beds F and B of 3823.29 m² and 4981.74 m² respectively. Thus, as in the operation of pile driving metal, depending on the placement of the equipment, the whole plot will be under the influence of occupational noise above the tolerance limits making the use of hearing protection since entering the construction site.

4　CONCLUSIONS

Through the maps spread was possible to analyze the behavior of the noise within the limits of the site and determine the locations as potential worker exposure above the tolerance limit provided by law.

The radius of coverage stake geoprocessing was inferior to the type of metal cutting and pre-shaped, as the radius of coverage of the metal stake was inferior compared to the radius of the pile preformed.

With the determination of the areas was possible to check the range of the noise generated and thereby delimit aas areas that need attention.

This study was conducted as a pilot, only two samples were collected from each types of cuttings ara future work the amount of samples should be increased so that we can perform a statistical analysis of the different variables in the environment and the various types of existing stakes.

For detailed studies regarding the magnitude of the coverage area of noise by equipment type crimped metal stakes and preformed it is necessary to extend the area of study well beyond the construction site, considering that its reach exceeded in the cases studied the boundaries of the work.

With the future of the technical detail will be possible to determine parameters for selection of pile driving equipment based on noise levels allowed in the area where the project is install.

REFERENCES

Bistafa, S.R. (2006) Acústica aplicada ao controle do ruído. São Paulo: Edgard Blücher.

Christofoletti, A. (1999). Modelagem de sistemas ambientais. São Paulo: Edgar Blücher, 1ª Ed, 200p.

Caumo, R.B. (2006). Estatística Espacial em Dados de Área: Uma Modelagem Inteiramente Bayesiana Para o Mapeamento de Doenças Aplicada à Dados Relacionados com a Natalidade em Mulheres Jovens de Porto Alegre. Porto Alegre-RS. (Monografia), Universidade do Rio Grande do Sul. 87p.

Do rio, Pires. R., Licínia. (2001) Ergonomia: Fundamentos da prática ergonômica. 3ª Ed., São Paulo: LTR, 225 p.

Gerges, S.N.Y. (1995) Efeitos do Ruído e das Vibrações no Homem sem Proteção. Porto Alegre-RS: MPF Publicações Ltda, 65p.

Gerges, S.N.Y. (2011) Tecnologia Para Atenuação de Ruído. Caderno Informativo de Prevenção de Acidentes—CIPA, Ano:32, Agosto, n° 318, São Paulo-SP.

Lida, I. (2005) Ergonomia—Projeto e produção. 2ª Edição revisada e ampliada, São Paulo: Edgard Blücher.

Maia, P.A. Estimativa de exposições não contínuas a ruído: Desenvolvimento de um método e validação na Construção Civil. Campinas: 2001. Tese (Doutorado em Engenharia Civil) Universidade Estadual de Campinas. Disponível em: http://www.fundacentro.gov.br/CTN/teses_conteudo.asp?retorno = 137. Acesso em: 06 set. 2010.

Medeiros, L.B. (1999) Ruído: Efeito extra Auditivo no Corpo Humano.(Monografia), Centro de Especialização em Fonoaudiologia Clínica – CEFC, Porto Alegre—RS.

Ministério da Previdência Social-MPS (2011) Acompanhamento Mensal dos Benefícios Auxílios-Doença Acidentários Concedidos, segundo os códigos da CID-10—Janeiro a Setembro.

Rodrigues, M. (1990) Introdução ao geoprocessamento. In: Simpósio Brasileiro de Geoprocessamento. São Paulo: Sagres Editora.

Sirvinskas, L.P. (2007) Manual de Direito Ambiental. 5ª Ed., São Paulo: Saraiva, 548.p.

World Health Organization—WHO (2010). Résumé d'Orientation des Directives de I'OMS Relatives au Bruit dans l'Environmental. Disponível em: <http://www.who.int/home.page/>. Acesso em: 5 abr. 2010.

Wu, Y., Ding, C. (1998) Effect of fighter cockpit noise on pilot hearing. Space Med Eng (Beijing) 11: 52–5.

Occupational Safety and Hygiene – Arezes et al. (eds)
© *2013 Taylor & Francis Group, London, ISBN 978-1-138-00047-6*

Indoor air quality in community health centers: A preliminary study

L. Santos, J. Santos, A. Rebelo & M. Vieira da Silva
Research Centre on Environment and Health, Allied Health Sciences School of Polytechnic of Porto,
Vila Nova de Gaia, Portugal

ABSTRACT: Indoor air quality (IAQ) in healthcare facilities is of great concern for both patients and staff. The aim of this study was to evaluate the IAQ in Community Health Centers. Total aerobic count and fungal load were assessed using a microbiological air sampler. The determination of the other parameters was performed using real-time equipments. The average concentrations ranged between 161 CFU/m^3 and 1423 CFU/m^3 for fungi and between 147 CFU/m^3 and 999 CFU/m^3 for total aerobic count. Only three waiting rooms exceeded the concentration reference value for carbon dioxide. The concentration of particulate matter and carbon monoxide were above the limit value. In general, physical parameters are outside the comfort zones. These results demonstrate the need for on-going efforts to ensure adequate ventilation in Community Health Centers to maintain an acceptable IAQ. Therefore, it is essential to accomplish a long-term surveillance of IAQ in this type of buildings.

1 INTRODUCTION

The indoor air quality (IAQ) is a determinant factor of health and well-being (WHO, 2009; Cabral, 2010). Many studies developed by the U.S. Environmental Protection Agency (US-EPA) indicate that concentration of pollutant inside buildings can be two or five times—and occasionally, more than 100 times – higher than outdoors. These contamination levels are especially important when it is known that modern society spend 80–90% of our lives indoors (Wang et al., 2007; Herberger et al., 2010). The indicators of IAQ are a combination of physical, chemical and biological factors and also ventilation rate. The factors that directly affect IAQ are building materials, cleaning products, the occupants' activities and HVAC systems. Outdoor air also influences the IAQ and the principal sources are smoke from burning vehicles, industrial emissions, pollen and waste disposal (Marínez et al., 2006).

Hospitals and healthcare facilities have complex indoor environments due the different end uses of indoor spaces and functions (Balaras et al., 2007; Yau et al., 2011). In these environments, the most studied agents are the airborne bacteria and fungi. Fungi such as *Penicillium sp.* and *Aspergillus sp.* have often been detected as dominant in the air within clinical units (Araújo et al., 2008). Other microorganisms such as *Staphylococcus aureus*, *Pseudomonas aeruginosa* and *Micobacterium tuberculosis* are also associated with such infections (Cole et al., 1998; Ekhaise et al., 2008). Poor IAQ in

health care facilities is responsible for health care-associated infections (HAIs) and occupational diseases (Bartley et al.,2010; Wan et al., 2011).

Although numerous studies investigated the health outcomes of indoor air exposure in different types of buildings (commercial and public buildings), few data exist with respect to the IAQ in Community Health Centers.

The Portuguese Primary Health Care are provided in called Health Centers Groups (ACES). The ACES are administrative structures that include a multiple functional units and include one or more Community Health Centers. These units ensure the provision of primary health care to the population of a specific geographic area and have crucial importance in health promotion and disease prevention. Therefore, the control of IAQ plays an important role in the prevention of infection and others diseases associated to this type of occupation.

The aim of this study was to evaluate IAQ in four Community Health Centers, by indoor monitoring of chemical, microbiological and thermal parameters. In addition, potential emission sources of indoor/outdoor air pollutants were investigated.

2 MATERIAL AND METHODS

2.1 Sampling sites and descriptions

This study was conducted in four Community Health Centers (CHC) located in Porto, Portugal, during the summer of 2012. Environmental conditions were monitored during a work day. A total of 9 waiting rooms were analyzed regarding their structural and operational conditions (cleaning of spaces, type of ventilation, type of construction and occupancy). Indoor and outdoor air sampling were carry out during work time. The monitoring evolved the measurement of chemical parameters (CO_2, CO and particulate matter (PM_{10})), physical parameters (air temperature, relative humidity and air velocity) and microbiological parameters (Total Aerobic Count (TAC) and fungi). TAC reflects bacterial load and is an indicator of the microbiological quality of an environment (Ortiz et al., 2009). In order to maintain the confidentiality of data, the four CHC were coded as: CHC_1, CHC_2, CHC_3 and CHC_4.

The characterization of the structural and operational of the building (Table 1) was made with a checklist, which include the fields: general setup (coating materials, furniture, etc.), activities, ventilation systems and identification of potential pollution sources, indoor and outdoor.

2.2 Sampling collection and data analysis

Indoor samplers were placed in representative locations (at least 0,6 m above the floor, below the ceiling, and away from windows and doors). It was taken into account the room's layout, the doors and the windows location and the existence of inner or outer contamination sources. The measurement of chemical and physical parameters was based on the recommendations outlined in the Technical Note-NT-SCE-02 and the "Technical Guide for Indoor Air Quality", published by the Portuguese Environment Agency.

The microbiological sampling was carried out with a microbiological air sampler, working at a constant air flow rate of 100 litters per minute and using the following culture media: *Triptycase Soy Agar* (TSA) for TAC and *Malt Extract Agar* (MEA) for

fungi quantification. It was followed the National for Occupational Safety and Health (NIOSH) 0800 Method—Bioaerosol Sampling (Indoor Air) (NIOSH, 1998). All the plates were incubated at 37°C for TAC and at 25°C for fungi, during two days and five days, respectively. After the incubation period, the colonies were quantified and the Colony Forming Units per cubic meter (CFU/m^3) were calculated. The Table 2, indicate the equipment used to measure the environmental parameters.

2.3 Evaluation criteria

The values of the monitored parameters were compared with the current Portuguese evaluation criteria: Decree-law n.° 79/2006 of April 4th (Annex VII—concentration reference value of pollutants within the existing buildings) and the Decree-Law n.° 80/2006 of April 4th (article n.°14—Interior conditions reference). The reference values are presented in Table 3.

2.4 Statistical analysis

All statistical analyses were performed using IBM SPSS Statistics 20. Descriptive statistics were performed to describe the levels of indoor air parameters. The difference in pollutants levels between CHC was tested using non-parametric test of Kruskal-Wallis.

Spearman's rank correlation analyses were performed to determine the relationship between

Table 2. Parameter and equipment of measure.

Parameter	Equipment
CO	IAQ-Calc 8760
CO_2	IAQ-Calc 8760
PM_{10}	Dust Track
Viable microorganisms	MAS 100
Air's Temperature and Relative Humidity	IAQ-Calc 8760
Air's Velocity	VelociCalc 8345

Table 1. Characterization of waiting rooms.

		Floor	Occupants	Room capacity*	Ventilation
CHC_1	R_1	2nd	7	15	N**
	R_2	3rd	6	24	N and M**
	R_3	4th	16	30	M
CHC_2	R_4	−1st	20	11	N and M
	R_5	0	2	30	M
CHC_3	R_6	1st	12	20	N
	R_7	0	18	15	N
CHC_5	R_8	1st	6	30	N
	R_9	0	1	16	N

* Taking into account the number of seats;
** N—Natural; M—Mechanical.

Table 3. Reference values for indoor air quality.

Decree-Law n.° 79/2006 of April 4th

CO	CO_2	PM_{10}	TAC (37°C)	Fungi (25°C)	Air's velocity
10,7 ppm	984 ppm	0,15 mg/m³	500 CFU/m^3	500 CFU/m^3	<0,2 m/s

Decree-Law n.° 80/2006 of April 4th

Air's Temperature	Relative Humidity
20°C—Heating Season 25°C—Cooling Season	50%—Cooling Season

indoor and outdoor levels and between air quality variables.

An association was considered statistically significant for a p-value less than or equal to 0.05.

3 RESULTS AND DISCUSSION

3.1 Total Aerobic Count and fungi

During the surveillance period, a total of 78 air samples (indoor and outdoor) were collected. Of all waiting rooms evaluated 44% had concentrations of TAC and fungi above the concentration reference value (500 CFU/m³) given by Portuguese Decree-Law n.°79/2006. Higher concentration of fungi where observed in indoor environment as compared with results obtained for the same group in outdoor environment. The same result was obtained for TAC on 33% of the waiting rooms studied, in spaces where the CO_2 concentration was lower than concentration reference value (R_4, R_7 and R_9).

The Figure 1 show the TAC and fungi counts for four CHC studied corresponding a total of nine different waiting rooms. According to the figure, the average concentrations of fungi and TAC ranged between 161 CFU/m³ and 1423 CFU/m³ and between 147 CFU/m³ and 999 CFU/m³, respectively. Considering the levels of microorganisms present in the indoor environment, no significant differences has been found for TAC ($p = 0,086$) and fungi ($p = 0,435$) among the four CHC studied. The highest bacteria concentration was found in the waiting room R_7, located on the ground floor of the CHC_3. This result may be associated with the high number of occupants in this room (18 occupants for a seating capacity of 15). Another similar situation was observed in room R_4 (20 occupants for a seating capacity of 11) which explains the high concentration of bacteria (847 CFU/m³). Human presence is a potential source of airborne indoor bacteria (Loftness et al., 2007; Stanley et al., 2008; Wan et al., 2011) which contributes to the quality of indoor air in terms of microbial contaminations

(Jaffal, 1997; Ekhaise et al., 2008; Ekhaise et al., 2010; Awosika et al., 2012).

Despite scientific evidence, the results obtained showed that there was no significant statistical evidence ($p = 0.814$) to state that indoor bacteria is related to the number of occupants. The ventilation system is also a factor that can influence indoor's microorganism concentration. The absence and/or non-operation of mechanical and natural ventilation in several rooms may have also contributed to the high concentration of bacteria in these spaces, particularly in R_4, R_7 and R_9. Heating, ventilating and air-conditioning (HVAC) installations in health care facilities control indoor environmental quality and aseptic conditions, and a healthy, safe and suitable indoor air quality for both patients and staff (Dascalaki et al., 2008; Giuli et al., 2012). In addition, no significant statistical correlation was found between bacteria concentration and relative humidity (RH). These results are consistent with Wan et al. (2011).

Regarding the fungi indoor concentration the highest value was obtained in the waiting room R_3, located on −1st ground of CHC_1, exceeding more than two times the concentration reference value. This result could be related to the surrounding of this building, characterized by the proximity of green zones. Indoor fungal contamination depends on numerous factors, including moisture, ventilation, temperature, organic matter present in building materials and also on outdoor fungal load (Medrela-Kuder, 2003). There was no significant correlations between indoor and outdoor fungi concentration ($p = 0.259$), indoor fungi concentration and relative humidity ($p = 0,145$) and indoor fungi concentration and indoor air temperature ($p = 0.366$). Fungi concentrations obtained in outdoor air were similar to those reported by Satour et al. (2009), which one achieved concentrations for viable fungi between 8 and 496 CFU/m³ in outdoor environments at a French Hospital. However, the concentrations of fungi obtained in indoor environments at this hospital were much lower (between 4 and 26 CFU/m³) than those recorded in the present study.

In general the outdoor concentration of airborne bacteria and fungi were lower than those obtained in indoor environments. According to Mandal et al. (2011) the main factors to affect the level of airborne microorganisms might be not the cleanliness of this type of buildings but the activity of people, organic materials derived from the outdoors and ventilation efficiency applied to healthcare centers.

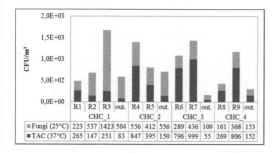

Figure 1. Indoor and outdoor fungi and TAC concentration.

3.2 Chemical pollutants and thermal parameters

The levels of CO_2 obtained for different CHC can be seen in Figure 2. There was no significant difference between indoor CO_2 levels measured in CHC ($p = 0.809$).

Figure 2 shows the mean concentration of CO_2 and the number of occupants by waiting room. In fact, only three waiting rooms showed levels above the limit value given by Portuguese legislation (984 ppm). Considering that CO_2 is a good indicator of the adequacy and efficiency of ventilation (Heudorf et al., 2009), the concentrations obtained may indicate that these areas have an insufficient ventilation rate. Rooms provided with a combined ventilation system (natural/mechanical system) had a lower concentration of CO_2 (527.5 ± 144.9) than natural system (867.7 ± 88.2) and mechanical system (1003.0 ± 230.4). However, the mechanical system was non-operational at the moment of measurements. Quian et al. (2010) referred that the major advantage of natural ventilation in healthcare facilities is the significant ventilation rates that can reduce cross-infection of airborne diseases.

Higher levels of CO_2 were positively correlated with number of occupants ($r = 0.697$; $p = 0.037$). These results are consistent with others authors (Heudorf et al., 2009; Zuraimi et al., 2008).

Figure 3 presents the mean concentration of CO and PM_{10} by waiting room. For indoor CO and PM_{10} concentration, there was no significant difference between the CHC ($p = 0.572$ and $p = 0,070$, respectively). The indoor CO concentrations ranged between 2.2 ppm and 10.0 ppm. The high levels of CO obtained in CHC_4 can be originated in outdoor. According to Chaloulakou et al.

(2002), in the absence of internal sources related to the emission of CO it may be suggested that the CO peaks found within the building are related to traffic. However, to understand the source of this agent, it would be necessary to study the variations inside and outside. Regarding PM_{10}, concentrations measured in the CHC were lower than the limit value of given by Portuguese legislation.

The air temperature in waiting rooms ranged between 23.4°C and 31.2°C and 44% of waiting room were above the limit value recommended (25°C). The relative humidity (RH) ranged between 26.6% and 62.4%. In 78% of the waiting rooms the RH was above the recommend limit value (50%).

In general, the air's velocity values obtained are below the threshold set in the Portuguese law (0.2 m/s), however, they are also below to the minimum value proposed by the IAQA (0.05 m/s), which can lead to the conclusion that the air renewal is not properly done.

4 CONCLUSIONS

The indoor airborne bacteria and fungi were higher than outdoor levels in most of the waiting rooms, which can be related with poor ventilation conditions. This potential IAQ-problem can be a source of HAIs due the enclosed spaces that can confine aerosols and allow them to build up to infectious levels. In general, the concentrations of others indoor air pollutants were below the recommended limit value. However, it is essential to ensure the existence and the correct operation of mechanical ventilation systems.

This study provides for the first time data about IAQ in the Portuguese Community Health Centers. Further to this work, it would be important to accomplish a long-term surveillance of IAQ in this type of buildings. In addition, there is also a need to compare different type of ventilation systems and their impact on IAQ parameters.

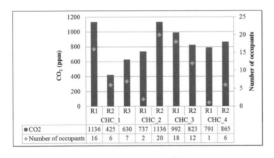

Figure 2. CO_2 mean concentration by waiting room.

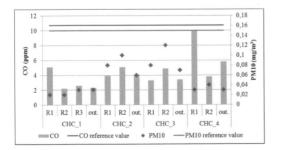

Figure 3.   CO and PM_{10} mean concentration by waiting room.

REFERENCES

Araújo, R., Cabral, J.P. & Rodrigues, A.G. 2008. Air filtration systems and restrictive access conditions improve indoor air quality in clinical units: Penicillium as a general indicator of hospital indoor fungal levels. *American Journal of Infection Control* 36: 129–134.

Awosika, S., Olajubu, F. & Amusa, N. 2012. Microbiological assessment of indoor air of a teaching hospital in Nigeria. *Asian Pacific Journal of Tropical Biomedicine* 2: 465–468.

Balaras, C.A., Dascalaki, E. & Gaglia, A. 2007. HVAC and indoor thermal conditions in hospital operating rooms. *Energy and Buildings* 39: 454–470.

Bartley, J.M., Olmsted, R.N. & Haas, J. 2010. Current views of health care design and construction: Practical implications for safer, cleaner environments. *American Journal of Infection Control* 38: S1–12.

Cabral, J.P.S. 2010. Can we use indoor fungi as bioindicators of indoor air quality? Historical perspectives and open questions. *Science of the Total Environment* 408: 4285–4295.

Chaloulakoua, A. & Mavroidis, I. (2002). Comparison of indoor and outdoor concentrations of CO at a public school. Evaluation of an indoor air quality model. *Atmospheric Environment* 36:1769–1781.

Cole, E.C. & Cook, C.E. (1998). Characterization of infectious aerosols in health care facilities: an aid to effective engineering controls and preventive strategies. *Association for Professionals in Infection Control and Epidemiology* 26: 453–464.

Dascalaki, E.G., Lagoudi, A., Balaras, C.A. & Gaglia, A.G. 2008. Air quality in hospital operating rooms. *Building and Environment* 43: 1945–1952.

Ekhaise, F.O., Ighosewe, O.U. & Ajakpovi, O.D. 2008. Hospital Indoor Airborne Microflora in Private and Government Owned Hospitals in Benin City, Nigeria. *World Journal of Medical Sciences* 3: 19–23.

Ekhaise, F.O., Isitor, E., Idehen, O. & Emogbene, O. 2010. Airborne microflora in the atmosphere of an hospital environment of University of Benin Teaching Hospital (UBTH), Benin City, Niger. *World Journal Agriculture Science* 6: 166–170.

Giuli, V., Zecchin, R., Salmaso, L., Corain, L. & Carli, M. 2012. Measured and perceived indoor environmental quality: Padua Hospital case study. *Building and Environment* 1–16.

Herberger, S., Herold, M., Ulmer, H., Burdack-Freitag, A. & Mayer, F. 2010. Detection of human effluents by a MOS gas sensor in correlation to VOC quantification by GC/MS. *Building and Environment* 45: 2430–2439.

Heudorf, U., Neitzert, V. & Spark, J. 2009. Particulate matter and carbon dioxide in classrooms—The impact of cleaning and ventilation. *International Journal of Hygiene and Environmental Health* 212: 45–55.

Jaffal, A.A., Banat, I.M., El Mogheth, A.A., Nsanze, H., Bener, A. & Ameen, A.S. 1997. Residential indoor airborne microbial populations in the United Arab Emirates. *Environment International* 23(4): 529–533.

Loftness, V., Hakkinen, B., Adan, O. & Nevalainen, A. 2007. Elements that contribute to healthy building design. *Environmental Health Perspectives* 115: 965–970.

Mandal, J. & Brandl, H. 2011. Bioaerosols in Indoor Environments—A Review with Special Reference to Residential and Occupational Locations. *The Open Environmental & Biological Monitoring Journal* 4: 83–96.

Marínez, F.J.R. & Callejo, R.C. 2006. Edificios saludables para trabajadores sanos: calidad de ambientes interiores. Available from: http://www.google.com/search?hl = pt-PT&q = Edificios+saludables+para+tr abajadores+sanos%3 Acaldad+de+ambientes+interio res&btnG = Pesquisar&aq = f&aqi = &aql = &oq = &-gs_rfai [Accessed September 2011].

Medrela-Kuder, E. 2003. Seasonal variations in the occurrence of culturable airborne fungi in outdoor and indoor air in Cracow. *International Biodeterioration & Biodegradation* 52: 203–205.

NIOSH *Manual of Analytical Methods* (NMAM). 1998. Bioaerosol Sampling (Indoor Air) 0800 Method., National Institute for Occupational Safety and Health (NIOSH). Fourth Edition, Issue 1.

Ortiz, G., Yagüe, G., Segovia, M. & Catalán, V. 2009. A Study of Air Microbe Levels in Different Areas of a Hospital. *Current Microbiology* 59: 53–58.

Qian, H., Li, Y., Seto, W.H., Ching, P., Ching, W.H. & Sun, H.Q. 2010. Natural ventilation for reducing airborne infection in hospitals. *Building and Environment* 45: 559–565.

Sautour, M., Sixt, N., Dalle, F., L'Ollivier, C., Fourquenet, V., Calinon, C., Paul, K., Valvin, S., Maurel, A., Aho, S., Couillault, G., Cachia, C., Vagner, O., Cuisenier, B., Caillot, D. & Bonnin, A. 2009. Profiles and seasonal distribution of airborne fungi in indoor and outdoor environments at a French hospital. *Science of the Total Environment* 407: 3766–3771.

Srikanth, P., Sudharsanam, S. & Steinberg, R. 2008. Bioaerosols in indoor environment: Composition, health effects and analysis. *Indian Journal of Medical Microbiology* 26(4): 302–312.

Stanley, N.J., Kuehn, T.H., Kim, S.W, Raynor, P.C, Anantharaman, S., Ramakrishnan, M.A. & Goyal, S.M. 2008. Background culturable bacteria aerosol in two large public buildings using HVAC filters as long term, passive, high-volume air samplers. *Journal of Environmental Monitoring* 10: 474–481.

Wan, G.H., Chung, F.F. & Tang, C.S. 2011. Long-term surveillance of air quality in medical center operating rooms. *American Journal of Infection Control* 39: 302–308.

Wang, S., Ang, H.M. & Tade, M.O. 2007. Volatic organic compounds in indoor environment and photocatalytic oxidation: State of the art. *Environment International* 33: 694–705.

WHO (World Health Organization), 2009. WHO Guidelines for Indoor Air Quality – Dampness and Mould. Available from: http://www.euro.who.int/__data/assets/pdf_file/0017/43325/E92645.pdf [Accessed January 2010].

Yau, Y.H., Chandrasegaran, D. & Badarudin, A. 2011. The ventilation of multiple-bed hospital wards in the tropics: A review. *Building and Environment* 46: 1125–1132.

Zuraimi, M.S. e Tham, K.W. 2008. Indoor air quality and its determinants in tropical child care centers. *Atmospheric Environment* 42: 2225–2239.

Occupational Safety and Hygiene – Arezes et al. (eds)
© 2013 Taylor & Francis Group, London, ISBN 978-1-138-00047-6

Miner's exposure to carbon monoxide and nitrogen dioxide in underground metallic mines in Macedonia

D. Mirakovski, M. Hadzi-Nikolova, Z. Panov, Z. Despodov & S. Mijalkovski
Faculty of Natural and Technical Sciences (FTNS), Goce Delcev University Stip, Macedonia

G. Vezenkovski
Sasa Mines, Makedonska Kamenica, Macedonia

ABSTRACT: The largest mining companies in Macedonia, in collaboration with Goce Delcev University and MOHSA, launched a campaign to measure the exposure of miners to CO and NO_2 while working in underground metallic mines. The aim of this campaign is to provide exposure data for risk assessment, and to develop an effective assessment program. This study took place in two mines, where the subjects were groups of workers that were involved in the operation of diesel powered equipment and blasting during a full shift of 8 hours of exposure. In each mine, two groups were assessed: workers in both production and development/service areas. Average exposures differentiate between groups and working positions, indicating diesel powered equipment as the main source of pollution. Ventilation efficiency played a significant role in the overall exposure levels, which is clearly indicated for all working positions in the development group primarily operating under local exhaust ventilation systems.

1 INTRODUCTION

The modern mining industry, due to an ever increasing intensity of production processes, including more powerful diesel equipment and increased blasting frequency and power, is experiencing an increase in the risk of miners' exposure to potentially harmful gases such as CO, CO_2, NO and NO_2.

With this in mind, as well as the lack of site-specific data which would reflect the actual exposure of miners to these gases, the largest national mining companies through MMA—Macedonian Mining Association, in collaboration with Mining Engineering Department at FTNS Goce Delcev University in Stip, and MOHSA Macedonian Occupational Health and Safety Association, launched an exposure assessment campaign to address the issue at hand. The campaign includes two hard rock metallic mines: Mine A (an underground operation having a total output of approximately 750.000 tons per year), and Mine B (a surface operation having a total output exceeding 8.000.000 tons per year).

The first step of the campaign was focused on determining the miners' exposure to CO and NO_2,

with the aim to provide solid exposure da-ta for a risk assessment, to develop efficient, cost effective and easily applicable assessment programs, as well as to recommend additional protection/control measures as needed.

A brief description of the methodology deployed, and results gained, from Mine A are discussed below.

2 DESCRIPTION OF THE MINING PROCESS

Mine A, which is an underground operation, is a lead and zinc mine producing about 750.000 tons of raw ore per year. This mine operates a fleet of about 50 diesel powered vehicles, as well as various other diesel powered equipment, which results in the level of installed diesel power reaching a combined total of about 3000 kW. Due to the high power and utilization time, the loaders and haulage trucks are the most significant source of diesel emissions, while production blasting is the most significant source of blasting fumes. Most other production machines, i.e. those used in blast-hole drilling, roof bolting and ex-plosive loading, are powered by electricity. Thus, the diesel engines are primarily used for the transfer of workers, equipment, and materials from one work-place to the other.

These vehicles, being primarily comprised of off-road vehicles equipped with four-wheel-drive, have been modified to suit their intended purposes. Other service machines in operation, such as fork lifts, cranes, and road graders are relatively smaller diesel powered multipurpose machines.

Due to specific geological conditions, a sublevel caving is applied. Ore body is assessed through horizontal drifts and access ramps, while shafts are used for ventilation and the transport of ore. Production and development processes are exclusively per-formed by drill and blast methods (Fig.1).

The quality of the underground atmosphere is controlled only by an exhaust ventilation system, which is powered by one main fan (500 kW) and two "buster" underground fans. Fresh air, having been drawn through drifts and intake shafts, is guided to the individual production areas where it is then distributed by local ventilation systems to the several working areas on different active sublevels. Used air is then returned back to the exhaust shafts in isolated drifts.

Air quality requirements are strictly defined in national mining regulation. Threshold limit values for carbon monoxide and nitrogen dioxide are given bellow (Tab. 1). The same regulations prescribe minimal control measures and assessment procedures.

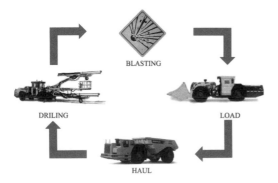

Figure 1. Production/development process.

Table 1. Treshold limit values.

	Threshold limit values*	
	TWA (8 hours) in ppm	STEL (15 minutes) ppm
CO	30	120
NO$_2$	5	/

* valid at the time of investigations.

3 MATERIALS AND METHODS

Within Mine A, an underground operation with 6 active production areas, a group of exposed workers includes operators of diesel powered equipment, blasting specialist and production supervisors. Due to a difference of working conditions, and a suspected level of exposure, two sub-groups where formed, one to include workers from the production areas under the general ventilation system, and another comprised of workers from development areas where auxiliary ventilation is usually applied. The group of workers from the production areas included two 2 diesel loader drivers, 2 jumbo drill operators, and 2 blasting specialists, while the group of workers from the underground construction areas consisted of 1 diesel loader driver, 1 jumbo drill operator, and 1 blasting specialist. The supervisor of each group was also included in the assessment. Real full shift exposure (spanning from the entry point, down into the mine itself, and back to the point of exit) to car-bon monoxide and nitrogen dioxide was determined. The workers were monitored for 3 consecutive shifts (I, II and III), with each worker wearing a Gastec direct-read dosimeter tube (Fig. 2) placed on the lapel within the breathing zone.

The tubes provide simple and reliable direct-read TWA (time-weighted average) monitoring of targeted chemicals. Since there is no need for user calibration, extra equipment, laboratory analysis or extensive training, the dosimeter tubes reduce costs, administrative and maintenance time, and the possibility of user error. Dosimeter tubes operate by direct diffusion exposure, and are highly sensitive and selective to the targeted chemicals, as opposed to other non-specific testing methods.

As a quality control, one of the workers in each shift wore a VRAE PGM 7800 hand-held monitor with built-in electrochemical CO and NO$_2$ sensors, sampling pump and data logging. The monitor was calibrated to clean air prior to each use, and span gases calibration was performed once a week.

The data collected were used to calculate TWA for 8 hour shifts and compared against the dositube readings.

In general, there was good correlation between TWA obtained from the tube (9,25 ppm) readings against TWA calculated from real-time concentration measurements (7,97 ppm) as shown in Figure 3.

Due to the small amount of data obtained in the first phase of the study, statistical processing was not possible.

All data were recorded in predefined measurement protocols, including photos of the tubes and raw readings from the handheld monitor.

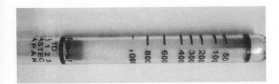

Figure 2. Carbon monoxide dosimeter tube after 8 hours usage.

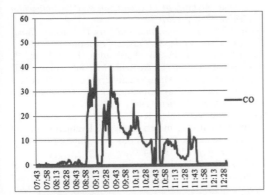

Figure 3. Carbon monoxide exposure during the shift of a supervisor.

4 RESULTS AND DISCUSSION

Compiled assessment data, including 36 readings for each pollutant from Mine A, are given in the Table 2.

Although exposure above national regulations was not noted, the average exposures determined could be regarded as significant.

We have found that the average exposure levels differentiate between the groups and working positions, indicating that diesel powered equipment was the main source of pollution, thus the loader drivers were shown to have the highest levels of exposure during their 8 hour shifts (15,84 ppm for carbon monoxide and 1,52 ppm for nitrogen dioxide) relative to employees working in other positions within the same timeframe (Tab. 3).

The efficiency of underground ventilation also plays a significant role in overall exposures, which is clearly indicated for all working positions in the development group usually operating under local exhaust ventilation systems. Workers in the development group, on average, were exposed at a rate of 10–48% higher than corresponding positions in the production group (Tab. 3).

Average exposures of different working positions obtained in this study are generally higher, compared to data from the extensive study in German potash mines (Dahman, Monz, Sönksen 2007).

The full shift exposures of LHD drivers obtained in our study are significantly higher compared to the same in German potash mines as shown in Figure 4.

Table 2. 8 hour's TWA exposure in Mine A.

Working position	Shift I CO ppm	Shift I NO₂ ppm	Shift II CO ppm	Shift II NO₂ ppm	Shift III CO ppm	Shift III NO₂ ppm
Production group						
LHD Driver 1	11,85	1,325	15,5	1,425	13,25	1,55
LHD Driver 2	9,75	1,075	12,25	1,05	11,53	0,95
Drilling operator 1	10,55	0,75	9,25	0,75	8,51	0,25
Drilling operator 2	7,50	0,25	8,50	0,50	6,52	0,25
Blasting operator 1	8,20	0,55	8,75	0,95	11,25	2,15
Blasting operator 2	4,50	0,25	7,325	0,75	9,75	1,85
Supervisor 1	10,25	0,87	7,85	0,25	7,85	0,50
Supervisor 2	9,25	0,55	5,25	0,25	5,55	0,25
Development group						
LHD Driver	22,5	2,50	25,80	2,25	19,85	1,55
Drilling operator	16,37	1,85	14,75	2,15	12,25	1,25
Blasting operator	11,25	1,55	10,05	1,85	11,85	2,05
Supervisor	12,50	1,25	12,5	1,50	9,85	1,15

Table 3. Average exposure of working positions in different groups.

Working position	Average exposure (8 hour's TWA) Production group CO ppm	Production group NO₂ ppm	Development group CO ppm	Development group NO₂ ppm
LHD drivers	12,41	1,23	15,84	1,52
Drilling operator	8,47	0,46	10,47	0,89
Blasting operator	8,30	1,08	9,21	1,33
Supervisor	7,67	0,45	8,98	0,73

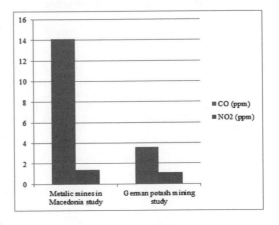

Figure 4. Average exposure of LHD drivers.

Drilling operators exposures also show similar relations, while other working positions are not comparable between studies.

The difference could be explained with different mining conditions and methods applied, but measures for improvement of the exposure situation in Macedonian mines are still necessary to achieve state of the art conditions as those found in German potash mines.

5 CONCLUSIONS

The results of this study clearly indicate that diesel equipment is the main source of carbon monoxide and nitrogen dioxide in underground metal mines, and although blasting can contribute to nitrogen di-oxide generation, if proper ventilation is applied, exposure to blasting fumes is rather insignificant. Of all the working positions that were studied, diesel loader drivers had the highest average expo-sure compared with other working positions. In general, the average exposures determined could be regarded as significant, although exposure above national regulations was not noted. Proper ventilation measures are the most efficient method for con-trolling underground atmosphere quality, thus con-trolling workers exposure to harmful gases. As such, every effort should be made to implement appropriate ventilation in all underground work areas. As part of the measurement campaign plan, 3 measurement sessions, conducted according to the pattern and methodology mentioned above, should be completed in each of the mines. The data collected should allow for the proper evaluation of miners' exposure in hard rock metal mines, and comparative analysis of exposures between different groups of workers at surface and underground operations. Additionally, quantum data will support the use of direct-read dosimeter tubes as an effective assessment method within specific terms of usage.

REFERENCES

Coble, J.B., Stewart, P.A., Vermeulen, R., Yerebm, D., Stanevich, R., Blair, A., Silverman, D.T., Attfield, M. 2010. The Diesel Exhaust in Miners Study: II. Exposure monitoring surveys and development of exposure groups. *The Annals of Occupational Hygiene*: 54(7): 747–61.

Dabill, D.W. 2004. Controlling and monitoring exposure to diesel engine exhaust emissions in non-coal mines. *HSE Books*: Sudbury: HSE Information Services.

Dahmann, D., Monz, C., Sönksen, H. 2007. Exposure assessment in German potash mining. *International Archives of Occupational and Environmental Health*: 81(1): 95–107.

Dahmann, D., Morfeld, P., Monz, C., Noll, B., Gast, F. 2009. Exposure assessment for nitrogen oxides and carbon monoxide in German hard coal mining. *International Archives of Occupational and Environmental Health*: 82(10): 1267–79.

Gastec. 2012. *Environmental Analysis Technology Handbook* (12th ed.). Japan: Gastec Corporation.

Stewart, P.A., Coble, J.B., Vermeulen, R., Schleiff, P., Blair, A., Lubin, J., Attfield, M., Silverman, D.T. 2010. The diesel exhaust in miners study: I. Overview of the exposure assessment process. *The Annals of Occupational Hygiene*: 54(7): 728–46.

Stewart P.A, Vermeulen R, Coble JB et al. (2012) The diesel exhaust in miners study: V. Evaluation of the exposure assessment methods. *The Annals of Occupational Hygiene*: 56: 389–400.

US EPA. 2002. Health assessment document for diesel engine exhaust. *National Center for Environmental Assessment EPA/600/8–90/057F*: Washington, DC: US EPA.

Vermeulen R, Coble J.B., Lubin, J.H. et al. 2010. The diesel exhaust in miners study: IV. Estimating historical exposures to diesel exhaust in underground non-metal mining facilities. *The Annals of Occupational Hygiene*: 54: 774–88.

Vermeulen, R., Coble, J.B., Yereb, D. et al. 2010. The diesel exhaust in miners study: III. Interrelations between respirable elementary carbon and gaseous and particulate components of diesel exhaust derived from area sampling in underground non-metal mining facilities. *The Annals of Occupational Hygiene*: 54: 762–73.

Yanowitz, J., McCormick, R.L., Graboski, M.S. 2000. In-use emissions from heavy-duty vehicle emissions. *Environmental Science & Technology*: 34: 729–40.

Zey, J.N., Stewart, P.A., Hornung, R. et al. 2002, Evaluation of side-by-side pairs of acrylonitrile personal air samples collected using different sampling techniques. *Applied Occupational and Environmental Hygiene*: 17: 88–95.

Occupational Safety and Hygiene – Arezes et al. (eds)
© 2013 Taylor & Francis Group, London, ISBN 978-1-138-00047-6

On the nature of hearing protection devices usage prediction

S. Costa & P.M. Arezes

Human Engineering group, Center for Industrial and Technology Management,
School of Engineering of the University of Minho, Guimarães, Portugal

ABSTRACT: Throughout the years, several investigators have placed the burden of their work on prediction, a fact that is explained by the desirability of such researches. Researchers focused in occupational settings matters, for instance, benefit greatly from the ability to predict, inasmuch as they acquire knowledge of particular features of human behaviour in advance. In the occupational field, factors that encourage or inhibit some type of behaviour, influence exposure and convey a perception are named predictors. This paper aims at analysing published studies on probable predictors of the use of Hearing Protection Devices in the occupational arena, through searches in databases relevant to the occupational field. It is concluded that the search and determination of predictors has indeed motivated the work of many researchers in many science disciplines. These studies have, in turn, impelled the emergence of many successful models in the occupational matters, which will grant more anticipated knowledge and render a more effective health and safety practitioners' work.

1 INTRODUCTION

On what concerns to occupational health and safety, assessing occupational risks, exposure to hazards and workers' health status are routine to health and safety practitioners. On the front line of this science, as on many other fields, are those who want to 'think ahead', i.e., those who privilege knowledge on beforehand. The ability to predict, given a certain context, has inspired many researchers throughout a multiplicity of science fields. This is understandable, given the convenience of such studies. Once a researcher has figured out what the key issues involving a certain context are, holding knowledge on the variables and their effect in a desirable result, the disclosure of the systematization of the problem will provide readily available solutions to cases that resemble the one studied. Moreover, the identification of those key factors (also known as predictors, determinants, indicators, etc. and, from now on, referred to as predictors) will allow for the construction of models that will theorize the problematic to a wider range of problems of the occupational settings. The most obvious advantage for researchers focused in occupational settings matters, is the attainment of particular traits of human behaviour (objectives, activities, likes and dislikes) in advance (Zukerman & Albrecht 2001). In the occupational arena, predictors may refer to factors that stimulate or inhibit some kind of (positive or negative) behav-

iour (Arezes & Miguel 2002b, 2003, 2005a, b, 2006, 2008, 2012, Edelson et al. 2009, Griffin et al., 2009, Kerr et al., 2002, Kushnir et al., 2006, Lusk et al., 1995, 1997a, b, 1998, 1999, 2003, 2004, Lusk & Kelemen 1993, McCullagh et al., 2002, Melamed et al., 1994, 1996, Sbihi et al. 2010, Suter 2002), that instil a perception (Alayrac et al., 2010), or even have influence on exposure (Abel et al., 1985, Burstyn et al. 2000a, b, Cavallari et al., 2012a, b).

According to NIOSH (National Institute for Occupational Safety and Health), the number of workers that are exposed to dangerous occupational noise levels in the US exceed 30 million, and hover around the 500 million all over the world (Arezes & Miguel 2002a, McReynolds 2005). Arezes and Miguel (2002a) estimated about 780 thousand workers in Portugal to have been exposed to occupational noise in the years 1998, 2000 and 2001.

The incurable occupational illness noise-induced hearing loss (NIHL) has a high prevalence among construction workers, even though it is liable to be completely prevented with the implementation of a proper hearing conservation program (Alberti 1992, Edelson et al., 2009, Goelzer et al., 2001, Rabinowitz, 2000).

In cases where the use of hearing protection devices (HPDs) is compulsory, their efficacy relies on their correct use, and only the use of HPDs during the total amount of time when exposed to high-noise levels grants effective protection (Brady, 1999).

The consequences of occupational noise exposure to the worker's health, performance and intercommunication has been the focus of studies of many researchers which, in turn, have enhanced the importance of hearing conservation programs (Arezes & Miguel 2002a, Concha-Barrientos et al. 2004, Dias et al. 2006, Kaczmarska et al. 2004, Kotabinska & Kozlowski 2005, McReynolds 2005, Popescu 2005).

The recourse to HPDs is not, by far, the best choice of a strategy to prevent NIHL, but the last alternative to protect the hearing of workers exposed to occupational noise, when all other protective measures, including legal measures of protection, are neither efficient nor enforceable in a given context. Moreover, in general, HPDs are preferred to technical and engineering controls, since the latter are more laborious and onerous measures (Arezes & Miguel 2005, Riko & Alberti 1983). Therefore, HPDs are still nowadays the main method of prevention of NIHL regardless of their constraints, very much so like almost three decades ago (Seixas et al. 2010, Malchaire & Piette 1997, Riko & Alberti 1983).

Even though the burden of NIHL prevention be placed in HPDs, these tools may not be effective in protecting the workers (Malchaire & Piette, 1997). In fact, the role of HPDs in protecting the workers' hearing against exposure to occupational noise is determined by the "real" attenuation provided by HPDs worn by workers exposed to occupational noise. The search for the predictors of the use of HPDs can give origin to more meaningful data regarding noise exposure, clarify noise exposure itself and, consequently, allow for better NIHL prevention strategies be developed.

This paper aims at analysing published studies that focused the discovery and testing of probable predictors in the occupational context of the use of Hearing Protection Devices (HPDs).

2 MATERIALS AND METHOD

Materials used for the pursuit of this paper's objective were obtained through searches in databases relevant to the occupational field (see References).

3 RESULTS

Early in 1994, Damongeot (1994) presented a range of HPDs, arranging them into several types according to their classification, and added recommendations to take into account in their selection (including when to avoid them), based on the noise exposure type, nature of the noise, environment conditions, nature of the task, other individual protective equipment and human factors, displaying some predictors of the use of HPDs.

Arezes and Miguel have engaged on the pursuit of the predictors of the use of HPDs for several occasions, and found that individual risk perception, subjective opinion on the company's safety climate and judgment on the complex concept of comfort for the HPD must be considered as important predictors of the workers' protective behaviour (Arezes & Miguel 2002b, 2003, 2005a, b, 2006, 2008, 2012). Education through interventions, age of the worker, trade group, seniority, perceived barriers to the use of HPDs and self-efficacy in their usage (the latter identified as the main predictor of the use of HPDs in Arezes and Miguel 2005b), acknowledged value of the use of HPDs and stemming benefits were found by Lusk, Kerr, and other colleagues, to be important predictors of the use of HPDs (Kerr et al, 2002, Lusk et al. 1995, 1997, 1998, 1999, 2003, Lusk & Kelemen 1993). Nevertheless, the researchers find that some of these variables are mediated by gender, education level and ethnicity (Kerr et al., 2002, Lusk et al. 1997). The results obtained by these authors meet the results of Kushnir et al. (2006) where self-efficacy and perceived barriers were found to be highly correlated with HPDs' use, as were the dysfunctional thinking patterns, meaning that the use of HPD is governed not only by rational, but also by irrational motivations (Kushnir et al. 2006).

Other researchers also discovered that perceived self-efficacy and self-perceived susceptibility to hearing loss were important predictors of the use of HPDs. These add up to noise annoyance and education, also found to be factors to take into account (Melamed et al. 1996).

The Pender's Health Promotion Model (HPM) has boosted the knowledge of HPDs' usage predictors. This model relies on social learning theory to explain workers health-promoting behaviours. It establishes factors of diverse nature that likely affect the actions of workers. The author has ever since revised the model, and several other authors have suggested amendments to that model and adapted to their own contexts, perpetuating the cycle of the development of the HPM—discovery of HPDs usage predictors (Lusk & Kelemen 1993, Lusk et al. 1995, 1997, Pender et al. 2011, Ronis et al. 2006). In short, the HPM rests on two spheres to assess major determinants of health behaviors. The first sphere encompasses the individual characteristics and experiences of the subject, in which prior related behaviour and, biological, psychological and sociocultural factors will set the grounds for a subject to tend to behave in a certain manner. The second sphere covers behavior-specific cognitions and affect, which includes perceived benefits of action, perceived barriers to action, perceived self-efficacy, activity-related affect, interpersonal influences of peers, family and providers, situational influences (such as options, demand characteristics and aesthetics), commitment to a plan of action and immediate competing demands and preferences), which will ultimately influence the subject's demeanor (Pender et al. 2011).

4 CONCLUSIONS

The literature review allowed the identification and compilation of the most important predictors of the use of HPDs. Given that the peers' results support each other, it seems reasonable to declare that, the factors of interest to the assessment of the use of HPDs are: individual risk perception, subjective opinion on the company's safety climate, individual judgment on comfort (which is closely related to another predictor—noise annoyance), education, age, trade group, seniority, perceived barriers to the use of HPDs and self-efficacy in their usage, acknowledged value of the use of HPDs and deriving benefits, gender, ethnicity, dysfunctional thinking patterns and self-perceived susceptibility to hearing loss.

The search and determination of predictors, also called determinants, has stimulated the work of many researchers in many science disciplines. It has driven the development of several successful models in the occupational matters and with great acceptability among the scientific community. These in turn, will provide more anticipated knowledge and, hence, make more effective the health and safety practitioners actions, by diminishing the time needed for assessments and by providing a bulk of early information.

ACKNOWLEDGMENTS

The author S. Costa expresses her acknowledgments for the support given by the Portuguese Foundation for Science and Technology (FCT) through the PhD grant SFRH/BD/69161/2010. This work was also partially funded with Portuguese national funds through Portuguese Foundation for Science and Technology (FCT), in the scope of the R&D project PEst-OE/EME/UI0252/2011.

REFERENCES

Abel, S.M., Kunov, H., Pichora-Fuller, M.K. & Alberti, P.W. 1985. Signal detection in industrial noise: Effects of noise exposure history, hearing loss, and the use of ear protection. Scandinavian Audiology 14(3): 161–173.

Alayrac, M., Marquis-Favre, C., Viollon, S., Morel, J. & Le Nost, G. 2010. Annoyance from industrial noise: indicators for a wide variety of industrial sources. The Journal of the Acoustical Society of America 128: 1128.

Alberti, P.W. 1992. Noise induced hearing loss. British Medical Journal 304(6826): 522.

Arezes, P.M. & Miguel, A.S. 2002a. A exposição ocupacional ao ruído em Portugal. Revista Portuguesa de Saúde Pública 20(1): 61–69.

Arezes, P.M. & Miguel, A.S. 2002b. Hearing protectors acceptability in noisy environments. Annals of Occupational Hygiene 46(6): 531–536.

Arezes, P.M. & Miguel, A.S. 2003. Risk perception and hearing protection use. Proceedings of the 8th International Congress on Noise as a Public Health Problem: Rotterdam, the Netherlands, 29 June-3 July 2003: 30.

Arezes, P.M. & Miguel, A.S. 2005a. Individual perception of noise exposure and hearing protection in industry. Human Factors: The Journal of the Human Factors and Ergonomics Society 47(4): 683–692.

Arezes, P.M. & Miguel, A.S. 2005b. Hearing protection use in industry: The role of risk perception. Safety Science 43(4): 253–267.

Arezes, P.M. & Miguel, A.S. 2006. Does risk recognition affect workers' hearing protection utilisation rate? International Journal of Industrial Ergonomics 36(12): 1037–1043.

Arezes, P.M. & Miguel, A.S. 2008. Risk perception and safety behaviour: A study in an occupational environment. Safety Science 46(6): 900–907.

Arezes, P.M. & Miguel, A.S. 2012. Assessing the use of hearing protection in industrial settings: A comparison between methods. International Journal of Industrial Ergonomics. Retrieved August 15, 2012, from http://www.sciencedirect.com/science/article/pii/S0169814112000637.

Brady, J. 1999. Training to promote worker's use of hearing protection: The influence of work climate factors on training effectiveness. PhD Thesis, Michigan State University: Michigan, 274 pgs.

Burstyn, I., Kromhout, H. & Boffetta, P. 2000a. Literature review of levels and determinants of exposure to potential carcinogens and other agents in the road construction industry. American Industrial Hygiene Association Journal 61(5): 715–726.

Burstyn, I., Kromhout, H., Kauppinen, T., Heikkilä, P. & Boffetta, P. 2000b. Statistical modelling of the determinants of historical exposure to bitumen and polycyclic aromatic hydrocarbons among paving workers. Annals of Occupational Hygiene 44(1): 43–56.

Cavallari, J.M., Osborn, L.V., Snawder, J.E., Kriech, A.J., Olsen, L.D., Herrick, R.F. & Mcclean, M.D. 2012a. Predictors of airborne exposures to polycyclic aromatic compounds and total organic matter among hot-mix asphalt paving workers and influence of work conditions and practices. Annals of Occupational Hygiene 56(2): 138–147.

Cavallari, J.M., Osborn, L.V., Snawder, J.E., Kriech, A.J., Olsen, L.D., Herrick, R.F. & Mcclean, M.D. 2012b. Predictors of dermal exposures to polycyclic aromatic compounds among hot-mix asphalt paving workers. Annals of Occupational Hygiene 56(2): 125–137.

Concha-Barrientos, M., Campbell-Lendrum, D., Steenland, K. & others. 2004. Occupational noise:

assessing the burden of disease from work-related hearing impairment at national and local levels (Vol. 9). Oms.

Damongeot, A. 1994. Les protecteurs individuels contre le bruit (PICB). Performances, choix, utilisation. Cahiers de Notes Documentaires-Hygiène et Sécurité du Travail 155: 169–179.

Dias, A., Cordeiro, R. & Gonçalves, C.G.O. 2006. Exposição ocupacional ao ruído e acidentes do trabalho. Cadernos de Saúde Pública 22(10): 2125–2130.

Edelson, J., Neitzel, R., Meischke, H., Daniell, W., Sheppard, L., Stover, B. & Seixas, N. 2009. Predictors of hearing protection use in construction workers. Annals of Occupational Hygiene 53(6): 605.

Goelzer, B., Hansen, C.H. & Sehrndt, G.A. 2001. Occupational exposure to noise: evaluation, prevention and control. Federal Institute for Occupational Safety and Health.

Griffin, S.C., Neitzel, R., Daniell, W.E. & Seixas, N.S. 2009. Indicators of hearing protection use: self-report and researcher observation. Journal of Occupational and Environmental Hygiene 6(10): 639–647.

Kaczmarska, A., Mikulski, W. & Smagowska, B. 2004. Noise in office rooms, Proceedings from the Eleventh International Congress on Sound and Vibration, St. Petersburg, Russia: 1299–1306.

Kerr, M.J., Lusk, S.L. & Ronis, D.L. 2002. Explaining Mexican American workers' hearing protection use with the health promotion model. Nursing Research 51(2): 100–109.

Kotabinska, E. & Kozlowski, E. 2005. Speech intelligibility in noise when hearing protectors are used. Magazine of the European Agency for Safety and Health at Work 8: 29–31.

Kushnir, T., Avin, L., Neck, A., Sviatochevski, A., Polak, S. & Peretz, C. 2006. Dysfunctional thinking patterns and immigration status as predictors of hearing protection device usage. Annals of Behavioral Medicine 32(2): 162–167.

Lusk, S.L. & Kelemen, M.J. 1993. Predicting use of hearing protection: A preliminary study. Public Health Nursing 10(3): 189–196.

Lusk, S.L., Ronis, D.L. & Kerr, M.J. 1995. Predictors of hearing protection use among workers: Implications for training programs. Human Factors: The Journal of the Human Factors and Ergonomics Society 37(3): 635–640.

Lusk, S.L., Ronis, D.L., & Baer, L.M. 1997a. Gender differences in blue collar workers' use of hearing protection. Women & Health 25(4): 69–89.

Lusk, S.L., Ronis, D.L. & Hogan, M.M. 1997b. Test of the health promotion model as a causal model of construction workers' use of hearing protection. Research in Nursing & Health 20(3):183–94.

Lusk, S.L., Kerr, M.J. & Kauffman, S.A. 1998. Use of hearing protection and perceptions of noise exposure and hearing loss among construction workers. American Industrial Hygiene Association 59(7): 466–470.

Lusk, S.L., Hong, O.S., Ronis, D.L., Eakin, B.L., Kerr, M.J., & Early, M.R. 1999. Effectiveness of an inter-

vention to increase construction workers' use of hearing protection. Human Factors: The Journal of the Human Factors and Ergonomics Society 41(3): 487–494.

Lusk, S.L., Ronis, D.L., Kazanis, A.S., Eakin, B.L., Hong, O.S. & Raymond, D.M. 2003. Effectiveness of a tailored intervention to increase factory workers' use of hearing protection. Nursing Research 52(5): 289.

Lusk, S.L., Eakin, B.L., Kazanis, A.S., & McCullagh, M.C. 2004. Effects of booster interventions on factory workers' use of hearing protection. Nursing Research 53(1): 53–58.

Malchaire, J. & Piette, A. 1997. A comprehensive strategy for the assessment of noise exposure and risk of hearing impairment. Annals of Occupational Hygiene 41(4): 467.

McCullagh, M., Lusk, S.L. & Ronis, D.L. 2002. Factors influencing use of hearing protection among farmers: a test of the Pender health promotion model. Nursing Research 51(1): 33.

McReynolds, M.C. 2005. Noise-induced hearing loss. Air Medical Journal, 24(2): 73–78.

Melamed, S., Rabinowitz, S., Feiner, M., Weisberg, E. & Ribak, J. 1996. Usefulness of the protection motivation theory in explaining hearing protection device use among male industrial workers. Health Psychology 15(3): 209.

Melamed, S., Rabinowitz, S., & Green, M.S. 1994. Noise exposure, noise annoyance, use of hearing protection devices and distress among blue-collar workers. Scandinavian Journal of Work, Environment & Health: 294–300.

Pender, N.J., Murdaugh, C. & Parsons, M.A. 2011. Health Promotion in Nursing Practice (6th ed.). Boston, MA: Pearson.

Popescu, D.I. 2005. Occupational exposure to noise in the machine building industry. Proceedings from the Twelfth International Congress on Sound and Vibration, Lisboa, Portugal.

Rabinowitz, P.M. 2000. Noise-induced hearing loss. American Family Physician 61(9): 2759–2760.

Riko, K. & Alberti, P. 1983. Hearing protectors: a review of recent observations. Journal of Occupational and Environmental Medicine 25(7): 523.

Ronis, D.L., Hong, O.S. & Lusk, S.L. 2006. Comparison of the original and revised structures of the health promotion model in predicting construction workers' use of hearing protection. Research in Nursing & Health 29(1): 3–17.

Sbihi, H., Teschke, K., Macnab, Y.C. & Davies, H.W. 2010. Determinants of use of hearing protection devices in Canadian lumber mill workers. Annals of Occupational Hygiene 54(3): 319–328.

Suter, A.H. 2002. Construction noise: exposure, effects, and the potential for remediation; a review and analysis. American Industrial Hygiene Association Journal 63(6): 768–789.

Zukerman, I. & Albrecht, D.W. 2001. Predictive statistical models for user modeling. User Modeling and User-Adapted Interaction 11(1): 5–18.

Occupational Safety and Hygiene – Arezes et al. (eds)
© 2013 Taylor & Francis Group, London, ISBN 978-1-138-00047-6

Exposure of children to VOCs in European schools and homes environments: A systematic review

J. Madureira, I. Paciência, G. Ventura & E. de Oliveira Fernandes
Institute of Mechanical Engineering, Faculty of Engineering of University of Porto, Porto, Portugal

ABSTRACT: The deterioration of the indoor air quality is currently due to the occurrence of a large number of chemicals. This paper summarizes recent data on the occurrence of most relevant volatile organic compounds in the indoor air of European schools and homes as prioritary indoor environments for children due to their particular susceptibility to air pollutants exposures and the long time spent in those spaces. The review process shows that significant differences do exist within and among countries, corresponding to the wide spectrum of differences in sources and emissions strength of airborne chemicals. This study highlights the need for harmonized sampling or auditing protocols that shall allow for a higher comparability and better conditions for the interpretation of the data. Nevertheless, it results clear the message for the need to consider the regulation of sources and other indoor air pollution determinants in indoor environments, including ambient air.

1 INTRODUCTION

Indoor air quality (IAQ) deterioration can be due to the occurrence of a number of chemicals and particulate matter. Volatile organic compounds (VOCs) are probably the most important categories of chemicals found indoors among which, the most common are benzene, toluene, xylenes (BTX), terpenes (e.g. α-pinene and limonene); and formaldehyde and acetaldehyde (ECA, 1995). The number of VOCs is quite high and their toxicity varies significantly. The reasons for the broad occurrence of those chemicals are their volatile character and the fact that they have been associated with a large number of products like paints, adhesives, glues, varnishes, waxes, solvents, detergents or cleaning products, carpets and personal care products. It has been also proven that they are emitted from (upholstered) furnishings as desks, shelves and chairs as well as by the use of electronic devices like photocopiers or printers, etc. VOCs were found to be related to building dampness (Yu et al., 2006, Katsoyiannis et al., 2008, Oliveira Fernandes et al., 2008). According to Kotzias D. et al. (2005), benzene and formaldehyde were classified in the Group 1 (high priority chemicals), acetaldehyde, toluene and xylenes, constituted the Group 2 (second priority chemicals) and, α-pinene and limonene were represented in the Group 3 (chemicals requiring further research with regard to human exposure or dose-response). For many of those chemicals, the impact on human health is almost totally unknown and difficult to predict because of the lack of toxi-cological and/or epidemiological data. Besides, a full toxicological testing as requested by the "existing chemicals" legislation is difficult to accomplish for those compounds in the air, because of their in general low concentrations. In December 2010, WHO produced for the first time the "WHO guidelines for indoor air pollution: selected pollutants" with guidelines for the protection of public health from the health risks associated to a number of chemicals more common in the indoor air. The selected substances considered in the guidelines are benzene, carbon monoxide, formaldehyde, naphthalene, nitrogen dioxide, polycyclic aromatic hydrocarbons (especially benzo[a]pyrene), radon, trichloroethylene and tetrachloroethylene. The selection of substances was done considering information on the existence of indoor sources of those pollutants and on the availability of toxicological and epidemiological data and on the exposure levels causing health concerns (WHO, 2010). To note that the occupational exposure limits are around 1000 times higher than the guidelines derived for non-occupational indoor environments, as they have been developed for healthy adult populations under a more or less controlled exposure in about 40 over the 168 hours a week.

The object of the present paper is to gathering and summarizing the existing data published, focusing on the occurrence of the most relevant VOCs in indoor air in schools and homes the latter being prioritary indoor environments for children due to their particular susceptibility and extended time of air exposure.

2 MATERIAL AND METHODS

The scientific literature research used in this review has covered studies published from 1993 to 2011 in PubMed and Scopus and was focused on at least one of the following compounds in European homes and schools environments: BTX, formaldehyde and acetaldehyde, as the most prominent and important carbonyl compounds, and chemicals like trichloroethylene, tetrachloroethylene, naphthalene, α-pinene, limonene as they are emitted in high rates from products that are used widely in homes and/ or because they are reactive and, under certain conditions, can give birth to secondary emissions.

3 RESULTS AND DISCUSSION

The concentration values of VOCs in Table 1 are as reported in the literature and for this reason some inconsistencies, e.g., in significant digits and in the presentation of statistical metrics, are evident.

Conducting a review for EU, on VOCs in the indoor air at schools and homes is a demanding task due to the relative paucity of data in some countries, namely, in Portugal, and to the inconsistencies in the respective datasets.

From the literature review, when compared to the WHO guideline values (WHO, 2010) and Kotzias D. et al. (2005), it can be underlined that the mean concentration levels found for some VOCs such formaldehyde, tetrachloroethylene, toluene and xylenes are generally low while some compounds such as benzene and naphthalene (in one study) tend to present higher concentration levels. Concerning the levels of acetaldehyde, α-pinene and limonene, for which there is no limit established, further research is needed with regard to the dose-response indoors.

Large variation in the measured data might be expected due to socio-economical differences, affecting both consumer products use and building materials/emission sources, as well as the effects of environment differences and urban locations which may imply with the outdoor/indoor air exchange. The difference in the aldehydes concentrations between countries could be explained by the differences on the current use or not of wood products, combustion processes and of environmental tobacco smoke. For other contaminants such as aromatic hydrocarbons, outdoor sources might have a more significant contribution than the indoor sources. In addition, pollutants like naphthalene are more linked to specific consumer products and behaviors. Then, their presence is particularly related to the socio-economic conditions and consumer behavior.

The overview of the results indicates that for some contaminants a wide variability is observed

mainly due to the different source characteristics. A realistic and representative view to indoor exposure to VOCs would beneficiate and be greatly facilitated by a sampling or auditing harmonization protocol.

4 FINAL REMARKS

The overview of the results indicates that for some contaminants a wide variability is observed, due to the different source characteristics and suggests that there is a need to consider priority to regulate sources nature and strength and other indoor air pollution determinants in indoor environments mostly by undertaking:

– Emission characterization and labeling schemes for building materials, household appliances and consumer goods;
– Publicizing good practices in handling newly acquired consumer goods (destined for primarily indoor use), maintenance of older products, and substitution of toxic chemicals in articles with less toxic ones.

This study also highlights the need for harmonized sampling and auditing protocols that will allow better data interpretation.

REFERENCES

Alexopoulos, E.C., Chatzis, C. & Linos, A. 2006. An analysis of factors that influence personal exposure to toluene and xylene in residents of Athens, Greece. *BMC Public Health*, 6, 50.

Clarisse, B., Laurent, A.M., Seta, N., Le Moullec, Y., El Hasnaoui, A. & Momas, I. 2003. Indoor aldehydes: measurement of contamination levels and identification of their determinants in Paris dwellings. *Environ Res*, 92, 245–53.

COMEAP 1997. Committee on the medical effects of air pollutants: handbook on air pollution and health. *In:* Department of Health, C.O.T.M.E.O. a. P.H.M.S.S.O. (ed.). Londom, UK.

D. Cavallo, D. Alcini, D. de Bortoli, D. Carrettoni, P. Carrer, M. Bersani & Maroni, M. Year. Chemical contamination of indoor air in schools and office buildings in Milan, Italy. *In:* Indoor Air 1993: The 6th International Conference on Indoor Air Quality and Climate, 1993 Helsinki. 45–49.

Dassonville, C., Demattei, C., Laurent, A.M., Le Moullec, Y., Seta, N. & Momas, I. 2009. Assessment and predictor determination of indoor aldehyde levels in Paris newborn babies' homes. *Indoor Air*, 19, 314–323.

ECA 1995. Report no 14, Sampling strategies for volatile organic compounds (VOCs) in indoor air. Brussels—Luxembourg.

Ergebnisse des repräsentativen Kinder-Umwelt-Surveys (KUS) des Umweltbundesamtes 2008. Vergleichswerte für flüchtige organische Verbindungen (VOC und

Aldehyde) in der Innenraumluft von Haushalten in Deutschland. *Bundesgesundheitsblatt—Gesundheitsforschung—Gesundheitsschutz,* 51, 109–112.

Fischer, P.H., Hoek, G., van Reeuwijk, H., Briggs, D.J., Lebret, E., van Wijnen, J.H., Kingham, S. & Elliott, P.E. 2000. Traffic-related differences in outdoor and indoor concentrations of particles and volatile organic compounds in Amsterdam. *Atmospheric Environment,* 34, 3713–3722.

Gustafson, P., Barregard, L., Strandberg, B. & Sallsten, G. 2007. The impact of domestic wood burning on personal, indoor and outdoor levels of 1,3-butadiene, benzene, formaldehyde and acetaldehyde. *Journal of Environmental Monitoring,* 9, 23–32.

Hutter, H.P., Moshammer, H., Wallner, P., Damberger, B., Tappler, P. & Kundi, M. 2002. Volatile Organic Compounds and Formaldehyde in Bedrooms: Results of a survey in Vienna, Austria. *Indoor Air 2002–9th International Conference on Indoor Air Quality and Climate.* Rotterdam (Netherlands): in-house publishing.

Janssen, N.A.H., van Vliet, P.H.N., Aarts, F., Harssema, H. & Brunekreef, B. 2001. Assessment of exposure to traffic related air pollution of children attending schools near motorways. *Atmospheric Environment,* 35, 3875–3884.

Järnström, H., Saarela, K., Kalliokoski, P. & Pasanen, A.L. 2006. Reference values for indoor air pollutant concentrations in new, residential buildings in Finland. *Atmospheric Environment,* 40, 7178–7191.

Katsoyiannis, A., Leva, P. & Kotzias, D. 2008. VOC and carbonyl emissions from carpets: A comparative study using four types of environmental chambers. *Journal of Hazardous Materials,* 152, 669–676.

Kirchner 2007. Etat de la qualite de l'air dans les logements francais. *Environnement, Risques & Santé,* 259–269.

Kotzias, D., Geiss, O., Tirendi, S., Barrero-Moreno, J., Reina, V. & Gotti, A. 2009. Exposure to multiple air contaminants in public buildings, schools and kindergartens: the European indoor air monitoring and exposure assessment (AIRMEX) study. *Fresenius Environmental Bulletin,* 18, 670–681.

Kotzias D., Koistinen K., Kephalopoulos S., Schlitt C., Carrer P., Maroni M., Jantunen M., Cochet C., Kirchner S., Lindvall T., McLaughlin J., Mølhave L., Oliveira Fernandes E. de & B., S. 2005. The INDEX project: Critical appraisal of the setting and implementation of indoor exposure limits in the EU. Ispra (VA) Italy: European Commission, JRC.

Lovreglio, P., Carrus, A., Iavicoli, S., Drago, I., Persechino, B. & Soleo, L. 2009. Indoor formaldehyde and acetaldehyde levels in the province of Bari, South Italy, and estimated health risk. *J Environ Monit,* 11, 955–61.

Marchand, C., Le Calvé, S., Mirabel, P., Glasser, N., Casset, A., Schneider, N. & de Blay, F. 2008. Concentrations and determinants of gaseous aldehydes in 162 homes in Strasbourg (France). *Atmospheric Environment,* 42, 505–516.

MJ. Jantunen, K. Katsouyanni, KnoppelH, KunzliN, E. Lebret & MaroninM 1999. Final report: air pollution exposure in European cities: the EXPOLIS study.

Oliveira Fernandes, E., Gustafsson, H., Seppanen, O., Crump, D., Ventura Silva, G., Madureira, J. & Martins, A. 2008. WP3 Final Report on Characterization of Spaces and Sources. EnVIE Project. European Comission 6th Framework Programme of Research, Brussels.

OO Hanninen, S. Alm, E. Kaarakainen & Jantunen, M. 2002. The EXPOLIS databases, B13/2002. *In:* National Public Health Institute, K., Ktl (ed.).

Otmar Geiss, Georgios Giannopoulos, Salvatore Tirendi, Josefa Barrero-Moreno, Bo R. Larsen & Dimitrios Kotzias 2011. The AIRMEX study—VOC measurements in public buildings and schools/kindergartens in eleven European cities: Statistical analysis of the data. *Elsevier—Atmospheric Environment,* 45, 3676–3684.

Pegas, P.N., Alves, C.A., Evtyugina, M.G., Nunes, T., Cerqueira, M., Franchi, M., Pio, C.A., Almeida, S.M. & Freitas, M.C. 2011. Indoor air quality in elementary schools of Lisbon in spring. *Environ Geochem Health,* 33, 455–68.

Preuss, R., Angerer, J. & Drexler, H. 2003. Naphthalene–an environmental and occupational toxicant. *Int Arch Occup Environ Health,* 76, 556–76.

Roda, C., Barral, S., Ravelomanantsoa, H., Dusseaux, M., Tribout, M., Le Moullec, Y. & Momas, I. 2011. Assessment of indoor environment in Paris child day care centers. *Environ Res.*

Sakai, K., Norbäck, D., Mi, Y., Shibata, E., Kamijima, M., Yamada, T. & Takeuchi, Y. 2004. A comparison of indoor air pollutants in Japan and Sweden: formaldehyde, nitrogen dioxide, and chlorinated volatile organic compounds. *Environmental Research,* 94, 75–85.

Schlink, U., Rehwagen, M., Damm, M., Richter, M., Borte, M. & Herbarth, O. 2004. Seasonal cycle of indoor-VOCs: comparison of apartments and cities. *Atmospheric Environment,* 38, 1181–1190.

Stranger, M., Potgieter-Vermaak, S.S. & Van Grieken, R. 2007. Comparative overview of indoor air quality in Antwerp, Belgium. *Environ Int,* 33, 789–97.

Stranger, M., Potgieter-Vermaak, S.S. & Van Grieken, R. 2008. Characterization of indoor air quality in primary schools in Antwerp, Belgium. *Indoor Air,* 18, 454–463.

Topp, R., Cyrys, J., Gebefugi, I., Schnelle-Kreis, J., Richter, K., Wichmann, H.E. & Heinrich, J. 2004. Indoor and outdoor air concentrations of BTEX and NO2: correlation of repeated measurements. *J Environ Monit,* 6, 807–12.

Ullrich, D., Gleue, C., Krause, C., Lusansky, C., Nagel, R., Schulz, C. & Seifert, B. 2002. German environmental survey of children and teenagers 2000 (Geres IV): A representative population study including indoor pollutants. *Indoor Air 2002–9th International Conference on Indoor Air Quality and Climate.* Rotterdam (Netherlands): in-house publishing.

VITO 2007. The influence of contaminants in ambient air on the indoor air quality—Part 1: exposure of children.

WHO 2010. Guidelines for indoor air quality—selected pollutants. Copenhagen, Denmark.

Yu, K.P., Lee, G.W.M., Huang, W.M., Wu, C.C. & Yang, S.H. 2006. The correlation between photocatalytic oxidation performance and chemical/physical properties of indoor volatile organic compounds. *Atmospheric Environment,* 40, 375–385.

Table 1. Concentrations (µg.m⁻³) of volatile organic compounds observed in homes and schools.

Volatile organic compounds	Health effects [1]	Homes, Concentrations (µg/m³)					Schools, Concentrations (µg/m³)				
		Mean	Min	Max	Country	Reference	Mean	Min	Max	Country	Reference
Formaldehyde	*Acute*- Respiratory Disorders *Chronic* - Nasal—and Pharynx Cancer *IARC* - Carcinogenic class 1[2]	19.4			France	Dassonville et al. (2009)		8.0	210.0	Italy	Cavallo et al. (1993)
		16	6	87	France	Lovreglio et al. (2009)		10.0	100.0	Europe	COMEAP (1997)
		33.5			France	Marchand et al. (2008)	16.7	1.5	49.7	11 cities over Europe	Geiss et al. (2011)
		19.6		86.3	France	Kirchner et al. (2007)	15.2	6.4	35.7	France	Roda et al. (2011)
		35			Belgium	VITO (2007)					
		26			Finland	Järnström et al. (2006)					
		35.0			Sweden	Gustafson et al. (2007)					
		8.3			Sweden	Sakai et al. (2004)					
		24.5			France	Clarisse et al. (2003)					
		24.3			France	-|-(3)					
		31.3	8.8	115.0	Austria	Hutter et al. (2002)					
Benzene	*Acute*- Neurotoxic/Immunotoxic *Chronic* – Leukaemia *IARC* - Carcinogenic class 1	32.2			Italy	WHO (2010)	1.54	0.30	3.08	Belgium	Stranger et al. (2008)
		18.9			Italy	-|-	0.41	0.14	0.75	Belgium	-|-
		11.0		91.7	Germany	-|-	3.2		8.1	Netherlands	Janssen et al. (2001)
		3.2			Germany	Topp et al. (2004)	4.4		63.7	11 cities over Europe	Geiss et al. (2011)
		1.6		3.3	Germany	Schlink et al. (2004)	2.1		4.5	France	Roda et al. (2011)
		10.1			Germany	Ullrich et al. (2002)					
		2.2			Greece	Hanninen et al. (2002)					
		17.0			Finland	-|-					
		8.0			Italy	-|-					
		7.7	2.2	18.8	Czech Republic	Fischer et al. (2000)					
		5.7	1.5	10.5	Netherlands	-|-					
					Netherlands	-|-					
Trichloroethylene	*Acute* - Neurotoxic *Chronic* - Disorders of Liver/Kidney/Endocrine Systems and Immunity/Testicle—Lymph and Esophageal Cancer *IARC* - Carcinogenic class 2 A	7.7			Italy	Jantunen et al. (1999)	3.3	0.2	44.3	11 cities over Europe	Geiss et al. (2011)
		8.2			Greece	-|-	1.5	0.4	9.5	France	Roda et al. (2011)
		13.6			Czech Republic	-|-	1.6	0.4	16.3	France	-|-
		n.d.			Finland	-|-					

Compound	Health effects			Country	Reference				Country	Reference
Naphthalene	*Acute* - Hemolytic anaemia	0.8		Germany	Schlink et al. (2004)					
		2.0	1.8	Germany	Preuss et al. (2003)					
	Chronic - Neurotoxic	83.5	14	Greece	Hanninen et al. (2002)					
	IARC - Carcinogenic class 2B									
Tetrachloroethylene	*Acute* - Kidney Disorders	n.d. [4]		Finland	Jantunen et al. (1999)					
	Chronic - Neurotoxic/ Cancer									
	IARC - Carcinogenic class 2 A									
Toluene	*Acute* - Neurotoxic	4		Greece	-|-	5.23	3.06	8.25	Belgium	Stranger et al. (2008)
	Chronic - Neurotoxic	7.4		Italy	-|-	3.64	2.14	5.91	Belgium	-|-
	IARC - Carcinogenic class 3	8.7		Czech Republic	-|-	12.6	1.0	103.8	11 cities over Europe	Geiss et al. (2011)
		37.3		Germany	-|-	6.0	3.3	25.5	France	Roda et al. (2011)
	Acute - Neurotoxic	4.25		Belgium	Stranger et al. (2007)	7.1	3.0	26.4	France	Roda et al. (2011)
	Chronic - Neurotoxic	39		Greece	Alexopoulos et al. (2006)	4.47			Portugal	Pegas et al. (2011)
	IARC - Carcinogenic class 3 [2]	20		Finland	Jarnström et al. (2006)					
		11		Finland	-|- [3]					
		29.5	43.6	Germany	Schlink et al. (2004)					
		20.1		Finland	Hanninen et al. (2002)					
		77.6		Italy	-|-					
		86.2		Czech Republic	-|-					
		10.62		Germany	-|-					
		15.63		Greece	-|-					
		14.88		Germany	-|-					
		6.61	7.13	Germany	-|-					
		9.4		Belgium	-|-					
		8.88		Finland	-|-					
		6.06		Germany	-|-					
Xylenes	*Acute* - Neurotoxic	1.36		Belgium	Stranger et al. (2007)	2.64	1.1	21.7	France	Stranger et al. (2007)
	Chronic - Neurotoxic	9.8	58.6	Germany	Schlink et al. (2004)	3.9	1.2	17.9	France	Roda et al. (2011)
	IARC - Carcinogenic class 3	1.95		Netherlands	Kotzias et al. (2009)	4.0				-|-
		2.44		Netherlands	-|-	19.7			Portugal	Pegas et al. (2011)
		6.65		Greece	-|-					
		5.28		Germany	-|-					
		2.99		Germany	-|-					
		2.53		Belgium	-|-					

(Continued)

461

Table 1. (Continued).

Volatile organic compounds	Health effects [1]	Homes, Concentrations (μg/m³)					Schools, Concentrations (μg/m³)				
		Mean	Min	Max	Country	Reference	Mean	Min	Max	Country	Reference
Acetaldehyde	*Acute* - Respiratory Disorders	14.3			France	Marchand *et al.* (2008)	8.5	1.4	29.1	11 cities over Europe	Geiss *et al.* (2011)
	Chronic - Nasal—and Larynx Cancer	23			Sweden	Gustafson *et al.* (2007)	5.3	4.5	6.3	France	Roda *et al.* (2011)
	IARC - Carcinogenic class 2B	11.25			Netherlands	Kotzias *et al.* (2009)	5.0	2.4	10.7	France	-|-
		11.68			Netherlands	-|-					
		9.81			Greece	-|-					
		20.61			Germany	-|-					
		12.42			Finland	-|-					
		9.59			Germany	-|-					
α-pinene	*Acute* - Headache/confusion	9.8		800	Germany	KUS (2008)	3.2	0.001	47.3	11 cities over Europe	Geiss *et al.* (2011)
	Chronic - Lung—and Kidney Disorders										
Limonene	*Acute* - Oxidised Forms are very allergic	71.5		400	Italy	Jantunen *et al.* (1999)	9.4	0.001	175.7	11 cities over Europe	Geiss *et al.* (2011)
	Chronic - Irritation of the skin, dermatitis, contact eczema due to air-oxidized limonene	11.5			Germany	KUS (2008)					
	IARC - Carcinogenic class 3										

[1] Broad references from literature; [2] IARC (International Agency for Research on Cancer) carcinogenic classes: class 1: Proven to cause cancer; class 2 A: Probably carcinogenic for humans; class 2B: Possibly carcinogenic for humans; [3] 3: Not Classifiable as carcinogenic for humans; [4] n.d. – not detected. -|- The same reference as above;

Occupational Safety and Hygiene – Arezes et al. (eds)
© *2013 Taylor & Francis Group, London, ISBN 978-1-138-00047-6*

Analysis of the combined exposure to noise and ototoxic substances—pilot study

M.M. Lopes, D. Tomé & P.C. Carmo
Audiology Department, School of Allied Health Technology, Vila Nova de Gaia, Portugal

M.A. Rodrigues
Polytechnic Institute of Porto, School of Allied Health Technology, Research Centre on Health and Environment, Vila Nova de Gaia, Portugal

P. Neves
Polytechnic Institute of Porto, School of Allied Health Technology, Research Centre on Health and Environment, Vila Nova de Gaia, Portugal
Environmental Health Department, National Health Institute Dr. Ricardo Jorge, Portugal

ABSTRACT: Ototoxic substances have been associated to damage of the auditory system, and its effects are potentiated by noise exposure. The present study aims at analyzing auditory changes from combined exposure to noise and organic solvents, through a pilot study in the furniture industry sector. Audiological tests were performed on 44 workers, their levels of exposure to toluene, xylene and ethylbenzene were determined and the levels of noise exposure were evaluated. The results showed that workers are generally exposed to high noise levels and cabin priming filler and varnish sector workers have high levels of exposure to toluene. However, no hearing loss was registered among the workers. Workers exposed simultaneously to noise and ototoxic substances do not have a higher degree of hearing loss than those workers exposed only to noise. Thus, the results of this study did not show that the combined exposure to noise and the organic solvent is associated with hearing disorders.

It is rather common to associate occupational noise to harmful effects on hearing. However, there are other agents in workplaces that may have an impact on the employees, such as vibrations and chemical agents (Bertoncello, 1999; Campbell, 2007; Campo et al., 2009). Certain chemicals defined as ototoxic, such as metals, asphyxiating substances and certain organic solvents, may have a negative effect on the auditory system, either due to their individual performance, or to synergistic effects with other agents, particularly noise (Campos & Santos, 2011; Costa et al., 2001; Campbell, 2007; Sliwinska-Kowalska et al., 2005).

It is acknowledged that noise above 85 dB causes damage to the auditory system, and in the early years, noise-induced hearing loss (NIHL) is usually observed at the level of high frequencies. With continued exposure, it could also reach the lower frequencies, depending mainly on the intensity, pitch and shape of the noise, as well individual susceptibility. The most common symptom is hearing loss in high frequencies, with a characteristic scotoma at 4000 Hz, which may be associated to other disorders of hearing dysfunction, including tinnitus and cochlear recruitment (Campbell, 2007; Reis, 2003; Portmann & Portmann, 1993). Exposure to intense noise can cause changes in the mechanical cochlear structures, as well as metabolic changes, mainly in vascular structures and organ of Corti where the outer hair cells are the most affected (Costa et al., 2001; Campbell, 2007; Campo et al., 2009; Cappaert et al., 2001).

So far, several studies performed both in humans and animals, have shown lesions of inner ear structures by exposure to some organic solvents, including toluene, ethylbenzene and xylene, chemicals characterized as ototoxic that can be found in a large number of chemicals in different industrial sectors (Costa et al., 2001; Campbell, 2007; Sliwinska-Kowalska et al., 2001; Sulkowski et al., 2002). Injuries such as these, caused by exposure to ototoxic, result in hearing loss initially verified at 3000 Hz level, followed by 4000 and 6000 Hz (Campbell, 2007; Sass-Kortsak et al., 1995; Sliwinska-Kowalska et al., 2005). Chronic exposure to solvent vapors, which normally are a mixture of different compounds, can also lead to a long term or even permanent functional pathology in the peripheral nervous system, and, in some cases, in the central nervous system, especially in workers exposed to high concentrations of these solvents (Campo et al., 2009; Sliwinska-Kowalska et al., 2001; Sulkowski et al., 2002). Both in the exposure to organic solvents or to noise, hearing loss is irreversible, sensorineural, predominantly cochlear and may stabilize when exposure to the triggering agent is eliminated.

Considering the recent association of exposure to certain organic solvents to the development of hearing loss, it becomes imperative to devote more attention to the issue of risks involving workers exposed to high levels of organic solvents and to analyze their synergistic effects combined with other agents, including noise (Bertoncello, 1999; Costa et al., 2001).

The wooden furniture industry is typically represented by a frame of tasks that are associated to risks that can be the potential causes of occupational diseases, mainly due to exposure to hazardous substances and noise. Among these are the constituent ototoxic substances of some chemicals that are widely handled, it being that exposure can occur through inhalation and/or dermal absorption. Due to the organic of the activities developed in this sector, there is a potential for a combined exposure to noise and ototoxic substance (Mayan et al., 2011). In Portugal, the exposure to these agents is regulated and this regulation is complemented by international standards that are aim at protecting workers and should be taken into account by organizations (Mayan et al., 2011). Given this situation and the scarcity of human studies, this study is aimed at analyzing the effects of combined exposure to noise and ototoxic substances through a pilot study in the furniture industry sector.

1 MATERIALS AND METHODS

1.1 Target population

This study involved a sector of the furniture industry, located in the North of the country. The target population is made up of workers exposed to organic solvents and the noise during working hours, randomly selected, based on group sampling according to the different sectors of the company.

1.2 Proceedings

To better define the sample under study, the anamnesis and otoscopy performed were directed to investigate the inclusion factors (exposure to organic solvents and/or noise) and the exclusion

factors (otologic surgery in the previous 6 months, ototoxic medication, suppurating otitis media and chronic, foreign bodies, cerumen plug, tympanic perforation). An air tone audiogram was then carried out, where hearing thresholds were tested at frequencies of 250, 500, 1000, 2000, 4000 and 8000 Hz and otoacoustic emissions by distortion product (DPOAE) were performed. Levels of noise exposure of workers taking part in the study were determined on the basis of ISO/DC 9612:31-05-2009, whereas the concentration levels of xylene, toluene and ethylbenzene were determined through personal samplers, using continuous air sampling, based on the methods described in the document NMAM-NIOSH-Manual of Analytical Methods, therefore recommended by the National Institute of Occupational Safety and Health (USA).

2 RESULTS AND DISCUSSION

The study involved a sample of 44 workers (5 were female and 39 were male) with the number of years of work corresponding to a mean ± standard deviation = 16 ± 12 years. The workers belonged to four different working sectors: finishing, storage, manufacturing and cutting. Among the sectors, in finishing only 6 workers were exposed to noise and ototoxic substances. The remaining workers, distributed by various sectors, were only exposed to noise.

For the limit value of daily personal exposure levels to noise (LEX,8h), Decree-Law n° 182/2006 is in force in Portugal. Taking into account the maximum values established in this document, and the values obtained after the measurement in factory subject to study, it was possible to observe that both in the cutting sector as well as in the manufacturing sector there are values that exceed the exposure limit value 87 dB (A) (Table 1). In the storage sector afforded a LEX,8h 83,6 dB (A) and finishing sector values between 79,5 to 82,8 dB (A), thus exceeding the lower action value of 80 dB (A). However, although the finishing sector workers are not the most exposed to noise in the factory under study, are simultaneously exposed to ototoxic substances, situation that enhances the risk of hearing loss.

Table 1. Distribution of daily exposure levels to noise, by work sector.

	<80 dB (A)	80–85 dB (A)	85–87 dB (A)	>87 dB (A)
Cutting sector	–	–	–	100%
Manufacturing sector	–	29,4%	11,8%	58,9%
Finishing sector	33,3%	33,3%	–	33,3%
Storage sector	–	100%	–	–

With regard to the exposure to xylene, toluene and ethylbenzene, workers of the finish sector, in particular cabin velatura, cabin priming filler and cabin varnish were selected as object of study. Statistically, 100% of these workers are exposed to low levels of concentration of ethylbenzene in the air (Table 2). According to the exposure limit value weighted average (TWA) for the concentration of xylene in the air, established in Decree-Law n° 24/2012, it was possible to verify that 83,3% of workers are exposed to low levels of xylene concentration, while the remaining 16,7%, corresponding only to the cabin priming filler, are exposed to moderate levels. As to levels of exposure to toluene, 50% of the workers (cabin priming filler and varnish cabin) are exposed to high levels and the remaining 50% (cabin velatura) to low levels of concentration of toluene in the air. In the cabin of priming filler, these levels coincide with the TWA established by ACGIH: 2012; in the varnish cabin they exceed the TWA established the same reference document. In regard to remaining values, both of toluene, ethylbenzene, and xylene as in any of the cabins, these are considered low, for any of the references followed in Portugal.

As to the use of protection by the workers (Table 3), it was found that in finishing sector, where there is both exposure to solvents and noise, 73,3% of employees claim to wear masks and only 33,3% wear hearing protectors. In the sectors of manufacturing and cutting, where noise levels are higher than in the finishing sector, the percentages of hearing protectors use increase, up to 38,9% in the manufacturing sector and up to 50% in the cutting sector.

According to literature, it is known that one of the characteristics of NIHL is having associated symptoms such as tinnitus (Campbell, 2007; Reis, 2003; Portmann & Portmann, 1993). As proved, there are at least two sectors that have excessive noise levels, it being more than enough reason for workers in these sectors to present different symptoms from other workers. However, through the results obtained, it is possible to state that in the sample under study there is no relationship between the symptoms presented and the sector of activity. Though statistically there is no relation, it is important to point out that 50% of workers engaged in cutting have tinnitus.

Table 2. Percentage of workers exposed to organic solvents.

	Toluene	Ethylbenzene	Xylene
Low level	50%	83,3%	100%
Moderate level	–	16,7%	–
High level	50%	–	–

Table 3. Percentage of workers that use protection, by work sector.

	Mask	H.P.*	M & H.P.**	No protection
Cutting sector	16,7%	50%	16,7%	50%
Manufacturing sector	5,6%	38,9%	5,6%	38,9%
Finishing sector	73,3%	33,3%	33,3%	13,3%
Storage sector	–	–	–	100%

* Hearing protection.
** Mask and hearing protection.

Considering the impact that the ototoxic agents under study have in the auditory system, the expected results in audiological exams would be the absence of DPOAE and in the audiogram, an increase of the thresholds in the frequencies affected by exposure to them. Analyzing the results obtained from the values presented in the audiogram and DPOAE at different frequencies, it was found that neither the frequency nor the DPOAE audiograms showed values significantly different between sectors. In contrast to the results of previous studies (Sliwinska-Kowalska et al., 2001; Sulkowski et al., 2002), finishing industry workers, who are exposed to noise as well as to organic solvent concentration levels above the allowed values, and cutting-sector workers who are exposed to noise levels higher than those established by law do not present results statistically different from workers exposed to noise levels under 80 dB (A). This evidence can be justified by adequate protection and by the number of years of exposure, since the effects of organic solvents may manifest themselves later, after several years of work, not being detectable at the time of the study.

3 CONCLUSION

The results of this study showed that neither the combined exposure to noise and ototoxic chemicals, nor the exposure to high noise levels were found associated with hearing losses, as would be expected in the factory under study. The reason for this may be the fact that many of the workers assessed are young and thus have not yet developed hearing loss and some of the workers use personal protective equipment, as well as the small size of the sample. In this way, this study shows the complexity of this analysis and the difficulty in finding a cause-and-effect relationship of combined exposure to noise and ototoxic substances in field studies. It is important to undertake more studies, involving broader samples, in order to better characterize this situation.

REFERENCES

Bertoncello, L. 1999. *Efeitos da Exposição Ocupacional a Solventes Orgânicos, no Sistema Auditivo.* Monografia de Especialização em Audiologia Clínica. Porto Alegre: CEFAC.

Campbell, K.C.M. 2007. *Pharmacology and Ototoxicity for Audiologists.* United States: Thomson Delmar Learning.

Campo, P, Maguin, K, Gabriel, S, Moller, A, Nies, E, Gómez, M.D.S., Toppila, E. 2009. Combined exposure to noise and ototoxic substances. Luxemburg: EU-OSHA—European.

Campos, C. & Santos, P. 2011. Combined exposure: noise and ototoxic substances. *International Symposion on Ocupational Safety and Hygiene.* 10 e 11 de Fevereiro, Guimarães, Portugal.

Cappaert, N.L.M., Klis, S.F.L., Muijser, H., Kuling, B.M., Smoorenburg, G.F. 2001. Simultaneous exposure to ethyl benzene and noise: synergistic effects on outer hair cells. *Hearing Research,* 162: 67–79.

Costa, E.A., Ibañez, R.N., Nudelmann, A.A., Seligman, J. 2001. *PAIR—Perda Auditiva Induzida pelo Ruído.* Rio de Janeiro: Livraria e Editora Revinter Ltda.

Mayan, O., Gonçalves, I., Trigo, L., Neves, P., Gonçalves, C., Guimarães, F., Rodrigues, S., Martins, M. 2011. *Condições de Trabalho na Indústria de Mobiliário de Madeira.* Lisboa: ACT—Autoridade para as Condições de Trabalho.

Portmann, M. & Portmann, C. 1993. Tratado de Audiometria Clínica. São Paulo: Livraria Roca Ltda.

Reis, J.L. 2003. *Surdez Diagnóstico e Reabilitação—volume II.* Lisboa: Servier.

Sass-Kortsak, A.M., Corey, P.N., Robertson, J.M. 1995. An investigation of the association between exposure to styrene and hearing loss. *AEP,* 5: 15–24.

Sliwinska-Kowalska, M., Zamyslowska-Szmytke, E., Szymczak, W., Kotylo, P., Fiszer, M., Dudarewicz, A., Wesolowski, W., Pawlaczyk-Luszczynska, M., Stolarek, R. 2001. Hearing loss among workers exposed to moderate concentrations of solvents. *Scandinavian Journal of Work, Environ & Health,* 27: 335–342.

Sliwinska-Kowalska, M., Zamyslowska-Szmytke, E., Szymczak, W., Kotylo, P., Fiszer, M., Wesolowski, W., Pawlaczyk-Luszczynska, M. 2005. Exacerbation of noise-induced hearing loss by co-exposure to workplace chemicals. *Environmental Toxicology and Pharmacology,* 19: 547–553.

Sulkowski, W.J., Kowalska, S., Matyia, W., Guzek, W., Wesolowski, W., Szymczak, W., Kostrzewski, P. 2002. Effects of Occupational Exposure to a Mixture of Solvents on the Inner Ear: a Field Study. *International Journal of Occupational Medicine and Environmental Health,* 15: 247–256.

Occupational safety

Occupational Safety and Hygiene – Arezes et al. (eds)
© 2013 Taylor & Francis Group, London, ISBN 978-1-138-00047-6

Mathematical relationship between age, working time and lost working days

E.G.S. de Medeiros, E.L. de Souza, F.B.R. de Brito, L.B. da Silva & M.B.F.V. de Melo
Federal University of Paraíba, João Pessoa, Paraíba, Brazil

ABSTRACT: Some individual's peculiarities such as age, lack of training and lack of experience have been studied as well as their mathematical relationship with the occurrence of occupational accidents. Specifically, this research sought to build negative binomial regression models, in order to evaluate the relationship between the variables age, working time of the employee in the company and lost days upon the occurrence of an accident considering the specificities the Brazilian electricity sector. As the materials and methods the negative binomial regression analysis was used to examine the relationship between the variables as it is suitable for counting data when over dispersion is observed resulting in a regression model wich relates the three variables involved. The conclusions show that workers with less time working in the company and aged 45 to 60 years are more likely to miss more days after an accident at work.

1 INTRODUCTION

Certain characteristics of the individuals were reported as factors that potentially increase the risk of accidents, such as age, lack of training and lack of experience (CHAU et al., 2002). The authors analyzed the case of workers in the construction industry in the region of Meurthe-et-Moselle, France. The sample consisted of 880 workers who suffered occupational accidents. The results showed that from the workers who had to be absent because of the occupational accident, 34.5% were aged 30 to 39 years and 31.2% between 40 and 49 years of age. Moreover, 90.3% of the victims who were away from work had been working for the company for more than 5 years.

According to Pransky et al. (2005), one of the consequences of the new millennium is the aging of the workforce. The US Bureau of Labor Statistics conducted a survey where it was shown that between the years 1995 and 2005 the number of workers over 55 years has grown at a rate of 2.5%, while the number of workers between 25 and 54 years of age grew at an rate of only 1.1%.

Pransky et al. (2005) also affirm that workers over 55 are more prone to the risk of occupational fatalities than younger individuals. The authors also claim that older workers may have major deficiencies and are less likely to return to work quickly after an accident. Moreover, as the resistance of the body decreases with age, the probability that the bodily trauma is severe or that the recovery time is slow, tends to increase.

About the consequences of accidents related to age, it was found that for miners of iron ore from Sweden, the highest average days lost due to occupational accidents was in the age group between 55 to 65 years (PRANSKY et al., 2005).

A cross-sectional study in the province of Quebec indicates a particular relationship between the accident and age in health care and social services in the region. Specifically, there has been an increasing rate of occurrence of accidents and average number of lost days concomitantly with the increasing age of workers (CSST, 1992).

Cloutier et al. (1998) also conducted research in the health sector and social services in Quebec, examining data from occupational accidents in the period between 1982 and 1991. The characteristics raised about the accidents and of accident victims were age, sex, occupation, accident characteristics (type of accident, injured body part, cause of accident, etc..) and the number of lost days. The results showed that for the group of individuals aged 55 or older, the mean number of days lost due to occupational accidents was larger than the other groups.

Studies on the relationship of age of workers and the working time in the company with the severity of accidents, as considering the days lost due to accident, can be found in the literature, such as the study by Blanch et al. (2009), which analyzed this relationship for 156 men working in shifts at a plant in Spain. The 156 individuals who participated in the study were employees of an industrial plant of manufacture, dedicated to the production, painting and coating of plastic film. The plant is located in the industrial area of Barcelona (Spain) and the work was carried out in three continuous shifts (morning, afternoon and evening).

In this study, it was found that the analysis of negative binomial regression can be used to examine the association between age, working time and days lost. It was possible to use the negative binomial regression due to the fact that the data show accentuated dispersions. The negative binomial regression model is suitable for counting data when over dispersion is observed. The phenomenon of over dispersion occurs when the dependent variable has a Poisson distribution, but the variance is greater than its average.

Poisson regression models have many of the desirable properties to describe the relationship between counting data, however, second and Miaou and Lum (1993), if the data presents significant over dispersion, using Poisson regression models may overestimate or underestimate the probability of occurrence of the dependent variable. Thus, yet according to the authors, more general probability distributions should be used as the Negative Binomial.

As a result, the regression indicated a significant relationship of age through interaction of working time to the days lost by accidents. However, this relationship was not significant when considering the age squared term in the regression equation, suggesting a nonlinear association between age and the days lost.

Similarly to what is adopted by the Brazilian legislation through the severity index (NBR 14208/2001), Blanch et al. (2009) suggests it to be used to count the days of work lost due to accidents at work. Thus, the more days lost due to an accident, the more severe will be the same. The same author conducted a survey about the influence of variables "age" and "working time in the company" in the amount of those lost days on the occasion of an accident at work, getting to build regressive models that demonstrated such a relationship.

The electricity sector wich is the object of this research, aims at generation, transmission and distribution of electricity. The work of electricians is characterized by the presence of physical and mental demands with prejudice to the health and safety because there are various risks, including those from the electrical, mechanical, biological, chemical, physical, psychosocial and biomechanical (GUIMARÃES et al, 2002; WHO, 2005).

The Hydroelectric Company of San Francisco-Chesf is a mixed capital company, a subsidiary of the Brazilian Electric Power S/A-Eletrobras. It was created in 1945 with the mission to produce, transmit and sell electricity to the Northeast region. Its headquarters is located in Recife since 1975, and the company currently has 5,635 employees (CHESF, 2012).

Thus, the present study aimed carry out a study similar to that of Blanch et al. (2009), aiming to build negative binomial regression models, to evaluate the relationship between age, working time of the employee in the company and lost days upon the occurrence of an accident, considering the specificities and the risks of sub-sectors of transmission lines maintenance, substations and centers of control and operations in the electricity sector.

2 MATERIALS AND METHODS

This study, documental and quantitative in nature, was conducted in Chesf. It was analyzed 500 forms filled by the company in the event of an accident at work, covering a period of four years between 2005 and 2008. As the forms were validated, the starting point was the statistical analysis, first by performing descriptive analyzes with calculations of means and variances, as well as construction of a frequency chart.

To construct the model, it was selected the dependent variable "number of days lost due to accidents at work" (DP), symbolizing the remoteness of the employee, and the independent variables "working time in the company" (TT) and "age" (ID).

Poisson regression models provide many of the desirable properties to describe the relationship between counting data, however, second Miaou and Lum (1993), if the data present significant over dispersion the use of Poisson regression models may overestimate or underestimate the probability of occurrence of the dependent variable. Thus, still according to the authors, more general probability distributions should be used as the Negative Binomial.

The negative binomial regression model is suitable for counting data when over dispersion is observed. The phenomenon of over dispersion occurs when the response variable has a Poisson distribution, but the variance is greater than its average.

After that, it was built a negative binomial regression model using the free domain software R, step performed in order to quantitatively analyze the relationships between the variables selected previously.

3 RESULTS AND DISCUSSION

To calculate the frequency of lost days, first the observations were separated into grades ranging from 1 to 232 days, with height of 21 days. Thus, in 92 of 114 cases observed, meaning about 81%, the lost days due to injury ranged from 1 to 21 days.

Then it was separated the cases into eleven classes of ages from [25.29[to [65.69[years. Table 1 shows the mean and variance of the variable DP

Table 1. Mean, standard deviation and variance of the variable DP with respect to eleven age classes.

ID	[25,29[	[29,33[	[33,37[	[37,41[	[41,45[	[45,49[	[49,53[	[53,57[	[57,61[	[61,65[	[65,69[
N	7	5	4	8	14	16	32	13	11	3	1
M_{DP}	10,29	5,20	9,25	33,13	19,64	16,75	33,13	8,00	11,45	18,33	–
VAR_{DP}	14,57	9,20	31,58	4194,13	701,48	468,73	2472,76	13,50	40,27	120,25	–

Table 2. Mean, standard deviation and variance of the variable DP with respect to nine classes working time.

TT	[1,5[	[5,9[	[9,13[	[13,17[	[17,21[	[21,25[	[25,29[	[29,33[	[33,37[
N	23	5	5	2	30	11	17	16	5
M_{TT}	17,61	15,20	14,80	9,50	16,03	15,00	18,71	44,69	7,00
VAR_{TT}	1480,98	500,70	151,70	40,50	368,72	349,60	780,85	4095,43	25,00

for them. The highlighted lines indicate the fact that the variance is greater than the average necessary prerequisite for using negative binomial regression.

It is observed that more working days were lost to the groups of ages between [37.41[and [49.53[both with average 33.13 days lost. The lowest average of lost days is in the age group between [29.33[years going against the other studies cited above, where the younger groups had higher rates of lost days. Table 2 shows the means and variances of the variable DP for nine classes of working time, from [1.5[to [33.37[years of work in the company. The class of [29.33[had the highest average with 44.69 lost days.

Observing the descriptive analysis above, a mathematical model based on the negative binomial regression models was constructed, where it was found thirteen possible models. Of the thirteen models, only one was selected, based on their significance with respect to the p-value and the SSR (sum of squared residuals).

Thus, the model that best represents the relationship between variables is m6_log with SSRm6_log = 127.1704. This model can be seen in Table 3.

This model assumes that the lost days can be predicted by an exponential function, considering the variable and the interaction between TT and TT ID, where, unlike the other models surveyed, in this the variable ID has impact on the variable DP, when combined with TT. To better visualize this model, Figure 1 shows how it behaves, where x = ID; y = TT and z = DP.

From Figure 1, analyzing the variable ID (from the track of 40 years of age) and TT together, it can be concluded that the less working time in the company, the greater the amount of lost workdays. A possible explanation for this lies on the fact that an employee with less labor time in his current role would be less adapted to it, and is therefore more

Table 3. Appropriate model.

Model	Equation
m6_log	$DP = e^{(2,4874 + 0,1381 + TT - 0,0022 + ID + TT)}$

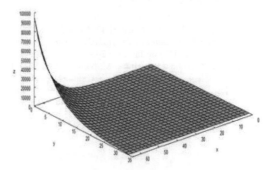

Figure 1. Model visualization.

prone to accidents and their consequences, such as lost workdays.

Moreover, still according to Figure 1, when keeping the variable TT constant, it is expected that the lost days decrease with decreasing age of the employee. The hypothesis is that younger employees would have greater ability to recover after an accident, requiring less time to return to work. (PRANSKY et al, 2005).

4 CONCLUSIONS

This study sought to establish the relationship between the independent variables "age" and "working time" in the behavior of the dependent variable "work days lost due to accidents."

This was possible through the construction of a negative binomial regression model, solely for the case of the Brazilian electric sector, specifically the company Chesf.

Thus, the relationship between age, working time in the company and days lost can be summarized as follows: Workers with less time working in the company and aged 45 to 60 years are more likely to miss more days after an accident at work.

As an interpretation of the results, management decisions can be taken to verify the adequacy of the model suggested in relation to the training for employees who were recently hired.

Moreover, the company can also establish security policies in order to consider the aging of the workforce, in order to control the risk of accidents to which employees are subjected.

Still as a suggestion for future works, it is indicated the possibility of using other independent variables such as time in the function performed, number of accidents suffered previously, training time for exercising of the function, schooling, among others.

REFERENCES

Associação Brasileira De Normas Técnicas. NBR 14280: *Cadastro de acidentes de trabalho—Procedimento e Classificação*. Rio de Janeiro: 2001.

Blanch, A.; Torrelles, B.; Aluja, A.; Salinas, J.A. (2009). Age and lost working days as a result occupational accident: A study in a shiftwork rotation of an system. *Safety Science Journal*, 47, 1359–1363.

Chau, N.; Mur, J.; Benamghar, L.; Siegfried, C.; Dangelzer, J.; Français, M.; Jacquin, R.; Sourdot, A. Relationships between some individual characteristics and occupation accidents in the construction industry: A case-control study on 880 victims of accidents occurred during a two-year period. *Journal of Occupational Health*, v. 44, p. 131–139, 2002.

Chesf. Available at <www.chesf.gov.br>. Access 03 Oct. 2011.

Cloutier, E.; David, H.; Duguay, P. Accident indicators and profiles as a Function of the age of female nurses and food services workers in the Québec health and social services sector. Safety Science, v. 28, n. 2, pp. 111–125, 1998.

Csst. Vieillissement de la population. Impact sur la sante' et la se'curite' du traĺail. Vice-présidence Planification et Programmation, v. 1, n. 3, p. 1–85, 1992.

Guimarães, L.B.M.; Fischer, D., Fae, C.S.; Salis, H.B.; Santos, J.A.S. Apreciação macroergonômica em uma concessionária de energia elétrica. In: *Anais do ABERGO 2002-VII Congresso Latino-Americano, XII Congresso Brasileiro de Ergonomia e I Seminário Brasileiro de Acessibilidade Integral*; 2002; Recife.

Miaou, S., Lum, H. Modeling Vehicle Accidents and Highway Geometric Design Relationships. *Accident Analysis and Prevention*, v. 25, n. 6, pp. 689–709, 1993.

Pransky, G.S.; Benjamin, K.L.; Savageau, J.A.; Currivan, D. Fletcher, K. Outcomes in work-related injuries: A comparison of older and younger workers. *American Journal of Industrial Medicine*, 47, 104–112, 2005.

Who, World Health Organization. *What are electromagnetic fields?* Geneva: WHO: 2005. Available at <http://www.who.int/pehemf/about/WhatisEMF/en>. Access 05 Oct 2012.

Occupational Safety and Hygiene – Arezes et al. (eds)
© 2013 Taylor & Francis Group, London, ISBN 978-1-138-00047-6

Evaluation of the thermal comfort in workplaces—a study in the metalworking industry

M. Talaia
DFIS, CIDTFF, University of Aveiro, Aveiro, Portugal

B. Meles
DEGEI, University of Aveiro, Aveiro, Portugal

L. Teixeira
DEGEI, GOVCOPP, IEETA, University of Aveiro, Aveiro, Portugal

ABSTRACT: Today, on the issue of the sustainability of countries in regards to economic aspects for the regulation of a balance in international trade, increasing the exportations volume is not enough. It is also needed that companies have standards to promote indoor levels of quality, productivity and satisfaction, being the Department of Medicine, Health and Safety the main advocate for this methodology. This paper presents a study, carried out in a multinational metalworking company, which aims to evaluate the thermal environment of a factory section characterized with a hot environment in order to enable the reorganization, change or the deployment of new equipment, improving the comfort of their occupants. After the data collection, using a pen data acquisition (*Easy Log USB*), the thermal index '*Temperature Humidity Index (THI)*' and the diagram of the *World Meteorological Organization (WMO)* were used. The results showed a consistency in the methods applied and suggest its applicability in the industrial context for the adoption of preventive and organizational measures, and/or for the replacement of equipment.

1 INTRODUCTION

Currently the industrial environment undergoes rapid global economic changes, fluctuations in demand and large technological developments. To survive with these constant changes, organizations are focusing on a more dynamic management process, where human resources are a crucial part. However, to overcome the constant changes it is important that workers are motivated and feel that the company is concerned with their well-being, ensuring that their working conditions are adequate (Talaia 2011). In this context, thermal comfort plays an important role in building sustainability (Yao *et al.* 2009) and it is a key factor for a healthy and productive workplace (Taylor *et al.* 2008, Wagner *et al.* 2007).

Thermal comfort has been defined by Hensen (1991) as "a state in which there are no driving impulses to correct the environment by the behavior". The American Society of Heating, Refrigerating and Air-Conditioning Engineers (ASHRAE) defined it as "the condition of the mind in which satisfaction is expressed with the thermal environment" (ANSI/ASHRAE 2004). According to ISO standard 7730 (ISO7730 2005), the thermal comfort can be defined

as "the satisfaction expressed when an individual is subjected to a given thermal environment".

The specialized literature presents some thermal indexes that aim to characterize a specific thermal environment (Gagge *et al.* 1986, Djongyang *et al.* 2010). According to Corleto (1998) and Djongyang *et al.* (2010), an index of thermal stress is a value that integrates the effect of various human indicators in the thermal environment, characterizing the comfort/discomfort to which an individual is subjected in a hot environment. It is in this context that the present project emerges, developed in an industrial environment in the scope of a multinational metalworking company.

This work aims to evaluate the thermal environment that is considered warm as a consequence of the presence of furnaces in a metalworking industry, in order to enable the reorganization, change or deployment of new equipment, improving the comfort of their occupants.

In order to accomplish this study, firstly we selected eight workstations, which we considered representative for the study due to the information provided by the Department of Health and Safety and by the section managers. After the data collection in workplaces selected, using a pen data acquisition (*Easy Log USB*), the thermal index '*Temperature Humidity Index (THI)*' and the *World Meteorological Organization* (WMO) were used. These indexes were used due to its easiness of application and the results allow the identification of the strategy of improvement of a thermal environment (Talaia *et al.* 2008, Meles 2012).

2 MATERIALS AND METHODS

As mentioned, the present experiment was performed using *Temperature Humidity Index* (*THI*), created by Thom (1959) and later modified by Nieuwolt (1977). This index was selected for its simplicity of application and because it uses the air temperature T (°C) and the relative humidity RH (%). The calculation formula for the *THI* index is:

$$THI = 0.7T + T(RH/500) \qquad (1)$$

where T = air temperature (°C); and RH = relative humidity (%).

Table 1 specifies the reference values for the index *THI*, delimiting situations of stress thermal and thermal comfort for human subjects.

In the present study, as shown by Talaia *et al.* (in press) and Meles (2012), the Table 1 was adapted to a wide range of values *THI*, based on the strategies depicted in the diagram of the *World Meteorological Organization* (WMO 1987). The resulting values are presented in Table 2.

In this work it was also used the diagram of the *World Meteorological Organization* (WMO) adapted by Talaia *et al.* (2006), by considering the air temperature (°C) and relative humidity RH, as well as by indicating intervention strategies for improving the thermal environment, as shown in the representation of Figure 1.

Table 1. Values of the comfort limits of the *THI* (Emmanuel 2005).

$26 \leq THI$	100% of the subjects felt uncomfortably hot
$24 \leq THI < 26$	50% of the subjects felt comfortable
$21 \leq THI < 24$	100% of the subjects felt comfortable

Table 2. Values of the comfort limits of the *THI* adapted [Talaia *et al.* (in press) and Meles (2012)].

$26 \leq THI$	too hot
$24 \leq THI < 26$	wind needed for comfort
$21 \leq THI < 24$	COMFORTABLE
$8 \leq THI < 21$	sun needed for comfort
$THI < 8$	too cool

Concerning the workplaces to be observed, we selected eight workstations in a specific section considered a hot environment. The choice of these observation posts was based on the criterion of criticalness of the workplace in terms of thermal comfort. This criterion was confirmed based on the observation *in situ*, also taking into account the orientation of the head of department of the manufacturing unit.

After identifying the workplaces more vulnerable to thermal discomfort, values of air temperature (°C) and relative humidity RH (%) were recorded in the daytime, having the measurements of the hygrometric variables been made using a pen data acquisition (*Easy Log USB*). In this work we also used a multi-function measuring instrument *Testo 435*, calibrated for validation of data recorded by the sensors through *Easy Log USB*.

Figure 2 presents the layout of the section, with the workplaces that were observed and where the data were collected.

Regarding the conditions of each operating position (workplaces) we have: 45—Manual welding booth. This compartment was installed in this section about one year ago, and has an adequate exhaust system; 78—Place of exit of products

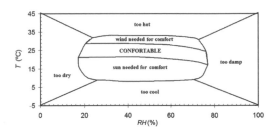

Figure 1. Diagram of the World Meteorological Organization adapted by Talaia *et al.* (2006).

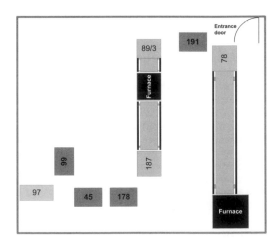

Figure 2. Layout of workstation in section.

474

from the furnace and where the quality inspection is made; 89/3—Place of entry of products (on the furnace), having a protective mechanism before the entrance; 97—Automatic welding booth. In this place there is a cooling system of products installed; 99—Manual welding booth. This station receives the product coming from the workstation 97, but it only works when it is necessary; 178—Manual welding booth. This station has an older system and consequently it has an inappropriate exhaust system; 187—Place of exit of products (from the furnace). The products from this station register a high temperature, being removed and placed in appropriate transport carriages; 191—Manual welding booth. Like workstation 178, this has an antique system and consequently an inappropriate exhaust system. Additionally, it is located close to workstation number 78 and near to the entrance door.

Related to the time of data collection, we chose the second shift that runs between 1:30 pm and 9:00 pm, so the data recording had into account the diurnal cycle of solar radiation and the greenhouse effect which is usually created within the section during the afternoon. Fifteen measurements of air temperature (°C) and relative humidity RH (%) were recorded during this work shift.

Data were collected in cloudless sky so that the received power of the sun was a condition for heat stress situations. Indeed it is known that the transfer of energy in the form of heat from the outside to the inside of the section is maximized after noon (Arhens 1999, McIntosh et al. 1981). Moreover the heat sources located within the section favor the increase of temperature of indoor air and a consequent decrease in relative humidity, causing thermal sensations on a hot environment for workplace occupants.

The results obtained by the application of the index THI and also with the use of the diagram WMO were compared. Therefore, whenever possible, intervention strategies were presented to improve the thermal comfort of the worker.

3 RESULTS AND DISCUSSION

We considered relevant to make a division between the obtained results by workstation location. We selected an area with the operating position near the door (entrance door) with three observation posts and other area farthest from the door with five observation posts.

Figure 3 shows the THI value during the work shifts to the stations 191, 78 and 89/3, near the entrance door.

The Figure 3 shows that the workplace 191 presents extreme conditions of thermal discomfort. This evaluation is justified by the circumstances of the workplace, as it is obsolete and has an inadequate

ventilation system. The results showed the need for intervention strategies (replacement or a strong improvement). The workstation 78 suggests that it should be occupied by persons with physical robustness. The acclimation of the worker is important.

Figure 4 shows an excellent concordance with the values indicated by THI showed in Figure 3.

The observation of diagram in Figure 4 shows that the post 191 requires intervention strategies, such as installing an adequate ventilation system and adjusting the operator times at the workstation. The workstation 78 requires an adequate ventilation system. The post 89/3 is appropriate.

Figure 5 shows the THI values throughout the work shift (farthest area from the entrance port) to the workstations 178, 187, 97, 45 and 99.

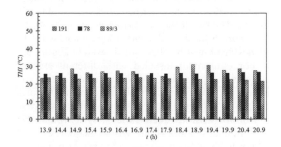

Figure 3. THI: stations 191, 78 and 89/3, near the entrance door.

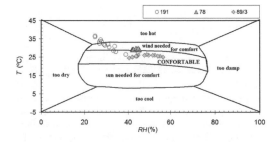

Figure 4. Diagram WMO: stations 191, 78 and 89/3, near the entrance door.

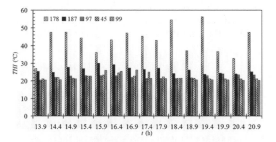

Figure 5. THI: stations 178, 187, 97, 45 and 99, farthest area from the entrance door.

475

The observation of the Figure 5 shows unequivocally that the job 178 demands immediate intervention. The registers throughout the day show that the workplace is too warm. The workstation 187 shows that *THI* is above 26 °C, over the major part of the work shift and that this situation demands intervention. These findings are consistent with the type of work of each workstation.

The remaining workstations, 97, 45 and 99, do not deserve discomfort valorization. At certain times an appropriate ventilation system solves the discomfort.

The representation of Figure 6 confirms the results of Figure 5 and shows that the workplace 178 requires immediate intervention strategies. The workstation 187 shows some concerns, but the use of an adequate ventilation system seems to be enough for a worker with physical robustness and acclimated. The remaining workstations are suitable.

Figure 6 shows an excellent agreement with the values displayed by *THI* and showed in Figure 5.

For equipment installed with different exhaust systems, it is possible to show how a workstation can have a very different thermal environment. Accordingly, we considered the workstation 45 (manual welding booth with adequate exhaust system) and 178 (old manual welding booth with inadequate exhaust system). These two workplaces are neighboring.

The results obtained for *THI* in the workstations 45 and 178 are shown in Figure 7.

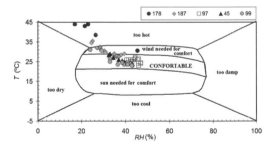

Figure 6. *Diagram WMO*: stations 178, 187, 97, 45 and 99, farthest area from the entrance door.

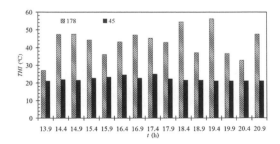

Figure 7. *THI*: stations 178 and 45.

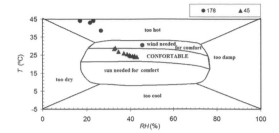

Figure 8. *Diagram WMO*: stations 178 and 45.

The observation of the chart in Figure 7 shows unequivocally that the workstation 178 requires immediate action due to being to warm, while the workstation 45 is appropriate. Being two similar workstations the results suggest that the major difference is in the exhaust system.

Figure 8 shows an excellent agreement with the values indicated by *THI* and showed in Figure 7.

The excellent agreement obtained by using the two methods suggests that both methods can be used by the Department of Health and Safety in order to improve the workstation conditions, which would lead to an improvement of the workplace occupant satisfaction.

4 CONCLUSION

In general, the results obtained showed a consistency in the methodology applied.

This work also shows that the index *THI*, for its simplicity of application, must be valorized by the industry. An interesting alternative is the diagram of the *WMO*.

The study shows that from the results obtained, it is possible to determine thermal environment places vulnerable to thermal stress and therefore adopt intervention strategies for improving workplaces. An example was the comparison of workstations 45 and 178. Accordingly, we must consider that these two methods should be valued by the Department of Health and Safety within any manufacturing unit.

In this sense it is possible, through the listed strategies, to modify the conditions of humid air as well as the organizational conditions in order to improve the operator's workstation and, thus, favor the indexes of productivity and quality, reducing fatigue risks, heat stress, and cardiovascular diseases, among others.

Today, on the issue of the sustainability of countries in regards to economic aspects for the regulation of a balance in international trade, increasing the exportations volume is not enough.

From the point of view of the authors, it is necessary to provide companies with a pattern of "climate" that not only favors the indexes of quality and productivity, but also the indexes of workplace occupant satisfaction. It is up to the departments of Medicine, Health and Safety, the promotion of this great paradigm shift.

REFERENCES

ANSI/ASHRAE 55. 2004. Thermal environmental conditions for human occupancy. American Society of Heating, Refrigerating and Air—Conditioning Engineers. *AINSI/ASHRAE Standard* 55: 1–26.

Arhens, C.D. 1999. *Essentials of meteorology—an invitation to the atmosphere.* Editor Thomson.

Corleto, R. 1998. *The evaluating of Heat Stress indices using physiological comparisons in an alumina refinery in a sub-tropical climate.* (Master), Deakin University, Geelong, Australia.

Djongyang, N., Tchinda, R. & Njomo, D. 2010. Thermal comfort: A review paper. *Renewable and Sustainable Energy Reviews* 14(9): 2626–2640.

Emmanuel, R. 2005. Thermal comfort implications of urbanization in warm—humid city: the Colombo Metropolitan Region (CMR). *Building and Environment* 40, 1591–1601.

Gagge, A.P., Fobelets, AP. & Berglund, L.G. 1986. A standard predictive index of human response to the thermal environment. *ASHRAE Transactions* 92(2B): 709–31.

Hensen, J.L.M. 1991. *On the thermal interaction of building structure and heating and ventilating system.* PhD thesis. Technische Universiteit Eindhoven.

ISO 7730. 2005. *Ergonomie des ambiances thermiques—Détermination analytique et interprétation du confort thermique par le calcul des indices PMV et PPD et par des critères de confort thermique local.* Switzerland, International Standardisation Organisation, Geneva, Suisse.

McIntosh, D.H. & Thom, A.S. 1981. *Essentials of meteorology.* The Wykeham Sciences Series. Taylor and Francis, Ltd. London.

Meles, B. 2012. *Ergonomia Industrial e Conforto Térmico em postos de trabalho.* Dissertação apresentada à Universidade de Aveiro (publicada). Aveiro *(in Portuguese).*

Nieuwolt, S. 1977. *Tropical climatology.* London: Wiley.

Talaia, M.A.R. 2011. Local de Trabalho e Ambiente Térmico—uma Avaliação de Conforto e Desconforto. *Proceedings VI Congresso Cubano de Meteorologia* (artigo CLI-21, 15 páginas). Havana, Cuba: Hotel Tryp Habana Libre, 29 de Novembro a 2 de Dezembro *(in Portuguese).*

Talaia, M.A.R. & Rodrigues, F.A.G. 2006. O Organismo Humano num Ambiente de Stress Térmico—caso de uma área com fornos. *Proceedings XXIX Jornadas Científicas AME & 7° Encuentro Hispano-Luso de Meteorología.* Ed. AMEspañola. Madrid, paper A25, 5 páginas *(in Portuguese).*

Talaia, M.A.R. & Rodrigues, F. 2008. Conforto e Stress Térmico: Uma Avaliação em Ambiente Laboral. *Proceedings em CD-ROM da CLME'2008/II CEM. 5° Congresso Luso—Moçambicano Engenharia—2° Congresso de Engenharia de Moçambique.* Maputo. Editores Gomes et al. Edições INEGI. Artigo 11 A020 (2008) 15 páginas *(in Portuguese).*

Talaia, M.A.R., Meles, B. & Teixeira, L. (in press). Worker perception in relation to workplace comfort—a study in the metalworking industry. *International Symposium on Occupational safety and hygiene, Sho13.* Guimarães, Feb 14–15.

Taylor, P., Fuller, R.J. & Luther, M.B. 2008. Energy and thermal comfort in a rammed earth office building. *Energy and Buildings* 40: 793–800.

Thom, E.C. 1959. The Discomfort Index., *Weatherwise* 57–60.

W.M.O. 1987. *World Climate Program Applications,* Climate and Human Health World Meteorological Organization.

Wagner, A., Gossauer, E., Moosmann, C., Gropp, T. & Leonhart, R. 2007. Thermal comfort and workplace occupant satisfaction—Results of field studies in German low energy office buildings. *Energy and Buildings* 39(7): 758–769.

Yao R., Li B. & Liu J. 2009. A theoretical adaptive model of thermal comfort adaptive predicted mean vote (aPMV). *Building and Environment* 44: 2089–96.

Occupational Safety and Hygiene – Arezes et al. (eds)
© *2013 Taylor & Francis Group, London, ISBN 978-1-138-00047-6*

Study of workers' perception on the risk of WRMSDs in the traditional bakery industry

J.M. Lima & M.E. Pinho
Faculdade de Engenharia da Universidade do Porto, Porto, Portugal

P.M. Arezes
DPS, Escola de Engenharia da Universidade do Minho, Guimarães, Portugal

ABSTRACT: This study aims at to analyse the perception of the workers of the traditional bakery on its exposure to the risk of Musculoskeletal Disorders and to verify the relationship between this perception and the reported symptomatology. An inquiry was carried out, based on the Nordic Musculoskeletal Questionnaire. 25 interviews, involving 96% of the workers of the 6 bakeries participating in the study, were performed. The workers were characterized concerning their profession and tasks executed. Further, the musculoskeletal symptomatology was assessed, thus contributing to explain some of the identified pathologies. The prevalence of symptoms, remaining for at least in 4 consecutive days, recorded in the last 12 months, was higher in knees (64%), and in the dorsal (52%), and lumbar (48%) body regions. Results seem to indicate that workers with higher prevalence of musculoskeletal symptoms show a greater perception on the exposure to the risk of Work related Musculoskeletal Disorders (WRMSD).

1   INTRODUCTION

The International Commission on Occupational Health (ICOH) designates the pathologies related to the work by "work-related disorders" and defines them as the injuries and illnesses of the musculoskeletal system. These pathologies can be caused or exacerbated by work conditions and that include a large number of inflammatory and degenerative injuries that have as consequences pain and/or loss of functional capacity (Armstrong, 1993). Nowadays, these pathologies constitute an important problem of health and threat workers well-being. The adoption and maintenance of extreme work positions, the intense work pace, the direct mechanical pressure on certain areas of the body, the excessive force, the handling of heavy loads, the static work and the extended states of stress are among the most important risk factors (Nunes, 2006).

However, the risk of developing musculoskeletal injuries is very often ignored by workers themselves until it is too late, probably due to the significant gap between the cause and the effect (Robertson & Stewart 2004). Often, it is also difficult to determine the underlying causes and to distinguish those of professional origin (related to the work) from those of a different nature. Therefore, in the area of health and safety, the perception of the risk is very important, because it can be used to influence workers' behaviour in their own benefit, what means that the adequate communication of the risk is fundamental.

In general, the literature on safety indicates that the safety behaviours not only influence the occurrence of work accidents, but also are influenced by perceptions of workers about the safety of the environment. If the objective is to decrease the accidents rates and the occurrence of work accidents, investments on the development of safety behaviours should be a priority. Moreover, research on risk perception has been focused on work accidents, thus there are some gaps in understanding how workers perceive and react to long-term risks, such as the risk of WRMSD development (Arezes and Miguel, 2005 and 2006).

The WRMSDS are currently one of the most studied subjects at the level of the occupational safety and health, mainly in the most developed countries. However, the micro and/or family companies in the manufacturing industry, due to either their dimension, or their adoption of an artisanal production process, as is the case in the traditional production of bread, are almost always forgotten and relegated for a second plan.

In Portugal, in terms of occupational diseases without disability, the data of the reports of 2007 and 2008 of the Ministry of the Work and Social

Solidarity (Social Reporting 2007/2008) show that the WRMSDs are the most prevalent, followed by the lung illnesses and the auditory disorders. Women are the most affected by the certified occupational diseases, representing 55.2%, 60.2% and 68.2% of the total of occupational diseases reported in 2006, 2007 and 2008, respectively. Throughout the years, the profile of the workers with occupational diseases remained unchanged, being the majority females, with ages between 50–54 years (Perista & Quintal, 2009).

The average rates of absenteeism at work in Europe vary between 3% and 6% of the work time has an estimated cost of about 2.5% of the GNP. In Portugal, in the period between 2003 and 2007, it represented almost 7%. Moreover, recent OCDE figures suggest that the costs of benefits for disability and disease are 2.5 times higher than unemployment, and show an increasing trend Cabrita (2010).

The main objective of this study is to know if there is a relationship between workers risk perception and the reported musculoskeletal symptoms. With this purpose, the following aspects were taken into account:

a. To understand how workers perception affects their attitude towards the risk;
b. To assess workers' musculoskeletal symptomatology;
c. To assess workers' perception on the exposure to work-related musculoskeletal disorders.

Thereby, this study aims to analyse the musculoskeletal symptoms in workers of the traditional bakery and better understand their perception on the exposure to the risk of work-related musculoskeletal disorders, henceforth referred to as WRMSDs, which is commonly used in Europe, and in particular by the European Agency for Safety and Health at Work.

2 MATERIAL AND METHODS

The 25 workers who agreed to participate in the study were surveyed by the method of direct interview, based on a questionnaire consisting of two parts, which comprised a total of 141 variables. The first part, concerning the musculoskeletal symptomatology, consists of the Nordic musculoskeletal questionnaire (NMQ), adapted by the Portuguese Society of Occupational Medicine (Serranheira et al., 2008), and consists of 52 questions, while the second part, on the study of risk perception, comprises a total of 37 questions adapted from the Health and Safety Executive questionnaire (Robertson & Stewart 2004).

A descriptive statistics analysis for all the variables under study was performed. Fisher's Exact test by Monte Carlo simulation was used, since it is the most appropriate to analyse small samples (n < 20) or whenever more than 20% of the expected frequencies are <5. The Qui-square test was also used in the analysis of the relationship between workers body mass index (BMI) and the prevalence of WRMSDs. The correlation coefficients between age and the variables frequency and intensity of pain were calculated. They were interpreted according to the classification proposed by Dancey and Reidy (2008), which states that the correlation is weak, moderate or strong, depending on if the absolute value of the coefficients are included in the ranges $r = 0.10$ to 0.30, $r = 0.40$ to 0.6 or $r = 0.70$ to 1. Two-tailed tests and a significance level of $\alpha = 0.05$ were used.

3 RESULTS AND DISCUSSION

The population analysed in the present study is mostly female (76%) and this is probably due to the type of work traditionally performed by women.

25 questionnaires were fulfilled and considered valid, representing about 96% of the total number of workers of the bakeries involved in the study.

The 25 individuals participating in the study are between 26 and 67 years old (mean = 47.3; SD = 11.5 years). The great majority (72%) are over 40 years old, attended only primary education (52%) and have been working for more than 20 years in this profession (60%). Most of them neither smoke (68%), nor drink alcoholic regularly (88%). On the contrary, more than 4 out of 5 workers drink coffee regularly (84%), and more than half of them visited a doctor in the past year (64%) while more than 2 out of 3 take regularly medicines, including contraceptives (72%).

Among the workers taking medicines regularly, 82% use the self-medication. The medicines most commonly used are analgesics, such as Ben-u-Ron, Aspegic and Aulin. The medicines intake occurs during the periods wherein they feel pains and about 23% of those who take the medication do it, at least, once or twice times a week.

44% of the workers declared to suffer from, at least, one illness. Among these, 72.7% reported suffering from varicose veins. This situation is in accordance with the literature found (Stoia, 2008), which refers to the risk of varicose veins in these professionals. In a study carried out in Poland (Szopińska et al., 2007), the authors compare the prevalence of this disease in bakers and office employees, concluding that the bakers are more susceptible to this pathology.

On the other hand, the prevalence of musculoskeletal symptoms in the last 12 months (table 1), present at least in 4 consecutive days, was higher in

Table 1. Prevalence of pain in the different regions.

Region	Prevalence in last the 12 months	Prevalence in last the 7 days
Neck	28,0%	20%
Dorsal zone	52,0%	44%
Lumbar zone	48,0%	44%
Shoulders	32,0%	20%
Elbows	00,0%	0,0%
Fist/Hand	36,0%	28%
Thighs	08,0%	04%
Knees	64,0%	40%
Ankles/Feet	40,0%	28%

knees (64%), then followed by the dorsal (52%), and the lumbar (48%) areas. These figures reveal a high prevalence in knees and dorsal region, because they were reported by more than half of the workers.

In a study carried out in Iranian bakeries (Ghamari et al., 2009), the prevalence of symptoms in the knees achieved 62.2% and was followed by 58.8% in the lumbar zone. These values are consistent with those found in the current study because they relate to the same regions and the figures are very close. The values obtained for these regions also reinforce the opinion of the workers on the activities carried out in the main work, wherein 32% of participants classified the standing work as "Totally related with the reported symptoms".

Concerning the prevalence of pain in the last 7 days (table 1), pain was more prevalent in the lumbar and dorsal areas. Both areas recorded 44% of responses, contrary to the prevalence in the past 12 months, where the knees were the most prevalent area. The analysis of the prevalence of the musculoskeletal symptoms in the different body regions during the last 7 days and during the last 12 months, in the present study, shows that often pain remained through at least a year and for a period of seven consecutive days. According to Uva et al. (2001), situations of this type indicate the presence of symptomatic cases.

The absenteeism rate was low. The total number of the absenteeism at work in the past 12 months for the totality of the workers was 22 days. This may be due to the inexistence of a labour contract, which means that the payment of the contributions for the social security system is a workers' responsibility.

In this study it was found that the prevalence of pain in relation to the age of workers, in the last 12 months, shows statistical significance in the lumbar region ($p = 0.017$) and in ankles/feet ($p = 0.044$). Regarding BMI statistical significance ($p = 0.003$) was found, indicating that the higher this ratio, the greater the likelihood of pain at the dorsal region.

It was still achieved statistic significance ($p = 0.022$) for the knees in relation to gender, revealing accordance with the regions previously identified, as the most prevalent. These values are in agreement with studies (Hagberg et al., 1995) that refer the gender as a factor influencing WRMSDs symptomatology.

The intensity of pain in the lumbar region shows a strong positive correlation with the age of the participants ($r = 0.707$, $p < 0.05$), indicating that older workers reported higher pain intensities in this region than the younger workers.

Furthermore, the age of the participants shows moderate positive correlations with the frequency of pain in the lumbar area ($r = 0.589$, $P < 0.05$), in the left wrist/hand ($r = 0.677$, $p < 0.05$) and in the left knee ($r = 0.630$, $p < 0.01$), indicating that the older workers reported higher frequencies of pain in these body regions than the younger workers.

Some studies (Torner et al., 1988) corroborate these differences in the symptomatology reported by workers in accordance with age. These results are consistent with previously reported symptoms in these areas.

Concerning the relationship between the developed activities and the related symptoms, it was found that the activity "to lift loads between 1 and 4 kg" is more often indicated as "without relation with the mentioned symptoms". This may be explained by the fact that few loads in this weight range are lifted by workers and, therefore, they may attribute more importance to other tasks. These values are in agreement with the results of a study conducted in Spain (Castaño et al., 2011), in which the manipulated weight is under 5 kg in 25% of the cases examined, while the tasks involving the handling of loads between 10 and 20 kg are more valued (31%). In third place, with 19%, it appears tasks related with the handling of loads greater than 20 kg. According to Buckle & Devereux (1999), strength as occupational risk factor is related to its intensity, with the time duration in which it is applied and respective periods of recovery, particularly in actions of predominantly static work. At the level of the upper limbs, the manipulation of weights (or loads) with more than 4 kg is understood as high force exertions.

The indication that the weight of the load may influence this perception is in the activity "Handling of loads above 4 kg", which was the only reported by all the workers in all the items. Standing work was the most reported risk factor found "Totally related with the reported symptoms" (32%). These values are in accordance with the symptomatology reported in spine and legs/knees.

Regarding the second part of the questionnaire—study of the workers' perception on risk exposure to WRMSDs—respondents indicated the repetitiveness of movements in only one task. In what concerns work conditions and postures, workers highlighted the environmental conditions (temperature and dusts) and standing work.

When inquired about their knowledge on WRMSDS, 24% of participants in the present study had previous knowledge of WRMSDs, most of which pointed out tendonitis, spinal disc herniation's and sprains. This figure shows that the number of workers knowing about WRMSDs is lower than that found in a study conducted by HSE (Robertson & Stewart 2004), involving workers of SME's, in the United Kingdom, and in which 38% of participants reported having already heard about them.

When asked about the likelihood of, in the future, feeling pain in the back and/or upper limbs due to the work they currently perform, 92% of the respondents believe they can suffer from back pain while 80% are aware that they may suffer from pain in the upper limbs, associated to the work. These figures are significantly higher than those obtained in the study of HSE for workers of SMEs, wherein 60% and 52% of respondents declared that they likely will suffer from pain in back and in the upper limbs, respectively. These differences may probably be explained by the type of activity performed. The same HSE study indicates that workers perception may be higher when risks are higher or when they suffered or witnessed an accident.

Regarding the likelihood of, in the future, workers feeling pain in back or upper limbs, the results show that it is higher for back than for the upper limbs, since 24% of respondents declared to have no doubts about the back pain, whereas only 16% indicated "No doubt" for the pain in upper limbs. The most frequent answer is "Maybe" with 40% and 44% of the participants expressing to be aware of the possibility of, in the future, to feeling pain in back and in upper limbs, respectively. In order to assess risk perception, the possibility of workers who have already had this symptomatology or those who had never had it, feeling pain in back and/or upper limbs, in the future was analysed (Figure 1). The results indicate that workers who have already felt pain are most likely to feel it in the future, because about 40% of these declared to have "no doubts", whereas workers who have never felt pain only admit the possibility of feeling it in the future.

On the other hand, workers who felt pain in the last 12 months are those who declared to have "no doubts" about the likelihood of having back pain in the future. Conversely, those who have not had symptoms in these regions in the last 12 months, only admit the possibility of feeling back pain in the future. Regarding the assessment of the perception of pain (gravity/intensity), workers were asked to rank it, on a scale from 0 to 5, (from the mildest,

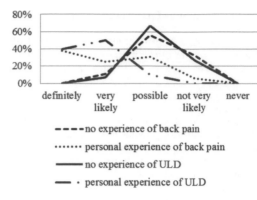

Figure 1. Probability to come to feel pain in the future versus already felt pain.

to the most severe). The results indicate that workers who have ever felt pain assess it more severely than those who have never felt it. It was also found the higher the intensity of pain experienced in these regions, the greater the perceived probability of feeling back pain in the future. Similar conclusion was made in the study developed by HSE (1999), although their values are slightly lower than those show in the present study. Likewise, the probability of the pain severity increasing is higher for those who have ever had back pain (p = 0.013). On the other hand, concerning the upper limbs, the higher the intensity of pain in the right shoulder, the higher the intensity/gravity of the perceived pain, that workers believe to have ever felt (p = 0.018).

4 CONCLUSIONS

The results of the present study allowed concluding that, in the analysed tasks, the risk of WRMSDS is significant and that the most frequently reported symptom is pain.

In the 12 months preceding the study, knees, and the dorsal and the lumbar regions were those that showed a higher prevalence of pain, with more than half of the workers declaring this symptom. On the other hand, the pain felt in the last 7 days mainly affected the lumbar and dorsal regions, similarly to what happened in the symptoms reported in the previous 12 months, except for the region of the knees which looses the position of most prevalent region.

The main mechanism used by most workers for the control of pain consists in the intake of medicines obtained through self-medication.

Most workers (92%) believe that, due to the current work, it is very likely that they feel back pain in the future, while 80% declared to believe that it is very likely that the same occurs in upper limbs, revealing a high perception on the risk factors for WRMSDs they are exposed at in their current work.

The study results provide useful information in order to characterize the perception on the risk of WRMSDs for the workers of the traditional bakery industry. In general, the results suggest a significant perception of risk for WRMSDs, which is as much high as higher the reported musculoskeletal symptoms.

The study has some limitations, namely regarding the reduced dimension of the sample, which was directly influenced by the extent of the study population—the traditional bakery workers-, whose number is reduced and widely dispersed throughout the country.

However, these limitations neither obstructed the validation of the study results nor hampered the conclusions achieved. However, further research is needed to explore the role of risk perception in the prevention of WRMSDs and, on the other hand, to validate the results of the current study. To investigate if a lower prevalence of musculoskeletal symptoms is associated to a lower perception of risk of WRMSDs was not included in the scope of this study.

REFERENCES

Arezes, P.M., Miguel, A.S. 2005. Individual Perception of Noise Exposure and Hearing Protection in Industry. *Human Factors,* 47(4): 683–692.

Arezes, P.M., Miguel, A.S. 2006. Does risk recognition affect workers' hearing protection utilisation rate? *International Journal of Industrial Ergonomics,* 36: 1037–1043.

Armstrong, T.J., Buckle, P., Fine, L.J., Hagberg, M., Jonsson, B., Kilbom, Å. 1993. A conceptual model for work-related neck and upper-limb musculoskeletal disorders. *Scand J Work Environ Health* 19(2): 73–84.

Buckle, P. Devereux, J. 1999. Work-related neck and upper limb musculoskeletal disorders. Luxembourg: European Agency for Safety and Health at Work. 92-828-8174-1.

Cabrita, J. 2010. Absentismo na UE27 e na Noruega. European Foundation for the Improvement of Living and Working Conditions. Available: http://www.eurofound.europa.eu/publications/htmlfiles/ef10431_pt.htm Accessed 9 September 2012.

Castaño, P.P., Cuesta, P.A., Mercé, C.P., Remesal, F.A., Folgado, R.R., Pastor, O.A., Ureña, L.A. 2011. *Manual para el asesoramiento técnico en prevención de riesgos ergonómicos en el sector de la panadería.* Edición Fundación para la Prevención de Riesgos Laborales, Valencia, La Gráfica ISG.

Dancey, C., & Reidy, J. 2008. *Statistics without Maths for Psychology: using SPSS for Windows,* 4ª ed, Edinburgh,Pearson Education.

Ghamari, F., Mohammad Beygi A., Tajik R. 2009. Ergonomic assessment of working postures in Arak bakery workers by the OWAS method, *Journal of School of Public Health and Institute of Public Health Research,* 7(1): 47–55.

Hagberg, M., Silverstein, B., Wells, R., Smith, M.J., Hendrick, H.W., Cayron, P., Pérusse, M. 1995. *L.A.T.R., les lesions atribuables au travail répétitif,* Paris, Éditions Multimondes.

HSE, 1999. Risk perception and Risk Communication: a Review of the Literature.

Nunes, I., 2006. *Lesões Músculo-Esqueléticas Relacionadas com o Trabalho—Guia para Avaliação do Risco.* V. Dashofer.

Perista, H. & Quintal, E. 2009. Ministério do Trabalho e da Solidariedade Social (MTSS), Balanço Social 2007, CESIS, Coleção Estatísticas Available: http://www.eurofound.europa.eu/ewco/studies/tn0911039s/pt0911039q.htm Accessed 9 September 2012.

Robertson, V. & Stewart, T. 2004. Risk perception in relation to musculoskeletal disorders. HSE—research report 284.

Serranheira F., Uva A., Lopes F. 2008. *Lesões Músculo-Esqueléticas e Trabalho—Alguns métodos de avaliação do risco.* Lisboa: Sociedade Portuguesa de Medicina do Trabalho, Cadernos avulso n.º 5.

Stoia, M., Oancea, S., 2008. Occupational risk assessment in a bakery unit from the District of Sibiu. Acta Universitatis Cibiniensis Series E: *Food Technology.* 12(2): 11–6.

Szopinska, I., Panorska, A.K., Kozinski, P., Blachowiak, K., 2007. Work related chronic venous disease in office and bakery workers, *Occupational Ergonomics,* 7(2): 125–137.

Torner M., Zetterberg C., Anden U., Hansson T., Lindell V., 1998. Musculoskeletal symptoms as related to working conditions among Swedish professional fishermen. *Applied Ergonomics.* 19(3): 191–201.

Uva, A., Lopes, M., Ferreira, L., 2001. *Critérios de avaliação das lesões músculo-esqueléticas do membro superior relacionadas com trabalho (LMEMSRT).* Lisboa: Sociedade Portuguesa de Medicina do Trabalho, Cadernos avulso n.º3.

Occupational Safety and Hygiene – Arezes et al. (eds)
© *2013 Taylor & Francis Group, London, ISBN 978-1-138-00047-6*

The identification of occupational hazards to administrative and technical staff of an educational institution

S.R. Benka
Associação Franciscana de Ensino Senhor Bom Jesus, Vacaria, Brazil

R.P. Maia
Associação Franciscana de Ensino Senhor Bom Jesus, Curitiba, Brazil

L.M.W. Prado
FAE Centro Universitário, Curitiba, Brazil

G.D.S. Ulbrich
Associação Franciscana de Ensino Senhor Bom Jesus, Curitiba, Brazil

ABSTRACT. The professional techno-administrative of an educational institution are exposed to several hazards in which primarily didn't seem to be. Sectors like janitorial, maintenance, inspection, health services, among others, are exposed to situations and risks into their schools, like the determined schedule, be in touch with the children, exposed to the noise, chemical products, assist parents and many others, they can be work accident generators.

This work will present some of these situations and point ways to minimize these events.

1 INTRODUCTION

When you talk about school, you picture a place full of children, who are there to learn or even to get something to eat, working as a company which has objectives, like many others, but it is an institution with peculiar characteristics, established according to the age range, which transfers knowledge to people through its masters.

The school in its totality considers that everybody from the school ambience must be considered educators, because the responsibility in the social ambience of an Educational Institution are much wider when comparing to a factory or another kind of services in businesses, since one of the goals of a school is just to prepare the student today for becoming the professional of tomorrow, there is no way of doing that without showing them their responsibilities in the world.

Accident prevention in most of the Educational Institutions, especially concern to health and worker safety, seems to reach only the minimum required by law. Considering Education an economic activity, it appears that rates of work accidents have been increasing in the last years, from 0.98% of the national total in 2000 (YEARBOOK..., 2004) to 1.16% in 2010 (YEARBOOK..., 2012). In 2004 alone, there were 14 deaths in the education activity according to the Yearbook 2006. Compared to other economic activities the education sector should show a minimum rate of deaths. There is a lack of information and investment to make the environment safe which compromises the health of workers.

2 METHOD

Due to the virtual absence of literature involving the occupational hazards of administrative personnel of IE, a survey was conducted in Environmental Risk Prevention Programs-ERPP from eleven schools, the disclosure of names was not allowed. According to NR-9-ERPP, "This norm establishes the obligation of drawing up and implementation by all employers and institutions that admit workers as employees (...) in order to preserve the integrity and health of workers, by anticipation, recognition, evaluation and control of the occurrence of consequent environmental hazards that exist or will exist in the workplace, taking into consideration the protection of the environment and natural resources" (BRASIL. 2012).

With this uprising was possible to identify the main risks to these professionals. Furthermore, existing literature were consulted regarding other economic activity, but that risks could be similar to the teaching, for example, the janitorial function. With the scarcity of Brazilian research in this area, we decided to investigate this matter on foreign sources, but without much success.

Due to this lack of reference, the motive for this work, having as main objective to point out the main risks to this category, in order to draw attention to this reality, especially of managers and public authorities and thus minimize through continuing education, these indexes. Whereas the importance of school is not to teach only on paper, but serving as a reference center for research related to a safe environment.

Although the faculty typically represent more than half of the staff of an IE and also be exposed to numerous risks, it will not be the focus of this work, since much (but not enough) has been written about them compared to the technical and administrative staff.

3 RESULTS AND DISCUSSIONS

According to PRADO (2007), the period between 2002 and 2005 in Brazil, there was a reduction of nearly 7,000 primary schools, although the number of students has increased. Regardless of the causes, in 2002 there were almost 256 students per school, in 2005, this number increased to about 273, facilitating the increase of accidents, due to the large number of people in the same physical space. While in higher education, the situation is the opposite, there was a large increase in the number of these institutions. Between 1980 and 1991, on average opened a higher EI per year, but between 1998 and 2005, this number was almost 192, considering only the last 3 years, this number jumps to 226.

In Brazil, there is little concern about the health and safety of the faculty and especially the technical and administrative employees who are exposed to several risks, according to the characteristics of each school.

Faced with the scenario presented, it appears that the concern about health and safety is highly important in any organization. In the case of EIs, this is even more relevant by several factors. The first one is the amount of people involved with the institutions, including faculty, students and staff. It is really necessary that school knows all the risks everybody is exposed to achieve this broader goal.

It is important to highlight that the risks mentioned in this article are common in EIs health and safety area. There are others which will not be considered here, because they are specific to particular courses (mainly technical level and higher level), such as the use of machinery; biohazards related

to courses in medicine and veterinary medicine among others. Below there are some risks that certain Education professionals may be exposed.

3.1 Administrative staff

Many accidents involving this category are due to the pressure suffered during the workday, mostly by the students' parents who are upset for many different reasons. The administrative staff is fundamental to good working of any school, regardless the institution size. Without them, there would be no office, data processing, parents assistance, library and many other services, and because of the type of activity they work, one of the biggest risks to which they are exposed is ergonomic.

The constant work of documentation and archiving, since there will always be areas with computer terminals in schools and colleges, it opens the possibility for the rising of Work-Related Musculoskeletal Disorders—WMSD. One of the ways to minimize the appearance of these disorders is the adequacy of the workplace, which can be done through the use of desks and ergonomic chairs, computer screens with appropriate distances in relation to depth and height, ergonomic keyboard and mouse pad, among other measurements.

But none of this will help if the work process is not compatible with human limitations, there must have adequate working hours (many people have double shifts) including regular overtime; constant pressure from bosses to reach deadlines, no breaks or micro breaks, taking 10 minutes every 50 worked minutes, among other causes is the lack of pleasure on what they do.

Although underestimated by being practically "invisible", there is another risk that can act on this category that is biological. There is a higher concentration of microorganisms in poorly ventilated rooms, archives, places where there are many plants and dim light, and can even be framed in Sick Building Syndrome—SED (NETO Kulcsar, 2000).

3.2 Janitorial employees

This is the category that works with just nobody wants to deal with, the waste. Despite this, it is unimaginable to think a week without these professional in a company. Even knowing their importance, it is one of the categories which are most at risk.

The biologics are in any kind of waste when cleaning toilets, waste (especially from small children) outpatient, among other places. Therefore, there is a need (must) use of PPE (such as gloves, safety shoes, masks semi-facial respiratory protection and special glasses against splashes and accidental impacts.) There is also the waste from the outpatient department (bandages, medicines and drill-cutting), which although normally are in short supply, there is a whole process of internal collection, transportation, storage and gathering outside, beyond the use of PPE.

With relation to chemical hazards, handling cleaning products require specific training, as some products (heavy cleaning) differ from household products due to their chemical concentration. The training necessity is due to the fact that to perform in this function the people from household chores are not aware of the high concentration of products used mainly in large quantities, plus the lack of use of PPE such as gloves and masks can be really harmful. These unknowns can cause intoxication and especially dermatitis, and other consequences, because intoxication can occur through the skin, lungs or by the swallowing indirectly these products.

Regard to the ergonomic hazards, there must be special care when transporting materials such as buckets and cleaning accessories, portable ladders, industrial polishers, moving furniture and other equipment for cleaning, because in some cases there may be an excessive physical exertion due to the weight. These ergonomic risks bring the possibility of major diseases such as tendinitis, bursitis which can be considered MSDs caused by sweeping steadily, without alternating arms and routine work. Therefore, some measures can reduce the possibility of these lesions, which are the constant training on the posture to stand during the workday as: use of ergonomic equipment (brooms, shovels, floor polishers); turnover in the positioning of the broom (or similar equipment) when such activity is undertaken, a work schedule with rotating employees to perform different types of tasks and this professional must be appreciated by the work system and himself. Another important measure would be to practice heating gymnastics.

The other type of risk to which these professionals are constantly exposed are the accidents. Aside from slips and falls on the same level, falls from ladders are one of the most common, because again the lack of information may be a generator fact, when it creates situations like chairs placed on tables for cleaning windows, defective ladders with missing steps, inadequate lean angles, use of flip-flops to climb this equipment, among many other situations.

It is important to mention the pressure on the schedule to be done (for example when they must clean classrooms in a limited amount of time) It contributes to the generation of diseases and especially accidents.

3.3 Maintenance staff

Maintenance is a sector which is always busy with unusual work and often pressured by unplanned ones, because it deals with more corrective than preventive with. Without them, the physical structure and pedagogical quality of a school would be threatened by broken desks, damaged windows, burned light bulbs and many other situations, exposing the students to accident risks. And it is due to this fact, of unexpected and/or immediate, it is often difficult planning tasks, and it may even result in accidents. It is usually found maintenance electricians, painters, gardeners, plumbers, masons and gardeners within the maintenance sector, many of them are outsourced professionals.

Despite being less exposure, biological hazards are present in this type of activity, especially in the maintenance of toilets, grease trap an when handling plants. Therefore, in these tasks the use of PPE is mandatory.

Chemical risks as already mentioned are also of concern, as they may occur in handling with petroleum products (including paints), the "(...) Prolonged contact with oil and grease cause a skin lesion known as elaioconiosis. (...)" (FUNDACENTRO, 1983). The gardening service is exposed to a unique risk, which is the use of pesticides (herbicides are included in this classification) in the treatment of gardens and can infect others even indirectly, for "The danger threatening the worker and his family also in direct or indirect contact with such powerful poisons" (TORREIRA, 1999), because the clothes arrive soaked at home, the worker threatens to contaminate with the wastes poison his home. Therefore, there is need for the company to take responsibility for cleaning this equipment, but also empower employees to be handling such products.

The constant movement of materials and improper posture in many cases can be considered ergonomic risk factor generators.

The physical risk is also present, especially when using drills, trimmers and other noisy equipment, requiring the use of PPE, because according to NR-15-Unhealthy, "It is not permitted exposure to noise levels above 115 dB (A) for individuals who are not adequately protected" (BRAZIL, 2012). It demands training besides adopting collective protective measures which are actions to be taken to minimize the risks.

Due to the type of activity performed, the most common risks of these professionals are accidents, such as falls from ladders; pressed fingers (when transporting materials), burns (using flammable liquids such as gasoline, kerosene and alcohol); cuts, electric shocks and accidents with sharp objects like glass. There are also many other risks (asphyxiation, explosion, poisoning, drowning, falls and other situations) services when in confined spaces, such as cleaning and maintenance on tanks, water tanks, elevator shafts and other locations that have limited exit space. Among the professional maintenance electricians are the ones exclusive to have a legislation in support of their security which is the NR-10-Security Equipment and Services in Electricity. This legislation, also fits the IEs, because according to this same item 10.1.2 NR "NR This applies to the phases of (...) consumption, including the stages of design, construction, (...) maintenance of electrical installations (...)" (BRAZIL, 2012). Some of the requirements of NR concern: the skills required, the collective protective measures, security projects and especially the development of a handbook of electrical installations.

3.4 Health professionals

It is impossible to measure the importance of health professionals in a school, because these are people who can immediately save the life of a child or adult in case of an event. Even so, few IEs have clinics on their facilities, although this group of professional are of extreme importance to be in any kind of economic activity. Professionals working in this area must be prepared to meet both students and employees, as well as become aware of accident prevention measures.

The professionals of this sector are subject to certain risks, such as biological, which according Bouillon (1998), quoting Neisson-Vernant (1986)," Nursing is particularly exposed to microbiological risks due to close contact (...) with infected patients."

Concern to chemical risks, the possibility of professionals exposure in a school is much smaller than in a hospital or a clinic. There are few existing chemicals in the ambulatory school, the majority consists on medicines.

The other types of risk in these places are accidents. One of the most serious, although not frequent in ambulatory IE is represented by sharp objects, especially needles.

3.5 Inspection professionals

In almost all EIs there is this professional, always walking in the yards and corridors of classrooms and because of this intense movement, some cannot sit for most of their working hours, and it may cause health problems such as varicose veins and sprains.

Another situation that they may be exposed, depending on EI, are the high levels of noise, as the noise generated by hundreds of children on a covered patio normally exceeds the amounts stipulated by the NR 15. In many cases, it is necessary to use earplugs. There are also accidents originated from falls and accidents caused by the children's "plays", which can cause temporary absence from work.

For all functions, the main strategy is the practice of continuing education through training, where language must be according to the public. They should approach the risks, the use of PPE, use of safety equipment, swimming rescue, fire escape and mainly first aid.

4 THE ROLE OF THE LEADER

Historically can be seen, especially in Brazil, there is little concern with the formal health and safety of faculty and especially the technical and administrative employees. Unfortunately many leaders are unaware of the need for policies and strategies for health and safety to employees that meet their own organization. He or she disagrees with it due to the lack of knowledge on the subject issued. Therefore, the lack of appropriate actions due to lack of information or even negligence can literally be fatal in an organization.

But in many cases there is negligence when the manager is unaware of the true working conditions in their company. He ignores them because he did not know he had to be aware of them. They do not know what are the various prevention programs on the use of PPE, functioning of the Internal Accident Prevention—CIPA, among other topics, because there are few basic teaching on these areas passed to these professionals, especially during their academic training.

However, the leader of an EI is not only the administrator of existing resources, but also an educator with a key role in the training of students, demanding and providing security conditions for its employees in the existing environment. The role of the manager is not managing people, but to manage people.

5 CONCLUSION

Educational activities are the ones which involve the participation of people, because there are million circulating around the world in EIs daily. While the school strives to be egalitarian, including all social classes, all races, all religions and so many other differences, it also segregates between those who can and cannot afford to study. In both cases, the quality is often questionable. This is noticeable in the reduction of almost seven thousand primary schools in eight years in Brazil, taking into account the increasing number of students and consequently, more students per class! Already in higher education, the opposite happens, an increase in the opening of new faculties never seen in Brazilian history, creating a competition that has forced many to offer their courses for really low prices, directly affecting the quality of education.

Currently when it comes to quality of education, it is impossible to know in a short term whether it is acceptable or not, because unlike most other branches of activities, these "results" will not be known years later and without the right to "complaints", unless states have developed effective evaluation mechanisms. It is a "journey" where there is no possibility of return. The main indicator of teaching quality will emerge with professional attitudes taken in the future when today's students are integrated into the labor market and active participation in their society.

Because of these future attitudes, there is the need for all professionals an EI develop their activities safely, showing the school community the right way to work and minimize the possibility of any accident at work, because in some occasions, the employee is questioned by students "what is it for?", "what is it?" etc, and what better teaching way about the prevention of accidents than the user himself (of PPE, for example) to explain to students why it is used for.

The daily pressure often suffered by employees of EI as the fulfillment of certain times and lean; direct contact with hundreds of children, exposure to noise, contact with chemicals; attending to students' parents, the risk of accidents among others usually goes unnoticed and underappreciated by some managers and for many parents who think the school as a totally safe place, like a medieval castle. But this pressure (which most often is impossible to be reduced, since it is intrinsic to the activity) should be studied and minimized to the maximum.

The institution must have, for technical and administrative employees, a corporate education policy with encouragement for studying, professional training programs and internal training. In these moments (the inmates) must have space for the explanation of the role of these professionals inside the school for health and safety, or remind them they are being watched constantly by students (especially in education) and that any action taken (positive or not) can be seen as an example or action to be followed in the future, this future includes their own children.

REFERENCES

ANUÁRIO brasileiro de proteção. Revista Proteção, Novo Hamburgo. Edição especial. 2004.

ANUÁRIO brasileiro de proteção. Revista Proteção, Novo Hamburgo. Edição especial. 2006.

ANUÁRIO brasileiro de proteção. Revista Proteção, Novo Hamburgo. Edição especial. 2012.

BRASIL. Lei Nº 6.514, de 22 de dezembro de 1977. 70. ed. São Paulo: Atlas. (Manuais de legislação) Altera o Capítulo V do Título II da Consolidação das Leis do Trabalho, relativo à Segurança e Medicina do Trabalho. 2012.

BULHÕES, I. Riscos do trabalho de enfermagem. 2. ed. Rio de Janeiro: Folha Carioca. 1998.

FUNDACENTRO. Curso de supervisores de segurança do trabalho. 2. ed. São Paulo: FUNDACENTRO. 1983.

KULCSAR NETO, F. Síndrome dos edifícios doentes. São Paulo: SENAC. 2000.

PRADO, L.M.W. Desafios às implementações de estratégias para as gestões de segurança e meio ambiente: estudo de caso de uma instituição de ensino do Paraná. Dissertação de Mestrado, Centro Universitário SENAC-SP. São Paulo, SP. 2007.

TORREIRA, R.P. Manual de segurança industrial. [s.l.]: MCT Produções Gráficas. 1999.

Occupational Safety and Hygiene – Arezes et al. (eds)
© 2013 Taylor & Francis Group, London, ISBN 978-1-138-00047-6

Using cloud computing in occupational risks

M. Izvercian, L. Ivascu & A. Radu
Politehnica University of Timisoara, Romania

ABSTRACT: Cloud Computing is a new way to develop new concepts, new tools for sustainable development of the organization. Organisations are using cloud services and technology now to develop innovative new products, improve operations, share information with customers, partners and suppliers, and run important organization applications. The purpose of this article is to highlight the usefulness of cloud computing in occupational risk, optimize business processes by adopting this approach, and in the final part, authors present an expert system for the evaluation of occupational risks. The authors perform a review of the specialized literature using recent sources, highlighting the advantages, disadvantages and directions of Cloud Computing. Also it shows various implemented platforms and the vision of various researchers in the field. Finally the authors' approach is presented using expert system generator. This approach considers Cloud computing environment as its primary hosting infrastructure.

1 INTRODUCTION

Human resource is essential to any organization, occupational risk assessment is imminent (Hunag, 2011). Work accidents and occupational diseases costs represent around 3% of the GDP in developed countries and about 20% in the less developed countries, according to statistic data (Draghici, Ivascu, Vacarescu, & Dragoi, 2011). Recurrent on Occupational Safety and Health (OSH), risk assessment is still characterized by technical deficiencies, due to ignorance or by lack of operational credibility, i.e., it's not recognized its potential as an effective technical tool, to improve OSH goals. Occupational risk assessment is the activity that identifies existing risk factors in workplaces and quantifies the risk dimension (Costa, Oliveira, & Fujao, 2011).

Correct identification and optimal evaluation of occupational risks contribute to the stability of the company, being a pillar in sustainable development of the organization. And taking into account the costs associated with these actions it can be considered that using cloud computing, an optimal process quality/price is done in the concerned organization. This emerging approach reduces costs primarily related to a standard process (in which there is at least one server, a computer and other peripherals) and then adds value to the assessed organization.

Information technologies (ITs) are systems of hardware and/or software that capture, process, exchange, and/or present information, using electrical, magnetic, and/or electromagnetic energy. It refers to anything related to computing technology, such as networking, hardware, software, the Internet, or the people that work with these technologies (Aksoy, & Denardis, 2011).

The new system for access to IT—Cloud Computing—significantly reduces costs, IT complexity and scope while increasing the optimization for workloads and delivery services. Cloud computing allows a very high degree of scalability, offering superior user experience and is based on the new Internet-based evaluation principles. The planet will be instrumented, interconnected and intelligent. Reducing costs is just the start. Organisations are using cloud services and technology now to develop innovative new products, improve operations, share information with customers, partners and suppliers, and run important organization applications. Despite security concerns and other challenges, executives believe cloud computing can provide their company with lasting competitive advantage.

2 CLOUD ARCHITECTURE FOR OCCUPATIONAL SAFETY

Thinking, planning, control, and working in the cloud require to cope with specific challenges of cloud environment (Bristow et al., 2010) such as uncertain definitions, privacy, contractual and jurisdictional issues, risk and non-performance, interoperability, network capacity, staff and perceptions. The technology allows a balance between the personal and professional life of the employees (Izvercianu, & Radu, 2011).

For the development of cloud architecture there must identified the needs and current state of the infrastructure (figure1). The service oriented architecture represents the base for understanding the data, services, processes and applications that may be migrated or need to be maintained within the occupational safety, so as to observe the security approach. With respect to the IT needs, their structure and usage, the analysis may start from the categories of users who interact with the present IT infrastructure (figure 2) and their necessities. Cloud computing includes 3 three fundamental models: Infrastructure as a service (IaaS), platform as a service (PaaS), and software as a service (SaaS) (Mu-Hsing Kuo, 2011).

- Software as a service (SaaS): The applications are hosted by a cloud service provider and made available to customers over a network, typically the Internet.
- Platform as a service (PaaS): The development tools are hosted in the cloud and accessed through a browser. The developers can build applications installing any tools on computer.
- Infrastructure as a service (IaaS): The cloud user outsources the equipment used to support operations, including storage, hardware, servers, and networking components.

The US National Institute of Standards and Technology (NIST) listed 4 models to deploy cloud computing (Mell, Grance, 2010) (figure 3).

- Public cloud: The cloud service provider makes resources (applications and storage) available to the general public over the Internet on a pay-as-you-go basis.
- Private cloud: The cloud infrastructure is operated solely for a single organization.

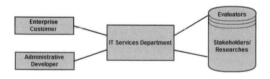

Figure 1.   Standard structure.

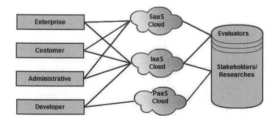

Figure 2.   The simplified structure of the processes in an organization using the service of cloud computing.

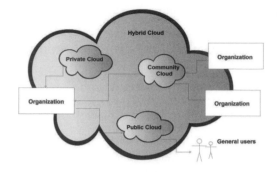

Figure 3.   Deployment models.

- Community cloud: The cloud infrastructure is shared by several organizations with common concerns.
- Hybrid cloud: The cloud infrastructure comprises 2 or more clouds (private, public, or community). In this infrastructure, an organization provides and manages some resources within its own data center and has others provided externally.

Developers may design, build and test applications that are executed on the infrastructure of the cloud provider and deliver those applications directly from the servers of the provider to the final users (Wyld, 2009).

3   THE PLATFORM SOLUTIONS

There are efforts to make the transition from traditional solutions for occupational risk assessment and e-health services to Cloud approaches. Until now there have been developed a number of approaches to this concept, so we can say that this opportunity is now increasingly addressed.

Many studies report the potential benefits of cloud computing and proposed different models or frameworks in an attempt to improve health service. Among them, Rolim et al. proposed a cloud-based system to automate the process of collecting patients' vital data via a network of sensors connected to legacy medical devices, and to deliver the data to a medical center's "cloud" for storage, processing, and distribution. The main benefits of the system are that it provides users with 7-days-a-week, real-time data collecting, eliminates manual collection work and the possibility of typing errors, and eases the deployment process (Rolim, Koch et al., 2010).

Some work shows the advantages and disadvantages of using trading platforms and Clouds (Such as Windows Azure, Google Apps) by organizations for their Cloud applications of e-health. For example, the benefits of rapid deployment and low cost are a compromise with security risks and issues of interruption of Cloud providers.

On the other hand, specialized e-health platforms provide a common set of services for the development and operation of various applications.

Data Capture and Auto Identification Reference (DACAR) is another example of using Cloud.

DACAR aims to develop, implement, validate and disseminate a novel, secure, "in-the-cloud" service platform for capture, storage and consumption of data within a health care domain. The objectives of the project include: development of novel distributed and secure infrastructures based on role and inter-domain security polices; smart device and system integration platform based on novel digital forensic security technology; generic risk assessment strategy for smart device and system integration; clinical evaluation, dissemination and commercialisation (Fan and et., 2011).

The other researchers developed an approach for a cloud platform called CyberHealth for Aggregation, Research, and Evaluation (CARE). CARE was mainly proposed to enable data integration, filtering, and processing for data mining in e-health. They identified a need for an infrastructure for data integration and languages, algorithms, and tools to analyze medical information to discover new medical patterns. With this approach, heterogeneous data from different sources can be integrated, processed and analyzed to improve health understanding and medical treatment effectiveness (Baru and et., 2012).

It can be affirmed that Cloud platform offers the possibility of optimal integration with other applications, while specialized e-health platforms provide better customized approaches for developing, implementing, integrating, and operating e-health applications.

4 A PROPOSED PLATFORM

This approach considers Cloud computing environment as its primary hosting infrastructure. In making the assessment tool it is used the expert system generator.

The importance of a professional risk assessment system is that for the sustainability of the organization it contributes the safe working environment and staff motivated and interested. These technologies, tools, emerging techniques can be implemented only with people and for helping them. According to legislation on health and safety, all employers must periodically assess the occupational risks. The main aim of occupational risk assessment is to protect the health and safety of employees, helping to maintain sustainability and productivity of the organization.

Occupational risk assessment treats the situations that can harm workers and environmental damage as a result of the activities in the organization. Likelihood and/or severity of an injury or occupational disease (at work) that occurs as a result of exposure to a hazard represents the professional risk. Anything that can cause harm is a danger.

A proper risk assessment to prevents and treats occupational injuries and/or illness, loss of production, damage to equipment, business interruption in one or more areas, environmental pollution, motivate staff by creating an optimal working environment within the organization.

The proposed platform is shown in Figure 4.

To identify hazards in the workplace it has been prepared a checklist which is included in PaaS. This list may be extended depending on the activity of the organization. Active involvement of all employees in the process of gathering information helps to correctly identify the hazards. In this list there is a given score (0—unidentified hazard, 1—hazard identified) then given the likelihood and severity of harm it is assessed the risk arising from a hazard that can be small, medium or large (high-unacceptable risk, low risk-acceptable). Using the expert systems generator based on a general hazard identification list it was created a knowledge base containing: award rules, rules for calculating the scores and probability assessment rules. After querying the knowledge base for a sector, you will see the result of the risk assessment, the assessment conclusion and methods of prevention, protection and proper treatment.

At the bottom are Security Mechanisms, which are used to meet the authentication, data integrity and confidentiality requirements.

Infrastructure as a Service includes the required base for developed application.

The applications are hosted by a cloud service provider and made available to customers over a network, typically the Internet, SaaS.

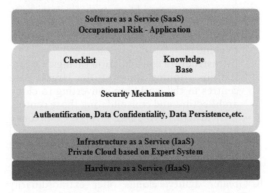

Figure 4. Conceptual structure of the proposed platform.

Cloud Computing—cloud services and storage are accessible from anywhere in the world over an Internet connection and occupational risk assessment is easily to achieve.

Cloud computing infrastructures are next generation platforms that can provide proper value to organizations of any size. They can help organizations achieve more efficient use of their IT hardware and software investments and provide a means to accelerate the adoption of innovations. Cloud Computing increases profitability by improving the tools utilization.

The proposed approach was tested on different companies in Romania and it was found that the implementation of a system using Cloud Computing leads to low investment in infrastructure and software, decreased time allocated for occupational risk assessment and the possibility for easily developed add-ons depending on the need and the business of each company. In future research, cost comparisons will be done between standard evaluation systems (online systems, static systems, etc.) and systems using the concept of Cloud Computing.

5 ADVANTAGES, RISKS AND LIMITATION

The defining advantages of using Cloud Computing can be summarized as (Harris, 2010):

a. Saving money
  − Energy savings: Cutting energy costs is a common motivator for pursuing clouds.
  − Efficiency and productivity improvements: Compared to conventional technologies, cloud services provide even more opportunities to save money when they are used to improve productivity and lower the cost of conducting business. Over half o executives say cloud computing has great potential for permanently and significantly lowering operating costs.
  − Savings on scarce IT resources: Cloud services are an attractive option when skilled IT labour or equipment is difficult and expensive to come by.
  − Infrastructure consolidation savings: Organisations running some of the largest data centres in the world are converting to cloud architecture and consolidating those centres.
  − Infrastructure avoidance savings: Companies, especially start-ups, are choosing to use cloud services to provide their IT infrastructure instead of buying equipment or licensing conventional software.
b. Greater agility: clouds offer extraordinarily flexible resources. They are scalable because of

their technical design. Clouds can be summoned quickly when needed, grow by assigning more servers to a job, then shrink or disappear when no longer needed. That makes clouds well suited for sporadic, seasonal or temporary work, for finishing tasks at lightning speed, processing vast amounts of data, and for software development and testing projects.

c. Integrating operations: The real surprise is that despite the widespread concern over how to secure and integrate clouds with older systems, and the risk of slow response times when data has to run over the Internet, the most commonly cited use for cloud computing is to integrate different organisations.

d. Better decisions: Most executives are interested in using clouds to make better decisions, but they appear divided over its importance compared with reducing costs and supporting processes.

By implementing the solution a gain that exceeds the capital costs and compensates the associated risks must be obtained. Many of the risks specific to cloud environment may be transferred to cloud providers. The important limitations are: not all applications run in cloud, security and protection of sensitive data, maturity of solutions, lack of confidence and risks related to data protection and security and accounts management.

As a conclusion of the strengths and weaknesses of the Cloud Computing in Occupational Safety, can say that the payment per use model and the management policies of risks and security represent positive factors in taking the decision of using this emergent approach.

6 FEATURES OF CLOUD COMPUTING

Cloud computing is the next big wave in computing. It has many benefits and more companies adopt this solution. The features of cloud computing are (Mohd and et., 2012):

− Cloud provides the resources which are on demand as there is isolation so no need to actual sharing;
− It is heterogeneous in nature;
− It adds the virtualization to the data and hardware resources;
− End user security;
− Up-to-date Clouds are operated by single companies;
− Clouds are easily usable hiding the deployment details from the user;
− User billing;
− Limited set of features exposed.

7 CONCLUSIONS

Despite its retentions and drawbacks, it seems that Cloud Computing is here to stay. Economic crisis will force more and more organizations at least to consider adopting a cloud solution. Cloud computing is the next big wave in computing.

A literature review of health Cloud issues was presented in this paper with emphasis on the importance of the concepts involved, implementations and challenges. Many companies have begun to adhere to this vision and there are proofs that indicate significant decreasing of expenses due to the implementation of cloud solutions. The aim of our work was to identify the particularities of using Cloud Computing within occupational safety.

Considering the risks and benefits of using Cloud Computing architecture, the authors have proposed a strategy of adopting this approach in occupational safety and a risk assessment system.

In future research we will test this possibility of risk assessment on various companies in Romania.

ACKNOWLEDGMENTS

This work was partially supported by the strategic grant POSDRU 107/1.5/S/77265, inside POSDRU Romania 2007–2013 co-financed by the European Social Fund—Investing in People.

REFERENCES

Aksoy, P., Denardis, L. 2011. *Information Technology*, Thomson Publisher.

Baru, C., Botts, N., Horan, T., Patrick, K. & Fedman, S.S. 2012. A Seeded Cloud Approach to Health Cyberinfrastructure: Preliminary Architecture Design and Case Applications, *Proceedings of the 45th Hawaii International Conference on System Sciences: 2727–2734*, USA.

Bristow, R., Dodds, T., Northam, R. & Plugge, L. 2010. *Cloud Computing and the Power to Choose*, EDUCAUSE, [Online], [Retrieved October 3, 2012], http://www.educause.edu/EDUCAUSE+Review/EDUCAUSEReviewMagazineVolume45/CloudComputingandthePowertoCho/205498.

Costa, R., Oliveira, C. & Fujao, C. 2011. Complimentarily of Risk Assessment Methods, International Symposium on Occupational Safety and Hygiene, Portuguese Society of Occupational Safety and Hygiene, Portugal, 186–193.

Draghici, A., Ivascu, L., Vacarescu, V. & Dragoi, G. 2011. Occupational Risk and Health System Design Process, *Proceedings of International Conference on Instrumentation, Measurement, Circuits and Systems:* 59–63, Hong Kong.

Fan, L., Buchanan, W., Thummler, C., Lo, O., Khedim, A., Uthmani, O., Lawson, A. & Bell, D. 2011. DACAR Platform for eHealth Services Cloud, *Proceedings of the 4th International Conference on Cloud Computing:* 219–226, USA.

Harris, J. & Alter A. 2010. Cloudrise: Rewards and Risks at the Dawn of cloud Computing, Institute for High Performance, *Research Report:* 10–21.

Huang, Y.H. 2011. *Assessment of Return on Human Resource Investments: Phillips, Stone and Phillips's ROI Process Model Perspective*(20): 443–451, European Journals Publisher, Londra.

Izvercianu, M. & Radu, A. 2011. The Role of Human Resources as Part of Corporate Social Responsibility in Increasing Competitiveness, *International Symposium on Occupational Safety and Hygiene, Portuguese Society of Occupational Safety and Hygiene (SPOSHO):* 209–304, Portugal.

Mell, P. & Grance, T. 2010. *The NIST definition of cloud computing. Commun ACM* 53(6): 50.

Mohd, R., Mohd, J. & Mohd J. 2012. *International Journal of Computer and Information Technology*, Impact of Cloud Computing on IT Industry: A Review & Analysis: 185–189.

Mu-Hsing Kuo, A. 2011, *Journal of Medical Internet Research*, Opportunities and Challenges of Cloud Computing to Improve Health Care Services.

Rolim, C.O, Koch, F.L, Westphall, C.B, Werner, J., Fracalossi, A. & Salvador, G.S. 2010. A cloud Computing Solution for Patient's Data Collection in Health Care Institutions. In: *Proceedings of the 2nd International Conference on eHealth, Telemedicine, and Social Medicine*, New York.

White, J. Cloud Computing in Healthcare: Is there a Silver Lining? Available online: http://www.aspenadvisors.net/results/whitepaper/cloud-computing-healthcare-there-silver-lining.

Wyld, D.C. 2009. *Cloud Computing 101: Universities are Migrating to The Cloud for Functionality and Savings*, Computer Sight.

Occupational Safety and Hygiene – Arezes et al. (eds)
© 2013 Taylor & Francis Group, London, ISBN 978-1-138-00047-6

Assessment of compliance with minimum safety requirements in machine operation: A case of assessing the control devices of a press

A. Górny

Faculty of Engineering Management, Poznań University of Technology, Poznań, Poland

ABSTRACT: Ensuring safe working conditions for workers and protecting their health is every employer's prime responsibility. Conformity with this requirement is particularly vital when operating machinery and equipment associated with high accident rates. Safety can be assured by complying with minimum requirements pertaining to technical solutions and work flows. To verify compliance, a certain risk level which the workers unavoidably need to take to be able to proceed with their work, has to be accepted. The paper describes key stages of risk assessment reflecting the minimum requirements set out in Directive 2009/104/EC. The assessment covers control devices of an eccentric press over all critical assessment stages from the identification of irregularities to the determination of risks and their potential impact to the identification of measures aimed at ensuring a safe working environment.

1 INTRODUCTION

1.1 *The principle behind the safety system*

A uniform safety system developed for work equipment in use has been based on the provisions of Directives 89/391/EEC and 2009/104/EC repealing Directive 89/655/EEC. The directives draw on article 153 of the Lisbon Treaty (formerly article 137 of the Amsterdam Treaty, and article 118a of the Treaty of Rome).

The requirements set out in the directives concern guidelines for the operation of work equipment and, in particular, equipment operated in an environment characterized by risks posed by hazardous factors. The basic improvement measures laid down in Directive 89/391/EEC rely on the principle that occupational safety, hygiene and health should not be subordinated to purely economic considerations. Every employer is obliged to learn about the latest available and practicable technologies and organizational arrangements, in particular those concerning the design of the work place, to secure proper health and safety conditions for its workers.

Due to substantial risks associated with mechanical factors, technical machinery and equipment are subject to specific provisions applicable where the available solutions are critical for ensuring work-related safety. Such requirements are described in Directive 2009/104/EC.

Directive 2009/104/EC addresses and is primarily concerned with employers as it defines measures which allow them to create safe conditions for operating machinery. Such requirements concern work equipment defined as any machine, apparatus, tool or installation used for work at heights. The directive concerns requirements applicable to working conditions wherever work equipment is used (regardless of its age) (Górny 2011, 2012). If complied with, the minimum requirements referred to in the directive will assure conformity with the required safety standards as attained by recognizing design requirements (i.e. by defining conditions for operational safety) and operational requirements (concerning day-to-day use of work equipment).

The European Union's approach to safety, as enshrined in legislation on machine operation, has been fully transposed into the Polish legal system.

1.2 *The meaning of minimum requirements*

Minimum requirements are defined as requirements which must be complied with during the operation of technical equipment. Stricter requirements which ensure a higher safety standard are not prohibited, especially where an employer finds them essential for e.g. minimizing losses caused by potential accidents.

Compliance with the minimum requirements set out in Directive 2009/104/WE, as envisioned to guarantee optimal safety and health standards when operating work equipment, is critical for ensuring the health and safety of workers. The requirements concern the operation of work equipment defined as any activity involving the day-to-day use of such equipment as well as the starting and stopping of such equipment and its use, transport, repair, modification, maintenance and servicing, including cleaning. The requirements concern any worker found wholly or partially in a danger zone.

The minimum requirements concern all areas in which hazards may arise during day-to-day operation, despite adherence to documentation, regular maintenance and repairs. This applies particularly to equipment control devices, normal as well as emergency starting and stopping, protection against hazards caused by the projection and emission of gases and vapours, prevention of contact with moving parts and protection against frostbites, burns, explosions and electric current (Bernard, Hasan 2002, Helander 2006, Górny 2012, Mrugalska & Kawecka-Endler 2012).

Compliance with minimum requirements in the context of controls and devices has been described by reference to requirements pertaining to visibility, markings, the work site, the assurance of the safe use of machinery and protection by preventing unauthorised (accidental or unintentional) start-up of machinery (Mrugalska & Kawecka-Endler 2011).

2 METHODOLOGY FOR ASSESSING SAFETY AGAINST MINIMUM REQUIREMENTS

Responsibility for the safety of workers during the operation of technical machinery and equipment rests with the employer.

The employer is therefore responsible for taking measures to ensure that the machines made available to workers are adequate for the performance of work or properly adapted for such a purpose. In choosing machinery, employers need to pay attention to the specific working conditions and work characteristics as well as any hazards involved in the operation which are essential for the attainment of particular worker health and safety standards. Regardless of the selection methodology, it is essential that assessments are carried out to ensure that work equipment can in fact be operated safely.

Such assessments must to be tailored to reflect any specific characteristics of the areas to be checked.

By inspecting machinery (as part of regular maintenance), employers ensure the safe operation of machinery in their workplace. Inspections help them plan any necessary improvements which, if adopted, will help their equipment comply with minimum safety requirements.

Machinery and technical equipment may be adapted by:

- inspecting and taking stock of work equipment (by, among other things, determining machine type, year of manufacture, wear and tear and estimated operating time),
- carrying out a general assessment to determine any discrepancies between the current state of machinery and any minimum requirements,
- assessing the time and expense required to have modifications made by a competent person,
- drawing up an adaptation plan and schedule,
- carrying out technical and organisational measures,
- supervising implementation and evaluating any measures taken.

The purpose of adapting control devices to meet minimum requirements is to ensure safe working conditions for operators and minimize the impact of potential hazards (Górny 2011). The intended outcome can be achieved by means of corrective, improving and preventive measures.

The measures have been taken in conformity with any relevant technical standards. They were assessed for effectiveness by verifying and validating the solutions in place. The purpose of validation was to confirm that all measures and solutions are consistent with client (worker, operator) needs and expectations.

3 SAFETY ASSESSMENT OF TECHNICAL EQUIPMENT AGAINST MINIMUM REQUIREMENTS

3.1 *PMS 40B eccentric press*

The equipment in question is a PMS40B eccentric press manufactured in 1963. The machine relies on solid coupling to prevent the press slide from stopping during the feed phase; hence, once its clutch is released, the slide completes a full stroke.

The press is operated in a manufacturing plant located in the Region of Wielkopolska. It is used primarily for stamping.

The press is operated by a single worker charged with all tasks involved in its operation. His responsibilities include ongoing press maintenance but exclude repairs requiring specialized tools or equipment.

3.2 *Outcome of safety assessment*

Employers may use safety assessments to check whether their existing measures comply with minimum requirements. It is critical output ensure that existing measures are be compared with those

Figure 1. The PMS40B eccentric press and its control system.

actually available for implementation and that the specific characteristics of measures are properly recognized. While minimum requirements define the objective to be achieved or a hazard to be eliminated, they fall short of defining or prescribing any specific technical solutions which will guarantee that such objectives are met. As a consequence, the user may choose the best way to satisfy requirements and adapt the design and working conditions to technological advances and the current state of the art. Safety assessments may draw on the findings of risk assessments seen as a tool to check the appropriateness of any solutions in place.

The assessment (verification) covered the control devices of a press.

The solutions in place were assessed for compliance with minimum requirements by means of a scorecard developed on the basis of the requirements laid down in Directive 2009/104/EC (Górny, Kowerski, Ostapczuk 2009). The list of the requirements is shown in Table 1.

Responses to the above survey have helped to determine the type, nature and scope of irregularities (incompliance with minimum requirements) and identify possible hazards.

The results are presented in Table 2.

Risk assessment findings can be assumed to provide an indication of the level of compliance with minimum requirements. In selecting measures, account needs to be taken of occupational risk assessment findings regarding the work in question and the technical risk inherent in any design solutions in place.

Risk can be defined as a likelihood and degree of any injuries or health impairments which may be sustained in dangerous situations. To ensure the required safety level and minimize burdens, ergonomic design principles need to be defined, incorporated and complied with for a given risk.

A risk mitigation strategy is deployed in five steps by:

- defining machinery operation rules which reflect its intended use and operating restrictions,
- identifying hazards and any related risks,
- assessing risks for each identified hazard or hazard situation,
- assessing risks and using risk assessment findings to decide whether risk mitigation is required,
- applying protection measures to eliminate hazards or mitigate hazard-related risks.

Risk assessment is a high priority for particularly dangerous machinery or machinery for which no technical specifications compliant with harmonized standards are available. Such machinery require special treatment which entails specific tests which, if not passed, will bar the use of such machinery.

In selecting machinery control devices, it is particularly important that the psychological and physical stress on the operator be reduced. Of key significance here are ergonomic considerations related to or derived from interactions between man (operator) and technical device.

3.3 Actions taken to improve operating safety

To ensure safe and hygienic working conditions for workers exposed to hazards associated with

Table 1. A checklist for assessing control devices.

Questions
Are control devices clearly visible?
Have control devices been identified appropriately? (Have they been described or marked clearly?)
Are control devices located outside hazard zones?
Is control device operation hazard-free?
Are control devices safeguarded against unintentional use?
Have control devices been chosen making due allowances for their circumstances of use?
Have control devices been chosen making due allowances for the circumstances of use of machinery?
Have control devices been chosen making due allowances for failures?
Are control devices the only system which starts work equipment?
Do control devices allow for stopping work equipment safely and completely?
Does the stop system have priority over the start system?
Is the machinery fitted with an emergency stop device?
If affirmative, responses to the above questions indicate that the measures in place comply with minimum requirements selected in view of their specific characteristics.

Table 2. Characteristics of irregularities identified through evaluation (examples).

Problem (1)

Identified irregularities

Controls vital for operational safety are out of the view, identifiable and adequately marked.

Characteristics of irregularities

Controls (STOP, START, EMERGENCY STOP) are insufficiently visible, there are no markings and the ones in place are insufficiently visible.

Hazards caused by irregularities

Worker reactions are slow during emergencies, what could lead to an accident.

Problem (2)

Identified irregularities

Control element creates threat in relationship with his accidental using.

Characteristics of irregularities

Press is fitted with a foot pedal which releases trip-dog activating plane movement of press slide. During standstill and after engaging EMERGENCY STOP controls the trip-dog releases making the press slide do a full operating cycle.

Hazards caused by irregularities

Unintentional use of foot pedal disengaging the clutch during standstill or after switch off could set into motion the slide, consequently crushing upper limbs.

Problem (3)

Identified irregularities

The control system for stopping the machinery does not have priority over start controls.

Characteristics of irregularities

Pressing the start button causes the machine to start, regardless of stop controls.

Hazards caused by irregularities

Pressing both controls at the same time could potentially start the press, causing an accident.

Problem (4)

Identified irregularities

The machine was not fitted with effective emergency braking device.

Characteristics of irregularities

Emergency stop device switches off the engine without blocking the clutch releasing the slide into motion.

Hazards caused by irregularities

No pedal brake or lock could cause unintentional slide movement posing additional hazard.

machinery and technical equipment, it is vital to apply safety precautions which will protect such workers from injuries, exposure to dangerous chemicals, electric shocks, excessive noise, mechanical vibrations and radiation as well as other deleterious factors in the work environment. It is equally critical to assure compliance with such ergonomic requirements. This can be done by guaranteeing design safety or, should that be impossible, fitting machinery and equipment with proper safety devices (means of collective protection).

Note that the safe operation of machinery and equipment relies on occupational safety as well as the operating reliability of technical equipment. The latter relates to technical risk with bearing on the safety of each component of installations and the overall man-machine system (technical equipment and workstation), as set in a particular environment. Complete safety is only achieved when all system components operate properly. Hence, it is the technical risk assessment that should become the prime duty of any employer who seeks to ascertain the adequacy of a particular set of measures designed to satisfy minimum requirements. It is only on that basis and, in particular, in view of any residual risks, that it becomes possible to assess occupational risk (to see whether work can be performed) and identify which organizational solutions (procedural measures) are necessary to allow the performance of work despite a high-risk environment, particularly for specific services (such as maintenance).

A hazard reduction (and risk mitigation) strategy requires that the ergonomic principles for machinery design are properly incorporated and perceived. The main effect of such efforts is to reduce the psychological and physical stress to which operators are exposed.

In keeping with the principles of ergonomics, stress in the selected operating circumstances should be limited to an acceptable level. In the case of control devices, such requirements include:

- recognizing that an operator's physical performance, including strength and endurance, may vary over time,
- ensuring that there is sufficient space for the movement of an operator's body parts,
- eliminating systems in which machines impose the work pace,
- avoiding monitoring which requires long-term concentration,
- ensuring that the man-machine interface is design to accommodate any foreseeable characteristics of operators.

Any related improvement actions can be adopted on a corrective, improving or preventive basis. To that end, one needs to recognize the time at which such actions are taken and the range of

Table 3. Actions taken to evaluate conformity with minimum requirements.

Corrective actions
– apply temporary markings, refurbish and clean controls,
– offer recommendations to increase operator focus whilst operating work equipment.
Improving actions
– fit machine with new controls,
– add permanent markings at controls,
– apply either an electronic or a two-handed control system.
Preventive actions
– carry out mandatory daily inspections to check the technical condition of control devices and report irregularities to superiors,
– carry out periodic checks to verify technical condition and adequate functionality of control devices.

any effects achieved. The actions can be classified into (EN ISO 9000:2005):

– corrective: actions taken to eliminate any identified nonconformities,
– improving: actions taken to eliminate the causes of any identified nonconformity,
– preventive: actions taken to eliminate the causes of any potential inconformity or other undesirable potential circumstances.

The above actions are means for the achievement of objectives which conform to the minimum requirements. The desirable and satisfactory outcome of the above measures is prescribed in advance. The related solutions are given in Table 3.

Many employers find the duties and requirements set out in the legislation highly burdensome as they are often capital-intensive. The greatest difficulties complying with new health and safety rules are experiences by SMEs. Such companies operate machinery and equipment which need to comply with minimum requirements (Górny 2011, 2012).

4 CONCLUSIONS

A compliance check of the control devices of an eccentric press has helped identify a range of serious irregularities. These can only be addressed by upgrading the machinery. Specific measures should focus on technical solutions as well as organizational arrangements.

An additional issue faced commonly in selecting measures are costly equipment adaptations placing a considerable financial burden on the plant owner. Note, however, that such an expense is a one-off

investment which will permanently improve worker safety and working comfort.

REFERENCES

Arezes P.M. & Miguel A.S. 2009. The role of ergonomics characteristics in the effective use of hearing protection devices. In: L.M. Pacholski, S. Trzcieliński S. (eds.), *Ergonomics in Contemporary Enterprise*. International Ergonomics Association Press, Madison.

Bernard A. & Hasan R. 2002. Working situation model for safety integration during design phase. *Manufacturing Technology*, 51 (1): 119–122.

Council Directive 89/391/EEC of 12 June 1989 on the introduction of measures to encourage improvements in the safety and health of workers at work; OJ L 183, 29.6.1989, pp. 1–8, as amended.

Council Directive 89/655/EEC of 30 November 1989 concerning the minimum safety and health requirements for the use of work equipment by workers at work (second individual Directive within the meaning of Article 16 (1) of Directive 89/391/EEC); OJ L 393, 30.12.1989, pp. 13–17, as amended.

Directive 2009/104/EC of the European Parliament and of the Council of 16 September 2009 concerning the minimum safety and health requirements for the use of work equipment by workers at work (second individual Directive within the meaning of Article 16(1) of Directive 89/391/EEC), OJ L 260, 3.10.2009, pp. 5–19.

EN ISO 9000:2005, Quality management systems. Fundamentals and vocabulary.

Górny A. 2011. Zastosowanie wymagań ergonomicznych w kształtowaniu bezpieczeństwa technicznego (Use of ergonomic requirements in a technical safety formation), In E. Tytyk (ed.), *Inżynieria ergonomiczna. Praktyka (Ergonomic engineering. Usage)*. Poznań: Publishing House of Poznan University of Technology.

Górny, A. 2012. Ergonomics in the formation of work condition quality. *Work: A Journal of Prevention, Assessment and Rehabilitation*, 41: 1708–1711.

Górny A., Kowerski A. & Ostapczuk M. 2009, Bezpieczeństwo i eksploatacja maszyn produkcyjnych (The safety and productive machine operation). Poznań: FORUM.

Helander M. 2006. *A Guide to Human Factors and Ergonomics*, 2nd ed. Boca Raton: CRC Press. Taylor and Francis Group.

Mrugalska, B. 2013. Design and Quality Control of Products Robust to Model Uncertainty and Disturbances, In K. Winth (ed.) *Robust Manufacturing Control, Lecture Notes in Production Engineering*. Berlin Heidelberg: Springer-Verlag (in print).

Mrugalska, B. & Kawecka-Endler, A. 2011. Machinery design for construction safety in practice. In C. Stephanidis (ed.), *Universal Access in HCI*, (part III, pp. 388–397). Berlin: Springer-Verlag.

Mrugalska, B. & Kawecka-Endler, A. 2012. Practical Application of Product Design Method Robust to Disturbances. *Human Factors and Ergonomics in Manufacturing & Service Industries*, 22:121–129.

Occupational Safety and Hygiene – Arezes et al. (eds)
© 2013 Taylor & Francis Group, London, ISBN 978-1-138-00047-6

Occupational safety and health: A comparative study between public policies in Brazil and Portugal

T.R. Ferreira & M.B.F.V. Melo
Federal University of Paraíba, João Pessoa, Paraíba, Brazil

ABSTRACT: The appropriate Occupational Safety and Health (OSH) conditions must be a target for various agents of society. These conditions can be attained throughout the adoption of public policies. This article aims at presenting the main OSH public policies in Brazil and Portugal, as well as comparing them. To do so, we accomplished an exploratory research to identify the main public policies adopted in the two countries. We also calculated the OSH rates in both countries. The encountered results show that the public policies adopted in Portugal and Brazil promoted a reduction in the rates of incidence of labor accidents and labor accidents with death.

1 INTRODUCTION

As a scientific and Professional area, Occupational Safety and Health (OSH) dedicates itself to the analysis of labor conditions and their impact on laborers' health, also proposing answers to eliminate, to reduce or neutralize the occupational risks, while turning the labor environment safer and promoting better quality of life for the laborers.

The search for the appropriate OSH conditions must be a target for various agents of society, because the absence of safety may cause labor accidents and diseases, which represent a high economic cost to companies and the State, further to causing inestimable damages to laborers, thus affecting society as a whole.

To attain the appropriate OSH conditions, each interested party (State, companies and laborers) has a fundamental role. For Neto (2011), the State's importance is on top because political will can generate better social conditions of life and labor, and the adoption of OSH public policies contributes to the country development.

Saraiva (2006) defines public policy as a system of public decisions, which aims at preventive or corrective actions destined to keep or change the reality of one or various sectors of social life, throughout definitions of performance targets and strategies and the allocation of the required resources to attain the established objectives.

Other authors emphasize that public policy only exists when the State gets into action, implanting a government project by means of programs and actions aiming at specific sectors of society (Höfling, 2001; Souza, 2006).

In the OSH area, the International Labor Organization (ILO) fundamentally contributes to governmental actions. Founded in 1919, it is the only one of the United Nations agencies with a tripartite structure, composed by representatives of the government, employers' and laborers' organizations, aiming at the promotion of social justice as a condition for the world's peace. ILO is responsible for the formulation and application of the International Labor Norms (OIT, 2012). These norms serve as a basis for the actions of the States, promote the respect to labor rights, also promoting productive and quality employment, the extension of social protection and the strengthening of the social dialogue.

In the community of the Portuguese-speaking countries, Portugal and Brazil stand out for their wide and solid OSH legislation (Santos et al. 2011). This article aims at presenting the main OSH public policies adopted by Portugal and Brazil, pointing out their differences and comparing them by means of OSH indexes.

2 MATERIALS AND METHODS

The study was accomplished by means of exploratory research, throughout the use of bibliographical and documental technical procedures.

To collect information on the OSH public policies adopted in each country, we carried out a bibliographical research in scientific articles and periodicals, as well as a documental one, on the theme, in the legislation of each country

To obtain the OSH rates, we first carried out a research with ILO, specifically in NORMLEX, which is an information system that gathers together information about International Labor Norms, wherein we obtained the quantity of the ratified conventions of each country.

Later on, we calculated the rates referring to the statistics of labor accidents in the period between 1992 and 2009, understood by the incidence rates of labor accidents and labor accidents with death of each country in the last few years. To calculate such rates, we used the Equations 1 and 2, and the results are expressed by the quantity of accidents per one hundred thousand laborers and the quantity of accidents with death per one hundred thousand laborers.

$$\frac{number\_of\_accidents}{quantity\_of\_laborers} \times 100,000 \quad (1)$$

$$\frac{number\_of\_lethal\_accidents}{quantity\_of\_laborers} \times 100,000 \quad (2)$$

To obtain the above-referred rates in Brazil, we consulted the Annual List of Social Information (RAIS) of the Ministry of Labor and Employment (MTE). The collection of data regarding the quantity of injured laborers was obtained from the Social Security Statistical Yearbook (AEPS) supplied by the Ministry of Social Security (MPS). In Portugal, the quantity of laborers was obtained from the National Institute of Statistics (INE), and the quantity of labor accidents between 1992 and 1999, from the Department of Labor, Employment and Professional Formation (DETEFP) described in the study of Macedo & Silva (2005), and from 2000 to 2009 from the Cabinet of Strategy and Planning (GEP)

3 RESULTS AND DISCUSSIONS

3.1 *OSH public policies in Portugal*

The main advances in the Portuguese OSH public policies were recorded in the 1990's. The 1991 Economic and Social Agreement on Safety, Hygiene and Health in Labor aimed at contributing to the modernization of the national economy to assure the competitiveness of the companies and, in a sustainable way, to improve the labor and life conditions of the laborers. This target turned into the improvement of the public policies toward the promotion of the laborers' health as regards OSH and the prevention of the professional risks, among which the demand to organize OSH activities stood out. The Agreement also allowed the transposition of the Directive 89/391/EEC into the Portuguese legislation (Neto, 2001; Sousa Uva, 2009).

With the Directive 89/391/EEC adhesion, there was a reorientation in the OSH section, which started marking the European strategies of the prevention. The prevention strongly needed to be endowed with a global and coherent management policy, capable to generate the required competences and resources to its appropriate and effective growth in the labor spaces (Santos, 2003). To turn this new scenario viable, the organizations needed to incorporate the OSH management systems, as is the case of OHSAS 18001:1999 – Occupational Health and Safety Management Systems, which was rendered into the Portuguese language and started to integrate the country's normative system. It is a management system that reinforces and certifies the systemic logics that the OSH area is looking to take over in Portugal (Neto, 2011).

Presently, the Ministry of the Economy and Employment is responsible for the OSH matter and the Authority for Labor Conditions (ACT), an organ under this Ministry, deals with the inspection toward the fulfillment of the labor norms and also the control of the fulfillment of the legislation related with safety and health in labor, as well as with the promotion of prevention policies against professional risks, either in the ambience of private labor relationships or in that of Public Administration.

The national strategy for the OSH (2008–2012), synchronized with the same European strategy, defines two fundamental targets to be achieved (Sousa Uva, 2009). The first one is the development of coherent and efficient public policies, which aim at:

– To develop a prevention culture;
– To improve the information system on OSH;
– To include OSH in the educational systems;
– To push dynamism into the National System of Prevention against Professional Risks;
– To improve the conditions of public services in the OSH area;
– To promulgate, to update and to simplify the OSH norms; and
– To establish a model for the Labor Conditions Authority.

The second target is the promotion of safety and health in the labor posts as presupposition of an effective improvement of the labor conditions. Its targets are:

– To promote the fulfillment of the OSH legislation, especially in small companies;
– To improve the quality of the OSH service rendering and to increase the competence of the professionals; and
– To deepen the role of the unions and entrepreneurial entities and to involve entrepreneurs and laborers in the improvement of labor conditions.

3.2 OSH public policies in Brazil

The international normative foundation of ILO pushed toward the emergence of a labor protection specific legislation throughout chapter V of the Consolidation of Labor Laws in 1943 and the emergence of several Regulating Norms as regards OSH. Among others, such norms deal with previous inspection (establishing requirements in OSH matter), dimensioning of medical and safety service in the companies, the organization of Internal Commissions against Accidents, specific norms o face physical, chemical and biological risks, further to the norms related with economic sectors of higher accident chances and the new diseases in the sector of services.

Presently in Brazil, there are three ministries with specific attributions in OSH: Ministry of Health (MS), the Ministry of Social Providence (MPS) and the Ministry of Labor and Employment (MTE). Until the mid 90's, regulation and inspection were the instruments that Brazil used to promote the OSH improvement. According to Chagas et al. (2011) and Santos (2011), Brazil has started a process of significant changes in OSH public policies through the 1990's and in the first years of the 21st century. For instance, the Prevention Academic Factor (FAP), implanted in 2010. For Silva & Fisher (2008) and Todeschini et al. (2011), Brazil adopts—with FAP—a public policy that explores the use of incentive in the OSH area. FAP consists in the flexibility of providential contribution rates. These rates can be reduced in up to fifty percent or be increased in up to a hundred percent, in accordance with the company performance in relation to its economic activity, measured by FAP. These governmental incentives for the promotion of OSH are understood as incentives that explore the differences between the organizational performances in OSH to proportionate advantage or disadvantage to the target public, with the intention of stimulating the accident prevention, decent labor and sustainable development.

Still there are records of OSH public policies in other spheres, as is the case of the city of Patos, in Paraiba State, where the City Hall started to bind up the concession of the building license with the presentation of collective protection projects and the electrical installations of the job camp, so stimulating the development of the promotion of health and safety in the civil construction labor in that city (MTE, 2011).

Another action of the Brazilian government was the creation of a Work Group composed of the MTE, MPS and MS for the elaboration of a base-document related with OSH. This group checked the continuity need of a Policy in the area, focusing it in a coherent way and contemplating the articulation between the actions of several organs. It also observed the need of the tripartite focusing, as per ILO's principles and direction lines, there resulting in the emergence of the National Policy of Safety and Health in Labor in November 2011. This policy aims at the promotion of health, the improvement in the laborer's quality of life, the prevention of accidents and health damages resulting from labor or that happen to occur during its course, by eliminating or cutting off the risks in the labor ambiences. Its implantation will be completed in up to 96 months.

3.3 Main differences between OSH public policies in Brazil and Portugal

Most government actions to promote the improvement in labor conditions, both in Brazil and Portugal, are centered in regulation and inspection, by means of labor inspection, which accomplished by the MTE in Brazil and the ACT in Portugal.

As to the inclusion of the norms in Portugal, these are based on the Directive 89/391/EEC and are applied to all sectors of activity, either private or public. In Brazil, the regulating norms just contemplate the laborers hired under the Consolidation of the Labor Laws, so excluding the remaining laborers (autonomous laborers, randomly-hired workers and public servants).

All in all, the Brazilian OSH legislation induces a "legalist" posture on the part of the companies, because most of them tend to orientate the programs of safety and health in labor to just meet the legislation accomplishment, thus turning them to be poor and of low performance. For Oliveira (2003), this problem is aggravated by the fact that there is not a full inspection coverage by the Ministry of Labor and Employment owing to the reduced number fiscal inspectors to cover the universe of companies where laborers are daily exposed to the risks of accidents and/or labor diseases.

On the other hand, the Brazilian OSH legislation stands out by the adoption of a public policy, which explores the use of incentive in the OSH area, as is the case of FAP. For Silva & Fisher (2008), these government incentives in the form of flexible rates in the collection of providential contributions are considered to be promising, because they influence the decision of the companies toward improving the OSH performance.

Following the European model, the Brazilian government created the National Policy of Safety and Health in Labor, in 2011. Among the strategic targets of this new policy, there emerges the implantation of OSH management systems in the public and private sectors.

In Portugal, after the adoption of the Directive 889/391/EEC, the need to incorporate OSH management system by the organizations becomes a differential, because a management system—chiefly when it is integrated—can bring advantages. Santos et al. (2012) detach that the main benefits are better labor conditions, guarantee of legislation fulfillment and better internal communication on risks and danger.

3.4 *Comparison between OSH rates—Brazil and Portugal*

In both countries, in general, the OSH norms are based on international references such as ILO conventions and recommendations. Brazil ratified 96 of these 82 in force. Portugal confirmed 78 conventions of these 70 in force, Although Brazil has more ratified conventions than Portugal, there is a group of conventions that ILO considers to be of higher priority and governance. Portugal has all these conventions ratified, whilst Brazil has not, as per the table below (ILO, 2012).

The ratification of the ILO conventions is of fundamental importance as they guide the State actions toward promoting productive and quality labor, in conditions of freedom, equity, human safety and dignity, being considered fundamental condition to overcome poverty, the reduction of social inequalities, the guarantee of democratic governability and sustainable development (ILO, 2012).

In Portugal, the reduction of labor accidents is a social priority. Since the Directive 89/391/EEC, the Portuguese policy of accident prevention and contribution to the improvement of labor conditions has resulted in a considerable reduction in labor accidents. Aires (2010) analyzed the effect of the European directives as for OSH in the countries integrating the European Economic Community (EEC), specifically of the civil construction sector. Among the results, Portugal is shown as presenting a reduction of 38.5% of this rate, compared to the years 1995 and 2005, one of the best performances among the EEC countries.

The rates calculated by the Equations 1 and 2 are shown in Table 2. Comparing 19992 and 2009, Portugal had a reduction of 30% in rate of incidence in labor accidents, while Brazil showed a reduction of 27%, a significant reduction in quantity of labor accidents in both countries. But in more recent years, Brazil registered a retrocession in this rate as can be seen in Figure 1. As from 2005, the quantity of labor accidents per a hundred thousand laborers has increased.

As for the data referring to accidents with death, Brazil and Portugal show expressive results. In the last few years, both countries show continuous falling rates as per Figure 2.

Table 1. Ratification of ILO conventions.

Conventions	Brazil	Portugal
Fundamental conventions	7 of 8	8 of 8
Governance conventions	3 of 4	4 of 4
Technical conventions	86 of 177	66 of 177

Table 2. Rates of incidence of accidents and accidents with death in Portugal and Brazil.

Year	Brazil		Portugal	
	Accidents	Death	Accidents	Death
1992	2,391	16	6,206	7
1993	1,780	13	5,599	6
1994	1,641	13	5,223	7
1995	1,785	17	4,714	8
1996	1,659	19	4,958	9
1997	1,748	14	4,828	6
1998	1,692	15	– *	5
1999	1,552	16	4,392	8
2000	1,387	12	4,765	7
2001	1,251	10	4,904	7
2002	1,370	10	4,828	7
2003	1,351	9	4,656	6
2004	1,483	9	4,574	6
2005	1,503	8	4,458	6
2006	1,457	8	4,616	5
2007	1,754	8	4,576	5
2008	1,917	7	4,637	4
2009	1,756	6	4,328	4

* Inconsistent statistic data.

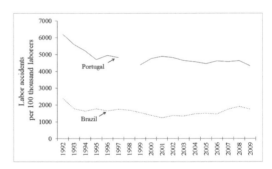

Figure 1. Rates of incidence of labor accidents in Portugal and Brazil.

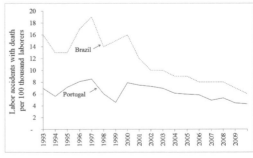

Figure 2. Rates of incidence of labor accidents with death in Portugal and Brazil.

Comparing with Brazil, one notices that the OSH public policies adopted by Portugal proportionate the humanization of life and labor conditions, as can be proved by the previously analyzed data.

4 CONCLUSIONS

The study shows that the governmental actions contribute to the improvement of labor conditions. With well-structured OSH public policies, it is possible to reduce the rate of labor accidents and related diseases, as is the case of Portugal who, since the new OSH national strategy, has attained expressive results. In Brazil, we notice that, despite the existence of well-founded laws and norms, the lack of a global and coherent management policy in this area makes a good performance impossible. In search of the continuous improvement of the labor conditions and influenced by the good results attained by countries that have adopted an OSH national policy based on management system, Brazil has launched the National Safety and Health in Labor policy in November 2011. Based on ILO principles and guidance, this policy starts a new era of approaching and solving problems related with the OSH issues in the country.

REFERENCES

Aires, M.D.M., Game, M.C.R. & Gibb, A. 2010. Prevention through design: The effect of European Directives on construction workplace accidents. *Safety Science.* 48 (2): 248–258.

Chagas, A.M.R.; Salim, C.A.; Servo, L.M.S. 2011. Saúde e segurança no trabalho no Brasil: os desafios e as possibilidades para atuação do executivo federal. In: _____ (eds) *Saúde e segurança no trabalho no Brasil: aspectos institucionais, sistemas de informação e indicadores.* Brasília: Ipea.

Cordeiro, R., Sakate, M., Clemente, A.P.G., Diniz, C.S. & Donalisio M.R. 2005. Subnotificação de acidentes do trabalho não fatais em Botucatu, SP. *Rev Saúde Pública.* 39: 254–60.

Höfling, E.M. 2001. Estado e políticas (públicas) sociais. *Cad. CEDES.* 21 (55): 30–41.

International Labour Organization (ILO). 2012. NORMLEX Information system on international labour standards. In: www.ilo.org/dyn/normlex/en

Macedo, A.C., & Silva, I.L. 2005. Analysis of occupational accidents in Portugal between 1992 and 2001. *Safety Science.* 43(5–6): 269–286.

Ministério do Trabalho e Emprego (MTE). 2011. Construção civil: um choque na cultura do improviso. *Revista Trabalho 11.* Brasília: MTE.

Neto, H.V. 2011. Segurança e saúde no trabalho em Portugal: um lugar na história e a história de um lugar. *Ricot Journal.* 2: 71–90.

Oliveira, J.C. 2003. Segurança e saúde no trabalho: uma questão mal compreendida. *São Paulo Perspec. [online].* 17(2): 03–12.

Organização Internacional do Trabalho (OIT). 2012. *A OIT no Brasil: trabalho decente para uma vida digna.* Brasília: OIT.

Santos, F.A. 2003. Custos e Benefícios na Segurança do Trabalho. *Eixo Atlántico: revista da Eurorrexión Galicia-Norte de Portugal.* 5, 61–74.

Santos, A.R.M. 2011. O Ministério do Trabalho e Emprego e a saúde e segurança no trabalho. In: Chagas, A.M.R.; Salim, C.A.; Servo, L.M.S.. *Saúde e segurança no trabalho no Brasil: aspectos institucionais, sistemas de informação e indicadores.* Brasília: IPEA.

Santos, B.H.F., Melo, A.L.P., Melo, M.B.F.V. & Borges, U.N. 2011. Aspectos legais de prevenção de riscos à saúde, higiene e segurança no ambiente laboral; um comparativo teórico entre os países que compõem a CPLP—comunidade dos países de língua portuguesa. In: Arezes, P.; Baptista, J.S.; Barroso, M.P.; Carneiro, P.; Cordeiro, P.; Costa, N.; Melo, R.; Miguel, A.S.; Perestelo, G.P. (eds.). *Segurança e Higiene Ocupacionais—SHO 2011. 1 ed. Guimarães: Sociedade Portuguesa de Segurança e Higiene Ocupacionais (SPOSHO).* 1, 570–573.

Santos, G., Barros, S, Mendes, F. & Lopes, N. 2012. The main benefits associated with health and safety management systems certification in Portuguese small and medium enterprises post quality management system certification. *Safety Science.* 51(2013): 29–36.

Saravia, E. 2006. Introdução à teoria da política pública. In E. Saravia & E. Ferrarezi (eds.). *Políticas públicas: coletânea (Vol. 1).* Brasília: ENAP.

Silva, R.G. & Fischer, F.M. 2008. Incentivos governamentais para promoção da segurança e saúde no trabalho: em busca de alternativas e possibilidades. *Saude soc. [online].* 17 (4): 11–21.

Sousa Uva, A. 2009. Salud y Seguridad del Trabajo en Portugal: apuntes diversos. *Med. segur. trab. [online].* 55(214): 12–25.

Sousa Uva, A. & Graça, L. 2004. Saúde e segurança do trabalho: da lógica do serviço à estratégia do sistema integrado de gestão. *Saúde e Trabalho 6:* 119–144.

Souza, C. 2006. Políticas públicas: uma revisão da literatura. *Sociologias [online].* 16: 20–45.

Todeschini, R., Lino, D. & Melo, L.E.A. (2011). O ministério da previdência social e a institucionalidade no campo da saúde do trabalhador. In: Chagas, A.M.R.; Salim, C.A.; Servo, L.M.S.. *Saúde e segurança no trabalho no Brasil: aspectos institucionais, sistemas de informação e indicadores.* Brasília: IPEA.

Occupational Safety and Hygiene – Arezes et al. (eds)
© *2013 Taylor & Francis Group, London, ISBN 978-1-138-00047-6*

Hearing protection: Selection factors and risks of excessive attenuation

M.L. Matos
LNEG/FEUP, Portugal

P. Santos
A.Ramalhão, Porto, Portugal

F. Barbosa
Cinfu, Porto, Portugal

ABSTRACT: Currently, noise exposure in the occupational setting continues to be a problem in industrialized countries, not only because this is the most common occupational disease in these countries, but also because it is transversal to all activity sectors, including leisure activities. Exposure to high noise levels can affect the hearing system. For this reason, the legislation requires that the employer implements a set of measures to protect the exposed worker, and as last solution the use of individual hearing protection. However, what occurs in practice and in the majority of cases is the adoption of the use of hearing protection, at the expense of collective protective measures. Nevertheless, not always their selection is made of the most efficient manner. Excessive attenuation of the protectors, as well as the existence of sectors where impulsive noise is produced, are currently the main concerns of safety technicians.

1 INTRODUCTION

Currently, noise exposure in the occupational setting continues to be a problem in industrialized countries, not only because this is the most common occupational disease in these countries, but also because it is transversal to all activity sectors, Abelenda (2006) including leisure activities. The perception of workers to hearing problems increased slightly. According to the results from the European survey, Eurogip, (2004), 7% of European workers consider that work affects their health in terms of hearing disorders. Workers who say they are subject to a higher level of noise exposure are also those who report more hearing problems.

The noise in the workplace is a global problem that affects a wide range of industrial sectors. Excessive exposure to noise can cause hearing problems, and they may arise from exposure to either a brief impulsive noise (over 140 decibels (dB (C)) or exposure to high intensity noise (more than 85 decibels (dB (A)) for several hours each day, during a long period of time (a few years) Within the EU-27, it is estimated that there are 60 million workers—30% of the workforce—exposed to noise, OSHA, (2009). Due to this reason, we list some important facts about the disturbances of hearing caused by noise in the workplace:

- The disorders are often accompanied by tinnitus, or buzz in the ears.
- The cost of hearing disorders caused by noise correspond to approximately 10% of the total cost of compensation for occupational diseases.
- The acknowledged incidence of hearing disorders differs, depending on the country and of the recognition policies. In 2005, the difference between the Member States was quite evident: 5.9% of EU-15 workers complained about hearing problems, against 13.5% in the 10 new Member States and 9.7% in Bulgaria and Romania.
- The largest numbers of cases are registered in the age groups of 40 to 54 years and of 55 to 60 years, OSHA, (2009).

Exposure to high noise levels can affect the hearing system. For this reason, the legislation requires that the employer implements a set of measures to protect the exposed worker. The purpose of these measures, both of technical and organizational nature, is to inform workers, medical surveillance and, as a last resort, the use of individual hearing protection, among other aspects. However, what occurs in practice and in the majority of cases is the adoption of the use of hearing protection, at the expense of collective protective measures, due to the direct economical cost that some of these measures entail. As a consequence, the use of hearing protection equipment has increased. However, not always it's selection is made of the most efficient manner done as efficiently. One of the most frequently-cited objections by workers is that they feel they are unable to hear speech and warning signals when they are wearing hearing protection equipment, Suter (1992), Wilkins & Martin, (1987). The excessive attenuation of hearing protectors, as well as the existence of sectors in which labor activity noises occur with specific characteristics of the impulsive type, are currently the main concerns of safety technicians.

2 RISKS AND SCENARIOS

2.1 Risks and factors for selection of individual hearing protection

Noise is not only a harm to the health, but being responsible for lowering the workers' attention, reduces the quality of the work environment. Particularly, noise makes difficult the communication between workers and the perception of sounds used by machines and working places, which can have serious consequences on workers safety that can lead to accidents. A personal protective equipment is a device designed to be used by a person; due to its characteristics in terms of acoustic attenuation, it reduces the harmful effects caused by noise on hearing.

The choice must be careful in order to assure that the hearing protectors supplied to the user don't cause an excessive attenuation. This situation can cause difficulties in communication, hearing and warning sounds differentiation, such as machinery and sound alarms for evacuation, in case of emergency. In this case, many workers state that they feel uncomfortable, that they do not hear or understand coworkers, do not hear the equipment noises and the warning sounds and that they feel isolated from the surrounding environment. As a result, they end up not using the hearing protectors thus endangering their hearing health.

Some studies on noise protection, Niquette, (2006) show that most of the protectors that are used cause an excessive attenuation at certain frequencies. The selection of hearing protectors involves preliminary analysis of several factors related to the noise in question (see Figure 1 diagram).

Probably the ideal protector does not exist. Therefore, on one hand, hearing protectors should be chosen based on their capability to attenuate the noise, in order to protect the worker, without restricting the hearing of warning sounds of working places

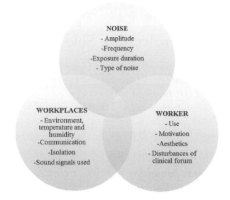

Figure 1. Diagram of the factors to be considered in selecting hearing protectors. INRS (2009).

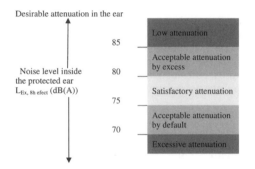

Figure 2. Evaluation of the hearing protector attenuation. Adapted from IPQ, (2006).

and without making difficult the communication between colleagues. On the other hand, the use of hearing protectors, during the whole exposure period, should be ensured. Due to these two reasons, the appropriate protector is the one that enables a reasonable a safety level and, at the same time, it is compatible with the comfort conditions.

The NP EN 458 contains recommendations regarding selection of hearing protectors, its use and usage care. According to this standard, a hearing protector must reduce noise to levels, below the action levels. This means that is the exposure level should be less than or equal to the action level set for each particular country, which requires the use of hearing protectors. According to the Portuguese Law Decree 182 of the year 2006, Decreto, (2006), there are two action levels, the lower: 80 dB (A) and the higher: 85 dB (A). Subparagraph b) of the 7th Article, establishes that the higher action level requires the mandatory use of hearing protectors, and the employer must assure the use of individual hearing protection. Table 1 shows an example

Table 1. List (exemplary) for risk assessment of exposure to noise and selection of hearing protectors. Adapted from INRS, (2009).

Characterization of the noise

	Yes	No	Comments
Continuous noise			Noise level dB(A) =
Intermittent noise			
Fluctuating noise			
Impulsive noise			
High frequency noise			
Low frequency noise			
Noisy industrial environment			—h
Exposure level $L_{Ex, 8h}$			$L_{Ex, 8h}$ = —dB(A)
The spectral analysis of the noise, is available?			Attach spectral analysis

Perception of word and acoustic signals

	Yes	No	Comments
Is necessary the perception of acoustic signals indicating a hazard?			
Oral commands or signals advertiment?			
Others information?			Indicate which:
Necessary to communicate verbally with others person at 1 meter?			
Necessary to communicate verbally with others person at 3 meters?			
There is a risk of falling objects?			
Is necessary climb stairs?			
Is necessary climb scaffolds?			

Electrical risks

	Yes	No	Comments
Electrical contacts?			Tension (Volts) =
Electrostatic discharge?			

Environmental conditions

	Yes	No	Comments
Cold			
Heat			
Humidity			
Rain or snow			

Table 1. (Continued).

Chemical risk

	Yes	No	Comments
Dust			
Liquid and droplets			
Gases			
Fumes			
Aerosols			
Ototoxic agents			Indicate which:
Safety cards are available?			

Hearing protection and PPE

	Yes	No	Comments
There removal and replacement of hearing protection is frequent?			Frequency of removal and replacement:
Simultaneous use of PPE?			
Helmet			
Glasses			
Masks			

Clinical Disorders

	Yes	No	Comments
There are workers on the job with clinical disorders?			Consulting the occupational doctor

Training and information of workers

	Yes	No	Comments
About the exposure risks?			
About the prevention measures?			
How to correctly use the hearing protectors?			

Accidents/incidents Investigation

	Yes	No	Comments
Accidents have occurred?			consulting workers:
Incidents have occurred?			
They are registed?			
Corrective measures were implemented?			Indicate which:
Preventive measures were implemented?			

of the evaluation of the level of attenuation provided by of a hearing protector, at a specific noise situation.

2.2 *Noise exposure scenarios*

Many or almost all industrial activities are responsible for workers exposure to very high noise levels, although there is concern in keeping workers protected. Some activities require more demanding protection measures due to their specific needs. Thus, Neves & Soalheiro, (2007), a study carried out in the School for the Improvement of Army Officers, tried to identify noise exposure and to evaluate if the hearing protection used by these workers was adequate for the exposure to impact noise during activities of Basic Shooting Instruction conducted in the Brazilian Army. In view of the conducted analysis, the author, Neves & Soalheiro, (2007), concluded that those workers were exposed to a minimum sound pressure level of 147.3 dB during shooting activities, and that this value could reach 171 dB. Hearing protection for this activity should provide not only noise attenuation, but also the possibility of understanding the orders and guidelines given during it. According to Berger, (2003), there are four paths ways in which sound can reach the ear, even when it is protected by an hearing protector: air gaps -, if there is no adequate adaptation of the hearing protector to the ear, this can lead to the occurrence of a attenuation reduction of about 5 to 15 dB, over a wide frequency band; protector vibration—when insert into ear protectors are used, vibration may occur, which is proportional to the hearing canal flexibility and can reduce the attenuation at low frequencies; transmission by the material—it is related to the mass, stiffness and sealing materials and transmission through bone and tissues. The protector usage period should also be considered when evaluating its attenuation effectiveness, since the intermittent use of that this device will cause a significant reduction of its efficiency, Berger, (2003).

3 ANALYSIS AND SOLUTIONS

3.1 *Analysis*

Given the mentioned selection factors, in particular when excessive attenuation is detected, it is necessary to select a protector with uniform attenuation which enables to provide a similar attenuation over a wide frequency range, thus supporting effective communications. When it is necessary that determined informative sounds are heard in the work environment or that they are recognized, such as the acknowledgment signals or verbal messages is advisable to use hearing protectors with a uniform attenuation feature across the frequency range. Risk assessment and characteristics of the job are crucial in the selection of suitable protector.Quantitative parameters such as noise level and spectral analysis as well as qualitative parameters such as environmental conditions must be analyzed. To aid the risk assessment, as an example we present in table 1, a checklist based in the original presented in INRS (2009). We suggested additional information, like as: clinical disorders, training and information of workers, contemplate de ototoxic agents and de accidents/incidents investigation. This checklist may be more or less comprehensive and detailed as the situation to apply. This working tool may also serve to promote further discussion among stakeholders: workers, employers, suppliers and manufacturers.For the choice of individual hearing protection, entities may also be able to appeal to the medicine at work to find the solution most adequate for workers with clinical problems, such as irritations, allergies and hearing aids.

3.2 *Solutions*

There are various solutions available in the market that solve problems of excessive attenuation on worker protection, that isolate the verbal communication and warning sounds of the environment that surrounds it. The passive hearing protectors are the type of hearing protector more known and used or alternatively the not passive hearing protectors equipped with electronic components that open new horizons in hearing protection. Electronic protectors use electronic microphone and speakers with built-in functions of dangerous noise attenuation with multiple levels of volume regulated by button.May allow amplifying lower ambient noises, in stereo, so that the user be aware of the direction of its source and reduce unwanted loud noises.

There are also with Bluetooth communication function, with microphone fitted with noise filtering, to allow clear communication with other workers even in very noisy environments. Protectors with radio may allow communicating with other radios or mobile phones or connect to CD or MP3 or FM radio for listening to music or news while meeting the priority input external signal communication for security reasons, as exemplified in Figure 3.

Electronic protectors may have electronic voice function to confirm the electronic volume and filtering adjustments without removing them, and active volume function to avoid unpleasant abrupt sound cuts when occur impulsive noise. The protection function with level dependence improves audibility of vital communications and emergency sounds in intermittent noise environments.

Figure 3. Electronic hearing protectors with MP3 or CD players and radio.

The latest hearing protectors are comfortable using for example liquid pads and ergonomic designs, and can even lead you to forget that you're using it, as a good protector should do.

Within the passive hearing protectors also already exist models with manual listening function, that pressing a button reduces attenuation and allow you to hear conversations and communicate verbally while maintaining some protection. The hearing protectors can also be used as a visual security, existing in bright colors for high visibility.

Given the range of solutions offered, when monitoring the use of hearing protectors by the safety technician and in situations where the employee has been provided a device that allows connection to CD players, MP3 or FM radio, should also be controlled its misuse. Although this option can work as a motivating factor for the use of hearing protectors, it can become a risk factor.

4 CONCLUSIONS

The hearing protector may not be the ideal, but currently, the market already offers several solutions to situations of excessive attenuation and exposure to impulsive noise. However, concern about these aspects, which are felt and expressed by workers, cannot be analyzed as an isolated event, but having to be a comprehensive analysis of various groups of factors that influence the selection of hearing protectors. Thus, using the risk assessment and analysis of global interaction of factors, we are on the way to find the ideal protector.

REFERENCES

Abelenda, C.S. 2006. Avaliação do Conforto de Protetores individuais auditivos. U. Minho, Ed. Guimarães.
Berger, E.H. 2003. Patologia do Trabalho.
Decreto, L. 2006. Prescrições mínimas de segurança e de saúde em matéria de exposição dos trabalhadores aos riscos devidos aos agentes físicos ruído.
Eurogip Agosto de 2004. Costs and funding of occupational diseases in Europe. Obtido em 9 de 11 de 2012, de Eurogip: http://www.eurogip.fr/pdf/Eurogip-08E-cost.pdf
Eswc. 2005. Inquérito Europeu sobre as Condições de Trabalho.
INRS 2009. Les équipements de protection individuelle de l' ouie. Choix et utilization.
IPQ 2006. NP EN 458. Protetores auditivos. Recomendações relativas à seleção, à utilização, aos cuidados na utilização e à manutenção. Documento guia.
Neves, E.B. & Soalheiro, M. 2007. A proteção auditiva utilizada pelos militares do Exercito brasileiro: há efectividade? Revista Temas Livres, pp. 889–898.
Niquette, P. 2006. Hearing protection for musicians. The Hearing Review.
OSHA 2009. Novos Riscos Emergentes para a Segurança e Saúde no trabalho. Luxemburgo: Serviço das Publicações Oficiais das Comunidades Europeias.
OSHA Janeiro de 2009. Perspectivas 1 – Novos riscos emergentes para segurança e saúde no trabalho. Agência Europeia para a Segurança e Saúde no Trabalho.
Suter, A.H. 1992. The effects of Hearing protectors on the perception of speech and warning signals. ASHA Monografs, 28.
Wilkins, P. & Martin, A.M. 1987. Hearing protection and warning sounds in industry: a review. Applied Acoustics, 24, pp. 267–293.

Occupational Safety and Hygiene – Arezes et al. (eds)
© *2013 Taylor & Francis Group, London, ISBN 978-1-138-00047-6*

Development of a multilevel safety climate measure for furniture industries

M.A. Rodrigues
Research Centre on Environment and Health, Allied Health Sciences School of Polytechnic Institute of Porto, Vila Nova de Gaia, Portugal
Department of Production and Systems, R&D Centre for Industrial and Technology Management, Engineering School of University of Minho, Guimarães, Portugal

P.M. Arezes
Centre for Industrial and Technology Management, Engineering School of University of Minho, Guimarães, Portugal

C.P. Leão
R&D Centro Algoritmi, Engineering School of University of Minho, Guimarães, Portugal

ABSTRACT: Safety climate is a relevant measure for monitoring safety conditions. However, there is still no consensus about safety climate measures. This study aims to develop and to analyse the suitability of an instrument to measure the safety climate in Portuguese furniture industries, using a multilevel structure, through a pilot survey. A questionnaire, called Safety Climate in Wood Industries, was developed. The first part comprised workers' demographic questions. The second part included 39 items for measuring safety climate, analysing three different levels: organizational, group and individual levels. The questionnaire was tested on a sample of 29 workers of a Portuguese furniture industry. The company safety conditions were also analysed. The analysis of the questionnaire results shows significant differences on safety climate among sectors, which may be related to differences in safety conditions among sectors. The study shows that the questionnaire allows identifying different safety climates in the same enterprise.

1 INTRODUCTION

The furniture sector represents one of the most important economic sectors of the Portuguese economy, particularly in the North of the country. However, it is a sector that has a reduced professionalization in terms of management, marketing and trade policies, and with low qualification and undifferentiated workers. Moreover, the number of occupational accidents in this sector remains high. Portugal does not have specific statistics on the number of accidents in the furniture industries. According to the Portuguese statistics, in 2007, a total of 77 423 accidents occurred in the manufacturing industries, where 6 128 refers to the wood, corks and related products sector (6 fatal and 6 122 non-fatal accidents) (GEP, 2010). This high number of accidents is a consequence of several factors, namely the specific risk of the sector, the stakeholders' low safety concerns, the lack of machine maintenance, and lower workers qualification. Contradicting this scenario, this is a sector that has attempted to increase its competitiveness.

Concurrently, it is expected its modernization, together with an increase of the safety concerns.

This actual scenario, on the Portuguese furniture industries, shows the need that enterprises have to analyse their safety state, as well as, the safety progress. With this purpose, quantitative measures can be used, as the accident rates (Cameron & Raman, 2005). However, in some cases, in particular for Small and Medium Sized Enterprises (SMEs), the data needed are not available or, in other cases, the number of accidents is under-reported, not reflecting the actual safety condition. Therefore, the use of other measures is important. In this context, safety climate have been referred as a relevant measure to monitor safety conditions (Flin *et al.*, 2000), overcoming some of the limitations of the traditional safety measures. Also Arezes & Miguel (2005) referred that workers' opinions about the company's safety climate play an important role as a safety predictor.

Zohar emphasized the safety climate concept in the eighties (Zohar, 1980). It is seen as a product/sub-component (Choudhry *et al.*, 2007) or an indicator of safety culture (Flin *et al.*, 2000; Tharaldsen *et al.*, 2008; Høivik *et al.*, 2009). According Flin *et al.* (2000), safety climate is a descriptive measure that "can be regarded as the surface features of the safety culture discerned from the workforce's attitudes and perceptions at a given point in time". In this context, safety climate can be used to describe the worker's perceptions and attitudes regarding safety (Flin *et al.*, 2000; Tharaldsen *et al.*, 2008).

Previous research has developed a considerable effort for the construction of a valid and reliable safety climate instrument (Guldenmund, 2007). However, there is still no consensus about safety climate measures, i.e., how many and what factors must be considered, and the number and which items need to be included (Rundmo, 2000; Lu & Yang, 2011). In this context, many researches led to different instruments to measure safety climate (see e.g. Zohar & Luria, 2005; Tharaldsen *et al.*, 2008; Håvold, 2010; Lu & Yang, 2011).

Most previous studies on safety climate only considered a single level of analysis. Recently, this practice has been contested. Different authors claim that, as the scores of the safety climate are aggregated in a single level, ignoring the hierarchical structure of the data, the climate relationships in an organization remain unwell specified (Zohar & Luria; 2005; Guldenmund, 2007). These criticisms are related with the companies' multilevel structure. This hierarchical structure has induced researchers recently to consider a multilevel analysis of the safety climate (see e.g. Brondino *et al.*, 2012). In this structure, and in accordance to Guldenmund (2007), it is possible to distinguish three key impact levels: organizational level, group level and individual level. Therefore, multi sub-climates can be founded in a specific organization (Clarke, 2006; Guldenmund, 2007; Zohar, 2008).

Faced to this problematic, the present study aims to develop and to analyse the suitability of an instrument to measure the safety climate in Portuguese furniture industries, using a multilevel structure, through a pilot survey. This survey attempts not only to get the feedback about the clarity of items, but also to analyse if the questionnaire identifies sub-climates, as expected. However, it is important to highlight that this corresponds to a first step of a study still in development, whose results will allow the tool to be applied to a set of furniture enterprises.

2 MATERIALS AND METHOD

2.1 Sample

The results of the study were based on data collected from a furniture enterprise with 69 workers, being classified as SMEs. The participants who accepted to take part in this study were 29 effective workers from five different sectors of the enterprise.

2.2 Instrument

An instrument for measuring the safety climate considering a multilevel structure, called Safety Climate in Wood Industries (SCWI), was developed and applied. The SCWI included two main parts. The first part included demographic questions about age, gender, department/sector, professional activity, number of years that works in the enterprise, number of years that develop the referred professional activity, and involvement in past accidents. The second part included 39 items for measuring safety climate, through the analysis of three different levels: organizational level, group level and individual level.

The items and scales of both organizational and group levels were adapted from Zohar & Luria (2005). The items were reworded and rephrased to adjust local working practices and culture. Seven items were eliminated because were double-barrelled or not suitable to the reality under study. Six new industry-specific items were included, assuming that this allowed to the instrument to reflect the most important organizational features in these companies and a better within-enterprises comparison (Zohar, 2008). The organizational level included 16 items and the group level 13 items (Table 1). The individual level, not considered in Zohar & Luria (2005) work', is considered an important one. The items included in group level are only related with supervisors' discretion, however, other factors related with workplace or co-workers may also have influence on individuals' safety climate perceptions/attitudes. In the specific case of this study, these factors were measured at an individual level. In this context, 10 items to measure individual level were included, based on literature review, in particularly Tharaldsen *et al.* (2008) work (Table 1). It is believed that these items allow a better distinction among sub-climates in this sector of activity, identifying perceptions/attitudes related with co-workers and activity specificities influences.

From all items, 26 items were phrased positively and 13 items negatively, in order to prevent any tendency on answers. The level of agreement on each item was assessed using a five-point Likert scale ranging from 1 = "Strongly disagree" to 5 = "Strongly agree". An alternative "not applicable" was also contemplated, to be used whenever appropriate. The SCWI questionnaire was delivered to five Occupational Safety & Health (OSH) experts in order to review, examine and test the questionnaire. Some improvements were suggested and taken into account. The final version was applied to the 29 effective workers of the company in analysis.

Table 1. Items included for each of the three considered levels.

Organizational level	Item
The management of this company...	reacts quickly when a dangerous situation is detected, or there is an accident/incident occurs.
	insists on thorough and regular safety audits and inspections.
	is not interested in to continually improve safety levels in each department.
	does not invest in the working machines modernization.
	invests in the implementation of measures to minimize the loads manual handling.
	provides all the equipment needed to do the job safely.
	is strict about working safely when we are working under pressure.
	considers a person's safety behaviour when moving–promoting people.
	requires each supervisor/team leader to help improve safety in his—her sector/department.
	invests a lot of time and money in safety training for workers.
	uses any available information to improve existing safety rules.
	promotes the development of appropriate work procedures to the tasks performed by workers.
	does not consider to workers' suggestions about improving safety.
	considers production before worrying about safety.
	provides workers with sufficient information on safety issues.
	do not gives safety personnel the power they need to do their job.

Group level	Item
My supervisor or team leader...	makes sure we receive all the equipment needed to do the job safely.
	do not check frequently if we are all obeying the safety rules.
	discusses how to improve safety with us.
	rather than to use explanations, compels us to act safely.
	worries that I fulfil with the regulations and work procedures.
	worries that I use all the machines protections
	lets be ignored safety rules and procedures when we are working under pressure.
	frequently tells us about the hazards in our work.
	makes sure we follow all the safety rules and not just the most important ones.
	praises the workers who pay special attention to safety.
	is strict about safety at the end of the shift, when we want to go home.
	spends time helping us learn to see problems before they arise.
	insists we wear our personal protective equipment even if it is uncomfortable.

Individual level	Item
I...	believe that safety is the main priority when I do my work.
	whenever check a dangerous situation, report it immediately to one of my superiors.
	when run my work, I try to always follow the rules and work procedures.
	do not use the personal protective equipment necessary for performing tasks
	not always use the machines protections.
	refuse to ignore safety rules, even when the work is delayed and it is needed to increase production.
	disregard safety rules at the end of the shift, when we want to go home.
	clarify all my questions about the risks to which I am exposed.
	do not call the attention of my colleagues when I see them violating some rule or some safety procedure.
	always participate in the safety training actions.

2.3 Safety conditions analysis

The analysis of the company safety conditions was performed. A checklist was formulated and applied. It included a set of items related with safety and health conditions of workplaces, tasks, equipment and machinery.

These items were evaluated based on a 5-point Likert scale adapted from Reese (2012), where 1 = "very deficient" and 5 = "excellent", in order to characterize the level of deficiency of each risk factor. A sixth level was contemplated, meaning "not applicable".

3 RESULTS AND DISCUSSION

The workers age vary from 19 to a maximum of 59 years old, with a mean around 35 years. Workers collaborate with the company in average for 7 years and exert such activity on average for 11, being considered experienced workers.

All inquired workers understood all the questions. Only three workers left one item in blank, despite the presence of the option "not apply". Four workers suggested restructuring the format of the questionnaire, so that the agreement scale also appears on the second page, facilitating the questionnaire fill.

An analysis of the safety climate total scores was performed by sector of activity, being the results presented on Table 2. The average of the obtained answers for each item by sector was estimated, after recoded the negative questions, and than summed to achieve the score for each level and sector. The total average score by sectors was also presented, representing the company safety climate. This procedure was adopted because the number of respondents was not the same among the five sectors.

As expected, the analysis of the questionnaire results shows significant differences on safety climate among sectors (based on the Friedman non parametric test, $p < 0.05$), indicating the presence of multi-subclimates. This is in accordance with the suggested by previous studies (Clarke, 2006; Guldenmund, 2007; Zohar, 2008). These differences can be related with sector work conditions, as suggested by Cooper & Phillips (2004) to justify the differences on safety climate among departments that they found. The relation between safety conditions and safety climate was previous observed by Varonen & Mattila (2000) in a study in wood-processing companies. Polishing sector presents the highest safety climate total average score (= 156) as opposed to Cut sector that has the lowest total average score (= 125). The results also show that these are more visible in the group and individual levels. This suggests that these differences can be related more with supervisors' performance, workplace features and co-workers influences, than with man-agement actuation. The safety conditions analysis shows that the company presents a great problem in the Cut sector related with machines without protection. Saws, drill and milling cutter were without any protection or the protection raised and workers non-use the driving-bar for cutting small pieces. This can be related with risk of contact with saws and blades, pinch, boring, projection of machine parts (for example saw or saw parts) and others objects. Also situations of excessive noise, loads manual handling (risk of excessive effort) and, materials and cables stored on passageways (risk of falls on the same level or collision with fixed objects) were identified in this sector. These risks, according Miguel et al. (2005), are common in this sector. Taking this into consideration, the Cut sector workers are in risk to suffer an accident, where the damages are immediate. This can have influence on safety climate, since previous studies found a relationship between accident rates and safety climate (see e.g. Varonen & Mattila, 2000; Arocena et al., 2008; Tharaldsen et al., 2008; Nielsen et al., 2008; Vinodkumar et al., 2009). Moreover, they need to fulfil with more rules and safety procedure than others sectors workers, being more likely not meet any of them. This is also an important sector in relation to production, because the following steps are dependent on the results of this sector. If the production on this sector has a delay, the others are left with no material to work. So, it is possible that workers can be subjected to a greater pressure and thus ignoring some safety rules/procedures. For other side, in the polishing sector, the workers exposition is more related with hygienic factors, as chemical agents and noise, as well as, excessive effort. The health damages take longer time to develop themselves, and may change workers' risk perceptions. Also, the procedures and safety rules that they need fulfil are lower. The supervisor actuation is also seen differently among sectors. However, the company only have two supervisors for all five sectors. Therefore, it is possible that the supervisors act differently among sectors, according to same risks and the production requirements.

Faced to these results, it is evident the importance of the inclusion of the individual level in the safety climate work questionnaire, because it can provide a better perspective on the influence of work conditions on subclimates. As also emphasized by Arezes & Miguel (2006), this also highlights the influence of attitudes and workplace culture over and beyond any regulations.

Table 2. Safety climate average scores by sector and level of analysis.

Sector	Organizational level	Group level	Individual level	Total average score
Production	65	49	37	150
Cut	57	39	29	125
Storage	60	52	38	150
Montage	62	45	34	141
Polishing	64	52	41	156

4 CONCLUSIONS

The present work describes the endeavour to develop and to analyse the suitability of an instrument to perform a multilevel analysis of the safety climate in furniture industries. The obtained results, although foreplay and despite the reduced sample size used on this pilot study, show that the questionnaire allows identifying different safety climates in the same enterprise. Furthermore, this work emphasized the use of three levels to perform a multilevel safety climate analysis: organizational level, group level and individual level. Using these three levels, differences among sectors were identified and sub-climates are highlighted, possibly due to differences on safety conditions.

This study was developed in a company with a good safety and health performance. The results cannot be generalized to the whole sub-sector. In the future, the questionnaire will be applied to more furniture industries, in the context of a PhD project, with different characteristics, in order not only to analyse its reliability, but also to find the main safety climate structure. A deep analysis of the relation between safety climate and safety performance will be performed in subsequent studies.

REFERENCES

Arezes, P.M., Miguel, A.S., 2005. Individual Perception of Noise Exposure and Hearing Protection in Industry. *Human Factors* 47 (4), 683–692.

Arezes, P.M., Miguel, A.S. 2006. Does risk recognition affect workers' hearing protection utilisation rate? *International Journal of Industrial Ergonomics* 36, 1037–1043.

Arocena, P. Núñez, I., Villanueva, M., 2008. The impact of prevention measures and organisational factors on occupational injuries. *Safety Science* 46, 1369–1384.

Brondino, M., Silva, S.A., Pasini, M. (2012). Multilevel approach to organizational and group safety climate and safety performance: Co-workers as the missing link. *Safety Science* 50, 1847–1856.

Cameron, I., Raman, R., 2005. *Process systems risk management*. San Diego, USA, Elsevier.

Choudhry, R.M., Fang, D., Mohamed, S., 2007. The nature of safety culture: A survey of the state-of-the-art. *Safety Science* 45, 993–1012.

Clarke, S., 2006. Contrasting perceptual, attitudinal and dispositional approaches to accident involvement in the workplace. *Safety Science* 44, 537–550.

Cooper, M.D., Phillips, R.A., 2004. Exploratory analysis of the safety climate and safety behavior relationship. *Journal of Safety Research* 35, 497–512.

Flin, R., Mearns, K., O'Connor, P., Bryden, R., 2000. Measuring safety climate: identifying the common features. *Safety Science* 34, 177–192.

Gabinete de Estratégia e Planeamento (GEP). (2010). *Séries Cronológicas acidentes de trabalho 2000–2007*. Portugal, Ministério do Trabalho e da Solidariedade Social. (in Portuguese).

Guldenmund, F.W., 2007. The use of questionnaires in safety culture research—an evaluation. *Safety Science* 45, 723–743.

Håvold, J.I., 2010. Safety culture aboard fishing vessels. *Safety Science* 48, 1054–1061.

Lu, C-S, Yang, C-S., 2011. Safety climate and safety behavior in the passenger ferry context. *Accident Analysis and Prevention* 43, 329–341.

Høivik, D., Tharaldsen, J.E., Baste, V., Moen, B.E., 2009. What is most important for safety climate: The company belonging or the local working environment?—A study from the Norwegian offshore industry. *Safety Science* 47, 1324–1331.

Miguel, A.S., Perestrelo, G., Machado, J.M., Freitas, M., Campelo, F. *et al.* (2005). *Manual de segurança higiene e saúde no trabalho para as industrias da fileira da madeira*. Associação das Indústrias de Madeira e Mobiliário de Portugal (AIMMP). Porto. (in Portuguese).

Nielsen, K.J., Rasmussen, K., Glasscock, D., Spangenberg, S., 2008. Changes in safety climate and accidents at two identical manufacturing plants. *Safety Science* 46, 440–449.

Rundmo, T., 2000. Safety climate, attitudes and risk perception in Norsk Hydro. *Safety Science* 34, 47–59.

Tharaldsen, J.E., Olsen, E., Rundmo, T., 2008. A longitudinal study of safety climate on the Norwegian continental shelf. *Safety Science* 46, 427–439.

Varonen, U., Mattila, M., 2000. The safety climate and its relationship to safety practices, safety of work environment and occupational accidents in eight wood-processing companies. *Accident Analysis and Prevention* 32, 761–769.

Vinodkumar, M.N., Bhasi M., 2009. Safety climate factors and its relationship with accidents and personal attributes in the chemical industry. *Safety Science* 47, 659–667.

Zohar, D., 1980. Safety climate in industrial organizations: theoretical and applied implications. Journal of Applied Psychology 65 (1), 96–102.

Zohar, D., 2008. Safety climate and beyond: A multi-level multi-climate framework. *Safety Science* 46, 376–387.

Zohar, D., Luria, G., 2005. A multilevel model of safety climate: cross-level relationships between organization and group-level climates. *Journal of Applied Psychology* 90 (4), 616–628.

Occupational Safety and Hygiene – Arezes et al. (eds)
© 2013 Taylor & Francis Group, London, ISBN 978-1-138-00047-6

Working conditions of firefighters: Physiological measurements, subjective assessments and thermal insulation of protective clothing

D.A. Quintela, A.R. Gaspar & A.M. Raimundo
ADAI-LAETA, Department of Mechanical Engineering, University of Coimbra, Pólo II, Coimbra, Portugal

A.V.M. Oliveira
Coimbra Institute of Engineering, Polytechnic Institute of Coimbra, Department of Mechanical Engineering, Quinta da Nora, Coimbra, Portugal

ABSTRACT: The present paper is dedicated to the study of working conditions of firefighters. Three main topics are considered: physiological measurements, subjective assessments and thermal insulation of protective clothing. The results show that all the physiological tests were carried out without any health risk, since the maximum heart rate was never achieved, that the total insulation of the three tested ensembles, calculated with the global method, is around 1.3 clo and that more than 50% of the firefighters report stress situations due to the firefighting activity and more than 75% declare that there is no institutional psychological support. This research represents one more contribution to the assessment of this high-risk activity and aims to provide the knowledge required to adopt preventive measures and good practices.

1 INTRODUCTION

Firefighting is a high-risk activity in which firefighters are exposed to a wide range of risks, diseases and even death. Unfortunately, fatal accidents still occur. In the United States, for instance, in 2009, a total of 82 on-duty firefighter deaths occurred. This represents a significant drop when compared to the 105 on-duty deaths that occurred in 2008, but is higher than the 79 deaths registered in 1993. A common scenario is that the largest share of deaths occurred while firefighters were operating on the fire ground. In Portugal, more than 180 deaths have occurred between 1980 and 2010 and 60 of them correspond to the last decade. Therefore, the firefighting activities are prone to the continuous attention of the scientific community and of several decision makers.

The thermal environments to which firefighters are exposed represent one of the areas that is frequently addressed by many researchers due to its effect on human beings. Depending on the level of exposure, working in extreme hot environments has several health effects that are related with changes in body heat storage, namely heat cramps, heat exhaustion by water depletion, heat exhaustion by salt depletion, heat syncope or even heat stroke.

Hence, the protective clothing is one of the most important parameters to be taken into account. The protective ensembles of firefighters add a signifi-cant physiological load and performance requirements related to design, mechanical and ergonomic characteristics have to be carefully considered. Whenever this is the goal, ISO 15384 (2003) should be adopted for reference. However, more specific standards on requirements of protective clothing are available. The thermal insulation and the water vapour resistance of the protective clothing represent other topics to be duly considered.

The present paper is directed to the assessment of physiological parameters of firefighters, to the measurement of the thermal insulation of the ensembles used by Portuguese firefighters in the field and to a subjective survey based on an individual questionnaire focused on a wide range of parameters with significant impact on the firefighter performance.

2 MATERIALS AND METHODS

2.1 Physiological measurements

In order to reproduce the exposure conditions of firefighters during a fire, several trials were carried out with 5 volunteers. These tests were performed in two consecutive days (18 and 19 of June), in a chamber used by firefighters during training. An ergometric bicycle (model Sprint Bike 3002) was placed inside the chamber and the volunteers were asked to simulate a high activity level (level 3 according to ISO 8996 (1990). During the tests the physiological changes, namely the tympanic temperature and cardiac rhythm, were recorded with a thermometer from Braun (model Thermoscan IRT 3520) and a heart rate monitor from Sigma Sport (PC 1600), respectively. In addition, a subjective assessment of the degree of thermal comfort and ergonomic effectiveness of the protective garments was performed through an individual questionnaire.

The tests took place under the supervision of a nurse, which is also a firefighter and one of the persons responsible for medical surveillance within the corporation. During the tests, the environmental conditions in the chamber were continuously measured with the heat stress monitor type 1219 from Bruel & Kjaer. The individual characteristics of the 5 volunteers, namely the Body Mass Index (*BMI*), the DuBois Body Surface Area (A_{DuBois}) and the Maximum Safety value of the Heart Rate ($HR_{saf,max}$), calculated according to equations 1, 2 and 3, respectively, are listed in Table 1.

$$BMI = m/h^2 \qquad (1)$$

$$A_{DuBois} = 0.202 \times m^{0.425} \times h^{0.725} \qquad (2)$$

$$HR_{saf,max} = 220\text{-Age} \qquad (3)$$

where *m* = body mass [kg] and *h* = height [m].

2.2 Subjective assessment

It is widely recognized that the activities developed in extreme thermal environments should be evaluated from multiple perspectives and subjective assessments are being looked at with growing interest. In fact, several researchers have been carrying out studies based on questionnaires and this methodology has, indeed, acquired an increasing importance in research studies. Thus, the working conditions of firefighters were also studied through a subjective assessment based on an individual questionnaire. The survey was carried out in Portugal and the sample presented here consists of 64 valid responses obtained in 4 firefighting corporations.

The questionnaire is divided in eight parts and has 44 questions, most of them with multiple choice answers. In the first part of the questionnaire a brief characterization of the firefighter is done, followed by assessments regarding medical surveillance, accidents, physical condition, working conditions during firefighting, hydration, protective clothing and cooling techniques.

2.3 Thermal insulation of protective clothing

The heat exchanges between the human body and the environment depend on various factors, namely the clothing thermal characteristics, the clothing fit, the surrounding thermal and aerodynamic fields and the body movement (Oliveira et al. 2011). The thermal insulation provided by clothing is thus affected by the activity and the environmental conditions. The measurement of clothing insulation should therefore be carried out under reference conditions (Oliveira et al. 2008). In the case of measurements with thermal manikins, the test specifications are referred in different standards, namely in ISO 9920 (2007). Such measurements are defined in terms of the heat loss and the mean skin temperature of the manikin and the environmental conditions within the test chamber (air velocity, operative temperature, difference between air and mean radiant temperatures and relative humidity).

The thermal insulation of the 3 clothing ensembles described in Table 2, was measured and the experiments were carried out with a thermal manikin in a climate chamber (CC). The CC (4.5 m × 4.5 m × 3 m) has four autonomous air-handling units with several capabilities to control air temperature, humidity and air velocity. The thermal manikin is divided into 16 parts and made of a fiberglass armed polyester shell covered with a thin nickel wire wound around all the body to ensure heating and temperature measurement.

The analysis of the thermal insulation of clothing when the measurements are carried out with different manikin regulation modes in the body parts, namely the constant skin temperature and the thermal comfort equation regulation modes, was the main goal of this part of the work. The

Table 1. Individual characteristics of the volunteers.

Code	Gender	*BMI* [kg/m²]	A_{DuBois} [m²]	$HR_{saf,max}$ [bpm]
1	Male	22	1.82	193
2	Male	26	1.94	186
3	Female	24	1.64	191
4	Male	24	2.03	186
5	Male	22	1.85	185

Table 2. Characteristics of the ensembles tested.

Ensemble	Description	State
1	49% Aramid, 50% viscose, 1% antistatic	New
2	100% cotton	New
3	100% cotton	Used

three thermal insulation calculation methods—the serial, the global and the parallel—are also considered and the results are discussed and presented for the total (I_T), the basic (I_{cl}) and the effective (I_{cle}) clothing insulations. The manikin regulation modes, the thermal insulation calculation methods and clothing insulation definitions can be found elsewhere (Oliveira et al. 2008). A comparative study is performed with the 3 clothing ensembles.

3 RESULTS AND DISCUSSION

The present study has gathered an extensive amount of data. However, due to space restrictions of the full paper only the most representative results of each of the three main topics of the research were selected.

3.1 Physiological measurements

In the case of the physiological measurements, the volunteers were mainly men (80%) and 67% have more than 30 years old. The remaining participants are between 25 and 29 years old. The body surface area ranged between 1.64 and 2.03 m², the body mass index (BMI) varied from 22 and 26 kg/m² and the lowest and highest safety values of the maximum heart rate calculated according to equation 3 (IAFF, 2006), were 185 and 193 beats per minute, respectively.

The tympanic temperature of the participants was measured moments before the beginning and immediately after each trial. The measurements carried out at the beginning show that the values varied from 36.1 to 37.2 °C, while the corresponding variation at the end was 36.9 to 37.7 °C. The mean variation was 0.5 ± 0.41 [°C] and the highest difference was 1.5 °C. This value is considered too high, since it doubles the second highest difference (0.7 °C), and is probably due to an error. The detailed results of these measurements are presented in Table 3.

The physical parameters of the environment, namely the air temperature, presented higher values in the second day of the trials (June, 19th), and that is probably the reason why the mean variation of the tympanic temperature is higher in June 19th. It is important to underline that these remarks are valid for the actual conditions in which the duration of the tests was 15 minutes. If longer exposures are adopted, higher values should be expected. In fact, during real firefighting operations, the body temperature can easily reach 39°C (Budd, 2001).

The cardiac rhythm was also measured and the details are listed in Table 4. The heart rate at rest, measured before the tests, the mean and the maximum heart rate values recorded during the tests,

Table 3. Tympanic temperature measurements.

Code	Measurement period [hour:min] Beginning	End	Tympanic temperature [°C] Beginning	End	ΔT
1_18	15:15	15:30	36.7	37.3	0.6
2_18	15:37	15:52	36.2	37.1	0.9
3_18	16:03	16:18	37.0	37.0	0.0
4_18	16:49	17:04	36.7	37.4	0.7
5_18	16:27	16:42	36.6	36.9	0.3
1_19	14:45	15:00	37.2	37.7	0.5
2_19	14:20	14:45	36.1	37.6	1.5
3_19	15:10	15:25	37.1	37.5	0.4
4_19	16:14	16:29	36.9	37.2	0.3
5_19	15:49	16:04	37.0	37.6	0.6
				Mean	0.5
				Standard deviation	0.41

Table 4. Heart rate measurements.

Code	At rest [bpm]	Trial values Mean [bpm]	Maximum [bpm]	End [bpm]
1_18	99	113	135	122
2_18	101	116	129	120
3_18	82	108	133	111
4_18	111	120	134	121
5_18	115	117	134	112
1_19	105	120	147	120
2_19	111	131	152	140
3_19	110	120	133	122
4_19	94	106	167	111
5_19	93	121	135	123

and the value obtained immediately after the end of each trial were registered. Considering the whole results, i.e., the tests carried out during the mentioned two days, the mean and standard deviation values of the five volunteers were 103± 10.32 bpm at rest, 119 ± 7.13 and 135 ± 11.85 bpm for the mean and maximum values recorded during the trials, respectively, and 121 ± 8.46 bpm for the case of the measurements performed at the end.

Most of the mean values of the heart rates were within the range of the third metabolic rate. All the tests were performed without any health risk since the corresponding maximum safety values were never attained.

These physiological evaluations showed that participants need a rest period of 30 minutes in a thermally controlled environment. The best ability to perform these tests was shown by participant 4, with a reduced increase of the tympanic

523

temperature and the highest heart rate. This individual reported that physical training was a daily practice, underlining the strong relevance of this issue.

3.2 *Subjective assessment*

The subjective survey carried out in 4 firefighting corporations shows that 85.7% of the firefighters are less than 40 years old and 47.6% are less than 30. The distribution of the 64 participants in the study by gender shows that the majority of firefighters are men (82.8%). The length in the activity is classified into 6 different categories and 29.7% of the firefighters have remained in this activity for less than 5 years and 17.2% are in this occupation for more than 20 years. The highest percentage corresponds to the 11 to 20 years class (26.6%) and the lowest fits in the class of less than 1 year (3.1%).

The periodicity of medical exams in the corporation was assessed through a multiple choice question with 5 possible answers. The majority of the respondents (51.6%) state that a medical exam was never carried out and among the 21.9% that mentioned the occurrence of medical exams, 20.3% point out that only one exam is performed per year. The results also highlight that 51.6% of the respondents report stress situations due to the firefighting activity and 76.6% state that there is no psychological support or surveillance within the corporation.

The physical condition of firefighters is an important matter, namely because their performance is directly linked to this issue. Thus, this topic was assessed by 2 multiple choice questions through which the weekly physical training and the duration of each session were quantified. The results show that 45.3% of the respondents performed at least one training session per week. However, 32.8% state that physical training is never performed and 26% did not answer. Regarding the duration of the sessions, considering only the firefighters that mentioned the existence of physical training, 41% state that its duration is between 30 and 60 minutes and for 7.7% the duration is higher than 2 hours.

A significant number of questions was based on a 10-level judgment scale. Representative examples are the questions related with the fire ground operations, namely those that quantified the hot feeling and the sweating and fatigue degrees. Regarding the feeling of heat, Figure 1 shows that the responses are spread along the 10-level judgment scale (level 1 – lower level; level 10 – higher level, i.e., the hot feeling is very high) but with a significant percentage (85.9%) in the 6 to 10 range.

The results of the assessment of the sweating level during firefighting also show an uneven distribution along the 10-level scale. In this case the mean value was 8.4 and the highest percentage corre-

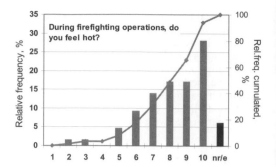

Figure 1. Results of question: During firefighting operations, do you feel hot?

sponds to level 10 with 35.9% of the votes. In terms of fatigue, the mean value was 6.9 and 25.0% of the votes are in the 1 to 5 range, and 67.2% between 6 and 10. These results clearly put in evidence the severe working conditions during firefighting.

The performance of the protective clothing was also assessed with the 10-degree judgment scale. Seven questions were asked, and an example is the subjective evaluation of the adequacy of the protective clothing. The majority of the votes (51.6%) are in the 6 to 10 range, but the highest percentage corresponds to level 5 (26.6%). The results regarding this matter highlight that a readjustment of the protective clothing is clearly needed to fulfill the desirable requirements and a special attention must be given to personal protective equipment since an important improvement in its performance may indeed be achieved.

3.3 *Thermal insulation of protective clothing*

The main objective of the thermal insulation tests was the assessment of the insulation of the forest fires protective clothing used by Portuguese firefighters. Ensemble 1 corresponds to the new protective clothing recently adopted by the Portuguese firefighters while ensembles 2 and 3 are the oldest ensembles. These last ensembles are identical but number 2 was brand new while number 3 was being regularly worn. In addition, the insulation differences between ensembles 1, 2 and 3 was studied. Two tests were performed for each ensemble, the first with the constant skin temperature (CST) regulation mode and the second with the thermal comfort equation (TCE) regulation mode. The thermal conditions of the tests are summarized in Table 5 where T_o is the operative temperature.

The trials were carried out according to ISO 9920 (2007) requirements and the recording period, that lasted for 30 minutes, was started only after achieving stationary conditions for the

Table 5. Thermal conditions of the tests.

Ensemble/ Regulation mode		T_o [°C]	Whole body mean value	
			T_{sk} [°C]	Q_s [W m^{-2}]
Ensemble 1	CST	21.4	33.0	56.94
	TCE	21.3	33.3	56.88
Ensemble 2	CST	21.4	33.1	60.35
	TCE	21.4	33.1	61.27
Ensemble 3	CST	21.4	33.1	61.55
	TCE	21.4	33.2	60.86
Nude	CST	21.4	33.0	98.33
	TCE	21.4	31.7	86.83

Table 6. Thermal insulation values.

Code	Regulation mode	Definition	Thermal insulation [clo] Calculation method		
			Serial	Global	Parallel
Ensemble 1	CST	I_{cle}	0.7	0.56	0.56
		I_{cl}	0.87	0.69	0.70
		I_T	1.49	1.33	1.33
	TCE	I_{cle}	0.72	0.61	0.57
		I_{cl}	0.88	0.75	0.71
		I_T	1.50	1.38	1.34
Ensemble 2	CST	I_{cle}	0.65	0.50	0.50
		I_{cl}	0.80	0.62	0.62
		I_T	1.43	1.27	1.26
	TCE	I_{cle}	0.56	0.48	0.45
		I_{cl}	0.70	0.60	0.56
		I_T	1.34	1.25	1.22
Ensemble 3	CST	I_{cle}	0.65	0.50	0.50
		I_{cl}	0.79	0.62	0.62
		I_T	1.43	1.27	1.26
	TCE	I_{cle}	0.55	0.47	0.44
		I_{cl}	0.69	0.59	0.55
		I_T	1.34	1.25	1.21

overall mean skin temperature (T_{sk}) and heat flux (Q_s). In order to assess the thermal insulation of the air layer (I_a), after the test performed with each ensemble, this was removed and a new trial under the same environmental conditions was carried out with the manikin nude.

Table 6 shows the effective (I_{cle}), the basic (I_{cl}) and the total thermal insulation (I_T) values obtained with the serial, the global and the parallel calculation methods for the mentioned CST and TCE manikin regulation modes.

An overall analysis shows that the total insulation (I_T) values are the highest, followed by the basic (I_{cl}) and effective (I_{cle}) thermal insulations, and that values obtained with the different calculation methods are quite different. The highest were obtained with the serial method, followed by the global and parallel calculation methods. The ensemble recently adopted by the Portuguese firefighters (ensemble 1) present the highest insulation values, despite the small differences to ensembles 2 and 3. This difference may be due to the single cotton layer of ensembles 2 and 3.

Considering the total insulation results calculated with the global method and operating the manikin in the CST regulation mode, the values varied between 1.27 and 1.33 clo. If the calculation methods are considered, assuming once more the I_T results for reference, the values varied from 1.34 and 1.50, from 1.25 and 1.38 and from 1.21 and 1.34 clo, respectively for the serial, global and parallel calculation methods. Often, protective clothing labels show lack of thermal insulation values. Furthermore, when they are displayed, there is no specification about the methods how such values have been obtained. This is a matter that should be duly considered since it may lead to erroneous choices between similar garments.

Ensembles 2 and 3 have similar thermal insulation values suggesting that, in this case, there is no evidence that washing cycles reduce insulation.

4 CONCLUSIONS

Besides of urban or industrial episodes, the activity of firefighters is mainly concerned with forest fires with a significant impact on the Portuguese economy. Therefore, any attempt to improve working conditions of firefighters should be welcomed. The preliminary results obtained in this research clearly demonstrate the need for further and more detailed studies like the choice of adequate protective clothing, which may ask for clarifying specific laboratory measurements. Otherwise, the awareness about the importance of the individual physical condition, the definition of medical protocols regarding admission tests, maintenance requirements or the consideration of psychosocial stressors are other preventive tools of great importance. The subjective assessments also suggested that firefighters are not familiar with active cooling techniques, a specific field where the civil protection hierarchy should promote formation actions.

REFERENCES

Budd, G.M. 2001. How do wildland firefighters cope? Physiological and behavioural temperature regulation in men suppressing Australian summer bushfires with hand tolls. *J. Thermal Biology* 26: 381–386.

IAFF 2006. Thermal heat stress protocol for firefighters and hazmat responders. http://www.iaff.org.hs.

ISO 15384 2003. Protective clothing for firefighters—Laboratory test methods and performance requirements for wildland firefighting clothing. International Standard, 1st edn. *International Organization for Standardization* (ISO), Geneva.

ISO 8996 1990. Ergonomics: Determination of the metabolic heat production. International Standard, 1st edn. *International Organization for Standardization* (ISO), Geneva.

ISO 9920 2007. Ergonomics of the thermal environment—Estimation of the thermal insulation and water vapour resistance of a clothing ensemble. International Standard, 2nd edn. *International Organization for Standardization* (ISO), Geneva.

Oliveira, A.V.M., Gaspar, A.R. & Quintela, D.A. 2008. Measurements of clothing insulation with a thermal manikin operating under the thermal comfort regulation mode: Comparative analysis of the calculation methods. *Eur J Appl Physiol* 104(4): 679–688. doi: 10.1007/s00421-008-0824-5.

Oliveira, A.V.M., Gaspar, A.R. & Quintela, D.A. 2011. Dynamic clothing insulation: Measurements with a thermal manikin operating under the thermal comfort regulation mode. *Applied Ergonomics* 42: 890–899. doi: 10.1016/j.apergo.2011.02.005.

Occupational health and safety management systems

Occupational Safety and Hygiene – Arezes et al. (eds)
© 2013 Taylor & Francis Group, London, ISBN 978-1-138-00047-6

Challenges in attending to OHS regulations in rice mills in southern of Brazil

I.G. Guimarães, A. Falcão, E.P. Ferreira & L.A.S. Franz
Federal University of Pampa, Rio Grande do Sul, Brazil

N. Costa
University of Minho, Guimarães, Portugal

ABSTRACT: This study aims to identify what are the biggest challenges that the rice mills encounter in respect the OHS, especially to the attendance to legal aspects. To reach the proposed objective, surveys were carried out on a sample of four rice mills located in southern Brazil, near the border with Uruguay. The research instruments consisted of interviews and questionnaires, transcripts of voice recordings and content analysis. The study showed that standards NR06 (referring to PPE) and NR33 (relating to confined spaces) are imposing major challenges in rice mills in southern Brazil. It is also important to note that the rice mills contain other risks beyond those surveyed as the most challenging in this study. After application of improvements over the PPE and confined spaces, sites under study must develop actions to adjust its installations to risks involving electrical shocks, exposure to noise, high falls, dust, vibrations and inadequate temperatures.

1 INTRODUCTION

The production of food, including the rice production, from storage to processing, sets itself as an economically vital aspect of human society. According to data from ABAG (2011), the Brazilian state of Rio Grande do Sul represents 64% of the national rice production in this country. The Rice Mills consist of industries specialized in receiving, storing and benefit grains, which are then distributed to the end customer or exported. These companies are common in the extreme south of Brazil and process annually approximately 650 tons of rice. However, despite of its strong connection with this product, the development of production technologies in rice processing is still not a recurring factor in these organizations (Weber 2005 & Palma 2005).

The rice mills are units with very peculiar characteristics although they not change significantly when observed in different countries. Generally these are local storage and processing grains units, which have greater or lesser importance according to the country in question. Thus, it can be inferred that under the subject safety aspects are also relevant and could have discussed their principles.

1.1 Understanding the rice mills and its operation

According to Weber (2005), commonly the rice crosses a sequence of operations very similar, independent of planted specie. Before the harvest, comes the improvement and multiplication of seeds, for example, genetically modified seeds. Also, it is performed the soil preparation and seeds planting. The cultural traits are specific for each species, seeking the necessary precautions to pest control. In turn, the harvest demands an additional care in the correct use of the equipment as well as the pursuit of elimination of rice wasting during transportation at the rice mills.

After the harvest process is necessary that the rice passes for a preliminary cleaning process in which impurities are removed, like soil, plant scraps or other organic materials. The reception and processing involves a large proportion of manpower that works in the mills. It is in this phase that the reception, separation and grain treatment occurs and subsequently is storage. Finally it gets shipped to the market using many different forms (road, rail, river or sea).

The rice mills usually have a layout similar to that shown in Figure 1, frequently having an office with laboratory, pathways for the circulation of trucks and points to collect samples of grains. The weighing is usually realized on a industrial truck scale, at the arrival of the truck loaded and after its finish the discharge completely.

Figure 2 depicts the flow of grain in a plant of a rice mill, which allows understanding in more detail the process of storing, processing and shipping.

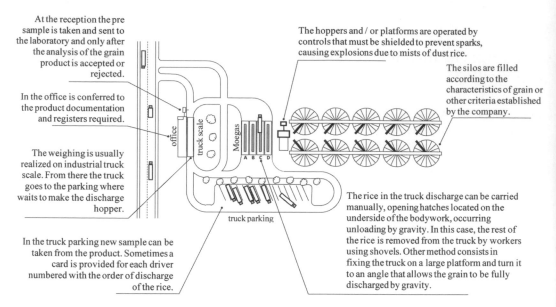

At the reception the pre sample is taken and sent to the laboratory and only after the analysis of the grain product is accepted or rejected.

In the office is conferred to the product documentation and registers required.

The weighing is usually realized on industrial truck scale. From there the truck goes to the parking where waits to make the discharge hopper.

In the truck parking new sample can be taken from the product. Sometimes a card is provided for each driver numbered with the order of discharge of the rice.

The hoppers and / or platforms are operated by controls that must be shielded to prevent sparks, causing explosions due to mists of dust rice.

The silos are filled according to the characteristics of grain or other criteria established by the company.

The rice in the truck discharge can be carried manually, opening hatches located on the underside of the bodywork, occurring unloading by gravity. In this case, the rest of the rice is removed from the truck by workers using shovels. Other method consists in fixing the truck on a large platform and turn it to an angle that allows the grain to be fully discharged by gravity.

Figure 1.   Example simplified layout of a rice mill. Font: (Weber 2005).

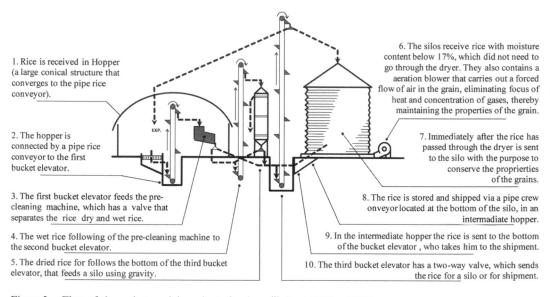

1. Rice is received in Hopper (a large conical structure that converges to the pipe rice conveyor).

2. The hopper is connected by a pipe rice conveyor to the first bucket elevator.

3. The first bucket elevator feeds the pre-cleaning machine, which has a valve that separates the rice dry and wet rice.

4. The wet rice following of the pre-cleaning machine to the second bucket elevator.

5. The dried rice for follows the bottom of the third bucket elevator, that feeds a silo using gravity.

6. The silos receive rice with moisture content below 17%, which did not need to go through the dryer. They also contains a aeration blower that carries out a forced flow of air in the grain, eliminating focus of heat and concentration of gases, thereby maintaining the properties of the grain.

7. Immediately after the rice has passed through the dryer is sent to the silo with the purpose to conserve the proprierties of the grains.

8. The rice is stored and shipped via a pipe crew onveyor located at the bottom of the silo, in an intermadiate hopper.

9. In the intermediate hopper the rice is sent to the bottom of the bucket elevator , who takes him to the shipment.

10. The third bucket elevator has a two-way valve, which sends the rice for a silo or for shipment.

Figure 2.   Flow of the grain trough in a plant of a rice mill. Font: (Weber 2005).

Weber (2005) also explains that the storage of rice can be performed in steel cylindrical silos or a deposit, and these sites must ensure that the basic properties of grain remain intact. In this case is possible to classify the silos in conventional warehouse, with bagged grain or bulk carrier (with horizontal compartments). In respect of the use of the silos, it depends on what the purpose is. Regarding the purpose they can be classified as storages silos, dryer silos, buffer silos and shipping silos. It is interesting to note that the silos can still be classified according to the building characteristics, either steel silos or concrete silos.

1.2   Occupational safety and rice mills panorama

The workers usually are exposed to different risks in their work environments, both in terms of physical and mental health. The incidents with injuries, in turn, can be fatal or disabling. In addition, it can be stated that the development of new technologies exposes workers to a growing possibilities of incidents (Zocchio 2002).

This author also defines that the concept of occupational hazard is based on the greater or lesser probability of an incident occurring with or without injury or illnesses. In turn, the OHSAS 18001-2007 defines risk as the combination of the probability of occurrence of a hazardous event or exposure with the severity of injury or illness resulting from this event. It defines hazard as the source, or situation with a potential for harm in terms of human injury or ill health, damage to property, damage to the workplace environment, or a combination of these. One possible way to control and improve in the working conditions comes from the Regulatory Standards (in Brazil it is used the acronyms NR), of the Ministry of Labour and Employment (MTE).

In Brazil, there are currently 35 regulatory standards, and will be addressed in the present study only the standards most directly related to the rice mills. Among these, it is possible to mention the standards NR01 (MLE 1993), NR02 (MLE 1983) and NR03 (MLE 2011) which include respectively the general provisions, preliminary facilities and embargo and interdiction.

Regarding to management structures, there is the NR04 standard, which deals with the implementation and maintenance of Specialized Services in Safety Engineering and Occupational Medicine (SSSEOM). The design of these services is associated with the National Classification of Economic Activities (NACE), where each NACE receives a degree of risk and from this index is determined the size of the teams that will compose the SSSEOM (MLE 2009). Another important standard is the NR05, which addresses the Internal Commission for Prevention at the Accidents (ICPA), which requires the establishment and maintenance of employees' group of eligible within the staff company and that plays encouraging initiatives aiming prevention of occupational accidents (MLE 2011). Still it is worth mentioning the NR07 standard, which addresses the Programme for Medical Control of Occupational Health (PMCOH) and presents the management structure responsible, among other things, to conduct the medical exams including: in the admission moment, the periodic exam, of return at the work, change of function and in the moment of demission (MLE 2011).

The NR06, in turn, deals with the Personal Protective Equipment (PPE), where the objective is to ensure the safety of employees through requirement for its use, and the employer is obliged to provide such equipment for free and according to the risk of each activity (MLE 2011).

Another important standard in this context is the NR09, Program for Prevention of Environmental Risks (PPER), which comes with an approach to environmental risks, which are considered due to physical, chemical and biological agents according with the available work environment (MLE 1994). The NR10 standard addresses the safety in electrical installations and services being that it defines guidelines for control measures, safety design, construction, installation, operation and maintenance (MLE 2004).

The unhealthy activities and operations are discussed in NR15 being that this standard establishes the Limits of Tolerance (LT) from which is not allowed that workers are exposed. The LT are suggested in several types of risk such as noise, heat exposure, and hyperbaric pressures, among others (MLE 2011). The NR16 refers to activities and hazardous operations, which are those involving explosives, subject to chemical degradation or autocatalytic. It also considers the action of external agents such as heat, humidity, sparks, fire, seismic phenomena, electrical shock and friction (MLE 2012). The NR17, focus on the theme ergonomics and aims to establish parameters that allow for adaptation to the working conditions at the psychophysiological characteristics of workers, so as to provide maximum comfort, safety and efficient performance (MLE 2007).

The fire protection is covered by NR23 consists in a regulate standard that determines the dimensioning of the exits, the required amount of fire fighting equipment, specifying what type of extinguisher to be used (MLE 2011).

Finally, NR33 discusses the safety and health at work in confined spaces, and sets minimum requirements for identification of these spaces, as well as recognition, monitoring and control of risks that may affect directly or indirectly the workers (MLE 2006). The confined spaces are particularly important in the case of rice mills. According Kulscar (2009a), workers are submitted to different risks during the work in confined spaces. In these places risks such as lack or excess of oxygen, the presence of atmosphere with toxic substances, biological agents to the possibility of electric shock can be found. Other incidents can occur, like burial, falls and drowning, which can lead employees to death or serious illness.

In this respect, these problems contribute to the increase of occupational risks, exposing workers to risks that affect both their physical and mental health (Zocchio 2002). Thus, the risks in the rice mills include: risk of explosion, especially in places like silos, mills and crushers; risk due to electrical shock; risk of burial and risk of suffocation due to emissions of gases especially in confined spaces (Sá 2007). The major attempts to reduce the incidents in Brazilian mills emerge from imposition of legal requirements, through Regulating Standards (NR) issued by the Ministry of Labour and Employment (MLE). In Brazil, there are currently 35 Regulating Standards, which allow the regulation of the conditions of Occupational Health and Safety (OHS) in different areas. However, despite the mandatory aspect of the NRs for private companies and public organisms, determined by the NR01 (MLE 1993), obedience to them is still precarious in several rice industries. Underlining the need for studies, that answers to the most immediate demands in terms of OHS in the productive sector. The biggest challenges for these work places must be identified and the way to adapt to the legal requirements must also be determined.

Thus, this study aims to identify what are the biggest challenges that the Rice Mills encounter in respect the OHS, especially with regard to the attendance to legal aspects.

2 MATERIALS AND METHOD

The content and method presented in this paper is part of a more comprehensive study, conducted between 2010 and 2011, according to Franco (2011). The partial results presented were collected and computed between February and March 2011. This work was characterized as a case study (Gil 2010), where the field research was initiated only after an exhaustive survey on the conceptual theme. The companies object of study are located in southern Brazil, in the state of Rio Grande do Sul, near the border with Uruguay. The sample, significant for the region under analysis, included four rice mills, called here by Company A, Company B, Company C and Company D. In Company A, the production manager was also the general manager of the company. At Company B, the production manager was one of the business owners. In company C the interviewee was the general manager of both the main and subsidiary units of the company, where this research was conducted. Finally, Company D, the production manager also works in sales.

In the field surveys the used resources were structured interviews and a questionnaire, applied to the production managers in each company. The questionnaire contained three categories, Sample Characteristics, Patterns of Risks and Potential Demands for Improvement in OHS. The first category, with eight questions, attempt to identify the company history, type of grains processed, sensitization courses in OHS, product shipping process, number of workers and existence of technicians in OHS, turnover, production capacity of the company and production workflow. The second category contained nine questions, which covered more specifics aspects like how the priority

of the risks was accessed, the presence of dust in the working environment, electrical risks, confined spaces, fire prevention and protection, presence of systems for immediate protection and orientation for works relatively to the safety care. At the final of the questionnaire, in third category, the interviewee was questioned about possible demands for improve the OHS conditions in your company.

Subsequently, the collected data were transcribed and submitted to content analysis, similar to that proposed by Bardin (2006). These analyses allowed us to identify which are the Regulatory Standards that present the greatest challenges to the company and therefore require greater attention in terms of improvement actions in OHS.

3 RESULTS AND DISCUSSION

In general, the biggest challenges in terms of OHS management found on the studied rice mills appear to be dependent of two elements, the safety culture and the compliance with Regulatory Standards. In this context, during the content analysis of the interviews, it was identified that in relation to safety culture, the efforts made in ensuring employee involvement appear to be affected by factors such as education level, outsourcing services and the absence of people directly involved with safety. With regard to standards, it was observed that even with the existence and concern for other risks such as fire, electric shocks, falls, confined spaces and the use of PPE (Personal Protective Equipment) proved to be a great challenge for the managers of rice mills.

In this sense, we found that these challenges are closely related standards: NR06 (MLE 2011) and NR33 (MLE 2006). The first rule focus on the use, control and provision of PPE while the second rule covers the access of workers to confined spaces, i.e., spaces not designed for continuous human occupation, and who have limited means of entry and exit. In these places there is insufficient ventilation to remove contaminants or there may be a deficiency of available oxygen. Regarding NR06, Company B showed no evidence regarding the awareness of its employees on the importance of using PPE. By their turn companies C and D, maintain good communication channels with their employees seeking thereby to make them aware about the importance of the PPE using. Never the less, to ensure that employees use PPE, both Company A and Company B, use the continuous monitoring as a means of imposing the use of equipment. By their turn in companies C and D there is the requirement of availability and use of the PPE, even though they don't persuade employees to use them. Generally, it is perceived that the four companies need to change their culture regarding the practice of safe acts, for this is a major challenge to meet the NR06.

Regarding the NR33 there were identified challenges for all companies since, companies A, B and C perform annual training and Company D only in 2011 conducted the first training on confined spaces for all employees. Another aggravating factor with respect to the NR33 compliance is that only the Company A, performs measurement of gases in confined spaces. These two points are serious violations of the regulations that are punishable and therefore would demand immediate action. Finally, it should be noted that during the content analysis, although it has indicated the use of PPE and access to confined spaces has the most challenging situations for the companies studied, there are still regulations and other practical aspects that are worth mentioning. It was found during the field surveys, safety deficiencies in the work at heights, electrical installations, rhythm of work, exposure to excessive noise, dust, vibrations and extreme temperatures. However, as can be inferred by the survey, an immediate action plan is required, under the OSH culture and the use of PPE, with the aim of acting in protection, and developing a plan of action that includes the urgent treatment of the pre-established requirements for access to confined spaces.

4 CONCLUSIONS

Rice production is important for the economy of the state and the local processing and storage facilities involve a large number of workers during the harvest period. Thus, actions to disseminate and maintain the OHS demands on the rice mills and the compliance to the Regulatory Norms are of extreme necessity. The study showed that standards NR06 (referring to PPE) and NR33 (relating to confined spaces) are imposing major challenges in rice mills in southern Brazil. These standards must be met immediately and its implementation mainly involves the change of the safety culture within companies.

It is also important to note that the rice mills contain other risks beyond those surveyed as the most challenging in this study. After application of improvements over the PPE and confined spaces, sites under study must develop actions to adjust its installations to risks involving electrical shocks, high falls, and exposure to noise, dust, vibrations and inadequate temperatures.

ACKNOWLEDGMENTS

The authors would like to thank Mirtô Fernades Morrudo Franco for the survey and the disposal of data used in this research.

REFERENCES

ABAG—Brazilian Agribusiness Association (2011). Retrieved April 18, from <http://www.abag.com.br/>.

Bardin, L. (2006). Análise de Conteúdo. Lisboa (Portugal): Edições 70 Publisher.

Franco, M.F.M. (2011). Identificação das normas regulamentadoras que apresentam maiores desafios para os engenhos de arroz. End of Course Dissertation in Production Engineering—Federal University of Pampa—Bagé—Brazil.

Gil, A.C. (2010). Como elaborar projetos de pesquisa (5th ed.). São Paulo (Brazil): Atlas Publisher.

MLE—Ministry of Labour and Employment (1983). Regulating Standard N° 02 (NR02)—Inspeção Prévia. Retrieved November 26, 2011, from <http://www.mte.gov.br/>.

MLE—Ministry of Labour and Employment (1993). Regulating Standard N° 01 (NR01)—Disposições Gerais. Retrieved October 15, 2010, from <http://www.mte.gov.br/legislacao/>.

MLE—Ministry of Labour and Employment (1994). Regulating Standard N° 09 (NR09)—Programas de Prevenção de Riscos Ambientais. Retrieved November 12, 2010, from <http://portal.mte.gov.br/>.

MLE—Ministry of Labour and Employment (2004). Regulating Standard N° 10 (NR10)—Segurança em Instalações e Serviços em Eletricidade. Retrieved November 26, 2011, from <http://www.mte.gov.br/>.

MLE—Ministry of Labour and Employment (2006). Regulating Standard N° 33 (NR33)—Segurança e Saúde no Trabalho em Espaços Confinados. Retrieved October 18, 2010, from <http://www.mte.gov.br/>.

MLE—Ministry of Labour and Employment (2007). Regulating Standard N° 17 (NR17)—Ergonomia. Retrieved October 22, 2012, from <http://portal.mte.gov.br/>.

MLE—Ministry of Labour and Employment (2009). Regulating Standard N° 04 (NR04)—Serviços Especializados em Engenharia de Segurança e em Medicina do Trabalho. Retrieved November 26, 2011, from <http://www.mte.gov.br/>.

MLE—Ministry of Labour and Employment (2011). Regulating Standard N° 03 (NR03)—Embargo ou Interdição. Retrieved November 24, 2011, from <http://www.mte.gov.br/>.

MLE—Ministry of Labour and Employment (2011). Regulating Standard N° 05 (NR05)—Comissão Interna de Prevenção de Acidentes. Retrieved November 22, 2011, from <http://portal.mte.gov.br/>.

MLE—Ministry of Labour and Employment (2011). Regulating Standard N° 06 (NR06)—Equipamentos de Proteção Individual. Retrieved October 16, 2010, from <http://www.mte.gov.br/legislacao/>.

MLE—Ministry of Labour and Employment (2011). Regulating Standard N° 07 (NR07)—Programas de Controle Médico de Saúde Ocupacional. Retrieved November 26, 2011, from <http://www.mte.gov.br/>.

MLE—Ministry of Labour and Employment (2011). Regulating Standard N° 15 (NR15)—Atividades e Operações Insalubres. Retrieved November 26, 2011, from <http://www.mte.gov.br/>.

MLE—Ministry of Labour and Employment (2011). Regulating Standard N° 23 (NR23)—Proteção Contra Incêndios. Retrieved November 12, 2012, from <http://portal.mte.gov.br/>.

MLE—Ministry of Labour and Employment (2012). Regulating Standard N° 16 (NR16)—Atividades e Operações Perigosas. Retrieved October 02, 2012, from <http://portal.mte.gov.br/>.

Palma, G. (2005). Pressões e fluxo em silos esbeltos (h/d ≥ 1,5). Retrieved August 28, 2010, from <http://www.set.eesc.usp.br/public/teses>.

Sá, A. (2007). Efeito devastador: explosões em locais onde existe muita poeira acumulada são ameaça constante. Revista Proteção, 181, 63–70.

Weber, E.A. (2005). Excelência em beneficiamento e armazenagem de grãos. Canoas (Brazil): Salles Publisher.

Zocchio, A. (2002). Prática da prevenção de acidentes: ABC da segurança do trabalho. São Paulo (Brazil): Atlas Publisher, 7th ed.

Occupational Safety and Hygiene – Arezes et al. (eds)
© *2013 Taylor & Francis Group, London, ISBN 978-1-138-00047-6*

Integrated Management Systems: A statistical analysis

J.P.T. Domingues, P. Sampaio & P.M. Arezes
Department of Industrial Engineering and Systems, University of Minho, Guimarães, Portugal

ABSTRACT: Designing a maturity model is a multi—methodological task. Maturity modelling had been reported in several research areas, namely, software development and inspection, e-Governance, knowledge management, networkability, health and safety enterprise culture, supplier relationship, project management, communicational levels on collaborative activities and management systems assessment. The present paper aims to report the efforts being made focusing the development of an integrated management system (IMS) maturity and efficiency assessment tool. It presents a survey statistical analysis, which is a methodology emphasized by several authors when developing a maturity model. The most appealing statistical conclusions, based on a survey focusing Portuguese IMS ruled companies, are now reported answering several questions, such as: Do initial IMS implementation motivation relates with the final achieved benefits? Do sub-systems standards integration relates to the integration level achieved? Which are the success factors to consider in order achieving a successful and high-level integration?

1 INTRODUCTION

1.1 *An IMS maturity model development*

Systems integration ranges from technical to organizational disciplines. Organizational management systems integration, outputting an integrated management system (IMS), should take into account several external features as macroergonomics and sustainable development. Internal cultural features, like corporate social responsibility, and proactive ones, like life cycle assessment, should be considered too (Domingues *et al.*, 2012c).

On the present project, two surveys have been carried out: the first one focusing organizations and the second one focusing an experts group. Domingues *et al.*, (2012a,b,c) reported results based on the first survey and based on both surveys (Domingues *et al.*, 2012b).

Koshgoftar and Gosman (2009), concerning the importance of organizational maturity models, stated that it allows the weaknesses and strengths identification of the organizational system were applied and, through benchmarking related methodologies, the information collection aiming an upper maturity level. Maturity models had been criticized too due to some shortcomings, namely, lack of empirical foundations, reality oversimplification and data reliability on its design process (Jia *et al.*, 2011). Becker *et al.*, (2009) defined

the main features to be considered prior to the development of a generic maturity model:

– Comparison with existing maturity models.
– Iterative procedure.
– Evaluation/assessment.
– Procedure sustained on a multiplicity of methodologies.
– Accurate problem identification.
– Results focused on the model dominium and on the users needs.
– Scientific documentation/methodologies sustaining the model development.

Taking into account this last item one may consider statistical analysis as an appropriate methodology to sustain the model development. Several authors had proposed this methodology when developing maturity models (Alessi, 2002) and the current paper intends to fulfil that assumption.

1.2 *State-of-the art*

The state-of-the art regarding IMS and maturity models has been reported in several previous publications (Sampaio *et al.*, 2012). Literature review regarding IMS implementation identified several features: motivation, obstacles, benefits, integration strategies, integration levels and audit typology.

Literature review regarding maturity models identified the following features: models typology and characteristics, application, evolution and maturity levels. This review, which was performed before the model development allowed the definition of the surveys carried out.

This paper is structured as follows: in section 2 the research methodologies adopted are described; section 3 presents the surveyed companies characterization and results derived from statistical analysis; section 4 discuss the available results. Finally, the paper presents the main conclusions answering several questions related to IMS ruled companies.

2 MATERIALS AND METHODS

A 30 Question/Statement (Q/St) online survey was carried out focusing on Portuguese organizations with more than one certified management subsystem according to the following standards: ISO 9001, ISO 14001 and OHSAS 18001/NP 4397. The survey was conceptually supported on a Likert type scale, for categorical and multiple option answers, being its structure reported in the appendix section (Table A1). The Q/St's were developed based on a deep bibliographic review and interviews with management systems managers.

A pre-test performed on three companies was used to validate the survey due to the reported limitations of using online surveys, such as sampling, representativeness, selection bias and response rate issues (Matsuo et al., 2004; Sackmary, 2012). The pre-test allowed also the refinement of Q/St's and answers scales minimizing the shortcomings when using Likert type scales which include the discrete nature of answers and the usual tendency for respondents for option the extreme values from the scale (Albaum, 1997; Clason and Dormody, 1994; Jamieson, 2004).

The following assumptions are supported on 53 valid answers given by the management systems manager during the period between 01–07–2011 and 01–11–2011. The response rate was of 15%. Statistical data analysis was performed with Portable IBM SPSS Statistics v19. String to numerical scale variable transformation was performed on St5 to St20, Q21, Q23, Q25 and Q28 to Q30.

Descriptive statistics, Kolmogorov-Smirnov with Lilliefors correction, Shapiro-Wilk, Kruskal-Wallis tests and dimension reduction were the statistical approaches performed on results dataset.

3 RESULTS

3.1 Sampled organizations characterization

Mainly higher than 50 workers organizations, located at North, Centre and Lisbon regions of Portugal with a QMS *and* EMS *and* OHSMS typology answered the survey. Construction, water supply, transport and logistics and other services were the most reported activity sectors in the sampled organizations matching with those reported by Sampaio and Saraiva (2012). These authors reported that QMS, EMS *and* OHSMS was the most reported IMS typology among certified organizations closely followed by QMS *and* EMS.

Regarding organization size (number of employees), mainly large enterprises were the respondents, suggesting that the assumption of Coelho and Matias (2010) that SMEs (Small and Medium Enterprises) are less interested in systems integration maybe accurate.

3.2 Organizations characterization parameters versus other surveyed parameters

Based on available results, no validated statistical relationship was found between companies' characterization features (Q1–Q4) and the other surveyed parameters.

3.3 Descriptive statistics

Descriptive statistics (Q/St5-20) included Kolmogorov-Smirnov with Lilliefors correction and Shapiro-Wilk tests, Normal Q-Q plot and detrended Normal Q-Q plot graphics in order to normal distribution assessment. Despite the fact that N = 53 (> 50) the tested results set were found to be non-normal distributed. This preliminary analysis implied that further statistical analysis would be performed through non-parametric tests. Data normalization was an alternate methodology to data analysis but it results' robustness is not comparable to that achieved by non-parametric methodologies.

3.4 One sample Kolmogorov-Smirnov (K-S)

One sample K-S test was performed (α = 0,05) on St5–20, St21, Q23, Q25 and Q28–30 in order to normal distribution assessment according to Table 1. Despite the fact that the N = 53 (> 50) the tested results set were found to be non-normal distributed.

Table 1. K-S test hypothesis and decision criteria.

Normality test hypothesis	Decision criteria
H_0: Q/St result set $\cap$ $N(\mu,\sigma)$	Accept H_0 if Sigma (p-value) > α = 0,05
H_a: Q/St result set do not $\cap$ $N(\mu,\sigma)$	Reject H_0 and accept H_a if Sigma (p-value) $\leq$ α = 0,05

3.5 Non-parametric Kruskal-Wallis test

Non-parametric Kruskal-Wallis test hypothesis (Table 2) was performed between Q28↔Q30, Q23↔Q25, St21↔Q23, St6↔Q23 and Q/St 14↔Q/St 23 ($\alpha = 0,05$). For all the tests described, group number (k) was four, so Kruskal-Wallis test hypothesis presents a three-freedom degrees (k–1) *Chi-square* distribution, validating the asymptotic sigma. String to numerical scale variables allows comparison according to the ranked means.

3.6 Dimension reduction and components identification

Variable reduction was performed considering St5–St20 in order to determine the enveloped concept behind the surveyed items determining the small set of factors/components that represent as much information as the entire 15 variables. The calculations were performed aiming an optimization of Cronbach's alpha coefficient (Christmann and Van Aelst, 2006; Lopez, 2007; Peterson, 1994; Prelog *et al.*, 2009; Santos, 1999; Shojima and Toyoda, 2002).

A preliminary rotated Varimax matrix identified four components among St5-St20 (Table 3).

The selection of items per components was performed considering the highest values from Table 3 and these are presented on Table 4. Considering table 4 results, component 3 was eliminated due to the scarce (2) items pertaining to.

A scale reliability analysis was performed per component considering only the items pertaining to each component. Table 5 presents the SPSS outputs considering component 1. A Cronbach's alpha coefficient of 0,852 was obtained when performing reliability analysis considering 6 items.

The same calculations were performed related to components two and four (not shown). Reliability statistics concerning the former component presented a Cronbach's alpha coefficient of 0,615 (4 items considered) and the reliability analysis concerning the latter component a Cronbach's alpha coefficient less than 0,5. Due to this fact component four was not considered.

Table 3. Rotated component matrix (Varimax).

| Q/St | Component | | | |
	1	2	3	4
5	0,216	0,054	0,857	0,089
6	0,772	0,038	0,176	0,009
7	0,770	− 0,021	0,397	− 0,110
8	0,039	− 0,107	− 0,004	0,655
9	0,101	0,733	0,421	− 0,083
10	0,428	0,718	0,144	− 0,229
11	0,459	0,491	0,409	− 0,295
12	0,796	0,200	− 0,072	0,055
13	0,397	0,217	0,233	0,564
14	− 0,662	− 0,139	− 0,074	0,158
15	− 0,333	0,001	− 0,083	0,730
16	0,633	0,364	− 0,215	0,158
17	0,665	0,215	0,477	− 0,017
18	− 0,057	0,224	0,689	− 0,036
19	0,076	0,844	0,050	0,141
20	0,475	0,405	0,471	0,043

Table 4. Selection of items per component.

| Q/St | Component | | | |
	1	2	3	4
5	–	–	0,857	–
6	0,772	–	–	–
7	0,770	–	–	–
8	–	–	–	0,655
9	–	0,733	–	–
10	–	0,718	–	–
11	–	0,491	–	–
12	0,796	–	–	–
13	–	–	–	0,564
14	–	–	–	0,158
15	–	–	–	0,730
16	0,633	–	–	–
17	0,665	–	–	–
18	–	–	0,689	–
19	–	0,844	–	–
20	0,475	–	–	–

Table 2. Non-parametric K-W test hypothesis and decision criteria.

K-W test hypothesis	Decision criteria
H_0: $M_{Q/St1} = M_{Q/St2} = ... = M_{Q/Stk}$.	Accept H_0 if Sigma > α^*
H_a: $M_{Q/Stk} \neq M_{Q/St1} = M_{Q/St2} .... = M_{Q/St(k-1)}$	Reject H_0 and accept H_a if Sigma ≤ α^*

* Sigma validation if k – 1 freedom degrees Chi-square distribution by Kruskal-Wallis statistics (≥ 3 (k) groups).

Table 5. Component 1—total statistics.

Q/St	Scale mean if item deleted	Scale variance if item deleted	Corrected item—Total correlation	Cronbach's alpha if item deleted
6	21,53	7,216	0,670	0,820
7	21,34	7,344	0,690	0,817
12	21,58	7,171	0,677	0,819
16	21,42	7,709	0,521	0,848
17	21,45	7,637	0,710	0,818
20	21,64	7,042	0,592	0,839

4 DISCUSSION

Regarding Q28↔Q30, SPSS outputs suggests enough evidence that motivation (internal, mainly internal, external or mainly external) do relate with benefits (internal, mainly internal, external or mainly external) achieved by IMS implementation. This relationship was predicted by Domingues *et al.*, (2012a), analysing results through an alternate methodology, concluding that IMS implementation motivation typology has a high probability to output the same benefits typology.

Regarding Q23↔Q25 SPSS outputs suggests enough evidence that the IMS organizational structure classification ((1- Documental/2- Management tools and 1)/3- Policies and objectives and 1) and 2)/Common organizational structure and 1), 2) and 3)) do relate with the management system manager perceived integration level (1- Minimum integration level/2- Low integration level/3- Medium integration level/4- High integration level/5- Total/Maximum integration level) achieved by IMS implementation.

Regarding St21↔Q23 SPSS outputs suggests no evidence that organizational structure IMS classification do relate with the perceived difficulty on integrating the management sub-systems standards. This result implies that a high or low achieved IMS integration level do not relates with the perceived difficulty on standards integration.

The organizational structure in an IMS context (Q23) had been tested too regarding St6 (Top management training) and St14 (Documental level integration). Results suggest that these features do relate with the IMS structure classification, reflecting they are required as success factors for a high-level integration achievement. Thus, top management training and not just a document-based integration, certainly among other features, guarantee a successful high integration level organization.

Dimension reduction and further analysis identified two valid components. Concerning the first valid component, we propose that the enveloped concept is: organizational awareness and monitoring. This concept is related to the six validated items (top management training, integrative concept taken into account, organizational interactions and IMS as an add value perceptions and integrated objectives promotion and application). These items would be difficult to identify in an organization without top management awareness and proper monitoring procedures.

Related to the second valid component we propose as the enveloped concept: organizational vision. Tools, methodologies and goals alignment, integrated management procedures and KPI´s, OPI´s and MPI´s implementation are items revealing (top management) organizational vision.

5 CONCLUSIONS

There seems to be enough evidence that IMS organizational structural classification does relate with the perceived integration level achieved and that initial motivation typology does relate with benefits typology achieved by IMS implementation.

No statistical relationship evidence was found when comparing difficulties on sub-systems management standards integration and the organizational IMS structure achieved suggesting that is not due to standards features that a higher integration level may or not be achieved.

Top management training and not just a documental-based integration were identified as required success factors on IMS implementation.

Dimension reduction analysis identified two components enveloped in the Q/St answers provided: organizational awareness and monitoring and organizational vision, being these also, success factors for a well succeeded integration.

ACKNOWLEDGEMENTS

Acknowledgements are due to all the companies that answered the survey and to CemPalavras.

REFERENCES

Albaum, G. 1997. The Likert scale revisited: An alternate version. *Journal of the Market Research Society* 39(2): 331–343.

Alessi, S. 2002. A simple statistic for use with capability maturity models. *Systems Engineering* 5(3): 242–252.

Becker, J., Knackstedt, R. and Pöppelbuβ, J. 2009. Developing maturity models for IT management— A procedure model and its application, *Business and Information Systems Engineering* 3: 213–222.

Christmann, A. and Van Aelst, S. 2006. Robust estimation of Cronbach's alpha. *Journal of Multivariate Analysis* 97(7): 1660–1674.

Clason, D.L. and Dormody, T.J. 1994. Analyzing data measured by individual Likert-type items. *Journal of Agricultural Education* 35(4): 31–35.

Coelho, D.A. and Matias, J.C.O. 2010. An empirical study on integration of the innovation management systems (MS) with other MSs within organizations. *In proc. of ERIMA 2010*, 11–12 June, Wiesbaden, Germany, 5–13.

Domingues, J.P.T., Sampaio, P., Arezes, P.M. and Ramos, G. 2012a. Integrated OHS management systems: Is it the *final frontier* regarding OHS?. *In proc. of ESREL 2012*, Helsinki, Finland, 1293–1302.

Domingues, J.P.T., Sampaio, P. and Arezes, P. 2012b. Latest developments aiming an integrated management systems tool focusing maturity assessment. *In proc of IEEM2012*, Hong Kong, China, 2063–2067.

Domingues, J.P.T., Sampaio, P. and Arezes, P.M. 2012c. New organizational issues and macroergonomics: Integrating management systems. *International Journal of Human Factors and Ergonomics,* in press.

Jamieson, S. 2004. Likert scales: How to (ab)use them. *Medical Education* 38: 1217–1218.

Jia, G., Chen, Y., Xue, X., Chen, J., Cao, J. and Tang, K. 2011. Program management organization maturity integrated model for mega construction programs in China. *International Journal of Project Management* 29: 834–845.

Koshgoftar, M. and Osman, O. 2009. Comparison between maturity models. *In proc. of 2nd IEE International Conference on Computer Science and Information Technology* 5: 297–301.

Lopez, M. 2007. Estimation of Cronbachs's alpha for sparse datasets. *Evaluation*: 151–156.

Matsuo, H., McIntyre, K.P., Tomazic, T. and Katz, B. 2004. The online survey: Its contributions and potential problems. *ASA sections on survey research methods*, pp. 3998–4000.

Peterson, R. 1994. Meta-analysis of alpha Cronbach's coefficient. *Journal of Consumer Research* 21(2): 381–391.

Prelog, A., Berry, K. and Mielke, P. 2009. Resampling permutation probability values for Cronbach's alpha. *Perceptual and Motor Skills* 108(2); 431–438

Sackmary, B. 2012. *Internet survey research: Practices, problems and prospects.* faculty.buffalostate.edu/.../ sackmary/Ama98.pdf (02/11/2012).

Sampaio, P. and Saraiva, P. 2012. *Barómetro da Certificação.* Edição 6 (CemPalavras).

Sampaio, P., Saraiva, P. and Domingues, P. 2012. Management systems: Integration or addition?. *International Journal of Quality and Reliability Management* 29(4): 402–424.

Santos, J.R.A. 1999. Cronbach's alpha: a tool for assessing the reliability of scales. *Journal of Extension* 37(2): 1–4.

Shojima, K. and Toyoda, H. 2002. Estimation of Cronbach's alpha coefficient in the context of item response theory. *The Japanese Journal of Psychology* 73(3): 227–233.

APPENDIX

Table A1. Survey questions/statements.

Q/St	Possible answers
St1-The company main activity is:	Unstructured
Q2-How many workers employ the company?	Unstructured
Q3-Where is geographically located the company?	North; Centre; Lisbon; Alentejo; Algarve; Madeira; Açores
St4-The management system is certified according the following standards:	ISO 9001 + ISO 14001; ISO 9001 + OHSAS 18001; ISO 14001 + OHSAS 18001; ISO 9001 + ISO14001 + OHSAS 18001; Other

(*Continued*)

Table A1. (Continued).

Q/St	Possible answers
St5-Quality, Environmental and Occupational Health and Safety policies are integrated.	Totally disagree; Disagree; Nor agree or disagree; Agree; Totally agree
St6-Training related to management systems integration had been provided to top management.	""
St7-Integration concept had been taken into account during IMS implementation.	""
St8-Management system is bureaucratized.	""
St9-The tools, methodologies and goals from each management sub-system are harmonized/aligned.	""
St10-Top management reveals integrated vision.	""
St11-Management procedures are integrated.	""
St12-Organizational interactions derived from IMS implementation are perceived by responsible and top management.	""
Q13-The implementation process was supported on a guideline or in a framework.	""
St14-Integration occurs at a documental level.	""
St15-Authority from Environmental and/or OHS responsible is residual.	""
St16-IMS is an add-value.	""
St17-Integrated objectives are defined.	""
St18-On the company organizational structure there is a clear responsible by the IMS.	""
St19-The company monitors their processes based on KPI´s, MPI´s and OPI´s.	""
St20-The company promoted the implementation of integrated indicators.	""
Q21-How do you classify the integration level of sub-systems standards?	Very easy; Easy; Reasonable; Difficult; Very difficult
Q22-If the company did not had implemented an IMS the overall performance comparing with the actual reality would be:	Lower than the present status; Equal to the present status; Higher than the present status

(*Continued*)

Table A1.   (Continued).

Q/St	Possible answers
Q23-How do you classify the management system integration level?	1-Documental/ 2-Management tools plus 1)/ 3-Policies and objectives plus 1) and 2)/ Common organizational structure plus 1), 2) and 3)
Q24-Audits performed to management sub-systems are?	Integrated; Simultaneous, Overlapped, Sequential
Q25-In a 1 to 5 scale how do you characterize the IMS?	1-Minimum integration level/ 2-Low integration level/ 3-Medium integration level/ 4-High integration level/ 5-Total/ Maximum integration level
Q26-The strategy followed during integration process was:	Sequential "All In"
Q27-Organizational items not susceptible of being integrated are identified?	Yes No
Q28-The main motivations to implement the IMS were.	Internal/ Mainly internal/ External/ Mainly external
Q29-The main benefits resulting from the integration of the management system were.	""
Q30-The main obstacles found during the implementation of the IMS were.	""

Occupational Safety and Hygiene – Arezes et al. (eds)
© 2013 Taylor & Francis Group, London, ISBN 978-1-138-00047-6

Options in managing hazards and risks of nanomaterials

Paul Swuste

Safety Science Group, Delft university of Technology, Delft, The Netherlands

David Zalk

Lawrence Livermore National Laboratory, CA, USA

ABSTRACT: Managing risks of manufactured nanomaterials includes managing scenarios leading to emission, and to exposure to these nanomaterials. Risk management of nanomaterials is like entering a field with many uncertainties, both related to relevant health endpoints, as to metrics of exposure. If no valid quantitative information is available yet, there is a preference for a qualitative tool, or method to assist risk management decisions. The Control Banding Nanotool and the method Design Analysis will be explained, discussed, and commented in their capacity to assess and predict scenarios, leading to emission and exposure to nanomaterials.

1 INTRODUCTION

Applications for nanomaterials seem endless and substantial efforts are being put forth by both government and private industries into the research and development of nanotechnologies. However, it is becoming increasingly clear that properties that make nanoparticles technologically beneficial may also make them hazardous to humans and the environment. Due to their size distribution, nano-materials can have an high degree of reactivity, an ability to deposit in various regions of the respiratory tract, and an ability to cross normally impenetrable barriers. Publications are fuelled by the fear of a second worldwide asbestos debacle, which nowadays, only in The Netherlands, has created an annual mortality exceeding that of traffic (Swuste et.al., 2004; Burdorf et.al., 2005; Ruers, 2012).

Journal articles on process conditions from companies that produce nanomaterials or nano-products are scarce. This "new technology" is referred to as the future industrial revolution. It is often assumed that the volume of production of nanomaterials is limited and automated process steps are carried out in closed systems. This is not always the case. Articles and media coverage from the United States have reported that in some companies 'the future looked a lot like the past' (Hanssen et.al., 2008). An example is reported of a company of the metallurgical sector, which used to produce powders, and had switched to the production of nanomaterials. Here the processes were open, large quantities of product were descended in open systems and in different places in the process there was a visible dust emissions (Weiss, 2006).

This contribution will focus on possibilities to reduce emission and exposure to nanomaterials, using more conventional occupational hygiene control measures during the production of nanomaterials, as well as options of (re)design its production processes. Two questions will be leading:

– Which methods or tools from the domains of occupational safety, hygiene, and health are suitable to reduce emission, or exposure to nanomaterials?
– Is it possible to incorporate these methods and tools, if proven successful, into a risk management approach?

2 RISK MANAGEMENT

The concept of risk management stems from the ability to define what may happen in the future and to choose among alternatives. This lies at the heart of contemporary societies. Risk management guides us over a range of decision-making, from allocating wealth to safeguarding occupational, public safety and health, from waging war to planning a family, from paying insurance premiums to wearing seatbelts, from planting corn to marketing cornflakes (Bernstein, 1998). In The Netherlands, risk and risk management plays a dominant role in the domain of external safety. This small country, with its major land mass below sea level, has a relative high level of industrialisation, including high hazard industries, and the associated transport of hazardous chemicals by road, trains, and boats. Risk management methods like Quantitative Risk Analysis are used to calculate risks related to flooding, and to processing, storage, and transport of these hazardous chemical (PSG, 1999; Lemkowitz and Pasman, 2012; Vlek, 2012).

Models for risk management are closely related to models for safety and health management. The first concepts and models were published in the professional literature in the period between the wars and just after World War II (DeBlois, 1926; Heinrich, 1950). These models describe general activities to ensure a safe and healthy production, and are variations of the well-known 'Plan—Do—Check—Adjust' model of Deming, which was introduced in Japan in the late 1940s to ensure the quality of their post-war production (Pindur et.al., 1995). In the last decades, academic research defined safety and health management as a set of problem solving activities and refined these models (for an overview see Hale et.al., 1997).

3   BOWTIE

Within the domain of safety science, metaphors have been developed depicting the process leading to consequences, like accidents and disasters, and also industrial diseases. One such metaphor, the bowtie (Visser, 1998), is rather well known in The Netherlands (figure 1). This metaphor is also useful to understand emission, and the exposure to nanomaterials, and the influence of managerial decisions. The focus of the metaphor is the 'central event'. A central event is a condition, when one or more hazards are becoming uncontrollable by following one or more scenario routes (left to right arrows). In case of nanomaterials, the central event represents a situation with emission to nanomaterials, and consequently leading to exposure in the presence of workers. Hazards of nanomaterials have become risks, following scenarios or pathways, or sequence of events represented by arrows on the left side of figure 1.

Barriers are physical entities, technical control measures that can stop or reduce the energy flow to and from the central event. Management is responsible for identifying risks, scenarios and central events, for selecting and identifying barriers and for the various activities undertaken. These managerial factors are the vertical arrow in figure 1. Apart from the barrier choice, these management factors will have no direct effect on scenarios, but they will determine the effectiveness of the barriers (Guldenmund et.al., 2006).

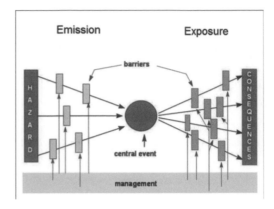

Figure 1.   The Bowtie metaphor.

4   OCCUPATIONAL HYGIENE: CONTROL BANDING NANOTOOL

Control Banding is a general term, referring to a qualitative risk assessment that stratifies chemical hazards at a given workplace across two sets of levels or bands, the hazards bands and exposure potentials. It follows a rather simple scheme of the occupational hygiene domain (figure 2), and of risk and safety management. The controls are the barriers in figure 1 and the risks are central events. The scheme in figure 2 is the general scheme in risk and safety management. The unique point of Control Banding approach is its relation to the selection of technical controls.

The qualitative tool is suitable when uncertainty exists on exposure metrics, or means are non-existing to measure exposure. Also, Control Banding operates without a lot of expert input. This is a strong point of the process, making it applicable in small and medium enterprises, both in the industrial and developing world (NIOSH, 2009; Swuste and Eijkemans, 2002). Various countries have adopted Control Banding, like the US, South Africa, countries in North-West Europe, South-East Asia, and supra national organisations like World Health Organisation (WHO) and the International Labour Office (ILO) (Zalk and Heussen, 2010, 2011).

Figure 2 represents the Control Banding Nanotool tool. Different from most Control Banding tools, the probability, and severity axes are scored, each using a pointing system. This pointing system combines nanomaterial composition parameters and toxicological information (surface activity, shape, size, solubility area and carcinogenicity, mutagenicity, reproductive, dermal, and asthmagen toxicity of both nanomaterials as parent materials) with their exposure availability (amount used, frequency and duration of operation, size of population at risk). Av detailed explanation is

HAZARDS + EXPOSURES/SCENARIOS →
RISKS + CONTROLS

Figure 2. Occupational hygiene scheme.

Probability

		extremely unlikely (11.25-25)	less likely (26-50)	likely (51-75)	probable (76-100)
Severity	very high (76-100)	RL 3	RL 3	RL 4	RL 4
	high (51-75)	RL 2	RL 2	RL 3	RL 4
	medium (26-50)	RL 1	RL 1	RL 2	RL 3
	low (5-25)	RL 1	RL 1	RL 1	RL 2

RL 1: general ventilation; RL 2: fume hoods or local exhaust ventilation; RL 3: containment; RL 4: seek specialist advice (RL: risk level).

Figure 3. Control banding nanotool.

given by Paik and collogues (2008) en Zalk and collogues (2009). These indices are linked to bands with four corresponding risk levels (RL), and control approaches. The control approaches are a grouping of three levels of engineering containment, based on conventional occupational hygiene principles.

Based on this success, the Control Banding Nanotool has also been implemented in many countries as a qualitative decision matrix for a risk assessment that leads to commensurate controls for nanomaterials (Zalk and Paik, 2010).

5 SAFETY SCIENCE: METHODICAL DESIGN, AND DESIGN ANALYSIS

Scenarios are the starting point for estimating the emissions at the workplaces. Scenarios consist of (process) conditions that specify how hazards become uncontrollable, and may become a risk. Scenarios are largely determined by the design of the installations for production of nanomaterials. Methodical design is a method to make design choices transparent. The method comes from the domain of mechanical engineering and is aimed to cover structurally the path from abstract definition of a problem to concrete solutions. Thereby the system sequence function-principle-form is observed.

The objective of a given design, of an installation or production process, is reflected in its function. This function is an abstract description. Functions can be fulfilled in different ways, according to various mechanical principles. Examples of production functions are: supply of raw materials, storage of raw materials, transportation of raw materials and (intermediate/final) products, processing of raw materials and final shaping, final finishing,

processing of waste, maintenance, and fixing process disturbances. In the second step principles are determined to fulfil the function. The production principle specifies the general principles of production process (continuous versus batch-wise), including the mechanical principle and the operating principle. Examples of mechanical principles of the production function 'processing raw materials' are: forming, mixing, separating, combining, surface treatment, etc. The second principle, the operating principle, is an indication of the distance of a worker to an installation. This distance is short for manual and mechanically operated functions and large at remote-controlled and automated functions. Production principles provide the variations possible for a corresponding production function. The final design is not yet determined by the production function and principle. Also, choices are possible on the level of production form.

The production form is the final design of the installation. This involves the detailed design of the installation by which the principle is implemented. This detailed design also determines whether an installation will operate as an open or a closed system and which aforementioned technical control measures are applied to reduce exposure. The form is thus a concrete elaboration of the above lying principles and function.

The power of methodical design lies in the development of as many possible alternative principles to perform a function. This prevents designers to think directly into concrete solutions and technical controls (Kroonenberg and Siers, 1978; Kroonenberg 1986, 1990; Swuste et al., 1993, 1997a). Every principle results in one or more forms. The final choice of a form is determined by a number of technical and business factors such as cost considerations, production efficiency of a specific form, gained technical experience with similar types of forms, etc.

With ten or less production functions the production of nanomaterials and nanoproducts can be described. It is estimated that for each production function five to ten different principles are possible and each principle will have several possible production forms. The emission and exposure to nanomaterials takes place at the level of production form. Logically, the focus on controlling exposures via add-on technical control measures is concentrated on the production form. But the source of emission, and related scenarios leading to a central event, is largely determined by the production principle: a batch wise, manually operated, open process of mixing raw materials will potentially give a higher level of emission and exposure than a process which is performed continuously, remotely controlled and in a closed system. In the latter case, emission and exposure occurs during maintenance and during fixing process disturbances.

In scientific literature of various domains, process data of the use and production of nanomaterials are mentioned. This information is discussed in detail in Swuste and Zalk (2013). In one of these publications from the Nanex project, work package 2 (Tongeren et.al., 2010), information of fieldwork studies and exposure assessments are combined with production functions and principles.

6 CONCLUSIONS

Both in risk and in safety management, scenarios are the starting points, which can lead to central events. These scenarios have to be managed. The Control Banding Nanotool addresses only limited information on scenarios, being the probability factors, the amounts used and the frequency and duration of operations. These process data, and especially the information on the production principle (the status of the process batch versus continuous, operating principle and the mechanical principle) in combination with production form information, open versus closed system, will determine scenarios that may occur. This opens possibilities to predict scenarios, even before a production line is built (Swuste et al., 1997b).

The second question dealt with the option of whether the Control Banding Nanotool and the Design Analysis can be incorporated into a risk management approach. The Control Banding Nanotool is an instrument that can facilitate risk management, because it can support decisions on risks and controls in areas with many uncertainties on safety, hygiene and health issues. The tool has been validated, and has been included into an occupational health and safety management system. The Design Analysis, as a classification of production processes, and installations has proven its use in predicting scenarios. Unfortunately, the method has not been tested yet for nanomaterials. But its future will be promising, if information from fieldwork is collected for production principles.

REFERENCES

Bernstein P (1998). Against the gods, the remarkable story of risk. John Wiley & Sons Inc., New York.

Burdorf A Järvholm B Englund A (2005). Explaining Differences in Incidence Rates of Pleural Mesothelioma Between Sweden and The Netherlands. International Journal of Cancer 113:298–301.

DeBlois L (1926). Industrial safety organisation for executives and engineers. McGraw-Hill Book Company, New York.

Guldenmund F Hale A Goossens L Betten J Duijm N (2006). The development of an audit technique to assess the quality of safety barrier management. Journal of Hazardous Materials 130:234–241.

Hale A Heming B Carthey J Kirwan B (1997) Modelling of safety management systems Safety Science 26(1–2):121–140.

Heinrich H (1950). Industrial accident prevention. McGraw-Hill Book Company, New York.

Kroonenberg H van den Siers F (1978). Methodic design (Methodisch ontwerpen—in Dutch). Vakgroep Ontwerp—en Constructieleer, Technische Universiteit Twente, Enschede.

Kroonenberg H van den (1986). Techniques of choise in methodic design (Keuzetechnieken bij methodisch ontwerpen—in Dutch). De Constructeur (3):24–32.

Kroonenberg H van den (1990). The impending demise of the mechanical design (De dreigende teloorgang van het werktuigbouwkundig ontwerp in Dutch). De Ingenieur (11):25–28.

Lemkowitz S Pasman H (2012). Chemical hazards and chemical risks (Chemische gevaren en chemische risico's—in Dutch). In: Ale B Muller E Ronner A (eds) (2012). Risk, risk and risk management in The Netherlands. Kluwer, Deventer.

NIOSH (National Institute for Occupational Safety and Health) (2009). Approaches to safe nanotechnology. DHHS (Department of Health and Human Services) (NIOSH), publication no 2009–125.

Paik S Zalk D Swuste P (2008). Application of a pilot control banding tool for risk level assessment and control of nanoparticles exposures. Annals of Occupational Hygiene 52(6):419–428.

Pindur W Rogers S Kim P (1995). The history of management: a global perspective. Journal of Management History 1(1):59–77.

PGS 3—CPR18E (1999), Guidelines for quantitative risk assessment, 'Purple book', http://www.publicatiereeksgevaarlijkestoffen.nl/publicaties/PGS3.html.

Ruers B (2012). Power and counter power in the Dutch asbestos regulation (Macht en tegenmacht in de Nederlandse asbestregulering—in Dutch). PhD thesis. Erasmus University Rotterdam. Boom Juridische uitgevers.

Swuste P Kromhout H Drown D (1993). Prevention and control of chemical exposures in the rubber manufacturing industry in The Netherlands. Annals of Occupational Hygiene 37(2):117–134.

Swuste P Drimmelen D van Burdorf A (1997a). Pneumatic chippers design analysis and solution generation. Safety Science 27:85–98.

Swuste P Goossens L Bakker F Schrover J (1997b). Evaluation of accident scenarios in a Dutch steel works using a hazard and operability study. Safety Science 26(12):63–74.

Swuste P Eijkemans G (2002). Occupational Safety Health and Hygiene in the urban informal sector of Sub-Saharan Africa. An application of the Prevention and Control Exchange (PACE) Program to the informal-sector workers in Healthy City Projects. International Journal of Occupational and Environmental Health 8(2):113–117.

Swuste P Burdorf A Ruers B (2004). Asbestos, asbestos related diseases, and compensation claims in The Netherlands. International Journal of Occupational and Environmental Health 10:159–165.

Swuste P Zalk D (2013). Risk management and nano-materials. IN: Fundamentals and nanomatyeials p. 156–173 (vol 1) Studium Press LLC, USA.

Visser K (1998). Developments in HSE Management in Oil and Gas Exploration and Production. In: Safety management, the challenge of change. Hale A Baram M (Eds.). Pergamon, Amsterdam, p. 43–66.

Vlek C (2012). Assessment, acceptation and control of risks: tradition, critique and innovation (Beoordeling, acceptatie en beheersing van risico's: traditie, kritiek en vernieuwing—in Dutch).). In: Ale B Muller E Ronner A (eds) (2012). Risk, risk and risk management in The Netherlands. Kluwer, Deventer.

Zalk D Paik S Swuste P (2009). Evaluating Control Banding Nanotool: a qualitative risk assessment method for controlling nanoparticle exposure. Journal of Nanoparticles Research 11:1685–1704.

Zalk D Paik S (2010). Control Banding and nanotechnology. The Synergist, 21(3):26–29.

Zalk D Heussen H (2010). 6th International Control Banding Workshop: Practical Primary Prevention. Workshop during the 8th International Scientific Conference van de International Occupational Hygiene Association, Rome 27 september–2 oktober (http://ioha.net/controlbanding.html). Results of earlier Control Banding Workshops can be found at http://www. bohs.org/eventDetails.aspx?event=42 (London, UK, 2002), http://www.acgih.org/events/controlband (Cincinnati, US, 2004), http://www.saioh.org/ioha2005/proceedings/SSI.htm (Pilanesburg, Zuid Afrika, 2005), http://ioha.net/controlbanding.html (Seoul, Zuid Korea, 2008), http://ioha.net/controlbanding.html (Cape Town, Zuid Afrika, 2009).

Zalk D Heussen H (2011). Banding the world together; the global growth of Control Banding and qualitative occupational risk management. Safety Health at Work; 2(4):375–9.

Occupational Safety and Hygiene – Arezes et al. (eds)
© *2013 Taylor & Francis Group, London, ISBN 978-1-138-00047-6*

Data quality assessment for performance measures in the area of safety and health at work

M.A. Cavallare, S.D. Sousa & E.P. Nunes
Centro Algoritmi, University of Minho, Campus de Gualtar, Braga, Portugal

ABSTRACT: Data quality issues of performance measures in the area of Safety and Health at Work are explored based on a set of uncertainty components previously proposed. A case study protocol is developed to ascertain the quality of such performance measures. Results provided by three case studies show that respondents have the perception that the performance measures used in companies are affected by the suggested uncertainty components. The results can lead to a revision of the existing data collection methods to reduce the uncertainty of such indicators. The characterization of data uncertainty can also be considered as a risk indicator of decisions based on these indicators. This work is part of a bigger project that aims to represent and reduce uncertainty in performance measurement systems.

1 INTRODUCTION

Generally organisations have a need to monitor their processes, and to measure, manage and express in a systematic way their performance at different levels of management, using performance indicators (PIs). In the Health & Safety domain companies select standard safety, hygiene and health indicators, to respond to legal requirements and to establish a reference framework with the ability to compare and evaluate their performance. In doing so consistently, companies can apply the resources needed to, either globally or locally, improve safety, hygiene and health of its employees. Most of the indicators used to evaluate the performance of organisations' management are "positive" indicators, i.e., they express gains rather than losses (company profits, number of customers, market share, etc.), but in the Health & Safety domain the used indicators are, traditionally, "negatives", i.e., they represent a specific issue that companies intend to minimize (accident rates, economical and material losses, costs, etc..) (Neto et al., 2008). The obligation to comply with legal requirements increasingly stringent, combined with the organizations' need to use appropriate PIs to improve human capital management have justified the research work in this field in recent years.

There are a large number of publications on design of performance assessment systems (PAS) and defining critical success factors (CSFs) for the development of PIs. These indicators are often associated with multi-dimensional concepts, presenting problems with the data quality and, therefore, may have a negative impact on decision-making based on these PIs.

The uncertainty is a quality measure of a PI, it is also an inevitable part of any measurement, and it becomes particularly relevant when the results are close to a specified limit. If the uncertainty is present in the data it will certainly be reflected in PAS and PIs.

In the formulation of a traditional PAS in the Occupational Health & Safety (OHS) area, most PIs are affected by non-probabilistic uncertainty like the imprecision, the indefiniteness or the ambiguity, however, they are usually represented by deterministic values. This is mainly due to the inability of current PASs to adequately represent this kind of uncertainty. It is considered, however, that a good PAS must be able to deal with the uncertainty since this uncertainty is part of the models used to obtain the PIs and also part of data that support them.

Generally, each PI is represented by a number that is not able to represent uncertainty. The problem is how to overcome this situation or how to deal with data uncertainty.

There is a wide variety of reasons why uncertainty is present in PMSs. Particularly, to reliability studies (Coolen 2004) presents three main reasons: (i) dependence on subjective information in the form of expert judgments; (ii) the relaxation of dependence on precise statistical models justified by physical arguments; (iii) the exact system structure and dependence relations between components are known, which may well be unrealistic. These relationships are conditioned by the system's environment and may generate contradictory information, vagueness, ambiguity data, randomness, etc. In reliability studies, the vagueness of the data have many different sources: it might be caused by subjective and imprecise perceptions of failures by a user, by imprecise records of reliability data, by imprecise records of the tools appropriate for modelling vague data, and suitable statistical methodology to handle these data as well (Nunes et al., 2006).

According to ISO 10012 (ISO_2003), section 7.3, the measurement uncertainty shall be estimated for each measurement process covered by the measurement management system and all known sources of measurement variability shall be documented. If these requirements are to be applied in all PMs of the organization there would be the need to identify all sources of variability. However, few works (Wazed et al., 2009, Herroelen 2005), report the inclusion of such variability in their studies.

Overall the studies published about data quality (Batini et al., 2009) mention the following problems on their quality: accuracy, completeness, timeliness and consistency. Other studies (Lee et al., 2002, Ge et al., 2008) show different categories/dimensions to classify data quality attributes: (i) intrinsic—contextual and reputation, (ii) internal—related internal data; (iii) external—system related objective and subjective, (iv) syntactic, (v) semantic, (vi) and pragmatic (vii) physical. This classification applicable to data or information, in general (Madnick et al., 2009), can also be applied for PIs in the OHS area.

Sousa et al. (2012) propose a set of seven uncertainty components (UCs) which may affect PIs. These UCs are classified into three main groups:

− *intrinsic* (development-related measurement system);
− *data collection* (refer to the problem of real time data quality introduced by method of data collection); and
− *PI definition* (difference between what is intended to be measured and what is really measured with the chosen PI).

This paper aims to make a contribution to the quality assessment of PIs in the OHS area.

2 MATERIALS AND METHODS

2.1 *Hypothesis and case study design*

The hypothesis of this study is that PIs in the OHS area are affected by uncertainty classified by seven uncertainty components (UC) previously defined (related with its definition, the data and contextual factors). Thus, case studies will be performed to understand how PIs are affected by such UCs.

There are different types of case studies (Yin 2003). The explanatory kind seemed appropriate for the present research, as they seek to explain how and why some events occurred. This method benefits from the prior existence of several complex theories. The identification of UCs was based upon different literature sources.

The choice between single-case and multiple-case studies depends on the research objectives and the availability of resources. Conducting multiple case studies is described as being similar to replication of experience. Evidence from multiple cases is often more compelling, and the overall study can therefore be regarded as being more robust (Yusof & Aspinwall 2001). Therefore in this research an explanatory multiple-case study was performed.

After choosing the appropriate type of case study, the researcher should define the unit of analysis, screen and select the case studies and define the case study protocol. The unit of analysis, in this research, are PIs related to OHS used by each organization.

This research proposed to identify how UCs influence the PIs.

Three companies were contacted from among the Companies that had expressed a willingness to cooperate with the Engineering School of the University of Minho. An initial contact was made to briefly describe the area of cooperation and a meeting scheduled with the Manager of the OHS area. The physical proximity to the University in most cases facilitated logistic issues and constraints such as time and money. The companies belonged to different sectors of activity and it was hoped that through this approach, the conclusions would be generic, rather than one that was limited to one specific activity (Yusof & Aspinwall 2001).

Finally, there have been repeated calls for more qualitative case-study-based research in operations management (Silvestro & Westley 2002), despite the clear difficulty of drawing generalised conclusions from a small number of instances.

2.2 *The case study protocol*

The case study a protocol, the interview instrument for conducting the case study, was defined and contains all the pertinent questions to be asked when investigating UCs in OHS PIs.

It is a major tool for increasing the reliability of case study research and is intended as a guide for the investigator in carrying out the study.

The initial section in the protocol concentrated on gathering general information about the respondent, such as its function in the company.

A description of the conceptual definition and classification of UCs would then be presented, and asked to give his comments on its overall structure.

A three page document was provided, including the proposed UCs definitions, scale of measurement and examples. The intention was to validate the UCs through open-ended questions without restriction to the above criteria. A pilot study was conducted with two respondents which suggested changes that were introduced in the used version of the protocol.

Finally, each respondent was asked to select, identify and define two PIs in the OHS area. For each one his/her perception about the influence of UCs on each PI was assessed with a three level scale (see Table 1).

3 CASE STUDY

3.1 Context

This work consists of, on a first step, in a study to identify relevant PIs in the area OHS. After identifying the PIs empirical studies are conducted to show which UCs are present in data and models that affect the quality of the PIs. Based on this study a relationship matrix PIs vs UCs is created evidencing the causes of uncertainty in existing PIs.

Quantitative methods for modelling the uncertainty in the data and PIs usually require more resources and information than qualitative methods. This work begins with an assessment of qualitative data and PIs based on analysis of UCs. To characterize data uncertainty in an OHS system firstly should be identified UCs that affects these data.

The next step is a classification of the uncertainty level of each UC. To be used in less structured systems a three levels scale is proposed to be applied to PIs. The scale for all UCs is presented in Table 1. For example, for *human evaluation uncertainty component*, the three levels of scale can be defined by:

a. Without uncertainty—person (s) responsible, which has objective and measurable criteria, and weighting is universally accepted?
b. Low uncertainty—one of the requirements in (a) is not satisfied;
c. High uncertainty—two or more of the requirements in (a) are not satisfied.

3.2 Data results

This study used four performance indicators proposed by the companies that participated in the study. These are indicators used by business manager in the area of OHS and, some of them, simultaneously, as elements to meet legal requirements.

Table 1. Measurement scale of uncertainty components.

Uncertainty components (UC)	Measurement scale
Definition/ Measurand	*Without Uncertainty* (a)– there is a clear, universally accepted, objective and aligned with what you want to measure. Is determined the method of measurement, context, responsible, frequency measurement. The PM does not depend on other PM with uncertainty; *Low Uncertainty*– one of the requirements in (a) is not satisfied; *High Uncertainty*– other cases.
Environmental	*Without Uncertainty* (a)– the physical environment (context) has not changed (processes, people, standards, products, equipment, etc..); *Low Uncertainty*– the context has changed slightly; *High Uncertainty*– other cases.
Aggregating of IDs	*Without Uncertainty* (a)– the formula to define a PM based on other is universally accepted (e.g. the weights are well-founded) and there is not uncertainty in the formula used in PI; *Low Uncertainty*– there is some sort of data validation; *High Uncertainty*– other cases.
Measurement method	*Without Uncertainty* (a)– there is a defined measuring method, and adequate mechanisms of training and verification of execution; *Low Uncertainty*– one of the requirements in (a) is not satisfied; *High Uncertainty*– two or more of the requirements in (a) are not satisfied.
Precision and accuracy of measurement tool	*Without Uncertainty* (a)– there are studies (eg R&R, PTTR) which demonstrate capability of the measurement system and measuring equipment is calibrated with error consistent with the decisions to be taken; *Low Uncertainty*– No studies but it is presumed that the measuring system is suitable; *High Uncertainty*– other cases.
Human evaluation	*Without Uncertainty* (a)– jury with jurisdiction, who decides based on objective and measurable criteria, whose weights are universally accepted; *Low Uncertainty*– one of the requirements in (a) is not satisfied; *High Uncertainty*– two or more of the requirements in (a) are not satisfied.
Data collection (equipment/ operator)	*Without Uncertainty* (a)– there are appropriate audit / validation methods and data filters to identify suspicious values; *Low Uncertainty*– there is some sort of data validation; *High Uncertainty*– other cases.

Then a brief discretion of indicators is presented, as well as analytical expressions for obtaining the respective numerical values.

3.2.1 Frequency index

The PI Frequency index, I_f, represents the number of injured per million hours of exposure to risk in a given period and is obtained by the Equation 1.

$$I_f = \frac{1000000 \times NI}{WH} \quad (1)$$

where NI = number of injured; and WH = number of work-hours of risk exposure, i.e., the sum of time during which each worker is available to the employer. The interpretation of this equation indicates how many accidents would occur if they have been working 1000000 hours in the time period considered (month, year, etc.). This indicator is used to compare companies in the same sector or sectors with the same risk, considering the time that workers were exposed to the risks (WH) and making a projection for 1000000 hours.

3.2.2 Number of days without accidents

This indicator represents the number of working days without accidents.

3.2.3 Lost workday cases

The lost-workday case, lw, is an indicator based on the occupational lost-workday injury/illness cases multiplied by 200000, then divided by the hours worked for the same time period in which the injury occurred. The "200000" used in this calculation is the equivalent number of work-hours for 100 employees working 40 hours per week for 50 weeks.

$$lw = \frac{2000000 \times NLW}{EH} \quad (2)$$

where NLW = number of lost-workday injury / illness cases; and EH = total hours worked by all employees during the calendar year. For example, if there was two lost-workdays injury/illness cases in a quarter and 50000 hours worked, the calculation would be: $2 \times 200000/50000 = 8.0$ lost-workday cases.

3.2.4 Illness rate

The illness rate, Ir, represents the number of injuries and illnesses per 100 full-time workers and is calculated as:

$$I_r = \frac{2000000 \times N}{EH} \quad (3)$$

where N = number of injuries and illnesses, EH = total hours worked by all employees during the calen-

dar year. 200000 = base for 100 equivalent full-time workers (40 hours per week, 50 weeks per year).

A high illness rate may be interpreted as an indicator of a heavy workload, bad working conditions, dangerous working environment, low employee satisfaction, and so on. As a simple key figure it can be used for planning purposes, for example, to shift resources from one area into an area with a high Illness rate. An analysis of the illness reasons or causes must include other factors as well. For example, a high overtime rate combined with a high number of accidents may indicate the reasons for an increase of the illness rate.

Based on the PIs and the UCs previously defined a matrix that relates with each PI with each UCs, according to scale on Table 1, is presented in Table 2. The resulting matrix identifies the most relevant to UCs and which PIs are more influenced by UCs.

3.3 Analysis and discussion of results

By analysing the results of Table 2 it can be seen that the UCs with the highest representation on the analysed PIs are related to the *measurement method* and *data collection*. A common feature of these indicators lies in their dependence on basic performance measures with a strong subjective component (assessments or judgments of experts such as physicians of occupational medicine among others). Moreover, with regard to *measurement methods*, it may be different interpretations of concepts influenced by social or cultural issues that introduce uncertainty and hinder the comparison of indicators.

The other component of uncertainty with a presence in all analysed indicators lies in *data collection*. This uncertainty relates to the lack of

Table 2. Uncertainty components of performance indicators.

Uncertainty components (UC)	Frequency index	Number of days without accidents	Lost workday cases	Illness rate
Measurement method	High	High		
Precision and accuracy of measurement		Low		
Human assessment	Low	Low		
Data collection	High	High	Low	Low
Definition/ Measuring	Low	Low		
Environmental	Low	Low		
PIs Aggregation	Low		Low	Low

appropriate audit/validation data and the lack of filters identifying suspicious values.

An aspect to note in these indicators is their weak direct dependence of measuring equipment. For this reason the UC associated with the *precision and accuracy of measurement* is not significant or does not apply to the PIs considered in this study.

There is also a reduced perception of uncertainty in UC *PIs aggregation*. This is justified by the low degree of aggregation of the indicators used in this study. This is indeed a common feature of the PIs in the OHS area. In general, results are obtained directly from basic measures of specialists diagnostics with a strong and subjective data collected over long work periods, by structured processes.

Table 2 represents the uncertainty level associated with the data and contextual factors in their PIs. It can also be seen as a support tool for management to suggest a revision of PIs to improve data quality.

4 CONCLUSIONS

This study aimed to test the hypothesis that PIs in the OHS area are affected by uncertainty classified by seven uncertainty components (UC) previously defined. Therefore, this paper begins by presenting a classification of Uncertainty Components (UCs) that may affect the quality of the performance indicators (PIs) in the area of OHS and establishes a theoretical framework for classification/modelling of the uncertainty of the PIs in this area.

Through the case studies clear evidence was obtained that suggests: i) the PIs in the OHS area are affected by several of the uncertainty components previously defined; ii) the main uncertainty components identified in the case studies are related to the measurement method and data collection; iii) any uncertainty component identified in the case studies is being considered in the assessment of PIs nor in decision making from these indicators. Reducing uncertainty in the main uncertainty components identified may provide less risk in decision making.

This work is part of a larger project that aims, through case studies, validate or refute the uncertainty components above proposed in different areas of performance management.

ACKNOWLEDGMENTS

This work was financed with FEDER Funds by Programa Operacional Fatores de Competitividade—COMPETE and by National Funds by FCT—Fundação para a Ciência e Tecnologia, Project: FCOMP-01-0124-FEDER.

REFERENCES

Batini, C., Cappiello, C., Francalanci, C. & Maurino, A. (2009). "Methodologies for data quality assessment and improvement ": J ACM Comput. Surv. 41(3): pp. 1–52.

Coolen, F.P.A. (2004). *On the Use of Imprecise Probabilities in Reliability.* Quality and Reliability Engineering International.

Ge, M., Helfert, M., Abramowicz, W. & Fensel, D. (2008). Data and Information Quality Assessment in Information Manufacturing Systems.

Herroelen, W. & Leus, R. (2005). Project scheduling under uncertainty: Survey and research potentials. *European Journal of Operational Research 165*: pp. 289–306.

ISO 10012. (2003). "ISO 10012 Measurement management systems—Requirements for measurement processes and measuring equipment": ISO.

Lee, Y.W., Strong, D.M., Beverly K. Kahn & Richard Y. Wang. (2002). "AIMQ: a methodology for information quality assessment.": Information & Management 40(2): pp. 133–146.

Madnick, S.E., Wang, R.Y., Lee, Y.W. & Hongwei Zhu. (2009). "Overview and Framework for Data and Information Quality Research ": J. Data and Information Quality. V. 1, n. 1: pp. 1–22.

Neto, H., Arezes, P.M. & Sousa, S.D. (2008). "New performance indicators for the Health and Safety domain: a benchmarking use perspective". Valencia—Spain: ESREL 2008 & 17th SRA-Europe Conference, 22–25 September, pp. 761–765.

Nunes, E., Faria, F. & Matos M. (2006). Using fuzzy sets to evaluate the performance of complex systems when parameters are uncertain.: Proceedings of Safety and Reliability for Managing Risk. Lisbon. v. 3: pp. 2351–2359.

Silvestro, R. & Westley, C. (2002). Challenging the paadigm of the process enterprise: a case-study analysis of BPR implementation, Omega, v. 30: pp. 215–225.

Sousa, S.D; Nunes, E.P. & Lopes, I. (2012). "Data Quality Assessment in Performance Measurement", Lecture Notes in Engineering and Computer Science: Proceedings of the World Congress on Engineering 2012, WCE 2012, London, UK 4–6 July, pp. 1530–1535, ISBN 978-988-19252-2-0.

Wazed, M.A., Ahmed, S. & Yusoff, N. (2009). Uncertainty Factors in Real Manufacturing Environment. *Australian Journal of Basic and Applied Sciences*, v. 3(2): pp. 342–351.

Yin, R.K. (2003). Applications of case study research. 2nd ed. Applied Social Research Methods Series, London: L. Bickman and D.J. Rog. v. 34.

Yusof, S.M. &. Aspinwall E. (2000). Total quality management implementation frameworks: comparison and review. *Total Quality Management*, 11(13): pp. 281–294.

Occupational Safety and Hygiene – Arezes et al. (eds)
© 2013 Taylor & Francis Group, London, ISBN 978-1-138-00047-6

Prevalence of work-related musculoskeletal symptoms in Portuguese volunteer firefighters

A. Seixas & F. Silva
Universidade Fernando Pessoa, Porto, Portugal

ABSTRACT: The prevalence of work-related musculoskeletal symptoms in firefighters has not been extensively investigated. The aim of this study was to identify the prevalence of work-related musculoskeletal symptoms in Portuguese volunteer firefighters, possible causes and activities behind their origin, the implications and preventive strategies reported by these professionals and find eventual associations between the symptoms and the type of duty. The results indicate a high prevalence of musculoskeletal symptoms, caused by physical exertion and mainly during fireground activities. Firefighters with musculoskeletal symptoms tend to reduce the techniques that aggravate or provoke the symptoms and to improve their physical condition as preventive strategy. Fireground and rescue activities were linked to higher prevalence of musculoskeletal symptoms.

1 INTRODUCTION

The term "Musculoskeletal Disorder" (MSD) has been extensively used in literature to identify several injuries that might affect muscles, ligaments, tendons, nerves, joints and blood vessels related to movement. It refers to a large group of injuries with sudden or insidious onset of symptoms that might affect an individual for a short period of time or for the whole life (Bernard, 1997, Sanders & Dillon, 2006, Sanders & Stricoff, 2006, Woolf & Pfleger, 2003, Mody & Brooks, 2012). This group of disorders is extremely common, affects strongly both individuals and society, is one of the major causes of disease burden in the world and has been a significant reason for the development of the Bone and Joint Decade, an initiative of the World Health Organization (Brooks, 2006) that has been prolonged to 2020 as it is yet to fulfil the goals that were proposed initially (Atik, 2010).

The subject has been widely studied and discussed in the literature but MSD remain a major source of negative economic impact on society, through both direct and indirect expenditures, treating the sequels of the disorders and loosing productivity. MSD conditions are related to age and affected by lifestyle factors and their burden is predicted to increase. The growing impact of these conditions on both individuals and society is not recognized at the level of health policy due to several reasons such as the diversity of disorders regarding to pathophysiology and the low mortality associated (Woolf et al., 2012).

Musculoskeletal pain (MSP) is extremely common and reported in the literature. McBeth & Jones (2007), in a literature review, stated that in adult populations nearly one-fifth reported widespread pain, one-third reported shoulder pain and up to one-half reported low back pain in a 1-month period.

Firefighters are exposed to biological, chemical, psychosocial, physical and ergonomic occupational risk factors. The working conditions of these professionals are often responsible for an acute risk of developing MSD.

Fire-fighting and rescue work in emergency scenarios leads to ergonomic problems that cannot be addressed normally as it is difficult to reduce the exposure without interfering with work equipment and methods (Punakallio et al., 2006). Fire-fighting activities require hard physical work, high metabolic and respiratory requirements and are not readily covered in ISO 8996 classification scheme of metabolic rates (International Organization for Standardization, 2004). Hólmer & Gavhed (2007) suggested two new classes to the classification table in ISO 8996 providing values for intensive and exhaustive short term work.

The majority of injuries suffered by firefighters in the United States occurs during fireground operations (43.5%) but performing non-fire emergency tasks such as emergency medical services leads to 21,3% of those injuries. Musculoskeletal injuries account for 56.6% of the injuries reported and are, by far, the most prevalent nature of injury (Karter & Molis, 2012).

Job tasks are physically challenging within these professionals and further research is needed to determine and address the effects of work-related physical demands on firefighters. Therefore, the objective of this study is to identify the prevalence of work-related musculoskeletal symptoms (WRMSS) among Portuguese volunteer firefighters in the last 7 days, in the last 12 months and during the entire career since no study, to our knowledge, addressed this problem in Portugal. The study aims to identify the possible causes involved in these disorders, the activities behind their origin, the implications to firefighter's work, the prevention strategies used by these professionals and to find possible associations between work-related complaints and type of duty.

2 MATERIALS AND METHOD

2.1 Research design and participants

A cross-sectional design was used for this study. Volunteer firefighters were recruited, using a convenience sampling method, from 16 corporations in the northern region of Portugal with potencial participants comprising 265 individuals.

2.2 Instrumentation

A two-part self administered instrument was selected for this study. One part was a demographic questionnaire, used to obtain information such as age, gender, working experience, daily working hours, type of duty, prevention strategies used, implications of symptoms in working experience and possible causes for musculoskeletal symptoms reported. The second part was a Portuguese version of the standardized Nordic Musculoskeletal Questionnaire used to access the occurrence of musculoskeletal complaints. The questionnaire divides the human body into nine anatomical regions (neck, shoulder, elbow, hand/wrist, upper back, lower back, hip/thigh, knee and ankle/foot) and includes a diagram with the anatomical regions clearly marked (Kuorinka et al., 1987). Binary choice questions (yes/no) were used to associate the musculoskeletal symptoms with the anatomical regions. The Nordic Musculoskeletal Questionnaire has been widely used in similar studies (e.g. Widanarko et al., 2011) and a Portuguese version has been adapted and published by Mesquita et al. (2010).

2.3 Procedures

After approval from the corporations, 265 questionnaires were delivered and all participants read and signed the informed consent for this study. Answers were provided to all questions the individuals thought necessary and the completed copies of the instrument were collected by the same researcher.

2.4 Data analysis

Statistical analysis was made using the software Statistical package for the Social Sciences, version 20 for Windows. Descriptive statistics were used to estimate the prevalence of WRMSS and demographic characteristics. Frequencies and cross-tabulations were used to compare the presence of musculoskeletal symptoms between demographics and type of duty. Chi-square tests were used to assess the relationship between the type of duty and the affected anatomical regions defined by the Nordic questionnaire. Statistical significance was set at $\alpha = 0.05$.

3 RESULTS

Out of the 265 questionnaires distributed 30 (11.3%) were not returned but the overall response rate was very high as 235 questionnaires were returned (88.7%). The respondents comprising 83.4% (n = 196) males and 16.6% (n = 39) females had a mean age of 32.5 years. The firefighters reported an average of 12.4 years of working experience and an average of 9.1 working hours per day. Socio-demographic characteristics of participants are presented in table 1.

As we were dealing with volunteer firefighters it was expected another professional activity and 60% of the respondents reported other professions. When asked about the workload as firefighter only 35.7% considered that it was excessive.

Table 1. Socio-demographic characteristics of participants.

Characteristics	
Gender (n = 235)	
male	196
female	39
Age (years) (n = 235)	
mean ± sd	32.5 ± 10.4
range	18–59
Experience (years) (n = 235)	
mean ± sd	12.4 ± 9.2
range	1–40
Daily working hours (hours) (n = 235)	
mean ± sd	9.1 ± 3.0
range	0–15

Out of the 235 respondents 67.7% (n = 159) have reported WRMSS at some point of their careers. When inquired about the last 12 months 65.9% (n = 155) reported WRMSS and 36.2% (n = 85) reported WRMSS in the last 7 days. Higher lifetime prevalence of WRMSS was observed in firefighters older than 40 years and with more professional experience as seen in table 2.

Without questioning the usefulness of lifetime prevalence of WRMSS, an accurate appreciation of the presence of musculoskeletal symptoms using a retrospective study is difficult so we focused on the impact of WRMSS within the last 12 months and within the last 7 days. The respondents' recall of the symptoms is more likely to be fresh, increasing the accuracy of the reported information. The results obtained in our study for the last 12 months and last 7 days prevalence of WRMSS in each anatomical region are shown in table 3.

The most affected anatomical region in the last 12 months was the lower back (46.0%), followed by the neck (26.8%) and the knees (22.1%) and the least affected anatomical regions were the elbows (6.8%) and the ankles/feet (8.9%). In the last 7 days

the most affected anatomical region was the lower back (21.3%), followed by the neck (8.5%) and the knees (7.2%) and the least affected anatomical regions were the ankles/feet (3.4%), the elbows (3.8%) and the wrists/hands (3.8%).

Firefighters in this study reported performing regularly several activities such as fireground activities, patient transport, rescue activities, driving, diving/aquatic rescue and training.

When performing the activities mentioned above the most reported perceived causes to WRMSS were physical exertion (30%), fatigue (23.3%) and unexpected sudden movement (21.9%), followed by sustained awkward postures (13.7%) and repeated movements (11.2%).

The most common coping strategies adopted among volunteer firefighters in this study were stop performing the tasks or using the techniques that aggravate or provoke the symptoms (41.8%) and asking for the help of colleagues (30.3%).

Preventive strategies were reported by 110 firefighters (46.8%) and the most common strategies adopted were the improvement in body mechanics (15.3%), the practice of regular physical activity to improve physical fitness (9.4%) and the avoidance

Table 2. Lifetime prevalence of WRMSS according to demographic characteristics.

Variables		Lifetime prevalence	
		Yes n (%)	No n (%)
Overall respondents		159 (67.7)	76 (32.3)
Gender	Male	133 (67.9)	63 (32.1)
	Female	26 (66.7)	13 (33.3)
Age	<40	116 (65.9)	60 (34.1)
	≥40	43 (72.9)	16 (27.1)
Work experience	≤5	36 (57.1)	27 (42.9)
	5 < 12	45 (69.2)	20 (30.8)
	≥12	78 (72.9)	29 (27.1)

Table 3. 12 month prevalence of WRMSS in the studied anatomical regions.

Anatomical region	12 months	7 days
	N(%)	N(%)
Neck	63 (26.8)	20 (8.5)
Shoulders	42 (17.9)	11 (4.7)
Elbows	16 (6.8)	9 (3.8)
Wrists/hands	31 (13.2)	9 (3.8)
Upper Back	38 (16.2)	12 (5.1)
Lower Back	108 (46.0)	50 (21.3)
Hips/thighs	20 (8.5)	10 (4.3)
Knees	52 (22.1)	17 (7.2)
Ankles/feet	21 (8.9)	8 (3.4)

Table 4. Type of duty believed to be in the origin of the symptoms.

Type of duty	12 months	7 days
	N(%)	N(%)
Fireground Activities	281 (37.6)	102 (37.2)
Patient Transport	232 (31.1)	72 (26.3)
Rescue Activities	145 (19.4)	65 (23.4)
Driving	70 (9.4)	31 (11.3)
Other	14 (1.9)	4 (1.5)

Table 5. Type of duty influence in anatomical regions affected by WRMSS in the last 12 months.

Anatomical regions	Fireground activities	Patient transfers	Rescue activities
p values (Chi-square) related to anatomical regions and type of duty analysis			
Neck	0.00*	0.00*	0.00*
Shoulders	0.00*	0.02*	0.00*
Elbows	0.01*	0.54	0.00*
Wrists/hands	0.00*	0.54	0.02*
Upper Back	0.00*	0.01*	0.01*
Lower Back	0.00*	0.00*	0.00*
Hips/thighs	0.00*	0.41	0.00*
Knees	0.00*	0.00*	0.12
Ankles/feet	0.00*	0.28	0.00*

* Significant at 95%, $p \leq 0.05$.

Table 6. Type of duty influence in anatomical regions affected by WRMSS in the last 7 days.

Anatomical regions	Fireground activities	Patient transfers	Rescue activities
p values (Chi-square) related to anatomical regions and type of duty analysis			
Neck	0.03*	0.01*	0.01*
Shoulders	0.54	0.14	0.00*
Elbows	0.03*	0.22	0.03*
Wrists/hands	0.05*	0.68	0.61
Upper Back	0.02*	0.23	0.00*
Lower Back	0.01*	0.00*	0.00*
Hips/thighs	0.00*	0.46	0.00*
Knees	0.00*	0.30	0.00*
Ankles/feet	0.01*	0.09	0.00*

* Significant at 95%, $p \leq 0.05$.

of unnecessary physical exertion (8.9%). No preventive strategies were reported by 125 (53.2%) firefighters and 53 (22.6%) took sick leave due to their complaints.

When asked about the type of duty that the respondents believed to be in the origin of the symptoms the most cited were fireground activities, patient transport and rescue activities in both 12 month and 7 days prevalence, as shown in table 4.

Looking at the relationship between the type of duty and the anatomical regions with WRMSS, performing fireground and rescue activities was found to be related with higher prevalence of WRMSS, followed by patient transfers (tables 5 and 6).

4 DISCUSSION

The prevalence of WRMSS among Portuguese volunteer firefighters was high, considering lifetime prevalence (67.7%), last 12 month prevalence (65.9%) and last 7 days prevalence (36.2%). Although high, our results were lower than those of Beaton et al. (1996) that reported a lifetime prevalence of 95%. We found a higher prevalence of WRMSS in firefighters older than 40 years that increased with the work experience of the respondents. The results of Szubert & Sobala (2002) however point that younger firefightes with less working experience reported higher values of work-related musculoskeletal disorders but, as stated by Widanarko et al. (2011) the association between age and musculoskeletal symptoms is rather inconsistent and in their study they found no significant differences in prevalence among age groups. Given the intense physical demands of this occupational group (Holmér & Gavhed, 2007) and the deleterious effect of age in the musculoskeletal

system it is not totally unexpected the older professionals report higher prevalence of symptoms.

The lower back, the neck and the knees were the anatomical regions with highest prevalence of WRMSS in both 12 months (46%, 26.8% and 22.1% respectively) and 7 days (21.3%, 8.5% and 7.1% respectively) period analysis. Other studies have found similar results (Bos et al., 2004) with 6 month prevalence values of 32% for low back disorders, 16% for neck disorders and 20% for knee disorders. The same authors identified several biomechanical demanding activities that were performed more often by firefighters like standing, lifting/carrying, pushing/pulling, kneeling/squatting, stooping, working in a twisted posture and jumping that might be responsible for the overloading of the mentioned anatomic regions. Musculoskeletal symptoms, particularly in the lower back are common among firefighters due to the physical requirements of the tasks performed (Beaton et al., 1996).

In our study the firefighters reported as perceived causes to their symptoms the physical exertion, fatigue, unexpected sudden movements and working in awkward postures. Heavy workload, sustained awkward postures, lack of attention, insufficient physical capacity, the lack of personal protective equipment and performing heavy tasks without warming up have also been pointed as perceived causes for WRMSS (Szubert & Sobala, 2002, Walton et al., 2003).

Coping strategies like stop performing the tasks or using the techniques that aggravate or provoke the symptoms and asking for the help of colleagues were identified but 22.6% of the respondents took sick leave due to their complaints. WRMSS may lead volunteer firefighters to quit and might be responsible as well for sick leaves in the main occupational activity (Gamble et al., 1991), with important economic impact on society through both direct health expenditure and indirect loss of productivity.

Preventive strategies include the improvement of physical capacity, the reduction in excessive load and the implementation of ergonomic programs (Walton et al., 2003, Szubert & Sobala, 2002, Gamble et al., 1991). The improvement in body mechanics, the improvement in physical capacity and the avoidance of unnecessary physical exertion were identified as preventive strategies adopted but no preventive strategies were reported by the majority of the respondents.

According to Karter & Molis (2012) 43.5% of all firefighter injuries occurred during fireground operations and the major types of injuries were strains, sprains and muscular pain (50.7%). During non-fireground activities, strains, sprains and muscular pain accounted for 61.1% of all injuries. Our results indicate that fireground activities are related with higher prevalence of WRMSS followed by non-fireground rescue activities and patient transfers and are in line with previous research (Karter & Molis, 2012, Beaton et al., 1996, Szubert & Sobala, 2002).

5 CONCLUSIONS

The present study has shown that the prevalence of WRMSS was high among Portuguese volunteer firefighters. The lower back, the neck and the knees were the most affected anatomical regions and the most reported causes were physical exertion, fatigue, unexpected sudden movements and working in awkward postures. The most commonly adopted coping strategies were stop performing the tasks or using the techniques that aggravate or provoke the symptoms and asking for the help of colleagues and the improvement in body mechanics, the improvement in physical capacity and the avoidance of unnecessary physical exertion were the most reported preventive strategies. Fireground and rescue activities and patient transfers were related to higher prevalence of WRMSS.

More satisfactory preventive and treatment measures are recommended to minimize the prevalence of WRMSS among Portuguese volunteer firefighters. Future research should contemplate professional firefighters and should address the implementation and evaluation of preventive strategies. Educational programs should be created and address the needs of these professionals.

REFERENCES

ATIK, O.S. (2010) Is the Bone and Joint Decade over? *Eklem Hastaliklari Ve Cerrahisi-Joint Diseases and Related Surgery*, 21, 123–123.

BEATON, R., MURPHY, S. & PIKE, K. (1996) Work and nonwork stressors, negative affective states, and pain complaints among firefighters and paramedics. *International Journal of Stress Management*, 3, 223–237.

BERNARD, B.P. (1997) Musculoskeletal Disorders and Workplace Factors—A Critical Review of Epidemiologic Evidence for Work-Related Musculoskeletal Disorders of the Neck, Upper Extremity, and Low Back. Cincinnati: U.S., National Institute for Occupational Safety and Health; Center for Disease Control and Prevention.

BOS, J., MOL, E., VISSER, B. & FRINGS-DRESEN, M. (2004) Risk of health complaints and disabilities among Dutch firefighters. *International archives of occupational and environmental health*, 77, 373–382.

BROOKS, P.M. (2006) The burden of musculoskeletal disease—a global perspective. *Clinical Rheumatology*, 25, 778–781.

GAMBLE, R., STEVENS, A., MCBRIEN, H., BLACK, A., CRAN, G. & BOREHAM, C. (1991) Physical fitness and occupational demands of the Belfast ambulance service. *British journal of industrial medicine*, 48, 592–596.

HOLMÉR, I. & GAVHED, D. (2007) Classification of metabolic and respiratory demands in fire fighting activity with extreme workloads. *Applied Ergonomics*, 38, 45–52.

INTERNATIONAL ORGANIZATION FOR STANDARDIZATION (2004) Ergonomics—determination of metabolic heat production. Geneva, International Organization for Standardization.

KARTER, M.J. & MOLIS, J.L. (2012) U.S. Firefighter Injuries-2011. Quincy, National Fire Protection Association.

KUORINKA, I., JONSSON, B., KILBOM, A., VINTERBERG, H., BIERING-SØRENSEN, F., ANDERSSON, G. & JØRGENSEN, K. (1987) Standardised Nordic questionnaires for the analysis of musculoskeletal symptoms. *Applied Ergonomics*, 18, 233–237.

MCBETH, J. & JONES, K. (2007) Epidemiology of chronic musculoskeletal pain. *Best Practice & Research Clinical Rheumatology*, 21, 403–425.

MESQUITA, C., RIBEIRO, J. & MOREIRA, P. (2010) Portuguese version of the standardized Nordic musculoskeletal questionnaire: cross cultural and reliability. *Journal of Public Health*, 18, 461–466.

MODY, G.M. & BROOKS, P.M. (2012) Improving musculoskeletal health: Global issues. *Best Practice & Research Clinical Rheumatology*, 26, 237–249.

PUNAKALLIO, A., LUSA-MOSER, S. & LOUHEVAARA, V. (2006) Fire-Fighting and Rescue Work in Emergency Situations and Ergonomics. *International Encyclopedia of Ergonomics and Human Factors, Second Edition-3 Volume Set*. CRC Press.

SANDERS, M. & DILLON, C. (2006) Diagnosis of Work-Related Musculoskeletal Disorders. *International Encyclopedia of Ergonomics and Human Factors, Second Edition-3 Volume Set*. CRC Press.

SANDERS, M. & STRICOFF, R. (2006) Rehabilitation of Musculoskeletal Disorders. *International Encyclopedia of Ergonomics and Human Factors, Second Edition-3 Volume Set*. CRC Press.

SZUBERT, Z. & SOBALA, W. (2002) Work-related injuries among firefighters: sites and circumstances of their occurrence. *International journal of occupational medicine and environmental health*, 15, 49–55.

WALTON, S.M., CONRAD, K.M., FURNER, S.E. & SAMO, D.G. (2003) Cause, type, and workers' compensation costs of injury to fire fighters. *American journal of industrial medicine*, 43, 454–458.

WIDANARKO, B., LEGG, S., STEVENSON, M., DEVEREUX, J., ENG, A., MANNETJE, A.T., CHENG, S., DOUWES, J., ELLISON-LOSCHMANN, L., MCLEAN, D. & PEARCE, N. (2011) Prevalence of musculoskeletal symptoms in relation to gender, age, and occupational/industrial group. *International Journal of Industrial Ergonomics*, 41, 561–572.

WOOLF, A.D., ERWIN, J. & MARCH, L. (2012) The need to address the burden of musculoskeletal conditions. *Best Practice & Research Clinical Rheumatology*, 26, 183–224.

WOOLF, A.D. & PFLEGER, B. (2003) Burden of major musculoskeletal conditions. *Bull World Health Organ*, 81, 646–56.

Risk assessment methods

Occupational Safety and Hygiene – Arezes et al. (eds)
© *2013 Taylor & Francis Group, London, ISBN 978-1-138-00047-6*

Safety function analysis in a manufacturing process of paper products

C. Jacinto & R. Beatriz
UNIDEMI, Mechanical and Industrial Engineering, Faculdade de Ciências e Tecnologia,
Universidade Nova de Lisboa, Portugal

L. Harms-Ringdahl
Institute for Risk Management and Safety Analysis, Stockholm, Sweden

ABSTRACT: This paper reports a study of the safety characteristics of a manufacturing line of paper products. The method Safety Function Analysis (SFA) was applied for this purpose. The line was divided in sections, which were studied consecutively. For one of them, 36 safety functions (SF) were identified and evaluated. In general, most SFs were found to have good performance but a few shortfalls were also unveiled during the analysis. Consequently, a number of specific recommendations were proposed to improve safety in the line.

1 INTRODUCTION

The study of safety at work can be supported by several analytical and practical approaches. Many methodological alternatives are available and most are thoroughly described in the safety literature. Most methods are based on hazard identification, risk reduction and subsequent safety improvement. An alternative approach is to focus directly on the safety features of the working system under analysis. In such cases the aim is to judge if they are adequate or if improvements are needed.

This study puts an emphasis on the approach associated with safety features. There are a number of related terms commonly found in the literature, such as, barrier function, defence or protection layer. Hollnagel (2004), for instance, characterises the term barrier as prevention or protection, depending on how they are used before or after the action takes place in time. Sklet (2006) proposes specific definitions for the terms Safety Barriers, Barrier Functions and Barrier Systems, all of which are relevant for the analysis of safety level and performance of a system.

A similar concept is Safety Function (SF), defined as *a technical or organizational function, a human action or a combination of these, that can reduce the probability and/or consequences of accidents and other unwanted events in a system* (Harms-Ringdahl, 2009, p. 361). This is the working definition used in the current study.

The objective of this work was twofold:

1. to study the safety characteristics of one particularly hazardous process in a manufacturing line of paper products, and

2. to test the application of a new version of the SFA method.

The manufacturing company is called Renova (Portugal) and is specialized in the production of paper tissue, printing paper and packaging. The new version of SFA (Harms-Ringdahl, 2011) is not published yet, but it introduces some improvements and new steps, as compared to its original version (Harms-Ringdahl, 2001, 2003).

2 METHODOLOGY

The general methodology of this work follows a case-study approach with a timeline of about five months. A starting point was a set of risk assessments previously done at the workplace. The major steps in this study were to select one part of the workplace, to apply safety function analysis on that, and finally to examine the results.

The choice of the case (production line) was based on the risk assessment (RA) corporate procedure currently used in Renova, which is essentially a simplified version of the W.T. Fine method (Fine, 1971); a very similar version of such method is recommended by the Spanish H&S Authorities (c.f. NTP-561, INSHT, Spain, Technical Norm for Prevention). According to Renova's records and experience, the production line selected (H4) is one that not only has more safety functions to assess, but it also is the most hazardous in the plant.

As mentioned, this study applied an approach to risk assessment that is based on the study of safety functions related to a specific hazardous workplace. This methodology is known as Safety Function Analysis (SFA) (Harms-Ringdahl, 2001, 2003). It is a general method that can be applied to most types of systems or events.

The SFA methodology can be briefly described in a series of 5 main steps (Fig. 1). The results of the first three stages produce a structured list of SFs, which follow the original manual. In step 4, the SFs are evaluated one by one in order to judge if improvements are needed. When needed, improvements will be developed in order to, for instance, increase the efficiency of SFs.

In this study, a fairly advanced evaluation procedure with six stages (a–f) has been applied, and therefore step 4 constitutes the core of the analysis in this approach. The first four parameters (a–d) are estimated, which was done in a group discussion. The characteristic **Intention** (a) is merely informative; it describes whether a certain function was designed intentionally for safety purposes (or if the main purpose is something else and safety is co-lateral). It is not an essential parameter, but it can give valuable information on how the function works and how it can be managed. The characteristic **Importance** (b) has four categories (Table 1 A) for describing the influence on safety. The characteristic **Efficiency** (c) gives a coarse estimate of "probability to function" or "error frequency"; the latter can be seen as the negative expression of efficiency (Table 1B).

The analyst can opt for the criterion that seems easier to apply. In addition, there are two kinds of efficiency (Estimated and Wanted); the *Wanted Efficiency* (WE) refers to the classification that the company wishes to have in that process (or system), so that one can give a score to the *Estimated Efficiency* (EE) by comparison.

The efficiency of a SF might degrade over time and, especially if latent failures could occur, this can increase the risks. This means that the monitoring of SFs is vital for safety. The characteristic Monitoring (d) is therefore included (Table 2).

Monitoring (d) (Table 2) distinguishes between the need for monitoring (MN) the SF under analysis and its present monitoring status (MS).

Table 1. Criteria for Importance and Efficiency (Harms-Ringdahl, 2011).

1 A—Importance of SF

Code	Description
0	SF has no or very small influence on safety
1	Small influence on safety
2	Rather large influence on safety
3	Large influence on safety

1B—Efficiency of SF

Code	Efficiency	Probability to function	Error Frequency
0	Very Low	<50%	–
1	Low	>50%	<100 times/year
2	Medium	>90%	<10 times/year
3	High	>99%	<1 time/year
4	Very High	≥99,99%	<0,01 time/year

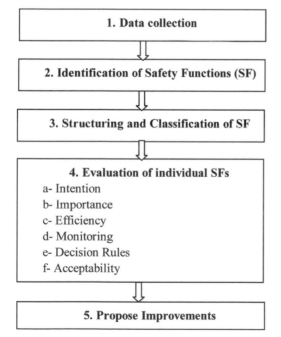

Figure 1. Flowchart of the SFA process (after Harms-Ringdahl, 2011).

Table 2. Need of monitoring (MN), and judgment of status (MS) (Harms-Ringdahl, 2011).

Code	Need	Status	Code
MN4	Monitoring is essential	Meets the requirements	MS2
MN3	Monitoring is necessary, at least periodically	Exists, but not fully meet the requirement	MS1
MN2	Monitoring is of interest, but not a critical issue	Monitoring function does not meet requirement	MS0
MN1	Of low interest	Ok, no need for monitoring	MS2
MN0	Not needed or irrelevant	Ok, no need for monitoring	MS2

Table 2 shows that, for instance, a critical SF in which monitoring is considered "essential" (MN4) can be classified within different status (MS2, MS1, or MS0) depending on how well (or not) the SF is being monitored in practice. The same reasoning applies for categories MN3 and MN2.

The aim of the evaluation stage (step 4; Fig.1) is to judge if the system and the individual SFs are acceptable as it is, or if improvements are needed. The characteristic **Acceptability** (f) gives a measure of this, as shown in Table 3.

Table 3. Evaluation scale for acceptability of safety function (Harms-Ringdahl, 2011).

Acceptability	Code	Description
Acceptable	0	*No need* for improvement
	1	Improving safety function can be *considered*
Not Acceptable	2	Improving safety function is *recommended*
	3	Improving safety function is *essential*
	4	*Intolerable*, work should not be started or continued until the risk is reduced

However, such judgments might be difficult to make for persons without a long experience, and the result could be quite unreliable and subjective. In order to support industry practitioners in these estimates, the authors saw the need for some kind of "guidance" or "rule of thumb".

Based on the estimates of Importance (b), Efficiency (c) and Monitoring (d) a set of **Decisions Rules** (e) (Harms-Ringdahl, 2011) have been applied. These Decision Rules are shown in Table 4. The aim was to improve the evaluation stage in order to make it more reliable and reproducible. The outputs are suggested acceptability values as in previous Table 3.

In its original version (Harms-Ringdahl, 2001, 2003), the evaluation stage used only three characteristics (intention, importance and efficiency) which made it more vulnerable to subjective influences. One feature of this study has been to develop a new modified version of SFA (Harms-Ringdahl, 2011), which brings a number of improvements and novelties. One important addition is that efficiency shall be regarded as a design parameter, where the management is asked to define the "wanted efficiency". The second addition is to systematically consider the monitoring status of the system and its subsystems. The third addition is to connect all characteristics in a set of decisions rules (Table 4).

Table 4. Decision rules for acceptability (Harms-Ringdahl, 2011).

Importance IMP	Efficiency *estimated versus wanted*	Monitor Status MS	Evaluation (acceptability)	Decision (needs improvement?)
0	EE ≥ WE	–	0	No need for improvement
Very Small	EE < WE	–	1	Can be considered
1	EE ≥ WE	–	0	No need for improvement
Small	EE < WE	0	2	Is recommended; prevent degrading SF
		1–2	1	Can be considered
2	EE ≥ WE	0–1	2	Is recommended; prevent degrading SF
Rather large		2	0	No need for improvement
	EE < WE	0	3	Is essential
		1–2	2	Is recommended; prevent degrading SF
	EE ≪ WE	–	3	Is essential
3	EE ≥ WE	2	1	Can be considered
Large		1	2	Is recommended; prevent degrading SF
		0	3	Is essential
	EE < WE	2	2	Is recommended; prevent degrading SF
		0–1	3	Is essential
	EE ≪ WE	0–1	4	Urgent improvement; intolerable situation
		2	3	Is essential

3 RESULTS AND DISCUSSION

The production line studied in this work (Line H4) transforms large reels of tissue paper (Fig. 2) into standard toilet paper rolls.

The entire production line (22 processes) was observed for a week after which one particular process was selected for the study for having the most hazardous operations. This partial process is internally called *transversal cut of logs*—it consists on the cutting of long logs into the final standard rolls, at very high speed. These logs are WIP (work in progress); they are around 2 m long but have small diameter. After the transversal cut, they become the final product (toilet paper rolls).

3.1 *Application to a specific SF*

The SFA methodology is illustrated next showing its application to this case and the evaluation of one particular SF.

Step 1: Data collection—the selection of the hazardous process to analyse resulted from different elements, such as: observations of the various processes, analyses of the existing "hazard identification and risk control map for line H4" provided by Renova, and discussions with workers of the line. Within the process chosen (transversal cut of logs), each equipment, infra-structure and task was meticulously analysed.

Step 2: Identification of safety functions (SFs)—the identification of safety functions was firstly made through observation of the work, and questions such as: how is the likelihood of an accident being kept low? or How are consequences kept to a low level?. Other means were the study of documents such as previous risk analysis protocols and the legal requirements for use of equipment and machinery. This exercise resulted in the identification of 36 safety functions.

Step 3: Structuring and classification of SFs—the safety functions aforementioned were divided into five categories: 1-containment, 2-automatic control, 3-reduction of consequences, 4-procedures and routines, and 5-management/organiza-tional. The category procedures/routines integrates both the formal and informal ones, because it was difficult to distinguish between one another in many instances. After this structuring, the safety functions were sorted out and ready to be evaluated.

Step 4: Evaluation of safety functions—this step was made in collaboration with two experienced plant engineers. The group went to the production line and watched the process once again, so that any question regarding more specific functions could be clarified at that moment and discussed at the place.

To give an example, a safety function related to *"barriers preventing walkthrough over conveyors"*, concerning a Trim Conveyor Belt, was identified as a legal requirement. However, the SF does not exist (absent SF) and therefore it cannot be seen in the photo (Fig. 3). It was classified as a "containment" type with relation to the risk of mechanical contact with moving parts (Fig. 3). Its evaluation was made as follows:

The **intention** of this barrier—if it existed—should be scored "**3**", i.e., specifically intended for safety purposes. The **importance** of this safety function should be considered "rather large", since it would prevent operators from losing their step (or balance) when crossing the conveyor, or even block the crossing (wrong action), which seems to occur quite frequently; so **importance** is "**2**". The overall **efficiency** was scored "**0**" because it does not exist (this is a missing barrier; EE < WE). As for the **monitoring status**, since this function does not exist yet, it does not meet any requirements and its classification was "**MS0**". In the light of the above, the **acceptance level** can be classified as *Not Acceptable* "**3**" (c.f. Table 4), and in this case the decision rule establishes that improving the safety function is essential.

Step 5: Propose Improvements—the acceptability criterion depends on the factors analysed previously (Importance, Efficiency and Monitoring Status); it leads to a plan of action with specific recommendations, which aims at improving safety.

In this example, the SF in question (barrier preventing walkthrough over the conveyor) does not

Figure 2. Raw material loading and transport (reels).

Figure 3. Trim Conveyor Belt.

exist and it should be implemented. The design of a physical barrier (new SF) might consider two alternative solutions:

1. to prevent the crossing at all (e.g. fixed metal grid); in such case, workers need to go around a couple of meters to reach the other side. In fact, this is what they are expected to do right now, but they tend to adopt a dangerous behaviour and simply jump or cross over the conveyer belt.
2. to build a crosswalk bridge over the conveyer; this gives operators the possibility to keep crossing it, but in a safer way. The plant managers preferred the second option and, in the meantime, it was already implemented.

3.2 Relevant results; synthesis of SF analysed

Within the safety analysis of the process "transversal cut of logs" a total of 36 SFs were assessed individually, after being structured and classified into five main groups: 1) Containment, 2) Automatic control, 3) Reduction of consequences, 4) Procedures & Routines, 5) Management/Organisational.

Of the 36 SFs analysed, 13 are related to "contact with sharp elements", 9 to "fire", 8 to "mechanical contact" and 6 to "particles and dust exposure". Most of them (22) were considered well maintained and working properly, whilst others (14) showed that improvement could either be considered or was necessary. A distinctive aspect in this case study was the identification of an "absent" SF, which needed to be implemented. By contrast, all SFs related to the (most dangerous) cutting operation, for which the analysts were expecting higher chance of problems, did not require further improvement.

3.3 Experiences of using the methodology

The SFA method has been used as a complement to a traditional hazard identification method. The method was fairly new for the people conducting the analysis, and especially in the beginning it was regarded as more time consuming than traditional methods.

On the other hand, it brought in some positive features. The most essential are:

– The methodology evaluates "safety" rather than potential "risk", bringing a different point of view into the analysis of the system.
– It supports the analyst to search for non-technical functions, e.g.: procedures or organisation.
– The focus on safety performance has resulted in a more comprehensive analysis than usual.
– The methodology was found useful by identifying problematic, inefficient and missing SFs.

The approach of evaluating safety is rather new, and the application of the method has two especially crucial stages. The first is the identification of SFs, which requires a shift in thinking and therefore is difficult at the beginning. However, after a short time, this was regarded as rather simple. The other crucial stage is the evaluation whether improvements are needed. The evaluation was constructed as a set of consecutive steps which can be followed systematically. In the practical application this was considered fairly easy to follow. Features that were perceived as especially useful were:

– The monitoring status; a SF may be accepted as good, but it can degrade easily with time if not adequately monitored.
– Table 4 with the decision rules provides guidance on acceptability and reduces subjectivity between different analysts.
– The method gives support in developing and improving existing or new safety features.

An improvement to make in future studies is to create a category "General SF", for classifying SFs that are common to several subsystems. For example, the function "First aid 24 hours" appears frequently associated with a variety of hazardous operations.

There is also guidance for an overall evaluation of the entire system. However, this ability was not tested in this study due to time constraints.

4 CONCLUSIONS

The experience is that the study of safety features is a valuable complement to traditional hazard identification. In this study, a developed version of SFA (Harms-Ringdahl, 2011) has been tested, and it was found clearly more useful than the earlier version (Harms-Ringdahl, 2001, 2003), especially considering less experienced analysts. Important positive features of the methodology were related to identification of safety barriers, and the procedure for evaluating them.

ACKNOWLEDGMENT

The authors are grateful to Renova for authorising this study, as well as for their vital contribution.

REFERENCES

Fine, W.T. 1971. Mathematical Evaluation for Controlling Hazards. Unclassified NOLTR 71-31, 8 March 1971. Naval Ordinance Laboratory, White Oak, Maryland.

Harms-Ringdahl, L. 2001. Safety Analysis—Principles and Practice in Occupational Safety. 2nd Edition. Taylor & Francis, London.

Harms-Ringdahl, L. 2003. Assessing Safety Functions—results from a case study at an industrial workplace. *Safety Science* 41(8): 701–720.

Harms-Ringdahl, L. 2009. Analysis of safety functions and barriers in accidents. *Safety Science* 47: 353–363.

Harms-Ringdahl, L. 2011, unpubl. Analysis of barriers and safety functions. Preliminary manuscript of *Guide to safety analysis for accident prevention* (book in prep.)

Hollnagel, E. 2004. Barriers and Accident Prevention. Ashgate Publishing Limited, England.

Sklet S. 2006. Safety barriers; definition, classification, and performance. *Journal of loss prevention in the process industries* 19: 494–506.

Occupational Safety and Hygiene – Arezes et al. (eds)
© *2013 Taylor & Francis Group, London, ISBN 978-1-138-00047-6*

Assessment of human factor in production engineering

I. Turekova & Z. Turnova
*Slovak University of Technology in Bratislava, Faculty of Materials Science and Technology in Trnava,
Department of Safety Engineering, Trnava, Slovak Republic*

ABSTRACT: The manual activities and attendance of smaller machines represent decisive human activities in small and medium enterprises. Individual workers are often qualified for attendance of several equipment. The effect of factors of working environment together with work organisation, combined with short delivery time and personal restrictions may exert unfavourable effects upon the workers. These may affect the work performance on one hand, but also the attention and capability for correct decision in a given instant on the other hand. Therefore it is important to assess the reliability of human factor in a given process. Aim of this study was to compare the human factor reliability of three working positions in a welding workplace by MIPS method.

1 INTRODUCTION

Selection of optimum analytic technique for human factor reliability is the basis for selection of a technique proper. Also own resources must be taken into account at method selection, i.e. what requirements are laid by the method on:

- number of persons needed for analysis,
- time amount necessary for performing the analysis,
- professionalism level of the assessing expert and/or entire team,
- sophistication of individual methods—number of values that are to be treated and the degree of work complexity which is necessary for a correct processing of analysis (Fiserova, 2009).

2 METHOD OF IDENTIFICATION OF FAILURE CAUSES

Method of identification of failure causes (MIPS) was created on the basis of system model—Work Process Analysis Model. This model involves the elements of task analysis, ergonomic elements and last but not least also the work psychology elements. Owing to this wide approach, the MIPS method allows to analyse majority of factors acting upon the workers causing the failure of human factor (Miklos & Solc 2011; Skrehot & Malcikova 2006).

For system characteristic (environment and processes) the organisational reliability factors (SOF) are introduced. These factors represent indices, characterising the effect of a part of system upon the appropriate worker, which exerted and/or could exert certain effect upon the occurrence and/or course of undesirable event. The SOF factors are divided to groups, where each SOF group characterises a wider circle of effects and is therefore subdivided to partial ones (DPSOF). The measure of SOF interaction with individual professions differs, therefore MIPS considers this fact via different approach at assessment of the collected answers. For this purpose it is necessary to identify the critical profession groups, which contribute most often at occurrence of undesirable event.

MIPS can be in simplified form illustrated via the flow chart, Figure 1.

Quantitative analysis of MIPS methodic is based on a controlled conversation with a selected worker of appropriate profession. He is subjected to questions with answer possibilities "yes" and/or "no". The questions are formulated in such a way that each of them would allow to reveal subsequently the possible cause of worker failure.

A workplace of fitter's fabrication in a company oriented to repair of trailers and semi-trailers was selected for analysis.

Quantification was performed via the human failure factor—F_p, which allows to determine the probability of a correct determination of human failure causes P_p. Each negative answer from the check lists was penalized by one point and

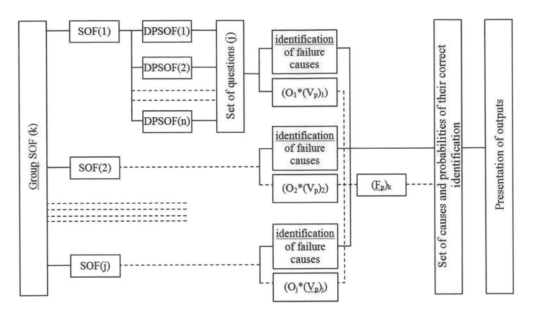

Figure 1. Flow chart of MIPS method for the (k) SOF group applied at work position (p), O_j-penalizing factor, V_p-weighting factor, F_p-factor of human failure (Havlikova, 2009).

subsequently calculated with the weight coefficient (value 1–3).

The factor of human element failure (F_p), to be determined for each SOF (k) group separately, is calculated by the formula (1).

$$(F_p)_k = \left[\frac{\Sigma_i o_i \cdot (V_p)_i}{\Sigma_i ((V_p)_i \cdot j)} \right]_k \quad (1)$$

where F_p = factor of human element failure; V_p = appropriate weight coefficient z; O_i = penalization coefficient (sum of penalization points for negative assessment of appropriate DPSOF—each negative answer = 1 point); i = question; j = the number of investigative questions for appropriate SOF; k = SOF group.

The dependence of P_p on F_p does not take into account just the general number of causes and probability of their acting leading to human error but it takes into account also the significance of possible consequences. This is considered by the penalization coefficient, weight coefficient and also the counting of cause occurrence in the investigated group. The task of MIPS is to offer the set of possible causes of human failure and the value of probability that the real cause of failure would occur in the presented output. The cause may not be just one, therefore we may speak just about the measure of relevance

of correct cause determination from the output (Skrehot, 2009).

Conditions of functional relationship (2):

$$F_p \in \langle 0;1 \rangle P_p \in (0;100\%), \lim_{F_p \to 0} P_p(F_p)$$
$$= 0\% \cap \lim_{F_p \to 1} P_p(F_p) = 100\% \quad (2)$$

The resultant P_p value expressed in per cents then shows the probability that there exists at least one from among the identified causes, which really caused the failure of a given person. The quantitative assessment shown in table 1 thus defines the measure of reliability of correct result determination.

An overview and arrangement of the attended equipment is shown in Figure 2.

The workers of appropriate profession, who exert decisive effect upon occurrence of undesired event were subjected to directed interview, namely two workers directly attending the cutting equipment and one managing worker from the top management. The first worker, attending the equipment responded to questions during the directed interview, other two workers responded via MIPS software application. The method is rather time demanding (about 2 hours), however, the fact that the software application may be interrupted any time allows a more responsible assessment, free from any time pressure.

Table 1. Assessment of probability of determination of human failure cause P_p by the values of factor of human element failure F_p (Malý, 2009).

F_p	P_p %	Probability of correct determination	Qualitative assessment
0–0.20	<10	Very low	Neither human failure, nor effect on other causes is supposed.
0.21–0.30	10–30	Low	Human failure is not supposed, however the identified cause might affect the acting of other causes belonging to another group.
0.31–0.40	30–50	Moderate	Cause acting might result in human failure at participation of causes belonging to another group.
0.41–0.54	50–80	High	Acting of cause in appropriate group might result in human failure and/ or significantly participate in it.
0.55–1.00	>80	Very high	Acting of cause in appropriate group caused the human failure and/or its effect on occurrence was decisive.

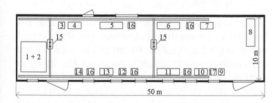

Figure 2. Scheme of machinery arrangement 1 = plasmacutting equipment type RS515B; 2 = cutting equipment type Plazma CUT 100; 3 = hydraulic presstype P 6324; 4 = roll bending machine type ZXP-50/7; 5 = versatile shears type NUD 500/20; 6 = bending machine type XOM-2000-6B; 7 = power braketype LOD 315; 8 = table shears typeNTE 3150/6,3; 9 = saw type Thomas; 10 = profile shears type HB 5221; 11 = hydraulic bending machine type XOM-2000/RA-4; 12 = manual press; 13 = hydraulic press; 14 = drilling machine type PH3271; 15 = crane-pulley (320 kg); 16 = welding machine (type MIG 350FW, WP 350); 17 = oxygen cutting rig (Kubovicova, 2012).

Actually, the activity of material cutting by use of plasma equipment was concerned. The workers of following professions were selected for analysis:

– two workers attending the equipment (a fitter and a shift foreman)—were subjected to 227 questions,
– operating manager—responded to 234 questions.

Differentiation of questions is essential; these must be formulated with regard to work position. Method also supposes that the managing workers should know the answers to all questions. Each negative answer allows to define the cause that might lead to failure of the appropriate worker.

3 RESULTS AND DISCUSSION

The results from analysis by MIPS method have shown, that the human failure factor on the workplace for plasma cutting of materials F_p varies in average from 0.17 to 0.30, what in quantitative assessment means that though human failure is not supposed, the identified cause might affect the behaviour of worker. In spite of the fact that the risk of human factor failure is rather low, the analysis has revealed several weak points actually in each of analysed fields.

Analysed worker judged the entire system and work team by his answers. Since he follows mainly from his own experience and subjective feelings, then he has revealed also the facts concerning himself. If contradiction in answers of several persons was observed, it is obvious that one side did not respond in accordance with real state. There was minimum subjective questions, mainly owing to the fact that such statements are not always trustworthy in practice.

Final step of analysis consisted in elaboration of protocol on investigation. This protocol contained the calculated values $(F_p)_k$ and $(P_p)_k$ and possible reasons of effect of human factor on the analysed work activity. In case of software processing of this method the protocol is generated automatically by computer.

Table 2 gives a comparison of results from analysis of individual work positions.

Imperfections that could exert the most significant effect upon the failure of human factor within the given SOF group are:

– There is not elaborated organisation order with clearly defined tasks and duties for duly performance of functions. Therefore the employees do not understand their appointed tasks always and at any circumstances.
– There were not issued organizational and management regulations in the plant. It happens that

Table 2. Comparison of estimation probability for the effect of human factor $(P_p)_k$ (Kubovicova, 2012).

k	Group name	Equipment operator %	Shift foreman %	Head of organisational unit %
1	Training	<10	<10	<10
2	Tasks and duties	>80	50–80	50–80
3	Decision making and control of processes	10–30	30–50	<10
4	Operations and manipulation	10–30	<10	<10
5	Work group	10–30	<10	<10
6	Attendance and supervision	<10	10–30	<10
7	Control and management	30–50	10–30	10–30
8	Personal features	30–50	30–50	<10
9	Risk factors of work environment	<10	<10	<10
10	Workplace	<10	<10	<10
11	Stress factors	<10	<10	<10

some employees adopt erratic decisions repeatedly. There are not stipulated any recourses in case of power excess. The occurred undesirable event is not always duly investigated.

– There are not always available appropriate work procedures and those existing are not updated regularly. There are not specified procedures for elimination of operational anomalies.

– There occur conflicts and discrepancies amongst the employees sometimes and they are not then willing to cooperate. The workers are not sufficiently attentive sometimes and they also neglect the importance of safety regulations.

– System allows that some operation need not be performed correctly. The worker may be affected by work stressors.

– Regarding the financial demands, the short-time projects have often priority against the long-term projects, what caused discrepancies between the instantaneous solutions and long-term needs. Performance of managing functions is not controlled. Employees are not subjected to any psychological analyses prior to entry. The plant does not provide the employees´ benefits, leading to relatively poor involvement of workers. The work risks are not regularly and continuously identified and assessed. This fact is caused by a low system pressure.

– Regarding absence of some work procedures, there occur improper work habits of some workers. The employees have not opportunity to utilize the medical care beyond the compulsory that would be organised by the employer.

– Risk factors are not regularly measured on the workplace, since the character of works does not suppose their increased values.

– Though order is currently kept on the workplace, there is also a material that is immediately or not at all needed and could be returned to the store. This activity is sometimes performed, but not with sufficient efficiency. There does not exist any systematic search for dangerous points on the workplace. Their revealing is thus rather incidental and non-systematic.

– The employees themselves do not realise the existence of all risks, connected with performance of technology they work on.

4 CONCLUSIONS

The SOF field Tasks and duties was identified as the weakest link from the viewpoint of human error in analyzed plant. It was found out that the tasks and duties for performance of functions were not clearly defined on all stages of management. Therefore the following corrective measures were suggested:

– to elaborate the organisation order for performance of necessary functions with clearly defined tasks and duties,
– to appoint unambiguous right powers and responsibilities on all stages of management.

Not only quantitative assessment of actual risk level brings about the greatest merit of this method, but also qualitative assessment in the form of verbal description of imperfections on the analysed workplace as well. It is just this qualitative assessment, based on which the corrective action can be then suggested. It can be concluded, that the studied method is a suitable instrument at application of reliability assessment for the human factor in case of cutting by use of plasma equipment.

The MIPS method is relatively new technique in the field of assessment of human factor failure. Therefore it would be inevitable for next development and application of this method to perform a statistical assessment of significant samples of mutually comparable plants. Unfortunately, only few data attained by application of this method were published up to now and therefore the expert's appraisals play an irreplaceable role in present practice.

ACKNOWLEDGMENTS

The submitted work was supported by the Slovak Grant agency VEGA MS VVS SR and SAV project No. 1/2594/12.

REFERENCES

Fiserova, S. 2009. Emerging physical, biologic, psycho-social and chemical risks related to safety and health protection at work. *Integrated safety 2009.* Trnava: STU Bratislava, ISBN 978-80-8096-107-7, pp. 26–34.

Havlikova, M. 2009. *Human factors in systems MMS.* IN JOSRA. No.1. ISSN 1803-3687.

Kubovicova, B. 2012. Analysis of the impact of human factors in production engineering. STU in Bratislava. pp. 82.

Maly, S. 2009. *Prevention of occupational hazards*, Vol. II. Praha: VUBP, ISBN 978-80-86973-79-1.

Miklos, V. & Solc, M. 2011. *The human factor- fundamental factor affecting the performance, quality and safety.* 16th International Scientific Conference: Quality and reliability of technical systems, May 2011, Nitra, ISBN 978-80-552-0595, pp. 46–50.

Skrehot, P. 2009. Errors of human factor and identification of their causes. In. JOSRA. No.1. ISSN 1803-3687.

Skrehot, P. & Malcikova, K. 2006. *Methodics for identification of failure causes*. User´s guide. Prague: VÚBP, ISSN 1803-3687.

Occupational Safety and Hygiene – Arezes et al. (eds)
© 2013 Taylor & Francis Group, London, ISBN 978-1-138-00047-6

Proposals for risk management in nanotechnology activities

L.R.B. Andrade
Fundacentro, Ministry of Labor and Employment, Porto Alegre, RS, Brazil

F.G. Amaral
Production Engineering Department, Federal University of Rio Grande do Sul UFRGS, Porto Alegre, RS, Brazil

ABSTRACT: Nanomaterials handling presents enormous challenges for the risk management in research and productions of new materials. However, there is a lack of data about the impacts of these new materials on human health and in the environment. In this scenario, a number of efforts have been made to mitigate the adversities and provide guidelines for the risk management associated with nanomaterials. This paper aims to give a broad overview and comparison between the main proposals in the literature. The methodology was a systematic analysis comprising 17 proposals for risks management with nanomaterials. The results indicate, although there is no consensus on the metrics used to characterize these risks, the adoption of the precautionary principle and the focus of the Control Banding stands out among the documents examined.

1 INTRODUCTION

The handling of nanomaterials presents great challenges for risk management. Although nanotechnologies are increasingly involved in research and in the production of new materials, data are lacking regarding the impacts of these new materials on human health and the environment.

In the context of this uncertainty, considerable effort has been devoted to mitigate the difficulties and to provide guidelines for managing the health risks associated with nanomaterials. In the literature, different approaches have been reported to assess and manage these risks (Ostiguy et al., 2009, Andrade & Amaral, 2012, Paik et al., 2008, Höck et al., 2012, among others).

Paik et al. (2008) notes that there are difficulties for the adoption of the traditional occupational hygiene (OH) approach to nanomaterials because of knowledge gaps regarding these materials.

Brouwer (2012) compared various approaches to understand the complexity of the problem and the various associated objectives and structures. In this study, it was deduced that the difficulties involved in working with nanomaterials are related to the metrics used to characterise the danger and the exposure limit.

The work of Brouwer (2012) was based primarily on proposals that employed the Control Banding (CB) method, which is an approach derived from an initiative of the UK Health and Safety Executive of 1999, titled Control of Substances Hazardous to Health (COSHH) Essentials Model.

The CB method is a plausible alternative to the traditional OH approach that allows the previously described barriers to be overcome. The CB can be used in situations in which the data about the dangers and exposure limit are scarce. Both the danger and exposure limit factors can be defined in a qualitative manner (not necessarily quantitative) to assign risk levels for each of which control actions would be suggested.

2 METHODOLOGY

It was analysed 17 studies with the common and basic goal of managing the safety risks and health at work associated with nanomaterials. Based on the examined proposals, it was developed a comprehensive list of strategies and actions that were components of the proposals. This list formed the basis for creating a table to compare the various reports, which indicates the presence or absence of these actions and strategies or, in certain cases, a reference to strategies described in a general or implied manner. At the same time, each proposal was briefly described, noting their main differences from the others.

The studies analysed were categorised into three groups according to their primary focus: 1) strategic approaches that define the general strategy and not the actions, 2) methodological approaches that provide strategies along with a practical set of actions to control the risks from nanomaterials and 3) pragmatic approaches that primarily define the actions, essentially the CB approaches (Brouwer, 2012).

Both strategies and actions were grouped according to the basic principles listed by the International Center for Technology Assessment— ICTA (2007) for overseeing nanotechnologies and are as follows: 1) the precautionary principle, 2) compulsory nano-specific regulation, 3) health and safety of the public and workers, 4) environmental protection, 5) transparency, 6) public participation, 7) inclusion of large impacts and 8) producer responsibility.

The strategies were grouped according to the principles that they most closely match without representing or accounting for the scope of the proposed principle. Thus, the principles in question are, in general, far more comprehensive than the set of strategies that were attributed to each principle. Several of the principles mentioned are not met by the proposed risk management methods because these principles are beyond the scope of the proposals, such as the principle of a compulsory nano-specific regulation. Likewise, none of the proposals includes strategies or actions with broader impacts (e.g., ethical or socioeconomic), bearing in mind that these impacts should be predicted or addressed using alternate tools with greater scope.

3 RESULTS AND DISCUSSION

The Table 1 presents a comparison of the main characteristics of the analysed proposals and tools, which are briefly described below. Each tool is associated with a letter in brackets that represents its reference within the table.

3.1 Strategic approaches

Among the strategic approaches, the study by Tyshenco & Krewski (2008) [A], "A risk management framework for the regulation of nanomaterials", can be cited; this study provides a general framework and proposes a set of strategies to regulate the handling of nanomaterials. As a comprehensive proposal, the study's primary aim is to provide an integrated and standardised focus to facilitate the future break of any trade barriers. Because the goal of the abovementioned study is the creation of a regulatory structure and not the specific control of a proposed activity, even in the specifications of the strategies, it is generic.

The nanoparticle risk assessment proposed by Tsuji et al. (2006) [B] ("Research Strategies for Safety Evaluation of Nanomaterials, Part IV: Risk Assessment of Nanoparticles") is part of a larger body of research strategies for the safe evaluation of nanomaterials. Specifically, this assessment contains richer details regarding the forms of exposure and possible adverse effects on human health.

The Nano Risk Framework (2007) [C] is intended to provide a generic structure for managing the risks associated with nanotechnologies, especially those related to the possible damage caused by products containing nanoparticles, which is more suited to large corporations. This framework contains the macro elements of a management system, including a description of nanotoxicology tests (most likely its great differential over the other proposals), and it incorporates concern for the concept of the "life cycle" of the product.

The Spanish contribution, "Evaluación de Riesgos de las Nanopartículas Artificiales—ERNA [Assessment of Artificial Nanoparticle Risks]" [D], proposed by Anton (2009), is basically a proposal supported by the conventional methods of risk assessment with the addition of an analysis of uncertainties as a way to mitigate the knowledge gaps regarding the effects of nanoparticles on the health.

3.2 Methodological approaches

Amoabediny et al. (2008) [E] presented the paper "Guidelines for Safe Handling, Use and Disposal of Nanoparticles". This document contains some general strategies and was therefore classified as a methodological approach. The strategies mentioned above were compiled from the literature available at the time the document was presented.

The British approach [F] (British Standard-BSI "Safe Handling Nanomaterials—PD 6699-2:2007") presents a set of strategies and specific actions for the control and management of the risk associated with nanomaterials.

Unlike the other options, which only presented qualitative evaluations, this standard describes the apparatus and methodologies that allow the quantitative analysis of nanoparticles and also notes some of the limits of exposure to these materials.

Another approach, from the German Federal Institute for Occupational Safety and Health [G] ("Bundesanstalt für Arbeitsschutz und Arbeitsmedizin/BAμA") from 2007, titled "Guidance for Handling and Use of Nanomaterials at the Workplace", is fairly generic and devotes special attention to the possible contamination with nanomaterials by inhalation. Although this proposal incorporates some quantitative assessment methods, it does not provide further evidence regarding the limits or methodologies to be applied.

The Quebec approach [H] is the "Best practices guide to synthetic nanoparticle risk management", presented by Osteguy et al. (2009). This guide proposes a comprehensive approach covering both general strategies for managing the risks associated with nanomaterials along with a CB approach,

Table 1 – Description and comparison among proposals

	A	B	C	D	E	F	G	H	I	J	K	L	M	N	O	P	Q
Reference of the proposal analysed in the text ->	A	B	C	D	E	F	G	H	I	J	K	L	M	N	O	P	Q
Grouping regarding CB methodology ->	Do not include CB						Include CB				Only CB						
Type of approach																	
Strategic approach (only strategies)	√	√	√	√													
Methodological approach (strategies and actions)					√	√	√	√	√	√	√						
Pragmatic approach (only actions) => CB tool												√	√	√	√	√	√
Type of evaluation																	
Only qualitative risk evaluation	√	√	√	√	√					√	√	√			√		√
Qualitative and quantitative risk evaluation						√	√	√	√				√	√		√	
Principles involved, strategies and associated actions																	
Transparency principle																	
Strategies for policy implementation	↕		↕														
Written, clear and transparent policy										↑	↔						↔
Policy developed with the participation of all							↔	↔	↑								
Principle of public participation and producer responsibility																	
Strategies focused in organisation	↕		↕														
Responsibility for accountability								↔		↑	↔						↑
Competence and training						↔	↑	↑	↑	↑			↔				↑
Documentation							↑	↑	↑	↑	↑		↑				↑
Broad communication								↔		↑							↑
Precaution principle																	
Strategy of hazard identification	↕	↕	↕	↕													
Nanomaterial characterisation						↑	↑	↑	↑	↑	↑	↑	↑	↑	↑	↑	↑
Principle of health protection and safety for the public and workers and environmental protection																	
Strategy of exposure evaluation	↕	↕	↕	↕													
Type of exposure (inhalation, dermal, ingestion)						↑	↑	↑	↑	↑	↔	↔	↑		↔	↔	↔
Monitoring of biological indicators							↑	↑	↑	↑		↔	↑				↑
Occupational and environmental monitoring							↑	↑	↑	↑	↔		↔	↑		↑	↔
Staff involved and possible exposures							↑	↑	↑	↔	↑		↑			↑	↑
Strategy for toxicity evaluation	↕	↕	↕	↕													
Toxicity studies		↑	↑														
Determination of the safe limits of exposure		↑	↑														
Strategy of risk characterisation	↕	↕	↕	↕													
Calculation of risk								↑							↔	↑	
Extrapolation of models								↑		↔	↔					↔	
Ranking of risks						↑		↑	↔	↔	↑	↑	↑	↑		↑	↑
Strategy of risk management	↕		↕	↕													
Technical actions						↑	↑	↔	↑	↑	↑	↑	↑	↑	↑	↑	↑
Organisational actions						↑	↑	↔	↑	↑	↑	↔	↑	↑	↑		↑
Labelling / storage						↑	↑		↔	↑	↑						↑
Cleanup / spill						↑			↔	↔	↔			↔			↑
Transport						↑			↔		↑			↔			↑
Destination / disposal of residues						↑	↑				↑			↔			↑
Personal protective equipment						↑	↑	↔	↑	↑	↑		↑	↑			↑
Risk of fire or explosion with nanoparticles							↑		↑	↑	↑						
Monitoring or surveillance strategy	↕		↕														
Monitoring							↑		↑	↑	↑	↑		↔		↑	↑
Research (accidents and incidents)							↑		↑		↑	↔					
Audit / review							↑	↑	↑	↑	↑						↑
Critical analysis of the administration								↑	↔								↔
Improvement strategy			↕														
Corrective and or preventive action							↑		↑	↔	↑	↑					↑
Continuous improvement							↑		↑	↔	↑	↑		↔			↑

√ = belongs to the category   ↕ = referred strategy   ↑ = referred action   ↔ = action with implicit or generic reference

Note: for the strategic proposals, only the presence of the strategy is indicated, except for the evaluation of toxicity, in which certain actions are also indicated.

which is referenced and based on the study by Paik et al. (2008), CB Nanotool.

The approach of the National Institute for Occupational Safety and Health (NIOSH) [I], called "General Safe Practices for Working with Engineered Nanomaterials in Research Laboratories" (2012), is quite comprehensive. This proposal assumes that there is a greater system for control and risk management in the organisation, into which the nanoparticle guidelines will be incorporated. In addition to generic guidelines, the proposal indicates the use of the CB approach as part of the nanomaterial risk control mechanism.

The study by Andrade & Amaral (2012) [J], "Methodological proposal for occupational health and safety actions in research laboratories with nanotechnologies activities", is also classified into the methodological approach group. This method not only presents a simplified flowchart for the characterisation of nanomaterials but also offers a number of suggestions for managing various specified and stratified operations, such as site clean-up, labelling and disposal. The great differential of this proposal is the inclusion of the OIT guidelines for management systems; thus, this method advocates the active participation of all stakeholders and not only the technical staff in addition to requiring the involvement of the administration in implementing the nanomaterials risk management.

The French ANSES Control Banding Tool for Nanoparticles [K] approach is considered by Brouwer (2012) to be only a CB tool; its implications are broader because it contains some elements of a management system (planning, implementation and operation, checking and corrective action and management review). Therefore, this approach can be characterised as methodological. Their authors notably indicate the need for specialised staff to implement this method.

3.3 Pragmatic approaches of the CB type

The CB method was developed as a pragmatic tool for performing risk management in situations involving potentially hazardous chemicals with almost no available toxicity data (Brouwer, 2012). In this type of approach, the risk levels are determined as a function of the exposure and danger. The focused situation is classified into a particular group, and specific actions to control risks are suggested for each group. Therefore, CB is a fully qualitative method in which risk is assessed instead of measured, and it is well-suited to be employed under conditions where there is much uncertainty, as in the case of nanomaterials.

By dispensing with quantitative surveys, which are usually more expensive, the CB approach is suitable for smaller operations, such as those carried out in research laboratories, or in micro and small companies. The CB approach has been used in the pharmaceutical industry, as highlighted by Brouwer (2012), and its use was expanded to the chemical industry in general and, more recently, to new technologies, especially nanotechnology. In most cases, these tools merely indicate the risk level of a given operation and the associated actions needed to mitigate the risks. Thus, as one would expect, these tools must be inserted into a larger framework of actions to produce effective risk management.

One of the first applications of the CB method to nanotechnologies (CB Nanotool) [L] was proposed by Paik et al. (2008), who classified a given operation involving nanomaterials into four risk levels. This classification is based on the integration of a severity score, obtained from certain physic-chemical characteristics of the nanoparticles and their toxicity, and a probability score that takes into account the amount of material used, the frequency and duration of the operations, the number of people involved and the dustiness of the material. Although the score compositions may be based on quantitative information, the tool can be used without performing any type of measurement.

The European Union guide [M] (2012), "Working Safely with Engineered Nanomaterials and Nanoproducts—A Guide for Employers and Employees", presents a methodology that, despite having a qualitative base, demonstrates features that enable the quantitative assessments of environmental work, including indications of the exposure limits. The activities are classified into three levels of control based on the integration of "exposure categories" and a "hazard category". The exposure category is determined by evaluating the possibility of nanoparticle emission, while the hazard category is defined by some of the nanomaterial characteristics, such as the bio persistence and shape.

The Stoffenmanager Nano 1.0 [N] tool presented by Duuren-Stuurman et al. (2012) is an application that is available on the Internet and, according to the authors, does not require specific knowledge concerning OH for its application. Stoffenmanager Nano 1.0 is an adaptation to nanoparticles of a generic system with the same name. In situations where no information on the nanoparticles is available, this system classifies the danger by data about the macro substance, thereby classifying the nanoparticles into danger ranges. The exposure, in turn, is defined by 14 multipliers, which when combined allow for the determination of the exposure range. These multipliers involve factors including the amount of material, dustiness, forms of manipulation, process types, collective protective equipment (CPE) and personal protective equipment (PPE). The interpolation between the ranges of danger and exposure allows situations to be classified into three groups of risk prioritisation.

Precautionary matrix [O] (Höck et al., 2011) is a tool for generating a score that determines two major classes of risk. The main parameters for defining the score are the relevance of the nanomaterial (based on the size and characteristics of the particle), the specific conditions of use and the potential effects of human exposure. The use of these parameters indicates the need for skilled staff to implement the tool. Moreover, a point that deserves mention is the use of the concept of half-life with respect to the stability of nanomaterials. Conversely, the NanoSafer (2011) [P] is a proposal focusing on nanoparticles dispersed in air, being based in corollary on the dustiness of the nanomaterials. This proposal also indicates the requirement for measurements in the workplace, including quantitative data. In contrast, the GoodNanoGuide (2009) [Q] is a tool with a greatly simplified focus, allowing its application at three progressive levels: basic, intermediate and advanced.

4 CONCLUSIONS

This set of 17 analysed proposals do not converge on a consensus approach, even though they each share the same theoretical basis, as detailed in the report by ANSES (2010). In general, all of the proposals refer to the process of hazard identification, exposure assessment and risk definition in addition to the elimination, substitution or control of risks through technical and organisational measures.

The solubility, lability, dustiness and shape of the nanoparticles are more critical factors than the amount of material involved, indicating that for nanomaterials, other metrics should be adopted. Although there is no consensus on which metrics should be used to characterise the risks associated with the nanomaterials, the adoption of a precautionary principle and the CB approach stand out among the documents examined.

There is still much to be undertaken to obtain a standard to define and characterise the risks arising from the manufacture and use of nanomaterials, starting with nanotoxicology studies and including social discussions of the impacts of these new technologies in society and on the work environment.

Thus, the inclusion of multiple stakeholders (industry, government, insurance, trade, academia, standardisation organisations, media, consumers and the general public) is cited by many as essential.

There is an urgent need to achieve this consensus not only for the occupational health safety but also for the legal and economic safety that is essential for progress and technological advancement.

REFERENCES

Amoabediny, Gh., Naderi, A., Malakootikhah, J., Koohi, MK., Mortazavi, SA., Naderi, M. and Rashedi, H. Guidelines for Safe Handling, Use and Disposal of Nanoparticles. 2008. *International Conference on safe production and use of nanomaterials* – Nanosafe 2008. *Journal of Physics: Conference Series* 170 (2009).

Andrade, L.R.B and Amaral, F.G. Methodological proposal for occupational health and safety actions in research laboratories with nanotechnologies activities. 2012. *Work: A Journal of Prevention, Assessment and Rehabilitation*, v.41, supplement 1/2012: 3174–3180.

Anton, J.M.N. La Nanotoxicología y la Evaluación del riesgo de las nanopartículas artificiales y la Salud 2009 [Nanotoxicology and Risk Assessment of the NA and Health]. *Seguridad y Medio Ambiente*; 114: 6–16.

British Standards. PD6699–2:2007. Part 2: Guide to safe handling and disposal of manufactured nanomaterials.

Brouwer, D.H. 2012. Control Banding Approaches for Nanomaterials. *British Occupational Hygiene Society. Ann. Occup. Hyg.*, 56, n. 5: 506–14.

DuPont and Environmental Defense. 2007. Nano Risk Framework.

Duuren-Stuurman, B.V.; Vink, S.R., Verbist, K.J.M., Heussen, H.G.A., Brouwer, D.H., Kroese, D.E.D., Niftrik, M.F.J.V., Tielemans, E. and Fransman, W. 2012. Stoffenmanager Nano Version 1.0. *British Occupational Hygiene Society. Ann. Occup. Hyg.*, vol. 56, no. 5: 525–41.

European Union. 2012. Working Safely with Engineered Nanomaterials and Nanoproducts - A Guide for Employers and Employees version 4.2. Retrieved Sep. 4, 2012 from http://www.rpaltd.co.uk/documents/ J771_NanoWorkSafetyGuidancev4.2_publ.pdf.

French agency for food, environmental and occupational health & safety (ANSES). 2010. Development of a Specific Control Banding Tool for Nanomaterials - Report. Retrieved September 5, 2012, from http:// www.anses.fr/Documents/AP2008sa0407RaEN.pdf.

Germany (Bundesanstalt für Arbeitsschutz und Arbeitsmedizin/BAµA). 2007. Guidance for Handling and Use of Nanomaterials at the Workplace.

GoodNanoGuide. 2009. Retrieved September 25, 2012 from www.goodnanoguide.org.

Höck J., Epprecht T., Furrer E., Hofmann H., Höhner K., Krug H., Lorenz C., Limbach L., Gehr P., Nowack B., Riediker M., Schirmer K., Schmid B., Som C., Stark W., Studer C., Ulrich A., von Götz N., Weber A., Wengert S. and Wick P. 2011. Guidelines on the Precautionary Matrix for Synthetic Nanomaterials. *Federal Office of Public Health and Federal Office for the Environment*, Berne (Swiss). Version 2.1. Retrieved September 5, 2012 from http://www.bag.admin.ch/na notechnologie/12171/12174/12175/index.html?lang=e n&download=NHzLpZeg7t, lnp6I0NTU042 l2Z6 ln lad1IZn4Z2qZpnO2Yuq2Z6gpJCHd3×9g2ym162ep Ybg2c_JjKbNoKSn6 A.

International Center for Technology Assessment (ICTA). 2007. Principles for the Oversight of Nanotechnologies and Nanomaterials. Retrieved October 2, 2012, from http://www.cleanproduction.org/library/Principles_Nano_finaldesign.pdf/.

NanoSafer. 2011. Retrieved September 5, 2012, from http://nanosafer.i-bar.dk/.

Ostiguy, C., Roberge, B., Ménard, L., and Endo, C. 2009. Best practices guide to synthetic nanoparticle risk management. *Institut de recherche Robert-Sauvé en santé et en sécurité du travail (IRSST),* Québec, Canada.

Paik, S.Y., Zalk, D.M. and Swuste, P. 2008. Application of a Pilot Control Banding Tool for Risk Level Assessment and Control of Nanoparticle Exposures. *British Occupational Hygiene Society. Ann. Occup. Hyg.,* vol. 52, no. 6: 419–28.

Tsuji, J.S., Maynard, A.D., Howard, P.C., James, J.T., Lam, C., Warheit, D.B., and Santamariak, A.B. 2006. Forum Series Research Strategies for Safety Evaluation of Nanomaterials, Part IV: Risk Assessment of Nanoparticles. *Toxicological sciences;* 89(1): 42–50.

Tyshenco, M.G. and Krewski, D.A. 2008. A risk management framework for the regulation of nanomaterials. *International Journal Nanotechnology,* vol 5, issue 1: 143–60.

US/National Institute for Occupational Safety and Health (NIOSH). 2012. General Safe Practices for Working with Engineered Nanomaterials in Research Laboratories. Retrieved September 24, 2012 from http://www.cdc.gov/niosh/docs/2012–147/pdfs/2012–147.pdf.

Occupational Safety and Hygiene – Arezes et al. (eds)
© *2013 Taylor & Francis Group, London, ISBN 978-1-138-00047-6*

Comparative study of risk evaluation methods for the development of worker related musculoskeletal disorders

G. Pereira & P.M. Arezes
Ergonomics Laboratory, School of Engineering, University of Minho, Guimarães, Portugal

ABSTRACT: The prevention of Work-Related Musculoskeletal Disorders (WRMSDs) requires an adequate risk assessment. A methodology that includes a risk assessment by body zone evaluation, known as RAMBoS, was previously proposed. Thus, it is pertinent to develop a comparison study between the evaluation by the proposed method and the workers answers obtained through the application of an appropriate questionnaire. The methodological process was based on a literature review, followed by the risk assessment of some workplaces using a previous questionnaire and the proposed method. Conclusion is that RAMBoS' approach to the workers' opinion has a high value of similarity, higher than 50%, between these responses and the level of risk, suggesting a convergence of results. By classifying several body areas with higher levels of risk, RAMBoS proves to be more sensitive than the symptomatology questionnaire and allows an immediate perception of body areas with higher risk, as well as to be more accurate in the definition of the associated levels of risk.

1 INTRODUCTION

When analysing the WRMSDs development, the continuous exposure to risk factors initially leads to intermittent symptoms that become gradually constant (Ranney, 2000). The correct and detailed knowledge of the workplaces associated with the risk assessment, allows the development of procedures to reduce the risk of developing this type of injury (Kuorinka & Forcier, 1995). Traditional methods of risk assessment report, usually, a general classification of the workplace (Uva, 2006). The specific body area in which fatigue is developed can be different when discussing different workplaces, or even when evaluating the same workplace by different methods. Therefore, it is pertinent to evaluate a workplace focusing on the most affected areas of the body. A Risk Assessment Methodology by Body Section (RAMBoS) is an observational method that is based on the range of movements and traditional methodologies for risk assessment of WRMSDs, in order to classify this risk specifically for each body zone, referencing it in a body chart (Carrelhas, 2010). This methodology considers a way to assess all movements, postures and risk factors present in a workplace with a weighting of additional described factors, such as repeatability and strength. Apart from that, it is also possible that there are other risk factors, such as vibration, skin compression by objects or tools or work rate (Carrelhas, 2010).

The evaluation performed must also consider the opinion of the workers involved. With this objective, it was applied an adapted version by Serranheira et al., 2003 of the Musculoskeletal Nordic Questionnaire (MSNQ), which was previously validate in its Portuguese version (Mesquita et al., 2010). The used version contains 36 questions, divided into two distinct parts. The first part contains nine questions of descriptive character. With these questions we intend to obtain information about the characteristics of the worker, such as age, gender, height, weight, working time per week and number of hours worked at the workplace in question. Apart from these, it also seeks information about the job, work shift and the materials and equipment used. The second part consists of 27 questions designed to assess the discomfort, pain and self-reported discomfort by workers in the past 12 months and if in this time period, there was absence from work. Each question is accompanied by a body chart with demarcation of the area of the body under evaluation. The self-assessment by the worker is based on a scale that assesses the intensity of discomfort. Considering both methods, the workplaces are classified into four risk levels, summarized in Table 1.

Considering the existing risk assessment methods it is important to perform this comparison between the evaluation of workplaces according to RAMBoS and workers responses obtained through MSNQ. This study aims, through this comparison, to check whether the results obtained in RAMBoS will meet the self-reported symptoms by workers. To accomplish our objective we

Table 1. Definition of risk assessment levels of used WRMSDs methods.

	Risk level			
	Green (1)	Yellow (2)	Orange (3)	Red (4)
MSNQ	Slight discomfort	Moderate discomfort	Intense discomfort	Unbearable discomfort
RAMBoS	Low risk of WRMSDs Acceptable workplace	Medium risk of WRMSDs Workplace needs investigation	High risk of WRMSDs Workplace needs quickly investigation	Very high risk of WRMSDs Workplace needs urgent investigation

intend to compare the results obtained in a perspective of convergence or divergence of results.

2 RESEARCH METHODOLOGY

In order to meet the objectives set in the first phase, it was carried out a detailed literature review on the topic, followed by a period of data collection. In this, we identified the workplaces under evaluation—Running Stitch, Cut-sewing, Inspection, Folding with no card (Folding WNC) and Folding with card (Folding WC), located in the area of confection (fig. 1) in a textile industry. The workplaces evaluated are divided into two samples, with five workplaces in each sample, constituted by the same workplaces, in two different shifts, in order to assess the workplace independently of the workers. Due to the absence of records of absenteeism rates specifically caused by WRMSDs, workplaces were selected considering the existence of symptomatic complaints by the workers.

In the third phase, we proceeded to image collection through photography (Canon Digital Ixus 50) and video (Sony Handycam DCR-SX30), in order to characterize the workplaces. In this phase, some characteristics of workplaces were recorded such as physical conditions, frequency of motion, strength, posture, unnecessary movements, static load and applied the evaluation methodology. The evaluation of the upper limb includes analysis of the shoulder, elbow and wrist joints follow by the score of neck, trunk and lower limbs postures, specifically the joints of the hip, knee and foot. After finding these values, the score from the additional factors is added on, according to the description of the methodology described in Carrelhas (2010). Also at this stage, and in order to consider the symptoms reported by workers themselves, it was applied an adapted version by Serranheira et al., (2003) of the MSNQ valid for the Portuguese population by Mesquita et al., (2010). In the last stage it was conducted a comparison between risk levels of the different workplaces in order to assess similarities / differences between RAMBoS and MSNQ.

Figure 1. Images of the five evaluated workplaces: A-Inspection; B-Cut-sewing; C-Running Stitch; D-Folding WC; E-Folding WNC.

RAMBoS was applied to both samples after the selection of the technical action of each workplace and this is what occupies the largest cycle time or providing more repetitive postures.

The duration of the work shift is on average 8 hours with a 30 minutes break for rest by work shift. In all the workplaces assessed there is a static posture or it is maintained throughout the working day, as well as a predominant use of upper limbs. In sewing machines, such as Running Stitch and Cut-sewing, the legs are also used to trigger the sewing pedal in a sitting position, not always in the proper position to the chair and thus keeping the column without support.

In the case of Inspection and Folding workplaces, the constant working position is in orthostatic position without support. At inspection there is hand predominance and an increased need for details verification—inspection action. In the Folding workplaces there is the need, at various stages of the work cycle and due to the size of the pieces, to use the shoulders and elbows in extreme positions.

3 RESULTS PRESENTATION AND DISCUSSION

In both samples, all elements are female. The characteristics of the samples under study are summarized in Table 2, such as age (years), height (in cm) and weight (in kilograms). The average age is very similar for both samples, varying in relation to height and weight.

As for the painful symptoms, it was possible to identify where it occurs and quantify the pain felt, according to the scale proposed in RAMBoS and MSNQ, for both samples. Table 3 shows the results obtained in a body chart summarizing risky body areas of WRMSDs through MSNQ and RAMBoS.

Considering the results obtained in MSNQ for Sample 1, the workplaces referring more areas of intense pain are Inspection, Cut-sewing and Folding WNC, indicating different areas corresponding to intense discomfort. Considering the results obtained in RAMBoS, it appears that the workplaces with the highest number of areas with high risk of WRMSDs are Inspection, Cut-sewing and Folding WC. Data analyses show that RAMBoS

considers more body areas with risk of injury than the areas of pain reported by workers.

Analysing the results in Sample 2, the data obtained in MSNQ reveal areas of more intense pain in Folding WNC and Inspection workplaces, like Sample 1. The result after applying RAMBoS shows that the workplaces with more areas with high risk of WRMSDs are the posts of Inspection and Folding WNC. Inspection and Folding WNC are the workplaces with more symptoms reported by workers and with greater risk of injury in RAMBoS assessment. This may be due to the characteristics of these workplaces. Inspection is characterized by a standing position, with a scissor in one hand, checking visually all stitches. Folding WNC consists of opening the piece and performs a series of maneuvers to fold the final product. In both workplaces, the operator is in standing position throughout the workday in which the upper limbs are required at the maximum range of movement, and without any support.

Table 4 presents the body chart corresponding to the differences found between the responses obtained in MSNQ and risk levels encountered

Table 2. Characterization of the samples analysed.

Average	Age (years)	Height (cm)	Weight (Kg)
Sample 1	49 ± 3	160 ± 3	74 ± 7
Sample 2	46 ± 4	156 ± 4	62 ± 10

Table 3. Comparison of the results obtained by MSNQ and RAMBoS in both samples.

WRMSDs Risk   ◯ No Risk   ☆ Low Risk (1)   ▢ Medium Risk (2)   ▲ High Risk (3)   ✚ Very High Risk (4)

Table 4. Comparison of results obtained in MSNQ and risk assessment according to RAMBoS.

Workplace	Inspection	Running stitch	Cut-sewing	Folding WC	Folding WNC
Sample 1 Differences					
% Similarity	63	88	69	69	88
Sample 2 Differences					
% Similarity	69	75	50	75	81

White: no difference / Light gray: difference of 1 level / Dark gray: difference of 2 levels / Black: difference of 3 levels

by RAMBoS. The differences are marked by gray, the lack of difference is indicated in white and the difference of three levels in black colour. The intermediate difference states are shown with light gray—for 1 level difference—and dark gray—for 2 levels of difference. Both samples show a percentage of similar body areas, ie how many zones were considered similar—no difference (white) and 1 level of difference (light gray) - in a total of 16 body areas analyzed for each workplace.

Regarding Sample 1, the workplaces with the highest percentage of similarity were Running Stitch and Folding WNC with 88%, while the workplace of Inspection presents the lowest value with 63%. For Sample 2, the workplace with the highest percentage of similarity is the Folding WNC with 81%, while the lowest with 50% is Cut-sewing. This is the lowest value for both samples, representing half of the body areas assessed as similar, being considered a quite remarkable similarity of results.

Each worker feels special difficulties in his workplace, often proposing changing's to it towards greater adaptability to its needs. To evaluate and consider the opinion of the workers, the MSNQ allowed to get results regarding pain symptoms related to their workplaces.

4 CONCLUSIONS

The prevention of WRMSDs in a workplace requires the study, description and analysis of the various risks. Each worker feels specific difficulties often proposing changes and / or suggestions for greater adaptability to their own needs. In order to evaluate and consider the opinion of each worker,

the MSNQ used in this study yielded results related to self-reported pain symptoms by workers. Considering the objective of comparing the results obtained in RAMBoS with the symptoms reported by workers at MSNQ, it was concluded that the RAMBoS approaches to the workers opinion with a high value of similarity between the degree of discomfort reported by workers and the level risk obtained by RAMBoS, for most body areas analyzed. All similarity values found are situated above 50%, which indicates a convergence of the results of RAMBoS with the response of workers. When compared both methods, the RAMBoS shown to be more sensitive than the MSNQ, since various body zones are ranked with higher levels of risk. The highest levels of difference found are due to a greater protection of workers from the RAMBoS than the actual symptoms reported by them. This can highlight the omission of symptoms experienced by workers in the face of a questionnaire and/or the thoroughness of the evaluation of workplaces by RAMBoS.

The high percentage of similarity between both analysed methodologies reveals to be highly beneficial for recruitment and collaboration of workers in improvement actions and active participation in prevention programs. Despite the similarities found between the methods, there are cases of significant differences, which could be explained more in detail through an increased number of workplaces analyzed.

REFERENCES

Carrelhas, V. 2010. Desenvolvimento de uma metodologia para avaliação do risco de LMERT por zona corporal. Tese de Mestrado. Guimarães: Universidade do Minho.

Kuorinka, I. & Forcier, L. 1995. Work-related musculoskeletal disorders (WMSDs). *A reference book for prevention*. London: Taylor & Francis.

Mesquita C., Ribeiro C. & Moreira P. 2010. Portuguese version of the standardized Nordic musculoskeletal questionnaire: cross cultural and reliability. *J Public Health* 18: 461–466.

Ranney, D. 2000. Distúrbios Osteomusculares crónicos relacionados com o trabalho. São Paulo: Editora Roca, Lda.

Serranheira, F., Pereira, M., Santos, C.S. & Cabrita, M. 2003. Auto-Referência de sintomas de lesões músculo-esqueléticas ligadas ao trabalho (LMELT) numa grande empresa em Portugal. *Saúde Ocupacional* (julho/dezembro). 21(2).

Uva, A. 2006. Diagnóstico e Gestão do Risco em Saúde Ocupacional: algumas vulnerabilidades. *Revista Portuguesa de Saúde Pública*. 6: 5–12.

Occupational Safety and Hygiene – Arezes et al. (eds)
© 2013 Taylor & Francis Group, London, ISBN 978-1-138-00047-6

Operational safety: Development of electronic system for dynamic balance evaluation of farm tractors

M.D.S. Luciano, J.E.G. Santos, A.G. Santos Filho, J.A. Cagnon & A.L. Andreoli
Universidade Estadual Paulista "Júlio de Mesquita Filho", Bauru, São Paulo, Brazil

ABSTRACT: The present study aimed at the development and evaluation of a low cost electronic device in order to provide safety for farm tractor users. The major accident occurrence in agricultural surroundings is from farm tractor side bending. Therefore, this sensor was designed to detect and alert about it. The results were satisfying.

1 INTRODUCTION

New farm tractors are acquired in order to look after its operators, offering higher comfort and safety. Due to the agricultural development in Brazil, new farm entrepreneurs are in search of new technologies in order to attend the productive demand. To improve the mechanical fleet prior in performance, comfort and safety of agricultural equipments and machinery, joined with appreciation of human capital, for improvement of human-machine relation. High-tech farm tractors are responsible for this change once that they are engaged in production enhance while adopt new electronic systems such as GPS (Global Positioning System) and several other embedded sensors (Anfavea 2010).

Despite these new high tech farm tractors there is not a sensor which is able to warn the operator about the risk they are submitted to on uneven ground. This can be corrected with the installation of an anti-tipping sensor which warns the farm tractor operators. These sensors will improve the machine reliability and will provide better safety conditions on task accomplishments over grounds with a high slope under operational risk, which means, out of dynamic balance and out of limit angle of balance.

Accidents with agricultural machinery happen major in rural surroundings. Although, there are few information and statistics about these accidents. There are several potential risk factors related to operators as unpreparedness, lack of attention on the activity, high speeds during opera-tion and the use of alcoholic beverages. There are also potential risks about machines themselves. For instance, some of them do not meet the ergonomic principles and/or are non-standard safety. Should be also highlighted the work in unsanitary conditions (heat/cold, sun exposure, dust, noise, and physical effort besides mechanical vibrations) as described by Santos (2012).

Brazilian farm tractors fleet is not new. Farm tractors still are the most used power source for the raise of agricultural activities and productivity. Also according to Anfavea (2010) of total number of farm tractors in Brazil, 59% of them is 10 or more years old and 37% have more than 20 years of use. Small family farmers use these machines for the accomplishment of difficulty tasks. Even though, these machines are old because they are cheaper when purchased. It is possible to determine through simple tests the center of gravity and its limit angles for tipping. Therefore the use of a inclination sensor device will provide the operator verifying the farm tractor potential risk of tipping during the operation.

Inclinometers (or tiltsensors) measure the inclination of an object (or a body part) and there are several ways of implementing it on a machine. These measure equipments can, for instance, be used in vehicles for measuring the acceleration and direction changes during a path as explained by Carmo et al. (2010). While operating farm tractors, the tipping or overturning risks can result in severe accidents with injuries for operators which, in most cases can results in death. For containing risks like these, there are low cost and low energy consumption sensors that can be implemented on these machines. With these agree Novais (2010), Leite et al. (2010) and Leite et al. (2011).

2 OBJECTIVE

The present study aimed at the development and evaluation of an inclinometer as a safety sensor device for farm tractors. Inasmuch as the high occurrences of accidents related with side bending of farm tractors. Also, there are not low cost equipments to warn the operator of the imminence of tipping of the farm tractor out of it dynamic balance.

3 MATERIALS AND METHODS

3.1 Materials

To construct this safety device, an accelerometer: ±1.5 g–6 g three axis low-g micromachined, LED lights (red, yellow and green), a controller circuit, and a 12 V battery 70 AH were used.

For the device evaluation was used the following farm tractors: Farm Tractors Massey Fergusson 290 PAVT, Ford 6610 and Agrale 4200 which belong to Unesp (Sao Paulo State University) Bauru. The evaluation happened on the Agricultural Machinery Laboratory belonging to Mechanical Engineering Department of the same university.

Also, for this evaluation was related to limit operational slope, was built a track in compact soil for simulate the inclination angles as described by Leite et al. (2007) and NBR 12567 (1992).

3.2 Methods

The side bending was determined considering that the entire weight of the farm tractor is exerting force on the rear wheels only on one side because it is 4 × 2 farm tractors, as recommended by Promersberger et al. (1962) and Leite (2007). In addition, the center of gravity (C.G.) of farm tractors was defined according to NBR 12567 (1992).

The inclinometer developed is presented with its circuits distributed in a board for tests. After field tests with machines and needed corrections made it was enclosed in order to allow a comfortable and appropriate use in machines.

The device was developed along with the Electrical Engineering Department, as can be seeing by Figures 1 and 2. On Figure 2 it is possible to see the lateral inclination variability through the warning lights which allows the adoption of an operational safety zone in function of operational slope.

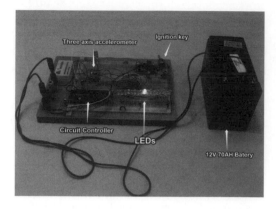

Figure 1. Inclinometer before being enclosed for field tests.

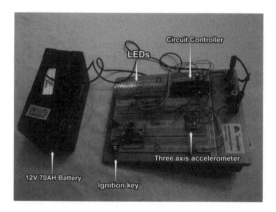

Figure 2. Top view of the inclinometer with warning lights on imminence of tipping.

4 RESULTS AND DISCUSSION

An inclinometer measures the instantaneous angle of farm tractors. This angle can be compared to limit angle of the farm tractor in order to not happening the side bending. From this, it is possible to manipulate the data acquired to control the farm tractor in case of excessive inclination, or even alert the operator that the machine can tip in case he/she submits it to higher inclination. However, the device configuration was the same for every farm tractor used because the limit angles were different according to C.G. of each machine.

4.1 Printed circuit board

The printed circuit board can be visualized in Figures 3 and 4. As the university has no specifically laboratory for circuit printing, this service was made by an outside company.

Figure 3.   Printed circuit board, top view.

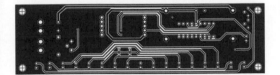

Figure 4.   Printed circuit board, bottom view.

The alert generated to the machine operator can be a sound advice or a luminous advice that priors a simple understanding. It was chosen these two types of warning.

It is possible to use three different types of color lights in order to inform the inclination of the farm tractor: green, yellow and red. When the inclination angle is safe related to the limit angle, the operator observes that there is no problem because green lights are shown. As long as the angle increases, it is possible to see besides the green lights, the yellow lights indicating that the limit angle is near, but not that much to point that tipping occurs. Finally, when the limit angle is near and the tipping risk is imminent, it indicates the high risk with the red lights.

5   CONCLUSIONS

On the anti-tipping device, the inclinometers will measure the inclination of farm tractors, while the lights will indicate how risk it is for the tractor suffers a tipping. Besides, the buzzer will reinforce that the machine is under risk or even advise the operator who is not looking at the display. As a result of this study, was obtained the prototype in Figure 5, with the electronic circuit enclosed, the warning lights and the sensor system.

The device was tested under methodology conditions of this study in a track as shown in Figure 6. Finally, the device presented satisfaction results related to operational safety during tests.

In addition, is being discussed on the research group "Design, Safety and Ergonomics applied in Agricultural Machinery" active ways to prevent accidents independently of operator. Also, ways to record risk situations submissions that an operator is submits him/her.

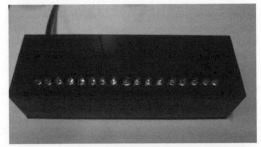

Figure 5.   Device prototype, enclosed and ready for field tests.

Figure 6.   Device ready and placed in a farm tractor during tests.

ACKNOWLEDGMENTS

CNPq—National Counsel of Technological and Scientific Development of Brazilian Government, for financial supporting and funding.

REFERENCES

Anfavea. 2010. *Anuário da Indústria Automobilística Brasileira*. São Paulo-SP, Brazil: ANFAVEA.
ABNT. Associação Brasileira de Normas Técnicas. 1992. Trator Agrícola—Determinação do centro de gravidade. CB-05—Comitê Brasileiro de Automóveis, Caminhões, Tratores, Veículos Similares e Autopeças. Rio de Janeiro-RJ, Brazil (NBR 12567).
Carmo, K.C., Santos, J.E.G., Gutierrez, A., Santos Filho, A.G., 2010. Desenvolvimento de sistema de segurança (inclinômetro) para minimizar acidentes com operadores de tratores agrícolas. *Anais do Congresso de Iniciação Científica da Unesp* XII ISSN 2178-860: p. 4537–4540.
Leite, F. 2007. Thesis: *Construção de um Inclinômetro para Avaliar o Efeito da Declividade Lateral no Desempenho*

de Tratores Agrícolas. Botucatu-SP, Brazil: Faculdade de Ciências Agronômicas, Universidade Estadual Paulista "Júlio de Mesquita Filho"

Leite, F., Santos, J.E.G., Lanças, K.P. 2010. Construção e calibração de um inclinômetro para uso em tratores agrícolas como instrumento de segurança. *Energia na agricultura*, v.25, p.40–52.

Leite, F., Santos, J.E.G., Lanças, K.P. et al. 2011. Evaluation of tractive performance of four agricultural tractors in laterally inclined terrain. *Engenharia Agrícola*, v.31, p.923–929.

Novais, I.R.W., Santos, J.E.G., Santos Filho, A.G. 2011. Desenvolvimento e avaliação de sensor de segurança como sistema de proteção visando à segurança operacional dos operadores de tratores agrícolas: estudo de caso. *Anais da 1ª fase do Congresso de Iniciação Científica da Unesp XIII* ISSN 2178–860.

Promersberger, W.J., Bishop, F.E. 1962. *Modern farm power*. New Jersey, USA: Prentice Hall.

Santos, J.E.G. 2012. *Curso de especialização em engenharia de segurança do trabalho, Módulo: Segurança Agropecuária.* Bauru-SP, Brazil: Faculdade de Engenharia Mecânica, Universidade Estadual Paulista "Júlio de Mesquita Filho".

Occupational Safety and Hygiene – Arezes et al. (eds)
© *2013 Taylor & Francis Group, London, ISBN 978-1-138-00047-6*

Analysis and risk assessment in a polymer manufacturing industry

Cláudia de Matos & Isabel L. Nunes

Faculdade de Ciências e Tecnologia, Universidade Nova de Lisboa, Caparica, Portugal

ABSTRACT: An analysis and risk assessment for work accidents and occupational diseases was performed in a Polymer Manufacturing Industry. The Company is a small company with a total of 23 workers. In order to perform such study the authors established an integrated methodology, where a preliminary analysis based on four activities (direct observation, interviews with workers and manager, detailed analysis of company work accidents reports and safety data sheets), followed by application of qualitative and quantitative methods were used to estimate the risk level of exposure to styrene, exposure to occupational noise, manual materials handling and contact with machinery. The combination of assessment methods contributed to improve the occupational safety and health conditions of the company; and thus the working conditions.

1 INTRODUCTION

Within the context of their general obligations, employers have to take the necessary measures for the safety and health protection of their workers, including prevention of occupational risks. Therefore, workers should be protected from occupational risks they could be exposed to. This can be achieved through a risk management process, involving risk analysis, risk assessment and risk control practices (Figure 1) (Nunes, 2010).

Risk assessment is a legal obligation in Europe, but it's also a good practice that contributes to keep companies competitive and effective. Risk assessment is a dynamic process that allows companies and organizations to put in place a proactive policy for managing occupational risks. Therefore, risk assessment constitutes the basis for implementation of appropriate preventive measures and, according to the Framework Directive (EU, 1989); it must be the starting point of any Occupational Safety and Health Management system (Nunes, 2011).

Work accidents and occupational diseases are a serious problem in the manufacturing industry of polymers (GEP, 2009), thus the aim of this paper is to present a case study regarding prevention of occupational risks performed in this type of industries.

2 MATERIALS AND METHODS

In order to perform an analysis and risk assessment for work accidents and occupational diseases in a

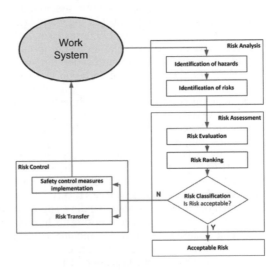

Figure 1. Risk management (Nunes, 2010).

manufacturing industry of polymers, the authors established an integrated methodology (Figure 2).

A preliminary analysis, based on direct observation, interviews, study of work accidents reports and safety data sheets regarding the chemical substances used in the company, was performed to identify the existing hazards and the occupational risks workers could be exposed to. The number of exposed workers was 7. To perform the assessment of the identified risks the appropriate methodologies were selected, as well as the applicable legislation.

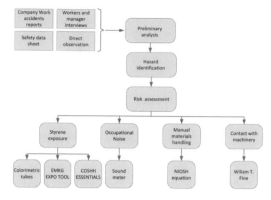

Figure 2.   Study methodology.

2.1   *Styrene exposure*

Regarding chemical contaminants, considering that styrene was the product used more frequently and in larger amounts by the company, it was assessed the level of exposure to this product in the workplace. The Portuguese standard (NP 1796) was taken into account. Styrene occupational exposure occurs mainly via inhalation. The quantification (approximate value) of the styrene concentration was performed using colorimetric tubes. A qualitative assessment was performed using two different methods. The COSHH ESSENTIALS, developed by the Health and Safety Executive (HSE, 2002) to help companies comply with the Control of Substances Hazardous to Health Regulations (COSHH), and the EMKG-Expo-Tool developed by the German Federal Institute for Occupational Safety and Health (BAuA), which is available online (BAuA, 2008). Both methods use the control banding approach in order to minimize worker exposure to hazardous chemicals and to assist small companies control occupational hazardous exposures, since it provides an easy-to-understand practical approach.

2.1.1   *Colorimetric tubes*
The colorimetric tubes are directly measuring (estimative) devices, where the collection and testing are done by the equipment used. They should be positioned in the breathing zone of the worker. These tubes consist of a glass vessel containing a chemical mixture which, when in contact with the analyzed substance, undergoes a reaction and changes color.

Most tubes contain a graduated scale and the length which reaches the coloring is an indicator of the concentration of the substance in the work atmosphere. As such, there should be a continuous observation during the measurement and evaluation immediately after the sampling period. If the tubes are not graduated, the presence of the analyzed substance is detected by comparing the obtained color

with the color of a tube that suffered no change (before and after effect) (Drager, 2006).

2.1.2   *COSHH ESSENTIALS method*
The COSHH ESSENTIALS method is based mainly on the analysis of the data available in safety data sheets (SDS) of chemicals and provides advice on how to control the use of chemicals in various tasks, such as their mixing and drying.

The methodology for tasks analysis involves five steps and several questions about the tasks must be answered, such as: chemicals compounds used, how hazardous are these chemicals, the hazard of which group they belong and how often they are used. As result the tool gives information about the safety measures to be implemented in the analyzed workplace, such as engineering measures, organizational measures and use of personal protective equipment.

2.1.3   *EMKG-Expo-Tool*
The EMKG-Expo-Tool is part of the "Easy-to-use workplace control scheme for hazardous substances" (EMKG "Einfaches Maßnahmenkonzept für Gefahrstoffe"). This is an IT-tool free of charge for a first exposure estimate (Tier 1 assessment) at the workplace. This tool was developed within the context of REACH by BAuA.

The EMKG-Expo-Tool is also based on data from safety data sheets. These data regarding styrene were introduced in a spreadsheet in excel and five steps (definition of volatility bands, scale of use bands, short term exposure, applications on surfaces >1 m^2 and control strategies) were performed until the final result—the risk assessment—is reached.

2.2   *Occupational noise*

Occupational noise was measured using a digital sound level meter, Center 322. In order to perform risk assessment the Portuguese legislation (DL n. 182/2006) was adopted.

2.3   *Manual materials handling*

The revised NIOSH equation developed by the National Institute for Occupational Safety and Health (Waters et al., 1993) and the Portuguese legislation (DL n. 330/1993) were used to evaluate manual materials handling tasks.

The assessment is based on the determination of the Recommended Weight Limit (RWL) through the following expression:

$$RWL = LC \times HM \times VM \times DM \times AM \times FM \times CM$$

where LC-Load Constant (23 Kg) is reduced by the following coefficients: HM-Horizontal Multiplier;

VM-Vertical Multiplier; DM-Distance Multiplier; AM-Asymmetry Multiplier; FM-Frequency Multiplier; and CM-Coupling Multiplier.

Then the LI—Lifting Index, computed as the ratio of Load Weight (L) by the RWL, provides a relative estimation of the risk level associated with a particular manual lifting task.

This empirical method is the most suitable tool to evaluate two-handed lifting tasks, as it does not interfere with the routine of workers and covers the majority of parameters that compose the Manual materials handling. Despite that, this method also presents some limitations such as: is not applicable to spaces where postures are unfavorable, where there are no good mechanical, thermal and visual conditions, and do not include unforeseen circumstances (Waters et al., 1993).

2.4 Contact with machinery

The William T. Fine method (Fine, 1971) was used to evaluate the risk of contact of the worker with equipment/machinery.

This method is frequently used for hazard identification, assessment, prioritization and control of risks associated with the activities and processes in order to determine what can or cannot be tolerated, and proposes the estimation of each risk—Magnitude of Risk (R)—based on three factors: Consequence (Fc), Exposure (Fe) and Probability (Fp). Each of these variables is analyzed using a 6 levels scale.

According to the method, the activities that require control measures to eliminate or reduce the potential risk for accidents can be identified.

For R values below 20 the risk is acceptable, but for values of R equal to or above 400 the risk is not acceptable (serious and imminent) and it is necessary to put into practice control measures with immediate suspension of activity.

3 RESULTS AND DISCUSSION

3.1 Styrene exposure

The results for the exposure to styrene are detailed in the next subsections.

3.1.1 Colorimetric tubes
Seven measurements were taken with time intervals between 15 and 30 minutes. The measurements were made in the workstation where the molds are manufactured and the resins that contain styrene are applied manually.

The color of the no-graduated colorimetric tubes, changed from white to yellow, an indicator for the presence of styrene in the local.

Using the graduate colorimetric tubes, the measured concentration of styrene in the work atmosphere was about 35 ppm, in a workplace located away from windows, and about 15 ppm in a workplace near a window.

The Exposure Limit Value (ELV) for styrene, during an 8h period is 20 ppm. According to a common criteria used, a safe atmosphere is one where the contaminant concentration is less than half of the ELV. In this case in both workplaces the styrene concentration is above the 'safety limit', therefore the implementation of safety measures is required.

3.1.2 COSHH ESSENTIALS method
In the analysis performed using the COSHH ESSENTIALS method, styrene was signed as a substance that can cause skin irritation and inhalation hazard. This tool also suggested control measures, such as engineering measures (e.g., local exhaust ventilation), organizational measures (e.g. restriction of entry to authorized personnel only in the areas where styrene is handled) and use of personal protective equipment (e.g., goggles, gloves and respiratory mask).

3.1.3 EMKG-Expo-Tool
Based on empirical and real data the EMKG-Expo-Tool assessed the exposure to styrene (for 8 hours' working days) as hazardous to health.

3.2 Occupational noise

Several noise measurements were taken along the day, representative of the work cycle. Usually workers are exposed to noise of the working machines (during six hours) and of work preparation and cleaning (during two hours). The calculated daily noise exposure level ($L_{EX,8h}$) for these workers was 90.1 dB(A). This value is above the legal exposure limit value, which is 87 dB(A). Thus, it was deemed necessary to implement control measures. Given the economic conditions and the company's current situation the suggested measures were: (1) a study of occupational noise on the basis of frequency bands, in order to select the adequate personnel hearing protection to be used by the workers, and, (2) organizational measures, such as reducing the exposure time of workers in each job and making smaller and rotating shifts. Maximum instantaneous sound pressure was also measured in each workstation. The values obtained along the day vary between 77.4 dB(C) and 98.7 dB(C). In this case the observed peak noise levels were lower than those defined by legislation.

3.3 Manual materials handling

Using the NIOSH equation, it was possible to verify that the risk involved in the 3 tasks of lifting loads was acceptable, since the computed LI values were less than 1 (0.64, 0.59 and 0.73, respectively). Therefore, it was not considered necessary to intervene or implement control measures for these activities.

3.4 Contact with machinery

According to the William T. Fine method, 40% of the assessed risks can be considered acceptable, i.e. the combination of the 3 factors Fc, Fe and Fp scored less than 20. The other 60% fall in the [70–200[interval; thus, the risk is considered remarkable, requiring urgent corrections such as: placing guard protections in the machines in order to delimit the scope of the hand to the cutting area; training workers on the machine's operation, and the use of personal protective equipment.

4 CONCLUSIONS

The use of an integrated methodology resulted from a combination of a preliminary analysis to identify the most critical hazards (based on direct observation, interviews, study of the company work accidents reports, and safety data sheets) and of a risk assessment methods contributed to improve the occupational safety and health conditions of this small company. This study allowed concluding that 7 workers (one third of the company workforce) were exposed to the following key risk factors: styrene inhalation, occupational noise and contact with dangerous machinery. Several safety measures were proposed in order to improve the working conditions.

Regarding the styrene exposure assessment the combination of the three methods used provided the means for identifying the presence and estimating the concentration of styrene and subsequently to advice on corrective safety measures. Nevertheless, users must be aware of the limitations inherent to each method. For instance, since the colorimetric tubes detect and give an indicative value of the substance concentration, the users must be aware

that this value is not precise; therefore its comparison with ELV is not totally reliable. On the other hand, the limitations of the methods based on the control banding approach relate with the need of professional knowledge and call for some experience to verify if the control measures specified are correctly installed, used and maintained.

REFERENCES

BAuA (2008). EMKG-Expo-Tool for Exposure Assessment. Available online: http://www.reach-clp-helpdesk.de/en/Exposure/Exposure.html (accessed Nov 2012).

DL n. 182/2006, September 6 (2006). Estabelece a protecção dos trabalhadores contra os riscos devido à exposição ao ruído durante o trabalho.

DL n. 330/1993, September 25 (1993). Prescrições mínimas de segurança e saúde na movimentação manual de cargas.

Drager (2006). Manual de Tubos Drager/CMS. Lubeck: Drager Safety AG & Co KGaA.

EU (1989). Directive 89/391/EEC of 12 June 1989 on the introduction of measures to encourage improvements in the safety and health of workers at work (Framework Directive). Official Journal L 183, 29/06/1989.

Fine, W.T. (1971). Mathematical evaluations for controlling hazards, Naval Ordnance Laboratory, USA.

GEP (2009). Acidentes de Trabalho 2009. Estatísticas em Síntese. Ministério do Trabalho e da Solidariedade Social. Gabinete de Estratégia e Planeamento.

HSE (2002). The technical basis for COSHH essentials: Easy steps to control chemicals. Retrieved from: www.coshh-essentials.org.uk/assets/live/CETB.pdf (April 2012).

NP 1796:2007, 26 de Março (2007). Segurança e Saúde do Trabalho. Valores limite de exposição profissional a agentes químicos.

Nunes, I.L. (2010). Risk Analysis for Work Accidents based on a Fuzzy Logics Model, 5th International Conference of Working on Safety- On the road to vision zero? Roros. Norway.

Nunes, I.L. (2011). Occupational safety and health risk assessment methodologies, unpublished working paper.

Waters, T., Putz-Anderson, V., Garg, A. & Fine, L. (1993). Revised NIOSH equation for the design and evaluation of manual lifting tasks. Ergonomics, 36, 749–776.

Other

Occupational Safety and Hygiene – Arezes et al. (eds)
© *2013 Taylor & Francis Group, London, ISBN 978-1-138-00047-6*

Evaluation of organizational psychosocial characteristics' effect on workers health: A systematic review

A. Cardoso, C. Neves, L. Afonso, P. Costa & S. Rosário[1]
Faculty of Engineering, University of Porto, Porto, Portugal

J. Torres da Costa
Faculty of Medicine, University of Porto, Al. Prof. Hernâni Monteiro, Porto, Portugal

ABSTRACT: The present systematic review aims to determine the evidence of the impact of psychosocial characteristics of organizations on workers health. Materials and Methods: The review of Literature was conducted using the systematic search of the database sources PUBMED and GOOGLE, using PRISMA Checklist and Flow Diagram. Results and Discussions: There were few studies concerning cause-effect relationship between psychosocial characteristics of organizations and workers health. Conclusions: Further studies are needed to assess the impact of psychosocial characteristics on workers health.

1 INTRODUCTION

Economic globalization based on free market principles, designed to create a flexible workforce, increase productivity and profitability (Schnall et al., 2009) and enabled by technological innovation, has profoundly changed the structure of the labor market and the nature of work in recent decades. Other manifestations of global trends include the increase of precarious contracts and job insecurity which raises the risk of occupational illnesses [(European Agency for Safety and Health at Work) (EASHW, 2002)] as lead to significant financial costs to organizations as well as to society in terms of both human distress and the impaired economic performance (EASHW, 2012).These changes have been associated with new and emerging types of risks to worker's health and safety referred as psychosocial risks (PSR)(ILO 2010). Stress is probably the most PSR mentioned. According to the European Commission, in 2002, the yearly cost of work-related stress and the related mental problems in the 15-EU of the pre-2004 EU was estimated to be on average between 3% and 4% of gross national product, amounting to €265 billion annually. Nearly one in three of Europe's workers (more than 40 million people) report they are affected by stress at work, which is responsible for millions of lost working days per year (EASHW, 2012). According to EASHW, in 2002, the economic cost of work-related stress in the EU-15 was estimated at about €20 billion, responsible for millions of lost working days per year, and because of it, the psychosocial risks and stress at work were the central theme of the European Week for Safety and Health at Work. Some studies suggest that changes in work conditions may be related to the fact that workers being more exposed to stress, burnout, anxiety, depression and musculoskeletal disorders (Leka & Jain, 2010). Recently, was given a special attention to PSR by the representatives of the various organizations in Europe and worldwide, which are regarded an important area of study (EASHW, 2012). In 2012, the Committee of Senior Labour Inspectors (SLIC) has launched its European Campaign on PSR, which the aim of the project was the development of an inspection toolkit for targeted interventions on occupational health and safety of PSR, in order to improve the quality of available risks evaluation. Psychosocial hazards are defined by SLIC (2012) as those aspects of work design and the organization and management of work, and their social and environmental contexts, which have the potential for causing psychological, social or physical harm. These risks are normally imperceptible when traditional methods of inspection work are used, therefore requires new forms of specific and validated assessment and intervention tools, and the development of specific rules. One of the methods used for the evaluation of PSR is the *Copenhagen Psychosocial Questionnaire* (COP-SOQ) which meets international consensus as to its

---

1. Author presenter.

validity in assessing and understanding the most relevant psychosocial dimensions inherent in the employment context, appearing in an attempt to better understand the influence of these factors on worker health. Some studies suggest a relationship between specific characteristics of organizations and the effects on health of workers (Laszlo et al., 2010; Schnall et al., 2009). There are others who do not support this relationship (Isepen & Langaa, 2012; Lindstrom, 2009). Although political and social community trend for recognize the role of working conditions "linked" to diseases, there is some difficulty in finding studies that establish a specific relationship of cause-effect relationship between the conditions the workplace and the problems that affect the health of workers. The systematic review aims to determine the evidence between psychosocial characteristics of organizations and their impact on workers' health.

2 MATERIALS AND METHODS

The Literature database was extracted by a systematic search by the "metalib exlibris" system of the PUBMED (www.ncbi.nlm.nih.gov/pubmed) and search engine (Google—http://scholar.google. com) using appropriate words—"mental health of workers", "occupational health", "psychosocial work environment", "psychosocial risks at work", for the period 2002 to 2012. A systematic review was carried out based on Statement for Reporting Systematic Reviews (PRISMA, 2009) checklist, flow diagram (Figure 1) and the review protocol was based on the information at http://www. prisma–statement.org/statement.htm.

This method was used to classify and analyze the articles with the consensus of all the authors of this study. Criteria used in this systematic review consisted in published articles in refereed journals, scientific quality, full text available online; related theme and suitable to the intended objectives; being

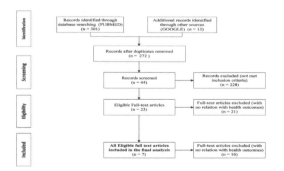

Figure 1. Identification, screening, eligibility and inclusion of data sources for the study.

written in accessible language (English; French; Portuguese; Spanish) and present validated instruments for the study population. Studies were selected according to the eligibility criteria and evaluated taking into consideration the following criteria: type of study (longitudinal studies, cross-sectional and case studies), publication year (2002[2] to 2012); country; psychosocial characteristics organizations mentioned in SLIC (eg. organizational culture and function; role in the organization; career development; decision control; interpersonal relationships at work; home-work interface; work equipment and work environment; organization and content of work tasks; workload/ pace of work; working hours), repercussions on health and worker behavior (eg. stress, burnout, violence, sleeps disorders, musculoskeletal disorders, among others not specified); confounding variables (eg. gender; age; marital status; number of children; educational attainment; numbers of professional experience in sector; total working hours per week); professional activity studied (health care, education, hotel industry; retail; services industry) and assessment methods (eg. questionnaires, interviews, medical evaluation registered, scales, self-report, socio-demographic data).

3 RESULTS AND DISCUSSIONS

The European Campaign of PSR promoted by the SLIC (2012) aimed to give special attention to the PSR at work, as they are seen as significant emerging risks (EASHW, 2007). This campaign has emphasized the reinforcing role of inspection at work as a way for promoting a holistic and preventive action in workplace. The top ten emerging PSR can be regrouped to the following five main topics: new forms of employment contracts and job insecurity; ageing workforce; work intensification; high emotional demands at work, and work interfering with family conflict (WIF) (EASHW, 2009). In recent years, there has been accumulating evidence that poor psychosocial working environment and work-related stress can have impact on worker's physical and mental well-being. Some studies indicate a link between the psychological working environment and impacts on workers health such as heart disease (Nabi et al, 2008), burnout (Maslach et al, 2001), depression (Cox et al., 2005) and musculoskeletal disorders (National Institute for Occupational Safety and Health, 1997). Stress factors are related to a poor performance, increase of absenteeism, a higher percentage of accidents

2. The year 2002 was chosen because it was the European Year's Week for Safety and Health at Work: Psychological Risk Prevention in Workplace.

(SLIC, 2012) and the desire to "leave the organization" (Schaufeli & Bakker, 2004). The main problems in the workplace, identified as major contemporary challenges to health and safety, consists on stress, violence, harassment and intimidation at work (ESENER, 2012). Several studies indicate there is consistent evidence that high strain at work, low control and the perceived imbalance between efforts and rewards are risk factors, and in turn, have negative consequences for physical and mental health (Melchoir et al., 2007). This systematic revision aims to explore how the PSR in organizations may be involved in the gradient of health. In this systematic review 7 eligible articles (Figure 1) indicate a relationship between PSR and the impact on the health of workers, of which only 1 indicates no specific effect to workers' health (László et al., 2009) and the remaining 6 revealed specific consequences to health (Fuß et al., 2008; Kozak et al., 2012; Eatougha et al., 2011; Heponieme et al., 2010; McNamara et al., 2010; Pisljar et al., 2011). These studies showed that low organizational justice and low job control were the cause of sleep disorders, high quantitative work demands have generated high work interfering with family (WIF) and in consequence have generated stress and burnout. Low safety-specific leadership and low job control were associated with increased employee strain which in turn generated musculoskeletal disorder symptoms. Workers with less control over their working hours were more likely to report increased levels of interpersonal conflicts and violence. Potential confounding factors in the 6 studies were assessed, such as socio-demographic (age, sex, marital status, number of children, education level) and characteristics of the work (the total number of hours worked per week, years of professional experience in the sector). The association of these variables with psychosocial factors in the workplace can take different contours and have been important to perform the analysis of the respective(s) influence(s) on the results. The health sector is the most represented professional sector in this field (Fuß et al., 2008; Heponieme et al., 2010; Pisljar et al., 2011). According to the European survey of enterprises on new and emerging risks (EASHW, 2010) the health sector (including social assistance) was considered one of the most problematic areas in relation to the PSR and for this reason, it was recently mentioned in SLIC (2012) as the target group of PSR assessments to meet the needs of all Member States of the EU. The health sector studies (Fuß et al., 2008; Pisljar et al., 2011) indicate that workers reporting high quantitative demands at work associated with work schedule irregularities (eg. unpredictable hours, long and unsocial hours) have been linked to health effect such as stress (Pisljar et al., 2011) and stress/burn-

out at work (Fuß et al., 2008). Some studies demonstrated that job control influences health status (Eatougha et al., 2011; Heponieme et al., 2010) while others clearly contradict the expected results in health (Pisljar et al., 2011). Prior studies have reported that high quantitative demands at work has been associated with stress (Cox, Griffiths & Rial-González, 2000) and that control on job may be a decisive factor in determining the level of health (Sauter et al., 1989). High work control were associated with less sleeping problems (Heponieme et al, 2010); less levels of work-related musculoskeletal disorder (WRMSD) symptoms (Eatougha et al., 2011), less levels of interpersonal conflicts/violence and less WIF (McNamara et al., 2010). In other study, work control is found to have a positive effect on the health of hospital employees in Western Europe, but a negative effect on the health of Eastern European hospital employees (Pisljar et al., 2011). Previous studies found that work control diminishes the negative impact of work stress on health only when employees cope actively with their work stress (De Jonge & Kompier, 1997). In this case, Eastern European employees have few opportunities to exercise active autonomy over their work and schedules. Fuß et al. (2008) refers high levels of WIF (Grzywacz et al., 2006) were correlated to higher rates to personal burnout, behavioral and cognitive stress symptoms and the intention to leave the job. On the contrary, low levels of WIF predicted higher job satisfaction, better self-judged general health status, better work ability, and higher satisfaction with life in general. Cognitive stress symptoms have in previous studies been associated with the psychosocial dimensions of psychological (quantitative) demands, lack of decision authority (influence), lack of meaning of work. Three of the studies have suggested that the intention to leave the job was related to high quantitative demands at work (Pisljar et al., 2011) and WIF (Fuß et al., 2008; Kozak et al., 2011), particularly in health and education sector. Only one study indicates a linkage between stress and poor job performance (Schaufeli & Bakker, 2004). According to our results, factors such as age and sex were positively associated with WIF, burnout and sleeping disorders. Female had significantly higher level of burnout, than their male counterparts (Fuß, et al., 2008; Kozak et al., 2011). Although, Maslach et al. (2001) identify that higher levels of burnout is more reported among young employees (<30 years), the result indicates that young employees between 30–39 years are more likely to be affected by high levels of burnout (Kozak et al., 2011). Consonant with others research findings our studies refers that young employees are more likely to report higher WIF (Fuß et al., 2008; Heponieme et al, 2010) comparing to older employees. Increased levels of interpersonal conflicts/violence, in hotel industry, were found to be associated with lower hour control of temporary or secure employment workers (McNamara et al., 2010). In addition, lower levels of hours control was significantly associate with WIF (Pisarski et al., 2002). Zohar (1994) study has demonstrated that the autonomy and work control are important mediating factors when considering stress in hotel industry. Additionally, Sprigg & Jackson (2006) reported that hotels have characteristics typical of lean service systems, which are significantly correlated with lower control, higher work demands and employee strain. Previous studies support the linkage between the vulnerability of hospitality workers to increased risks of violence (Gleeson, 2001). Literature indicates that low control over work hours leads to greater WIF (Pisarski et al., 2002). Heponieme et al. (2009) identified low justice levels in organisations as directly associated with more sleeping problems (Kivisto et al., 2008) and WIF, and low job control levels with more sleeping problems. Similar findings have been reported organizational justice might be considered as an important aspect in health care sector, given that earlier studies has suggested that WIF is maybe the most influencing factor on job burnout among medical staff (Chaoping et al., 2003). In a study to explore the associations between appraised safety-specific leadership, psychosocial work factors (control over work; role conflict) and musculoskeletal complaints among workers reported that high levels of role conflict, low job control and low safety-specific leadership were associated with increased employee strain, which was related to higher levels of WRMSD symptoms of the wrist/hand, shoulders and lower back, even considering the requirement control made of physical activity performed (Eatougha et al., 2011). The analyses showed that job control is related to reduced strain and subsequently, reduced muscle tension or other physiological reactions that put individuals at a greater risk for developing WRMSD´S. These results are consistent with existing literature (Swanson, 1996) which demonstrated how WRMSD´s are related to stressors and strains. Considerable evidence of effective leadership has been found to contribute to employees' safety and well being (Kelloway et al., 2006) as job control are too an important factor in job design and work organization (Leka & Jain, 2010).In methodological terms the results of 6 studies, 5 were obtained from cross-sectional studies (Fuß et al., 2009; Kozak et al., 2012; Eatougha et al., 2011; Heponieme et al., 2010; McNamara et al., 2010) and only 1 was from a case study (Pisljar et al., 2011). A major limitation recognized by most studies, refers to the fact that the cross-sectional may be an appropriate method at an early stage to establish relations of inference,

but is insufficient to allow establishing cause-effect relationships. One of the recommendations focused on the need for more longitudinal studies undertaken in order to contribute to a better demonstration of organizations psychosocial characteristics and their impact on workers' health. Furthermore, the assessment of the work environment has been dependent solely on quantitative evaluation methods, which some authors have recommended to adopt qualitative methods as well (eg. interviews or open semi structured, observation) to obtain more information allowing a more accurate analyses (Rugulies, 2012).With regard to the theoretical basis, it was found that five of the studies (Fuß et al., 2008; Kozak et al., 2012; Eatougha et al., 2011; McNamara et al., 2010; Pisljar et al., 2011) refer to model demand-control-social support (Karaseck & Theorell, 1990), one of the studies (Eatougha et al., 2011) mentions the Transactional Model Stress (Lazarus, 1991) and finally another one of these studies (McNamara et al., 2010) references the model stress-reward Siegrist (1996). Another important factor in the 6 studies consisted of the use of validated assessment tools. It is important the evaluation methodology of psychosocial factors to be congruent with the theoretical assumptions and the definition of the concept. Considering the evaluation instruments used in this consulted bibliography, two of the most used were COPSOQ and *Job Content Questionnaire* (JCQ) (Karasek, 1985). The instrument COPSOQ allowed assessing more accurately the PSR. The large number of studies that have been conducted in the context of psychosocial factors in organizations, has shown that although different methods can be adopted in research, it is advisable to adopt well-validated instruments, in order to more accurately assess the psychosocial dimensions inherent in the employment context. This review indicates that psychosocial characteristics of organizations may have a detrimental impact on workers health.

4 CONCLUSIONS

The most used instruments of evaluation in the consulted bibliography were the COPSOQ and the JCQ. The COPSOQ is an internationally recognized instrument as to its validity, allowing a more adequately evaluation the PSR. In this regard, it is suggested that this instrument should be systematically used on organizations, in order to proceed an effective PSR assessment on health of workers. It becomes clear the need for further studies in order to accurately confirm the complex relations between psychosocial characteristics in organizations and consequences to individual´s physical, mental and social health.

REFERENCES

Chaoping, L., Kan, S., Zhengxue, L. (2003). Work-family conflict and job burnout of doctors and nurses. *Chinese mental health journal*, 17, 807–809.

Committee of Senior Labour Inspection. (SLIC). (2012). European Campaign on Risk Assessment Psychosocial.

Cox, T., Griffiths, A. & Rial-Gonzalez, E. (2000). *Research on work-related stress*. Luxembourg: Office for Official Publications of the European Communities (OOPEC).

Cox, T. & Griffiths, A. (2005). The nature and measurement of work –related stress: theory and practice. In Wilson, J. & Corlett, N. (Eds.). *Evaluation of Human Work* (3rd ed.). London: CRS Press.

De Jonge, J. & Kompier, M. (1997). A critical examination of the demand-control-support model from a work psychological perspective. *International Journal of Stress Management*, 4:253–258.

Eatougha, E., Way, J. & Chang, C.-H. (2011). Understanding the link between psychosocial work stressors and work-related musculoskeletal complaints. *Applied Ergonomics*, 43:554–563.

European Agency for Safety and Health at Work. (EASHW) (2002). New forms of contractual relationships and the implications for occupational safety and health. Luxembourg: (OOPEC).

EAHSW. (2007). *Expert forecast on emerging psychosocial risks related to occupational safety and health*. Luxembourg: OOPEC.

EAHSW. (2009). *OSH in figures: Stress at work—facts and figures*. Luxembourg: OOPEC.

EAHSW. (2010). *European Survey of Enterprises on New and Emerging Risks: managing safety and health at work*. European Risk Observatory Report.

EASHW. (2012). Drivers and Barriers for psychological risk management: an analysis of the findings of the European Survey of Enterprises on New and Emerging Risks (ESENER) report. Luxembourg: OOPEC.

Fuß, I., Nübling, M., Hasselhorn, H-M., Scwappach, D. & Rieger, M. (2008). Working conditions and Work-Family Conflict in German hospital physicians: psychosocial and organizational predictors and consequences. *BMC Public Health*, 8:353.

Gleeson, D. (2001). Health and safety in the catering industry. Occupational Medicine 51(6):385–391.

Grzywacz, J., Frone, M., Brewer, C. & Kovner, C. (2006). Quantifying work–family conflict among registered nurses. *Research in Nursing & Health*, 29(5):414–426.

Heponieme, T., Kouvonen, A., Sinervo, T. & Elovainio, M. (2010). Do psychosocial factors moderate the association of fixed-term employment with work interference with family and sleeping problems in registered nurses: a cross-sectional questionnaire survey. *International Journal of Nursing Studies*, 47:1096–1104.

International Labour Office (ILO). (2010).Emerging risks and new patterns of prevention in a changing world of work.Geneva: ILO.

Isepen, C. & Langaa, J. (2012). Organizational options for preventing work-related stress in knowledge work. *International Journal of Industrial Ergonomics*, 42:325–334.

Karasek, R. (1985). *Job Content Questionnaire and User's Guide*. University of Massachusetts.

Karasek, R. & Theorell. (1990). Health work, stress, productivity and the reconstruction of working life. New York: Basic Books.

Kelloway, K., Mullen, J. & Francis, L. (2006). Divergent effects of transformational and passive leadership on employee safety. *Journal of Occupational Health. Psychology*, 11:76–86.

Kivistö, M., Harma, M., Sallinen, M. & Kalimo, R. (2008). Work-related factors, sleep debt, and insomnia in IT professionals. *Occupational Medicine*, 58:138–140.

Kozak, A., Kersten, M., Schillmoller, Z. & Nienhaus, A. (2012). Psychosocial work-related predictors and consequences of personal burnout among staff working with people with intellectual disabilities. *Research in Developmental Disabilities*, 34:102–115.

László, K., Pikhart, H., Kopp, M., Bobak, M., Pajak, A., Malyutina, S., Gyongyvér, S. & Marmot, M. (2009). Job insecurity and health: A study of 16 European countries. *Social Science & Medicine*, 70:867–874.

Lazarus, R. (1991). Progress on a cognitive-motivational-relational theory of emotion. Am. Psychol. 46, 819–834.

Leka, S. & Jain, A. (2010). *Health impact of psychosocial hazard at work: an overview*. Geneva: World Health Organization.

Lindstrom, M. (2009). Psychosocial work conditions, unemployment and generalized trust in other people: a population-based study of psychosocial health determinants. *The Social Science Journal*, (46):584–593.

Maslach, C., Schaufelli, W. & Leiter. (2001). Job burnout. *Annual Review of Psychology*, 52:397–422.

McNamara, M., Bohle, P. & Quinlan, M. (2010). Precarious employment, working hours, work-life conflict and health in hotel work. *Applied Ergonomics*, 42:225–232.

Melchior, M., Caspi, A., Milne, B., Danese, A., Poulton, R. & Moffitt, T. (2007). Work stress precipitates depression and anxiety in young, working women and men. *Psychological Medicine*, 37(8):1119–1129.

Nabi, H. *et al.* (2009). Differences between day and non day workers in exposure to physical and psychosocial work factors affect inflammation and incident coronary heart disease. The Whitehall II Study. *Arteriosclerosis, Thrombosis & Vascular Biology,* 28(7):1398–1406.

Pisarski, A., Bohle, P., Callan, V. (2002). Extended shifts in ambulance work: influences on health. *Stress and Health,* 18(3):119–1226.

Pisljar, T., Van der Lippe, T. & Dulk, L.(2011).Health among hospital employees in Europe: A cross-national study of the impact of work stress and work control. *Social Science & Medicine*, 72:899–906.

Rugulies, R. (2012). Studying the effect of psychosocial work environment on risk of ill-health: towards a more comprehensive assessment of working conditions. *Scandinavian Journal of Work, Environment and Health*, 38, 187–192.

Sauter, S., Hurrel, J. & Cooper, C. (1989). *Job Control & Worker Health*. Chichester: Wiley & Sons.

Schnall, P., Rosskam, E., Dobson, M., Gordon, D., Landsbergis, P. & Baker, D. (2009). *Unhealthy Work: Causes, Consequences and Cures*. Baywood Publishing: New York.

Shaufeli, W. & Bakker, A. (2004). Job demands, job resources, and their relationship with burnout and engagement: a multi-sample study. *Journal of Organizational Behavior*, 25(3):293–315

Siegrist, J. (1996). Adverse Health effects of high-effort/low-reward conditions. *Journal of Occupational Health Psychology*, 1:27–41.

Sprrig, C. & Jackson, P. (2006). Call centers as lean service environments: job-related strain and the mediating role of work design. *Journal of Occupational Health Psychology*, 11(2):197–212.

Swanson, N.G. & Sauter, S.L. (2006). A multivariate evaluation of an office ergonomic intervention using longitudinal data. *Theoretical Issues in Ergonomic Science*, 7:3–17.

The PRISMA Statement for Reporting Systematic Reviews and Meta-Analyses of Studies That Evaluate Health Care Interventions: Explanation and Elaboration. (2009). *Plos Medicine*: 6(7):1–28.

Zohar, D. (1994). Analysis of job stress profile in the hotel industry. *International Journal of Hospitality Management*, 13(3):219–231.

Occupational Safety and Hygiene – Arezes et al. (eds)
© 2013 Taylor & Francis Group, London, ISBN 978-1-138-00047-6

Working conditions in the urban transport sector in the metropolitan region of Recife: Risk analysis of the jobs of driver and conductor

Béda Barkokébas Jr.
University of Pernambuco, Brazil

Laura Bezerra Martins
Federal University of Pernambuco, Brazil

Eliane Maria Gorga Lago
University of Pernambuco, Brazil

Bruno Maia de Guimarães
Federal University of Pernambuco, Brazil

Felipe Mendes
University of Pernambuco, Brazil

ABSTRACT: The objective of this article was to analyze the conditions of the bus driver and the bus conductor's workstations in the urban passenger transport sector in the metropolitan area of Recife, in Pernambuco, Brazil. Analyzing five bus terminals and the bus environment from five different companies, an evaluation of the conditions of the driver and the conductor's workstations was made. Thus, we performed evaluations of the physical risks, chemical risks and ergonomic risks. In addition, bus terminals and the workstations of the drivers and the conductors were evaluated using checklists and questionnaires. We also analyzed the prevalence of pain in drivers and conductors. The bus terminals have good architectural conditions and the workers are exposed to noise, dust and carbon monoxide within permitted by Brazilian law. However, these workers are subject to an average temperature above the maximum limit of WBGT allowed by national law. Furthermore, it was found that both the workplaces of drivers and conductors had good conditions and allowed the adoption of good posture. Finally, the lumbar spine and hip were the regions where the workers had a higher prevalence of pain. Thus, one verifies that the research is in progress and proposes to discuss with the productive sector joint solutions to be implemented, aiming at the quality of life of workers and the safety of all citizens.

1 INTRODUCTION

Public transportation has a great importance in people's day-to-day lives, whose main representative in Brazil is the city bus. Drivers and conductors are agents that, in a brief review, make the interface between the passenger public transportation road organization and society, interfering with the feeling of security and well being of the entire municipal community. The behavior of these operators is of fundamental importance for the development of this activity, since failures in the work may result to losses, which may endanger both society and the workers themselves. Errors in the work of the conductor can cause conflicts with passengers and financial loss, while mistakes at the driver's work may cause accidents and endanger people's lives (Prange, 2011). According to Zanelato & Oliveira (2003), the bus driver is a professional who drives public and private company vehicles, which operates gear and steering commands by driving the car through a pre-established route, according to traffic laws, in order to transport passengers to their destinations.

Costa (2006) reports that the workstations of drivers and conductors began on trams and, over the years, public enterprises were being privatized and they started working for private companies. Thus, the author makes the criticism by stating that despite the labor achievements, these professionals are still subjected to a grueling work, with the presence of noise above the limit, the overload of the musculoskeletal system due to repetitive movements and maintaining sitting posture for long periods, high temperatures, heavy traffic and vehicle vibration which can lead to some kind of dysfunction of the organism.

The Brazilian Association of Occupational Hygienists (ABHO) defines the security risks that mechanical and ergonomic agents cause, due to the fact of being static or due to the inadequacy of the environment to man. Moreover, occupational hazards are the physical, chemical, biological, ergonomic and safety agents, if any of these are attacking workers' health. Therefore, the working conditions have a major influence on workers' productivity, which confirms the need for a study on working conditions with the purpose to point solutions and tools that contribute with the least risk of accidents and damage to health for these professionals to perform their tasks, with comfort and well-being, thus increasing work efficiency. The aim of this paper is to analyze the employment conditions of the bus driver and conductor of the urban passenger transportation sector in the metropolitan area of Recife.

2 MATERIALS AND METHOD

We conducted a survey of the work conditions of the driver and the conductor. We analyzed the bus terminals, the bus environment and the functions of the worker in a real work situation. For this, measurements were made of the physical risks: noise and heat; chemical risks: total dust and CO (carbon monoxide), and ergonomic risks, according to the normalization of 8 hours of activity. The survey and measurements were conducted between May-June of 2012 and took place between 6:30 am and 2:30 pm in five urban bus terminals corresponding to five different selected companies (company A, B, C, D and E), each with its route towards downtown Recife. The survey of the working conditions and the measurements were performed by the research team and monitored by the Regional Labor Attorney of the 6th Region of the Ministry of Labor (MPT).

Unsystematic interviews were conducted with the five coordinators/inspectors from each of the routes evaluated. Furthermore, researchers completed the checklists, developed for the evaluation of the bus terminals and of the driver and conductor's workstations. Twenty-four bus drivers and twenty-three bus conductors from the companies participating in this study completed the questionnaire developed to confirm the findings of checklists and identify the worker's main complaints about the conditions of the workplace environment. Finally, for the evaluation of postural discomfort the diagram of Corlett (1976) was applied to these forty-seven workers.

For the evaluation of the noise, dosimeters Edge 020 017/020 007 were installed in the lapel of drivers and conductors of the lines and it was monitored along the partial journey of 4 hours. Since they are cyclical activities, it will be standardized for 8 hours. The analysis of the temperature consisted in using the Heat Stress Monitor Heat Meter Questemp° 34, during the trip cycle at the hottest period, corresponding to the times between 10 am and 1:30 pm. The equipment was positioned close to the driver and the conductor's workstations. In addition to this, in order to collect dust, gravimetric pumps were installed on the waist of the professionals and attached to cassettes with PVC filters positioned near the breathing zone with flow rates of 1.7 L/min, and the sampling time of 4 hours. During the dust evaluations at the workstations of drivers and conductors of companies A and B, it was not possible to determine the parameters, thus it was not possible to determine the values. Finally, evaluations of carbon monoxide and oxygen were performed using the M40 multi-gas monitor. The equipment was placed in the mediations of the workstations of the conductor and the driver.

3 RESULTS AND DISCUSSION

During visits to the five bus terminals, the terminal and the workstations of the driver and the conductor were evaluated. The questionnaires used in this study were developed by the authors and were answered by employees during times of breaks or at the end of the workday. While checklists, which were also developed by the authors, were completed by researchers during visits to bus terminals.Regarding the conditions of maintenance of the terminal building, it was considered good. All the terminals had places for meals with table and chair and drinking water was available, but none of the terminals had disposable cups available. Furthermore, they all had male and female toilets. Cleaning was performed by outsourced companies at the integrated terminals, and performed by the employees of the bus companies in the case of neighborhood terminals.

3.1 Bus driver

In the evaluations through checklist of the driver's workplace, it was observed that the bus cabins of all companies surveyed did not allow easy access due to the presence of the vehicle engine. The height and distance from the steering wheel allow the driver a good posture. Of the five companies examined, four have seats and backs in good condition and with adjustments. The arrangement of the gears allows good posture and also the mirrors allow good visualization of the passengers on all the companies analyzed. While four of the five companies presented seatbelt on the bus, drivers reported to the researchers that they did not use this device, claiming that it was not required and that the competent agency did not inspect it.

Twenty four drivers, of which only one was female, answered questionnaires about the working conditions. 79% were married and 21% single. Furthermore, 55% were aged 25–40 years, 12% between 40–50 years and 33% over 50 years. Regarding dominance 84% were right-handed, 8% were left-handed and 8% did not answer. Regarding the opinion of these drivers on the work environment at the bus terminal we notice that 62.5% considered the maintenance of the terminal between regular and proper. Regarding the place to park the bus, 50% rated between very bad and poor and 37.5% as good or very good. Moreover, it was evident the lack of supply of disposable cups for water consumption.

In relation to the work environment of the bus, in the opinion of the drivers questioned, as for the cabin space 41.66% said they work in very bad or bad conditions and 58.34% between fair and good. The system of open/close the doors was rated as good by 96% of drivers and only 4% as regular. The seat was rated as good by 62.50% of drivers, only 33.34% rated as poor or regular and only 4.17% as very good. Regarding the adjustment of the seat, 70.83% reported it as good, 16.67% as bad and 12.5% as regular. As for the headrest, 54.17% of the drivers stated that there was no such accessory, 37.5% stated that the headrest was good and only 8.33% classified it as regular. Of the 24 drivers, 79.17% said they did not make use of seat belts, despite Brazilian law require the use of seat belts.

When asked about the workstation, 37.50% of drivers rated the location of the motor as very bad, 20.83% as bad, 20.83% as regular and only 16.67% as good. It is noteworthy that all five buses observed had frontal motors. Regarding the gearbox of the bus, 79.17% of those interviewed rated it as good and only 8.33% as regular. As for the pedals, 91.67% of the drivers rated it as very good or good. The steering wheel was considered as very good or good by 91.67%. Furthermore, 87.50% of them classified the panel devices as good, 8.33% as regular and only 4.17% as very good. As for the rearview mirrors, 91.67% classified them as good and only 4.17% as regular.

Regarding the feeling of one's own physical and cognitive conditions, only 12.5% answered that they feel good physically at the end of the workday and 87.5% responded that they feel physically tired or exhausted, while 95.84% feel mentally tired or exhausted at the end of the workday. From the evaluation of the prevalence of bodily pain, the locations that showed higher rates in the drivers were: lumbar spine (63%), hip (58%), neck, legs and right foot (54%).

Subsequently, researchers conducted evaluations of physical and chemical hazards at the workstation of the driver of the five companies evaluated. The results of these evaluations can be observed in Table 1.

From the analysis of the ergonomic risks of the workstation, we can identify that the driver, during the execution of his activities, spends a lot of time sitting and has little time for breaks, which can cause circulatory disorders of the lower limbs. The act of driving requires a lot of twisting of the trunk (side bending associated with trunk rotations), with repetitive movements of the upper limb, which can cause osteomuscular disorders in the lumbar spine. With the engine beside the workplace it is difficult to pass and it causes the driver to elevate his legs to get in or out of his workstation, showing a dimensional problem of the workstation.

3.2 Bus conductor

Regarding the workstation of a bus conductor, the checklist used by the researchers shows that, of the five bus companies evaluated, only one had a cabin with easy access. The height of the table, the design and space for opening the cash drawer allows good posture in all the companies evaluated. The seats were in good condition, with adjustments and armrests on all the buses evaluated, but only two of the five had movable armrests. The backrests of the seats were good and had adjustments. We observed the presence of footrest in four of the five companies analyzed, however, only in three companies the buses had space for changing position. In all five buses there is lack of seat belts for this workstation.

Twenty three conductors, of which 61% were male, 35% female and 4% did not answer, answered questionnaires about the working conditions. 57% were married and 39% were single. Furthermore, 78% were aged 25–45 years, 9% between 45–50

Table 1. Ratings of physical and chemical hazards of the workstation of the driver of the five companies evaluated.

Company	Noise	Temperature	Dust	CO
A	79,6 dB(A)	30,5°C	–	8 ppm
B	79,8 dB(A)	32,7°C	–	8 ppm
C	84,3 dB(A)	32,7°C	0,199 mg/m³	8 ppm
D	78,4 dB(A)	29,7°C	0,09 mg/m³	8 ppm
E	82,8 dB(A)	28,5°C	0,159 mg/m³	8 ppm
Average	80,9 dB(A)	30,8°C	0,149 mg/m³	8 ppm

years, 4% over 50 years and 9% did no answer. Regarding dominance 83% were right-handed, 13% were left-handed and 4% did not respond. 60.87% of the conductors who responded to the questionnaires, considered the conditions of the maintenance of the terminal between regular and good. Regarding the place to park the bus, 43.48% rated it between very bad and bad, 34.78% as good or very good and 21.74% as regular. Moreover, it was evident the lack of supply of disposable cups for water consumption.

In relation to the work environment of the bus, according to the conductors, as for the cabin space, 65.22% classified it as regular, 21.74% as good and 13.04% as bad. The seat was rated as very good or good by 47.83% of the conductors, 17.39% rated it as regular and 34.78% rated it as bad. Regarding the adjustments of the seats, 39.13% reported it as good or very good, 30.43% as regular, 17.39% as bad and 13.04% as very bad. Of the 23 conductors, 43.48% judged the backrest as good, 21.74% as regular, 17.39% as bad, 13.04% as very bad and only 4.35% as very good. As for the adjustment of the backrest, 30.4% classified it as regular, 26.04% as good and very good, 17.39% as very bad, 13.04% as bad and 8.7% as inexistent. The system of open/close the doors was rated as good or very good by 52.18% of the conductors, 26.09% rated it as regular and 8.70% as very bad.

When asked about the workstation, conductors evaluated the turnstile as good or very good at 60.90%, 17.4% as regular and 13% as very bad. Meanwhile, 52.17% considered the footrest good, 26.09% classified it as regular, 8.70% as bad, and 4.35% as very good. The cash drawer was considered by 60.87% as good or very good, as regular by 26.09% and by 13.04% as bad. Among the participants, 65.22% rated the table as good, 30.43% as regular and just 4.35% as very good.

Regarding the physical fatigue at the end of their working hours, 13.04% of the bus conductors said they felt good and 82.6% of the workers said they felt tired or exhausted. In regard to mental fatigue, 82.6% reported they felt tired or mentally exhausted at the end of their activities and only 13% felt well. From the evaluation of the prevalence of bodily pain, the regions that showed higher rates in the conductors were: lumbar spine (78%), hip (70%), legs and right foot (65%).

Subsequently, researchers conducted physical and chemical risk evaluations of the workstations of the bus conductor of the five companies evaluated. The results of these evaluations are shown in Table 2.

From the analysis of the ergonomic risks of the workstation, one can identify that at various times conductors adopt postures during rest with anterior tilt of the torso or arms raised above his shoulder

Table 2. Ratings of physical and chemical hazards of the conductor's workplace in the five companies evaluated.

Company	Noise	Temperature	Dust	CO
A	74,5 dB(A)	30,5°C	–	8 ppm
B	79,8 dB(A)	32,7°C	–	8 ppm
C	70,1 dB(A)	32,7°C	0,189 mg/m³	8 ppm
D	76,7 dB(A)	29,7°C	0,102 mg/m³	8 ppm
E	71,3 dB(A)	28,5°C	0,152 mg/m³	8 ppm
Average	74,5 dB(A)	30,8°C	0,147 mg/m³	8 ppm

to support himself and keep his balance. Often also adopt a relaxed position, with support sitting on top of the sacral region, which can lead to back pain and overload the musculoskeletal system and may lead to osteomuscular disorders in the lumbar region. During the execution of work activities, they use many repetitive movements of pinch and grip with wrist movements, besides trunk movements, which can lead to nervous disorders, as is the case of carpal tunnel syndrome. As these workers remain long hours with their legs dangling, in case there is no footrest or it is not in use, there may also be compression on the posterior thigh and generate circulatory disorders of the lower limbs. The workstation needs to allow easy access due to the position of the arms of the chair which, in some of the buses analyzed, are not retractable; and due to the position of the turnstile of the bus and the cash drawer. Some conductors of shorter height need to bend over the drawer so that he/she can register the journey in the electronic turnstile, showing dimensional problems. Moreover, the function of the conductor is an activity that has great mental burden since it deals with money, because the conductor is constantly required to receive the money and give the change correctly, and often has to give information to users. It is also a job that requires the employee to remain in a constant state of alert due to probable burglary/theft.

From the analysis of Tables 1 and 2, one can see that the bus drivers and conductors are exposed to average noise levels below the Maximum Daily Permissible Noise Level Exposure 85 dB (A) for a period of 8 working hours per day as recommends Norm No. 15 (Ministry of Labour and Employment, 1995). Furthermore, these workers are also exposed to the harmful agent dust, the result of which the total dust showed average concentration for the five companies below the tolerance limit set for total dust of 10 mg/m3, as recommended by Norm No. 15 (Ministry of Labour and Employment, 1995). Meanwhile, Carbon Monoxide showed up in its mean value and unit for the companies evaluated, a constant value below the maximum allowed 39 ppm, as recommended by

Norm No. 15 (Ministry of Labour and Employment, 1995). However, with regard to temperature, the bus drivers and conductors are subjected to an average WBGT above the maximum WBGT limit allowable for light activities of 30.5 ° C as recommended by Norm No. 15 (Ministry of Labour and Employment, 1995).

Additionally, organizational problems were found, both for drivers and for conductors that take into account breaks between trips, work rhythms, monotony, time to accomplish the route, which often comes all mixed, because the moment that traffic jams and travel delays occur, consequently the stress increases, the breaks tend to decrease, the pace of work becomes more pronounced and the demand by the terminal controllers becomes greater.

4 CONCLUSIONS

From the collection and analysis of data obtained, it was possible to analyze the working conditions of the driver and the conductor in the bus environment. The five bus terminals evaluated had good architectural conditions.

Concerning the workstations of drivers and conductors, we found that the exposure of these workers to noise, dust and carbon monoxide are within permitted by Brazilian law and do not endanger the health of these individuals. However, it was found that drivers and conductors are subject to an average temperature above the maximum limit of WBGT for activities allowed by national legislation, which can cause discomfort and affect the health of these workers.

Also regarding workstations, it was found that both the driver's and the conductor's workplace had good conditions and allowed the adoption of good posture, but both had problems with access and lack of seat belt, or of its use, by the workers. Finally, it was found that the lumbar spine and hip were the regions where drivers and conductors had a higher prevalence of pain.

In this sense, this research has relevance to the areas of work safety and ergonomics as it proposes to raise and examine the working conditions of these professionals so there can be interference in their safety and quality of life. It is worth noting that the research is in progress and has the proposal to discuss joint solutions to be implemented by the productive sector, aiming at the quality of life of workers and the safety of all citizens.

REFERENCES

Corlett, E.N., Bishop, R.P. (1976). A technique for assessing postural discomfort. *Ergonomics*, 19:175–182.

Costa, E.A.V.G. da. (2006). Estudo dos Constrangimentos Físicos e Mentais Sofridos pelos motoristas de ônibus urbano na cidade do Rio de Janeiro. Dissertação. Universidade Pontifícia Católica- PUC—RIO.

Ministério do Trabalho e Emprego. (1995). Norma Regulamentadora do Ministério do Trabalho—*NR 15 – atividades e operações insalubres*.

Prange, A.P.L. (2011). Quem dá mais, cobra mais! Uma análise das normas antecedentes do ofício de motorista de ônibus em um contexto específico. *Revista Estudos Pesquisas em Psicologia*. 11(2).

Zanelato, L.C., Oliveira, L.C. *Fatores Estressantes Presentes no Cotidiano dos Motoristas de Ônibus Urbano*. 2003 [Retrieved jan, 2012, from http://www.sepq.org.br/sitesipeq/pdf/poster1/08.pdf].

Occupational Safety and Hygiene – Arezes et al. (eds)
© *2013 Taylor & Francis Group, London, ISBN 978-1-138-00047-6*

Nurse practice to assist students with heart condition at school

R.P. Maia & G.D.S. Ulbrich
Associação Franciscana de Ensino Senhor Bom Jesus, Curitiba, Brazil

L.M.W. Prado
FAE Centro Universitário, Curitiba, Brazil

S.R. Benka
Associação Franciscana de Ensino Senhor Bom Jesus, Vacaria, Brazil

ABSTRACT: It is the nurse's job search inside the school, through the student's medical records, who are the students with a heart disease, let the parents know the importance on asking the child's doctor a cardiologic exam to practice exercises and special assistance at school. For this reason at school, the professional nurse can use all his knowledge to promote health to the students.

1 INTRODUCTION

Caring is part of human life since the beginning of the mankind, as an answer to serving their needs. To accomplish care, the nurse, as an integral member of a multidisciplinary team, uses a set of knowledge that enables the search for answers to solving the phenomena of health, defined by the International Council of Nurses as health aspects relevant to the practice of nursing.

The instrument for accomplishing care is the care process through an interactive action between the nurse and the patient, and in the school context this patient is a student, in this case with emphasis on patients with heart disease. In it, the professional activities are developed "to" and "with" the student, anchored in scientific knowledge, skill, intuition, critical thinking and creativity accompanied by behavior and attitudes to care, to promote, keep/or recover all and the human dignity of the child.

Developing care occurs in its different specialties and in this study, it will address the process of taking care of the student with heart disease whether congenital or acquired. According to Conceição (1994), Health scholar is an area of health that includes activities related to students' health. School can provide primary health care, where nurses are ideally prepared to provide care in these services. Nursing professional is adapting to meet the changing expectations and necessities of the school environment. Brunner & Suddarth (1999) state that one of these adaptations is through the expanded role of the nurse, which has been developed in response to the demand to improve the distribution of health services.

The nurse has a key role in the health care team, as through daily clinical evaluation of the student, may conduct a survey of various phenomena, whether in the external appearance or the subjectivity of the multidimensionality of human beings. Also could provide for the student to be served in many different segments of the multidisciplinary team or nursing. The World Health Organization—WHO, in the document "innovative care for chronic conditions," emphasizes that the patient/student with a chronic disease needs to have a planned care plan, which are capable of predicting their basic needs and provide integrated care. This kind of attention involves time and health scenario.

When performing nursing actions through a holistic approach, the nurse helps the client to purchase a health condition. However, to effectively perform these actions, the nurse must correctly identify the faults or deficiencies related to the client's health. In pursuit of this improvement, nursing has sought to direct and integrate knowledge with doing, to contribute to improving the quality of their care.

Daily practice with children and adolescents of school age in Cardio pediatrics presents unique responses that need to be improved and worked with a scientific character, through the intervention of nursing, in a systematic way, according to Dalcina (2000), one of which is the process of investigation and service to students with heart disease. It is essential strategies are implemented action oriented towards health to promote safety and quality of life of cardiac patients in school. The most common needs is information about their own heart, adjust eating habits, physical activity promotion, nursing care in specific symptoms and or risky situations. It is estimated that for every thousand Brazilian children, eight born with heart disease. The incidence of congenital heart disease in the general population ranges from 8 to 11 per thousand live births.

Due to the progressive improvement in diagnostic methods and treatment of heart diseases in childhood in recent decades, growing every year the number of patients with congenital heart disease who reach adolescence and adulthood.

However, compared to developed countries, Brazil is delayed in relation to the treatment of congenital and acquired diseases. There are insufficient financial resources, trained professionals and health services prepared to treat these children. Such as this scenario, the Cardiology Society of the State of São Paulo—SOCESP created the Reference Center for Congenital Heart Defects, aiming to empower health services from various cities to assist these children with heart disease, through training that can offer quality treatment. This initiative allows the school-age children to have an adequate cardiologist and nurse attention whose role is to pass medical guidelines for the teaching staff, guide the physical education teacher regarding medical clearance for physical exercise and or situations where there is restriction to sports.

2 METHODOLOGY

The survey was conducted by quality quantitative method, using direct observation in five Brazilian schools, through the literature associated to the authors with the person responsible for the student, addressing a request for an opinion and assessment experience in addition to the research protocol and monitoring of students with heart disease. The initial stage of this process is characterized by previous contact of the cardiac pathology report. Through a systematic roadmap of the nurse's actions in the school environment towards to cardiac disease, the data collected showed and proved the existence of heart disease in school.

The purpose of this process is to guide the school community, which is part of the student's daily life, how to act before a crisis. All cardiac requires a plan of care according to the needs presented by him and his family, for the purpose of effective guidelines to restore to its maximum potential in order to prevent complications and have the prospect to resume their activities previously performed. The vast majority of students who have heart disease need to understand the process that helps you to learn to incorporate healthy habits that will become part of their everyday life (SOCESP, 2012). At the same time the medical report contributed to the survey data in this research. This study aimed to clarify on the matter concern to the reference Conception (1994) and Brunner & Suddarth (1999), about the role of the school nurse. 8781students' health records were also checked, these were filled out by parents or guardians at the time of the enroll in the 2012 academic year. With the collected data we created a table with the proportion of students who have heart disease per school.

The schools that participated in the study allowed the dissemination of data, on the condition of not publishing the names of students and their institutions.

3 DISCUSSION AND RESULTS

Heart diseases in childhood, congenital or acquired, are not very widespread in the general population, prevalent cardiovascular disease as the leading cause of mortality in Brazil and around the world and represent the leading cause of mortality and disability in Brazil. Carvalho (sd), quotes data from the WHO figures, reporting that in 2002 there were 16.7 million deaths, of which 7.2 million were due to coronary artery disease. It is estimated, that by the year 2020, this number may rise to between 35 and 40 million. In Brazil, cardiovascular disease is the leading cause of morbidity and mortality, occurring at an early age. It is known that since the 60 s, these diseases have been showing progressive increase worldwide. We highlight the Brazilian cities of the South and Southeast, with a high incidence of ischemic diseases. According to Leite, 2007, according to the Department of the Health System of the Ministry of Health—DATASUS/MS, Brazil currently has about 30 million teenagers. Only in 2006 were held in the country 1857 pediatric heart surgery, with a mortality rate of about 14%. In 1950, only 20% of infants with complex congenital heart diseases survived the first year. Currently, 90% of them survive to adulthood in the U.S.

Regarding acquired heart diseases, the survival rate of these patients has also increased significantly, emphasizing the great need for their training so that they can reach adulthood in a position to become fulfilled people, implying therefore a good social adaptation. But nowadays has performed with some frequency, as can be seen in the table, in which approximately 1.25% of the records analyzed showed that diagnosis of heart disease in five Brazilian schools.

In suspected cases of heart disease, the diagnosis should be early, with a form and treatment correct, as they are often serious and life-threatening form. It is undisputed that the opinion of the cardiologist is the gold standard for the diagnosis of cardiopaties. Since heart disease can be detected in children soon after birth, by the doctor, also through taking exams at

Table 1. Percentage of students with heart diseases in five schools surveyed.

School	%
A	1,90
B	1,40
C	1,00
D	0,25
E	1,70
Average	1,25

school for physical exercises, in their adolescence. In most cases, the doctor suspects heart disease in children when auscultation heart murmur, detect cyanosis (purplish extremities) when the child gets very tired during breastfeeding, has difficulty in gaining weight, frequent colds or recurrent pneumonia. To Kobinger (2003), hypertrophic cardiomyopathy is genetic in 20% to 60% of cases may progress asymptomatic for years showing up only in special situations, for this reason the concern with children who have these characteristics and make an option for competitive physical activities. The sudden death in childhood or young adult is a precedent that should be valued. An example is the case of the 16-year-old boy in the city of São Paulo, who died after feeling bad while in his Physical Education class. He was helped by the school staff and taken to the city hospital where he arrived with cardiopulmonary arrest. The boy died two hours after receiving medical attention. According to information from the hospital, doctors tried to revive him, but he did not resist (Portal of Physical Education, 2012). Another case even more shocking was the student of 15 years old, who suffered from heart problems, who died after getting sick during a Physical Education class in a state school in São Paulo. According to family members, the 8th grade student, he used to go to the doctor often, but no one knew he had health problems. According to the student's aunt, he had recently gone to a health clinic. "The doctor prescribed a cough syrup." (Aluno que morreu... 2012).

The occurrence of sudden death from heart problems during exercise practice is a tragedy, it is not acceptable that an apparently healthy young is at risk of suffering from cardiovascular collapse. According to Rowland (1994), the risk of sudden death from heart problems in young people, in American medical literature, is reported about 10 to 13 cases annually.

Even assuming that this fact is undersized, chances of a tragedy occur in an athlete is probably less than 1:250,000. An innocuous impact on the anterior portion of the chest can trigger sudden cardiac arrest, known as commotio cordis, responsible for about 20 deaths per year in the United States, predominantly in children and adolescents. Most experts theorize that commotio cordis is caused by a relatively small non-penetrating impact on the precordium (area over the heart) that occurs in an electrically vulnerable portion of the cardiac cycle, while others believe that a coronary vasospasm may play a role in its development. Whatever the mechanism, the final result is arrhythmia leading to the sudden cardiac arrest. This condition occurs most often during amateur sports activities in which the victim is hit by a projectile such as a baseball (most common in the U.S.). However, the cardiac concussion was also reported in the body after impact, such blows of karate. After impact, the victims were able to give one or two steps, and then fell to the ground and had cardiac arrest.

The heart rhythm must be determined as quickly as possible, perform rapid defibrillation if ventricular fibrillation is identified. The prognosis is bad, with survival chances of 15% or less. Almost all the survivors of this condition received cardiopulmonary resuscitation and cardiac defibrillation in place, often with an Automatic External Defibrillator—AED. The analysis of these pathologies have been focused on the importance of health care providers, although those who have responsibility to identify risks that these students are taking and alert them from a sudden heart at school.

Children with heart disease may attend school normally, perform sports according to the type of heart disease they have, but it is important the role of the professional nurse with critical and active in the evaluation of complaints, signs and symptoms, the occurrence of pain chest, dizziness, fainting, chest wheezing or fatigue during exercise may be a "red flag" for any sportsman. This symptom indicates the possibility of a high risk of cardiovascular disease.

The historical occurrence of palpitation or heart murmurs are signs that indicate the need to see a cardiologist, guiding the leaders of orienting parents for the necessity of evaluation by the specialist due to the situation presented, it is very important and the role of the school nurse. When the diagnosis and treatment of heart disease in children are performed in time, it may allow the cure of heart disease or improve the quality of life in an attempt to avoid sudden illness.

It is part of the nurse's job to fetch inside the school environment who are the students with heart disease, make the parents aware of the importance of asking the child's physician a cardiology opinion for physical activity, the goal of health assessment with pre-sports participation is determine whether the child or adolescent can safely participate in an activity organized sports. It's mandatory to pay attention to the parts of the body which are most vulnerable to the stress of sports. The clinical history and physical examination needs to focus on the cardiovascular system (stenotic lesions, hypertension, surgery). It is a fact that any teenager who has desire for physical activity, especially competitive, should be evaluated by his physician previously (Leite, 2007).

The history and physical examination are essential in this appointment because through them the doctor can find teenagers with heart disease already diagnosed or suspected, who must be assessed and monitored. They should be evaluated also those who, although healthy, have risk factors for sudden death associated with physical activity, family history of sudden death or sudden death associated with heart disease (hypertrophic cardiomyopathy, long QT syndrome, arrhythmogenic right ventricular dysplasia, Marfan syndrome, Kawasaki, etc.).

Moreover, it is the time when asymptomatic diseases, like hypertension, can be diagnosed and/or symptoms such as chest pain, dyspnea and syncope associated with exercise, can be investigated (Leite, 2007). The request for additional examination as electrocardiogram—ECG and exercise stress test for healthy students, but no family history of sudden death risk, is a matter of discussion worldwide.

According to Rowland (1994), in countries like the United States, in which the number of athletes is significant and the incidence of sudden death is low, about 0.4%, it becomes incoherent from the point of view risk/benefit ratio, require laboratory exams for all subjects. In Europe, countries such as Italy, that has a high incidence of arrhythmogenic right ventricular dysplasia, prefer to adopt the holding of at least one ECG before starting the adolescent physical activity. On this topic, it is important to remember that the ECG is not faithful to diagnose 100% of patients with the possibility of sudden death. Thus, the history and physical examination are still the best weapons for this type of diagnosis, reserving the ECG for specific populations where the risk of sudden death is higher. Under this guideline, the teaching staff must take the necessary precautions against the crisis.

In the presence of symptoms, the teacher should ask the assistance of the school nurse, who will check the clinical complaints during or after physical activity, will observe the general state checking vital data with emphasis on heart rate, presence of pallor and tiredness for small efforts. So at school, the nurse practitioner can use all their knowledge to promote the health of these students.

4 CONCLUSIONS

The objective of this study is to show the role of the nurses inside the school, that this person has professional expertise to assist health problems, caring crises or sudden illness. To provide planned nursing care and skilled integration is necessary between the multidisciplinary team, nursing and a cardio pediatrician. The performance of the nursing staff at that time is critical in the prevention and early diagnosis of complications and maintaining the comfort of the student with heart disease, with careful observation, detailed and systematized. According to Azevedo (2008), students with heart disease have different needs, planning care and guidelines they must follow individual parameters.

Thus, the role of the nurse is to encourage students to externalize symptoms and needs as well as their way of life and emotions, to enable global customer support with a view to relieve the physical and psychological stress as a result of the process of the symptoms and pathology previously diagnosed heart condition.

In order to perform a plan of care, shared and effective, it is necessary to have an effective interpersonal relationship between the nurse and the student. By guiding the student with heart disease, this professional should be alert to the fact that the student is able to learn and promote self-care, prevent and seek help immediately after the perception of deviations from standards of health, reducing the risk of premature death. To develop knowledge and understanding of their cardiac condition, the student can change their health behavior, which enables the nurse to act as a facilitator of behavioral changes in health aspects. A student's education is a process that helps them learn and incorporate healthy habits that will become part of their everyday life.

The role of the professional nurse consists, among other actions, the health education of students with heart disease, family and team teaching for conservation and rehabilitation of health, beyond the implementation of measures to prevent the disease.

The promotion of education for pupils, aims to avoid cardiac complications in these cases, the nurse monitors, supports students and families to self-care, which requires knowledge of the conditions of each student. In an attempt to reduce modifiable risk factors, students should be instructed to make changes in lifestyle.

The nurses in the role of health educators need to be aware that the change in lifestyle of patients with heart diseases is essential for maintaining health and improving quality of life. These changes are summarized in controlling the ingestion of healthy foods, control of hypertension and systemic hypotension, diabetes and hypercholesterolemia, body weight loss, dehydration and stress. The nurse can be defined as a professional who has the care of the patient, based on meeting the basic human needs. It requires the ability to make accurate decisions, aimed primarily on the prevention and reduction or resolution of identified problems. Such actions are modified according to the individual characteristics of students and with advances in science, ranging from the simplest to those extremely complex activities. For this it is necessary that the professional school nurse is in constant updating of existing protocols regarding the resuscitation cardiopulmonary.

The participation of the nurse in the prevention of health problems has been very limited, given the emphasis on curative action focus of the profession. However, the educational activities are of fundamental importance for developing the consciousness of the individual to act and to prevent injuries in their disease process. Opportunities to exchange health information arise through contact with the student, in any scenario, but it will only happen if the school nurse is engaged to educate for prevention. It is highlighted as a fundamental aspect of care sharing decisions and orientations between cardio pediatrician, school nurse, the student and family, so that the joint planning proposals achieve the targets, developing the maximum potential of the student.

REFERENCES

Almeida, T.L.V. (2012). Cardiopatias congênitas para o leigo. Retirado 19 de agosto de 2012, de http://amigosdocoracao.wordpress.com/cardiopatias.

Aluno que morreu depois de passar mal em aula tinha problemas cardíacos. Retirado 4 de outubro de 2012, de http://m.g1.globo.com/sao-paulo/noticia/2012/10/aluno-que-morreu-depois-passar-mal-em-aula-tinha-problemas-cardiacos.html.

Atendimento pré-hospitalar ao traumatizado, PHTLS/NAEMT (2011). (7a ed.). Rio de Janeiro: Elsevier.

Azevedo, R.V.M., Moretão, D.I.C., Moretão, V.J. (2008). Prevenção de acidentes vascular cerebral em pacientes portadores de cardiopatia. Interseção-Vol. 1, n. 2, p. 82–90 Retirado 7 de novembro de 2012, de http://www.saocamilo-mg.br/publicacoes/edicao2/sao_camilo/artigo.

Brunner & Suddarth, S.C.S. (1999). Tratado de enfermagem médico-cirúrgica (8a ed). Rio de Janeiro: Guanabara Koogan.

Carvalho, T.M.W. (s.d.) Doenças cardíacas na infância. Retirado 11 de outubro de 2012, de http://www.ebah.com.br/content/abaaaeijoah/doencas-cardiacas-na-infancia.

Conceição, J.A.N. (1994). Assistência Integral à Saúde Escolar. São Paulo: Sarvier.

Dalcina, J., et al.(2000). Assistência de Enfermagem à criança portadora de cardiopatia. Revista SOCERJ—jan/fev/mar Vol.XIII, n.1, Retirado 11 de outubro de 2012, de http://sociedades.cardiol.br/socerj/revista/200001/.

Kobinger, M.E.B.A. (2003).Avaliação do sopro cardíaco na infância. Jornal de Pediatria - Vol.79, Supl.1. Retirado 13 de setembro de 2012, de http://www.scielo.br/pdf/jped/v79s1/v79s1a10.pdfMaria Elisabeth B.A. Kobinger.

Leite, M.F.M.P., Borges, M.S. (2007). Qualidade de vida do adolescente portador de cardiopatia: alguns aspectos práticos. Adolescência e Saúde – Revista oficial do núcleo de estudos da saúde do adolescente/UERJ. Vol. 4, n.3 (Jul./set.2007), Retirado 11 de outubro de 2012, de http://www.adolescenciaesaude.com/detalhe_artigo.asp.

Portal da Educação Física. (2012). Adolescente morre após passar mal em aula de educação física. Retirado 13 de setembro de 2012, de http://www.educacaofisica.com.br/index.php/escola/canais-escola/cotidiano.

Rowland, T.W. (1994). Diagnóstico do risco de morte súbita em atletas jovens com problemas cardiopáticos. Retirado 31 de agosto de 2012, de http://www.gssi.com.br/artigo/70/sse-74-diagnostico-do-risco-de-morte-subita-em-atletas-jovens-com-problemas-cardiacos.

Socesp (2012). Institucional. Retirado 15 de setembro de 2012, de http://socesp.org.br/publico/espaco_leigo/ebrasil_criancas_cardiopatias.asp.

Occupational Safety and Hygiene – Arezes et al. (eds)
© *2013 Taylor & Francis Group, London, ISBN 978-1-138-00047-6*

Terms and concepts: A reflection on occupational health and safety definitions and terminology

C. Gomes de Oliveira
Instituto Superior de Educação e Ciências, Portugal

F.O. Nunes
Instituto Superior de Engenharia de Lisboa, Portugal

A. Pinto
Independent researcher

ABSTRACT: Often occupational health and safety (OH&S) terminology is ambiguous and controversial. Different terms for the same concept or different definitions for the same term lead to misunderstanding even between OH&S's experts. The development of a methodology aim to construct a OH&S glossary containing the terms that are necessary to describe. characterize, analyze/assess, evaluate, manage and communicate risks, based on a better understanding of the underlying terms seems to be of paramount importance.

1 INTRODUCTION

Occupational health and safety (OH&S) is a relatively new and developing branch of knowledge composed by knowledge from different sciences. Its terminology is a source of ambiguity and at times even a source of controversy within experts from different sciences, because roots deeply in different views on the applied terms often derail a discussion from its core issue(s). This led to ambiguity in the use of terms, both between different OH&S sciences and between the different parties involved in OH&S debates.

It must be noted that some of the concepts used in OH&S are diffuse/uncertain by nature (such as: safety and risk), which means that this kind of concept cannot be defined with precision, certainty and rigor.

Several authors (Leeuwen and Hermens, 1995; Lewalle, 1999; Christensen et al., 2003) and institutions (NRC, 1983; IPCS, 1989; UNEP/ILO/WHO, 1994; EC, 1996, 2000; US-EPA, 1997; ISO, 2001; SRA, 2012) had regularly made suggestions on terminology in different branches of OH&S related sciences.

The main object on OH&S is the risk. In fact, risk and safety are linked both conceptually and pragmatically (Hollnagel, 2008). The conceptual link can be seen by comparing definitions of the two main concepts: risk and safety. Risk is usually defined as the likelihood that something unwanted can happen (Hollnagel, 2008). Safety is likewise defined as the absence of unwanted (health harmful) events, which essentially means as the absence of risk (Hollnagel, 2008).

From an historical point of view, the risk concept has been related with the culture and the language of those who use it (Oliveira, 2010).

In some way, all the other OH&S concepts are risk-related.

Consulting the legal and bibliographic sources, different terms related to the same concept can be found, according to the specific scientific/technological field where it is applied.

The elaboration of a coherent, consistent and efficient glossary needs the establishment of criteria in order to justify the chosen relationship between concept, definition and term.

The purpose of this paper is to explain and discuss a rational criteria for the development of a methodology aim to construct a OH&S glossary containing the terms that are necessary to describe, characterize, analyze/assess, evaluate, manage and communicate risks, based on a better understanding of the underlying physical (fundamental terms) and societal (action oriented, such as legal terms) terms. The approach will follow the criteria:

- Identification and/or assessment of terms widely used on current practices but taking into account the legal and normative terms;
- Minimization of interfaces between definitions;

- Rejection of redundancies;
- Systemic approach by fundamental terms and action oriented terms considering the significant levels.

2 METHODOLOGY

As early referred, in addressing this subject prima facie, a structural issue emerges: the object of safety is admittedly the risk, i.e., risk is the core concept of a whole set of limiting, derived and engaging notions. But if risk is, in itself, a concept which is intimately related to the concept of uncertainty, how to define uncertainty? How can one frame, relate, make consistent, different concepts that bind with it (upstream and downstream)?

And of course, the concept of risk is paradigmatic, not only by the different approaches found in the technical and scientific literature, but also by the specificity of legal definitions, often with restricted and restrictive scopes.

It is not intended, in this study, to prepare a glossary of terms useful in the area of Safety, though this goal will be necessarily a development of the theme. It is intended, firstly, to establish a set of rules and procedures allowing such elaboration, so that glossary can be a structured and coherent system of terms, linked to definitions and translating unambiguously the concepts used in this area of knowledge.

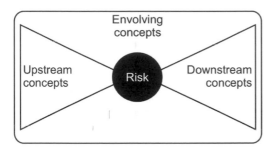

Figure 1.   Related concepts.

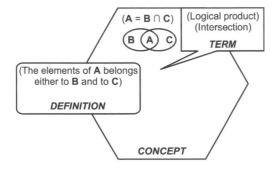

Figure 2.   Concept-definition-term.

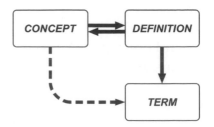

Figure 3.   Relationship between concept, definition and term.

The fundamental question is, then, not only to find a clear and justified definition of each term but also, for each concept, the term that best fits his definition. To clarify the relationship between concept, definition and term the example in Figure 2 may be used.

Naturally, the relationships between these entities are not easy to fit, given the interdependence that exists between them. However, one can establish a two-way relationship between concept and definition, so that for each definition there is a corresponding term that is, naturally, semantically conditioned by concept.

Consulting dictionaries and other reference sources, with a necessarily critical approach, is crucial as a starting point for proposing a set of concepts/definitions.

A number of glossaries, dictionaries and thesaurus of OH&S used terms can be found in the bibliography (Garcia, F. M. et al., 1993, Uva, A. S. & Graça, L. 2004, US-EPA, 2006, Society for Risk Analysis, 2012, as an examples) but most of them approaches the terminology in a sectorial point of view and, sometimes, using restricted sources (Nunes, 2010).

Essentially, two types of sources can be found: the regulations that correspond to various legal documents and other related literature where it is possible to obtain scientific and technical justification for defined terms. The use of either (or both) will require, mandatorily, a critical and comparative analysis, particularly in those cases where discrepancies or even contradictions between definitions may occur.

However, one must relativize these sources of information, given that the context of the presented definitions has a very comprehensive character, therefore, not consistent with specific interpretations, proper of a well delimited area of knowledge. They will always be, then, necessary but not sufficient sources on their own.

Options arising from the consultation of different sources involve the application of valid criteria that allows the selection of the one with the best characteristics—simplicity, consistency, scientific rigor, usability—present. These criteria will be based mainly on two key aspects:

- Priority—the obligation arising from law must of course be taken into account. However, one must consider the scope defined in the law and assess its comprehensiveness. Recommendations resulting from technical standards are necessarily important aspects to consider, overlapping, in terms of significance, the information contained in non-normative technical documentation.

The analysis of bibliographic sources presents major difficulties regarding the establishment of evaluation criteria. The context in which the settings are inserted, the type of development where they fall and, of course, technical and scientific credibility of authors and/or institutions, are aspects that determine the priority and significance of definitions analyzed. The extent to which the various definitions are applied, in terms of expertise and/or coverage, also contributes to determine the weight they may have.

- Synthesis—is expected that, in a search for sources, either regulatory or bibliographic, one will find different definitions for the same concepts, varying both the scope and form (syntactic and/or semantic). This means that there will probably be, complementarities and overlapping between the various definitions referenced (see diagram in figure 4).

It is then necessary for an effort of synthesis to be made, in order to obtain a single definition which is, on one hand, complete and comprehensive and, on the other, concise and not redundant.

The correspondence between the term and definition adopted has to, necessarily, take into account the cultural appropriateness and comprehensiveness of the latter by using a common language or a technical-professional "jargon", obviously while respecting their scientific relevance.

Thus, the establishment of normative and methodological basis for the preparation of a glossary implies a sequential process of research, either starting from the term to the concept (deductive method) or from the concept to the term (inductive method).

The important thing is, then, to establish a two-way relationship between these two concepts, which results in a two-column table: "term"/"definition".

The fundamental issue is, therefore, not only finding a clear and justified definition of each term but also to the term that best fits your definition for each concept. Naturally, the relationships between these entities are not easy to fit, given the interdependence that exists between them. However, you can establish a two-way relationship between concept and definition, so that to each definition there is a corresponding term that is naturally semantically conditioned by the concept.

For a better understanding among all stakeholders, a uniform glossary of the technical terms used within OH&S is crucial.

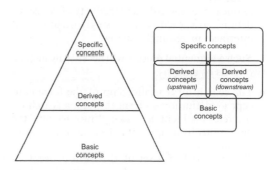

Figure 5.   Hierarchy of concepts.

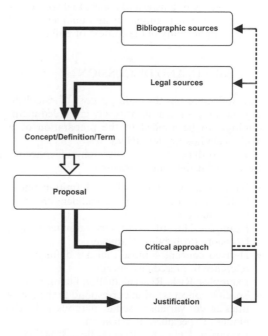

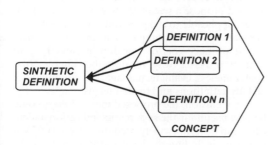

Figure 4.   Synthesis of definitions.

Figure 6.   Flowchart of the proposed methodology.

A consistent way of building definitions, implies a hierarchy of several concepts that shows, as noted earlier, the dependency ratio—and causality—connecting them.

Hence, a glossary will always sit in a tree structure, in which

- the definition of the basic concepts should be broad and based on a perceptive language to most participants in the process, without ever yielding to a generalization that will remove accuracy, and seeking, in particular, to minimize possible overlaps between definitions;
- there are, obviously, concepts which are derived from each other, so that for a particular concept, there might be several definitions of terms that result from it and that can be related through an inductive tree;
- as one moves towards the specificity of concepts, definitions should be increasingly accurate and interfaces increasingly narrow.

Such a sequential process will necessarily include two levels of analysis:

- The search for a relationship of the form concept ↔ definition ↔ term through a search on the relevant literature, taking into account the definitions from applicable legal and regulatory documents.
- The proposal of a set of the form concept ↔ definition ↔ term based on a critical analysis of the data obtained in the previous level, duly justified by linguistic and conceptual considerations, given their autonomy and uniqueness and their applicability in the context of Safety.

3 RESULTS AND DISCUSSION

This reflection has the main purpose of establishing and proposing a scientifically grounded methodology, to be applied to the preparation of a Technical Glossary for OH&S.

In accordance with the hierarchy of concepts, the following definitions are used:

- Basic concepts—those whose definition under OH&S, is absolute, i.e., not dependent on previous definitions.
  Examples: Hazard; Event; Operation technique; Decision; Probability; Damage.
- Derived concepts—those that are defined with reference to preceding concepts.
  Examples: Risk, Risk calculation factor; Cause.
- Specific concepts—that correspond to the quantities and/or variables used in methods and techniques of specialized application.
  Examples: Deficiency level (Belloví & Malagón, 1993); Index of safety conditions in the work-

place (Nunes, 2003); Training and employee participation (Kinney & Wiruth, 1976; Malchaire, 2003); Top event (Fault Tree Analysis).
Note: these concepts should not be included in the glossary of the technical terms of OH&S, but must be clearly defined in the description of the method/technique to which they relate.

A way to relate concepts/definitions/terms systemically coherent and comprehensive is proposed, emphasizing the semantic and conceptual relationship between the definition and the defined.

An example can be the term "risk" for which several definitions can be founded even in legal documents. US-EPA, (2006) states up to 19 different definitions related to different applications.

A structured methodology and a clear identification of criteria for the analysis of such cases will permit the proposal of a synthetic but comprehensive definition.

Searching for some legal definitions of "hazard" we can find: "source, situation, or act with a potential for harm in terms of human injury or ill health, or a combination of these" (BS OHSAS, 2007) or: "the intrinsic property of a facility, activity, equipment, agent or other material component of work with the potential to cause harm" (Portuguese Law no. 102/2009).

From those sources, it is possible to propose a definition of "hazard" rooted on the synthesis of legal and normative terms:

"Intrinsic property of any source, equipment, situation, or act with the potential to cause injury or/and ill health".

4 CONCLUSIONS

The diversity of disciplines that contribute to the Occupational Health and Safety reinforces the need to harmonize the terminology used.

The development of criteria to ensure a coherent, complete and synthetic relationship between concept, definition and term must be the preliminary but necessary step to achieve this harmonization.

This paper proposes a reflection aimed to establish and propose a scientifically grounded methodology, applicable to the preparation of a Technical Glossary on OH&S, preparing in the near future.

The primary goal of a technical glossary is to achieve a better way to describe, characterize, analyze/assess, evaluate, manage and communicate OH&S definitions and related terms.

In conclusion, the proposed methodology could be very useful for the systematic and rational identification of terms widely used on current (OH&S) practices, including the legal and normative terms and to propose a term associated to a definition rooted on the synthesis of the various definitions currently used.

REFERENCES

Assembleia da República, Lei no. 102/2009 de 10 de setembro.

Bell100, M.B. & Malagón, F.P. (1993). Sistema simplificado de evaluación de riesgos de accidente. Nota Técnica de Prevención 330. Instituto Nacional de Seguridad y Higiene en el Trabajo, Barcelona.

BS OHSAS 18001. (2007). Occupational Health and Safety Management Systems: Requirements. British Standard Institution.

European Commission, (1996). Technical Guidance Document in Support of Commission Directive 93/67/EECon Risk Assessment for New Notified Substances and Commission Regulation (EC) No. 1488/94 on RiskAssessment for Existing Substances. Part I. Office for Official Publications of the European Communities,Luxembourg.

European Commission, (2000). First Report on the Harmonisation of Risk Assessment Procedures. Part 2. Appendices, Appendix 1: glossary of terms, 26–27 October 2000 (published on the internet on 20–12–2000).

Garcia, F.M. ed al (1993). Diccionario MAPFRE de Seguridad Integral, Fundación MAPFRE Estudios, Madrid.

Health and Safety Executive (HSE), Generic Terms and Concepts in the Assessment and Regulation of Industrial Risks, London, 1995.

Hollnagel, (2008). E. Risk + barriers = safety?, Safety Science 46, 221–229.

International Programme on Chemical Safety (IPCS), (1989). Glossary of terms on chemical safety for use in IPCS publications, World Health Organization, Geneva.

International Organization for Standardization (ISO), (1999). Safety aspects – Guidelines for Their Inclusion in Standards, ISO/IEC Guide 51.

International Organization for Standardization (ISO), (2001). Risk Management—Vocabulary—Guidelines for Use in Standards, Draft ISO Guide 73.

Kinney, G.F. & Wiruth, A.D. (1976). Practical Risk Analysis for Safety Management. California, Naval Weapons Center. Technical Publication 5865.

Leeuwen, C.J., Hermens, J.L.M. (1995). Risk Assessment of Chemicals: An Introduction, Kluwer Academic Publishers, Dordrecht.

Lewalle, P. (1999).Terminology Standardization and Harmonisation (TSH), 11 (1–4).

Malchaire, J. (2003). Estratégia Sobane de Gestão de Riscos Profissionais., Available at: http://www.deparisnet. be/sobane/pt/Estrategia_SOBANE_Port_8-4-09.pdf (assessed oct. 2012).

National Research Council (NRC), Committee on the Institutional Means for Assessment of Risks to Public Health, (1983). Risk assessment in the federal government: Managing the process, National Academy Press, Washington, DC.

Nunes, F.O. (2003). Avaliação de Níveis de Segurança nos Locais de Trabalho: Uma Abordagem Quantitativa. 3.º Colóquio Internacional sobre Segurança e Higiene do Trabalho, Ordem dos Engenheiros, pp. 77–82.

Nunes, F.O. (2010). Sobre a utilização de termos e conceitos em avaliação de riscos profissionais, 1ª Parte, Segurança, nº 198, set/out, p. 3–5; 2ª Parte, Segurança, nº 199, nov/dez, p. 5–7.

Oliveira, C.G. (2010). Proposta de uma Metodologia Integrada de Avaliação de Riscos Profissionais, PhD Thesis, DepartamentodeCiênciasBiomédicas—Universidadde Léon, Available at: https://buleria.unileon.es/bitstream/ handle/10612/904/2010ON-CUNHA%20GOMES%20 DE%20OLIVEIRA%2c%20CARLOS%20A1. pdf?sequence=1 (assessed nov. 2012).

Society for Risk Analysis, (2012). Glossary of Risk Analysis Terms, Available at: http://www.sra.org/resources_ glossary.php (assessed nov. 2012).

United Nations Environment Programme, International Labour Organization, World Health Organization (UNEP/ILO/WHO), (1994). Assessing human health risks of chemicals: derivation of guidance values forhealth-based exposure limits, Environmental Health Criteria 170, World Health Organization, Geneva.

US-EPA, (1997). Risk assessment and risk management in regulatory decision-making, The Presidential/Congressional Commission on Risk Assessment and Risk Management, Final Report, vol. 2, Glossary (posted in RiskWorld on March 27, 1997), Available at: http:// www.riskworld.com (assessed aug. 2010).

US-EPA, (2006). Thesaurus Of Terms Used In Microbiological Risk Assessment, Available at: http://water.epa. gov/scitech/swguidance/standards/upload/2007_10_01_ criteria_humanhealth_microbial_thesaurus_microbialthesaurus.pdf (asses-sed nov. 2012).

Uva, A.S. & Graça, L. (2004). Saúde e Segurança: glossário de termos e expressões mais comuns cadernos avulso nº 4, Sociedade Portuguesa de Medicina do Trabalho.

Occupational Safety and Hygiene – Arezes et al. (eds)
© *2013 Taylor & Francis Group, London, ISBN 978-1-138-00047-6*

Study of the food waste of the canteens of the polytechnic institute of Coimbra

M. Gaspar, C. Santos, A. Ferreira & J.P. Figueiredo
Escola Superior de Tecnologia da Saúde de Coimbra, Coimbra, Portugal

ABSTRACT: Modern society is characterized by the high consumption of different products, including food. The school canteens are an example where this reality is verifiable, translating into an increase of food waste. In this study is intended to analyze the food waste produced in the canteens of the Polytechnic Institute of Coimbra, through the analysis and description of its percentages, assessing the level of satisfaction of users with the quality of the meals and services and the evaluation of level of good practices about the hygiene and food safety of the food handlers.

The results reveal satisfactory levels concerning the users' satisfaction with the meals and services of the canteens. The food handlers revealed good levels good practices, being these levels higher in the individuals with training in hygiene and food safety. The waste percentages, data obtained showed values above those recommended (3% the food surplus and 10% the leftovers).

1   INTRODUCTION

With the passing of time, our society is undergoing major technological development and consequently its habits and behaviors are naturally modified. This fact, allied with population growth, is predictably associated to excessive consumption, as a wider variety of products is made available, consequently generating a larger amount of waste. In this sense, this reality is also a consequence of uncontrolled and unorganized consumerism of food, as well as its subsequent organization and processing at our homes. However, the consumer isn't the only one to blame, as most mass catering industry establishments, such as school canteens, aren't accomplishing a good processing in food preparation and subsequent presentation of the dishes. Usually, these are served overfilled, contributing to the increase of the amount of food waste and predictably harming our daily life economically and in terms of the health status of the population, which is currently a concern of great importance (Campos 2010).

In association to the preparation of the meals in the canteens, as well as to their quality, emerges the increase of food waste production, which may have multiple origins, being the most common excessive meal production, flaws in the menu planning, unpredictable variations in the number of sales and even the lack of knowledge concerning the nutritional needs of the target population. Furthermore, the way the canteens' staff handles the food also contributes to food waste, regarding the inadequate use of machinery and essentially the absence of training in proper techniques of food handling after its harvest, that is, the conditions of storage and the management of stocks (Martins 2002; Farias 2002).

The importance of reducing food waste becomes then evident in all types of services that supply meals, since this fact leads to significant benefits, because there is a decrease in the production of organic waste which subsequently increases the profits of the services and the satisfaction of entrepreneurs, staff and users. The increase of food waste is thus related to financial, environmental, ethical and social matters, being in this manner considered a world problem (Campos 2010).

The research's key objective was the study of the food waste produced in the canteens of the Polytechnic Institute of Coimbra (PIC), through the analysis and description of its percentages, as well as, the assessment of the level of satisfaction of users concerning the quality of the meals and services and the evaluation of the food handlers' level of good practices about hygiene and food safety.

2   MATERIAL AND METHODS

The research study was conducted between October 2011 and July 2012 and developed in the school canteens of the PIC, where the collection of data took place between March and April 2012.

The study developed was observational and of cross-cutting nature, with non-probabilistic sample and convenience sampling.

The target population of the research study was constituted by 502 users and 31 food handlers, in a total of the PIC's 6 school canteens. The questionnaire for assessing the good practices of the food handlers was distributed to all the handlers of all the canteens. The sample to the application of the questionnaire to the users was calculated taking into account the number of students, teaching and non-teaching staff, that constitute each school, since hard data concerning the average number of meals served daily in each canteen don't exist and considering that these will be the individuals that more often eat at the respective school canteens. Thus, it was obtained a minimum number of questionnaires to take into consideration to apply in each canteen. However, during the collection of the data, it was possible to distribute and collect more questionnaires and so that sought to increase the reliability of results, once the sample became more meaningful. The sample calculation was taken into account to finite population (<100.000 cases). The interpretation of the statistical tests was based on the significance level of p-value = 0,05 with a Confidence Interval (C.I.) of 95%. To a p-value > 0,05, meaningful differences or associations between the analyzed variables weren't observed. The IBM SPSS Statistics 19 software was used to process the statistical data. The assessment and analysis of the variables in study was carried out by applying the following statistical tests: the Mann-Whitney Test, the Kruskal-Wallis Test, the Chi-squared Test of adherence and the Spearman's ordinal correlation coefficient Test.

The data collection was held in two separate occasions: first the application of two different questionnaires to two target-populations (users and food handlers of the PIC canteens), and then the weighing of food waste, that is, the food surplus and leftovers, produced in each canteen during the five business day of a week.

The users' questionnaire aimed at assessing of satisfaction concerning the quality of the meals served and of the services provided by the staff. These were distributed randomly when picking up the trays at the beginning of the meal. This questionnaire was divided into four essential parts: the first part concerning the social-biographical characterization of the respondents; the second part estimated the food waste produced by the users; the third part appraised the satisfaction about the quality of the meals; the forth part assessed the satisfaction related to the quality of the services of the canteens. For each aspect pertaining to the evaluation of the quality of the meals and services of the canteens, the following rating scale was applied: 1—Unsatisfactory; 2—Not Very Satisfactory; 3—Satisfactory; 4—Good; 5—Very Good.

The food handlers questionnaire, aimed at assessing the level of good practices in the canteen and they were distributed to all of the food handlers of each canteen. The questionnaire was divided into four fundamental parts: the first part was about the social-biographical data of the food handlers; the second part estimated the good practices in the reception and storage of foodstuffs; the third part assessed the good practices in the preparation of food and cooking; the fourth part was concerned with the evaluation of the management of the meal treatment and costumer care. Regarding the assessment of the food handlers' good practices, the scale applied was the following: 1—Yes; 2—No; 3—Sometimes; 4—Does not apply, with a larger number of "Yes" answers indicating a more positive evaluation in relation to the good practices that subsequently may be an indicator of a greater knowledge and training (Saurim and Basso 2008).

In the measurement of the quality of the meals and of the canteen services by the users and in the global assessment of the food handlers' good practices, the scores were adjusted to the percentages, that is, the higher the observed percentage, the better classified is the degree of the users' satisfaction and the level of the food handlers' good practices.

The weighing of the food waste was conducted during the five business days of one week in each of the PIC's canteens, since in this period of time one could observe their variation. The calculation of the food waste was carried out through the quantification of the food surplus and leftovers. In this sense, the weighing of three necessary data was executed, so that later on, the respective percentages could be calculated: the weight of the meals (weight of all cooked food in Kg), weight of the food surplus (weight of all the food remaining in the distribution line, in Kg, rejecting the weight of the containers) and the weight of the leftovers (weight of all the food remaining on the dishes, in Kg, including bones, fish spines and skin, since their weight was also taken into account initially).

The calculation of the food surplus percentage was done through the following expression, according to Augustini V., 2008 (Campos, 2010).

Surplus (%) = (Weight of the Surplus/Weight of the Cooked Meal) × 100.

The calculation of the percentage of the leftovers was conducted through the food surplus indicator, according to Teixeira S, 1999. (Campos, 2010).

Indicator of the Leftovers (%) = (Weight of the Leftovers/Weight of the Distributed meal) × 100.

In order to achieve these weighings the following material was used: bag to put waste (surplus and leftovers), personal protective equipment adequate to the data collection (cap, disposable gloves, protection feet and gown) and the scales existent in each canteen. The identification of the scales existent in each canteen and their characteristics are as follows: Coimbra Institute of Accounting and Administration: TKTB Industrial Weighing Scale—digital weighing scale with maximum capacity of 300 Kg and minimum of 0,10 kg; Coimbra College of Education, Oliveira do Hospital College of Technology and Management: MIC Mechanical Platform Weighing Scale—industrial mechanic weighing scale with maximum capacity of 300 Kg and minimum of 1 Kg; Coimbra College of Agriculture, Coimbra College of Health Technology, Coimbra Institute of Engineering: Dial Scale with 300 Kg capacity.

3 RESULTS

3.1 Sample description

From the 500 individuals' sample, 249 were female and 251 were male.

It was observed that 431 of the respondents were students, 34 teaching staff and 35 non-teaching staff. The average age was of approximately 22 years old in the student group, 43 in the teaching staff group and 37 in the non-teaching staff group.

In relation to how often they have been going to the canteen, the student group displayed an average period of 2 years, the teaching staff 8 years and the non-teaching staff 6 years. As for the number of times eating at the canteen each week, the average was of approximately 3 days in the student and teaching staff groups and of 4 days in the non-teaching staff group.

3.2 Analysis of the results

The obtained results about the reasons why the users eat at the PIC's canteens reveal that there are significant differences, observed $X2 = 649,055>$ critical $X2 = 16,919$. It was noted that among the several reasons for going to the canteen, the ones that highlighted the most were, first price (29,8%), followed by location (23,7%), and then the quality of the food (11,5%).

The table 1, it was aimed to assess the level of satisfaction with the quality of the meals and services of the canteens in relation to the social condition.

Statistically significant differences concerning the assessment of the satisfaction with the quality of meals and services of the canteens according to social condition were verified (p-value <0,05). On average, the levels of satisfaction regarding the quality of the meals were once more inferior to the level of satisfaction related to the quality of the services. In both assessments, the student group was the one showing less satisfaction and the non-teaching staff the group revealing more.

The satisfaction with the quality of the meals and services of the canteen was analyzed, according

Table 1. Assessment averages of the quality of the meals and services of the canteens according to social condition.

	S.C.	Av.	S.D.	N
Assessment of	students	66,98	11,95	431*
the quality	teaching staff	69,72	8,59	34*
of the meals	non-teaching staff	70,59	9,79	35*
(%)				
Total		67,42	11,65	500
Assessment of	students	72,21	14,40	431*
the quality	teaching staff	78,12	12,13	34*
of the services	non-teaching staff	80,86	13,16	35*
(%)				
Total		73,22	14,38	500

S.C.-Social Condition; Av.-Average; S.D.-Standard Deviation; t-Kruskal-Wallis Test; *p < 0,005.

to the number of times eating at the canteen each week, taking into account the total of the sample. It was verified that there is a statistically meaningful correlation between the satisfaction with the quality of the meals and the number of times eating at the canteen per week (p-value <0,05), nevertheless this variation was weak. However, a significant relation between the satisfaction with the quality of the services and the number of times eating at the canteen each week wasn't verified (p-value > 0,05).

The satisfaction with the quality of the meals and services of the canteens, according to the number of times eating at the canteen each week was also assessed taking the social condition into account. (Table 2).

Through the analysis of the table, referring to both assessments, it was noticed that it was the student group that registered a statistically meaningful correlation (p-value < 0,05), as the students who less use the canteen, better classify the quality of the served meals. On the contrary, the teaching staff and non-teaching staff groups registered a statically non-significant relation (p-value > 0,005).

It was intended to analyze the relation between the quality of the served meals and the services of the canteens regarding the users of each of the PIC's colleges and institutes. Through attachment 5, statistically significant differences were verified (p-value < 0,05). The CCA users revealed, on average, a greater satisfaction with the quality of the meals and services of the canteen. The lower levels of satisfaction with the quality of the meals were registered by the OHCTM users and the CCHT users revealed, on average, less satisfaction with the quality of the services of the canteen.

The relation between the levels of satisfaction with the quality of the meals and services in the canteens was analyzed, as well as the fact that the users try served food that they do not know.

Table 2. Relation between the assessment of the quality of the meals and services of the canteens and the number of times eating at the canteen each week, regarding social condition.

Number of times eating at the canteen per week

		S.C.	C.C	p-value	N
Assessment of the quality of the meals (%)	students		−0,146	**0,002**	430
	teaching staff		−0,113	0,523	34
	non-teaching staff		−0,123	0,480	35
Assessment of the quality of the services (%)	students		−0,102	**0,034**	430
	teaching staff		−0,142	0,424	34
	non-teaching staff		−0,217	0,212	35

S.C.-Social Condition; C.C-Correlation Coefficient; Spearman's ordinal correlation coefficient Test.

Statistically significant differences (p-value < 0,05) were observed in regard to the condition that evaluates the fact that the users try food that is served to them and that they do not know. It was verified that the users who answered that they "always" try the food they don't know, registered on average, a more positive assessment concerning the quality of the food and of the services of the canteens. On the other hand, the users who say that they "never" tried food they didn't know registered a more negative assessment of the quality of the meals and services.

It was analyzed the relation between the satisfaction with the quality of the meals and services of the canteens and the perception that the users have from each other regarding waste.

When comparing the levels of satisfaction concerning the quality of the meals and services of the canteens with the perception that the users have from each other regarding their food waste during the meal at the canteen, there weren't any statistically significant differences to register (p-value > 0,05). The results revealed that the users, who believe that the other users "rarely" waste food, assessed more positively the quality of the meals and services of the canteens where they usually eat. The users, who believe that the other users "always" reject parts of their meal, assessed more negatively the meals and services of the canteens.

4 DISCUSSION

Through the data obtained, it was verified that the averages of the levels of satisfaction of the users according to their gender, was similar. However, when regarding their social condition, it was observed that the students showed a lower assessment of the canteens' meals and services.

It was also determined that the assessment of the meals and services of the canteens, according to how often the users eat there each week, was statistically significant in the student group. These two previously mentioned facts may be justified by the fact that the students are one of the groups that eats more often in the canteens, as shown by the results, having thus a better perception/opinion regarding the meals and services rendered by the canteens.

Of the colleges involved in this research, it was noted that the users from CCA were the ones more satisfied with the quality of the meals and services of the canteen. The lower satisfaction rates, in regard to the quality of the meals, were registered by the HOCTM users and as for the quality of the services, it was the CCHT users who revealed less satisfaction.

The results obtained revealed that on average, the satisfaction rates concerning the quality of the meals and services of the canteens are higher in the users who claim to "always" try the food that is served to them and that they don't know.

It was also noticed that the most positive assessment of the quality of meals and services, belongs to the users who have an understanding that the others "rarely" leave a large amount of food on their dishes. However, the users who believe that the other users "always" leave food waste display the most negative assessment of the quality of the meals and services of the canteens.

The food handlers who mentioned having some kind of training in the areas of hygiene and food safety, or who received some at the beginning of their activities in the canteens, revealed a higher assessment average in regard to their good practices. It is pointed out that this association is meaningful in regard to the condition of the food handlers having some sort of training. Once more, as in the previously analyzed results, it is noticed that the level of educational qualification and the fact that the food handlers receiving training, will be a preponderant factor on the good performance of their activities. The handlers of the meal service units should receive permanent training.

In the present study, the total value obtained from the food surplus average was of approximately 10%, that is, it was also registered in all the colleges an average value higher than the reference value. This may be a synonym of flaws in the planning and preparation of the meals.

The total average value of the leftovers obtained in this study was of approximately 19%, that is, a high percentage of leftovers was registered in all the canteens with the exception of the CCA canteen. These results are distressing and may be justified by the inadequate size of the portions of the food served or even by the definition of the menu.

5 CONCLUSIONS

The food waste produced in the school canteens may have multiple origins, namely the satisfaction of the users and the good practices and level of training of the food handlers (7). The present research study intended, not only to analyze this fact, but also to describe the food waste percentage produced in the PIC's canteens (Saurim and Basso 2008).

Although the averages of the levels of the users' satisfaction were registered in a satisfactory way in all the canteens (superior to 50%), the results indicated that the food surplus, 10%, was higher than the recommended value (3%). It was also noticed that the percentage of leftovers, 19%, was above the desirable value (10%), except in the case of CCA. Thus, the importance of continuing to implement corrective measures in the planning and preparation of the meals becomes evident (Campos, 2010).

The failure concerning the leftovers shows how important it is to improve the planning, production and the distribution of the meals. Selling meal tickets the day before is a common practice that improves the management of the meals, but that can also be significant in the management of food production. It would be equally important if each canteen implemented a food waste control system, in order to develop corrective measures (Viana 2007; Saurim and Basso 2008; Reforço 2010).

The school canteens should offer a wider variety of food and in quantities fitting the nutritional needs of this type of users. The food handlers should also mind the presentation of the dishes, so that they may "please" the user and in this manner captivate him and avoid food waste. The responsible for each canteen should also pay attention to the organization of the kitchen and of the refectory space, reorganizing the self-service lines and for example, placing the bread at the end of the distribution line (Campos 2010).

Each college should promote public awareness and alert all the community to the importance and benefits of food waste reduction (Campos 2010).

Food waste is relevant in environmental, social and economic matters. It is thus fundamental to develop actions and measures that contribute to the decrease of waste and consequently that contribute in a positive way to the sustainability and environmental preservation. These measures may have to do with the raising of the users awareness through information or even with the development of campaigns encouraging the reduction of food waste (Campos 2010).

REFERENCES

Campos, V. (2010). Estudo dos Desperdícios Alimentares em Meio Escolar. Faculdade de Ciências da Nutrição e Alimentação. Universidade do Porto. p. 1–3.

Code of hygienic practice for precooked and cooked foods in mass catering. CAC/RCP 39-(1993). Rome: Codex Alimentarius Comission.

Martins, C., Farias, R. (2002). Produção de Alimentos × Desperdício: Tipos, causas e como Reduzir perdas na Produção Agrícola. Revista da FZVA. Uruguaiana. 20–32.

Oliveira, B. (2007). Qualidade e Segurança alimentar na Restauração colectiva. Segurança e Qualidade Alimentar, 2, 38–40. Editideias.

Pereira, F. (2009). Auditorias Internas aos Sistemas de segurança Alimentar Implementados em cantinas Universitárias [dissertação]. Lisboa. Universidade Técnica de Lisboa—Faculdade de Medicina Veterinária.

Reforço, A. (2010). Segurança Alimentar no Refeitório de uma escola secundária—estudo para implementação do HACCP [dissertação]. Universidade Aberta.

Saurim, I.; Basso, C. (2008). Avaliação do Desperdício de Alimentos de Bufê em Restaurante Comercial em Santa Maria, RS; Disc. Scientia. Série: Ciências da Saúde, Santa Maria, v. 9, n. 1, p. 115–120.

Sousa, A. (2010). Impacto Ambiental das Empresas do Canal Horeca. Faculdade de Ciências da Nutrição e Alimentação—Universidade do Porto; p. 1–2.

Viana, I. (2007). Estudo do desperdício nas refeições hospitalares na unidade CHAM—Viana do Castelo. Faculdade de Ciências da Nutrição e Alimentação—Universidade do Porto. p. 6–7.

Occupational Safety and Hygiene – Arezes et al. (eds)
© 2013 Taylor & Francis Group, London, ISBN 978-1-138-00047-6

25 years of ergonomics in Spain: Current status

Javier Llaneza & Gustavo Rosal López
Spanish Ergonomics Association, Spain

Elsa Peña Suarez & Julio Rodriguez Suarez
University of Oviedo, Spain

ABSTRACT: The development and advancement of ergonomics in Spain is linked to the Polytechnic University of Catalonia, some of the mutual work accidents and occupational diseases in industry and the framework of the European Coal and Steel Community (ECSC). There is a confluence of these actions with the creation of the Spanish Association of Ergonomics (AEE) in 1989. With the Law on Prevention of Occupational Risks and the entry into force of Royal Decree 39/1997, specialists have been trained according to the training program that gives official recognition by authorities and academic work in one of these three areas or specialties: Work Safety, Industrial Hygiene and Ergonomics and Applied Psychology, plus Occupational Medicine. The research shows different aspects related to daily activities, relationships among other professionals, the use of techniques and perspectives for new applications, etc. shaping a study that complements other European studies about the work of ergonomists.

1 INTRODUCTION

The origins of Ergonomics in Spain are in a certain way consistent with its development in the European Community. The construction of Europe after World War II was carried out in stages and unification efforts were limited initially to large industries: Mining and Steel, Luxembourg becoming the lead agency of the European Coal and Steel Community (ECSC). The ECSC Treaty (1951) was the legal framework under which ergonomic research developed, with funding coming from a fee imposed on the production of coal and steel.

Early research in 1954 was dedicated to deriving the pathological consequences of certain environmental conditions of labor under the heading of Occupational Medicine. The antecedents of Ergonomics' Programs can be found in the creation of a Programme for Work Physiology and Psychology, out of which came two programs on the Human Factor and Security, the first in 1957–1963 and the second from 1965–1970, and a First Ergonomics Research Program in 1964. As result of these last Programmes, 43 investigations were funded, 30 of which were directly developed in the mining and steel companies and laboratories, and only 13 were essentially laboratory investigations.

The characteristics of these researches were:

– The multinational dimension, they were addressed to the constituent states: France, West Germany, Italy and the Benelux (Belgium, Luxembourg and the Netherlands).
– The duration of six years.
– The methodological steering. This is a Community Research, in singular.
– The industrial research framework.
– The subject of the investigation was the workforce, which excluded studies of individual character.

Considering previous experiences, in the mid-seventies, the Commission of the European Community decided to develop, through the Community Ergonomics Action, a first Ergonomics Program, which was specific to the coal and steel sectors. This program arose from the need to bring Ergonomics closer to the workplace, from the requirement of Ergonomics being participatory and showing an adequate interrelation between co-researchers.

Several national Ergonomics teams in the steel and coal mines of the ECSC were constituted to carry out this program. Considering the need for coordination and support for these teams, a managing organization was established in 1980: The Bureau d'Information et de Coordination des Programmes de l'Action Communautaire Ergonomique of the ECSC. In addition to this organization, a full organic structure was created: the Ergonomics Community Action Network, with the aim of optimizing the development of ergonomic activity. This Network consisted of:

- Ergonomics National Teams, whose aim was to develop the Ergonomic Community Action projects in the industries of each country
- Expert Committee, made up of representatives from National Teams.
- Coordination Groups, who were responsible for coordinating the projects of each thematic area, maintaining a close relationship with all teams and Community projects.

Public Enterprises of iron and steel (ENSIDESA) and coal mining (HUNOSA), which were situated in the north of Spain, in the province of Asturias, created their own ergonomic teams and conducted various research projects in ergonomics The aim of these projects was not only to improve and adapt working conditions for workers, but also to create a body of knowledge on Ergonomics that could be applicable, through the Office of Information and Coordination of Ergonomics Community Action or through Community Legislation, to the rest of the industries in the EEC.

1.1 The creation of the spanish association of Ergonomics

In the late eighties, there began to emerge in Spain a clear presence of Ergonomics in industry and in University. At that time, the first university courses were delivered, in some cases as University degrees, either as Master or as Expert courses. A highlight of this were the first two editions of the Master of Ergonomics and Working Conditions in Oviedo organized by the University of Mining Engineering in Oviedo, with ENSIDESA and the Safety Commission in the iron and steel industry (CSIS) led by Jesus Portillo. Other outstanding actions focused on an attempt to establish a framework for training in the discipline, with courses at the Higher School of Management in the "Universidad Complutense" of Madrid, "Universidad Autónoma" of Madrid and "Universidad Politécnica" of Barcelona (UPC). These actions meant a great attempt to spread this discipline and to bring it closer to other types of professionals. Nowadays, only the course taught at UPC led by Pedro R. Mondelo remains, maintaining excellent reputations. In 1989, some of these pioneers of Spanish Ergonomics decided to create the Spanish Ergonomics Association (AEE).

Figure 1. Logo of Asociación Española de Ergonomía (www.ergonomos.es).

Since 1997, with the publication of the Royal Decree 39/1997, specialists have been trained under the training program given official recognition by labor and academic authorities—one of the four areas or specialties (including Occupational Health) contemplated under the preventative techniques to address occupational risks.

Therefore, Spanish ergonomists do not require to work as such anything other than that legal recognition regulated by the Royal Decree 39/1997, which states: "It will be necessary to have an official university degree and to have a minimum and appropriate training accredited by a university with a content specified in the program which is referred to in Annex VI, the development of which will last no less than six hundred hours with a proper time distribution for each training project, respecting the distribution established in that Annex ... ". The previous may explain certain skepticism about those "European ergonomists" degrees under the excuse of being necessary for ergonomists to work as such.

The training program for Ergonomics experts includes the following subjects:

- Ergonomics: concepts and objectives.
- Environmental conditions in Ergonomics.
- Workplace conception and design.
- Physical workload.
- Mental workload.
- Psychosocial Factors.
- Organizational structure.
- Characteristics of the company, the job and the individual.
- Stress and other psychosocial problems.
- Consequences of adverse psychosocial factors and their evaluation.
- Psychosocial intervention.

2 MATERIALS AND METHODS

A research about the practice of ergonomics experts in Spain was conducted, with a sampling framework of 356 ergonomists, which were associated to Regional Associations of Ergonomist professionals and to the Spanish Ergonomics Association. 356 questionnaires were sent, and

97 questionnaires were received, representing a 27.24% of the whole population of interest. This instrument is based on a conceptual model of the factors that may be relevant to the practice of the ergonomists. Specifically, it consists of 104 items that fall into the following areas: demographic, professional practice of Ergonomics in relation to university education and to working in other preventative disciplines, practice of ergonomics in relation to the used methodology, specific training received, actions for the promotion and acceptance of Ergonomics, and perception of the practice of Forensic Ergonomics. The average age of respondents was 41.62 years old (SD = 8.63). The youngest respondent was 25 years old, while the oldest was 64 years old. The gender distribution is shown in the following pie chart, Figure 2, showing 61.9% (N = 60) of men.

The Professional practice as Ergonomist for 70.5% of respondents was related to Occupational Risk Prevention functions, whereas for 8.4% of respondents works in Ergonomics together with Occupational Risk Prevention, in relation to other fields of work such as teaching, research and consulting, as shown in figure 3.

37.5% of the sample has been practicing Ergonomics professionally for more than 10 years. 57.40% of the participants work as Ergonomics specialist within an Internal Prevention Service, 40.40% work within External Prevention Services and 2.10% perform their work in the Public Administration. Generally, all Ergonomics and SocialPhsychology experts also have two other

Figure 2.   Gender distributton.

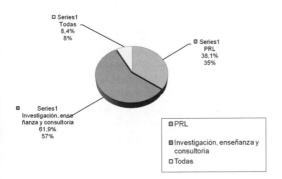

Figure 3.   Professional practice of Ergonomics.

technical specialties (Occupational Safety and Industrial Hygiene), as do those experts engaged in the Health Surveillance specialty (Occupational Medicine and Nursing). Analysis of the questionnaires showed that the fields of Ergonomics and of Applied Psychology occupied over 60% of the working day for graduates in Psychology and Sociology (68.42%). The health related specialties also show this trend very clearly, as these fields occupy over 60% of their working day, 50% for medical graduates, and 30% for nurses. In the Hygiene specialty this trend was clearly not observed, with chemistry and biology experts proving to be the most active in these fields. There are 36 people who spend more than 60% of their day working in Ergonomics, which represents 39.6% of the sample. In the Safety specialty it was observed that experts with Engineering, Law, Biology and Chemistry university degrees devote greater time percentages to these fields.

However, those Occupational Risk Prevention professionals who spend more than 60% of their work time working in Ergonomics, showed statistically significant differences (p < .05) with respect to those who spent less than 60% of their workday. They showed a higher frequency in tasks related to work and mental health, work organization, new projects and judicial reports.

Analyzing the use of instruments and metrological equipment, the most widely used are the camera and the video camera, followed by different devices for the evaluation of environmental conditions: light meter (lighting), hygrometer (humidity and temperature) and sound level meters (noise). More specific equipment such as the goniometer (angle measurement), the HRM (heart rate measurement) or electromyography (measuring muscle activity) was not used by most of participants. It is striking that the tape measure has never been used by 63.30%. Not so strikingly 61.50% of the participants have never used lighting contrast measures.

The techniques employed for evaluating physical load to (among other goals) objectivize biomechanical aspects do not show very high frequencies. For instance, NIOSH guide is used by 19.30% of the sample, and more than half of the respondents have never implemented the Strain Index (57.30%), nor the Cirello and Snock tables (50%). Paradoxically, the greater frequency of intervention is found on those tasks that are related to occupational health, in particular to muscle and bone risk assessment (load handling, stress positions, etc..).

Participation in new projects and the application of the principles of Ergonomics design has not happened for 39.50% of respondents. Only 30.60% of the sample has participated in the design and specification of physical devices and/or the functionality of systems. Both results show the small development of Design Ergonomics and the limited involvement of ergonomists in new projects or in the design or the elaboration of specifications for Request for Proposals (RfPs).

3 RESULTS AND DISCUSSION

The aim of this research has been to investigate the practice of Ergonomics in Spain from its origins. We must underline that international research in the practice of the ergonomist's profession is also still very scarce, with the exception of studies conducted in Europe by Breedveld and Dul (2005), in North America by Dempsey (2005) and more recently by Anceaux et al (2012) in France. Therefore we understand that this research represents a significant contribution to expanding knowledge on the way Ergonomics is applied. In Spain the specialty of Ergonomics and Psychosociology is linked to the prevention of occupational hazards and to the improvement of working conditions, but it gets diluted by other preventive areas that companies are obliged to apply in relation to the protection of safety and health at work. Consequently, the practice of Ergonomics in the prevention area is very limited, and its expansion depends not only on imposed actions based on new regulations or on a more rigorous enforcement of existing regulations, but also on the personal initiative of the Ergonomics specialist himself.

4 CONCLUSIONS

There are difficulties to extend the field of prevention of occupational risks and to expand its development in organizations. The reasons behind these difficulties are diverse and although the research points to some strategies, there are no clear ways to overcome them, recognizing the key role played by those who have decision making power, within the management of organizations and companies.

Strategies must be developed to give frame to the creation of new actions for the enhancement of the professional practice and its different applications, which can be contrasted both by future European research on ergonomists, and by the advance of new practices and fields of application, like for example Forensic Ergonomics or Ergonomics Project Management.

REFERENCES

Anceaux, F., Barcellini, F., Boccara, V., Forrierre, J., Gaillard, I., Nelson, J. y Toupin, C. (2012). Cartographie de la recherche en ergonomie. Collège des Enseignants-Chercheur (CE2) et les *Réseau des Jeunes Chercheurs en Ergonomie (RJCE)*. En www.rcje.fr/site/uploads/activites/ Cartographie_de_la recherche2012.pdf.

Breedveld, P., Dul, J., (2005). The Position and Success of Certified European Ergonomists. *Rotterdam School of Management, Erasmus* University, Rotterdam, 22 p.

Chung, A.Z.Q. y Shorrock, S. (2011). The research-practice relationship in ergonomics and human factors, surveying and bridging the gap. *Ergonomics*, vol. 54(5), pp. 413–419.

Comisión de las Comunidades Europeas. CEE (1994) Acción Ergonómica en la Siderurgia. Resultados del V Programa. *Oficina de Publicaciones de las Comunidades Europeas*. Luxemburgo.

Dempsey, P.G, McGorry, R.W. y Maynard, W.S. (2005). A survey of tools and methods used by certified professional ergonomists. *Applied Ergonomics* vol. 36, pp. 489–503.

Dul, J., Bruder, R., Buckle, P., Carayon, P., Falzon, P., Karras, W., Wilson, J.R. y Van der Doelen, B. (2012). A strategy for Human Factors/Ergonomics: Developing the discipline and profession. Final report of the IEA Future of Ergonomics Committee. *Ergonomics*, vol. 55(4), pp. 377–395.

Llaneza, F.J. (2009). Ergonomía y Psicosociología Aplicada. *Manual para la formación del especialista*. (15 ed.). Valladolid: Editorial Lex Nova.

Shorrock, S. y Murphy, D.J. (2005). The ergonomist as skilled helper. En: Bust, P.D. (Ed.), *Contemporary Ergonomics*. Taylor & Francis.

Author index